Transcriptome and Genome Analyses Applied to Aquaculture Research

Transcriptome and Genome Analyses Applied to Aquaculture Research

Editor

Patricia Pereiro

MDPI • Basel • Beijing • Wuhan • Barcelona • Belgrade • Manchester • Tokyo • Cluj • Tianjin

Editor
Patricia Pereiro
Immunology and Genomics
Institute of Marine Research
Vigo
Spain

Editorial Office
MDPI
St. Alban-Anlage 66
4052 Basel, Switzerland

This is a reprint of articles from the Special Issue published online in the open access journal *Biology* (ISSN 2079-7737) (available at: www.mdpi.com/journal/biology/special_issues/TAGAATAR).

For citation purposes, cite each article independently as indicated on the article page online and as indicated below:

LastName, A.A.; LastName, B.B.; LastName, C.C. Article Title. *Journal Name* **Year**, *Volume Number*, Page Range.

ISBN 978-3-0365-5922-3 (Hbk)
ISBN 978-3-0365-5921-6 (PDF)

Cover image courtesy of Patricia Pereiro

Contents

biology

Editorial

Transcriptome and Genome Analyses Applied to Aquaculture Research

Patricia Pereiro

Institute of Marine Research (IIM), National Research Council (CSIC), Eduardo Cabello, 6, 36208 Vigo, Spain; patriciapereiro@iim.csic.es; Tel.: +34-986231930

The Special Issue "Transcriptome and Genome Analyses Applied to Aquaculture Research" had great success among the researchers specialized in different fields of aquaculture. It reached 27 published manuscripts covering different interests (immunology, response to stressors, sexual dimorphism, development, etc.) in a variety of fish and shellfish, and even in turtles. Different transcriptome (mRNAs and non-coding RNAs –ncRNAs–), genome (Single Nucleotide Polymorphisms –SNPs–), and metatranscriptome analyses were conducted to unravel those different aspects of interest.

Understanding the immune response to different pathogens is pivotal for developing breeding selection strategies to improve the resistance to diseases or for formulating new treatments, among other objectives. In this sense, massive analyses of the immune response have attracted a lot of interest and are being conducted in different species of interest for the aquaculture industry.

Parasites, and especially endoparasites, are highly problematic for fish production due to the absence of treatments and vaccines on the market for most of them. Because of this, the study of the response to these pathogens could provide interesting clues aimed to improve the fish's resistance to endoparasites. Two works included in this Special Issue analyzed the response of the flatfish turbot (*Scophthalmus maximus*), a very valuable fish species both in Europe and China, to two different endoparasites causing important economic losses. Whereas Valle et al. [1] explored the transcriptome response of the turbot peritoneal cells at 1, 2, and 4 h post-infection (hpi) with the ciliate parasite *Philasterides dicentrarchi*, as well as the transcriptome of the own parasite after the challenge, Ronza et al. [2] carried out transcriptomic and histopathological analyses of the thymus after infection with the myxozoan parasite *Enteromyxum scophthalmi*.

But several fish viruses are also problematic for the same reasons: no vaccines nor therapeutic treatments commercially available. Different components of the antiviral response were analyzed in this Special Issue. Tripartite motif proteins (TRIMs) are increasingly well known for their antiviral immune functions in mammals, although their exact function in fish still needs to be fully elucidated, which is hindered by the massive gene expansion of the teleost TRIM family. Qin et al. [3] identified 42 *trim* genes in the grass carp (*Ctenopharyngodon idella*) genome and conducted different domain and phylogenetic analyses to classify them. The authors also found that 11 of these TRIMs were significantly induced during grass carp reovirus (GCRV) infection and, by co-expression analyses, they identified three *trim* genes that could be considered as a part of the type I interferon response [3]. On the other hand, Bello-Perez et al. [4] analyzed the expression of the seven C-reactive proteins (CRPs) previously identified in zebrafish (*Danio rerio*) under basal conditions and after spring viraemia of carp virus (SVCV) infection in different tissues. Most of these *crp* genes were altered after the viral challenge in different tissues, including the skin. This tissue also showed a higher expression of these genes in the recombinant activation gene 1 (*rag1*) mutant zebrafish compared to their wild-type counterparts, suggesting an overcompensation of the innate immune in the absence of the adaptive one [4]. Finally, Pereiro et al. [5] analyzed by RNA-Seq the repertoire of long non-coding RNAs (lncRNAs) in the brain and

Citation: Pereiro, P. Transcriptome and Genome Analyses Applied to Aquaculture Research. *Biology* **2022**, *11*, 1312. https://doi.org/10.3390/biology11091312

Received: 1 September 2022
Accepted: 2 September 2022
Published: 4 September 2022

head kidney from European sea bass (*Dicentrarchus labrax*) and identified those lncRNAs significantly modulated after infection with nodavirus. By identifying the neighboring coding genes of these differentially expressed lncRNAs, the potential functions affected by these lncRNAs were determined [5].

But, in addition to the study of the immune response to different pathogenic challenges, the knowledge of the immunological properties of the different immune cell types is also a cornerstone for fish immunologists. Perdiguero et al. [6] conducted single-cell RNA sequencing to analyze for the first time in teleost the transcriptional profile of single B cells from rainbow trout (*Oncorhynchus mykiss*) peripheral blood. Based on their transcriptome profiles, the authors identified ten different B cell subpopulations. This information provides valuable information to understand the biology of teleost B cells and could serve as a source of potential markers to be used to differentiate B cell subsets [6].

In addition to fish, several pathogens also impact the production of other economically relevant organisms produced in aquaculture systems. This is the case of the Chinese soft-shelled turtle (*Trionyx sinensis* or *Pelodiscus sinensis*), an important cultured reptile in East Asia taken as a food resource, especially in China and Japan. Lv et al. [7] conducted RNA-Seq analyses of liver and spleen from uninfected turtles and two groups of turtles with different susceptibility to *Aeromonas hydrophila* infection. Whereas the diseased turtles showed a high induction of genes encoding for pro-inflammatory cytokines, pattern recognition receptors (PRRs), and apoptosis-related molecules and low phagocytic response, asymptomatic turtles showed the opposite pattern. All these data are very useful to understand the molecular basis of the resistance to the hemorrhagic sepsis caused by *A. hydrophila* in Chinese soft-shelled turtles [7]. On the other hand, hepatopancreas necrosis disease of the Chinese mitten crab (*Eriocheir sinensis*) is causing huge economic losses in China, although the pathogens potentially involved in this disease have not yet been identified. To elucidate this question, Shen et al. [8] conducted metatranscriptome analyses of the hepatopancreatic flora of diseased crabs with mild symptoms, diseased crabs with severe symptoms, and crabs without visible symptoms. The authors observed a decrease in the prevalence of the bacteria *Absidia glauca* and *Candidatus Synechococcus spongiarum*, whereas the prevalence of *Spiroplasma eriocheiris* increased in the hepatopancreatic flora of diseased crabs. Moreover, the flora from diseased crabs showed a higher expression of genes involved in certain functions that could be also involved in the pathological mechanism [8].

Chronic or acute inflammation is an undesirable condition in fish aquaculture, which could be a consequence of inadequate aquaculture rearing conditions (husbandry, transportation, crowding densities, water parameters, etc.) or infections. These stressful environments can compromise fish growth, welfare, and immunity. The neuroendocrine-immune axis is a bi-directional network; for instance, the immune response conditions the stress response and vice versa. It is known that opioid receptors are involved in regulatory mechanisms of both the immune and the stress responses in mammals. Nevertheless, their exact roles and responsiveness to immune stimulation in the brain are still not known, especially in non-mammalian species. To better understand the link between the peripheral immune signaling and the central neuroendocrine responses, Azeredo et al. [9] and Peixoto et al. [10] analyzed the effect of acute and chronic inflammation, respectively, induced by the intraperitoneal injection of Freund's Incomplete Adjuvant in European sea bass. Azeredo et al. observed that different neuroendocrine and opioid receptors were altered during acute inflammation in different brain regions, which demonstrates the interconnection between peritoneal inflammation and brain neuroendocrine response, as well the existence of potential site-specific functions for these receptors in the different regions of the brain [9]. On the other hand, Peixoto et al. analyzed different immune parameters after chronic inflammation, which was characterized by higher leukocyte numbers in the blood accompanied by an intensification of the respiratory burst and higher transcription levels of *cxcr4* and *il34* in the head kidney [10]. Nevertheless, contrary to that observed by Azeredo et al. [9] in the European sea bass brain, opioid receptors seem not to be affected by inflammation in head kidney cells [10].

Stress can significantly impact different physiological aspects of the animals, which could result in a low growth rate, higher susceptibility to diseases, and even a low meat quality. There are a variety of stressors that can induce a pernicious effect in the aquaculture species (changes in temperature, pH, salinity, water oxygen saturation, handling, etc.) The impact of different stressors was analyzed in different works included in this Special Issue. Dettleff et al. [11] evaluated the effects of high-temperature stress on the liver of red cusk-eel (*Genypterus chilensis*) through different approaches, including RNA-Seq analysis. The results revealed that this stressor induced oxidative damage in the liver and affected the expression of a multitude of genes, which are mainly involved in the unfolded protein response, heat shock response, and oxidative stress, among others, providing a rich source of information to be considered for the aquaculture and fisheries industry of this species under a climate change scenario [11]. Interestingly, Naya-Catalá et al. [12] found that gilthead sea bream (*Sparus aurata*) juveniles acclimated for 45 days to mild-hypoxia and then returned to normoxia showed an increased contribution of lipid metabolism to the whole energy supply to preserve the aerobic energy production, a better swimming performance regardless of changes in feed intake, as well as reduced protein turnover and improved anaerobic fitness with the restoration of normoxia. These results indicate that mild-hypoxia conditioning could serve as a promising prophylactic measure to prepare these fish for predictable stressful events [12].

Salinity alteration is also a very frequent stressor that aquatic animals need to cope with. Gills are one of the most important tissues in osmoregulation and therefore, to known the mechanisms affected in this tissue could help to better understand the adaptation to salinity. Malik and Kim [13] conducted a meta-analysis of publicly available RNA-Seq datasets to identify differentially expressed genes in the Chinese mitten crab gills under different salinity conditions. The authors found a great modulation of many different types of ion transporters and channels (transportome) under different salinity conditions that may serve as biomarkers for osmoregulation. On the other hand, Li et al. [14] conducted a systematic analysis of 95 heat shock protein (*hsp*) genes in spotted sea bass (*Lateolabrax maculatus*) gills, a fish species that can widely adapt to diverse salinities from fresh to seawater, and moderately adapt to high alkaline water. The study revealed that these genes were mainly affected in the gills by alkalinity stress, responding with a coordinated modulation of *hsp*40-70-90 families [14]. But changes in salinity are also a part of the life cycle of the diadromous (catadromous and anadromous) fish species. Smoltification is a complex physiological process that prepares freshwater fish for successful osmoregulation in seawater, and it has been mainly studied in salmonid species. Nevertheless, how non-coding RNAs can contribute to these complex physiological alterations remained practically unexplored. Whereas Valenzuela-Muñoz et al. [15] analyzed the mRNA and miRNA profiles in gills, intestine, and head kidney from Atlantic salmon (*Salmo salar*) exposed to a gradual salinity change or to a salinity shock, Shwe et al. [16] studied their modulation in the liver during smoltification. The results revealed a set of miRNAs associated with smoltification and allowed us to predict their highly confident target genes [15,16]. Additionally, Valenzuela-Muñoz et al. [15] also found that salinity changes induced a broad modulation of transposable elements, which suggests a relevant role of these molecules in the smoltification process.

Gender dimorphism has drawn a lot of attention due to the different growth rates and body sizes of males and females depending on the species. This is the case of the Chinese soft-shelled turtles, with males exhibiting the best growth pattern. Consequently, to understand the sex determination- and differentiation-related genes could help to easily achieve single-sex breeding populations. In this sense, Zhu et al. [17] carried out transcriptome analyses of the ovaries and testes of these turtles using RNA-Seq to identify the genes, lncRNAs, miRNAs, and circRNAs that could be involved in gender dimorphism and well as to determine the relationship between genes and ncRNAs in the regulatory mechanism of sex differentiation. Pseudo-female turtles, which have a female phenotype and male genotype after estradiol administration, can be used to obtain an all-male offspring.

Nevertheless, the genes involved in the gonad differentiation and sex reversal of Chinese soft-shelled turtles were poorly known. Wang et al. [18], by analyzing the transcriptome profile of male, female and pseudo-female gonads found that differences between males and pseudo-females were mainly related to steroid hormone synthesis at the transcriptome level. Moreover, the SRY-related high-mobility group (HMG)-box (SOX) family of genes seems to be playing an important role in the process of sex reversal from male to pseudo-female, being *sox3* induced by exogenous estrogen and *sox8* and *sox9* inhibited [18]. But, in addition to gender, growth dimorphism could be also observed in tetraploid individuals compared to their diploid wild-type counterparts. Because of this, Luo et al. [19] compared the transcriptome profiles through RNA-Seq of different tissues (brain, gonad, liver, and muscle) from diploids and tetraploids of both sexes of the freshwater fish loach (*Misgurnus anguillicaudatus*). Their results suggested that energy metabolism levels, steroid hormone synthesis, and fatty acid degradation abilities might be important reasons for the sexual and polyploidy dimorphisms in loaches, three processes that were clearly differentiated between fast-growing loaches (tetraploids, females) and slow-growing loaches (diploids, males) [19].

For several fish species, interspecific hybrids also show higher feed conversion, disease resistance, carcass yield, and harvestability compared to both parental species, which is known as heterosis or hybrid vigor. This is the case of the hybrid between female channel catfish (*Ictalurus punctatus*) and male blue catfish (*Ictalurus furcatus*). Interestingly, heterosis for these hybrids only occurs in small culture units (pond culture) but not in tank culture, being therefore environment-dependent. Wang et al. [20] conducted different morphometric, immune, and metabolic measurements and analyzed through RNA-Seq the differences in the liver transcriptome among channel catfish, blue catfish, and their reciprocal F1s reared in tanks. The results showed that the tank environment resulted in a higher growth performance for channel catfish, which could be related to a lower immune function and a higher expression of genes involved in fatty acid metabolism/transport [20]. On the other hand, hybrids showed a higher level of blood glucose, which can be explained by the phenomenon known as transgressive expression (overexpression/underexpression in F1s than the parental species); indeed, a total of 1140 transgressive genes were identified in F1 hybrids, showing an enrichment in glycan degradation function [20]. This gene expression and physiological differences could be contributing to the lack of heterosis in the tank culture environment.

Polyunsaturated fatty acids (PUFAs) content is a desirable trait to be selected in fish aquaculture. Zhang et al. [21] identified single nucleotide polymorphisms (SNPs) in the coding regions of the fatty acid elongase 5 genes, *elovl5a* and *elovl5b,* in common carp (*Cyprinus carpio*) and correlated them with the PUFAs content. Three SNPs, one in *elovl5a* and two in *elovl5b,* were found to be associated with higher PUFA levels, and could be used in the future as markers for selective breeding in common carp [21]. But, in addition, to improve the PUFAs content in this way, optimal fish nutrition can also significantly impact the lipid composition, among others aspects. Katan et al. [22] used microarrays to examine transcriptome differences in the liver of Atlantic salmon fed with a diet containing a high ratio of omega-6 to omega-3 (ω6:ω3) fatty acids and those fed with a low ω6:ω3 ratio. The results revealed differences in the expression of genes involved in the immune response, lipid metabolism, cell proliferation, and translation, and determined that two helicases with zinc finger 2 (*helz2*) paralogues were positively correlated with ω3 and negatively with ω6 fatty acids both in liver and muscle, indicating their potential as biomarkers of tissue ω6:ω3 variation [22].

Fish development has been extensively studied since it could help to improve different production bottlenecks, such as massive mortality episodes or malformations. Flatfishes have the particularity that undergoes a dramatic metamorphosis to change their body pattern from a pelagic bilateral symmetric larva to a benthonic flat asymmetric juvenile. Since this process is mainly controlled by the brain, Guerrero-Peña et al. [23] conducted RNA-Seq analyses of the brain from different developmental stages of turbot (pre-metamorphic,

metamorphic climax, and post-metamorphic) to understand the transcriptome modulations occurring during the metamorphosis. The authors found 1570 genes differentially expressed in the three developmental stages, being the climax stage of metamorphosis characterized by a high expression of genes involved in inflammation and apoptosis, which could be related to processes of larval tissue inflammation, resorption, and replacement, as occurs in other vertebrates [23]. Moreover, their results confirmed the relevance of the thyroid stimulating hormone in the induction of the teleost metamorphosis process [23].

But not only does the expression pattern of the protein-coding genes changes during development, the expression of ncRNAs, as regulators of the gene expression, is also fine-tuned during development. To obtain the complete miRNAome of lumpfish (*Cyclopterus lumpus*), Chakraborty et al. [24] conducted miRNA analysis of different organs from adult individuals and entire embryos and larvae. They obtained 443 unique mature miRNAs from 391 conserved and eight novel miRNA genes, which represent an essential reference for future expression analyses. Additionally, the expression patterns of specific conserved miRNAs in particular tissues and developmental stages indicated that they are involved in organ- and developmental stage-specific functions reported in other teleost [24]. To obtain as most complete as possible a first transcriptome of edible red sea urchin, *Loxechinus albus* was the objective of Antiqueo et al. [25]. For this, the authors sequenced gonads, intestines, and coelomocytes of juvenile urchins. After de novo assembly, a total of 185,239 transcripts were obtained, of which 57,106 (31%) were successfully annotated. Comparative transcriptome analyses revealed that differentially expressed transcripts among tissues were specialized in different biological processes according to the specific functions of these tissues [25].

Finally, this Special Issue also included research about the development of the intermuscular bones (IBs), which negatively affect the edibleness and economic value of fish. Comparing species with and without epineural and epipleural IBs or fish strains with different abundances of them could provide interesting clues for reducing the number of IBs. Zhou et al. [26] analyzed different stages of myosepta development using morphological and transcriptomics analyses in blunt snout bream (*Megalobrama amblycephala*) and Nile tilapia (*Oreochromis niloticus*), fish species with and without epineural and epipleural IBs, respectively. Among other results, the authors identified a repertoire of genes that may play important roles in IB development, with could provide interesting information for further studies [26]. On the other hand, Cui et al. [27] conducted whole-genome resequencing and bulked segregant analysis in three common carp strains with a different abundance of IBs to identify SNPs that could be associated with the number of IBs. The 21 SNPs significantly associated with the abundance of IBs could serve as potential markers for selective breeding to generate IB-reduced common carps [27].

Funding: This research received no external funding. Patricia wishes to thank the Spanish Ministerio de Ciencia e Innovación for her postdoctoral contract IJC2020-042682-I.

Conflicts of Interest: The author declares no conflict of interest.

References

1. Valle, A.; Leiro, J.M.; Pereiro, P.; Figueras, A.; Novoa, B.; Dirks, R.P.H.; Lamas, J. Interactions between the *Parasite Philasterides dicentrarchi* and the Immune System of the Turbot *Scophthalmus maximus*. A Transcriptomic Analysis. *Biology* **2020**, *9*, 337. [CrossRef]
2. Ronza, P.; Robledo, D.; Losada, A.P.; Bermúdez, R.; Pardo, B.G.; Martínez, P.; Quiroga, M.I. The Teleost Thymus in Health and Disease: New Insights from Transcriptomic and Histopathological Analyses of Turbot, *Scophthalmus maximus*. *Biology* **2020**, *9*, 221. [CrossRef]
3. Qin, B.; Xiao, T.; Ding, C.; Deng, Y.; Lv, Z.; Su, J. Genome-Wide Identification and Expression Analysis of Potential Antiviral Tripartite Motif Proteins (TRIMs) in Grass Carp (*Ctenopharyngodon idella*). *Biology* **2021**, *10*, 1252. [CrossRef] [PubMed]
4. Bello-Perez, M.; Adamek, M.; Coll, J.; Figueras, A.; Novoa, B.; Falco, A. Modulation of the Tissue Expression Pattern of Zebrafish CRP-Like Molecules Suggests a Relevant Antiviral Role in Fish Skin. *Biology* **2021**, *10*, 78. [CrossRef] [PubMed]

5. Pereiro, P.; Lama, R.; Moreira, R.; Valenzuela-Muñoz, V.; Gallardo-Escárate, C.; Novoa, B.; Figueras, A. Potential Involvement of lncRNAs in the Modulation of the Transcriptome Response to Nodavirus Challenge in European Sea Bass (*Dicentrarchus labrax* L.). *Biology* **2020**, *9*, 165. [CrossRef] [PubMed]
6. Perdiguero, P.; Morel, E.; Tafalla, C. Diversity of Rainbow Trout Blood B Cells Revealed by Single Cell RNA Sequencing. *Biology* **2021**, *10*, 511. [CrossRef]
7. Lv, Z.; Hu, Y.; Tan, J.; Wang, X.; Liu, X.; Zeng, C. Comparative Transcriptome Analysis Reveals the Molecular Immunopathogenesis of Chinese Soft-Shelled Turtle (*Trionyx sinensis*) Infected with *Aeromonas hydrophila*. *Biology* **2021**, *10*, 1218. [CrossRef]
8. Shen, Z.; Kumar, D.; Liu, X.; Yan, B.; Fang, P.; Gu, Y.; Li, M.; Xie, M.; Yuan, R.; Feng, Y.; et al. Metatranscriptomic Analysis Reveals an Imbalance of Hepatopancreatic Flora of Chinese Mitten Crab *Eriocheir sinensis* with Hepatopancreatic Necrosis Disease. *Biology* **2021**, *10*, 462. [CrossRef]
9. Azeredo, R.; Machado, M.; Pereiro, P.; Barany, A.; Mancera, J.M.; Costas, B. Acute Inflammation Induces Neuroendocrine and Opioid Receptor Genes Responses in the Seabass *Dicentrarchus labrax* Brain. *Biology* **2022**, *11*, 364. [CrossRef]
10. Peixoto, D.; Machado, M.; Azeredo, R.; Costas, B. Chronic Inflammation Modulates Opioid Receptor Gene Expression and Triggers Respiratory Burst in a Teleost Model. *Biology* **2022**, *11*, 764. [CrossRef]
11. Dettleff, P.; Zuloaga, R.; Fuentes, M.; Gonzalez, P.; Aedo, J.; Estrada, J.M.; Molina, A.; Valdés, J.A. High-Temperature Stress Effect on the Red Cusk-Eel (*Geypterus chilensis*) Liver: Transcriptional Modulation and Oxidative Stress Damage. *Biology* **2022**, *11*, 990. [CrossRef]
12. Naya-Català, F.; Martos-Sitcha, J.A.; de las Heras, V.; Simó-Mirabet, P.; Calduch-Giner, J.À.; Pérez-Sánchez, J. Targeting the Mild-Hypoxia Driving Force for Metabolic and Muscle Transcriptional Reprogramming of Gilthead Sea Bream (*Sparus aurata*) Juveniles. *Biology* **2021**, *10*, 416. [CrossRef]
13. Malik, A.; Kim, C.-B. Role of Transportome in the Gills of Chinese Mitten Crabs in Response to Salinity Change: A Meta-Analysis of RNA-Seq Datasets. *Biology* **2021**, *10*, 39. [CrossRef] [PubMed]
14. Li, X.; Liu, S.; Wang, Y.; Lu, W.; Zhang, Q.; Cheng, J. Genomic and Transcriptomic Landscape and Evolutionary Dynamics of Heat Shock Proteins in Spotted Sea Bass (*Lateolabrax maculatus*) under Salinity Change and Alkalinity Stress. *Biology* **2022**, *11*, 353. [CrossRef]
15. Valenzuela-Muñoz, V.; Gallardo-Escárate, C.; Benavente, B.P.; Valenzuela-Miranda, D.; Núñez-Acuña, G.; Escobar-Sepulbeda, H.; Váldes, J.A. Whole-Genome Transcript Expression Profiling Reveals Novel Insights into Transposon Genes and Non-Coding RNAs during Atlantic Salmon Seawater Adaptation. *Biology* **2022**, *11*, 1. [CrossRef] [PubMed]
16. Shwe, A.; Krasnov, A.; Visnovska, T.; Ramberg, S.; Østbye, T.-K.K.; Andreassen, R. Expression Analysis in Atlantic Salmon Liver Reveals miRNAs Associated with Smoltification and Seawater Adaptation. *Biology* **2022**, *11*, 688. [CrossRef]
17. Zhu, J.; Lei, L.; Chen, C.; Wang, Y.; Liu, X.; Geng, L.; Li, R.; Chen, H.; Hong, X.; Yu, L.; et al. Whole-Transcriptome Analysis Identifies Gender Dimorphic Expressions of mRNAs and Non-Coding RNAs in Chinese Soft-Shell Turtle (*Pelodiscus sinensis*). *Biology* **2022**, *11*, 834. [CrossRef]
18. Wang, Y.; Luo, X.; Qu, C.; Xu, T.; Zou, G.; Liang, H. The Important Role of Sex-Related Sox Family Genes in the Sex Reversal of the Chinese Soft-Shelled Turtle (*Pelodiscus sinensis*). *Biology* **2022**, *11*, 83. [CrossRef]
19. Luo, L.-F.; Xu, Z.-S.; Elgazzar, E.A.B.E.; Du, H.; Li, D.-Y.; Zhou, X.-Y.; Gao, Z.-X. Comparative Transcriptome Analysis Revealed Genes Involved in Sexual and Polyploid Growth Dimorphisms in Loach (*Misgurnus anguillicaudatus*). *Biology* **2021**, *10*, 935. [CrossRef]
20. Wang, H.; Bruce, T.J.; Su, B.; Li, S.; Dunham, R.A.; Wang, X. Environment-Dependent Heterosis and Transgressive Gene Expression in Reciprocal Hybrids between the Channel Catfish *Ictalurus punctatus* and the Blue Catfish *Ictalurus furcatus*. *Biology* **2022**, *11*, 117. [CrossRef]
21. Zhang, Y.; Li, Q-S.; Ye, Y.-Q.; Wang, Q.; Sun, X.-Q.; Zhao, R.; Li, J.-T. Association Analysis between Genetic Variants of *elovl5a* and *elovl5b* and Poly-Unsaturated Fatty Acids in Common Carp (*Cyprinus carpio*). *Biology* **2022**, *11*, 466. [CrossRef] [PubMed]
22. Katan, T.; Xue, X.; Caballero-Solares, A.; Taylor, R.G.; Parrish, C.C.; Rise, M.L. Influence of Varying Dietary ω6 to ω3 Fatty Acid Ratios on the Hepatic Transcriptome, and Association with Phenotypic Traits (Growth, Somatic Indices, and Tissue Lipid Composition), in Atlantic Salmon (*Salmo salar*). *Biology* **2021**, *10*, 578. [CrossRef] [PubMed]
23. Guerrero-Peña, L.; Suarez-Bregua, P.; Méndez-Martínez, L.; García-Fernández, P.; Tur, R.; Rubiolo, J.A.; Tena, J.J.; Rotllant, J. Brains in Metamorphosis: Temporal Transcriptome Dynamics in Hatchery-Reared Flatfishes. *Biology* **2021**, *10*, 1256. [CrossRef]
24. Chakraborty, S.; Woldemariam, N.T.; Visnovska, T.; Rise, M.L.; Boyce, D.; Santander, J.; Andreassen, R. Characterization of miRNAs in Embryonic, Larval, and Adult Lumpfish Provides a Reference miRNAome for *Cyclopterus lumpus*. *Biology* **2022**, *11*, 130. [CrossRef] [PubMed]
25. Antiqueo, P.; Zuloaga, R.; Bastias-Molina, M.; Meneses, C.; Estrada, J.M.; Molina, A.; Valdés, J.A. De novo Assembly and Analysis of Tissue-Specific Transcriptomes of the Edible Red Sea Urchin *Loxechinus albus* Using RNA-Seq. *Biology* **2021**, *10*, 995. [CrossRef]
26. Zhou, J.-J.; Chang, Y.-J.; Chen, Y.-L.; Wang, X.-D.; Liao, Q.; Shi, R.-H.; Gao, Z.-X. Comparison of Myosepta Development and Transcriptome Profiling between Blunt Snout Bream with and Tilapia without Intermuscular Bones. *Biology* **2021**, *10*, 1311. [CrossRef]
27. Cui, M.-S.; Zhao, R.; Wang, Q.; Zhang, Y.; Li, Q.-S.; Huang Yang, M.-D.; Sun, X.-Q.; Li, J.-T. Bulked Segregant Analysis and Association Analysis Identified the Polymorphisms Related to the Intermuscular Bones in Common Carp (*Cyprinus carpio*). *Biology* **2022**, *11*, 477. [CrossRef]

biology 2020, 9, 337

MDPI

Article

Interactions between the Parasite *Philasterides dicentrarchi* and the Immune System of the Turbot *Scophthalmus maximus*. A Transcriptomic Analysis

Alejandra Valle [1], José Manuel Leiro [2], Patricia Pereiro [3], Antonio Figueras [3], Beatriz Novoa [3], Ron P. H. Dirks [4] and Jesús Lamas [1,*]

1 Department of Fundamental Biology, Institute of Aquaculture, Campus Vida, University of Santiago de Compostela, 15782 Santiago de Compostela, Spain; alejandra.valle.cao@usc.es
2 Department of Microbiology and Parasitology, Laboratory of Parasitology, Institute of Research on Chemical and Biological Analysis, Campus Vida, University of Santiago de Compostela, 15782 Santiago de Compostela, Spain; josemanuel.leiro@usc.es
3 Institute of Marine Research, Consejo Superior de Investigaciones Científicas-CSIC, 36208 Vigo, Spain; patriciapereiro@iim.csic.es (P.P.); antoniofigueras@iim.csic.es (A.F.); beatriznovoa@iim.csic.es (B.N.)
4 Future Genomics Technologies, Leiden BioScience Park, 2333 BE Leiden, The Netherlands; dirks@futuregenomics.tech
* Correspondence: jesus.lamas@usc.es; Tel.: +34-88-181-6951; Fax: +34-88-159-6904

Received: 4 September 2020; Accepted: 14 October 2020; Published: 15 October 2020

Simple Summary: *Philasterides dicentrarchi* is a free-living ciliate that causes high mortality in marine cultured fish, particularly flatfish, and in fish kept in aquaria. At present, there is still no clear picture of what makes this ciliate a fish pathogen and what makes fish resistant to this ciliate. In the present study, we used transcriptomic techniques to evaluate the interactions between *P. dicentrarchi* and turbot leucocytes during the early stages of infection. The findings enabled us to identify some parasite genes/proteins that may be involved in virulence and host resistance, some of which may be good candidates for inclusion in fish vaccines. Infected fish responded to infection by generating a very potent inflammatory response, indicating that the fish use all of the protective mechanisms available to prevent entry of the parasite. The findings also provide some valuable insight into how the acute inflammatory response occurs in fish.

Abstract: The present study analyses the interactions between *Philasterides dicentrarchi* (a ciliate parasite that causes high mortalities in cultured flatfish) and the peritoneal cells of the turbot *Scophthalmus maximus* during an experimental infection. The transcriptomic response was evaluated in the parasites and in the fish peritoneal cells, at 1, 2 and 4 h post-infection (hpi) in turbot injected intraperitoneally (ip) with 10^7 ciliates and at 12 and 48 hpi in turbot injected ip with 10^5 ciliates. Numerous genes were differentially expressed (DE) in *P. dicentrarchi*, relative to their expression in control ciliates (0 hpi): 407 (369 were up-regulated) at 1 hpi, 769 (415 were up-regulated) at 2 hpi and 507 (119 were up-regulated) at 4 hpi. Gene ontology (GO) analysis of the DE genes showed that the most representative categories of biological processes affected at 1, 2 and 4 hpi were biosynthetic processes, catabolic processes, biogenesis, proteolysis and transmembrane transport. Twelve genes of the ABC transporter family and eight genes of the leishmanolysin family were DE at 1, 2 and 4 hpi. Most of these genes were strongly up-regulated (UR), suggesting that they are involved in *P. dicentrarchi* infection. A third group of UR genes included several genes related to ribosome biogenesis, DNA transcription and RNA translation. However, expression of tubulins and tubulin associated proteins, such as kinesins or dyneins, which play key roles in ciliate division and movement, was down-regulated (DR). Similarly, genes that coded for lysosomal proteins or that participate in the cell cycle mitotic control, glycolysis, the Krebs cycle and/or in the electron transport chain were also DR. The transcriptomic analysis also revealed that in contrast to many parasites, which passively evade the host immune system, *P. dicentrarchi* strongly stimulated turbot peritoneal cells. Many genes related to inflammation were DE in peritoneal cells at 1, 2 and 4 hpi. However, the response was much lower at 12 hpi and almost disappeared completely at 48 hpi in fish that were able to kill *P. dicentrarchi* during the first few hpi. The genes that were DE at 1, 2 and 4 hpi were mainly related to the apoptotic process, the immune response, the Fc-epsilon receptor signalling pathway, the innate immune response, cell adhesion, cell surface receptors, the NF-kappaB signalling pathway and the MAPK cascade. Expression of toll-like receptors 2, 5 and 13 and of several components of NF-κB, MAPK and JAK/STAT signalling pathways was UR in the turbot peritoneal cells. Genes expressing chemokines and chemokine receptors, genes involved in prostaglandin and leukotriene synthesis, prostaglandins, leukotriene receptors, proinflammatory cytokines and genes involved in apoptosis were strongly UR during the first four hours of infection. However, expression of anti-inflammatory cytokines such as Il-10 and lipoxygenases with anti-inflammatory activity (i.e., *arachidonate 15-lipoxygenase*) were only UR at 12 and/or 48 hpi, indicating an anti-inflammatory state in these groups of fish. In conclusion, the present study shows the regulation of several genes in *P. dicentrarchi* during the early stages of infection, some of which probably play important roles in this process. The infection induced a potent acute inflammatory response, and many inflammatory genes were regulated in peritoneal cells, showing that the turbot uses all the protective mechanisms it has available to prevent the entry of the parasite.

Keywords: *Philasterides dicentrarchi*; turbot; immune response; transcriptomics; infection

---

1. Introduction

The subclass Scuticociliatia (Protozoa, Ciliophora) includes several families of free-living ciliates that are abundant in marine habitats. Some scuticociliate species, including *P. dicentrarchi*, have acquired the capacity to infect fish, causing high mortalities in some cultured fish species, including flatfish [1–5]. However, there is no clear picture of what makes this scuticociliate a fish pathogen. It is generally thought that *P. dicentrarchi* enters the fish through external lesions and then proliferates in the blood and in most internal organs where it causes important histopathological changes [1,3,6–9]. Although the mechanisms that *P. dicentrarchi* uses to invade fish tissues are not known, it has been suggested that during infection the ciliate can release proteases, which are considered virulence factors [10]. Ciliate proteases can destroy components of the turbot humoral immune response [11,12] and modify fish leukocyte functions [13,14], thus providing a mechanism for circumventing the fish immune system. In addition, *P. dicentrarchi* has developed other mechanisms that appear to be important in relation to parasitism. For example, it possesses mitochondria with two respiratory pathways: the cytochrome pathway, which exists in all aerobic organisms, and an alternative, cyanide-insensitive oxidase pathway, which enables the ciliate to survive and proliferate under normoxic and hypoxic conditions [15] and thus to become adapted to differences in oxygen levels in the host and seawater. In addition, *P. dicentrarchi* has potent antioxidant defence mechanisms that may be important during infection, including several superoxide dismutases that help the ciliate resist the reactive oxygen species released by the host [16]. Finally, it has been suggested that *P. dicentrarchi* can release its extrusome contents to create a protective barrier against soluble factors of the host immune system [17].

Several components of *P. dicentrarchi* stimulate fish leucocytes, thereby increasing respiratory burst, degranulation and the expression of pro-inflammatory cytokines. Despite the high level of stimulation, the toxic substances produced by the fish leucocytes do not seem to be sufficient to kill the parasite [11,18]. However, a turbot NK-lysin (an effector molecule of cytotoxic lymphocytes) has been reported to have very high antiparasitic activity, thus directly affecting the viability of *P. dicentrarchi* [19]. The innate humoral and adaptive immune responses appear to be crucial in defending the fish against this parasite. In this respect, the complement system plays a key role in the fish defence against this pathogen, as it displays potent antiparasitic activity, especially when activated by the classical pathway [11,20]. In addition to complement, the fish coagulation system also plays an important role in immobilizing and killing the parasite [21].

Little is known about how the fish immune system provides protection against parasites or even how it recognises the parasites (particularly ciliates) or which immune signalling pathways are stimulated. Pardo et al. [22,23] investigated how *P. dicentrarchi* stimulates the immune system of fish, by analysing the changes in gene expression in organs of the turbot immune system (spleen, liver and kidney) during a *P. dicentrarchi* infection. These researchers found that many genes regulating substances involved in the immune response, including chemokines, chemotaxins, complement, immunoglobulins, major histocompatibility complex, interferon, lectins, cytochrome P450 and lysozyme, were differentially expressed and that the response was strongest in the spleen at 1- and 3- days post-infection. Most transcriptomic studies of the immune response in fish infected with ciliates have been carried out with ectoparasites such as *Cryptocaryon irritans* and *Ichthyophthirius multifiliis* [24–27], usually with an infection time longer than 12 h. In addition, previous transcriptomic studies have analysed the response in either the parasite or in the cells of the fish immune system, but not in both. The scuticociliate *P. dicentrarchi* can overcome the external barriers of fish and proliferate in the blood and in internal organs, causing a systemic infection. The initial interaction between the ciliate and fish immune system may be crucial to the success of infection. In the present study, we used transcriptomic technologies to analyse the interactions between *P. dicentrarchi* and the turbot immune system during the very early stages of infection. Parasites were injected intraperitoneally and the transcriptomic response was evaluated in both fish cells and parasites found in the peritoneal cavity at several different times. This infection model enabled us to obtain sufficient numbers of host cells and parasites for the analysis.

2. Material and Methods

2.1. Fish and Ethical Statement

Healthy turbot, *Scophthalmus maximus* (L.), weighing about 50 g, were obtained from a local fish farm and maintained in 250 L tanks with recirculating and aerated seawater at 18 °C and fed daily with commercial pellets. All experimental protocols carried out in the present study followed the European legislation (Directive 2010/63/EU) and the Spanish legislative requirements related to the use of animals for experimentation (RD 53/2013) and were approved by the Institutional Animal Care and Use Committee of the University of Santiago de Compostela (Spain) (Ethic code: AGL2017-83577) on 01-02-2019. Before any experimental manipulation, fish were anaesthetized by immersion in a 100 mg/L solution of MS-222 (tricaine methane sulfonate; Sigma-Aldrich, Madrid, Spain) in seawater. At the end of the experiments, the fish were fully anaesthetized before being killed by pithing.

2.2. Parasites

Specimens of *Philasterides dicentrarchi* (isolate I1), obtained from experimentally infected turbot, were cultured as previously indicated [11]. Briefly, parasites were cultured at 18 °C in flasks containing L-15 Leibovitz medium with 10% heat-inactivated foetal bovine serum, lipids (lecithin and Tween 80), nucleosides and glucose. To prepare the parasites, the ciliates were collected from the flasks, centrifuged at 700× *g* for 5 min, washed twice in phosphate-buffered saline (PBS) and resuspended in the same buffer.

2.3. Experimental Infection with *P. dicentrarchi*

Immediately prior to the present study, and using the same groups of fish, we found that with an inoculum dose of 10^7 parasites per fish, a high percentage of the ciliates remained alive in the peritoneal cavity. However, with an inoculum dose of 10^5 parasites per fish, all the ciliates died in the peritoneal cavity. Both concentrations of parasites were used in the present study. The higher concentration (10^7 ciliates per fish) was used to investigate the gene expression in the parasites and in turbot peritoneal cells at 0, 1, 2, and 4 post-infection (hpi). The lower concentration was used to analyse the gene expression in turbot peritoneal cells at 12 and 48 hpi, comparing the response between fish injected with ciliates and fish injected with PBS. This ciliate concentration was selected because most fish cells would be engulfed and killed at 12 and 48 hpi if the higher concentrations were used.

Two parallel experiments were carried out to evaluate the changes in gene expression in P. dicentrarchi and turbot peritoneal cells during experimental infection. In one experiment, 36 fish were each injected intraperitoneally with 1×10^7 ciliates (3 replicates of three fish each per time point were used; 0, 1, 2 and 4 hpi) (Table S1). Ciliates and peritoneal cells were obtained by washing the peritoneal cavity with 5 mL of PBS. The ciliates and peritoneal cells were then counted in a Neubauer counting chamber and concentrated by centrifugation (400× *g*, 10 min). The pellet thus obtained was then diluted in 1 mL of RNAlater solution and incubated overnight at 4 oC. The samples were diluted in cold PBS (50%) and concentrated by centrifugation at 12,000× *g* for 10 min. The pellets were then frozen in liquid nitrogen and sent on dry ice to Future Genomics Technologies (Leiden, The Netherlands) for analysis.

In the second experiment, 42 fish were injected with 1×10^5 ciliates and 42 fish were injected with the corresponding volume of PBS (as controls) (Table S1). Three replicates of 7 fish each were analysed at each time point (12 and 48 hpi). The peritoneal cells were obtained and processed as described above.

In addition, a parallel experiment with the same groups of fish (3 fish per sampling time) and the same concentration of ciliates described above was used to validate the RNAseq.

2.4. RNA Extraction and RNAseq Analysis

Total RNA was extracted from samples preserved in RNA*later*, by using the TissueRuptor homogenizer (Qiagen Iberia, Madrid, Spain) and the miRNeasy mini kit (Qiagen Iberia, Madrid, Spain) according to the manufacturer's instructions. RNA was eluted in RNAse-free water and quantified on a total RNA 6000 Nano series II chip (Agilent, Santa Clara, CA, USA) in an Agilent Bioanalyzer 2100 device. Illumina RNAseq libraries (150–750 bp inserts) were prepared, from 2 mg total RNA, using the TruSeq stranded mRNA library prep kit according to the manufacturer's instructions (Illumina Inc., San Diego, CA, USA). The quality of the RNAseq libraries was checked using a DNA 1000 chip (Agilent, Santa Clara, CA, USA) and a Bioanalyzer (Agilent 2100, Santa Clara, CA, USA). RNAseq libraries were sequenced on an Illumina HiSeq2500 sequencer, as 2×150 nucleotides paired-end reads, according to the manufacturer's protocol. Image analysis and base calling were performed using the standard Illumina pipeline. The read sequences were deposited in the Sequence Read Archive (SRA) (http://www.ncbi.nlm.nih.gov/sra); BioProject accession number: PRJNA648859.

2.5. Raw Data Cleaning, De Novo Assembly, and Gene Annotation

CLC Genomics Workbench v. 10.0.1 (CLC Bio, Aarhus, Denmark) was used for filtration and assembly and to perform the RNAseq and statistical analyses. Prior to assembly, the raw data from each sample were trimmed to remove adapter sequences and low-quality reads (quality score limit 0.05). One de novo assembly was conducted for each challenge experiment (1×10^7 ciliates/fish or 1×10^5 ciliates / fish). Although the turbot genome is available [28], a de novo assembly strategy was applied to analyse the *P. dicentrarchi* transcriptome. Therefore, for each experiment, all the high-quality reads were assembled in a unique file using default parameters (mismatch cost = 2, insert cost = 3, minimum contig length = 200 bp, and similarity = 0.8). The contigs yielded by these assemblies were annotated with the Blast2GO program (https://www.blast2go.com/) against the UniProtKB/SwissProt database (http://UniProt.org), applying a cut-off E-value of 1×10^{-3}.

2.6. RNAseq and Differential Gene Expression Analysis

The transcriptome database generated for each experiment was used as a reference for the RNAseq analysis. Expression levels were calculated as transcripts per million (TPM) values. Finally, a differential expression analysis test (a Robinson and Smyth's Exact Test, which assumes a Negative Binomial distribution of the data and considers the overdispersion caused by biological variability) was used to identify the DE genes. In the first experiment, the parasites at time 0 h were considered control samples, and the following comparisons were conducted: 1 hpi vs. 0 hpi, 2 hpi vs. 0 hpi and 4 hpi vs. 0 hpi. In the second experiment, the samples obtained from the infected fish were compared with the samples from the turbot injected with PBS (controls) at the corresponding sampling points. Contigs showing a >2 fold change in the absolute value relative to the control group and a Bonferroni corrected p-value < 0.05 were considered DE genes.

The first experiment included a mixture of contigs belonging to the parasite and contigs belonging to turbot. Two strategies were used to differentiate the contigs: (1) the DE contigs were mapped against the turbot genome [28] and those contigs with a high identity value (above 85%) were considered of piscine origin; (2) the remaining contigs were analysed by multiple BLASTx alignment (https://blast.ncbi.nlm.nih.gov/Blast.cgi) and were identified as *P. dicentrarchi* or turbot contigs on the basis of the number of hits for the preferred annotation (invertebrates vs. vertebrates).

2.7. Gene Ontology (GO) Assignment and Enrichment Analysis

The GO term assignments of the contig lists were obtained from the Uniprot/Swissprot BLASTx results, with Blast2GO software [29]. GO categorization of the *P. dicentrarchi* DE contigs was conducted on the basis of the biological process terms at level 2. For enrichment analysis of the turbot DE contigs (for biological processes), a Fisher's exact test was run with default values and an FDR cut-off of 0.05 was applied.

2.8. RNAseq Validation by qPCR

Transcript abundance in the samples was estimated by reverse transcription-quantitative PCR (qPCR). RNA was extracted from fish tissue with TRI Reagent® (Sigma-Aldrich, Madrid, Spain), according to the manufacturer's instructions. The resulting RNA was dried, dissolved in RNase-free water and quantified in a NanoDrop ND-1000 (Thermo Fisher Scientific Inc., Wilmington, DE, USA) spectrophotometer. To prevent genomic DNA contamination, the total RNA was treated with DNAase I (Thermo Scientific, Surrey, UK). cDNA was synthesized using the cDNA synthesis kit (NZYTech, Lisboa, Portugal), with 1 µg of sample RNA. cDNA was amplified with a qPCR reaction mixture (NZYTech, Lisboa, Portugal) and 0.3 µM of each specific primer (Table S2). For each gene, primer efficiency was determined as indicated by [30]; the expression of each gene was determined in each fish, in triplicate. A total of 5 genes in *P. dicentrarchi* (*cathepsin B*, *leishmanolysin 1589*, *leishmanolysin 17670*, *ABC transporter G family member 10* and *ABC transporter G family member 11*) and 5 genes in turbot (*tumor necrosis factor alpha*, *cd11b*, *stat6*, *c-x-c motif chemokine ligand 8* and *interleukin 1 beta*) were used to validate the RNAseq. In *P. dicentrarchi* and turbot, respectively, two genes (*β-actin* and *elongation factor 1-alpha* (*ef1α*)) and three genes (*β-actin*, *glyceraldehyde-3-phosphate dehydrogenase* and *ef1α*) were tested as candidate housekeeping genes. *Ef1-α* was identified as the most stable in both cases and was therefore selected as the reference gene for qPCR analysis. The qPCR analysis was carried out as previously described [31]. Relative gene expression was quantified by the $2^{-\Delta\Delta C}$t method [32] applied with software conforming to minimum information for publication of qRT-PCR experiments (MIQE) guidelines [33]. The results are presented as the normalized expression in experimental groups, divided by the normalized expression in the control group. The concordance between RNAseq and qPCR data was evaluated using Spearman's correlation coefficient and the null hypothesis of no differences between both methods was checked by a Wilcoxon–Mann–Whitney U test. The critical value was adjusted to $p \leq 0.05$, as described by [30].

2.9. Statistics

The values shown in the figures are means ± standard deviation (SD). Significant differences ($p \leq 0.05$) were determined by analysis of variance (ANOVA) followed by Tukey–Kramer multiple comparisons test.

3. Results

3.1. RNAseq Analysis and Validation

In the first experiment, in which the fish were injected with 10^7 ciliates, about 2–2.5 × 10^6 ciliates were recovered from each fish at all sampling times. The number of free peritoneal cells was between 2 and 3 × 10^5 cells per fish at 0, 1 and 2 hpi and was slightly higher at 4 hpi (5 × 10^5 cells) (Figure S1). At the time points considered (1, 2 and 4 hpi), the ciliates contained numerous engulfed peritoneal cells, particularly at 4 hpi (Figure S2). In the second experiment, no live ciliates were found in the peritoneal cavity of turbot at the sampling times (12 and 48 hpi).

Two of the replicates, corresponding to "ciliates 4 h, replicate 1" and "ciliates 4 h, replicate 2" did not produce sufficient Illumina RNAseq library for a proper run. These samples were therefore excluded from further analysis. A total of 357.3 million reads were obtained from the 22 libraries, and the number of reads obtained per sample was between 11.8 and 22.1 million, with a mean value of 16.2 million reads per library. Samples obtained at 0, 1, 2 and 4 hpi were assembled and analysed separately from samples obtained at 12 and 48 hpi. In the former, 99,821 contigs with an N50 of ~1 kb and an average length of 691 bp were obtained, and of these, 34,034 (34.1%) contigs were successfully annotated. The numbers of differentially expressed (DE) contigs (Bonferroni ≤ 0.05; FC ≥ 2), relative to the 0 h samples, were 3101, 4376 and 3499 at 1, 2 and 4 hpi respectively. About 65% of the DE contigs were annotated, and 407, 769 and 507 of these contigs were DE in *P. dicentrarchi* at 1, 2, 4 hpi (File S1) and the others (1627, 2067 and 1822) were DE in turbot peritoneal cells (File S2), respectively. In the samples obtained at 12 and 48 hpi, in which fish injected with ciliates were compared with fish injected with PBS, 167,321 contigs with an N50 of ~1.1 kb and an average length of 740 bp were obtained; of these, 31,238 (18.87%) contigs were successfully annotated. The numbers of DE contigs (Bonferroni ≤ 0.05; FC ≥ 2) in peritoneal cells of fish injected with *P. dicentrarchi*, relative to fish injected with PBS, were 464 (370 up-regulated) at 12 hpi and 57 (15 up-regulated) at 48 hpi (File S3). In this experiment, about 61% of the DE contigs were annotated against the UniProtKB/SwissProt database.

The expression of five genes in *P. dicentrarchi* and five genes in turbot peritoneal cells was evaluated by qPCR (Figure S3), to validate the findings of the RNAseq experiment. The RNAseq and qPCR values were strongly correlated for *P. dicentrarchi* (Spearman's correlation $\rho = 0.885$) and turbot peritoneal cells (Spearman's correlation $\rho = 0.846$).

3.2. Changes in Gene Expression in P. dicentrarchi during an Experimental Infection

In order to identify which parasite genes may be important during the early stages of infection and which may play a role in virulence or in resistance against the fish immune response, the changes in gene expression in the scuticociliate were analysed during infection. Differential expression of 407 (369 up-regulated (UR)), 769 (415 UR) and 507 (119 UR) genes was observed at 1, 2 and 4 hpi, respectively.

GO analysis was carried out to explore the biological processes that were best represented during infection (Figure 1). Five categories were well represented at 1 and 2 hpi, including biosynthetic processes, catabolic processes, biogenesis, proteolysis and transmembrane transport, with a similar distribution at both times. At 4 hpi, there was a substantial increase in the percentage of genes involved in catabolic processes and proteolysis (Figure 1).

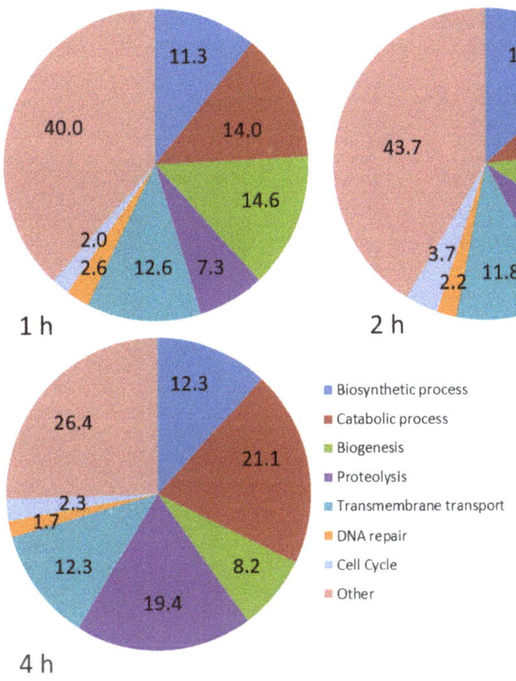

Figure 1. Pie charts showing the proportion (%) of differentially expressed genes involved in various biological processes in *P. dicentrarchi* at 1, 2 and 4 hpi, relative to the number at 0 h.

The most strongly regulated genes in *P. dicentrachi* during infection were those related to the ATP-Binding Cassette (ABC) transporter gene family, which included 12 DE genes (Table 1). Of these, *ABC transporter G family member 10* (*abcg10*) was strongly UR at 1, 2 and 4 hpi. Other genes that were UR at the three sampling times were *abca4* and *abcb4*. Six genes were DE only at 1 and 2 hpi, including three members of the G family, with *abcg14* and *abcg11* being strongly UR. Overall, the highest expression of ABC transporter family genes was found at 1 hpi. Finally, *abca3* and *abcb1* were down-regulated (DR) at 4 or at 2 and 4 hpi, respectively (Table 1).

Several leishmanolysin-like genes were also DE. Four genes of that family were UR and two genes were DR (Table 1). As *P. dicentrarchi* proteases are considered virulence factors, we were interested in determining how the genes of lysosomal enzymes would behave during the early stages of infection. The expression of several cathepsins, especially *ctsb, ctsl* and *ctsd*, was DR at 2 and 4 hpi (Table 1). In addition, several other genes that coded for lysosomal proteins, such as *lysosomal acid phosphatase* and *lysosomal alpha-mannosidase* were also DR at 2 and 4 hpi (Table 1).

Tubulins and tubulin associated proteins, such as kinesins and dyneins, play key roles in, e.g., ciliate division and movement. Apart from *kinesin family member 1*, expression of alpha and beta tubulins and many kinesins and dyneins (including the *outer dynein arm 1* gene, which is required for ciliary motion) was DR at 2 and 4 hpi (Table 1).

Table 1. Heat map showing a group of differentially expressed (DE) genes in *P. dicentrarchi*, including ABC transporters, leishmanolysins and genes related to microtubules and to lysosomes, at 1, 2 and 4 hpi. Results are expressed as mean expression ratios of groups 1, 2 and 4 hpi vs. 0 h. Red indicates increased gene expression levels; green indicates decreased levels.

Gene Name	Gen Abbrev.	1 h	2 h	4 h
ABC transporter G family member 10	*abcg10*	1547.333077	1525.204333	135.9495633
ABC transporter G family member 14	*abcg14*	155.981555	66.32506772	
ABC transporter G family member 11	*abcg11*	117.2851641	31.01838328	
ABC transporter G family member 22	*abcg22*	9.047137204	6.428493341	
ABC transporter F family member 3	*abcf3*	5.149616627	5.972621581	
ABC transporter A family member 2	*abca2*	17.05398277	7.274235391	
ABC transporter A family member 4	*abca4*	17.12815109	26.00467446	39.15870736
ABC transporter B family member 4	*abcb4*	8.243947246	3.745339799	6.402422485
ABC transporter A family member 7	*abca7*	3.043246918	3.532536485	
ABC transporter B family member 3	*abcb3*	3.36028248		
ABC transporter A family member 3	*abca3*			−7.218505508
ABC transporter B family member 1	*abcb1*		−2.872487246	−4.36433424
leishmanolysin 17670	*lmln17670*	13.02222148	56.71743416	13.79572431
leishmanolysin 4908	*lmln4908*	4.636533774	6.397241791	12.09714031
leishmanolysin 6891	*lmln6891*	6.660199714	15.9960053	
leishmanolysin 906	*lmln906*	2.789707738	3.478134169	
leishmanolysin 8301	*lmln8301*	−3.341460004	−5.425240786	
leishmanolysin 1589	*lmln1589*	−7.499602753	−19.34465739	−29.97419775
tubulin alpha chain	*tba*		−3.541692478	
tubulin beta chain	*tbb1*		−3.494382907	
dynein alpha flagellar outer arm	*dyha*			−4.485994554
dynein heavy chain axonemal heavy chain 7	*dyh7*		−2.931765009	−5.770286234
dynein heavy cytoplasmic	*dyhc1*		−3.566132897	−13.96776319
dynein light chain cytoplasmic	*dyl2*		−2.781891083	
dynein regulatory complex 1	*drc1*			−4.298212013
outer dynein arm 1	*oda1*			−4.407810139
kinesin 7l	*kn7l*		−5.659610804	
kinesin family member 1	*kif1*	15.69290132	13.68478782	
kinesin fla10	*fla10*		−3.064471017	−4.729088437
kinesin kif15	*kif15*	−2.943783569	−6.688399519	−11.43932373
kinesin-ii 95 kda subunit	*krp95*		−8.465050495	
cathepsin d	*ctsd*	3.7	−2.45	−5.7
cathepsin l	*ctsl*		−4	−18.8
cathepsin z	*ctsz*		−3.3	
cathepsin b	*ctsb*		−5.3	−41.49
lysosomal acid phosphatase	*ppal*		−5.19988665	−11.71489074
lysosomal alpha-mannosidase	*mana*		−4.216965484	−9.663142631

Ribosome synthesis plays a central role in regulating cell growth [34]. Many genes involved in ribosome biogenesis, DNA transcription and RNA translation were UR (Table 2). The strongest regulation occurred in the gene *regulator of rDNA transcription 15*. For most genes, UR occurred at 1, 2, and 4 hpi, peaking at 2 hpi (Table 2).

Table 2. Heat map showing a group of DE genes in *P. dicentrarchi*, including genes related to ribosome biogenesis, DNA transcription, cell cycle, metabolism, mitochondrial respiratory chain and detoxification, at 1, 2 and 4 hpi. Results are expressed as mean expression ratios of groups 1, 2, and 4 hpi vs. 0 h. Red indicates increased gene expression levels; green indicates decreased levels.

Gene Name	Gen Abbrev.	1 h	2 h	4 h
regulator of rdna transcription 15	*rrt15*	−19.2921801	305.2526145	14961.40172
pre-rrna-processing tsr1 homolog	*tsr1*	8.39214507	17.7662126	12.40935626
ribosome biogenesis regulatory homolog	*rrs1*	8.397288759	17.61314827	10.18017877
60s ribosome subunit biogenesis nip7	*nip7*	11.22783554	17.445352	
ribosome biogenesis bop1	*bop1*	7.627651039	13.73086419	5.8194392
ribosome production factor 1	*rpf1*	6.256019845	11.74039515	5.947145101
u3 small nucleolar rna-associated 6	*utp6*	7.9733096	17.52126323	7.607224123
eukaryotic translation initiation factor 6	*eif6*	3.03261779	4.997917367	
eukaryotic translation initiation factor 3 subunit l	*eif-3*	2.695718872	4.426628843	
atp-dependent rna helicase has1	*has1*	11.52254521	16.62575106	5.879612164
rna polymerase i subunit 1	*rpa1*	6.376072721	9.0470461	5.741738543
rna polymerase i subunit 2	*rpa2*	4.419921248	7.480794029	
cyclin-dependent kinase 2	*cdk2*	179.5992883	233.6336546	119.049648
cyclin-dependent kinase 1	*cdk1*	−3.206564057	−9.082843233	−8.784823344
cyclin-b2-3	*ccnb23*		−2.98467857	−5.526397608
cyclin-dependent kinases regulatory subunit 2	*ppp2r1a*		−4.068280735	−4.813378142
centrosomal of 78 kda	*cep78*		−4.499861387	−7.279573252
triose-phosphate isomerase	*tpi1*			−5.037605277
isocitrate dehydrogenase	*idhp*		−3.619776764	−18.74348161
citrate synthase	*cs*		−3.677561365	−12.28402905
acetyl-coenzyme a synthetase	*acsa*		−4.375876016	−8.810778878
cytochrome c mitochondrial	*ccpr*		−3.056139192	−7.538025415
cytochrome c oxidase subunit 1	*cox1*		−14.4260752	
cytochrome c oxidase subunit 2	*cox2*	−27.83893378		
hypoxia up-regulated 1	*hyou1*	−3.4739462	−8.629031534	−8.645030215
cytochrome p450 3a19	*cyp3a19*			−22.82674295
cytochrome p450 4b1	*cyp4b1*		−3.684326503	
cytochrome p450 4e3	*cyp4e3*			−28.32575971
glutathione s-transferase 2	*gstm2*		−5.124694687	−15.81386515
glutathione s-transferase 3	*gst3*		−5.175570202	−24.89756053
glutathione s-transferase theta-2b	*gst*		−2.786019803	
trichocyst matrix t1-b	*t1-b*		−13.95	−14.54
trichocyst matrix t2-a	*t2-a*	−3.67	−12.13	−13.88
trichocyst matrix t4-b	*t4-b*	−3.92	−8	−14.84

Analysis of the genes involved in the cell cycle showed that the *cyclin-dependent kinase 1* (*cdk1*) gene, which is involved in mitotic control, was DR. Similarly, *cyclin-B2-3* and *cyclin-dependent kinase regulatory subunit 2*, which participate in the cell cycle, were also DR. By contrast, *cyclin-dependent kinase 2* (*cdk2*), which participates in G1-S transition and the DNA damage response, was strongly UR at 1, 2 and 4 hpi (Table 2).

Genes involved in glycolysis, such as *triose-phosphate isomerase*, or in the Krebs cycle, such as *isocitrate dehydrogenase*, *citrate synthase* and *acetyl-coenzyme A synthetase*, and components of the electron transport chain, such as mitochondrial *cytochrome c*, were DR at 2 and 4 hpi. Other genes involved in the mitochondrial electron transport chain, such as *cytochrome c oxidase subunit 1* and *2*, were also DR at 2 and 1 hpi, respectively (Table 2). Finally, genes coding for enzymes of the microsomal compartment possibly involved in detoxification processes were also DR, including several cytochrome p450 (*cyp3a19*, *cyp4b1*, *cyp4e3*) and *glutathione S-transferase 2* and *3* genes (Table 2).

3.3. *P. dicentrarchi Generates an Intense Inflammatory Response in Turbot during the Aarly Stages of Infection*

We observed intense regulation of many genes of peritoneal cells at 1, 2 and 4 hpi. The response was much lower at 12 hpi and almost disappeared completely at 48 hpi in those fish that were able to kill *P. dicentrarchi* during the first few hours after injection. To analyse their function, DE genes were subjected to gene ontology and enrichment analysis for biological processes, molecular function and cellular components. The proportion of genes involved in different biological processes was very similar across groups at 1, 2 and 4 hpi (Figure 2). DEGs at 1, 2 and 4 hpi were associated with the apoptotic process (between 7.9 and 8.4%), inflammatory response (between 4.3 and 4.8%), immune response (3.4 to 5.8%), Fc-epsilon receptor signalling pathway (4.3 to 4.6%), innate immune response (3.5 to 3.6%), cell adhesion (3.3 to 3.8%), phagocytosis (2.8 to 3.1%) and the cell surface receptor signalling pathway (2.5–2.7%) (Figure 2). Many genes of signalling pathways or cascades were also DE, such as those involved in the MAPK cascade (3.6 to 3.8%), the NIK/NF-kappaB signalling pathway (1.7 to 1.9%) and the I-kappaB kinase/NF-kappaB signalling pathway (1 to 1.4%). However, the proportion of genes involved in the defence response to bacterium (2.3 to 2.9%) was higher than the number of genes involved in the defence response to protozoan (0.5 to 0.6%) (Figure 2).

Injection of ciliates into the peritoneal cavity generated a very potent inflammatory response in turbot, with the peritoneal cells expressing genes related to this response, even though many of them were already phagocytosed by the ciliate. One of the first events that occur during the host response to pathogens is recognition by host pattern recognition receptors. In peritoneal cells, *toll-like receptor* (*tlr*) *2* was UR at 1 and 2 hpi, *tlr13* was UR at 1, 2, 4 and 12 hpi, and *tlr5* was UR at 2, 4 and 12 hpi. Of these three genes, *tlr5* was the most strongly regulated (Table 3). Adaptor *myd88*, which is involved in almost all TLRs signalling pathways, was UR. Other genes that were DE included *interleukin-1 receptor-associated kinase 4* (*irak4*), several other components of the NF-κB pathway, including *nuclear factor nf-kappa-b p100 subunit* (*nfkb2*) or *nf-kappa-b inhibitor alpha* (*nfkbia*) or *map kinase kinase kinase 14* (*map3k14*). In addition, several components of the MAPK and JAK/STAT pathways were also DE (Table 3). Except for *mapk6*, which was also UR at 12 hpi, the other components of the MAPK pathway were differentially expressed at 1, 2 and 4 hpi, and most of them peaked at 4 hpi. In the genes related to the JAK/STAT pathway, *stat1*, *stat3*, *stat4* and *stat6* or *jak1* and *jak3* were UR at 1, 2 and 4 hpi, peaking at 4 hpi. However, negative regulators of the pathway, *socs1* and *socs3*, were also UR at 1, 2, 4 and 12 hpi (Table 3).

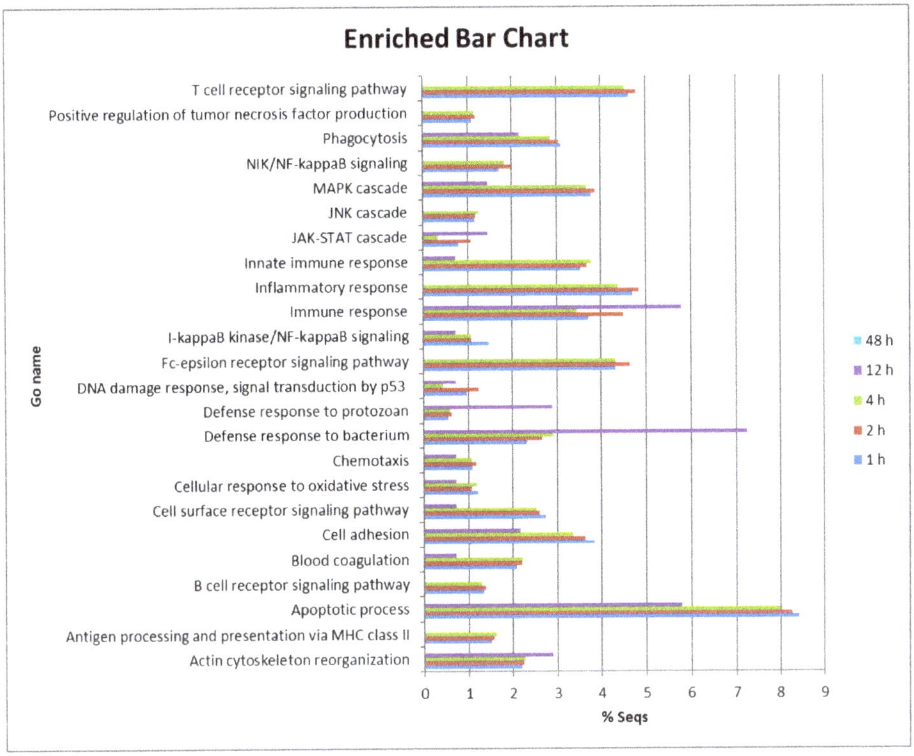

Figure 2. Bar chart showing the proportion (%) of DE genes involved in various biological processes in turbot peritoneal cells at 1, 2 and 4 hpi.

During infection by *P. dicentrarchi*, several chemokines were DE, including *c-x-c motif chemokine ligand 8* (*cxcl8*) and *c-c motif chemokines* (*ccl*) *11*, *3*, *4*, *2* and *20*. *Cxcl8* was strongly expressed at 1, 2 and 4 hpi, but was DR at 48 hpi, and *ccl20* was only UR at 2 hpi. In addition, several chemokine receptors were also UR, including *c-x-c chemokine receptor type 4*, *1*, *3*, *3-2*, and *2* and *chemokine-like receptor 1* (*cmklr1*). *Cxcr4* was most strongly expressed at 1, 2 and 4 hpi, and *cxcr1* and *cmklr1* were UR at 1, 2, 4 and 12 hpi (Table 4).

The genes *prostaglandin g h synthase 2* and *prostaglandin e synthase 3*, which code for enzymes involved in the synthesis of prostaglandin E2, were up-regulated, the former at 1, 2 and 4 hpi, and the latter at 1 hpi (Table 4). Genes coding for enzymes involved in the synthesis of leukotrienes, i.e., *12s-lipoxygenase* and *leukotriene a-4 hydrolase*, were UR at 1, 2 and 4 hpi. Positive differential expression of prostaglandin and leukotriene receptors, i.e., *leukotriene b4 receptor 1* and *prostaglandin i2 receptor*, was also observed. Finally, *arachidonate 5-lipoxygenase* was only UR at 12 and 48 hpi and *arachidonate 15-lipoxygenase* only at 48 hpi (Table 4).

Table 3. Heat map showing a group of DE genes in turbot peritoneal cells at 1, 2, 4 and 12 hpi, including genes coding for toll-like receptors and genes involved in different signalling pathways. Results are expressed as mean expression ratios of groups 1, 2 and 4 hpi vs. 0 h, or 12 hpi vs. phosphate-buffered saline (PBS). There were no DE genes at 48 hpi. Red indicates increased gene expression levels; green indicates decreased levels.

Gene Name	Gene Abbrev.	1 h	2 h	4 h	12 h
toll-like receptor 13	*tlr13*	33.5730	84.0101	252.8270	23.7867
toll-like receptor 2 type-1	*tlr2-1*	26.4006	33.7079		
toll-like receptor 5	*tlr5*		8963.1958	13,251.6757	27.6041
myeloid differentiation primary response 88	*myd88*	62.1266	170.5665	380.9293	
interleukin-1 receptor-associated kinase 4	*irak4*		283.4938	349.8690	
nuclear factor nf-kappa-b p100 subunit	*nfkb2*		349.8690	142.5494	3.5853
nf-kappa-b inhibitor alpha	*nfkbia*	46.3156	554.3636	72.7335	20.6977
nf-kappa-b inhibitor epsilon	*nfkbie*	38.6016	54.1131	72.7335	
map kinase-activated kinase 2	*mapkapk2*	78.9174	312.7962	458.7292	
map kinase 3	*mapk3*		268.6767	444.6294	
map kinase kinase kinase 4	*map2k4*	113.5394	171.4468		
map kinase 12	*mapk12*	100.5704	169.6868	176.4579	
map kinase 14	*mapk14*	69.3530	150.9215	115.8456	10.5737
map kinase 6	*mapk6*	36.5802	123.2329	464.0316	
map kinase 2	*mapk2*		65.5898		
map kinase kinase 3	*map3k3*		32.1681		
map kinase kinase kinase 4	*map3k4*		0.0000	5291.6003	
mitogen-activated kinase kinase kinase 14	*map3k14*	45.4755	63.9774	190.0614	
mitogen-activated kinase kinase kinase 8	*map3k8*	79.1107	137.7203	292.0333	4.8744
signal transducer and activator of transcription 1	*stat1*	44.2231	76.8213	145.9930	
signal transducer and activator of transcription 3	*stat3*		33.8213	70.3976	
signal transducer and activator of transcription 4	*stat4*	16.6058	37.3109	33.8135	
signal transducer and activator of transcription 6	*stat6*	46.6821	67.9804	168.4172	
janus kinase 1	*jak1*	29.7269	75.7555	238.0845	
janus kinase 2	*jak2*	52.6834	33.4907	133.9884	
suppressor of cytokine signaling 1	*socs1*	30.3920	54.9888	358.7853	26.6144
suppressor of cytokine signaling 3	*socs3*	76.5795	194.8081	787.6507	29.6728

Table 4. Heat map showing a group of DE genes in turbot peritoneal cells at 1, 2, 4, 12 and 48 hpi, including several genes coding for chemokines, chemokine receptors and enzymes involved in prostaglandin synthesis. Results are expressed as mean expression ratios of groups 1, 2 and 4 hpi vs. 0 h, or 12 and 48 hpi vs. PBS. Red indicates increased gene expression levels; green indicates decreased levels.

Gene Name	Gene Abbrev.	1 h	2 h	4 h	12 h	48 h
c-x-c motif chemokine ligand 8	*cxcl8*	696.184767	4267.68045	6208.1874		−4.0735920
c-c motif chemokine 11	*ccl11*	126.694565	162.000922	55.8971143		
c-c motif chemokine 3-like 1	*ccl3l1*	33.0565132	281.708153	177.056028		
c-c motif chemokine 4	*ccl4*	24.6363172	22.0159395		5.04909594	
c-c motif chemokine 2	*ccl2*	45.3990169	46.1844289			
c-c motif chemokine 20	*ccl20*		448.823876			
c-x-c chemokine receptor type 4	*cxcr4*	3607.59607	8791.52888	4637.55324		
chemokine-like receptor 1	*cmklr1*	990.041628	1608.67403	3857.3838	6.59418415	
c-x-c chemokine receptor type 1	*cxcr1*	835.451583	3172.20243	5777.64259	11.6779436	
c-x-c chemokine receptor type 2	*cxcr2*	75.7457614	163.539151	575.328912	27.9524211	
c-x-c chemokine receptor type 3	*cxcr3*	249.193869	1113.09829	906.716695		
c-x-c chemokine receptor type 3-2	*cxr3-2*	103.928423	81.3886854	26.2114885		
prostaglandin g h synthase 2	*ptgs2*	81.9664047	624.749387	418.015344		
prostaglandin e synthase 3	*ptges3*	19.3125976				
prostaglandin i2 synthase	*ptgis*		80.9131124			
arachidonate 12s-lipoxygenase	*alox12*	93.5362347	145.035988	386.109999		
arachidonate 5-lipoxygenase	*alox5*				9.32644124	56.2132685
arachidonate 15-lipoxygenase	*alox15b*					164.506767
leukotriene a-4 hydrolase	*lkha4*	19.7350686	59.5675737	155.71424		
leukotriene b4 receptor 1	*ltb4r*	67.2372589	510.252678	866.318345	5.73528823	
prostaglandin i2 receptor	*pi2r*	47.2490711				

Several interleukins (IL) and IL receptors were also DE during *P. dicentrarchi* infection (Table 5). For most of these, the response peaked at 4 hpi. The pro-inflammatory *il1β* was strongly UR at 1, 2 and 4 hpi and, similarly to *cxcl8*, was DR at 48 hpi. Other interleukins that were DE at 1, 2 and 4 hpi included *il16*, *il27β* and *myeloid-derived growth factor* (also *il27*). However, *il10* was only UR at 12 hpi and *il12β* was UR at 12 and 48 hpi. Numerous IL receptors were also UR, including *il6ra*, *il6rb*, *il22r2*, *il1r1*, *il1r2*, *il10r1*, *il3b2*, *il7ra*, *il2rb*, *il2rg* and *il31r*. Of these, *il6ra*, *il22r2* and *il1r2* were most strongly regulated. In addition, *il1r1*, *il1r2*, *il10r1* and *il3b2* were DE at 12 hpi (Table 5).

Among all the cytokines, up-regulation was strongest in the pro-inflammatory cytokine *tumour necrosis factor-alpha* (*tnfa*) at 2 and 4 hpi, peaking at 4 hpi. Many other genes related to TNF, including TNF receptors, were also differentially expressed, including *tnf alpha-induced protein 8 like 2*, *tnf ligand superfamily member 6*, *tnf superfamily member 13b*, *tnf receptor superfamily member 26*, *tnf alpha-induced 2*, *tnf receptor superfamily member 1b*, *tnf receptor-associated factor 2*, *tnf receptor superfamily member 5* and *tnf receptor superfamily member 11b*. In almost all cases, expression of TNF and TNF-related genes peaked at 4 hpi (Table 5).

Table 5. Heat map showing a group of DE genes in turbot peritoneal cells at 1, 2, 4, 12 and 48 hpi, including genes coding for interleukins, interleukin receptors, genes of the tumour necrosis factor (TNF) family and genes involved in different signalling pathways. Results are expressed as mean expression ratios of groups 1, 2 and 4 hpi vs. 0 h, or 12 and 48 hpi vs. PBS. Red indicates increased gene expression levels; green indicates decreased levels.

Gene Name	Gene Abbrev.	1 h	2 h	4 h	12 h	48 h
interleukin-1 beta	*il1b*	327.182241	1589.55462	1847.60352		−6.7387485
interleukin-16	*il16*	19.2507463	35.3562406	59.9248328		
interleukin-27 subunit beta	*il27b*	13.7300327	242.32254	179.231112		
myeloid-derived growth factor	*mydgf*		95.9450925	122.26906		
interleukin-10	*il10*				62.3687203	
interleukin-12 subunit beta	*Il12b*				11.4185297	23.9473534
interleukin-6 receptor subunit alpha	*il6ra*	262.635667	669.959144	1017.60006		
interleukin-6 receptor subunit beta	*il6rb*	85.5323549	50.6037764	193.061244		
interleukin-22 receptor subunit alpha-2	*il22r2*	122.220327	297.897713	1069.37689		
interleukin-1 receptor type 1	*il1r1*	106.471491	238.123796	283.029024	4.98247239	
interleukin-1 receptor type 2	*il1r2*	83.0593856	1201.91688	2250.71899	25.847296	
interleukin-10 receptor subunit alpha	*il10r1*	101.62706	240.910175	664.168172	9.10643463	
interleukin-3 receptor class 2 subunit beta	*il3b2*	66.3168627	197.186399	800.67187	6.3092039	
interleukin-7 receptor subunit alpha	*il7ra*	31.0267486				
interleukin-2 receptor subunit beta	*il2rb*	28.098746	35.2061352	47.6837501		
interleukin-2 receptor subunit gamma	*il2rg*	19.6387686	38.542965	60.4108006		
interleukin-31 receptor subunit alpha	*il31r*				3.92933661	
tumor necrosis factor alpha	*tnfa*		3614.33852	5953.71784		
tumor necrosis alpha-induced 8 2 b	*tnfaip8l2b*	333.798683	408.811225	737.275001		
tumor necrosis factor ligand superfamily member 6	*tnfl6*	113.72684	163.681789			
tumor necrosis factor ligand superfamily member 13b	*tnfsf13b*	110.928512	52.0480829	271.529023		
tumor necrosis factor receptor superfamily member 26	*tnfrsf26*	105.956279	215.164494	196.925793		
tumor necrosis factor alpha-induced 2	*tnfaip2*	86.1712612	364.257117	384.174929		
tumor necrosis factor receptor superfamily member 1b	*tnfr1b*	65.5284083	57.8161671	64.5415275		
tnf receptor-associated factor 2	*traf2*	57.1024509	101.996426	195.25474		
tumor necrosis factor receptor superfamily member 5	*tnfrsf5*			41.613748	21.2114576	
tumor necrosis factor receptor superfamily member 11b	*tnfrsf11b*				24.2204824	

Another group of genes that were DE, but at a much lower level than the previously mentioned genes, was the group of interferon (IFN) related genes, including several interferon regulatory factors and other IFN related genes. Most of these were up-regulated at 1, 2 and 4 hpi, but some of them were also up-regulated 12 h after intraperitoneal injection. Among this group of genes, *interferon-double-stranded rna-activated kinase* and *interferon-induced 44-like* were the most strongly regulated (Table 6).

Table 6. Heat map showing a group of DE genes in turbot peritoneal cells at 1, 2, 4 and 12 hpi, including several interferon regulatory factor genes. Results are expressed as mean expression ratios of groups 1, 2 and 4 hpi vs. 0 h, or 12 hpi vs. PBS. There were no DE genes at 48 hpi. Red indicates increased gene expression levels; green indicates decreased levels.

Gene Name	Gene Abbrev.	1 h	2 h	4 h	12 h
interferon-induced 44-like	*if44l*	71.9994179	111.460711	150.105919	12.8949636
interferon-induced helicase c domain-containing 1	*ifih1*	27.3095849	27.3256044	32.2526372	
interferon- double-stranded rna-activated kinase	*eif2ak2*	49.9656668	106.466172	196.446601	196.446601
interferon-related developmental regulator 1	*ifrd1*	22.9845565	31.2624166	75.4855289	
interferon-induced 35 kda homolog	*ifi35*	18.9428499	34.0369582	47.5850614	
interferon alpha beta receptor 2	*ifnar2*	74.1505791	71.2747429	76.2223874	
interferon regulatory factor 1	*irf1*	14.0519125	18.1839762		10.9989039
interferon regulatory factor 2	*irf2*	24.2396496	43.3669569	81.4920354	
interferon regulatory factor 3	*irf3*	22.1667324	54.0067235	79.6835222	
interferon regulatory factor 8	*irf8*	10.1543906	24.2065428	63.6000421	4.01348333
interferon-induced gtp-binding mx	*mx*	15.4202932	37.2305381	68.2243287	
interferon-induced with tetratricopeptide repeats 1	*ifit1*		18.5057029	57.1941474	
interferon regulatory factor 5	*irf5*		21.2670525	31.5542826	
interferon-induced very large gtpase 1	*gvinp1*			95.658683	5.53328268

Several genes associated with cell death were DE at 1, 2 and 4 hpi, but many were also DE at 12 hpi (Table 7). Some members of the *bcl-2 gene family*, which are regulators of cell apoptosis, were strongly UR at 1, 2 and 4 hpi, including *apoptosis facilitator bcl-2 14*, *bcl2 associated agonist of cell death*, *apoptosis regulator bax*. Other DE genes were involved in processes that occurred during apoptosis, such as DNA fragmentation, including *dna fragmentation factor subunit beta*, or the extrinsic pathway of apoptosis, including *fas cell surface death receptor* and *fas-associated death domain*, which are associated with Fas ligand cell death. We also observed up-regulation of *caspases*, such as *caspase 3* and *8*, which were UR at 2 and 4 hpi or at 12 hpi, respectively (Table 7). Finally, several apoptotic regulator genes were also differentially expressed, including *programmed cell death 1 ligand 1* and *programmed cell death 4, 6* and *10*.

Several genes related to T and B lymphocytes or other immune cells were DE (Table 8). Up-regulation of components of IgM and IgD was observed at 1, 2 and 4 hpi, such as *immunoglobulin heavy constant mu* and *immunoglobulin delta heavy chain*, and several genes involved in B cell development and function in mammals, including *b-cell antigen receptor complex-associated alpha chain* and *b-cell receptor cd22*. Several genes associated with cytotoxic T lymphocytes and natural killer cells, such as *granzyme a*, *granzyme b* and *perforin-1*, were also UR. (Table 8). In addition, genes that are expressed by activated B and T lymphocytes in mammals, but that are also markers of dendritic cells, e.g., *cd83 antigen*, were also UR (Table 8). Finally, genes that can be expressed in mast cells, such as *high-affinity immunoglobulin epsilon receptor subunit gamma* and *mast cell protease 1a*, were DE at 1, 2 and 4 hpi.

Table 7. Heat map showing a group of DE genes in turbot peritoneal cells at 1, 2, 4 and 12 hpi, including many genes involved in cell death. Results are expressed as mean expression ratios of groups 1, 2 and 4 hpi vs. 0 h, or 12 hpi vs. PBS. There were no DE genes at 48 hpi. Red indicates increased gene expression levels; green indicates decreased levels.

Gene Name	Gene Abbrev.	1 h	2 h	4 h	12 h
apoptosis facilitator bcl-2 14	*bcl2l14*	113.684	898.176	1577.930	
bcl2 associated agonist of cell death	*bad*	189.050	352.048	452.689	
apoptosis regulator bax	*bax*	32.616	72.452	176.064	4.928
caspase recruitment domain-containing 11	*card11*	16.892	32.920	40.394	
cysteine serine-rich nuclear 1	*csrn1*	223.940	974.630	2534.320	
DNA fragmentation factor subunit beta	*dffb*	45.668	44.024	110.253	
programmed cell death 10	*pdcd10*	25.266	167.783	67.591	
programmed cell death 6	*pdcd6*	41.736	95.197	162.973	
programmed cell death 6-interacting	*pdcd6i*	30.355	84.740	67.733	
programmed cell death 4	*pdcd4*	68.163	110.907	162.973	−3.435
programmed cell death 1 ligand 1	*pd1-l1*			503.383	64.837
serine threonine- kinase 17a	*st17a*	31.071	93.784	201.549	
serine threonine- kinase 17b	*st17b*	56.748	56.748	251.152	
fas cell surface death receptor	*faslgr*	129.441	119.114	339.304	
fas-associated death domain	*fadd*	174.716	313.210	515.828	132.765
apoptosis-associated speck containing a card	*pycard*		16.474		
baculoviral iap repeat-containing 2	*birc2*		122.763	233.354	3.976
caspase recruitment domain-containing 9	*card9*			6092.587	
caspase-3	*casp3*		158.260	327.980	
casp8 and fadd-like apoptosis regulator	*cflar*				6.554
caspase-8	*casp8*				5.508
fas apoptotic inhibitory molecule 3	*faim3*				132.765
b-cell lymphoma leukemia 10	*bcl10*	67.237	278.829	504.966	
pyrin	*mefv*		102.595	208.050	
receptor-interacting serine threonine- kinase 2	*ripk2*	16.727	95.564	103.176	
receptor-interacting serine threonine- kinase 3	*ripk3*	31.089	64.526	110.020	

Several genes related to the complement and the coagulation systems were DE. Some components of the complement system, including *complement c1q subcomponent subunit b, complement c1s subcomponent, complement factor d, complement factor h, c3* and *c4* were UR at 1, 2 or 4 hpi, depending on the gene. Similarly, some components of the coagulation system were also DE at 1, 2 and 4 hpi, including *coagulation factor VIII, coagulation factor XIII a chain* and *tissue factor pathway inhibitor*. However, *tissue factor* was only UR at 12 hpi (Table 9).

Table 8. Heat map showing a group of DE genes in turbot peritoneal cells at 1, 2, 4 and 12 hpi, including several genes related to B cells and cytotoxic T cells. Results are expressed as mean expression ratios of groups 1, 2 and 4 hpi vs. 0 h, or 12 hpi vs. PBS. There were no DE genes at 48 hpi. Red indicates increased gene expression levels; green indicates decreased levels.

Gene Name	Gene Abbrev.	1 h	2 h	4 h	12 h
high affinity immunoglobulin epsilon receptor subunit gamma	*fcerg*	176.817257	311.445965	598.32369	
immunoglobulin heavy constant mu	*ighm*	44.659176	56.1328349	52.6153499	
immunoglobulin lambda constant 6	*iglc6*	109.625362	108.139186	95.911698	
immunoglobulin lambda variable 7-46	*iglv7-46*	49.9266366	61.4466313		
v-type immunoglobulin domain-containing suppressor of t-cell activation	*vsir*	169.315053	314.679434	756.121188	
granzyme a	*graa*	49.9266366	116.08526	60.9648276	
granzyme b	*grab*	169.315053			
perforin-1	*perf*	8033.37216			
plastin-2	*lcp1*	26.9806006	60.8179271	112.085751	
immunoglobulin delta heavy chain	*igd*	106.748406	166.358035		
b lymphocyte-induced maturation 1	*blimp*		375.643759	652.345739	
lymphocyte cytosolic 2	*lcp2*		42.2394554		
immunoglobulin superfamily member 3	*igsf3*			4813.39137	
b-cell receptor cd22	*cd22*				5.21002715
cd83 antigen	*cd83*	33.9279207	157.493124	189.841453	
b-cell antigen receptor complex-associated alpha chain	*cd79a*		38.4830819	58.4676507	

Table 9. Heat map showing a group of DE genes in turbot peritoneal cells at 1, 2, 4 and 12 hpi, including complement and coagulation related genes. Results are expressed as mean expression ratios of groups 1, 2 and 4 hpi vs. 0 h, or 12 hpi vs. PBS. There were no DE genes at 48 hpi. Red indicates increased gene expression levels; green indicates decreased levels.

Gene Name	Gene Abbrev.	1 h	2 h	4 h	12 h
coagulation factor viii	*f8*	24.9906392	126.629374	84.4903238	33.3428858
coagulation factor xiii a chain	*f13a1*	144.605804	117.464967	111.905425	
tissue factor pathway inhibitor	*tfpi1*	32.3367549	59.5824002	52.4958644	
tissue factor	*tf*				106.435028
complement c1q subcomponent subunit b	*c1qb*	27.8372678	54.1585349	31.7288673	
complement c1s subcomponent	*c1s*	76.2220545	76.8459537		
complement factor d	*cafd*	115.641444	434.049445		
complement factor h	*cafh*	68.4428124	112.403061	39.3661006	
complement c3	*c3*			51.9446282	
complement c4	*c4*	28.5701165	52.4312679	73.1753333	

4. Discussion

In the present study, we evaluated gene expression in the ciliate parasite *P. dicentrarchi* at 1, 2 and 4 hpi, relative to the gene expression in samples processed at 0 h, during experimental infection in the turbot *S. maximus*, to identify the genes participating in this process, including those potentially involved in virulence or in resistance to the host immune system. The most strongly regulated genes in *P. dicentrarchi* during infection were the ABC transporters, which were mainly UR. Protozoan ABC transporters are involved in nutrient transport but they also protect cells from both internally produced and exogenous toxins, and some have been associated with antiparasitic drug resistance as well virulence and oxidative stress [35–37]. These transporters are very abundant in ciliates, and 165 ABC transporter genes were identified in the macronuclear genome of the ciliate *Tetrahymena thermophila* [38]. In the present study, the *P. dicentrarchi* ABC transporter G family genes showed a strong, early response during infection, indicating that some ABC transporters are very important in this process. It is not known whether they are involved in ciliate resistance against the attack of fish immune system or whether they have other roles in infection.

Very few free-living ciliate species can invade the fish, resist attack by the immune system and cause systemic infection. However, *P. dicentrarchi* is capable of infecting cultured fish and causing massive mortalities. Thus, the ciliates must be able to resist the attack from the fish immune system and also to destroy fish tissue, for which proteases may be important. In the present study, we observed an increase in the expression of several genes of the leishmanolysin family, which are membrane-bound metalloproteases capable of degrading and cleaving many biological molecules [39]. Leishmanolysin GP63 is a major surface protein and is considered one of the main virulence factors in the human pathogen *Leishmania* [40]. GP63 degrades a large number of proteins, including complement [41], and down-regulation of gene expression makes the parasites more susceptible to complement-mediated lysis [42]. The complement system is considered important in defence against *P. dicentrarchi* [11,20], and although complement levels are lowered in the serum of infected turbot [12], it is not known whether leishmanolysins participate in the process. Our results agree with those obtained in previous studies, which reported high level of leishmanolysin expression in a *P. dicentrarchi* isolate (I1) obtained from turbot infections [43], and UR of several leishmanolysin genes in *Miamiensis avidus* fed a cell line [44], suggesting that these molecules play a role in cell degradation. Thus, although the role of each regulated leishmanolysin gene in the success of *P. dicentrachi* infection remains to be established, it appears that leishmanolysins are probably involved in ciliate virulence.

Ciliate proteases, other than leishmanolysins, are considered virulence factors in *P. dicentrarchi* [10,12–14]. However, several members of the cathepsin family were DR during the early stages of *P. dicentrarchi* infection. Interestingly, other genes whose proteins are also located in the lysosome were also DR. Because some ciliates contained phagocytosed turbot cells, release of lysosomal enzymes probably occurred during the early stages of infection. However, the expression of several cathepsin genes also did not increase in the scuticociliate *Anophryoides haemophila* during an infection in American lobster [45]. Although the expression of these genes will probably be enhanced at later stages in order to degrade the endocytosed material, the expression of these enzymes may be delayed during infection due to other priorities of the cells.

Many genes associated with rDNA transcription, ribosome biogenesis or rRNA synthesis were UR in *P. dicentrarchi* during the early stages of infection, indicating a focus on ribosome production and protein synthesis in the ciliate. However, except for *kinesin family member 1*, the *tubulin alpha* and *beta* genes and genes coding for several dyneins and kinesins were DR in *P. dicentrarchi* during infection at 2 and/or 4 hpi. Tubulins and the associated motor proteins play many roles in eukaryotic cells, and particularly in ciliates, including cell division or cell movement. Genes involved in cell division, such as *cyclin-dependent kinase 1* (*cdk1*) and *cyclin B*, were also DR in *P. dicentrarchi* during infection. *Cdk1* is a key protein that regulates mitotic entry and spindle assembly and its activation depends on *cyclin B* [46]. Our findings suggest that cell division is not activated during the early stages of infection. However, other genes associated with the cell cycle, such as *cdk2*, were strongly UR. *Cdk2* activation leads to DNA replication in the cell, and it is also involved in DNA damage and the DNA repair response [47]. Some ciliates probably suffered DNA damage because of attack from the turbot immune system, thus explaining the increase in *cdk2* expression. Several genes involved in glycolysis, the tricarboxylic acid (TCA) cycle and the mitochondrial respiratory chain were DR at 2 and (particularly) 4 hpi. If the *P. dicentrarchi* mitochondrial genome is similar to that of *Tetrahymena pyriformis* and *Paramecium aurelia* [48], the genes of the respiratory chain will be located in the mitochondrial genome, while those of the TCA cycle and glycolysis will be located in the nucleus. Unfortunately, little is known about how the expression of these genes is regulated in ciliates during infection. Many ciliate mitochondria may be altered because of interactions between the ciliate and the fish immune system. Previous studies have shown that turbot phagocytes are strongly stimulated by contact with *P. dicentrarchi*, releasing a high amount of reactive oxygen species (ROS) [11]. Leucocytes do not seem to produce sufficiently high concentrations of toxic substances to kill the parasite, but some ciliate components may be affected, as an increase in ROS levels is associated with alterations in the *P. dicentrachi* mitochondria [49]. Due to retrograde regulation, mitochondrial dysfunctions may affect the expression of enzymes involved in the TCA cycle, as observed in cells in other organisms [50], thus explaining the DR of these genes.

The present study shows that intraperitoneal injection of turbot with *P. dicentrarchi* induced an acute inflammatory response and enhanced the expression of genes involved in inflammatory pathways, inflammatory cytokines or genes coding for enzymes involved in the synthesis of prostaglandins and leukotrienes. Recruitment of inflammatory cells, which is part of the inflammatory response [51], was not evaluated in the present study but has been demonstrated in previous studies [21]. Recognition of the pathogen by pattern recognition receptors is an important event in the inflammatory process. High levels of expression of *tlr5* and, to a lesser extent, of *tlr13* and *tlr2*, occurred in turbot peritoneal cells although *tlr2* and *tlr13* expression began earlier. *Tlr2* expression increased in some organs of channel catfish (*Ictalurus punctatus*) and orange-spotted grouper (*Epinephelus coioides*) during infection with the ciliate ectoparasites *I. multifiliis* and *C. irritans* respectively [24,52,53]. Studies in mammals have shown that glycosylphosphatidylinositol (GPI) is a potent activator of TLR2 in several protozoan parasites and that GPI is recognised by this receptor [54,55]. It has also been suggested that TLR2 mediates activation of the MAPK and NF-κB pathways [56]. An increase in the expression of genes involved in these pathways was also observed in the present study, indicating that turbot TLR2 is activated during *P. dicentrarchi* infection and that it may be important in the recognition of some ciliate molecules. However, because *P. dicentrarchi* causes extensive damage in tissues and TLR2 can also be activated by several damage-associated molecular patterns [57], an increase in the expression of this receptor associated with any of these molecules cannot be ruled out. Expression of *tlr13* and particularly *tlr5* was strongly enhanced in turbot peritoneal cells, peaking at 4 hpi and lasting until at least 12 hpi. TLR13 is located in endolysosomes and recognises bacterial 23s ribosomal RNA in mice, while TLR5 is located at the plasma membrane and recognises bacterial flagellin [58]. Although TRL13 has been associated with the resistance of some insects to parasites [59], it is not known whether it plays a role in fish parasite infections. Fish *tlr5* was also strongly up-regulated in some tissues of *E. coioides* and in the Tibetan highland fish (*Gymnocypris przewalskii*) in response to infection with the ciliates *C. irritans* [24,60] and *I. multifiliis* [61] respectively, although, as far as we know, no particular role for this receptor in parasite recognition has been reported. The UR of *tlr5* found in the present study may be a consequence of prior stimulation of other cell receptors. In this respect, we cannot rule out unspecific stimulation of TLRs generated by a substance released by the parasite or even by a mechanical effect generated by the contact between the parasite membrane (which is moving all the time due to the movement of the cilia) and the leukocyte membranes.

Regarding the stimulation of inflammatory pathways, many components of NF-κB, MAPK and JAK/STAT pathways were UR. Some of those genes, such as *myd88*, which was UR at 1, 2 and 4 hpi, have a positive effect on the pathways; however, others, such as *nfkbia*, *socs1* and *socs3*, which were UR at 1, 2, 4 and 12 hpi, exert a negative effect in order to limit the inflammatory response [62]. NF-κB, MAPK and JAK/STAT are involved in inflammatory processes, inducing expression of genes that regulate the inflammatory response, although they can also participate in other processes [63–65]. Several chemokines and their receptors were UR in turbot peritoneal cells during *P. dicentrarchi* infection. Among these, the strongest regulation was found in *cxcl8* and their receptors *cxcr1* and *cxcr2*, which can be expressed by several types of leukocytes, especially neutrophils; *cxcl8* expression has been associated with neutrophil recruitment [66]. This chemokine also induces chemotaxis of neutrophils and, to a lesser extent, lymphocytes and macrophages in turbot [24,67]. These results support those demonstrating intense migration of neutrophils to the turbot peritoneal cavity during *P. dicentrarchi* infection [21]. Several interleukins and their receptors, as well as several members of the TNF family and their receptors, were also strongly UR in turbot peritoneal cells, particularly the proinflammatory cytokines IL1b and TNFa. These and other inflammatory cytokines are produced by leukocytes, due to TLR stimulation, and promote inflammation [58]. In addition, molecules such as TNFa can modulate multiple signalling pathways, some of which are related to the immune response during inflammation and also other responses, such as cell death [68]. The role of both IL1b and TNFa in regulating the inflammatory response is well conserved in fish [69]. Several other interleukins, interleukin receptors and members of the TNF superfamily that participate in the regulation of the inflammatory response are also strongly expressed in turbot peritoneal cells, including *il27b, il6ra, il6rb, il22r2, il3b2, il3b2, tnfaip8l2b, tnfaip2* and others, some of which, such as *tnfaip8l2b*, are negative regulators of the immune response [70], indicating the complexity of the response generated. Some of the interleukin genes such as *il10* and *il10r* were UR at 12 hpi and at 1, 2, 4 and 12 hpi, respectively. IL10 displays potent anti-inflammatory and regulatory activity in most immune processes during infection [71], suggesting an anti-inflammatory state in fish capable of controlling the infection. Increased expression of *il12b* was observed at 12 and 48 hpi, although to a lower extent than for *il10*. However, there was no increase in the expression of the complementary subunit or the receptors. In mammals, IL12B is secreted by phagocytic cells and acts on T and NK cells, inducing IFN-γ production and generating a proinflammatory state [72]. An increase in IFN-γ has been observed in leukocytes of other flatfish species stimulated with recombinant Il12 [73]. However, *ifn-γ* expression did not increase in turbot peritoneal leukocytes in the present study, suggesting that in this case, an increase in *il12b* expression will not generate a proinflammatory state.

Several genes of the interferon family were UR, although to a much lesser extent than those related to other cytokines. *Interferon-induced double-stranded RNA-activated kinase* was one of the genes in which DE was highest, being up-regulated at 1, 2, 4 and 12 hpi. This gene is a key component of host innate immunity that restricts viral replication and propagation but also participates in the stress response, being crucial for cell survival and proliferation, functions that are beyond the viral response [74]. Most of the other DE genes were interferon regulatory factors (IRFs). Some of these factors are involved in TLR signalling pathways, inducing the expression of *type 1 ifn* or proinflammatory cytokine genes [75]. IRFs have been shown to play several roles in parasite infection, particularly *irf8*, which appeared UR at 1, 2, 4 and 12 hpi in turbot peritoneal cells during *P. dicentrarchi* infection and which regulates the production of proinflammatory cytokines during malarial infections [76]. However, IRFs are also involved in differentiation and apoptosis of immune cells in mammals [75] and in fish [77]. All of these events seem to occur in the turbot peritoneal cavity, and different IRFs may participate in any of them.

Many genes associated with apoptosis were highly UR in turbot peritoneal cells during initial infection with *P. dicentrachi*. Regulation of most genes occurred at 1, 2, 4 hpi, peaking at 4 hpi, as well as at 12 hpi in some. Previous studies have shown that *P. dicentrarchi* proteases induce apoptosis in turbot cells [14]. However, increased expression of genes related to apoptosis or in the number of apoptotic cells in turbot peritoneal cells after injection with other stimuli have also been observed [31,78]. Other studies have also shown that large numbers of neutrophils migrate to the peritoneal cavity after injection of *P. dicentrarchi* [21] and many of these cells seem to initiate the expression of genes leading to apoptosis. In later stages, these cells can be phagocytosed by macrophages, contributing to the resolution of inflammation [79]. On the basis of these findings, it appears that most of the gene expression associated with apoptosis in turbot peritoneal cells occurs in neutrophils. However, it has been shown that other stimuli also cause apoptosis in fish peritoneal macrophages [80], indicating that other cell types may also be involved in this process.

Increased expression of several T and B lymphocyte related genes was also observed in the turbot peritoneal cells. Thus, e.g., *igm* and *igd heavy chains*, *cd79a* and *cd22*, which are expressed in B lymphocytes, were UR. This regulation may be related to the recruitment of B cells to the peritoneal cavity, as shown in other fish species after injection with different stimuli [81]. Several genes related to cytotoxicity were also DE. *Granzyme A* was UR at 1, 2 and 4 hpi and *granzyme b* and *perforin-1* were UR only at 1 hpi. These genes are produced by cytotoxic lymphocytes in mammals and fish [82,83].

Information about how these molecules interact with parasites in fish is scarce. However, in mammals, it has been observed that in addition to killing parasite-infected cells, these cytotoxic granule effectors can kill parasites [84]. It is not known whether these molecules can kill *P. dicentrarchi*. However, a turbot NK-lysin, an effector molecule of cytotoxic T lymphocytes and NK-cells, has been shown to be toxic for *P. dicentrarchi*, although infection did not clearly affect mRNA expression [19]. On the basis of these observations, these molecules from cytotoxic cells appear highly likely to play a role in *P. dicentrarchi* infection.

Finally, several genes related to the coagulation and the complement systems were also DE in turbot during *P. dicentrarchi* infection. Several complement genes have been found to be expressed in immune organs of turbot during a *P. dicentrarchi* infection [23], as well as locally in *E. coioides* infected with the ciliate *C. irritans* [24] and also in the skin or liver of common carp (*Cyprinus carpio*) infected with *I. multifiliis* [85]. The findings of the present study are consistent with the aforementioned findings in suggesting that peritoneal cells express complement and that the expression is regulated during infection. Similar observations have been made for genes involved in coagulation. Both the complement and the coagulation systems play important roles in defending turbot against *P. dicentrarchi* [20,21], and the complement or coagulation proteins released by leukocytes at the site of infection may represent an important additional source of these molecules.

How does the immune response generated by *P. dicentrarchi* in turbot compare with that induced by other parasites in fish? The immune response induced in fish by different parasites varies widely. Thus, some parasites do not elicit a host response, while others induce a more or less intense inflammatory response [86,87]. Previous transcriptomic studies carried out in fish infected with ciliate parasites have shown that these parasites usually generate an inflammatory response. However, the sampling times were longer than used in the present study, and the responses are therefore not necessarily comparable. Transcriptomic analysis, involving a microarray enriched in immune genes, of spleen, kidney and liver of turbot infected with *P. dicentrarchi*, revealed strong up-regulation of many immune response genes [22,23] at 1 dpi, but a less intense response than observed in turbot peritoneal cavity in the present study. Sampling times after experimental infection with ciliates were 12 h [88], 24 h [25] or 3 dpi [24] in fish infected experimentally with *C. irritans*. In the ectoparasite *I. multifiliis*, sampling was conducted 2 dpi [61] or 8 dpi [27]. However, regardless of the differences between the sampling times in the present and previous studies, ectoparasite ciliates such as *C. irritants* also induced a potent inflammatory response on the skin of infected fish, with up-regulation of genes involved in innate immunity, genes coding for TLRs, TLR signalling pathways, chemokines and chemokine receptors and genes associated with complement activation [24]. Similarly, *I. multifiliis* generated a massive immune response in the gills of rainbow trout (*Oncorhynchus mykiss*) [27].

5. Conclusions

In conclusion, many genes of the ABC transporter and leishmanolysin gene families were strongly regulated in the ciliate parasite *P. dicentrarchi* during the early stages of infection in turbot, suggesting that they play an important role in this process. During infection, the ciliate generated a very high inflammatory response at the site of injection, with the regulation of most of the genes known to be involved in the inflammatory response in mammals, indicating that turbot use all the protective mechanisms they have available to prevent entry of the parasite. The findings also provide some valuable insight into how the acute inflammatory response occurs in fish.

Supplementary Materials: The following are available online at http://www.mdpi.com/2079-7737/9/10/337/s1, Figure S1: Number of turbot cells and ciliates in the peritoneal cavity, Figure S2: Micrographs of ciliates at different stages of infection, Figure S3: qPCR validation of RNA-seq findings, File S1: Excel file showing the contigs that were DE in *P. dicentrarchi* at 1, 2, 4 hpi, the fold change, *p*-value, and the gene names, File S2: Excel file showing the contigs that were DE in turbot peritoneal cells at 1, 2, 4 hpi, the fold change, *p*-value, and the gene names, File S3: Excel file showing the contigs that were DE in turbot peritoneal cells at 12 and 48 hpi, the fold change, *p*-value, and the gene names, Table S1: Sample information, Table S2: Primer sequences used in the qPCR analysis.

Author Contributions: Conceptualization, J.M.L. and J.L.; Data curation, P.P. and A.F.; Formal analysis, P.P., A.F., B.N. and R.P.H.D.; Investigation, A.V.; Methodology, R.P.H.D. Resources, J.M.L. and J.L.; Software, P.P., A.F. and B.N.; Supervision, J.L.; Writing—original draft, J.L.; Writing—review & editing, J.M.L., P.P., A.F., A.V. and R.P.H.D. All authors have read and agreed to the published version of the manuscript.

Funding: This study was financially supported by grant AGL2017-83577-R awarded by the Ministerio de Economía y Competitividad (Spain) and Fondo Europeo de Desarrollo Regional -FEDER- (European Union), by grant ED431C2017/31 from the Xunta de Galicia (Spain), and by PARAFISHCONTROL project, via funding from the European Union's Horizon 2020 research and innovation programme under grant agreement no. 634429. This publication only reflects the views of the authors, and the European Commission cannot be held responsible for any use which may be made of the information contained herein.

Acknowledgments: We thank I. Folgueira and A.P. De Felipe for helping in ciliate culture.

Conflicts of Interest: The authors declare no conflict of interest.

References

1. Iglesias, R.; Paramá, A.; Alvarez, M.F.; Leiro, J.; Fernández, J.; Sanmartín, M.L. *Philasterides dicentrarchi* (Ciliophora, Scuticociliatida) as the causative agent of scuticociliatosis in farmed turbot *Scophthalmus maximus* in Galicia (NW Spain). *Dis. Aquat. Organ.* **2001**, *46*, 47–55. [CrossRef] [PubMed]
2. Kim, S.M.; Cho, J.B.; Kim, S.K.; Nam, Y.K.; Kim, K.H. Occurrence of scuticociliatosis in olive flounder *Paralichthys olivaceus* by *Philasterides dicentrarchi* (Ciliophora: Scuticociliatida). *Dis. Aquat. Organ.* **2004**, *62*, 233–238. [CrossRef] [PubMed]
3. Rossteuscher, S.; Wenker, C.; Jermann, T.; Wahli, T.; Oldenberg, E.; Schmidt-Posthaus, H. Severe scuticociliate (*Philasterides dicentrarchi*) infection in a population of sea dragons (*Phycodurus eques* and *Phyllopteryx taeniolatus*). *Vet. Pathol.* **2008**, *45*, 546–550. [CrossRef] [PubMed]
4. Shin, P.S.; Han, J.E.; Gómez, D.K.; Kim, J.H.; Choresca, C.H., Jr.; Jun, J.W.; Park, S.C. Identification of scuticociliate *Philasterides dicentrarchi* from Indo-Pacific seahorses *Hippocampus kuda*. *Afr. J. Microbiol. Res.* **2011**, *5*, 738–741.
5. De Felipe, A.P.; Lamas, J.; Sueiro, R.A.; Folgueira, I.; Leiro, J.M. New data on flatfish scuticociliatosis reveal that *Miamiensis avidus* and *Philasterides dicentrarchi* are different species. *Parasitology* **2017**, *144*, 1394. [CrossRef]
6. Dragesco, A.; Dragesco, J.; Coste, F.; Gasc, C.; Romestand, B.; Raymond, J.-C.; Bouix, G. *Philasterides dicentrarchi*, n. Sp., (Ciliophora, scuticociliatida), a histophagous opportunistic parasite of *Dicentrarchus labrax* (Linnaeus, 1758), a reared marine fish. *Eur. J. Protistol.* **1995**, *31*, 327–340. [CrossRef]
7. Paramá, A.; Iglesias, R.; Álvarez, M.F.; Leiro, J.; Aja, C.; Sanmartín, M.L. *Philasterides dicentrarchi* (Ciliophora, Scuticociliatida): Experimental infection and possible routes of entry in farmed turbot (*Scophthalmus maximus*). *Aquaculture* **2003**, *217*, 73–80. [CrossRef]
8. Jin, C.N.; Harikrishnan, R.; Moon, Y.G.; Kim, M.C.; Kim, J.S.; Balasundaram, C.; Azad, I.S.; Heo, M.S. Histopathological changes of Korea cultured olive flounder, *Paralichthys olivaceus* due to scuticociliatosis caused by histophagous scuticociliate, *Philasterides dicentrarachi*. *Vet. Parasitol.* **2009**, *161*, 292–301. [CrossRef]
9. Harikrishnan, R.; Jin, C.N.; Kim, J.S.; Balasundaram, C.; Heo, M.S. *Philasterides dicentrarchi*, a histophagous ciliate causing scuticociliatosis in olive flounder, *Philasterides dicentrarchi*-histopathology investigations. *Exp. Parasitol.* **2012**, *130*, 239–245. [CrossRef]
10. Paramá, A.; Iglesias, R.; Álvarez, M.F.; Leiro, J.; Ubeira, F.M.; Sanmartín, M.L. Cysteine proteinase activities in the fish pathogen *Philasterides dicentrarchi* (Ciliophora: Scuticociliatida). *Parasitology* **2004**, *128*, 541–548. [CrossRef]
11. Piazzon, M.C.; Wiegertjes, G.F.; Leiro, J.; Lamas, J. Turbot resistance to *Philasterides dicentrarchi* is more dependent on humoral than on cellular immune responses. *Fish Shellfish Immunol.* **2011**, *30*, 1339–1347. [CrossRef] [PubMed]
12. Piazzon, C.; Lamas, J.; Leiro, J.M. Role of scuticociliate proteinases in infection success in turbot, *Psetta maxima* (L.). *Parasite Immunol.* **2011**, *33*, 535–544. [CrossRef]
13. Paramá, A.; Castro, R.; Arranz, J.A.; Sanmartín, M.L.; Lamas, J.; Leiro, J. Scuticociliate cysteine proteinases modulate turbot leucocyte functions. *Fish Shellfish Immunol.* **2007**, *23*, 945–956. [CrossRef] [PubMed]

14. Paramá, A.; Castro, R.; Lamas, J.; Sanmartín, M.L.; Santamarina, M.T.; Leiro, J. Scuticociliate proteinases may modulate turbot immune response by inducing apoptosis in pronephric leucocytes. *Int. J. Parasitol.* **2007**, *37*, 87–95. [CrossRef] [PubMed]
15. Mallo, N.; Lamas, J.; Leiro, J.M. Evidence of an alternative oxidase pathway for mitochondrial respiration in the scuticociliate *Philasterides dicentrarchi*. *Protist* **2013**, *164*, 824–836. [CrossRef]
16. Folgueira, I.; Lamas, J.; de Felipe, A.P.; Sueiro, R.A.; Leiro, J.M. Identification and molecular characterization of superoxide dismutases isolated from a scuticociliate parasite: Physiological role in oxidative stress. *Sci. Rep.* **2019**, *9*, 13329. [CrossRef]
17. Folgueira, I.; Lamas, J.; De Felipe, A.P.; Sueiro, R.A.; Leiro, J.M. Evidence for the role of extrusomes in evading attack by the host immune system in a scuticociliate parasite. *Fish Shellfish Immunol.* **2019**, *92*, 802–812. [CrossRef]
18. Piazzon, M.C.; Leiro, J.; Lamas, J. Fish immunity to scuticociliate parasites. *Dev. Comp. Immunol.* **2013**, *41*, 248–256. [CrossRef]
19. Lama, R.; Pereiro, P.; Costa, M.M.; Encinar, J.A.; Medina-Gali, R.M.; Pérez, L.; Lamas, J.; Leiro, J.; Figueras, A.; Novoa, B. Turbot (*Scophthalmus maximus*) Nk-lysin induces protection against the pathogenic parasite *Philasterides dicentrarchi* via membrane disruption. *Fish Shellfish Immunol.* **2018**, *82*, 190–199. [CrossRef]
20. Leiro, J.; Piazzon, M.C.; Budiño, B.; Sanmartín, M.L.; Lamas, J. Complement-mediated killing of *Philasterides dicentrarchi* (Ciliophora) by turbot serum: Relative importance of alternative and classical pathways. *Parasite Immunol.* **2008**, *30*, 535–543. [CrossRef]
21. Blanco-Abad, V.; Noia, M.; Valle, A.; Fontenla, F.; Folgueira, I.; de Felipe, A.P.; Pereiro, P.; Leiro, J.; Lamas, J. The coagulation system helps control infection caused by the ciliate parasite *Philasterides dicentrarchi* in the turbot *Scophthalmus maximus* (L.). *Dev. Comp. Immunol.* **2018**, *87*, 147–156. [CrossRef] [PubMed]
22. Pardo, B.G.; Fernández, C.; Millán, A.; Bouza, C.; Vázquez-López, A.; Vera, M.; Álvarez-Dios, J.A.; Calaza, M.; Gómez-Tato, A.; Vázquez, M.; et al. Expressed sequence tags (ESTs) from immune tissues of turbot (*Scophthalmus maximus*) challenged with pathogens. *BMC Vet. Res.* **2008**, *25*, 4–37. [CrossRef] [PubMed]
23. Pardo, B.G.; Millán, A.; Gómez-Tato, A.; Fernández, C.; Bouza, C.; Álvarez-Dios, J.A.; Cabaleiro, S.; Lamas, J.; Leiro, J.M.; Martínez, P. Gene expression profiles of spleen, liver, and head kidney in turbot (*Scophthalmus maximus*) along the infection process with *Philasterides dicentrarchi* using an immune-enriched oligo-microarray. *Mar. Biotechnol.* **2012**, *14*, 570–582. [CrossRef] [PubMed]
24. Hu, Y.; Li, A.; Xu, Y.; Jiang, B.; Lu, G.; Luo, X. Transcriptomic variation of locally-infected skin of *Epinephelus coioides* reveals the mucosal immune mechanism against *Cryptocaryon irritans*. *Fish Shellfish Immunol.* **2017**, *66*, 398–410. [CrossRef]
25. Yin, F.; Qian, D. Transcriptomic analysis reveals the key immune-related signalling pathways of *Sebastiscus marmoratus* in response to infection with the parasitic ciliate *Cryptocaryon irritans*. *Parasites Vectors* **2017**, *10*, 576. [CrossRef]
26. Zhang, X.; Ding, L.; Yu, Y.; Kong, W.; Yin, Y.; Huang, Z.; Zhang, X.; Xu, Z. The Change of Teleost Skin Commensal Microbiota Is Associated with Skin Mucosal Transcriptomic Responses during Parasitic Infection by *Ichthyophthirius multifillis*. *Front. Immunol.* **2018**, *9*, 2972. [CrossRef]
27. Syahputra, K.; Kania, P.W.; Al-Jubury, A.; Jafaar, R.M.; Dirks, R.P.; Buchmann, K. Transcriptomic analysis of immunity in rainbow trout (*Oncorhynchus mykiss*) gills infected by *Ichthyophthirius multifiliis*. *Fish Shellfish Immunol.* **2019**, *86*, 486–496. [CrossRef]
28. Figueras, A.; Robledo, D.; Corvelo, A.; Hermida, M.; Pereiro, P.; Rubiolo, J.A.; Gómez-Garrido, J.; Carreté, L.; Bello, X.; Gut, M.; et al. Whole genome sequencing of turbot (*Scophthalmus maximus*; Pleuronectiformes): A fish adapted to demersal life. *DNA Res.* **2016**, *23*, 181–192. [CrossRef]
29. Conesa, A.; Götz, S.; García-Gómez, J.M.; Terol, J.; Talón, M.; Robles, M. Blast2GO, a universal tool for annotation, visualization and analysis in functional genomics research. *Bioinformatics* **2005**, *21*, 3674–3676. [CrossRef]
30. Millán, A.; Gómez-Tato, A.; Pardo, B.G.; Fernández, C.; Bouza, C.; Vera, M.; Alvarez-Dios, J.A.; Cabaleiro, S.; Lamas, J.; Lemos, M.L.; et al. Gene expression profiles of the spleen, liver, and head kidney in turbot (*Scophthalmus maximus*) along the infection process with *Aeromonas salmonicida* using an immune-enriched oligo-microarray. *Mar. Biotechnol.* **2011**, *13*, 1099–1114. [CrossRef]

31. Fontenla, F.; Blanco-Abad, V.; Pardo, B.G.; Folgueira, I.; Noia, M.; Gómez-Tato, A.; Martínez, P.; Leiro, J.M.; Lamas, J. Vaccine-induced modulation of gene expression in turbot peritoneal cells. A microarray approach. *Mol. Immunol.* **2016**, *75*, 188–199. [CrossRef] [PubMed]
32. Livak, K.J.; Schmittgen, T.D. Analysis of relative gene expression data using real-time quantitative PCR and the 2(-Delta Delta C(T)) Method. *Methods* **2001**, *25*, 402–408. [CrossRef] [PubMed]
33. Bustin, S.A.; Benes, V.; Garson, J.A.; Hellemans, J.; Huggett, J.; Kubista, M.; Mueller, R.; Nolan, T.; Pfaffl, M.W.; Shipley, G.L.; et al. The MIQE guidelines: Minimum information for publication of quantitative real-time PCR experiments. *Clin. Chem.* **2009**, *55*, 611–622. [CrossRef]
34. Rudra, D.; Warner, J.R. What better measure than ribosome synthesis? *Genes Dev.* **2004**, *18*, 2431–2436. [CrossRef] [PubMed]
35. Sauvage, V.; Aubert, D.; Escotte-Binet, S.; Villena, I. The role of ATP-binding cassette (ABC) proteins in protozoan parasites. *Mol. Biochem. Parasitol.* **2009**, *167*, 81–94. [CrossRef] [PubMed]
36. Kos, V.; Ford, R.C. The ATP-binding cassette family: A structural perspective. *Cell. Mol. Life Sci.* **2009**, *66*, 3111–3126. [CrossRef]
37. Manzano, J.I.; Perea, A.; León-Guerrero, D.; Campos-Salinas, J.; Piacenza, L.; Castanys, S.; Gamarro, F. *Leishmania* LABCG1 and LABCG2 transporters are involved in virulence and oxidative stress: Functional linkage with autophagy. *Parasites Vectors* **2017**, *10*, 267. [CrossRef]
38. Xiong, J.; Feng, L.; Yuan, D.; Fu, C.; Miao, W. Genome-wide identification and evolution of ATP-binding cassette transporters in the ciliate *Tetrahymena thermophila*: A case of functional divergence in a multigene family. *BMC Evol. Biol.* **2010**, *10*, 330. [CrossRef]
39. Murase, L.S.; de Souza, J.V.P.; de Lima Neto, Q.A.; de Mello, T.F.P.; Cardoso, B.M.; Lera-Nonose, D.S.S.L.; Teixeira, J.J.V.; Lonardoni, M.V.C.; Demarchi, I.G. The role of metalloproteases in *Leishmania* species infection in the New World: A systematic review. *Parasitology* **2018**, *145*, 1499–1509. [CrossRef]
40. Olivier, M.; Atayde, V.D.; Isnard, A.; Hassani, K.; Shio, M.T. *Leishmania* virulence factors: Focus on the metalloprotease GP63. *Microbes Infect.* **2012**, *14*, 1377–1389. [CrossRef]
41. Brittingham, A.; Morrison, C.J.; McMaster, W.R.; McGwire, B.S.; Chang, K.P.; Mosser, D.M. Role of the *Leishmania* surface protease gp63 in complement fixation, cell adhesion, and resistance to complement-mediated lysis. *J. Immunol.* **1995**, *155*, 3102–3111. [CrossRef]
42. Thiakaki, M.; Kolli, B.; Chang, K.P.; Soteriadou, K. Down-regulation of gp63 level in *Leishmania* amazonensis promastigotes reduces their infectivity in BALB/c mice. *Microbes Infect.* **2006**, *8*, 1455–1463. [CrossRef] [PubMed]
43. Mallo, N.; De Felipe, A.P.; Folgueira, I.; Sueiro, R.A.; Lamas, J.; Leiro, J.M. Combined antiparasitic and anti-inflammatory effects of the natural polyphenol curcumin on turbot scuticociliatosis. *J. Fish Dis.* **2017**, *40*, 205–217. [CrossRef]
44. Seo, J.S.; Jeon, E.J.; Jung, S.H.; Park, M.A.; Kim, J.W.; Kim, K.H.; Woo, S.H.; Lee, E.H. Molecular cloning and expression analysis of peptidase genes in the fish-pathogenic scuticociliate *Miamiensis avidus*. *BMC Vet. Res.* **2013**, *9*, 10. [CrossRef] [PubMed]
45. Acorn, A.R.; Clark, K.F.; Jones, S.; Després, B.M.; Munro, S.; Cawthorn, R.J.; Greenwood, S.J. Analysis of expressed sequence tags (ESTs) and gene expression changes under different growth conditions for the ciliate *Anophryoides haemophila*, the causative agent of bumper car disease in the American lobster (*Homarus americanus*). *J. Invertebr. Pathol.* **2011**, *107*, 146–154. [CrossRef] [PubMed]
46. Hayward, D.; Alfonso-Pérez, T.; Gruneberg, U. Orchestration of the spindle assembly checkpoint by CDK1-cyclin B1. *FEBS Lett.* **2019**, *593*, 2889–2907. [CrossRef] [PubMed]
47. Liu, Q.; Gao, J.; Zhao, C.; Guo, Y.; Wang, S.; Shen, F.; Xing, X.; Luo, Y. To control or to be controlled? Dual roles of CDK2 in DNA damage and DNA damage response. *DNA Repair* **2020**, *85*, 102702. [CrossRef] [PubMed]
48. Burger, G.; Zhu, Y.; Littlejohn, T.G.; Greenwood, S.J.; Schnare, M.N.; Lang, B.F.; Gray, M.W. Complete sequence of the mitochondrial genome of *Tetrahymena pyriformis* and comparison with *Paramecium aurelia* mitochondrial DNA. *J. Mol. Biol.* **2000**, *297*, 365–380. [CrossRef]
49. Morais, P.; Piazzon, C.; Lamas, J.; Mallo, N.; Leiro, J.M. Effect of resveratrol on oxygen consumption by *Philasterides dicentrarchi*, a scuticociliate parasite of turbot. *Protist* **2013**, *164*, 206–217. [CrossRef]
50. Liao, X.S.; Small, W.C.; Srere, P.A.; Butow, R.A. Intramitochondrial functions regulate nonmitochondrial citrate synthase (CIT2) expression in *Saccharomyces cerevisiae*. *Mol. Cell Biol.* **1991**, *11*, 38–46. [CrossRef] [PubMed]

51. Chen, L.; Deng, H.; Cui, H.; Fang, J.; Zuo, Z.; Deng, J.; Li, Y.; Wang, X.; Zhao, L. Inflammatory responses and inflammation-associated diseases in organs. *Oncotarget* **2017**, *9*, 7204–7218. [CrossRef] [PubMed]
52. Li, Y.W.; Luo, X.C.; Dan, X.M.; Huang, X.Z.; Qiao, W.; Zhong, Z.P.; Li, A.X. Orange-spotted grouper (*Epinephelus coioides*) TLR2, MyD88 and IL-1β involved in anti-*Cryptocaryon irritans* response. *Fish Shellfish Immunol.* **2011**, *30*, 1230–1240. [CrossRef] [PubMed]
53. Zhao, F.; Li, Y.W.; Pan, H.J.; Shi, C.B.; Luo, X.C.; Li, A.X.; Wu, S.Q. Expression profiles of toll-like receptors in channel catfish (*Ictalurus punctatus*) after infection with *Ichthyophthirius multifiliis*. *Fish Shellfish Immunol.* **2013**, *35*, 993–997. [CrossRef] [PubMed]
54. Campos, M.A.; Almeida, I.C.; Takeuchi, O.; Akira, S.; Valente, E.P.; Procópio, D.O.; Travassos, L.R.; Smith, J.A.; Golenbock, D.T.; Gazzinelli, R.T. Activation of Toll-like receptor-2 by glycosylphosphatidylinositol anchors from a protozoan parasite. *J. Immunol.* **2001**, *167*, 416–423. [CrossRef] [PubMed]
55. Durai, P.; Govindaraj, R.G.; Choi, S. Structure and dynamic behavior of Toll-like receptor 2 subfamily triggered by malarial glycosylphosphatidylinositols of *Plasmodium falciparum*. *FEBS J.* **2013**, *280*, 6196–6212. [CrossRef] [PubMed]
56. Li, L.; Li, X.; Gong, P.; Zhang, X.; Yang, Z.; Yang, J.; Li, J. *Trichomonas vaginalis* Induces Production of Proinflammatory Cytokines in Mouse Macrophages Through Activation of MAPK and NF-κB Pathways Partially Mediated by TLR2. *Front. Microbiol.* **2018**, *9*, 712. [CrossRef] [PubMed]
57. Roh, J.S.; Sohn, D.H. Damage-Associated Molecular Patterns in Inflammatory Diseases. *Immune Netw.* **2018**, *18*, e27. [CrossRef]
58. Vijay, K. Toll-like receptors in immunity and inflammatory diseases: Past, present, and future. *Int. Immunopharmacol.* **2018**, *59*, 391–412. [CrossRef]
59. Zhong, D.; Pai, A.; Wang, M.H.; Keech, N.; Yan, G. Fine-scale analysis of parasite resistance genes in the red flour beetle, *Tribolium castaneum*. *Genetics* **2013**, *195*, 253–261. [CrossRef]
60. Bai, J.S.; Li, Y.W.; Deng, Y.; Huang, Y.Q.; He, S.H.; Dai, J.; Zhao, S.Z.; Dan, X.M.; Luo, X.C. Molecular identification and expression analysis of TLR5M and TLR5S from orange-spotted grouper (*Epinephelus coioides*). *Fish Shellfish Immunol.* **2017**, *63*, 97–102. [CrossRef]
61. Tian, F.; Tong, C.; Feng, C.; Wanghe, K.; Zhao, K. Transcriptomic profiling of Tibetan highland fish (*Gymnocypris przewalskii*) in response to the infection of parasite ciliate *Ichthyophthirius multifiliis*. *Fish Shellfish Immunol.* **2017**, *70*, 524–535. [CrossRef] [PubMed]
62. Newton, K.; Dixit, V.M. Signaling in innate immunity and inflammation. *Cold Spring Harb. Perspect. Biol.* **2012**, *4*, a006049. [CrossRef] [PubMed]
63. Lawrence, T. The Nuclear Factor NF-κB Pathway in Inflammation. *Cold Spring Harb. Perspect. Biol.* **2009**, *1*, a001651. [CrossRef] [PubMed]
64. Kiu, H.; Nicholson, S.E. Biology and significance of the JAK/STAT signalling pathways. *Growth Factors* **2012**, *30*, 88–106. [CrossRef]
65. Arthur, J.S.; Ley, S.C. Mitogen-activated protein kinases in innate immunity. *Nat. Rev. Immunol.* **2013**, *13*, 679–692. [CrossRef]
66. Russo, R.C.; García, C.C.; Teixeira, M.M.; Amaral, F.A.; Kiu, H.; Nicholson, S.E. The CXCL8/IL-8 chemokine family and its receptors in inflammatory diseases. *Expert Rev. Clin. Immunol.* **2014**, *10*, 593–619. [CrossRef]
67. Hu, Y.H.; Chen, L.; Sun, L. CXCL8 of *Scophthalmus maximus*: Expression, biological activity and immunoregulatory effect. *Dev. Comp. Immunol.* **2011**, *35*, 1032–1039. [CrossRef]
68. Holbrook, J.; Lara-Reyna, S.; Jarosz-Griffiths, H.; McDermott, M. Tumour necrosis factor signalling in health and disease. *F1000Research* **2019**, *8*. [CrossRef]
69. Zou, J.; Secombes, C.J. The Function of Fish Cytokines. *Biology* **2016**, *5*, 23. [CrossRef]
70. Sun, H.; Gong, S.; Carmody, R.J.; Hilliard, A.; Li, L.; Sun, J.; Kong, L.; Xu, L.; Hilliard, B.; Hu, S.; et al. TIPE2, a negative regulator of innate and adaptive immunity that maintains immune homeostasis. *Cell* **2008**, *133*, 415–426. [CrossRef]
71. Piazzon, M.C.; Lutfalla, G.; Forlenza, M. IL10, A Tale of an Evolutionarily Conserved Cytokine across Vertebrates. *Crit. Rev. Immunol.* **2016**, *36*, 99–129. [CrossRef]
72. Zundler, S.; Neurath, M.F. Interleukin-12: Functional activities and implications for disease. *Cytokine Growth Factor Rev.* **2015**, *26*, 559–568. [CrossRef] [PubMed]
73. Kim, D.; Lee, S.H.; Lee, H.; Kim, S.J.; Lee, K.H.; Song, S.K. Analyses of the gene structure and function of olive flounder (*Paralichthys olivaceus*) interleukin 12 (IL-12). *Fish Shellfish Immunol.* **2019**, *92*, 151–164. [CrossRef]

74. Bou-Nader, C.; Gordon, J.M.; Henderson, F.E.; Zhang, J. The search for a PKR code-differential regulation of protein kinase R activity by diverse RNA and protein regulators. *RNA* **2019**, *25*, 539–556. [CrossRef] [PubMed]
75. Zhao, G.N.; Jiang, D.S.; Li, H. Interferon regulatory factors: At the crossroads of immunity, metabolism, and disease. *Biochim. Biophys. Acta* **2015**, *1852*, 365–378. [CrossRef]
76. Gun, S.Y.; Claser, C.; Tan, K.S.; Rénia, L. Interferons and interferon regulatory factors in malaria. *Mediat. Inflamm.* **2014**, *2014*, 243713. [CrossRef]
77. Holland, J.W.; Karim, A.; Wang, T.; Alnabulsi, A.; Scott, J.; Collet, B.; Mughal, M.S.; Secombes, C.J.; Bird, S. Molecular cloning and characterization of interferon regulatory factors 4 and 8 (IRF-4 and IRF-8) in rainbow trout, *Oncorhynchus mykiss*. *Fish Shellfish Immunol.* **2010**, *29*, 157–166. [CrossRef]
78. Noia, M.; Domínguez, B.; Leiro, J.; Blanco-Méndez, J.; Luzardo-Álvarez, A.; Lamas, J. Inflammatory responses and side effects generated by several adjuvant-containing vaccines in turbot. *Fish Shellfish Immunol.* **2014**, *38*, 244–254. [CrossRef] [PubMed]
79. Havixbeck, J.J.; Rieger, A.M.; Wong, M.E.; Hodgkinson, J.W.; Barreda, D.R. Neutrophil contributions to the induction and regulation of the acute inflammatory response in teleost fish. *J. Leukoc. Biol.* **2016**, *99*, 241–252. [CrossRef] [PubMed]
80. Costa-Ramos, C.; Vale, A.D.; Ludovico, P.; Dos Santos, N.M.; Silva, M.T. The bacterial exotoxin AIP56 induces fish macrophage and neutrophil apoptosis using mechanisms of the extrinsic and intrinsic pathways. *Fish Shellfish Immunol.* **2011**, *30*, 173–181. [CrossRef]
81. Castro, R.; Abós, B.; González, L.; Granja, A.G.; Tafalla, C. Expansion and differentiation of IgM(+) B cells in the rainbow trout peritoneal cavity in response to different antigens. *Dev. Comp. Immunol.* **2017**, *70*, 119–127. [CrossRef]
82. Voskoboinik, I.; Whisstock, J.C.; Trapani, J.A. Perforin and granzymes: Function, dysfunction and human pathology. *Nat. Rev. Immunol.* **2015**, *15*, 388–400. [CrossRef] [PubMed]
83. Chaves-Pozo, E.; Valero, Y.; Lozano, M.T.; Rodríguez-Cerezo, P.; Miao, L.; Campo, V.; Esteban, M.A.; Cuesta, A. Fish Granzyme A Shows a Greater Role Than Granzyme B in Fish Innate Cell-Mediated Cytotoxicity. *Front. Immunol.* **2019**, *10*, 2579. [CrossRef] [PubMed]
84. Dotiwala, F.; Mulik, S.; Polidoro, R.B.; Ansara, J.A.; Burleigh, B.A.; Walch, M.; Gazzinelli, R.T.; Lieberman, J. Killer lymphocytes use granulysin, perforin and granzymes to kill intracellular parasites. *Nat. Med.* **2016**, *22*, 210–216. [CrossRef]
85. González, S.F.; Buchmann, K.; Nielsen, M.E. Complement expression in common carp (*Cyprinus carpio* L.) during infection with *Ichthyophthirius multifiliis*. *Dev. Comp. Immunol.* **2007**, *31*, 576–586. [CrossRef] [PubMed]
86. Sitjà-Bobadilla, A. Living off a fish: A trade-off between parasites and the immune system. *Fish Shellfish Immunol.* **2008**, *25*, 358–372. [CrossRef]
87. Gómez, D.; Bartholomew, J.; Sunyer, J.O. Biology and mucosal immunity to myxozoans. *Dev. Comp. Immunol.* **2014**, *43*, 243–256. [CrossRef]
88. Jiang, B.; Du, J.J.; Li, Y.W.; Ma, P.; Hu, Y.Z.; Li, A.X. Transcriptome analysis provides insights into molecular immune mechanisms of rabbitfish, *Siganus oramin* against *Cryptocaryon irritans* infection. *Fish Shellfish Immunol.* **2019**, *88*, 111–116. [CrossRef]

Publisher's Note: MDPI stays neutral with regard to jurisdictional claims in published maps and institutional affiliations.

biology

MDPI

Article

The Teleost Thymus in Health and Disease: New Insights from Transcriptomic and Histopathological Analyses of Turbot, *Scophthalmus maximus*

Paolo Ronza [1,*], Diego Robledo [2], Ana Paula Losada [1], Roberto Bermúdez [1,3], Belén G. Pardo [3,4], Paulino Martínez [3,4] and María Isabel Quiroga [1,3]

1 Departamento de Anatomía, Producción Animal y Ciencias Clínicas Veterinarias, Universidade de Santiago de Compostela, 27002 Lugo, Spain; anapaula.losada@usc.es (A.P.L.); roberto.bermudez@usc.es (R.B.); misabel.quiroga@usc.es (M.I.Q.)
2 The Roslin Institute and Royal (Dick) School of Veterinary Studies, University of Edinburgh, Midlothian EH25 9RG, UK; diego.robledo@roslin.ed.ac.uk
3 Instituto de Acuicultura, Universidade de Santiago de Compostela, 15782 Santiago de Compostela, Spain; belen.gomez@usc.es (B.G.P.); paulino.martinez@usc.es (P.M.)
4 Departamento de Zoología, Genética y Antropología Física, Universidade de Santiago de Compostela, 27002 Lugo, Spain
* Correspondence: paolo.ronza@usc.es; Tel.: +34-982822306

Received: 15 July 2020; Accepted: 11 August 2020; Published: 13 August 2020

Abstract: The thymus is a primary lymphoid organ that plays a pivotal role in the adaptive immune system. The immunobiology of the thymus in fish is considered to be similar to that of mammals, but it is actually poorly characterized in several cultured teleost species. In particular, while investigations in human and veterinary medicine have highlighted that the thymus can be affected by different pathological conditions, little is known about its response during disease in fish. To better understand the role of the thymus under physiological and pathological conditions, we conducted a study in turbot (*Scophthalmus maximus*), a commercially valuable flatfish species, combining transcriptomic and histopathological analyses. The myxozoan parasite *Enteromyxum scophthalmi*, which represents a major challenge to turbot production, was used as a model of infection. The thymus tissues of healthy fish showed overrepresented functions related to its immunological role in T-cell development and maturation. Large differences were observed between the transcriptomes of control and severely infected fish. Evidence of inflammatory response, apoptosis modulation, and declined thymic function associated with loss of cellularity was revealed by both genomic and morphopathological analyses. This study presents the first description of the turbot thymus transcriptome and provides novel insights into the role of this organ in teleosts' immune responses.

Keywords: fish; T lymphocytes; infection; malnutrition; inflammation; aquaculture; RNA-Seq; histopathology; immunohistochemistry; enteromyxosis

1. Introduction

The thymus is a primary lymphoid organ that plays a critical and unique role in the development of T cells. This organ provides the necessary microenvironment for the precursors of lymphocytes to proliferate, rearrange their receptors, and finally differentiate into mature T lymphocytes [1–3]. Once these cells have gone through the processes of selection and differentiation, they leave the thymus and migrate to the periphery to perform their major role in the body's immune response [4,5].

The immunobiology of this organ, particularly its pivotal role in the adaptive immune response as the main site of T-cell development, is considered to be similar in fish and mammals [6,7]. Despite its relevance as a primary lymphoid organ and its potential importance in the response to different diseases affecting aquaculture production, the characterization of the thymus in several cultured teleost species has scarcely been addressed to date. In particular, little is known about its behavior in pathological conditions. This is not surprising, since it is still considered an underexplored organ even in human medicine, mainly due to the difficulty of accessing in [8]. Similarly, in fish the thymus is usually not included in routine tissue sampling, probably also because of the difficulties due to its appearance and localization.

In terrestrial species, it has been demonstrated that different pathological conditions, such as malnutrition and infectious diseases, can affect the thymus, disrupting its architecture and function and having a detrimental effect on the ongoing immune response [9–11]. The underlying pathogenic mechanisms are beginning to be understood, mainly through studies in human and murine models [9,12–14].

The thymus in fish usually appears as a paired organ located in the dorsal region of each gill chamber [6,15]. In turbot, *Scophthalmus maximus*, a commercially important flatfish species, the gross anatomy and histology of the organ in juvenile fish was described by Vigliano et al. [16]. In this species, a zonation of thymus resembling the division in the cortex and medulla of mammals was reported: lymphoblasts and macrophagic cells were mainly observed in the inner part, whereas thymocytes were mostly found in the outer region [16,17]. Nevertheless, information on the function of this organ is still scarce, starting from the lack of high-throughput gene expression profiling, and its integration with morphological analysis to further understand the function of the thymus in health and disease. In this sense, the role of the thymus in response to infectious diseases has not been studied in turbot.

Enteromyxosis, caused by the myxozoan *Enteromyxum scophthalmi*, is a fast-spreading parasitic disease with a great impact on turbot aquaculture. *E. scophthalmi* is transmitted directly between animals, triggering a cachectic syndrome associated with poor growth performance and high mortality rates [18]. The main target organ is the gastrointestinal tract, where the parasite localizes into the lining epithelium, causing catarrhal enteritis of increasing severity throughout the disease. Further, the main lymphohematopoietic organs, kidney and spleen, develop severe cell depletion with the progression of the infection [19]. Recently, significant advances in our understanding of the pathogenesis of this parasitosis have been made using a multidisciplinary approach combining histopathology and RNA-seq transcriptomics [20,21]. The available data strongly indicate that turbot show a dysregulated and ineffective immune response against this parasite [18]. Nevertheless, the role of the thymus has never been evaluated in enteromyxosis.

Here, we employed a multidisciplinary approach to improve the knowledge of this primary lymphoid organ in teleosts under physiological and pathological conditions. We analyzed the transcriptomic profiles and performed histopathological studies of the thymuses of healthy and experimentally infected turbot, using enteromyxosis as the disease model.

2. Materials and Methods

2.1. Experimental Design and Histopathology

The experimental design was described in a previous paper (see Reference [20]). Briefly, 120 juvenile turbot (150 g mean weight) were divided into 55 receptor and 65 control fish and kept in two 500 L tanks per group with 5 µm filtered and UV-irradiated open-flow seawater (37.5‰ salinity) at 19 ± 1 °C at the facilities of the Instituto de Acuicultura de Torre la Sal (IATS, Cabanes, Castellón, Spain).

The experimental infection was carried out by oral route; receptor fish received 1 mL of intestinal scraping homogenates in Hank's balanced salt solution (HBSS) from 20 donor fish containing live *E. scophthalmi* parasites, whereas control fish were inoculated with the same amount of HBSS alone. Donor turbot came from an experimentally infected stock maintained at IATS (Castellón, Spain).

Tissue samples from control and infected fish were collected at 7, 24, and 42 days postinoculation (dpi) and were preserved in Bouin's fluid and RNAlater for histopathological and transcriptomic analyses, respectively.

The health status of control and infected fish was assessed by light microscopy on H&E- and toluidine-blue-stained sections. Challenged fish were classified into three groups (slightly, moderately, and severely infected) according to the histopathological grading described by Bermúdez et al., which considers the lesional degree and parasite burden [19]. For the immunohistochemical study, five slightly infected turbot and five control turbot were selected at 24 dpi, and five severely infected and five control turbot at 42 dpi. Of those animals, the thymus samples from four control (two pools) and three infected fish from each time point were also RNA-sequenced.

The experiment was carried out following international (Directive 2010/63/EU, on the protection of animals used for scientific purposes), national (Royal Decree RD1201/2005, for the protection of animals used in scientific experiments), and institutional regulations (CSIC, IATS Review Board and Institutional Animal Care and Use Committee of the University of Santiago de Compostela).

2.2. Immunohistochemistry

Thin sections (3 μm thick) were placed on slides treated with silane to improve section adherence and dried overnight at 37 °C. Immunohistochemical markers of cell proliferation (proliferating cell nuclear antigen, PCNA), apoptosis (active caspase-3), and T cells (CD3ε) were used, together with antibodies against the proinflammatory molecules tumor necrosis factor alpha (TNF-α) and inducible nitric oxide synthase (iNOS) (Table 1). The anti-CD3ε antibody, generously donated by Dr. Erin Bromage (University of Massachusetts Dartmouth, USA), was raised using as an immunogen a synthetic peptide consisting of 14 amino acids from the cytoplasmic tail region of rainbow trout CD3ε [22]. The sequence alignment for the immunogenic peptide and its turbot homologue showed 85% identity (Supplementary File 1). All the other antibodies employed were commercial and were previously used in turbot tissues (see references in Table 1).

Table 1. List of the immunohistochemical markers used in this work. AR = antigen retrieval, HUP = heating under pressure.

Marker	Description	Working Conditions	Reference
CD3ε (Donated by Dr. Erin Bromage)	Chain of T-cell co-receptor	1:500 No AR required	[22]
PCNA (M0879, Dako)	Cofactor of DNA polymerase delta	1:500 dilution AR: HUP in pH 6 buffer	[23]
Active caspase-3 (G7481, Promega)	Effector protease in apoptosis	1:200 dilution AR: HUP in pH 8 buffer	[24]
Inos (RB-1605, Thermo Fisher Scientific)	Nitric oxide synthase	1:5000 dilution No AR required	[25]
TNF-α (ab6671, Abcam)	Proinflammatory cytokine	1:600 dilution AR: HUP in pH 6 buffer	[26]

Endogenous peroxidase activity was quenched by incubation with a peroxidase-blocking solution (Dako, Glostrup, Denmark). Background prevention and antigen retrieval were performed as required. After incubation with primary antisera, slides were incubated with peroxidase (HRP)-labeled polymer conjugated to rabbit secondary antibody and peroxidase reaction was developed using a diaminobenzidine-positive chromogen (EnVision+ System-HRP kit, No. K 4011; Dako, Glostrup, Denmark). All incubations were performed in a humid chamber at room temperature. Finally, sections were counterstained with hematoxylin, dehydrated, and coverslipped with DePeX mounting medium. The sections were washed three times for 5 min in 0.1 M phosphate-buffered saline containing 0.05%

Tween-20 between all subsequent steps, and positive and negative controls were included in each assay to assess its specificity.

2.3. Transcriptome Analysis

RNA extraction was performed using the RNeasy mini kit (Qiagen, Hilden, Germany) with DNase treatment following the manufacturer's instructions. RNA quality and quantity were evaluated in a Bioanalyzer (Bonsai Technologies, Madrid, Spain) and in a NanoDrop® ND-1000 spectrophotometer (NanoDrop® Technologies Inc., Wilmington, DE, USA), respectively. Samples were barcoded and prepared for sequencing by the Wellcome Trust Centre for Human Genetics (Oxford, UK) on an Illumina HiSeq 4000 (Illumina Inc., San Diego, CA, USA) as 150 bp paired-end reads.

Raw sequencing data were deposited in NCBI's Short Read Archive (SRA) under BioProject ID PRJNA623212. The quality of the sequencing output was assessed using FastQC v.0.11.5 (http://www.bioinformatics.babraham.ac.uk/projects/fastqc/). Quality filtering and removal of residual adaptor sequences were conducted on read pairs using Trimmomatic v.0.38 [27]. Specifically, residual Illumina-specific adaptors were clipped from the reads, leading and trailing bases with a Phred score less than 20 were removed, and the reads were trimmed if a sliding window average Phred score over four bases was less than 20. Only reads where both pairs had a length greater than 36 bp after filtering were retained.

Filtered reads were mapped to the most recent turbot genome assembly (GenBank accession No. GCA_003186165.1; [28]) using STAR v.2.7.0e [29] two-pass mode and the following parameters: the maximum number of mismatches for each read pair was set to 10% of trimmed read length, and minimum and maximum intron lengths were set as 20 bases and 1 Mb, respectively. Alignment files were assembled into potential transcripts using Cufflinks v.2.2.1 [30]. Uniquely mapped paired-end reads were counted and assigned to genes using FeatureCounts v.1.6.4 [31] and statistical analyses related to differential expression were performed using R v.3.5.2 (http://www.R-project.org/). Gene count data were used to estimate differential gene expression using the Bioconductor package DESeq2 v.3.4 [32]. Transcript abundances were calculated as transcripts per million (TPM). Gene ontology (GO) enrichment analyses were performed using Blast2GO v.4.1 [33] and Kyoto Encyclopedia of Genes and Genomes (KEGG) enrichment analyses were performed using KOBAS v3.0.3 [34]. GO and KEGG enrichments for specific gene lists were tested by comparison to the whole transcriptome of the turbot using Fisher's exact test, and those terms or pathways showing a Benjamini–Hochberg FDR-corrected p-value < 0.05 were considered to be enriched.

3. Results

3.1. Histopathological Evaluation of Thymuses

The histological evaluation of the thymuses did not reveal any significant lesion in controls or in experimentally infected fish graded as slightly or moderately infected. Conversely, most fish with severe infections showed some degree of lymphocyte depletion in the outer zone of the organ, sometimes accompanied by an alteration of tissue architecture, namely a loss of differentiation between the inner and outer zones (Figure 1A,B).

3.2. Immunohistochemistry

The immunohistochemistry for CD3ε and PCNA was strongly positive in healthy (control) thymus, especially in the outer zone, where most labeled cells were consistent with thymocytes (Figure 1C,E). No differences were found between slightly infected and control fish, while reduced numbers of thymocyte-like cells immunoreactive to both markers were observed in turbot with severe enteromyxosis (Figure 1D,F).

The immunochemical staining pattern of TNF-α in turbot thymus under physiological conditions has been previously described [26]. Accordingly, we observed some TNF-α-immunoreactive macrophage-like cells randomly distributed in the stroma of the organ in control (Figure 2A) and slightly infected fish. Conversely, severely infected fish showed increased immunoreactivity to the cytokine, with higher number of positive macrophage-like cells and/or enhanced staining of the reticulo-epithelial cellular network of the thymic stroma (Figure 2B).

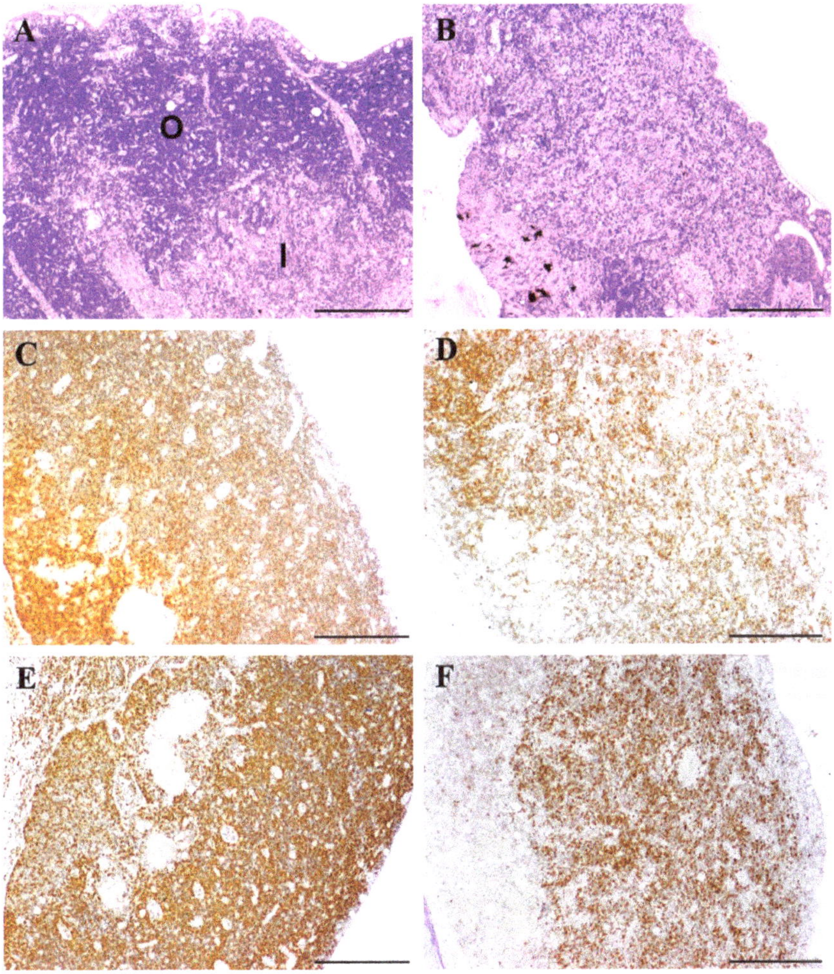

Figure 1. Comparative photomicrographs of thymus from healthy turbot (**A**,**C**,**E**) and turbot with severe infection by *Enteromyxum scophthalmi* (**B**,**D**,**F**). Scale bars = 200 µm. (**A**,**B**). Stained with H&E. Note that the zonation of the thymus into inner (I) and outer (O) parts was indistinguishable in the infected specimen, which also showed a reduced density of deeply basophilic thymocyte-like cells. (**C**,**D**) Immunostained for the T-cell marker CD3ε. The thymus tissues of diseased fish showed a reduced immunoreactivity, indicative of a loss of cellularity. (**E**,**F**) Immunostained for the cell proliferation marker PCNA (proliferating cell nuclear antigen). The strong immunolabel observed in the healthy fish was largely reduced during severe enteromyxosis.

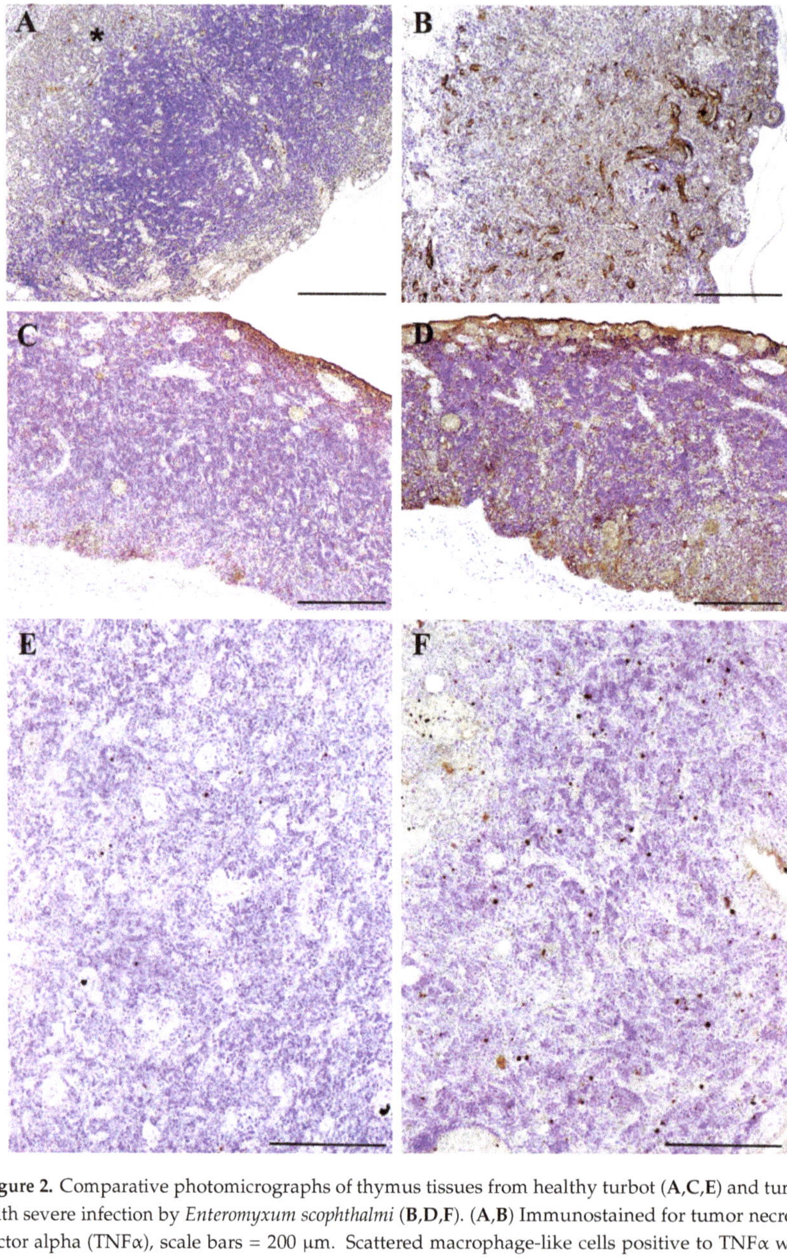

Figure 2. Comparative photomicrographs of thymus tissues from healthy turbot (**A**,**C**,**E**) and turbot with severe infection by *Enteromyxum scophthalmi* (**B**,**D**,**F**). (**A**,**B**) Immunostained for tumor necrosis factor alpha (TNFα), scale bars = 200 μm. Scattered macrophage-like cells positive to TNFα were present in the inner region of the thymus tissues (star) of healthy fish, while the severely infected turbot showed broader immunoreactivity to the cytokine, also found in the reticulo-epithelial network of the thymic stroma. (**C**,**D**) Immunostained for inducible nitric oxide synthase (iNOS), scale bars = 100 μm. Note that the immunostaining to this proinflammatory molecule was more diffuse in the thymus of the diseased specimen. (**E**,**F**) Immunostained for active caspase-3, scale bars = 100 μm. Higher numbers of thymocyte-like cells positive to this marker of apoptosis were noted in the outer part of the thymus of the *E.-scophthalmi*-infected fish. * shows the location of the inner region of the thymus: Scattered macrophage-like cells positive to TNFα were present in the inner region of the thymus tissues (star).

Similarly, few iNOS-immunoreactive cells were found in the thymus tissues of control and slightly infected turbot. The immunoreactivity was mainly observed in the epithelial layer and in the cytoplasm of some mucous and macrophage-like cells (Figure 2C). Severely infected fish displayed a stronger and more diffuse immunostaining for iNOS, also observed in the thymic stroma, with a greater number of labeled macrophage-like and mucous cells (Figure 2D).

Regarding the apoptotic marker active caspase-3, the immunolabeling was mainly detected in scattered cells consistent with thymocytes in the outer zone of the organ. The result was similar when comparing control and slightly infected turbot, while a mild increase of caspase-3 was detected in severely infected fish, which showed a higher density of positive thymocyte-like cells (Figure 2E,F).

3.3. *Thymus Transcriptome of Healthy Fish*

To characterize the turbot thymus transcriptome, we studied different subsets of highly expressed genes, selected based on their expression levels in control animals (TPM > 1000, 100, and 10), and performed GO term and KEGG pathway enrichment analyses (Supplementary File 2). The enrichment analyses revealed several overrepresented functions related to energy production, protein synthesis, and cell proliferation (Figure 3). Different GO terms and KEGG pathways related to immunological functions were found, such as chemokine signaling, antigen processing and presentation, T-cell receptor signaling pathways, and natural-killer-mediated cytotoxicity. Programmed cell death (i.e., apoptosis) was also overrepresented in both enrichment analyses.

3.4. *Enteromyxosis-Induced Transcriptomic Changes*

Only four genes showed differential expression between slightly infected and control fish, whereas a total of 4640 differentially expressed genes (DEGs) were detected between the thymus transcriptomes of control and severely infected turbot: 2150 upregulated and 2490 downregulated (Figure 4, Supplementary File 3).

GO term and KEGG pathway enrichment analyses were performed to study these DEGs between control and severely infected turbot (Figure 5, Supplementary File 4). Both sets of up- and downregulated genes showed enriched GO terms and KEGG pathways related to immune response. These were mainly associated with inflammatory reaction in the case of upregulated genes (i.e., macrophage chemotaxis, leukocyte transendothelial migration, cytokine activity, TNF signaling pathway), whereas some functions related to immunological role of the organ, such as antigen processing and presentation or lymphocyte differentiation, were found in the enrichment analysis of downregulated genes. The functional analysis also revealed overrepresentation of genes related to protein synthesis and cell proliferation among downregulated genes. Several DEGs related to apoptosis were also found, both up- and downregulated; a GO term related to apoptosis was found to be enriched in the set of downregulated genes ("Apoptotic signaling pathway"), while the "Apoptosis" KEGG pathway was marginally enriched in the upregulated gene set (Benjamini-Hochberg FDR-corrected *p*-value: 0.087, *p*-value: 0.0055; Supplementary File 4).

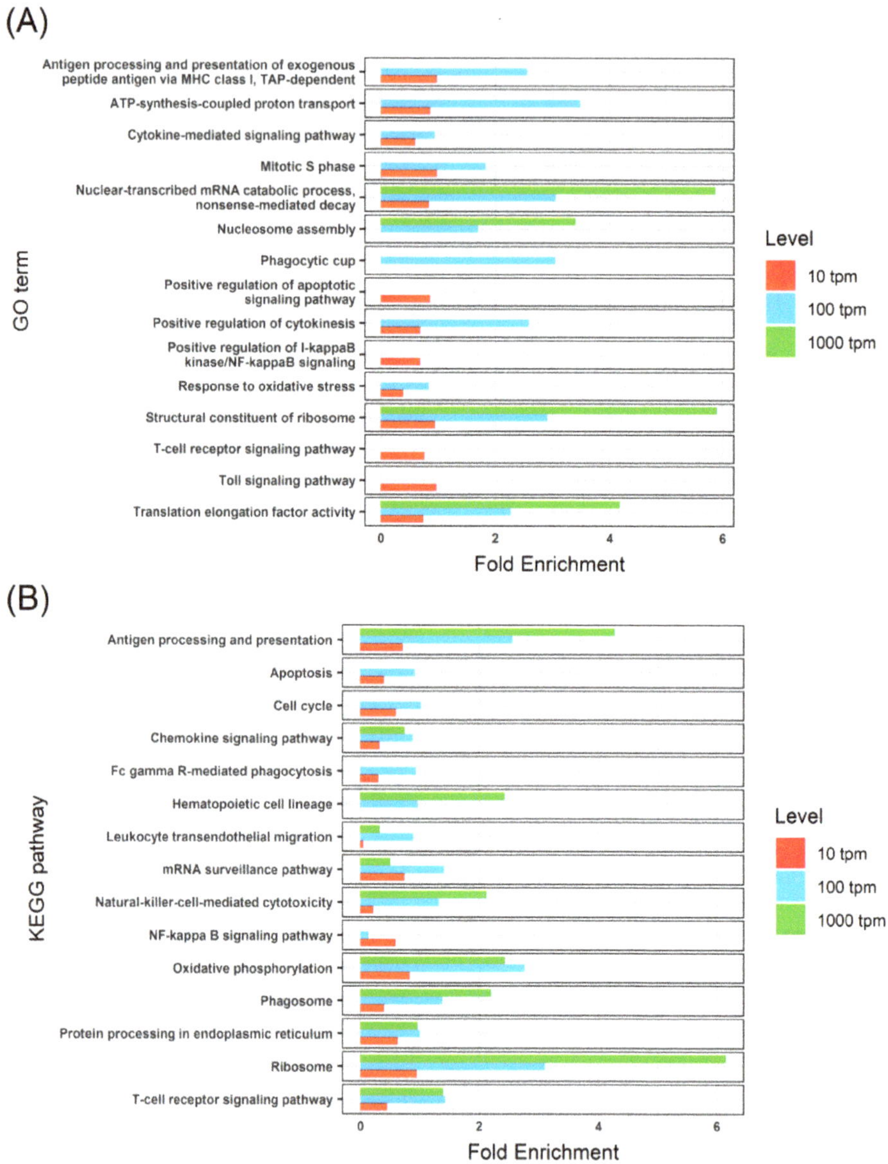

Figure 3. Bar graphs of representative enriched gene ontology (GO) terms (**A**) and KEGG pathways (**B**) found in three subsets of the most expressed genes in the thymus tissues of healthy turbot, tested by comparison to the whole transcriptome of the turbot. The subsets were selected based on different expression thresholds: TPM (transcripts per million) > 1000 (red, 134 genes), TPM > 100 (blue, 1322 genes) and TPM > 10 (green, 10371 genes). GO terms or pathways showing a Benjamini–Hochberg FDR-corrected p-value < 0.05 were considered to be enriched.

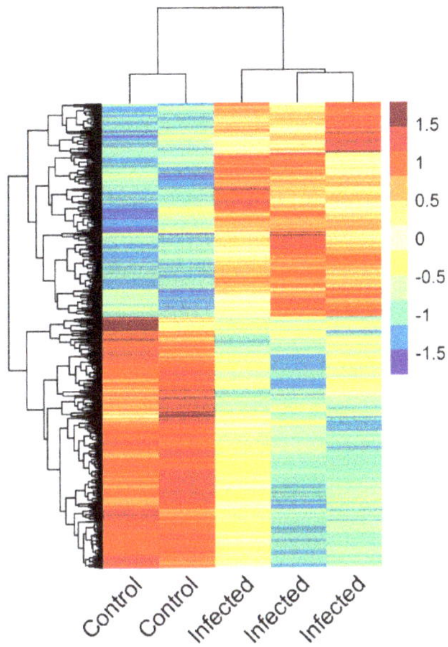

Figure 4. Heat map of the 4640 differentially expressed genes found in thymus tissues of turbot severely infected by *Enteromyxum scophthalmi* compared to controls. Expression values for each gene were scaled from −1.5 to 1.5 by subtracting the mean and dividing by the standard deviation. Samples from infected fish were analyzed individually, while the thymus tissues from four control fish were pooled into two samples for the transcriptomic analysis.

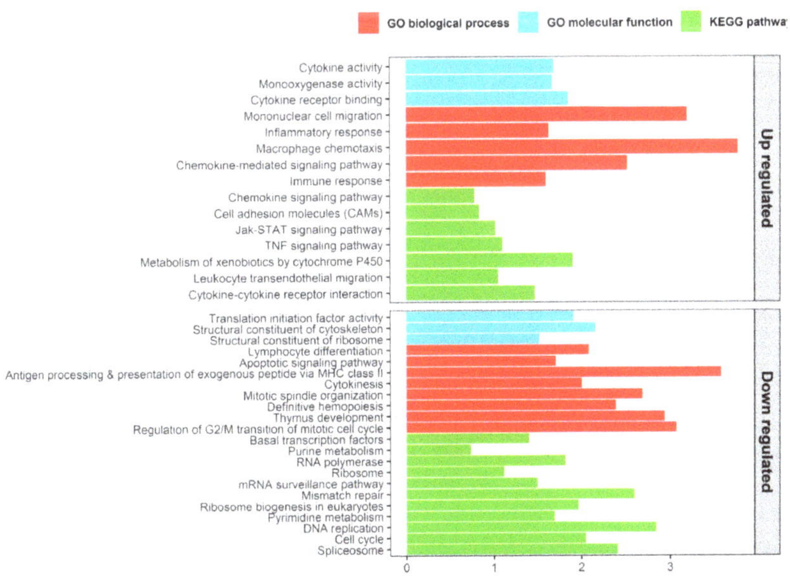

Figure 5. Bar plots of representative enriched gene ontology (GO) biological processes (red), GO molecular functions (blue), and KEGG pathways (green) in the thymus tissues of turbot at advanced stages of infection by *Enteromyxum scophthalmi*. The enrichment analyses were conducted on the sets of up- and downregulated genes found by comparison with healthy specimens. GO terms or pathways showing a Benjamini-Hochberg FDR-corrected *p*-value < 0.05 were considered to be enriched.

4. Discussion

The structural and functional characterization of the thymus has scarcely been addressed in many teleost species to date. In particular, the response of the thymus during diseases has barely been studied in cultured species, although several diseases of relevance in aquaculture are known to compromise the immune response and/or cause malnutrition. Infectious diseases and malnourishment are conditions markedly connected to thymus alterations in mammals [9–11]. In this study, transcriptomics and histological approaches were employed to contribute to the knowledge of this primary lymphoid organ in turbot, a commercially important species. The histological study ensured the healthy status of control specimens and served to classify *E.-scophthalmi*-infected turbot according to a histopathological grading system [19]. This approach allowed the selection of fish with similar lesions along the time course of experiment for the transcriptomic and immunohistochemical analyses, ensuring the evaluation of the same stage of disease and reducing the intragroup variance and consequently the number of animals needed to attain significant results.

Most of the functions enriched in the thymus tissues of control turbot were related to energy production, protein synthesis, and cell proliferation. This reflects the high anabolic environment of this organ in physiological conditions, also described for terrestrial vertebrates and associated with lymphoblastic proliferation [35]. Accordingly, the immunohistochemical assay using the cell proliferation marker PCNA showed a broad immunoreactivity in the thymus tissues of control fish. As expected, we also found overrepresented functions related to the molecular mechanisms involved in T-cell development (reviewed in Reference [4]): chemokine signaling defines the microenvironments for thymocyte development; antigen processing and presentation and apoptosis act in positive and negative selection of these cells; and T-cell receptor signaling pathways or natural-killer-mediated cytotoxicity can be linked to thymocytes' definitive maturation as different types of T lymphocytes.

This study was focused on turbot juveniles, which we expected to have fully functional thymuses. In future studies, it would be interesting to assess the potential histological and transcriptomic changes related to aging in turbot. Age-associated thymic involution is a process that apparently occurs in all terrestrial vertebrates [36,37], but it is still unclear to what extent it occurs in fish. The degree of involution seems to be variable among teleosts; some fish species undergo thymic involution in a similar way to mammals, whereas some other species do not show any sign of age-dependent thymic atrophy [6,38].

The lesions associated with disease-related effects on thymus tissue and the underlying pathogenic mechanisms are beginning to be understood from studies in human and murine models. Malnutrition and infection by a variety of pathogens, including parasites, have been shown to induce similar atrophy-related alterations in the thymus, consisting of cell depletion, reduction of the cortex:medulla ratio and loss of corticomedullary limit distinction [10,11,39]. This evidence is mainly due to an increased apoptotic rate of immature thymocytes and/or their abnormal egress caused by an impaired selection and export as a consequence of changes in the thymic microenvironment [9,11]. These changes primarily affect chemokines and extracellular matrix (ECM) proteins [11–14], which are essential components of this microenvironment, with a major role driving thymocyte migration [2,3].

The thymus tissues of turbot infected by *E. scophthalmi* showed intense transcriptomic changes at advanced stages of infection, whereas hardly any changes were noticed in fish with incipient infection, whether with morphological or with transcriptomic techniques. This might be related to a delayed response of the organ to the infection, which is characterized by a long pre-patent period and circumscribed to the gastrointestinal tract [18]. Further, potential parasite-induced silencing of the immune response was postulated at the first stages of infection [21].

Turbot enteromyxosis is a serious disease against which the host is incapable of triggering an effective immune response. Previous studies (reviewed in Reference [18]) have shown that an exacerbated and dysfunctional inflammatory response occurs once the parasitic load in the gut increases and the intestinal lesions start to become important. Specifically, high levels of proinflammatory mediators such as TNF-α and iNOS were found in the digestive tract as well as in the immune-related kidney and spleen organs, whereas a poor activation of protective anti-inflammatory mechanisms was observed [20,25,40]. Here, an increased presence of these proinflammatory molecules was observed by immunohistochemistry in the thymus tissues of severely infected fish, and, simultaneously, the transcriptomic profile showed evident signs of activation of the inflammatory response.

In mammals, pro-inflammatory mediators have been suggested to be responsible for the deleterious effects of infection on the architecture and function of the thymus, inducing apoptosis or alterations of the thymic microenvironment [9,12]. Nitric oxide locally produced through the iNOS signaling pathway has been implicated in thymic atrophy during infection by *Mycobacterium avium* [41], and TNF-α has also been associated with thymic atrophy in Chagas disease, caused by the parasite *Trypanosoma cruzi* [42]. It has also been suggested that TNF-α is involved in the enhanced activity of matrix metalloproteinases (MMPs) in the thymus during infection by *Plasmodium berghei* in mice, which have been related to thymus dysfunction via the alteration of extracellular matrix (ECM) proteins [43]. In that study, an enhanced expression of MMP-2 and MMP-9 and the tissue inhibitors TIMP-1 and TIMP-2 was observed. Here, all the DEGs related to MMPs were upregulated in the thymus tissues of heavily infected turbot, including MMP-2 and TIMP-2 (Supplementary File 3). In general, these results suggest that the thymus is exposed to the deleterious effects of the exacerbated immune response of turbot during enteromyxosis.

On the other hand, enteromyxosis induces a cachectic syndrome, which is particularly evident in advanced stages of infection, during which fish show severe weight loss and anorexia [18]. Lymphoid atrophy is a well-recognized consequence of nutritional deprivation in humans and animals [10,11,44]. In particular, given the pronounced loss of lymphoid tissue induced by starvation in this organ, the thymus was designated a "barometer of malnutrition" by Prentice [45].

The adipocyte-secreted hormone leptin has been indicated as a possible mediator of malnutrition-induced thymic atrophy in mammals [11,12]. Decreased leptin serum levels during malnutrition would compromise its immunoregulatory and anti-apoptotic functions in the thymus, where it acts through the leptin receptors present on thymocytes and reticulo-epithelial cells [39,46,47]. These aspects need to be further studied in teleosts, but most leptin functions appear to be evolutionarily conserved [47]. Interestingly, a gene encoding a leptin receptor was upregulated in the thymus tissues of turbot infected by *E. scophthalmi* (Supplementary File 3).

Both infectious diseases and malnutrition have been shown to ultimately provoke alterations of chemokines and ECM proteins of the thymic microenvironment [9,12–14]. In this study, several genes related to chemokines and ECM proteins governing thymocyte development and migration in mammals [2,3] showed modulated expression in heavily infected turbot (Supplementary File 3). Among these, changes in the expression of C-X-C chemokine receptor type 4, C-C chemokine receptors type 5 and 9, C-C chemokine ligand 25, and C-X-C chemokine ligand 9, and of the ECM components fibronectin and laminin, have been previously related to alteration of the thymic microenvironment in malnourished subject and during infection by *T. cruzi* and *P. berghei* [9,11,13].

Regarding apoptosis, the transcriptomic data revealed a modulation of this pathway at advanced stages of enteromyxosis. Moreover, the immunohistochemical technique for the detection of active caspase-3, an effector protease in the apoptotic process, indicated an increased presence of this protein. Thymocytes physiologically undergo apoptosis during the processes of positive and negative selection [48], and our results point towards an increased apoptosis rate in turbot infected by *E. scophthalmi*. However, the mechanisms by which this process is activated and the extent of its role in the development of thymus atrophy should be carefully investigated. A recent study on the thymus tissues of malnourished mice infected by *Leishmania infantum* showed that the alteration of the thymic microenvironment was the main cause underlying thymic atrophy, rather than an increased thymocyte apoptotic rate [13]. Furthermore, these authors demonstrated that the effect of malnutrition on the thymus deeply affects the immune response to this parasite by disturbing T-lymphocyte migration [13,14]. Indeed, both infection and malnutrition can independently give rise to a decline in thymic function that affects the peripheral T-cell pool and determines a negative impact on the ongoing immune response [9,11,12]. In the case of turbot enteromyxosis, we could imagine a scenario where the effect of the proinflammatory mediators and oxidative stress might be enhanced by starvation, preventing a proper immunological response by the thymus and causing the development of atrophy.

Although further investigations on the underlying mechanisms are necessary, strong evidence of declined thymic function during enteromyxosis was found in our study. Many genes found to be highly expressed in the thymus tissues of control fish, and putatively associated with thymic function, were found to be downregulated in infected fish. These genes were related to the immunological role of the organ, such as antigen presentation or lymphocyte differentiation, as well as to the processes of protein synthesis and cell proliferation. The transcriptomic profile was concordant with immunohistochemical and histopathological findings. The immunohistochemistry to PCNA confirmed a minor cell proliferation activity, which is a common feature observed in the thymus during infection and malnutrition [11,14,39]. Furthermore, reduced numbers of cells expressing the T-cell marker CD3ε were found. Most of the severely infected turbot showed cell depletion and altered thymus architecture upon histological examination. All in all, the results suggest a scenario of loss of cellularity and impaired thymic function, consistent with the findings associated with the development of thymus atrophy also observed in other species as a consequence of infectious diseases or dietary deficiencies [10,11].

Interestingly, cell depletion is also typically observed in spleen and kidney tissues at advanced stages of enteromyxosis [19]. This observation has been linked to an exacerbated leukocyte recruitment to the intestine, where these cells suffer high apoptotic rates [24,49], as well as to the deleterious effect of a prolonged exposure to proinflammatory cytokines [40]. RNA-seq analysis of those organs in heavily infected turbot also showed a decline in immunological function, as demonstrated by the downregulation of numerous genes related to adaptive immunity and involved in the coordination between innate and adaptive immune responses [20]. The failure in development of a coordinated immune response against *E. scophthalmi* in turbot has been previously associated with the cell depletion suffered by the lymphohematopoietic organs [20,49,50], and this study highlighted that the thymus might also be affected by this phenomenon, which could contribute to the inefficacy of the turbot response against *E. scophthalmi*. It is plausible that common pathogenic mechanisms of cell depletion occur in all lymphoid organs during enteromyxosis, with a probable synergistic effect of inflammatory mediators and malnutrition. A comprehensive understanding of these mechanisms might help to implement therapeutic measures [12,39].

5. Conclusions

The present investigation provides the first description of the turbot thymus transcriptome and gene expression profiling integrated with morphological observations. The results indicate the functional relatedness of this organ with that of mammals in physiological conditions. The significant changes found in advanced enteromyxosis suggest that thymic function is strongly affected, pointing towards the development of thymus atrophy as a consequence of the disease. This situation might contribute to the failure of the turbot immune response in fighting the disease. More studies are needed to increase our knowledge of the physiological and pathological thymus microenvironment and to investigate thymus atrophy during diseases in fish. This future research will benefit from combining genomic and proteomic approaches as well as from the identification of further immunohistochemical markers for the characterization of the thymus microenvironment and T-cell subsets. Overall, the present findings highlight that the role of the thymus in pathological conditions should not be overlooked, as it can contribute to a deeper understanding of host–pathogen interactions in infectious diseases.

Supplementary Materials: The following are available online at http://www.mdpi.com/2079-7737/9/8/221/s1, Supplementary File 1: Sequence alignment between the cytoplasmic tail of the CD3ε proteins of rainbow trout (*Oncorhynchus mykiss*) and turbot (*Scophthalmus maximus*). The peptide used by Boardman et al. (2012) as immunogen to produce an anti-rainbow trout CD3ε monoclonal antibody is highlighted in red, and the statistics of the local BLASTp comparison with its turbot homologue are shown. The reference sequences used were NP_001182103.1 (NCBI Reference Sequence) for rainbow trout and ENSSMAT00000000938.1 (Ensembl Transcript ID) for turbot. Supplementary File 2. Healthy thymus GO and KEGG enrichment. Three overlapping subsets of thymus genes were selected based on their normalised levels of expression (genes with TPM > 1000, 100, or 10), and GO term and KEGG pathway enrichment analyses were performed using the whole turbot transcriptome as background. Supplementary File 3. Differentially expressed genes in the thymus of turbot during enteromyxosis. List of genes showing significant differential expression (FDR p-value < 0.05) in the thymus at 24 and 42 days postinfection when compared to controls. Supplementary File 4. Enriched GO terms and KEGG pathways in turbot thymus 42 dpi after infection with *Enteromyxum scophthalmi*. List of GO terms and KEGG pathways showing significant enrichment (FDR p-value < 0.05) in lists of up- and downregulated genes observed in the thymus at 42 days post infection with *Enteromyxum scophthalmi* when compared to control.

Author Contributions: Conceptualization, M.I.Q., P.M., B.G.P., R.B. and P.R.; methodology, D.R., P.R., R.B. and A.P.L.; validation, P.M., M.I.Q. and B.G.P.; formal analysis, D.R.; investigation, P.R., D.R., R.B., A.P.L., B.G.P. and M.I.Q.; data curation, P.R., D.R. and A.P.L.; writing—original draft preparation, P.R.; writing—review and editing, P.R., D.R., R.B., B.G.P., P.M. and M.I.Q.; visualization, P.R. and D.R.; supervision, M.I.Q. and P.M.; project administration, M.I.Q., P.M., B.G.P. and P.R.; funding acquisition, M.I.Q., P.M., B.G.P. and R.B. All authors have read and agreed to the published version of the manuscript.

Funding: This research was funded by THE SPANISH MINISTRY OF ECONOMY, INDUSTRY AND COMPETITIVENESS AND THE EUROPEAN REGIONAL DEVELOPMENT FUND (ERDF) through the projects AGL2015–67039–C3–1–R and AGL2015–67039–C3–3–R. DR is supported by BBSRC INSTITUTE STRATEGIC PROGRAM GRANTS to the Roslin Institute (BB/P013732/1, BB/P013740/1, BB/P013759/1).

Acknowledgments: The authors would like to thank Sandra Maceiras and Lucía Insua for technical assistance and they acknowledge the support of the Centro de Supercomputación de Galicia (CESGA) in the completion of this work.

Conflicts of Interest: The authors declare no conflict of interest.

References

1. Miller, J.F.A.P. The golden anniversary of the thymus. *Nat. Rev. Immunol.* **2011**, *11*, 489–495. [CrossRef]
2. Lancaster, J.N.; Li, Y.; Ehrlich, L.I. Chemokine-mediated choreography of thymocyte development and selection. *Trends Immunol.* **2018**, *39*, 86–98. [CrossRef]
3. Savino, W.; Mendes-da-Cruz, D.A.; Smaniotto, S.; Silva-Monteiro, E.; Villa-Verde, D.M.S. Molecular mechanisms governing thymocyte migration: Combined role of chemokines and extracellular matrix. *J. Leukoc. Biol.* **2004**, *75*, 951–961. [CrossRef]
4. Takaba, H.; Takayanagi, H. The mechanisms of T cell selection in the thymus. *Trends Immunol.* **2017**, *38*, 805–816. [CrossRef]
5. Klein, L.; Kyewski, B.; Allen, P.M.; Hogquist, K.A. Positive and negative selection of the T cell repertoire: What thymocytes see (and don't see). *Nat. Rev. Immunol.* **2014**, *14*, 377–391. [CrossRef]
6. Bowden, T.J.; Cook, P.; Rombout, J.H. Development and function of the thymus in teleosts. *Fish Shellfish Immunol.* **2005**, *19*, 413–427. [CrossRef]
7. Nakanishi, T.; Shibasaki, Y.; Matsuura, Y. T cells in fish. *Biology* **2015**, *4*, 640–663. [CrossRef]
8. Carneiro-Sampaio, M. Thymus: Still an underexplored organ in medical practice. *Rev. Med. São Paulo* **2016**, *95*, 76–83. [CrossRef]
9. Nunes-Alves, C.; Nobrega, C.; Behar, S.M.; Correia-Neves, M. Tolerance has its limits: How the thymus copes with infection. *Trends Immunol.* **2013**, *34*, 502–510. [CrossRef]
10. Valli, V.E.O.; Kiupel, M.; Bienzle, D.; Wood, R.D. Chapter 2—Hematopoietic System. In *Jubb, Kennedy & Palmer's Pathology of Domestic Animals*, 6th ed.; Maxie, M.G., Ed.; W.B. Saunders: London, UK, 2016; Volume 3, pp. 102–268.e1.
11. Savino, W.; Dardenne, M.; Velloso, L.A.; Silva-Barbosa, S.D. The thymus is a common target in malnutrition and infection. *Br. J. Nutr.* **2007**, *98*, S11–S16. [CrossRef]
12. Majumdar, S.; Nandi, D. Thymic atrophy: Experimental studies and therapeutic interventions. *Scand. J. Immunol.* **2018**, *87*, 4–14. [CrossRef] [PubMed]

13. Losada-Barragán, M.; Umaña-Pérez, A.; Cuervo-Escobar, S.; Berbert, L.R.; Porrozzi, R.; Morgado, F.N.; Mendes-da-Cruz, D.A.; Savino, W.; Sánchez-Gómez, M.; Cuervo, P. Protein malnutrition promotes dysregulation of molecules involved in T cell migration in the thymus of mice infected with *Leishmania infantum*. *Sci. Rep.* **2017**, *7*, 45991. [CrossRef]
14. Losada-Barragán, M.; Umaña-Pérez, A.; Durães, J.; Cuervo-Escobar, S.; Rodríguez-Vega, A.; Ribeiro-Gomes, F.L.; Berbert, L.R.; Morgado, F.; Porrozzi, R.; Mendes-da-Cruz, D.A.; et al. Thymic microenvironment is modified by malnutrition and *Leishmania infantum* infection. *Front. Cell. Infect. Microbiol.* **2019**, *9*, 252. [CrossRef] [PubMed]
15. Langenau, D.M.; Zon, L.I. The zebrafish: A new model of T-cell and thymic development. *Nat. Rev. Immunol.* **2005**, *5*, 307–317. [CrossRef]
16. Vigliano, F.A.; Losada, A.P.; Castello, M.; Bermúdez, R.; Quiroga, M.I. Morphological and immunohistochemical characterisation of the thymus in juvenile turbot (*Psetta maxima*, L.). *Cell Tissue Res.* **2011**, *346*, 407–416. [CrossRef]
17. Padrós, F.; Crespo, S. Ontogeny of the lymphoid organs in the turbot *Scophthalmus maximus*: A light and electron microscope study. *Aquaculture* **1996**, *144*, 1–16. [CrossRef]
18. Ronza, P.; Robledo, D.; Bermúdez, R.; Losada, A.P.; Pardo, B.G.; Martínez, P.; Quiroga, M.I. Integrating genomic and morphological approaches in fish pathology research: The case of turbot (*Scophthalmus maximus*) enteromyxosis. *Front. Genet.* **2019**, *10*, 26. [CrossRef]
19. Bermúdez, R.; Losada, A.P.; Vázquez, S.; Redondo, M.J.; Álvarez-Pellitero, P.; Quiroga, M.I. Light and electron microscopic studies on turbot *Psetta maxima* infected with *Enteromyxum scophthalmi*: Histopathology of turbot enteromyxosis. *Dis. Aquat. Organ.* **2010**, *89*, 209–221. [CrossRef]
20. Robledo, D.; Ronza, P.; Harrison, P.W.; Losada, A.P.; Bermúdez, R.; Pardo, B.G.; Redondo, M.J.; Sitjà-Bobadilla, A.; Quiroga, M.I.; Martínez, P. RNA-seq analysis reveals significant transcriptome changes in turbot (*Scophthalmus maximus*) suffering severe enteromyxosis. *BMC Genom.* **2014**, *15*, 1149. [CrossRef]
21. Ronza, P.; Robledo, D.; Bermudez, R.; Losada, A.P.; Pardo, B.G.; Sitja-Bobadilla, A.; Quiroga, M.I.; Martinez, P. RNA-seq analysis of early enteromyxosis in turbot (*Scophthalmus maximus*): New insights into parasite invasion and immune evasion strategies. *Int. J. Parasitol.* **2016**, *46*, 507–517. [CrossRef]
22. Boardman, T.; Warner, C.; Ramirez-Gomez, F.; Matrisciano, J.; Bromage, E. Characterization of an anti-rainbow trout (*Oncorhynchus mykiss*) CD3ε monoclonal antibody. *Vet. Immunol. Immunop.* **2012**, *145*, 511–515. [CrossRef]
23. Losada, A.P.; Ronza, P.; Bermúdez, R.; Castrillo, A.; de Azevedo, A.M.; Quiroga, M.I. Proliferative Cell Nuclear Antigen (PCNA) Expression in the Intestine of *Scophthalmus maximus* Experimentally Infected with *Enteromyxum scophthalmi* (Myxozoa). In Proceedings of the 18th International Conference on Diseases of Fish and Shellfish, Belfast, UK, 4–8 September 2017.
24. Losada, A.P.; Bermúdez, R.; Faílde, L.D.; de Ocenda, M.V.R.; Quiroga, M.I. Study of the distribution of active caspase-3-positive cells in turbot, *Scophthalmus maximus* (L.), enteromyxosis. *J. Fish Dis.* **2014**, *37*, 21–32. [CrossRef]
25. Losada, A.P.; Bermúdez, R.; Faílde, L.D.; Quiroga, M.I. Quantitative and qualitative evaluation of iNOS expression in turbot (*Psetta maxima*) infected with *Enteromyxum scophthalmi*. *Fish Shellfish Immunol.* **2012**, *32*, 243–248. [CrossRef]
26. Ronza, P.; Losada, A.P.; Villamarín, A.; Bermúdez, R.; Quiroga, M.I. Immunolocalisation of tumor necrosis factor alpha in turbot (*Scophthalmus maximus*, L.) tissues. *Fish Shellfish Immunol.* **2015**, *45*, 470–476. [CrossRef]
27. Bolger, A.M.; Lohse, M.; Usadel, B. Trimmomatic: A flexible trimmer for Illumina sequence data. *Bioinformatics* **2014**, *30*, 2114–2120. [CrossRef]
28. Figueras, A.; Robledo, D.; Corvelo, A.; Hermida, M.; Pereiro, P.; Rubiolo, J.A.; Gomez-Garrido, J.; Carrete, L.; Bello, X.; Gut, M.; et al. Whole genome sequencing of turbot (*Scophthalmus maximus*; Pleuronectiformes): A fish adapted to demersal life. *DNA Res.* **2016**, *23*, 181–192. [CrossRef]
29. Dobin, A.; Davis, C.A.; Schlesinger, F.; Drenkow, J.; Zaleski, C.; Jha, S.; Batut, P.; Chaisson, M.; Gingeras, T.R. STAR: Ultrafast universal RNA-seq aligner. *Bioinformatics* **2013**, *29*, 15–21. [CrossRef]
30. Trapnell, C.; Williams, B.A.; Pertea, G.; Mortazavi, A.; Kwan, G.; van Baren, M.J.; Salzberg, S.L.; Wold, B.J.; Pachter, L. Transcript assembly and quantification by RNA-Seq reveals unannotated transcripts and isoform switching during cell differentiation. *Nat. Biotechnol.* **2010**, *28*, 511–515. [CrossRef]
31. Liao, Y.; Smyth, G.K.; Shi, W. featureCounts: An efficient general purpose program for assigning sequence reads to genomic features. *Bioinformatics* **2013**, *30*, 923–930. [CrossRef]
32. Love, M.I.; Huber, W.; Anders, S. Moderated estimation of fold change and dispersion for RNA-seq data with DESeq2. *Genome Biol.* **2014**, *15*, 550. [CrossRef]

33. Conesa, A.; Gotz, S.; Garcia-Gomez, J.M.; Terol, J.; Talon, M.; Robles, M. Blast2GO: A universal tool for annotation, visualization and analysis in functional genomics research. *Bioinformatics* **2005**, *21*, 3674–3676. [CrossRef]
34. Wu, J.; Mao, X.; Cai, T.; Luo, J.; Wei, L. KOBAS server: A web-based platform for automated annotation and pathway identification. *Nucleic Acids Res.* **2006**, *34*, W720–W724. [CrossRef]
35. Griffith, A.V.; Venables, T.; Shi, J.; Farr, A.; van Remmen, H.; Szweda, L.; Fallahi, M.; Rabinovitch, P.; Petrie, H.T. Metabolic damage and premature thymus aging caused by stromal catalase deficiency. *Cell Rep.* **2015**, *12*, 1071–1079. [CrossRef]
36. Palmer, D. The effect of age on thymic function. *Front. Immunol.* **2013**, *4*, 316. [CrossRef]
37. Shanley, D.P.; Aw, D.; Manley, N.R.; Palmer, D.B. An evolutionary perspective on the mechanisms of immunosenescence. *Trends Immunol.* **2009**, *30*, 374–381. [CrossRef]
38. Torroba, M.; Zapata, A.G. Aging of the vertebrate immune system. *Microsc. Res. Techniq.* **2003**, *62*, 477–481. [CrossRef]
39. Ansari, A.R.; Liu, H. Acute thymic involution and mechanisms for recovery. *Arch. Immunol. Ther. Exp.* **2017**, *65*, 401–420. [CrossRef]
40. Ronza, P.; Bermudez, R.; Losada, A.P.; Sitja-Bobadilla, A.; Pardo, B.G.; Quiroga, M.I. Immunohistochemical detection and gene expression of TNFalpha in turbot (*Scophthalmus maximus*) enteromyxosis. *Fish Shellfish Immunol.* **2015**, *47*, 368–376. [CrossRef]
41. Borges, M.; Barreira-Silva, P.; Flórido, M.; Jordan, M.B.; Correia-Neves, M.; Appelberg, R. Molecular and cellular mechanisms of *Mycobacterium avium*-induced thymic atrophy. *J. Immunol.* **2012**, *189*, 3600–3608. [CrossRef]
42. Pérez, A.R.; Berbert, L.R.; Lepletier, A.; Revelli, S.; Bottasso, O.; Silva-Barbosa, S.D.; Savino, W. TNF-α is involved in the abnormal thymocyte migration during experimental *Trypanosoma cruzi* infection and favors the export of immature cells. *PLoS ONE* **2012**, *7*, e34360. [CrossRef]
43. Lima, A.C.D.; Francelin, C.; Ferrucci, D.L.; Stach-Machado, D.R.; Verinaud, L. Thymic alterations induced by *Plasmodium berghei*: Expression of matrix metalloproteinases and their tissue inhibitors. *Cell. Immunol.* **2012**, *279*, 53–59. [CrossRef] [PubMed]
44. Tanaka, M.; Suganami, T.; Kim-Saijo, M.; Toda, C.; Tsuiji, M.; Ochi, K.; Kamei, Y.; Minokoshi, Y.; Ogawa, Y. Role of central leptin signaling in the starvation-induced alteration of B-cell development. *J. Neurosci.* **2011**, *31*, 8373–8380. [CrossRef] [PubMed]
45. Prentice, A.M. The thymus: A barometer of malnutrition. *Br. J. Nutr.* **1999**, *81*, 345–347. [CrossRef] [PubMed]
46. Velloso, L.A.; Savino, W.; Mansour, E. Leptin action in the thymus. *Ann. N. Y. Acad. Sci.* **2009**, *1153*, 29–34. [CrossRef] [PubMed]
47. Procaccini, C.; La Rocca, C.; Carbone, F.; De Rosa, V.; Galgani, M.; Matarese, G. Leptin as immune mediator: Interaction between neuroendocrine and immune system. *Dev. Comp. Immunol.* **2017**, *66*, 120–129. [CrossRef] [PubMed]
48. Romano, N.; Ceccarelli, G.; Caprera, C.; Caccia, E.; Baldassini, M.R.; Marino, G. Apoptosis in thymus of teleost fish. *Fish Shellfish Immunol.* **2013**, *35*, 589–594. [CrossRef]
49. Bermúdez, R.; Vigliano, F.; Marcaccini, A.; Sitjà-Bobadilla, A.; Quiroga, M.I.; Nieto, J.M. Response of Ig-positive cells to *Enteromyxum scophthalmi* (Myxozoa) experimental infection in turbot, *Scophthalmus maximus* (L.): A histopathological and immunohistochemical study. *Fish Shellfish Immunol.* **2006**, *21*, 501–512. [CrossRef]
50. Sitjà-Bobadilla, A.; Redondo, M.J.; Bermúdez, R.; Palenzuela, O.; Ferreiro, I.; Riaza, A.; Quiroga, I.; Nieto, J.M.; Alvarez-Pellitero, P. Innate and adaptive immune responses of turbot, *Scophthalmus maximus* (L.), following experimental infection with *Enteromyxum scophthalmi* (Myxosporea: Myxozoa). *Fish Shellfish Immunol.* **2006**, *21*, 485–500. [CrossRef]

biology

MDPI

Article

Genome-Wide Identification and Expression Analysis of Potential Antiviral Tripartite Motif Proteins (TRIMs) in Grass Carp (*Ctenopharyngodon idella*)

Beibei Qin †, Tiaoyi Xiao †, Chunhua Ding, Yadong Deng, Zhao Lv *,‡ and Jianming Su *,‡

Hunan Engineering Technology Research Center of Featured Aquatic Resources Utilization, Hunan Agricultural University, Changsha 410128, China; beibeiqin@stu.hunau.edu.cn (B.Q.); tiaoyixiao@hunau.edu.cn (T.X.); chunhuading@stu.hunau.edu.cn (C.D.); ya-dong.deng@stu.hunau.edu.cn (Y.D.)
* Correspondence: lvzhao_0320@hunau.edu.cn (Z.L.); sjmauhn@hunau.edu.cn (J.S.)
† Beibei Qin and Tiaoyi Xiao should be considered joint first authors.
‡ Zhao Lv and Jianming Su should be considered joint last author.

Simple Summary: Grass carp, *Ctenopharyngodon idellus*, is an important freshwater cultured teleost in China, and its annual production has reached 5,533,083 tons. However, its aquaculture is severely restricted by hemorrhagic disease caused by the grass carp reovirus (GCRV). For the better control of grass carp hemorrhagic disease, the breeding of resistant grass carp strains based on antiviral immune molecule markers is a potential solution. However, the molecular basis of grass carp's resistance to GCRV infection remains largely unknown, greatly limiting the breeding of grass carp resistant to hemorrhagic disease. Given the importance of tripartite motif proteins (TRIMs) in animal antiviral immunity, we used the Hidden Markov Model Biological Sequence Analysis software (HMMER) and SMART to identify TRIMs in the grass carp genome and analyze their gene loci, as well as structural and evolutionary features. We also tried to uncover antiviral TRIMs and their mediated immune processes based on two sets of transcriptomes during GCRV infection in grass carp. This study provides information for the understanding of TRIMs and antiviral immunity in grass carp.

Abstract: Tripartite motif proteins (TRIMs), especially B30.2 domain-containing TRIMs (TRIMs-B30.2), are increasingly well known for their antiviral immune functions in mammals, while antiviral TRIMs are far from being identified in teleosts. In the present study, we identified a total of 42 *Ci*TRIMs from the genome of grass carp, *Ctenopharyngodon idella*, an important cultured teleost in China, based on hmmsearch and SMART analysis. Among these *Ci*TRIMs, the gene loci of 37 *Ci*TRIMs were located on different chromosomes and shared gene collinearities with homologous counterparts from human and zebrafish genomes. They possessed intact conserved RBCC or RB domain assemblies at their N-termini and eight different domains, including the B30.2 domain, at their C-termini. A total of 19 TRIMs-B30.2 were identified, and most of them were clustered into a large branch of *Ci*TRIMs in the dendrogram. Tissue expression analysis showed that 42 *Ci*TRIMs were universally expressed in various grass carp tissues. A total of 11 significantly differentially expressed *Ci*TRIMs were found in two sets of grass carp transcriptomes during grass carp reovirus (GCRV) infection. Three of them, including *Ci*btr40, *Ci*TRIM103 and *Ci*TRIM109, which all belonged to TRIMs-B30.2, were associated with the type I interferon response during GCRV infection by weighted network co-expression and gene expression trend analyses, suggesting their involvement in antiviral immunity. These findings may offer useful information for understanding the structure, evolution, and function of TRIMs in teleosts and provide potential antiviral immune molecule markers for grass carp.

Keywords: tripartite motif proteins; B30.2 domain; antiviral immunity; *Ctenopharyngodon idella*; grass carp reovirus

Citation: Qin, B.; Xiao, T.; Ding, C.; Deng, Y.; Lv, Z.; Su, J. Genome-Wide Identification and Expression Analysis of Potential Antiviral Tripartite Motif Proteins (TRIMs) in Grass Carp (*Ctenopharyngodon idella*). *Biology* **2021**, *10*, 1252. https://doi.org/10.3390/biology10121252

Academic Editors: Patricia Pereiro and Oswaldo Palenzuela

Received: 27 September 2021
Accepted: 30 November 2021
Published: 1 December 2021

1. Introduction

Tripartite motif proteins (TRIMs) are generally characterized by three domains at the N-terminus, including a RING finger domain, one or two B-box domains and a coiled-coil domain, and are also known as the RING finger/B-box/coiled-coil (RBCC) domain-containing proteins [1,2]. The B-box domains represent a very ancient domain family that can be traced back to a common ancestor in protozoa, metazoa and even plants [3,4]. Meanwhile, the complete RBCC domain assemblies have only been discovered in TRIM family proteins from metazoa, such as arthropods, teleosts, amphibians, birds and mammals [4–7]. It has recently been confirmed, in mammals and teleosts, that proteins only containing the RING finger domain and B-box domains (RB domain assemblies) also exert biological functions similar to those containing RBCC domain assemblies, which further expands the TRIM family of proteins [8,9]. Although the RBCC or RB domain assemblies appear to be conserved in animals, almost every species has more than one TRIM or a specific repertoire of TRIMs, all of which, together, constitute a large protein family with highly variable sequences [4,10]. These TRIM family proteins play multiple roles in animal tissue development [11], metabolism and autophagy [12], transcriptional regulation [13], tumor suppression [14] and viral restriction [15].

Most TRIM family proteins possess an additional distinct domain at the C-terminus, including at least 11 categories of domains such as the Plant Homeo Domain (PHD), meprin and TRAF homology domain (Math), bromodomain (BROMO) and B30.2 domain (constituted by the juxtaposition of a PRY and a SPRY domain, alternatively called the PRY/SPRY domain) [16]. The C-terminal domains often determine the specificity of the interactions of TRIMs with other proteins [12,17]. Hence, TRIM proteins' RING-dependent E3 ubiquitin ligase activity is associated with the capacity to build multiprotein complexes though interactions with C-terminal domains. According to the categories of the C-terminal domains, the mammalian TRIM family proteins can be classified into nine main subsets, further extended to eleven subsets, according to Ozato's nomenclature [15]. Among the identified C-terminal domains, B30.2 domains are the most frequent in TRIMs. In humans, 35 TRIMs containing the B30.2 domain (TRIMs-B30.2) have been found at the C-termini of 80 TRIM family proteins [16]. In addition, these TRIMs-B30.2 evolve significantly faster than other TRIMs based on the calculated ratio of the non-synonymous substitution rate (*Ka*) to the synonymous substitution rate (*Ks*) [4]. Recently, several studies have identified a cluster of TRIM-B30.2 genes flanking the human major histocompatibility complex (MHC) gene locus, a well-known immune gene [18]. Functional experiments have also demonstrated that TRIMs-B30.2 tend to be involved in host immune defense against viral infections in mammals. For example, human TRIM5a recognizes human immunodeficiency virus-1 via the B30.2 domain and inhibits viral replication by promoting the degradation of the viral outer capsids [19]. TRIM21 can bind to hepatitis B virus through the B30.2 domain and participate in the degradation of the viral DNA polymerase [20]. Evidence indicates that the B30.2 domain of TRIMs is critical for antiviral immunity in mammals.

In teleosts, TRIMs-B30.2 have undergone a massive expansion under positive selection pressure and duplicated to develop into three subfamilies: bloodthirsty-like (btr), hematopoietic lineage switch-5 (hltr) and fintrim (ftr) [10,21,22]. In the zebrafish genome, the TRIMs-B30.2 consist of 33 btrs, 43 hltrs and 88 ftrs [10]. The results of a molecular evolution analysis indicate that teleost hltrs and TRIM35 can be traced back to a common ancestor, while teleost btrs and TRIM39 share a common ancestor [10,21]. Mammalian TRIM35 and TRIM39 are well known for their immune-defense roles in viral infections [23,24]. In teleosts, a new btr gene has been identified in Atlantic cod, *Gadus morhua*, and detected in a poly I:C subtractive library [25]. In zebrafish, the btr subfamily TRIM gene btr20 shows high expression levels in immune tissues, including the intestines, gills, kidneys, and spleen [26]. Therefore, teleost btrs and hltrs are also believed to participate in antiviral immunity. To date, no common ancestor of ftrs has been found in the TRIM family proteins from other species, and ftrs are considered to be teleost-specific TRIMs [22]. Zebrafish ftr83 has been confirmed to regulate the expression of interferon (IFN) and IFN-stimulating genes and is involved in the immune defense against infectious hematopoietic necrosis virus, viral hemorrhagic septicemia virus (VHSV) and spring viremia of carp virus infection [27]. Scattered evidence suggests conserved roles for TRIMs-B30.2, including btrs, hltrs and ftrs, in antiviral immunity in teleosts. However, teleost antiviral TRIMs are far from being identified due to the gap in research on fish immunology and many TRIM family proteins.

Transcriptome analyses, including weighted network co-expression analysis (WGCNA) and gene expression trend analysis, represent useful methods with which to survey host anti-infection immune molecules because they provide genome-wide profiles of gene expression [28–30]. In humans, a set of key immune genes, including IFN, interferon regulatory factor (IRF) 1, IRF 7, etc., interacting with human immunodeficiency virus (HIV)-1 have been defined by using WGCNA to construct gene co-expression networks based on transcriptome data from 52 patients [31]. Several antiviral immune genes, including laboratory of genetics and physiology 2 (LGP2), transforming growth factor-β-activated kinase 1 (TAK1) and zinc finger protein 36 (ZFP36), activated during encephalomyocarditis virus infection, have also been identified through gene expression trend analysis based on transcriptome data [32]. In teleosts, Ning et al. employed gene expression trend analysis combined with WGCNA to identify the differentially expressed gene clusters associated with anti-infection immune processes, including cytokine–cytokine receptor signaling, Toll-like receptor signaling and other immune-related pathways, from the transcriptomes of Japanese flounder (*Paralichthys olivacrus*) infected with *Vibrio anguillarum* [33].Recently, 10 hltrs associated with the type I IFN response significantly upregulated during VHSV infection were also identified as antiviral TRIMs by transcriptome analysis in rainbow trout [34].

Grass carp, *Ctenopharyngodon idellus*, is an important freshwater cultured teleost species in China, and its annual production has reached 5,533,083 tons [35]. However, the aquaculture of grass carp is severely restricted by grass carp hemorrhagic disease, which is caused by a double-stranded RNA virus known as the grass carp reovirus (GCRV) [36]. To better control grass carp hemorrhagic disease, it is urgent to investigate the molecular basis of grass carp's ability to resist GCRV infection, and the breeding of resistant grass carp strains based on antiviral immune molecule markers is a potential solution [37]. However, the molecular basis of grass carp resistance to GCRV infection remains largely unknown, greatly limiting the breeding of grass carp resistant to hemorrhagic disease [38]. Therefore, the genome-wide identification of antiviral immune molecules could uncover the molecular basis of GCRV resistance in grass carp and contribute to disease resistance breeding. Given the importance of TRIMs, especially TRIMs-B30.2, in animal antiviral immunity, we used the Hidden Markov Model Biological Sequence Analysis software (HMMER) in the present study to screen TRIM family genes in the grass carp genome and identify TRIMs-B30.2 in line with their structural and evolutionary features. We also tried to identify potential antiviral TRIMs by analyzing the gene expression patterns during GCRV infection in two sets of transcriptome data in grass carp by using WGCNA and gene expression trend analysis. This study may not only find potential antiviral immune molecule markers for disease resistance breeding in grass carp, but also provide useful information for understanding the structure, evolution, and function of TRIMs in teleosts.

2. Materials and Methods

2.1. Identification of TRIMs in the Grass Carp Genome

The grass carp genome data were downloaded from the Grass Carp Genome Database (GCGD, http://bioinfo.ihb.ac.cn/gcgd/php/index.php, accessed on 20 November 2020) [39]. Putative TRIMs were first retrieved from the grass carp genome using the HMMER3.1 software with a multi-sequence alignment algorithm and with an E-value of 1E-5, using three models from the Pfam database (http://pfam.xfam.org/, accessed on 23 November 2020) [40], including two RING finger domains (PF13445 and PF14634) and one B-box domain (PF00643) as templates, respectively. The intersection of the three produced hmm search result files were then extracted with a shell script and submitted to the Simple Modular Architecture Research Tool (SMART, http://smart.embl-heidelberg.de, accessed on 27 November 2020) [41] for domain analysis. According to the domain architecture results, redundant TRIMs were manually filtered out. Finally, grass carp TRIMs (*Ci*TRIMs) were identified according to the criterion of whether they possessed conserved RBCC or RB domain assemblies.

2.2. Gene Structure Analysis and Subcellular Localization Prediction of CiTRIMs

The information of the amino acid sequence length and number of introns and exons for *Ci*TRIMs was extracted from the grass carp genome annotation file by using a shell script. The protein molecular weights and isoelectric points of the *Ci*TRIMs were predicted through Expasy (https://www.expasy.org/, accessed on 10 December 2020) [42]. The amino acid sequences of the *Ci*TRIMs were submitted to the online platform Euk-mPLoc 2.0 Server (http://www.csbio.sjtu.edu.cn/bioinf/Cell-PLoc-2/, accessed on 11 December 2020) [43] for subcellular localization prediction.

2.3. Domain/Motif Architecture and the Dendrogram of CiTRIMs

The domain architecture results for the *Ci*TRIMs were collected from SMART and then visualized using Adobe Illustrator 2020 (version 24.1.0). The protein sequences of the *Ci*TRIMs were submitted to the MEME Suite 5.3.3 (http://meme-suite.org/tools/meme, accessed on 21 December 2020) [44] with the application of Motif Discovery for motif analysis and with the parameter of 10 for selecting the number of motifs. The dendrogram of the *Ci*TRIMs was constructed as previously described [45]. Multiple amino acid sequence alignments of the *Ci*TRIMs were conducted using the ClustalW 1.81 software with the default parameters. The MEGA 6.06 software [45] was then used to construct a dendrogram with the neighbor-joining algorithm and with the parameters including the p-distance, complete deletion and gap setting; the results were tested for reliability over 1000 bootstrap replicates, after which the editing was carried out online by using EVOLVIEW (https://evolgenius.info, accessed on 23 December 2020) [46].

2.4. Chromosomal Localization and Collinearity Analysis

The chromosomal localization analysis of *Ci*TRIMs was conducted according to the previous method with slight modifications [47]. In brief, the chromosome map draft was redrawn by mapping the assembled 301 scaffolds (with an average length of >179,941 bp) from the published grass carp genome into chromosomes. The number and localization information for the *Ci*TRIMs on the chromosomes was obtained using a shell script and then visualized using Mapgene2Chromosome (version 2.1) [48]. The gene collinearity analysis of the *Ci*TRIMs was also performed according to the previous methods. The human genome (GRCh38) and zebrafish genome (GRCz11) data were downloaded from the Ensembl Animal Genome database (http://www.ensembl.org/index.html, accessed on 3 January 2021) [49]. The TBtools software was used to handle the redundancies in the grass carp, zebrafish, and human genomes [50]. The longest transcript sequence for each gene in these three de-redundant genomes was extracted as the representative sequence using TBtools. The gene collinearities of TRIMs from grass carps, zebrafish and humans were analyzed by using multiple Collinear Scanning Toolkits (MCScanX) [51].

2.5. Expression Analysis of CiTRIMs in Uninfected Grass Carp Tissues

To investigate the tissue expression patterns of the *Ci*TRIMs, the published transcriptome data for uninfected grass carp tissues, including the kidneys, liver, head kidneys, spleen, brain, and embryo, were downloaded from GCGD. The protein sequences of the *Ci*TRIMs from the grass carp genome were submitted to the online platform eggNOG-MAPPER (http://eggnog-mapper.embl.de/, accessed on 4 January 2021) [52] with default parameters for genome-wide functional annotation. The RPKM (reads per kilobase per million mapped reads) values for the *Ci*TRIMs were obtained from the transcriptome data and then submitted to TBtools for normalization and the production of a heatmap.

2.6. Expression Analysis of CiTRIMs in Spleen Tissue during GCRV Infection

The published transcriptome raw data (SRP095827) [53] for the spleens from grass carp infected with GCRV on days 1, 3, 5 and 7 were downloaded from the Sequence Read Archive (SRA, https://www.ncbi.nlm.nih.gov/sra/?term=, accessed on 17 January 2021) database for identifying *Ci*TRIMs differentially expressed during GCRV infection. These raw data were reanalyzed with the following workflow: TrimGalore (version 0.6.4) was first used to eliminate adapter and low-quality sequences from the raw reads with the parameters of -q 20, -phred 33, -stringency 2, -length 20 and -e 0.1. FastQC (version 0.11.8) was also adopted to assess whether the cleaned reads met the requirements for subsequent analyses. Then, Hisat2 (version 2.1.0) [54] was employed to align cleaned reads to the grass carp genome with default parameters, followed by the counting of transcripts using FeatureCount (version 1.6.4) [55]. A differential expression analysis was performed using the DESeq R package (version 1.30.1) [56] with default parameters. The Benjamini and Hochberg approach was used to control the false discovery rate (FDR) through the adjustment of the resulting *p*-values. The average RPKM values for the *Ci*TRIMs from three biological replicates were obtained from the transcriptome data and submitted to TBtools for normalization and the production of a heatmap. The differentially expressed *Ci*TRIMs were identified in terms of fold changes > 2 and FDRs (or adjusted *p*-values) < 0.05.

WGCNA (version 1.70-3) was used to further explore if the differentially expressed *Ci*TRIMs were associated with immune processes based on the transcriptomes of spleens from grass carp infected with GCRV on Days 1, 3, 5 and 7, using the previous method with slightly modifications [57]. In brief, the function of genefilter's varFilter (version 1.72.1) [58] in the R package was used to exclude the genes with low expression variation within samples, with a var.cutoff of 0.3. The soft-thresholding power was selected by using the function pickSoftThreshold; then, the function blockwiseModules was adopted for gene network construction and module identification, with the parameters of corType = pearson, power = 6, networkType = unsigned, TOMType = unsigned, maxBlockSize = 100,000 and other default parameters, followed by the calculation of the coefficients of the correlation between module and trait (infection time points) by using the function cor. The Student asymptotic *p*-value was determined using the function of corPvalueStudent, with an Student asymptotic *p*-value < 0.05, marking a significant difference. The targeted module was exported from the Cytoscape software using the function of exportNetworkToCytoscape and with a threshold of 0.415. Finally, highly interconnected gene networks including differentially expressed *Ci*TRIMs were obtained by the application of MCODE within Cytoscape, followed by a GO enrichment analysis of functional annotations including Biological Process, Cellular Component and Molecular Function with Metascape (https://metascape.org/gp/index.html#/main, accessed on 3 February 2021) [59].

2.7. Expression Analysis for CiTRIMs in Kidney Cell Line during GCRV Infection

The published transcriptome raw data (PRJNA597582 and PRJNA597542) [60] of the grass carp kidney cell line (CIK) after GCRV challenge at 0 h (control), 6 h, 12 h and 24 h were also downloaded for identifying *Ci*TRIMs differentially expressed during GCRV infection. These raw data were reanalyzed with the same workflow as described above. The average RPKM values for the *Ci*TRIMs from three biological repeats at each infection time point were obtained from the transcriptome data and then submitted to TBtools for normalization and the production of a heatmap. The differentially expressed *Ci*TRIMs were identified in terms of fold changes > 2 and adjusted *p*-values (or FDRs) < 0.05.

To further explore if the differentially expressed *Ci*TRIMs were associated with immune processes in the CIK transcriptomes, a gene expression trend analysis was conducted using Short Time-series Expression Miner (STEM, version 1.3.13), with reference to the method previously described [61]. Briefly, the medians of the differentially expressed genes'

RPKM values from the CIK transcriptomes were first taken and imported into STEM to analyze the gene expression trends with the parameter of log normalize data. Different profiles where specific gene clusters showed similar expression trends were produced. The significantly similar gene expression trends were indicated as p-values < 0.05 by STEM. Targeted profiles containing differentially expressed *Ci*TRIMs were exported and submitted to Metascape, followed by a GO enrichment analysis with the zebrafish annotation database as the reference [59].

2.8. The Verification of Differentially Expressed CiTRIMs by qPCR

CIK cells were cultured in an incubator (Thermo Fisher Scientific, Waltham, MA, USA) at 28 °C with 5% CO_2 and with Medium 199 (Gibco, Grand Island, NY, USA) liquid medium containing a 1% penicillin–streptomycin mixture and 10% fetal bovine serum. When covering 80% of the bottom of the culture flask (Corning, NY, USA), the cells were detached using trypsin and transferred into 6-well plates (Corning, NY, USA). For the GCRV challenge experiment, GCRV (GCRV JX-01 strain, kindly provided by Professor Zeng Lingbing from the Yangtze River Fisheries Research Institute of the Chinese Academy of Fishery Sciences) suspension was added into the 6-well plates. The cell samples were collected after the GCRV challenge at 0 h, 6 h, 12 h and 24 h. Three biological replicate samples were taken for each infection time point.

Total RNA from the cell samples was extracted using an RNA-easy™ Isolation Reagent Kit (Vazyme, Nanjing, China), according to the manufacturer's instructions, followed by cDNA synthesis with a RevertAid First Strand cDNA Synthesis Kit (Thermo Fisher Scientific, Waltham, MA, USA). Specific primers (Table 1) were designed to detect the mRNA expression levels of genes, including eleven differentially expressed *Ci*TRIMs and interferon regulatory factor 3 (*Ci*IRF3) identified in transcriptomes, and VP2, which represented a GCRV protein component, by quantitative real-time polymerase chain reaction (qPCR) using a CFX96 Touch Real-Time PCR Detection System (Bio-Rad, Hercules, CA, USA). The grass carp β-actin gene was employed as the internal control. The amplifications were performed in triplicate in a total volume of 10 μL, containing 5 μL of ChamQ™ Universal SYBR qPCR Master Mix (Vazyme, Nanjing, China), 1 μL of diluted cDNA, 0.4 μL of each primer and 3.2 μL of ddH_2O. The cycle conditions were as follows: 1 cycle at 95 °C for 3 min, 40 cycles at 95 °C for 15 s, 60 °C for 15 s and 72 °C for 15 s. The relative expression levels of the genes were analyzed with the Ct method ($2^{-\Delta\Delta Ct}$ method) [62]. The data are expressed as means ± standard deviations and were analyzed with the Statistical Package for Social Sciences Version 25.0 (SPSS Inc., Chicago, IL, USA). The significance of the differences in expression levels was tested by one-way analysis of variance (ANOVA) and multiple comparisons. Statistically significant differences were represented by $p < 0.05$.

Table 1. The qPCR primers used in this study.

Primer Name	Primer Sequence (5′–3′)
*Ci*TRIM2-F	TGGTGCGTCAGATCGACAAA
*Ci*TRIM2-R	CTGTGGGCGGGAATGTAGTT
*Ci*TRIM35-16-F	TCTGGTTCCTGTCCTCAATGC
*Ci*TRIM35-16-R	TGTTAGCCACAATGCGGTTG
*Ci*TRIM-35-50-F	CCTCCAGTCAATCAGGCTCT
*Ci*TRIM-35-50-F	ATTTCCTTTGTTGCCTCTGCT
*Ci*btr40-F	AAAAGACAGCAGTGCAGCAG
*Ci*btr40-R	CGATCTCCTTCTCTTTGGCTTG
*Ci*TRIM46b-F	TAGAAAGCGGCATTGCTCAG
*Ci*TRIM46-R	ACCACGCAATTCACTCACAC
*Ci*TRIM5-like-F	ACGCCATTGATGCTCTTGTG
*Ci*TRIM54-like-R	TTGGCACGTTGAGCATTGTC
*Ci*TRIM71-F	ACCATCGCATTCAGGTGTTCG
*Ci*TRIM71-R	TCATTCCATCTGGGGTAACCGCTA
*Ci*TRIM103-F	CCACCTTCATTGCCCCATCT

Table 1. *Cont.*

Primer Name	Primer Sequence (5′–3′)
*Ci*TRIM103-R	GCGTCTGGTAAAATTCCCGC
*Ci*TRIM109-F	AACAGATCCAGTGCTCCGTG
*Ci*TRIM109-R	CTGCATTCCGGACACAGTCT
*Ci*TRIM110-F	TGCACAATTTCAGCACCAGC
*Ci*TRIM110-R	GATGGTGACCCTGCTGTTCA
*Ci*TRIM112-F	TCCAGAACCACCCGCTTGTGA
*Ci*TRIM112-R	CCCCTTGTGCGACCCAACCAG
IRF3-F	ACTTCAGCAGTTTAGCATTCCC
IRF3-R	GCAGCATCGTTCTTGTTGTCA
VP2-F	ATCAAGGATCCCATTCCGCCTTCA
VP2-R	TTAGAGGATCGTGCCATTGAGGGT
β-actin-F	GCTATGTGGCTCTTGACTTCG
β-actin-R	GGGCACCTGAACCTCTCATT

Note: F, forward primer; R, reverse primer.

3. Results

3.1. Genome-Wide Identification of *Ci*TRIMs

TRIMs are characterized by RBCC or RB domain assemblies [9]. A total of 42 *Ci*TRIMs were identified in the grass carp genome with hmmsearch and SMART analysis according to this criterion (Table 2). Among them, 37 *Ci*TRIMs were named and numbered with reference to their homologous counterparts from the genomes of zebrafish and humans based on sequence similarity and identity, while five *Ci*TRIMs, including *Ci*TRIM35-50, *Ci*TRIM39-like, *CiTRIM*111, *Ci*TRIM112 and *Ci*btr40, whose homologs were not identified in the genomes from zebrafish and humans, and could be found in the teleost genomes of Pimephales promelas and Sinocyclocheilus anshuiensis by BLAST search (Table 2), were given names referring to the nomenclature previously described [10].

Forty-two *Ci*TRIMs were structured with different numbers of introns and exons (Table 2). Their coding sequence and encoded amino acid sequence lengths were 813~3978 bp and 271~1326 aa, respectively (Table 2). The proteins of the *Ci*TRIMs were predicted with molecular weights (MWs) ranging from 14.64 to 89.11 KDa (Table 2). A total of 26 *Ci*TRIMs were acidic, with isoelectric points (PIs) ranging from 4.87 to 6.65, and 16 *Ci*TRIMs were alkaline, with PIs ranging from 7.53 to 8.73 (Table 2). Subcellular localization prediction showed that *Ci*TRIMs tended to be located in the cytoplasm, cytoskeleton and nucleus, with 35 *Ci*TRIMs in the cytoplasm and 17 *Ci*TRIMs in other multiple regions of the cell, among which only two *Ci*TRIMs (*Ci*TRIM18 and *Ci*TRIM55b) were located in the cytoskeleton (Table 2).

Table 2. Overall information for the 42 *Ci*TRIMs identified in this study.

Gene Name	Genome ID	PL (aa)	MW (KDa)	PI	EN	PSL
*Ci*TRIM1	CI01000000_14975127_14983931	404	45.89	8.63	4	cytoplasm
*Ci*TRIM2	CI01000300_10176172_10189470	812	89.11	6.20	12	cytoplasm
*Ci*TRIM3	CI01000304_12076818_12085420	784	86.13	8.11	12	cytoplasm
*Ci*TRIM3a	CI01000095_00776224_00788366	770	84.11	7.53	13	cytoplasm
*Ci*TRIM13	CI01000009_00343750_00344964	404	45.63	5.92	1	cytoplasm, nucleus
*Ci*TRIM18	CI01000349_00034813_00052670	676	75.74	6.32	9	cytoplasm, cytoskeleton
*Ci*TRIM23	CI01000304_04650911_04659754	579	64.74	6.03	11	cytoplasm, nucleus
*Ci*TRIM25	CI01000112_00810157_00821637	473	53.60	8.65	5	cytoplasm
*Ci*TRIM25-like	CI01000354_01204555_01213381	405	46.89	6.65	6	nucleus
*Ci*TRIM32	CI01000059_09706218_09708197	659	72.50	6.58	1	nucleus
*Ci*TRIM33-like	CI01000027_07545610_07559939	1326	14.64	8.00	19	nucleus
*Ci*TRIM35-1	CI01000258_00137907_00154325	525	59.04	8.27	8	nucleus
*Ci*TRIM35-13	CI01000113_01547153_01553082	387	44.07	8.25	5	cytoplasm
*Ci*TRIM35-16	CI01000158_00190147_00195261	401	45.70	8.51	6	cytoplasm, nucleus
*Ci*TRIM35-17	CI01000087_02092947_02096033	544	61.76	6.31	6	cytoplasm

Table 2. *Cont.*

Gene Name	Genome ID	PL (aa)	MW (KDa)	PI	EN	PSL
*Ci*TRIM35-29	CI01000027_06329025_06333271	406	46.45	8.43	6	cytoplasm, nucleus
*Ci*TRIM35-30	CI01000013_11271292_11274155	271	31.64	8.80	3	nucleus
*Ci*TRIM35-50	CI01000013_03911586_03916513	411	47.45	8.29	6	cytoplasm, nucleus
*Ci*TRIM39-like	CI01000196_00233827_00236332	541	59.59	6.39	2	cytoplasm
*Ci*TRIM45	CI01000009_10250936_10259244	568	62.21	7.97	7	cytoplasm
*Ci*TRIM46b	CI01000027_04750483_04762663	753	83.61	7.65	11	cytoplasm, nucleus
*Ci*TRIM47	CI01000304_12065676_12071391	491	56.31	5.89	8	cytoplasm, nucleus
*Ci*TRIM54	CI01000029_01758347_01773935	380	43.11	5.18	9	cytoskeleton
*Ci*TRIM54-like	CI01000009_08283637_08293491	575	64.21	4.93	7	nucleus
*Ci*TRIM55a	CI01000098_02956318_02961605	419	47.23	4.95	9	cytoplasm, nucleus
*Ci*TRIM55b	CI01000018_06417005_06423657	379	43.18	5.06	8	cytoplasm, cytoskeleton, nucleus
*Ci*TRIM59	CI01000092_04869145_04870424	425	47.95	6.03	1	cytoplasm
*Ci*TRIM63	CI01000024_00581310_00583174	371	41.93	5.34	2	cytoplasm, nucleus
*Ci*TRIM67	CI01000051_06837978_06881158	713	79.43	6.58	12	cytoplasm, cytoskeleton
*Ci*TRIM71	CI01000016_10600154_10633808	934	10.27	6.61	5	cytoplasm
*Ci*TRIM101	CI01000001_04746153_04754209	465	52.81	4.87	10	cytoplasm, nucleus
*Ci*TRIM103	CI01000354_01417265_01426361	491	55.39	6.01	4	cytoplasm, nucleus
*Ci*TRIM109	CI01000016_05986400_05994821	501	56.96	6.29	7	cytoplasm, nucleus
*Ci*TRIM110	CI01000004_15999644_16005705	469	53.57	6.58	7	cytoplasm, extracellular
*Ci*TRIM111	CI01000012_13551086_13554035	545	62.27	5.77	6	cytoplasm
*Ci*TRIM112	CI01000180_01434810_01440307	493	54.31	5.98	4	nucleus
*Ci*RNF207	CI01000001_06965170_06976281	633	72.34	5.96	17	nucleus
*Ci*btr1	CI01000119_00092461_00099753	592	66.61	7.82	6	cytoplasm
*Ci*btr11	CI01000344_01402598_01409449	468	52.78	6.17	6	cytoplasm, nucleus
*Ci*btr12	CI01000354_01450329_01466394	645	73.13	8.73	8	cytoplasm
*Ci*btr40	CI01000344_01289427_01295915	520	60.11	6.65	5	cytoplasm
*Ci*ftr83-like	CI01000339_06060658_06063401	299	34.30	8.21	4	cytoplasm

Note: PL, protein length; MW, molecular weight; PI, isoelectric point; EN, exon numbers; PSL, predicted subcellular localization.

3.2. Dendrogram and Structural Features of CiTRIMs

According to the topological structure of the dendrogram, 42 *Ci*TRIMs could be divided into two major branches, with 25 *Ci*TRIMs in Group 1 and 17 *Ci*TRIMs in Group 2 (Figure 1A). A total of 24 *Ci*TRIMs harbored conserved RBCC domain assemblies and the other *Ci*TRIMs harbored conserved RB domain assemblies at their N-terminal; on the other hand, the C-terminal domains of the *Ci*TRIMs, especially those in Group 1, were quite varied (Figure 1B); they consisted of eight categories of domains, including the COS (C-terminal subgroup one signature) domain, TM (transmembrane) domain, FN3 (fibronectin type 3) domain, B30.2 domain, ARF3 (ADP-ribosylation factor 3) domain, AIP3 (actin interacting protein 3) domain, filamin (filamin-type immunoglobulin) domain and PHD (plant homeodomain) domain. Most of the *Ci*TRIMs from Group 2 harbored only the B30.2 domain at the C-terminus, except for *Ci*TRIM112, *Ci*ftr83-like and *Ci*TRIM35-30 (Figure 1B). In total, 19 TRIMs-B30.2 were identified in the grass carp genome (Figure 1B). Meanwhile, 10 conserved motifs in *Ci*TRIMs were predicted by MEME Suite 5.3.3. The results showed that all the *Ci*TRIMs had the two motifs Znf-RING_LisH and Znf-B-box at their N-terminal motif architectures (Figure 1C), while the C-terminal motif architectures of the *Ci*TRIMs, especially those in Group 1, were significantly diversified (Figure 1C).

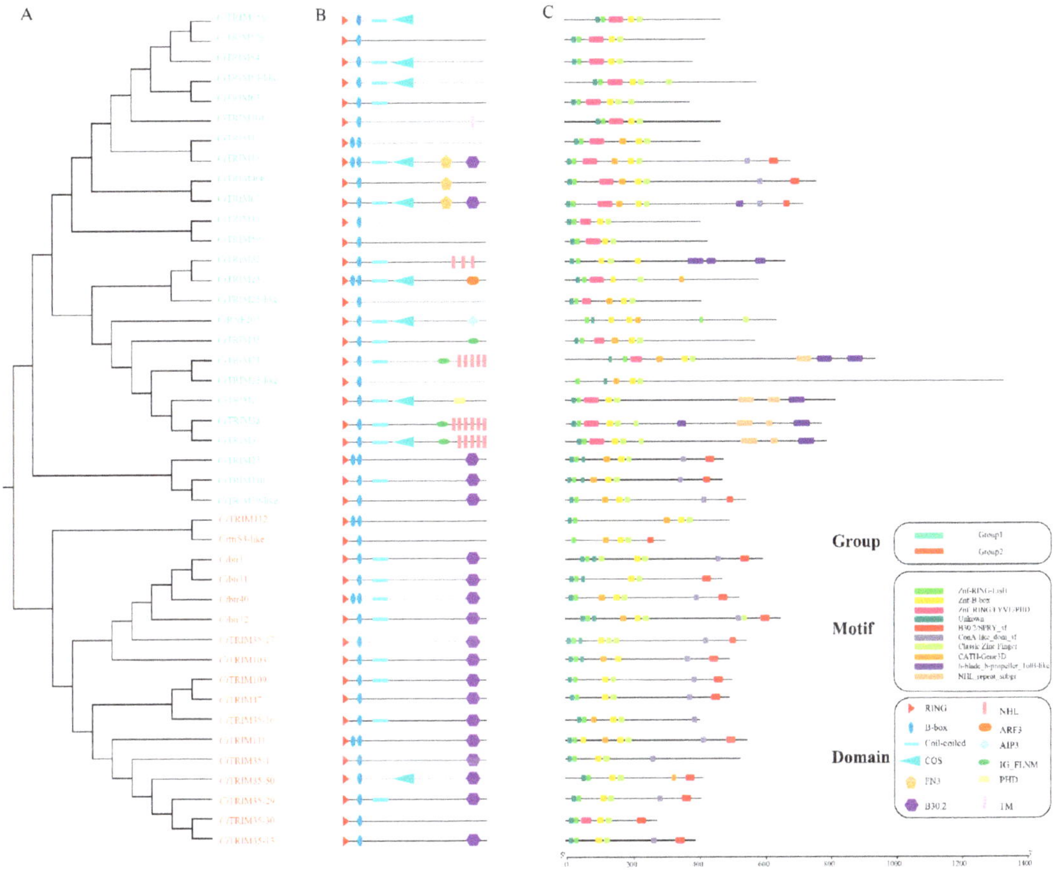

Figure 1. Dendrogram, domains and motifs of *Ci*TRIMs. (**A**) The dendrogram of *Ci*TRIMs was divided into two large branches, indicated in green and red. (**B**) The schematic diagram of domain architectures of *Ci*TRIMs. (**C**) Conserved motifs of *Ci*TRIMs predicted by MEME. The scale on the bottom margin was measured by amino acid number.

3.3. Chromosomal Location of *Ci*TRIMs

A total of 301 scaffolds (with an average length of >179,941 bp) from the grass carp genome were assembled into 24 chromosomes by Mapgene2Chromosome (V2.1), with 114 scaffolds anchored on linkage groups (Figure 2). Only 31 out of 42 *Ci*TRIMs were discovered on 16 chromosomes. The other 11 *Ci*TRIMs might have been lost due to low-quality chromosome assembly. No *Ci*TRIM was found on chromosomes 3, 4, 9, 10, 11, 20, 21 and 23, while five *Ci*TRIMs were located on chromosome 12, which possessed the largest number of *Ci*TRIMs (Figure 2). It has been reported that teleost TRIMs undergo massive expansion mainly though tandem repeats to adapt to environmental changes during evolution [22]. The possible presence of tandem repeats in the *Ci*TRIMs was investigated with MCScanX, but none were found (data not shown).

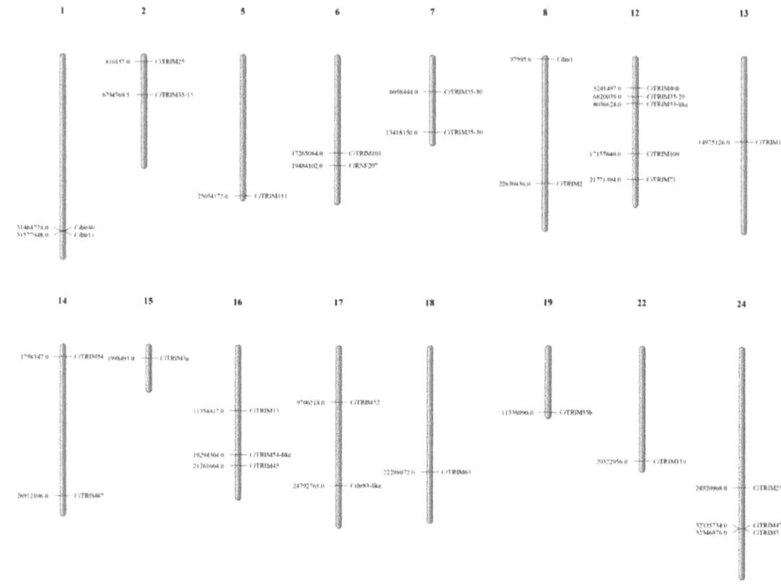

Figure 2. Chromosomal locations of *Ci*TRIMs. The number on the left represents the position of the *Ci*TRIM gene on the chromosome.

3.4. Gene Collinearities of *Ci*TRIMs with TRIMs from Zebrafish and Humans

To further understand the evolutionary features of *Ci*TRIMs, the gene collinearities of *Ci*TRIMs with TRIMs from zebrafish and humans were analyzed using MCScanX. A total of 42 *Ci*TRIMs were distributed on 30 different scaffolds of the grass carp genome (Figure 3). According to the results of the collinearity analysis, 30 pairs of homologous TRIMs were identified between the grass carp genome and zebrafish genome, while 10 pairs of homologous TRIMs were identified between the grass carp genome and human genome (Figure 3). No homologous TRIMs were detected in the zebrafish and human genomes for five *Ci*TRIMs, including *Ci*TRIM35-50, *Ci*TRIM39-like, *Ci*TRIM111, *Ci*TRIM112 and *Ci*btr40 (Figure 3).

3.5. Tissue Expression Patterns of *Ci*TRIMs

To clarify the tissue expression patterns of *Ci*TRIMs, the transcriptomes of the kidneys, liver, head kidneys, spleen, brain, and embryo from uninfected grass carp were analyzed. As presented in Figure 4, 21 *Ci*TRIMs, including *Ci*TRIM103, *Ci*btr40 and *Ci*TRIM55a, showed high mRNA expression levels in the embryos, and nine *Ci*TRIMs, including *Ci*TRIM13, *Ci*TRIM18 and *Ci*TRIM23, were highly expressed in the brain. Six *Ci*TRIMs (including *Ci*TRIM13, *Ci*TRIM18 and *Ci*TRIM23), three *Ci*TRIMs (*Ci*TRIM101, *Ci*TRIM110 and *Ci*TRIM33-like), two *Ci*TRIMs (*Ci*TRIM35-1 and *Ci*TRIM54) and *Ci*TRIM112 were mainly expressed in the spleen, head kidneys, kidneys, and liver, respectively, which indicated the specific and high expression of certain *Ci*TRIMs that occurred in the immune tissues of uninfected grass carp.

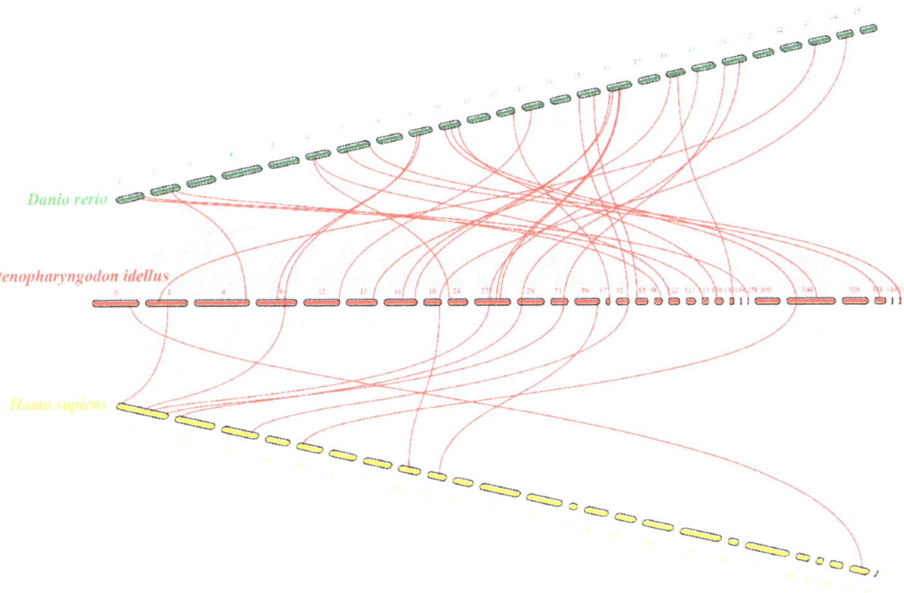

Figure 3. Gene collinearity analysis of *Ci*TRIMs with those from the zebrafish and human genomes. Red, green, and yellow short sticks represent scaffolds from the grass carp genome, zebrafish genome and human genome, respectively. The red lines indicate TRIM gene pairs between species.

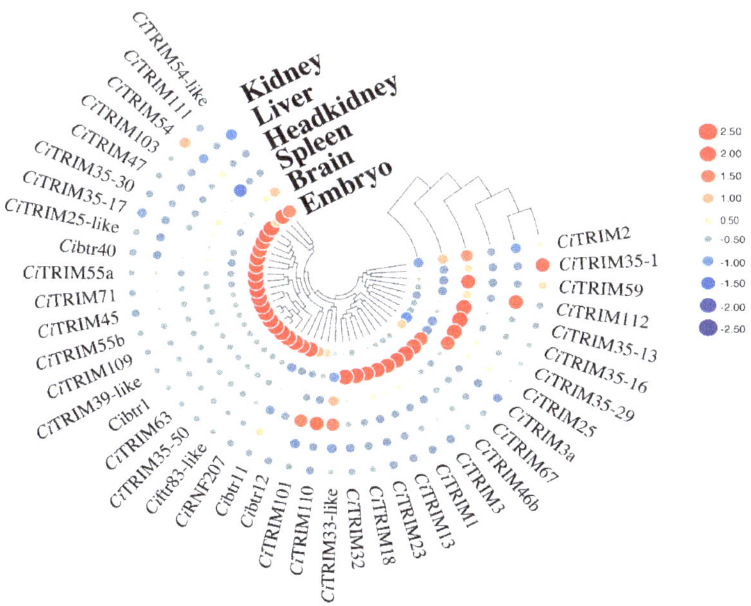

Figure 4. Heatmap based on expression profiles of *Ci*TRIMs in six tissues from uninfected grass carp including kidneys, liver, head kidneys, spleen, brain, and embryo. The sizes of red or blue dots in the heatmap reflect gene expression levels, and the color shades represent the cluster correlations of genes in terms of their expression levels.

3.6. Expression Patterns of Potential Antiviral CiTRIMs in Grass Carp Spleen Tissue during GCRV Infection

To identify *Ci*TRIMs differentially expressed during GCRV infection, the transcriptomes of the spleens in grass carp after GCRV challenge on Days 1, 3, 5 and 7 were analyzed. Although various *Ci*TRIMs were slightly upregulated after GCRV challenge at each time point, most of their expression levels showed no significant difference ($p > 0.05$), compared to those in the control (Figure 5A). Only three *Ci*TRIMs, namely, *Ci*btr40, *Ci*TRIM103 and *Ci*TRIM112, were significantly differentially expressed in the transcriptome of the spleen in grass carp after GCRV challenge on the fifth day, compared to the expression in the control ($p < 0.05$; Figure 5A).

WGCNA was used to explore the related biological functions of three significantly differentially expressed *Ci*TRIMs based on the spleen transcriptomes during GCRV infection. A soft-threshold power that was the most suitable for the construction of a gene co-expression network was chosen from a list of 1–20 candidate powers (Figure 5B). Then, the dynamic hybrid cleavage method was adopted to merge gene clusters with co-expression into the specific modules on the same branch (Figure 5C). After removing the genes with low gene expression variation, a total of 22,924 genes were obtained and clustered into 26 modules (Figure 5D). Except for the gray module, the turquoise module possessed the largest number of genes, with a total of 2883 co-expressed genes, while the dark-gray module harbored the fewest genes, with a total of 31 co-expressed genes (data not shown). After the Pearson correlation analysis of 26 modules with trait (the infection time points), it was found that the gene cluster in the turquoise module showed the highest positive correlation coefficient (0.83), followed by that in the brown module (0.69), while the gene cluster in the magenta module represented the highest negative correlation coefficient (−0.82; Figure 5D). Since all the three differentially expressed *Ci*TRIMs were clustered in the brown module, this module was selected for subsequent analyses. When filtering the undirected network with an adjacency threshold > 0.45 in the brown module using Cytoscape, *Ci*TRIM112 was excluded. The filtered network showed that both *Ci*TRIM103 and *Ci*btr40 were linked with the hub gene DEAD (Asp-Glu-Ala-Asp) Box Polypeptide 58 (DDX58) (alternatively named retinoic acid-inducible gene I (RIG-I)-like receptor), as well as several vital type I IFN response pathway genes, such as melanoma differentiation associated gene 5 (MDA5), signal transducer and activator of transcription 1 (STAT1), IFN stimulating gene 58 (ISG58), double-stranded RNA-dependent protein kinase (PKR), myxovirus 1 (MX1) and viperin (Figure 5E). The GO enrichment analysis of this gene network revealed that the top three enriched GO terms were associated with immune defense processes, including defensive responses to other organisms, cellular responses to cytokine stimulation and cysteine-type endopeptidase activity involved in apoptosis (Figure 5F).

3.7. Expression Patterns of Potential Antiviral CiTRIMs in CIK during GCRV Infection

The transcriptomes of CIK after GCRV challenge at 6 h, 12 h and 24 h were also analyzed to survey the *Ci*TRIMs differentially expressed during GCRV infection. A total of 4167 differentially expressed genes were identified. Among the 42 identified *Ci*TRIMs, eight whose expression levels showed significant differences at 6 h, 12 h or 24 h compared to the control (0 h) were determined as the differentially expressed ones ($p < 0.05$; Figure 6A). Among these differentially expressed *Ci*TRIMs, *Ci*TRIM46b was significantly down-regulated at 6 h; *Ci*TRIM35-16 was significantly upregulated at 12 h and 24 h; *Ci*TRIM109 was significantly upregulated at 6 h and 24 h; *Ci*TRIM2, *Ci*TRIM71 and *Ci*TRIM110 were significantly upregulated at 24 h; and *Ci*TRIM35-50 and *Ci*TRIM54-like were significantly down-regulated at 24 h after GCRV challenge, compared to control ($p < 0.05$; Figure 6A).

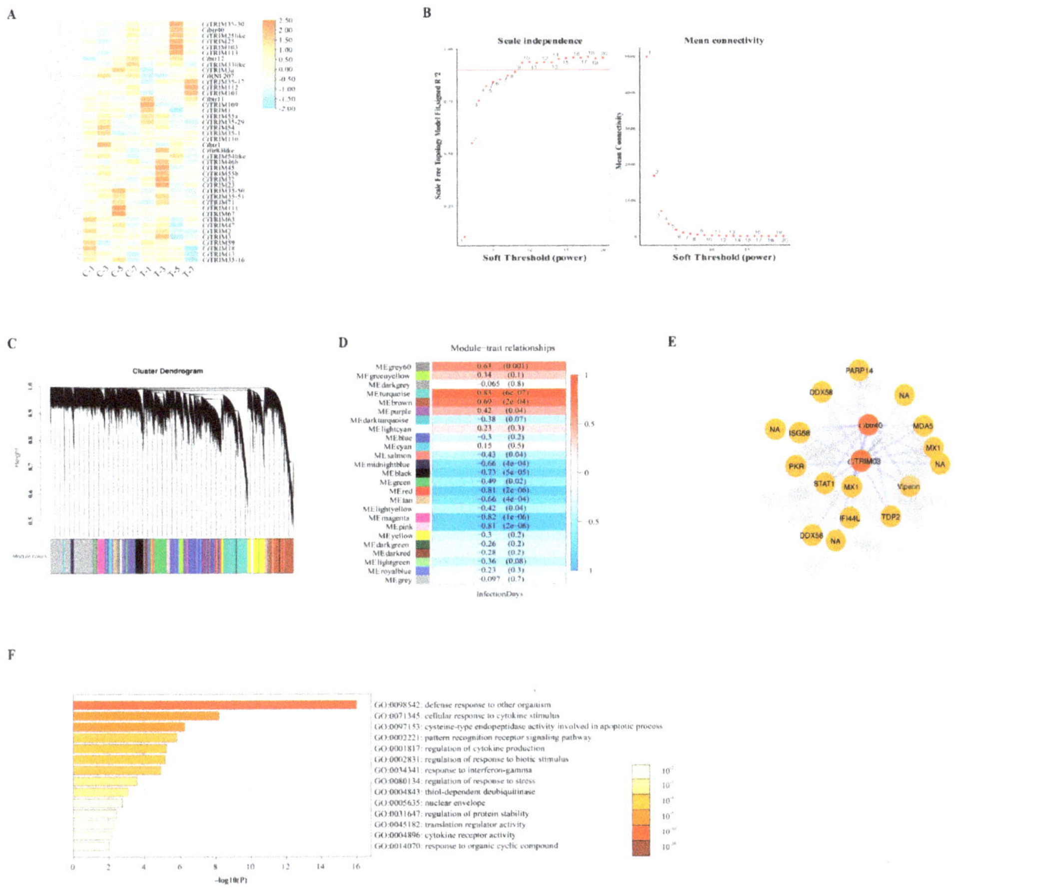

Figure 5. Transcriptome analyses of the spleens from grass carp infected with GCRV within 7 days. (**A**) Heatmap of 42 *Ci*TRIMs' expression changes in spleen after GCRV infection on days 1, 3, 5 and 7. C-1: control group, day 1; C-3: control group, day 3; C-5: control group, day 5; C-7:control group, day 7; T-1: treatment group, day 1; T-3: treatment group, day 3; T-5: treatment group, day 5; T-7: treatment group, day 7. (**B**) Analysis of network topology for various soft-thresholding powers. The left panel shows the scale-free fit index (y-axis) as a function of the soft-thresholding power (x-axis). The right panel displays the mean connectivity (degree, y-axis) as a function of the soft-thresholding power (x-axis). (**C**) Clustering dendrogram of genes, with dissimilarity based on topological overlap, together with assigned module colors, that contain a cluster of genes with similar biological functions. (**D**) Pearson correlation analysis of module and trait (infection time points). Each cell contains the Pearson correlation coefficient and Student asymptotic *p*-value. The correlation intensity color is illustrated by the legend on the right. (**E**) Gene co-expression network in brown module. *Ci*btr40, *Ci*TRIM103 and 12 genes, including DDX58 (RIG-I-like receptor), PKR, Viperin, MDA5, STAT1, MX1 and ISG58, with the highest node degree distribution value were highlighted in the network. (**F**) GO enrichment analysis of genes in the brown module using Metascape. Only top 14 enriched GO terms are shown. The color shade reflects the significance (*p*-value) of the GO enrichment analysis, which is represented as –log10(P) on the bottom coordinate ruler (x-axis).

Gene expression trend analysis using STEM was performed to further reveal the biological functions of these eight differentially expressed *Ci*TRIMs based on the CIK transcriptomes during GCRV infection. All the differentially expressed genes from the CIK transcriptomes were classified into fifty clusters and formed fifty profiles (numbered from 1 to 50), where the gene cluster exhibited a similar expression trend after GCRV challenge at 0 h, 6 h, 12 h and 24 h (Figure 6B). The expression trends of gene clusters in 12 profiles with colored backgrounds were significantly similar when analyzed by STEM (p-value < 0.05; Figure 6B). Profile 40 included the largest number of genes, with a total of 775 genes (Figure 6B). Six out of eight differentially expressed *Ci*TRIMs were, respectively clustered in five profiles where the gene expression trend was significantly similar (p-value < 0.05; Figure 6B). In detail, *Ci*TRIM35-50 and *Ci*TRIM54-like were clustered in Profile 11, *Ci*TRIM109 was clustered in Profile 40, *Ci*TRIM46b was clustered in Profile 4, *Ci*TRIM35-16 was clustered in Profile 29 and *Ci*TRIM71 was clustered in Profile 42 (Figure 6B). All the genes within these five profiles were imported into Metascape for GO enrichment analysis. The results reveal that the gene clusters in these five profiles were annotated with different GO terms (Figure 6C–F). Notably, the gene cluster including *Ci*TRIM109 in Profile 40 was enriched with a total of 20 GO terms, two of which were tightly associated with antiviral immune defense processes, including the NOD-like receptor signaling pathway and regulation of type I interferon production (Figure 6F).

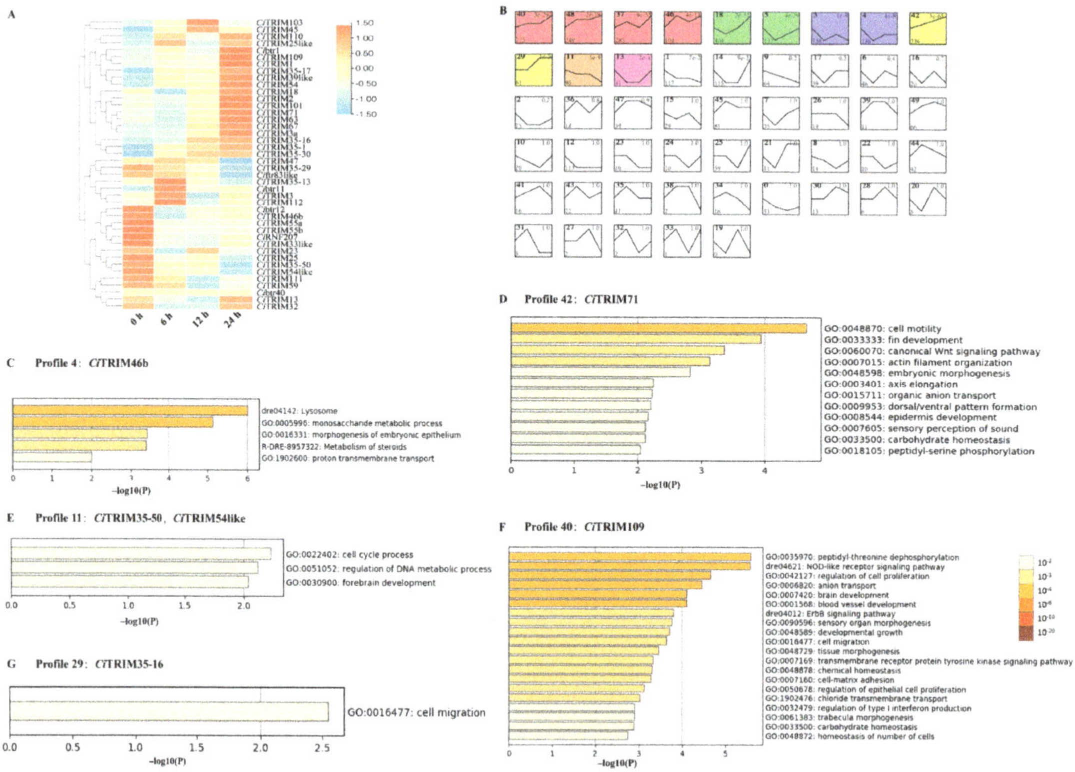

Figure 6. Transcriptome analyses of CIK after GCRV challenge. (**A**) Heatmap of 42 *Ci*TRIMs' expression changes in CIK after GCRV infection at 0 h, 6 h, 12 h and 24 h. (**B**) Expression trend for genes in CIK transcriptomes during GCRV infection. A total of 50 profiles where the gene cluster exhibited a similar expression trend after GCRV challenge at 0 h, 6 h, 12 h and 24 h were produced using STEM. The data in the top-left corner, top-right corner and top-right corner represent the profile ID, the significance of cluster correlation and the number of genes in profiles, respectively. GO enrichment analysis for genes including differentially expressed *Ci*TRIMs in Profile 4 (**C**), Profile 42 (**D**), Profile 11 (**E**), Profile 40 (**F**) and Profile 29 (**G**). Coordinate ruler on the bottom (x-axis) of profiles represents the significance (p-value) from GO enrichment analysis, which is shown as –log10(P).

3.8. The Verification of Differentially Expressed CiTRIMs during GCRV Infection

Eleven differentially expressed *Ci*TRIMs as well as several type I IFN response pathway genes were identified in the two sets of transcriptomes above. To verify their gene expression, an experiment in which CIK cells were challenged with GCRV over 24 h was conducted. The sequential expression changes of *Ci*IRF3 (an important IFN regulatory factor) [63], VP2 (a protein component of GCRV) [64] and these 11 differentially expressed *Ci*TRIMs were detected by qPCR to explore the correlation of the gene expression trends between *Ci*IRF3/VP2 and differentially expressed *Ci*TRIMs and to reveal the process of host–pathogen interaction during GCRV infection. The results show that the mRNA expression level of VP2, which indicated GCRV replication [64], was significantly upregulated after challenge with GCRV at 24 h, compared to that at 0 h ($p < 0.05$; Figure 7). Meanwhile, *Ci*IRF3, along with eleven differentially expressed *Ci*TRIMs identified in the two sets of transcriptomes above, was also significantly upregulated after GCRV challenge at 24 h ($p < 0.05$; Figure 7). In addition, three *Ci*TRIMs, including *Ci*btr40, *Ci*TRIM46b and *Ci*TRIM109, showed similar expression trends to VP2, all of which were little expressed after GCRV challenge at 0 h, 6 h and 12 h and sharply upregulated at 24 h ($p < 0.05$; Figure 7). The expression trends for eight other *Ci*TRIMs were similar to the trend for *Ci*IRF3 after GCRV challenge at 0 h, 6 h, 12 h and 24 h, all of which first decreased and then increased, with peak expression at 24 h (Figure 7). Moreover, the expression levels of *Ci*btr40, *Ci*TRIM103 and *Ci*TRIM109, which all belonged to TRIMs-B30.2 and were linked with the type I IFN response pathway by WGCNA and the gene expression trend analysis based on the transcriptome data, were significantly upregulated by 235-fold, 2-fold, and 916-fold after GCRV challenge at 24 h, respectively, compared to at 0 h ($p < 0.05$; Figure 7).

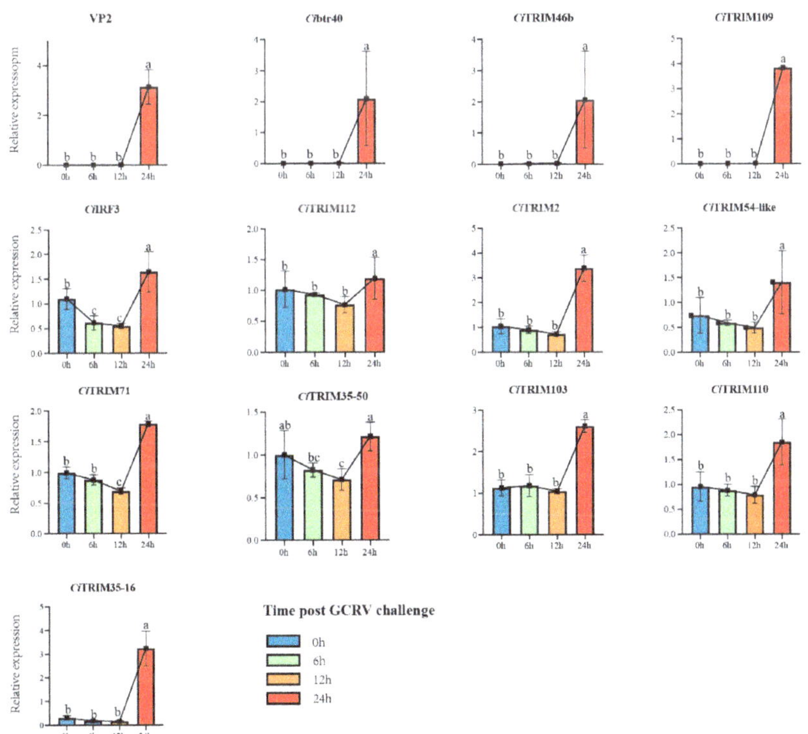

Figure 7. Expression profiles of VP2, *Ci*IRF3 and 11 *Ci*TRIMs in response to GCRV challenge in CIK. Expression data were normalized using β-actin as internal control; error bars indicate standard deviations among three biological replicates. The letters (a, b and c) indicate significant differences among expression levels at different time points after GCRV challenge ($p < 0.05$).

4. Discussion

The TRIM protein family is a class of proteins that possess conserved RBCC or RB domain assemblies at their N-termini and a variety of domains at their C-termini, widely existing in the majority of metazoa, which play multiple roles in physiological or pathological processes such as growth and development [65], metabolism and autophagy [12], transcriptional regulation [13], carcinogenesis [14] and antiviral immunity [15]. During the long-term evolution of animals, large differences in TRIM family gene numbers have arisen among species. Compared to invertebrates, vertebrates seem to possess, overall, a larger number of TRIMs. For example, humans have a total of 69 TRIMs, which is more than three times the number of TRIMs in nematodes, *Caenorhabditis elegans* [4]. In teleosts, two or more rounds of whole genome duplication can be observed [66,67], resulting in more universal variation in the numbers of TRIMs among species. The Ballan wrasse (*Labrus bergylta*) from Perciformes seems to have the largest number, with 369 extant TRIMs; the Red-bellied piranha (*Pygocentrus nattereri*) from Characiformes possesses 229 TRIMs, and the tiger tail seahorse (*Hippocampus comes*) from Gasterosteiformes only harbors 62 TRIMs [68]. The number of TRIMs may reflect the evolutionary processes or divergence times of species, making the identification of TRIMs an attractive topic for teleosts.

Recently, different software and methods have been employed to identify TRIMs in teleosts. Sardiello et al. combined the software packages PHI-BLAS and TBLASTN to identify a total of 240 TRIMs in zebrafish based on NR databases [4]. Zhang et al. used the blast method to identify a total of 196 zebrafish TRIMs in the NCBI and Ensembl databases, with the sequence encoding the RBCC domain assemblies as the bait [69], while Boudinot et al. refined the number of zebrafish TRIMs to 208 in the Ensembl databases by using the Hidden Markov Model with RING finger and B-box domains as the templates and with the domain architectures of human TRIMs as the reference to exclude redundancy [10]. These studies indicate that using different methods and databases may identify different numbers of TRIMs even among the same species. In the present study, we adopted the Hidden Markov Model to identify 42 *Ci*TRIMs in the grass carp genome with a conserved B-box domain and two conserved RING finger domains as the templates and according to the criterion of whether they possessed RBCC or RB domain assemblies. The authenticity of the identified *Ci*TRIMs was further verified by gene collinearity and chromosomal location analyses. On the other hand, we also tried to use our method in other species to test its specificity and sensitivity. In humans, for example, we found 267 TRIMs containing the RBCC or RB domain assemblies from 354 TRIM transcripts in genome annotation files (GRCh38.p13), which represented results consistent with those of a previous study [9]. A total of 106 TRIM genes were annotated in the grass carp genome [39]. Our results implied that 64 TRIM genes annotated in the grass carp genome did not harbor the N-terminal conserved RBCC or RB domain assemblies. To our knowledge, both the annotations for 354 TRIM transcripts in humans and 106 TRIM genes in grass carp tend to be obtained through sequence alignments against multiple databases, which would cause false positive annotations [9,39]. Although several *Ci*TRIMs identified in the previous study were not found in our study, two reasons may explain this difference based on reviewing multiple studies on the identification of zebrafish TRIMs. One reason is that different template sequences and search methods were used in the present and previous studies. Luo et al. utilized a set of zebrafish ftr gene sequences as templates to search all the putative ftrs in grass carp by using the blast method, while our study sought to identify all the *Ci*TRIMs based on conserved domains across species using the Hidden Markov Model [70]. Using the domain architectures of zebrafish and human TRIMs as the reference to exclude redundancy, we identified at least seven putative *Ci*ftrs, fewer than the previous study identified [70]. Additionally, the databases used in the identification of TRIMs also differ between the present and previous studies [70]. Luo et al. chose the NCBI and Ensembl databases, as well as the non-referenced transcriptomes, for retrieving the TRIMs in grass carp [70], while, in this study, we only downloaded the grass carp genome as the search database for the identification of *Ci*TRIMs, which could effectively reduce false positive results. Nevertheless, we realize that several *Ci*TRIMs have still not been identified through our subsequent gene tandem repeat analysis, probably due to the assembly of the extant genome (only with the contigs N50 of 40,781 bp). It is believed that a higher quality genome is required for identifying all the *Ci*TRIMs containing intact RBCC or RB domain assemblies, since Boudinot et al. found more zebrafish TRIMs in the genome Version Z9 than that in the genome Version Z8 [10].

The C-terminal domains help to build interactions with other proteins and often determine the functional specificity in TRIMs [16]. To date, more than 11 C-terminal domains have been identified in TRIMs, which perform a variety of biological functions [16,71]. For example, a deficiency of the C-terminal domains of COS, FN3 and B30.2 in TRIM18 weakens the activation of the mechanistic target of rapamycin complex 1 (mTORC1) signaling, causing abdominal midline dysplasia in the fetus [72]. The C-terminal domains of PHD and BROMO in TRIM24 can bind to chromatin and serve as potential therapeutic targets for breast cancer [73]. The filamin and NHL domains of TRIMs play important roles in neuronal differentiation [11]. The ARF domain at the C-terminus in TRIM23 appears to trigger autophagy by activating TBK1 through the non-traditional ubiquitinated GTPase [74]. In vertebrates, the B30.2 domain is the most frequent at the C-termini of TRIMs [4]. In humans, there are a total of 35 TRIM-B30.2s [16], 30 of which have been proved to participate in immune responses [16,75]. Distinct tissue expression patterns also reflect the functional differentiation of TRIMs. For instance, when TRIM4 is highly expressed in nervous tissues, neural tube defects are detected [76], while the overexpression or knockdown of TRIM4 in immune tissues significantly affects the expression of IRF3, nuclear factor-kappa B (NF-κB) and IFN, demonstrating its involvement in antiviral immunity in humans [77]. Such evidence indicates that different tissue expression patterns and C-terminal domains confer TRIMs with functional diversity. In the present study, the expression analysis showed that 42 *Ci*TRIMs were obviously expressed in different tissues. Specifically, nine, six, three, two and one *Ci*TRIMs were mainly expressed in the brain, spleen, head kidneys, kidneys, and liver, respectively. The structural analyses also identified eight C-terminal domains, including the COS domain, B30.2 domain, FN3 domain, TM domain, NHL domain, ARF3 domain, AIP3 domain, filamin domain and PHD domain in *Ci*TRIMs, implying the functional differentiation of grass carp TRIMs. In addition, 19 TRIMs-B30.2 were found, with the largest number in *Ci*TRIMs (accounting for ~45.2%), and they were clustered into a distinct branch in the dendrogram, suggesting that grass carps have also evolved with a cluster of TRIMs linked to immune defense, similar to other vertebrates.

Although the expansion of the B30.2 domain is common in vertebrates, TRIM-B30.2 family members differ among various species [4]. Thus, almost every species has specific TRIM-B30.2 family members; for example, 44% of TRIMs contain the B30.2 domain in humans; 55%, in puffer fish; and 83%, in zebrafish [10,16]. In teleosts, three major subfamilies of TRIMs-B30.2 have developed, including hltrs, btrs and ftrs, via the massive duplication of three ancestor TRIM-B30.2 genes, while no corresponding duplication phenomena are found in other vertebrate species [10]. In fact, teleost hltrs and btrs are produced by the duplication of TRIM35 and TRIM39, respectively, both of which play important roles in immune responses [10,21]. TRIM35 can directly mediate polyubiquitination by invading virus particles and degrade them or polyubiquitinate TNF receptor-associated factor (TRAF) 3 to activate the IFN response against the virus [23], while TRIM39 regulates the NF-κB signaling pathway to participate in immune defense [24]. Ftrs are teleost-specific TRIM-B30.2 genes, with no homologs identified in other vertebrates, while two evolutionarily close TRIM genes, TRIM16 and TRIM25, are found in a phylogenetic tree, which have also been reported to exert multiple antiviral functions [22]. These studies suggest specific selective pressure derived from virus–host interactions for the massive duplication of teleost TRIMs-B30.2 [68]. In the present study, we also identified seven hltrs, four btrs and one ftr in the grass carp genome, and we speculate that they might be involved in the immune defense against GCRV. We also analyzed the tandem repeat of grass carp TRIMs-B30.2 and tried to further reveal the mechanism for gene expansion, referring to previous hints from other teleost species [22]. Unexpectedly, no tandem repeat was found in the identified *Ci*TRIMs by MCScanX. We think that the fewer *Ci*TRIMs identified due to the low-quality assembly of the extant grass carp genome may partly explain this, because genome assembly with DNA fragment shifting or mutation can mask gene tandem repeats [47]. In addition, the MCScanX analysis of gene tandem repeats requires the genome to be assembled to the chromosome level [51], while the extant genome of grass carp in this study was only assembled to the draft level and with only 114 out of 301 scaffolds anchored on linkage groups, which may also have led to biased results for the gene tandem repeats. However, the real reason should be further explored in the future.

It has been reported, in teleosts, that WGCNA and gene expression trend analysis represent two effective methods for identifying anti-infection immune gene co-expression networks or temporal gene expression profiles and revealing their related immune processes based on transcriptome data [29,30]. For example, Ning et al. employed a gene expression trend analysis combined with WGCNA to identify the differentially expressed gene clusters associated with immune processes, including cytokine–cytokine receptor signaling, Toll-like receptor signaling and other immune-related pathways from the transcriptomes of Japanese flounder (*Paralichthys olivacrus*) infected with *Vibrio anguillarum* [33]. Given the importance of specific TRIMs in animal antiviral immunity [2,9], we reanalyzed two published grass carp transcriptomes during GCRV infection to seek potential antiviral *Ci*TRIMs based on their expression patterns. In the spleen transcriptomes after GCRV challenge at 1, 3, 5 and 7 days, we identified three significantly differentially expressed *Ci*TRIMs. A further WGCNA showed two out of these three *Ci*TRIMs, *Ci*btr40 and *Ci*TRIM103, which both contain the B30.2 domain, were clustered into the co-expression network that contains vital type I IFN response pathway genes such as RIG-I-like receptor, MDA5, STAT1, ISG58, PKR, viperin and MX1. The genes in this co-expression network have mainly been associated, by GO enrichment analysis, with immune processes including defense responses to other organisms, cellular responses to cytokine stimulation and cysteine-type endopeptidase activity involved in apoptosis. To find more potential antiviral *Ci*TRIMs, the CIK transcriptomes after GCRV challenge at 0 h, 6 h, 12 h and 24 h were also reanalyzed, and a total of eight significantly differentially expressed *Ci*TRIMs were identified. It has been reported that gene expression trend analysis is more suitable than WGCNA for the transciptome data with less than 15 samples [61]. A further gene expression trend analysis was conducted for the CIK transcriptomes and showed that one of these eight *Ci*TRIMs, *Ci*TRIM109, which is also a TRIM-B30.2, shows an expression trend significantly similar to that of the gene cluster that is enriched in the regulation of type I IFN production and the NOD-like receptor pathway. Expectedly, our qPCR results verify that 11 differentially expressed *Ci*TRIMs identified from the transcriptomes, including *Ci*btr40, *Ci*TRIM103 and *Ci*TRIM109, were significantly upregulated along with the upregulation of *Ci*IRF3, an important IFN regulatory factor [63], after GCRV challenge at 24 h. Increasing evidence has shown that certain TRIMs, especially TRIM-B30.2, can regulate RIG-I-like receptor, NOD-like receptor and the MDA5-mediated type I IFN response and promote the production of antiviral molecules, including ISGs, PKR, viperin and MX1, in various species [13,27]. We admit that the results for the expression profiles obtained by the qPCR analysis are not sufficient for the functional identification of actual antiviral TRIMs in grass carp. Nevertheless, our results strongly indicate that *Ci*btr40, *Ci*TRIM103 and *Ci*TRIM109, all containing the B30.2 domain, are associated with the type I IFN response during GCRV infection according to WGCNA and gene expression trend analysis, suggesting their involvement in the antiviral immunity of grass carp.

5. Conclusions

In conclusion, this study identified a total of 42 *Ci*TRIMs from the grass carp genome, which possessed conserved RBCC or RB domain assemblies at their N-termini and eight different domains at their C-termini. Among them, 19 *Ci*TRIMs contained the B30.2 domain at their C-termini, which has previously been proved to be fast-evolving and to play important roles in the antiviral immune defense. We also found a total of 11 significantly differentially expressed *Ci*TRIMs in two transcriptomes during GCRV infection. Despite the lack of further functional verification, three of them, including *Ci*btr40, *Ci*TRIM103 and *Ci*TRIM109, all belonging to TRIMs-B30.2, are associated with the type I IFN response during GCRV infection and deduced as potential antiviral TRIMs in grass carp (Figure 8). These findings may offer useful information for understanding the structure, evolution, and function of TRIMs in teleosts and provide potential antiviral immune molecule markers for the disease resistance breeding of grass carp.

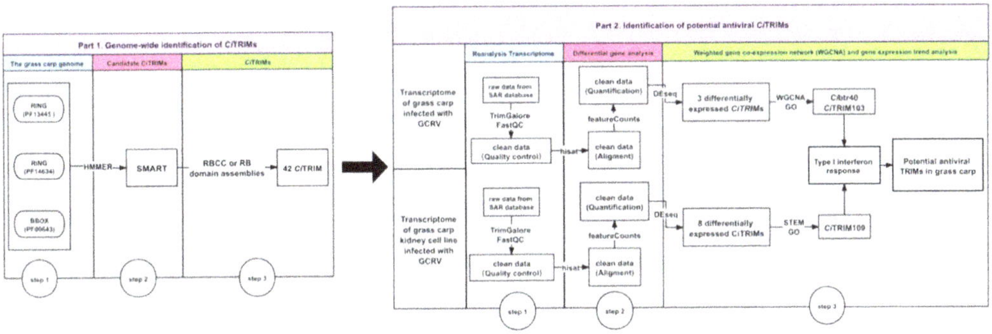

Figure 8. The overall framework for genome-wide identification of potential antiviral TRIMs in grass carp using transcriptomes in this study. First, the candidate *Ci*TRIMs containing RBCC, or RB domain assemblies were searched from the grass carp genome using the HMMER software, followed by removing redundancy through SMART analysis; finally, 42 *Ci*TRIMs were identified. Secondly, two published transcriptomic datasets based on experiments with grass carp individuals and cells infected with GCRV were downloaded from the SRA database (SRP095827, PRJNA597582 and PRJNA597542), followed by reanalysis. Eleven differentially expressed *Ci*TRIMs were identified through fold changes > 2 and FDRs (or adjusted p-values) < 0.05 using the DESeq R package. Through WGCNA and gene expression trend analysis, 3 TRIMs-B30.2, including *Ci*TRIM103, *Ci*btr40 and *Ci*TRIM109, associated with the type I interferon response during GCRV infection, were deduced as potential antiviral TRIMs in grass carp.

Author Contributions: Conceptualization, B.Q., J.S., T.X. and Z.L.; methodology, B.Q.; validation, B.Q.; formal analysis, B.Q., Y.D. and C.D.; investigation, B.Q., Y.D. and C.D.; data curation, B.Q.; writing—original draft preparation, B.Q.; writing—review and editing, Z.L.; visualization, B.Q. and Z.L.; supervision, J.S.; funding acquisition, J.S., T.X. and Z.L. All authors have read and agreed to the published version of the manuscript.

Funding: This work was supported by the Outstanding Youth Fund of Hunan Provincial Department of Education (10B046), Postdoctoral Research Foundation of China (2019M662783) and National Natural Science Foundation of China (U20A2063 and 31972787).

Institutional Review Board Statement: This study was approved by the Animal Care and Use Committee of Hunan Agricultural University (Changsha, China; Approval Code: 20200733; Approval Date: 9 October 2020).

Informed Consent Statement: Not applicable.

Data Availability Statement: The grass carp genome and tissue expression data were downloaded from the Grass Carp Genome Database (GCGD, http://bioinfo.ihb.ac.cn/gcgd/php/index.php, accessed on 20 November 2020) [39]. The transcriptome raw data (accession number: SRP095827) [53] for the spleens from grass carp infected with GCRV and the raw transcriptome data (accession numbers: PRJNA597582 and PRJNA597542) [60] for the grass carp kidney cell line (CIK) after GCRV challenge are available in the public NCBI database.

Acknowledgments: We appreciate Li.Z. from the Yangtze River Fisheries Research Institute of the Chinese Academy of Fishery Sciences for his gift of the GCRV JX-01 strain.

Conflicts of Interest: The authors declare no conflict of interest. The funding sponsors had no role in the design of the study; in the collection, analyses or interpretation of data; in the writing of the manuscript; or in the decision to publish the results.

References

1. Andrew, J.S.; Katherine, L.B.B.; Michael, N.B.; Paul, S. Freemont Does This Have a Familiar RING? *Trends Biochem. Sci.* **1996**, *21*, 208–214. [CrossRef]
2. Nisole, S.; Stoye, J.P.; Saïb, A. TRIM Family Proteins: Retroviral Restriction and Antiviral Defence. *Nat. Rev. Microbiol.* **2005**, *3*, 799–808. [CrossRef] [PubMed]
3. Zhao, J.; Li, H.; Huang, J.; Shi, T.; Meng, Z.; Chen, Q.; Deng, J. Genome-Wide Analysis of BBX Gene Family in Tartary Buckwheat (*Fagopyrum Tataricum*). *PeerJ* **2021**, *9*, e11939. [CrossRef] [PubMed]
4. Sardiello, M.; Cairo, S.; Fontanella, B.; Ballabio, A.; Meroni, G. Genomic Analysis of the TRIM Family Reveals Two Groups of Genes with Distinct Evolutionary Properties. *BMC Evol. Biol.* **2008**, *8*, 225. [CrossRef] [PubMed]
5. Yoshigai, E.; Kawamura, S.; Kuhara, S.; Tashiro, K. Trim36/Haprin Plays a Critical Role in the Arrangement of Somites during Xenopus Embryogenesis. *Biochem. Biophys. Res. Commun.* **2009**, *378*, 428–432. [CrossRef]
6. Wei, Y.; Zhou, H.; Wang, A.; Sun, L.; Wang, M.; Jia, R.; Zhu, D.; Liu, M.; Yang, Q.; Wu, Y.; et al. TRIM25 Identification in the Chinese Goose: Gene Structure, Tissue Expression Profiles, and Antiviral Immune Responses In Vivo and In Vitro. *BioMed Res. Int.* **2016**, *2016*, 1–14. [CrossRef]
7. Shi, Y.; Hu, S.; Duan, W.; Ding, T.; Zhao, Z. The Distinct Evolutionary Properties of the Tripartite Motif-Containing Protein 39 in the Chinese Softshell Turtle Based on Its Structural and Functional Characterization. *Dev. Comp. Immunol.* **2019**, *99*, 103407. [CrossRef]
8. Hirata, Y.; Katagiri, K.; Nagaoka, K.; Morishita, T.; Kudoh, Y.; Hatta, T.; Naguro, I.; Kano, K.; Udagawa, T.; Natsume, T.; et al. TRIM48 Promotes ASK1 Activation and Cell Death through Ubiquitination-Dependent Degradation of the ASK1-Negative Regulator PRMT1. *Cell Rep.* **2017**, *21*, 2447–2457. [CrossRef]
9. Wang, H.; Hur, S. Substrate Recognition by TRIM and TRIM-like Proteins in Innate Immunity. *Semin. Cell Dev. Biol.* **2020**, *11*, 76–85. [CrossRef]
10. Boudinot, P.; van der Aa, L.M.; Jouneau, L.; Du Pasquier, L.; Pontarotti, P.; Briolat, V.; Benmansour, A.; Levraud, J.-P. Origin and Evolution of TRIM Proteins: New Insights from the Complete TRIM Repertoire of Zebrafish and Pufferfish. *PLoS ONE* **2011**, *6*, e22022. [CrossRef]
11. Goyani, S.; Roy, M.; Singh, R. TRIM-NHL as RNA Binding Ubiquitin E3 Ligase (RBUL): Implication in Development and Disease Pathogenesis. *Biochim. Et Biophys. Acta (BBA)-Mol. Basis Dis.* **2021**, *1867*, 7. [CrossRef]
12. Hatakeyama, S. TRIM Family Proteins: Roles in Autophagy, Immunity, and Carcinogenesis. *Trends Biochem. Sci.* **2017**, *42*, 297–311. [CrossRef]
13. Jin, Z.; Zhu, Z. The Role of TRIM Proteins in PRR Signaling Pathways and Immune-Related Diseases. *Int. Immunopharmacol.* **2021**, *98*, 107813. [CrossRef] [PubMed]
14. Liu, J.; Zhang, C.; Wang, X.; Hu, W.; Feng, Z. Tumor Suppressor P53 Cross-Talks with TRIM Family Proteins. *Genes Dis.* **2021**, *8*, 463–474. [CrossRef]
15. Ozato, K.; Shin, D.; Chang, T.; Morse, H.C. TRIM Family Proteins and Their Emerging Roles in Innate Immunity. *Nat. Rev. Immunol.* **2008**, *8*, 849–860. [CrossRef] [PubMed]
16. Koepke, L.; Gack, M.U.; Sparrer, K.M. The Antiviral Activities of TRIM Proteins. *Curr. Opin. Microbiol.* **2021**, *59*, 50–57. [CrossRef]
17. Giraldo, M.; Hage, A.; van Tol, S.; Rajsbaum, R. TRIM Proteins in Host Defense and Viral Pathogenesis. *Curr. Clin. Microbiol. Rep.* **2020**, 1–14. [CrossRef]
18. Jia, X.; Zhao, C.; Zhao, W. Emerging Roles of MHC Class I Region-Encoded E3 Ubiquitin Ligases in Innate Immunity. *Front. Immunol.* **2021**, *12*, 687102. [CrossRef] [PubMed]
19. Stremlau, M.; Owens, C.M.; Perron, M.J.; Kiessling, M.; Autissier, P.; Sodroski, J. The Cytoplasmic Body Component TRIM5a Restricts HIV-1 Infection in Old World Monkeys. *Nature* **2004**, *427*, 6. [CrossRef]
20. Mu, T.; Zhao, X.; Zhu, Y.; Fan, H.; Tang, H. The E3 Ubiquitin Ligase TRIM21 Promotes HBV DNA Polymerase Degradation. *Viruses* **2020**, *12*, 346. [CrossRef]
21. Yergeau, D.A.; Cornell, C.N.; Parker, S.K.; Zhou, Y.; Detrich, H.W. Bloodthirsty, an RBCC/TRIM Gene Required for Erythropoiesis in Zebrafish. *Dev. Biol.* **2005**, *283*, 97–112. [CrossRef] [PubMed]
22. van der Aa, L.M.; Levraud, J.; Yahmi, M.; Lauret, E.; Briolat, V.; Herbomel, P.; Benmansour, A.; Boudinot, P. A large new subset of TRIM genes highly diversified by duplication and positive selection in teleost fish. *BMC Biol.* **2009**, *7*, 7. [CrossRef]
23. Sun, N.; Jiang, L.; Ye, M.; Wang, Y.; Wang, G.; Wan, X.; Zhao, Y.; Wen, X.; Liang, L.; Ma, S.; et al. TRIM35 Mediates Protection against Influenza Infection by Activating TRAF3 and Degrading Viral PB2. *Protein Cell* **2020**, *11*, 12. [CrossRef] [PubMed]
24. Suzuki, M.; Watanabe, M.; Nakamaru, Y.; Takagi, D.; Takahashi, H.; Fukuda, S.; Hatakeyama, S. TRIM39 Negatively Regulates the NFκB-Mediated Signaling Pathway through Stabilization of Cactin. *Celluarl. Mol. Life Sci.* **2016**, *73*, 1085–1101. [CrossRef]
25. Furnes, C.; Robertsen, B. Molecular cloning and characterization of bloodthirsty from Atlantic cod (*Gadus morhua*). *Fish Shellfish. Immunol.* **2010**, *29*, 903–909. [CrossRef] [PubMed]

26. Zhang, X.; Zhao, H.; Chen, Y.; Luo, H.; Yang, P.; Yao, B. A Zebrafish (*Danio Rerio*) Bloodthirsty Member 20 with E3 Ubiquitin Ligase Activity Involved in Immune Response against Bacterial Infection. *Biochem. Biophys. Res. Commun.* **2015**, *457*, 83–89. [CrossRef]
27. Langevin, C.; Boudinot, P. FTR83, a Member of the Large Fish-Specific FinTRIM Family, Triggers IFN Pathway and Counters Viral Infection. *Front. Immunol.* **2017**, *8*, 16. [CrossRef]
28. Wentzel, A.S.; Petit, J.; van Veen, W.G.; Fink, I.R.; Scheer, M.H.; Piazzon, M.C.; Forlenza, M.; Spaink, H.P.; Wiegertjes, G.F. Transcriptome Sequencing Supports a Conservation of Macrophage Polarization in Fish. *Sci. Rep.* **2020**, *10*, 13470. [CrossRef]
29. Savino, A.; Provero, P.; Poli, V. Differential Co-Expression Analyses Allow the Identification of Critical Signalling Pathways Altered during Tumour Transformation and Progression. *Int. J. Mol. Sci.* **2020**, *21*, 9461. [CrossRef]
30. Cruz, A.; Arrais, J.P.; Machado, P. Interactive and Coordinated Visualization Approaches for Biological Data Analysis. *Brief. Bioinform.* **2019**, *20*, 1513–1523. [CrossRef]
31. Hossain, S.M.M.; Khatun, L.; Ray, S.; Mukhopadhyay, A. Identification of Key Immune Regulatory Genes in HIV-1 Progression. *Gene* **2021**, *792*, 145735. [CrossRef]
32. Wei, J.; Zhang, H.; Li, X.; Li, Q.; Ma, Z.; Bai, J.; Qiao, Z.; Feng, R. Transcriptional Profiling of Host Cell Responses to Encephalomyocarditis Virus (EMCV). *Virol. J.* **2017**, *14*, 45. [CrossRef]
33. Tang, Y.; Xin, G.; Zhao, L.-M.; Huang, L.-X.; Qin, Y.-X.; Su, Y.-Q.; Zheng, W.-Q.; Wu, B.; Lin, N.; Yan, Q.-P.; et al. Novel Insights into Host-Pathogen Interactions of Large Yellow Croakers (*Larimichthys Crocea*) and Pathogenic Bacterium *Pseudomonas Plecoglossicida* Using Time-Resolved Dual RNA-Seq of Infected Spleens. *Zool. Res.* **2020**, *41*, 314–327. [CrossRef]
34. Verrier, E.R.; Genet, C.; Laloë, D.; Jaffrezic, F.; Rau, A.; Esquerre, D.; Dechamp, N.; Ciobotaru, C.; Hervet, C.; Krieg, F.; et al. Genetic and Transcriptomic Analyses Provide New Insights on the Early Antiviral Response to VHSV in Resistant and Susceptible Rainbow Trout. *BMC Genom.* **2018**, *19*, 482. [CrossRef]
35. China National Knowledge Infrastructure. Available online: https://data.cnki.net/Trade/yearbook/single/N2021020168?z=Z009 (accessed on 1 August 2021).
36. Zeng, W.; Wang, Q.; Wang, Y.; Zhao, C.; Li, Y.; Shi, C.; Wu, S.; Song, X.; Huang, Q.; Li, S. Immunogenicity of a Cell Culture-Derived Inactivated Vaccine against a Common Virulent Isolate of Grass Carp Reovirus. *Fish Shellfish Immunol.* **2016**, *54*, 473–480. [CrossRef] [PubMed]
37. Rao, Y.; Su, J. Insights into the Antiviral Immunity against Grass Carp (*Ctenopharyngodon idella*) Reovirus (GCRV) in Grass Carp. *J. Immunol. Res.* **2015**, *2015*, 670437. [CrossRef]
38. Zhao, X.; Xiao, T.; Jin, S.; Wang, J.; Wang, J.; Luo, H.; Li, R.; Sun, T.; Zou, J.; Li, Y. Characterization and Immune Function of the Interferon-β Promoter Stimulator-1 in the Barbel Chub, *Squaliobarbus Curriculus*. *Dev. Comp. Immunol.* **2020**, *104*, 103571. [CrossRef]
39. Wang, Y.; Lu, Y.; Zhang, Y.; Ning, Z.; Li, Y.; Zhao, Q.; Lu, H.; Huang, R.; Xia, X.; Feng, Q.; et al. The Draft Genome of the Grass Carp (*Ctenopharyngodon Idellus*) Provides Insights into Its Evolution and Vegetarian Adaptation. *Nat. Genet.* **2015**, *47*, 625–631. [CrossRef] [PubMed]
40. Mistry, J.; Chuguransky, S.; Williams, L.; Qureshi, M.; Salazar, G.A.; Sonnhammer, E.L.L.; Tosatto, S.C.E.; Paladin, L.; Raj, S.; Richardson, L.J.; et al. Pfam: The Protein Families Database in 2021. *Nucleic Acids Res.* **2021**, *49*, D412–D419. [CrossRef]
41. Letunic, I.; Khedkar, S.; Bork, P. SMART: Recent Updates, New Developments and Status in 2020. *Nucleic Acids Res.* **2021**, *49*, D458–D460. [CrossRef] [PubMed]
42. Duvaud, S.; Gabella, C.; Lisacek, F.; Stockinger, H.; Ioannidis, V.; Durinx, C. Expasy, the Swiss Bioinformatics Resource Portal, as Designed by Its Users. *Nucleic Acids Res.* **2021**, *49*, W216–W227. [CrossRef]
43. Chou, K.; Shen, H. Cell-PLoc 2.0: An Improved Package of Web-Servers for Predicting Subcellular Localization of Proteins in Various Organisms. *Nat. Protoc.* **2008**, *3*, 153–162. [CrossRef] [PubMed]
44. Bailey, T.L.; Boden, M.; Buske, F.A.; Frith, M.; Grant, C.E.; Clementi, L.; Ren, J.; Li, W.W.; Noble, W.S. MEME SUITE: Tools for Motif Discovery and Searching. *Nucleic Acids Res.* **2009**, *37*, W202–W208. [CrossRef]
45. Tamura, K.; Stecher, G.; Peterson, D.; Filipski, A.; Kumar, S. MEGA6: Molecular Evolutionary Genetics Analysis Version 6.0. *Mol. Biol. Evol.* **2013**, *30*, 2725–2729. [CrossRef]
46. Subramanian, B.; Gao, S.; Lercher, M.J.; Hu, S.; Chen, W.-H. Evolview v3: A Webserver for Visualization, Annotation, and Management of Phylogenetic Trees. *Nucleic Acids Res.* **2019**, *47*, W270–W275. [CrossRef] [PubMed]
47. Xia, J.; Liu, F.; Zhu, Z.; Fu, J.; Feng, J.; Li, J.; Yue, G. A Consensus Linkage Map of the Grass Carp (*Ctenopharyngodon Idella*) Based on Microsatellites and SNPs. *BMC Genom.* **2010**, *11*, 135. [CrossRef] [PubMed]
48. Chao, J.; Kong, Y.; Wang, Q.; Sun, Y.; Gong, D.; Lv, J.; Liu, G. MapGene2Chrom, a Tool to Draw Gene Physical Map Based on Perl and SVG Languages. *Yi Chuan* **2015**, *37*, 91–97. [CrossRef]
49. Howe, K.L.; Achuthan, P.; Allen, J.; Allen, J.; Alvarez-Jarreta, J.; Amode, M.R.; Armean, I.M.; Azov, A.G.; Bennett, R.; Bhai, J.; et al. Ensembl 2021. *Nucleic Acids Res.* **2021**, *49*, D884–D891. [CrossRef]
50. Chen, C.; Chen, H.; Zhang, Y.; Thomas, H.R.; Frank, M.H.; He, Y.; Xia, R. TBtools: An Integrative Toolkit Developed for Interactive Analyses of Big Biological Data. *Mol. Plant* **2020**, *13*, 1194–1202. [CrossRef]
51. Wang, Y.; Tang, H.; DeBarry, J.D.; Tan, X.; Li, J.; Wang, X.; Lee, T.; Jin, H.; Marler, B.; Guo, H.; et al. MCScanX: A Toolkit for Detection and Evolutionary Analysis of Gene Synteny and Collinearity. *Nucleic Acids Res.* **2012**, *40*, e49. [CrossRef]

52. Cantalapiedra, C.P.; Hernández-Plaza, A.; Letunic, I.; Bork, P.; Huerta-Cepas, J. EggNOG-Mapper v2: Functional Annotation, Orthology Assignments, and Domain Prediction at the Metagenomic Scale. *BioRxiv* **2021**. [CrossRef]
53. He, L.; Zhang, A.; Pei, Y.; Chu, P.; Li, Y.; Huang, R.; Liao, L.; Zhu, Z.; Wang, Y. Differences in Responses of Grass Carp to Different Types of Grass Carp Reovirus (GCRV) and the Mechanism of Hemorrhage Revealed by Transcriptome Sequencing. *BMC Genom.* **2017**, *18*, 452. [CrossRef] [PubMed]
54. Kim, D.; Langmead, B.; Salzberg, S.L. HISAT: A Fast Spliced Aligner with Low Memory Requirements. *Nat. Methods* **2015**, *12*, 357–360. [CrossRef]
55. Liao, Y.; Smyth, G.K.; Shi, W. FeatureCounts: An Efficient General Purpose Program for Assigning Sequence Reads to Genomic Features. *Bioinformatics* **2014**, *30*, 923–930. [CrossRef] [PubMed]
56. Love, M.I.; Huber, W.; Anders, S. Moderated Estimation of Fold Change and Dispersion for RNA-Seq Data with DESeq2. *Genome Biol.* **2014**, *15*, 550. [CrossRef] [PubMed]
57. Langfelder, P.; Horvath, S. WGCNA: An R Package for Weighted Correlation Network Analysis. *BMC Bioinform.* **2008**, *9*, 559. [CrossRef] [PubMed]
58. Bourgon, R.; Gentleman, R.; Huber, W. Independent filtering increases detection power for high-throughput experiments. *Proc. Natl. Acad. Sci. USA* **2010**, *107*, 9546–9551. [CrossRef]
59. Zhou, Y.; Zhou, B.; Pache, L.; Chang, M.; Khodabakhshi, A.H.; Tanaseichuk, O.; Benner, C.; Chanda, S.K. Metascape Provides a Biologist-Oriented Resource for the Analysis of Systems-Level Datasets. *Nat. Commun.* **2019**, *10*, 1523. [CrossRef]
60. Chu, P.; He, L.; Huang, R.; Liao, L.; Li, Y.; Zhu, Z.; Hu, W.; Wang, Y. Autophagy Inhibits Grass Carp Reovirus (GCRV) Replication and Protects *Ctenopharyngodon Idella* Kidney (CIK) Cells from Excessive Inflammatory Responses after GCRV Infection. *Biomolecules* **2020**, *10*, 1296. [CrossRef]
61. Ernst, J.; Bar-Joseph, Z. STEM: A Tool for the Analysis of Short Time Series Gene Expression Data. *BMC Bioinform.* **2006**, *7*, 191. [CrossRef]
62. Livak, K.J.; Schmittgen, T.D. Analysis of Relative Gene Expression Data Using Real-Time Quantitative PCR and the 2(-Delta Delta C(T)) Method. *Methods* **2001**, *25*, 402–408. [CrossRef] [PubMed]
63. Liu, S.; Cai, X.; Wu, J.; Cong, Q.; Chen, X.; Li, T.; Du, F.; Ren, J.; Wu, Y.-T.; Grishin, N.V.; et al. Phosphorylation of Innate Immune Adaptor Proteins MAVS, STING, and TRIF Induces IRF3 Activation. *Science* **2015**, *347*, aaa2630. [CrossRef] [PubMed]
64. Yan, L.; Liu, H.; Li, X.; Fang, Q. The VP2 Protein of Grass Carp Reovirus (GCRV) Expressed in a Baculovirus Exhibits RNA Polymerase Activity. *Virol. Sin.* **2014**, *29*, 86–93. [CrossRef] [PubMed]
65. Tocchini, C.; Ciosk, R. TRIM-NHL Proteins in Development and Disease. *Semin. Cell Dev. Biol.* **2015**, *47–48*, 52–59. [CrossRef] [PubMed]
66. Jaillon, O.; Aury, J.; Brunet, F.; Petit, J.; Stange-Thomann, N.; Mauceli, E.; Bouneau, L.; Fischer, C.; Ozouf-Costaz, C.; Bernot, A.; et al. Genome Duplication in the Teleost Fish *Tetraodon Nigroviridis* Reveals the Early Vertebrate Proto-Karyotype. *Nature* **2004**, *431*, 946–957. [CrossRef]
67. Xu, P.; Zhang, X.; Wang, X.; Li, J.; Liu, G.; Kuang, Y.; Xu, J.; Zheng, X.; Ren, L.; Wang, G.; et al. Genome Sequence and Genetic Diversity of the Common Carp, *Cyprinus Carpio*. *Nat. Genet.* **2014**, *46*, 1212–1219. [CrossRef]
68. Langevin, C.; Levraud, J.; Boudinot, P. Fish Antiviral Tripartite Motif (TRIM) Proteins. *Fish Shellfish Immunol.* **2019**, *86*, 724–733. [CrossRef]
69. Zhang, X. Phylogenetic Analysis of Fish TRIM Genes and Their Functional Roles in Innate Immune System. Ph.D. Dissertation, Chinese Academy of Agricultural Sciences, Beijing, China, 2015. (In Chinese)
70. Luo, K.; Li, Y.; Ai, K.; Xia, L.; Zhang, J.; Hu, W.; Gao, W.; Guo, L.; Qi, Z.; Yuan, H.; et al. Bioinformatics and Expression Analysis of FinTRIM Genes in Grass Carp, *Ctenopharyngodon idella*. *Fish Shellfish Immunol.* **2017**, *66*, 217–223. [CrossRef]
71. Reymond, A. The Tripartite Motif Family Identifies Cell Compartments. *EMBO J.* **2001**, *20*, 2140–2151. [CrossRef]
72. Cainarca, S.; Messali, S.; Ballabio, A.; Meroni, G. Functional Characterization of the Opitz Syndrome Gene Product (Midin): Evidence for Homodimerization and Association with Microtubules throughout the Cell Cycle. *Hum. Mol. Genet.* **1999**, *8*, 1387–1396. [CrossRef]
73. Tsai, W.; Wang, Z.; Yiu, T.T.; Akdemir, K.C.; Xia, W.; Winter, S.; Tsai, C.; Shi, X.; Schwarzer, D.; Plunkett, W.; et al. TRIM24 Links a Non-Canonical Histone Signature to Breast Cancer. *Nature* **2010**, *468*, 927–932. [CrossRef]
74. Sparrer, K.M.J.; Gableske, S.; Zurenski, M.A.; Parker, Z.M.; Full, F.; Baumgart, G.J.; Kato, J.; Pacheco-Rodriguez, G.; Liang, C.; Pornillos, O.; et al. TRIM23 Mediates Virus-Induced Autophagy via Activation of TBK1. *Nat. Microbiol.* **2017**, *2*, 1543–1557. [CrossRef] [PubMed]
75. Yang, W.; Gu, Z.; Zhang, H.; Hu, H. To TRIM the Immunity: From Innate to Adaptive Immunity. *Front. Immunol.* **2020**, *11*, 02157. [CrossRef] [PubMed]
76. Stelzer, G.; Rosen, N.; Plaschkes, I.; Zimmerman, S.; Twik, M.; Fishilevich, S.; Stein, T.I.; Nudel, R.; Lieder, I.; Mazor, Y.; et al. The GeneCards Suite: From Gene Data Mining to Disease Genome Sequence Analyses. *Curr. Protoc. Bioinform.* **2016**, *54*. [CrossRef] [PubMed]
77. Yan, J.; Li, Q.; Mao, A.; Hu, M.; Shu, H. TRIM4 Modulates Type I Interferon Induction and Cellular Antiviral Response by Targeting RIG-I for K63-Linked Ubiquitination. *J. Mol. Cell Biol.* **2014**, *6*, 154–163. [CrossRef]

biology MDPI

Communication

Modulation of the Tissue Expression Pattern of Zebrafish CRP-Like Molecules Suggests a Relevant Antiviral Role in Fish Skin

Melissa Bello-Perez [1], Mikolaj Adamek [2], Julio Coll [3], Antonio Figueras [4], Beatriz Novoa [4] and Alberto Falco [1,*]

1 Institute of Research, Development, and Innovation in Healthcare Biotechnology in Elche (IDiBE), Miguel Hernández University (UMH), 03202 Elche, Spain; melissa.bello@alu.umh.es
2 Fish Disease Research Unit, Institute for Parasitology, University of Veterinary Medicine, 30559 Hannover, Germany; mikolaj.adamek@tiho-hannover.de
3 Department of Biotechnology, National Agricultural and Food Research and Technology Institute (INIA), 28040 Madrid, Spain; juliocoll@inia.es
4 Institute of Marine Research, Consejo Superior de Investigaciones Científicas (IIM-CSIC), 36208 Vigo, Spain; antoniofigueras@iim.csic.es (A.F.); beatriznovoa@iim.csic.es (B.N.)
* Correspondence: alber.falco@umh.es; Tel.: +34-966-658-744

Simple Summary: The clinical use of the human short pentraxin C-reactive protein as a health biomarker is expanded worldwide. The acute increase of the serum levels of short pentraxins in response to bacterial infections is evolutionarily conserved, as are the main functions of pentraxins. Interestingly, fish orthologs have been found to increase similarly after bacterial and viral stimuli, thus becoming promising candidates for health biomarkers of both types of infection in this group of vertebrates. To preliminarily assess their adequacy for this application, zebrafish and a fish rhabdovirus were chosen as infection model systems for the analysis of the levels of gene expression of all short pentraxins in healthy and infected animals in a wide range of tissues. Because some significant increases were found in skin (a very suitable sampling source for testing purposes), further transcript analyses were carried out in this tissue. Due to the functional similarities between pentraxins and antibodies, it was also checked whether short pentraxins can compensate for the deficiencies in adaptive immunity by using mutant zebrafish lacking this system. In conclusion, the obtained results suggest that short pentraxins are highly reactant against viruses in skin and their overexpression seems to reflect a mechanism to compensate for the loss of adaptive immunity.

Abstract: Recent studies suggest that short pentraxins in fish might serve as biomarkers for not only bacterial infections, as in higher vertebrates including humans, but also for viral ones. These fish orthologs of mammalian short pentraxins are currently attracting interest because of their newly discovered antiviral activity. In the present work, the modulation of the gene expression of all zebrafish short pentraxins (CRP-like proteins, CRP1-7) was extensively analyzed by quantitative polymerase chain reaction. Initially, the tissue distribution of *crp1-7* transcripts and how the transcripts varied in response to a bath infection with the spring viremia of carp virus, were determined. The expression of *crp1-7* was widely distributed and generally increased after infection (mostly at 5 days post infection), except for *crp1* (downregulated). Interestingly, several *crp* transcription levels significantly increased in skin. Further assays in mutant zebrafish of recombinant activation gene 1 (*rag1*) showed that all *crps* (except for *crp2*, downregulated) were already constitutively highly expressed in skin from *rag1* knockouts and only increased moderately after viral infection. Similar results were obtained for most *mx* isoforms (a reporter gene of the interferon response), suggesting a general overcompensation of the innate immunity in the absence of the adaptive one.

Keywords: short pentraxins; c-reactive protein; zebrafish; transcript expression; antiviral; SVCV; rag1 mutants; skin; mucosal immunity

Citation: Bello-Perez, M.; Adamek, M.; Coll, J.; Figueras, A.; Novoa, B.; Falco, A. Modulation of the Tissue Expression Pattern of Zebrafish CRP-Like Molecules Suggests a Relevant Antiviral Role in Fish Skin. *Biology* **2021**, *10*, 78. https://doi.org/10.3390/biology10020078

Academic Editor: Herman Spaink
Received: 28 December 2020
Accepted: 19 January 2021
Published: 22 January 2021

1. Introduction

Circulating pentraxins are considered pattern-recognition molecules that contribute to innate immunity by mainly facilitating the clearance of damaged cells and bacterial pathogens [1–4]. The representative forms of these molecules show an annular pentameric structural symmetry in humans [5]. Their monomers are characterized by the presence of a C-terminal domain of approximately 200 amino acid residues, containing the so-called "pentraxin signature", a conserved 8 amino acid residue sequence (HxCxS/TWxS) [2,6,7]. When an additional N-terminal region is present, they are termed fusion or long pentraxins, and their prototype is pentraxin 3 (PTX3) [8]. In contrast, those soluble pentraxins consisting of just the C-terminal domain are termed classical or short pentraxins, and they include C-reactive protein (CRP) and serum-amyloid P component (SAP) [9].

The classification into CRPs and SAPs is mostly based on their differential ligand affinities in humans, although in other species this feature shows overlapping and even total reversion between the two groups [3,10–15]. Likewise, several oligomerization forms, other than pentameric, have also been identified across evolution, even in humans [16,17]. Nevertheless, because of shared homologous sequences corresponding to functionally important motifs and analogous molecular structures, fundamental activities associated with short pentraxins in humans are evolutionarily conserved [3,4,12–15]. One such activity is their role as reactive plasma proteins during the acute phase response (APR), an immediate and systemic physiological feedback of the innate immune system to trauma, injury, and infection [1–4]. In this regard, CRP is the predominant acute phase protein (APP) in most mammals and thus its clinical use as a health biomarker is common [1–4].

Although APRs are usually triggered during both bacterial and viral infections [18], increases of serum CRP levels in mammals are more characteristic of bacterial rather than viral infections [1,18,19]. In fish, short pentraxins are common ancestors of both mammalian CRP and SAP counterparts, which diverged very early in mammalian evolution after the separation of mammalian and avian lineages [20], and thus arose independently from a homologous differentiation in arthropods [7]. The existing studies analyzing the serum levels of fish short pentraxins (hereafter termed CRP-like proteins) show milder reactions than mammalian ones with similar response levels to bacterial and viral stimuli [21,22], which makes them potential biomarkers for both types of infection in fish. Interestingly, recent comparative studies of these ancestral orthologs have also revealed new interactions and activities such as affinity for different cholesterol metabolites, the modulation of autophagy, and antiviral activity [15,23–25].

In the present study, the transcript expression of all CRP-like protein (CRP1–7) encoding genes was analyzed extensively in zebrafish, which offers a great variety of scientific tools [26,27]. In addition to the determination of the tissue distribution of all *crp* transcripts and their modulation after infection with the spring viremia of carp virus (SVCV), they were also analyzed on the skin from zebrafish that are mutants of recombinant activation gene 1 (*rag1*$^{+/+}$ and *rag1*$^{-/-}$). The *rag1*$^{-/-}$ mutants present a point mutation that generates a premature stop codon in the catalytic domain of this protein, at the endonuclease responsible for the V(D) J recombination [28]. This process results in functional macrophages, natural killer cells, and neutrophils and in a complete block of immunoglobulin gene assembly and T cell receptors (TCR) formation [29]. Moreover, it leads to the absence of adaptative immunity, providing a living platform to elucidate mechanisms of the innate immune responses [30]. Since CRPs have been previously referred to by other authors as ancient antibodies, because of their several analogous activities [2], this work aimed at investigating the behavior of these molecules in the absence of antibodies and at exploring their potential as health biomarkers of viral infections in fish.

2. Materials and Methods

2.1. Cell Lines and Virus

Epithelioma papulosum cyprinid (EPC) cells from fat-head minnow (*Pimephales promelas*) were purchased from the American Type Culture Collection (ATCC, Manassas, VI, USA, ref. no. CRL-2872). EPC cell monolayers were grown in Dutch-modified Roswell Park Memorial Institute (RPMI)-1640 culture medium (Sigma, St. Louis, MO, USA), supplemented with 10% fetal bovine serum (FBS) (Sigma), 2 mM glutamine, 1 mM sodium pyruvate, 50 µg/mL gentamicin, and 2 µg/mL of fungizone (Gibco BRL-Invitrogen, Carlsbad, CA, USA), at 28 °C in a 5% CO_2 atmosphere.

SVCV isolate 56/70 from carp (*Cyprinus carpio*) was replicated in EPC cells at 22 °C without CO_2 supply and by using previously described cell culture media except for 2% FBS (infection media). After 7 days post-infection (dpi), infective supernatants were harvested, clarified by centrifugation at 4 kg and 4 °C for 30 min, aliquoted, and stored at −80 °C until use. Virus titers in plaque forming units (pfu) were determined by the focus forming assay as described elsewhere [31].

2.2. Animals

The adult wild-type zebrafish (*Danio rerio*) were obtained by natural spawning from mating adults at one of the host institutions (Instituto de Investigaciones Marinas-CSIC, Vigo, Spain). Likewise, adult zebrafish that are mutants of recombinant activation gene 1 (*rag1*$^{+/+}$ and *rag1*$^{-/-}$) were obtained at the University of Murcia (Murcia, Spain), and genotyped when they reached ~1 g (~6 months of age). Fish were maintained at 28 °C in 30-L aquaria, following established protocols [26], and fed daily with commercial food (Vipan BioVip, Berlin, Germany). Before the infection experiments, fish were acclimatized to 22 °C for 2 weeks. Prior to methodological handling, fish were anaesthetized by immersion in 100 mg/L tricaine methanesulfonate (MS-222) (Sigma). End-point fish euthanasia was performed by overdosing with tricaine methanesulfonate (500 mg/L).

2.3. Ethical Statement of Zebrafish Handling

All experiments with zebrafish complied with the Spanish Law for Animal Experimentation (Royal Decree-Law, 53/2013) and the European Union Council Directive 2010/63/UE. Animal trial procedures were approved by the local government ethics committee on animal experimentation (Dirección General de Agricultura, Ganadería y Pesca, Generalitat Valenciana), the Project Evaluation Board of Miguel Hernández University (permit no. UMH.IBM.JFG.01.14), and the CSIC National Committee on Bioethics (permit no. ES360570202001/16/FUN01/PAT.05/tipoE/BNG).

2.4. In Vivo Viral Infection

For each infection experiment, zebrafish were exposed to 10^4 ffu/mL of SVCV by bath immersion for 90 min at 22 °C (optimal temperature for SVCV replication). Mock-infected zebrafish were incubated with equivalent volumes of cell culture medium in parallel experiments. Fish were then transferred to tanks with clean water and kept at 22 °C to allow for the progress of SVCV infection. Tissues were harvested at specific time points after infection for each experimental design. All tissues were dissected under a binocular loupe for transcript expression analysis.

For a general analysis of the tissue-specific transcription levels (tissue distribution assays) and their modulation in response to SVCV infection, four wild-type zebrafish were used for each condition; that is, four for the mock-infection, four for the 2-day infection, and four for the 5-day infection. Tissue samples from the mock-infected fish were collected at 2 dpi and used to determine the basal (and reference) expression. The tissues collected comprised head kidney (HK), liver, skin, gills, gut, muscle, and spleen.

To analyze the modulation of the transcription levels in skin from *rag1*$^{+/+}$ and *rag1*$^{-/-}$ zebrafish in response to SVCV, four individuals from each genotype were mock- and SVCV-infected. Skin samples from these fish were collected at 2 dpi to measure the early immune responses evoked by the virus entry and replication in one of its initial target tissues.

2.5. RNA Isolation, cDNA Synthesis, and qPCR

Total RNA was extracted by using EZNA HP Tissue RNA kits (Omega Bio-tek, Norcross, GA, USA) and subsequently treating the samples with DNase (Turbo DNA-free™ Kit, Ambion Inc., Austin, TX, USA) by following the manufacturer's instructions. RNA concentrations were estimated with a Nanodrop 1000 spectrophotometer (Thermo-fisher Scientific, Waltham, MA, USA). Isolated RNA samples were stored at −80 °C until use.

For the synthesis of cDNA, 0.5 µg of isolated RNA from each sample and the Moloney murine leukemia virus (M-MuLV) reverse transcriptase were used (Gibco BRL-Invitrogen), as previously described elsewhere [22].

qPCR was performed by using an ABI PRISM 7300 thermocycler (Applied Biosystems, Branchburg, NJ, USA). Reactions were conducted in 20-µL-volume reactions comprising: 2 µL of cDNA, 900 nM of each corresponding forward and reverse primer (Sigma) (primer sequences are shown in Table S1), and 10 µL of SYBR Green PCR master mix (Life Technologies, Paisley, UK). Non-template controls were added for each gene analysis. All reactions were performed using technical duplicates. Cycling conditions were an initial denaturing step (10 min at 95 °C), followed by 40 cycles comprising 1 min at 65 °C and 1 min at 95 °C, and finally an extension step of 10 min at 65 °C. Melting curves were checked for inconstancies in each reaction. Cell and viral gene expression results were obtained and represented as relative transcription levels, by normalizing the expression level of each target gene to endogenous elongation factor 1-α (*ef1a*) levels by using a variation of Livak and Schmittgen's method [32] by the formula $2^{\mathrm{Ct\ ref.} - \mathrm{Ct\ target}}$. For the analysis of the transcriptional responses to SVCV infection, the data are represented as fold changes relative to the corresponding samples from mock-infected individuals (*ef1a*-normalized expression in samples from infected individuals/*ef1a*-normalized expression in samples from mock-infected individuals), and the data lower than 1-fold were inverted and represented as negative values to show downregulation.

2.6. Statistical Analysis and Graphics

Data are shown as mean and standard deviation (SD). Resulting datasets were subjected to the most appropriate statistical analysis depending on each particular experimental design. Significant differences were determined by one-way ANOVA and Tukey's multiple comparison (datasets in Section 3.1) or two-way ANOVA and Sidak's multiple comparison test (datasets in Sections 3.2 and 3.3). Prism v7 (Graphpad software, La Jolla, CA, USA) was used for creating the graphs and statistical analysis. $p < 0.05$, $p < 0.01$, and $p < 0.001$ statistical differences with respect to control groups are indicated in each graph by the letters a, b, and c, respectively. Further details are specified accordingly in the figure captions.

3. Results and Discussion

3.1. Broad Tissue Distribution of Zebrafish crp1–7 Expression

All seven *crp* isoforms were expressed in all the tested tissues (HK, liver, skin, gills, gut, muscle, and spleen) from healthy adult zebrafish, albeit with quite different expression patterns (Figure 1). Among all the isoforms, *crp3* and *crp5* showed the most consistent and broad expression patterns, since the site of expression significantly affected the transcription levels reached by the others ($p < 0.05$).

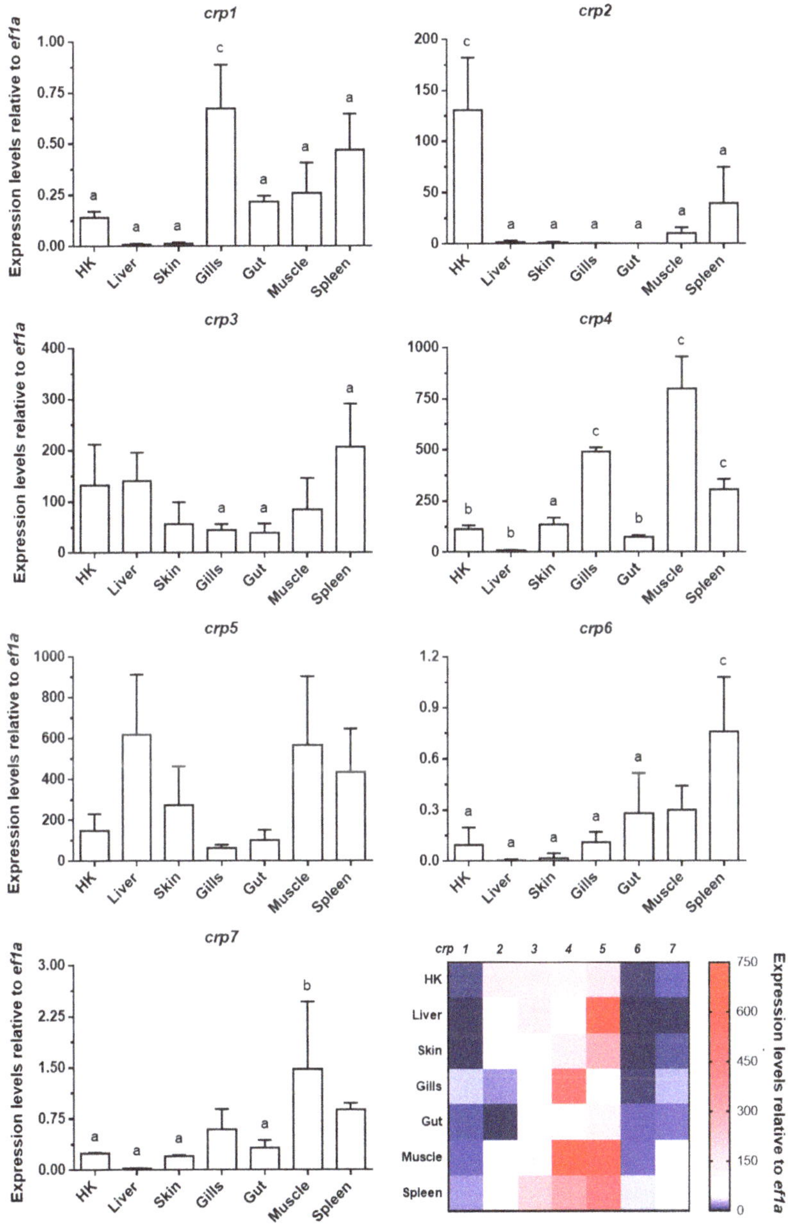

Figure 1. Gene expression analysis of *crp1–7* in tissues of healthy zebrafish. The expression of *crp1–7* was determined by RT-qPCR by using specific primers for each isoform (Table S1). *ef1a* mRNA was used as the endogenous control to normalize data, which are represented as the mean relative expression level $\times 10^3 \pm$ SD of four different individuals. Statistical differences ($p < 0.05$, one-way ANOVA) between tissues are represented by: a (different from up to 1–2 tissues), b (different from up to 3–4 tissues), and c (different from up to 5–6 tissues). Data in bar graphs are summarized in a final double gradient colormap (descending blue gradient for values from 0 to 1 and ascending red gradient from values from 1 to $\geq$750).

According to our results, the most pronounced expression level was found for *crp4* in muscle (799.95 ± 155.24), and the lowest one for *crp1* in liver (0.009 ± 0.003). The spleen and muscle tissues appeared to be among the major basal expression sites for all *crps*. In general terms, our results confirm the broad extrahepatic expression of zebrafish *crps*, in line with previous observations in wild-type zebrafish [25], as well as in other fish species such as *Salmo salar* [33], *Carassius auratus* [34], *Cyprinus carpio* [35], *Oplegnathus fasciatus* [36], *Plecoglossus altivelis* [37], *Sebastes schlegelii* [38], and *Cynoglossus semilaevis* [39,40]. Interestingly, the skin also showed notable expression levels for some isoforms, particularly *crp3–5*. A few studies previously reported the expression of *crps* in skin from zebrafish [25], *Cyprinus carpio* [35,41], *Oplegnathus fasciatus* [36], and *Sebastes schlegelii* [38] and its presence in skin mucus from *Gadus morhua* [42], *Raja kenojei* [43], *Cyclopterus lumpus* [44], and *Tilapia mossambica* [45].

3.2. *Systemic Modulation of Zebrafish crp1–7 Expression Levels in Response to SVCV Infection*

The modulation of the *crp1–7* expression profiles at each tissue in response to SVCV was also analyzed at 2 and 5 dpi by qPCR (Figure 2), and it revealed that both the infection stage and the tissue tested significantly altered the expression levels of each *crp* ($p < 0.001$). As can be observed in Figure 2, SVCV infection resulted mostly in the upregulation of all *crp* isoforms, except for *crp1*, which was mostly downregulated. The highest increases occurred at 5 dpi, mainly for *crp6* in liver (253 ± 18 folds), *crp2* in gut (65 ± 18 folds), and *crp7* in skin (62 ± 8 folds), matching with the highest viral loads, as observed from the expression pattern of the SVCV *n* gene (Figure 2), which was similar to previously reported levels [46,47]. The SVCV infection levels at 2 dpi (Figure 2) were high in HK, spleen (the two most relevant lymphoid organs in fish [48]), and skin. In contrast, the lowest levels of SVCV *n* expression were found in the liver at both time points. The modulation of the expression of certain *crp* genes in response to viral infections has been previously described, in some particular tissues, for zebrafish infected with SVCV [49] and viral hemorrhagic septicemia virus (VHSV) [15,25], for *Cyprinus carpio* infected with cyprinid herpesvirus 3 (CyHV-3) [50] and injected with polyinosinic:polycytidylic acid (poly(I:C), which mimics viral infections) [22] and for *Cynoglossus semilaevis* [40] and *Oplegnathus fasciatus* [36] infected with red seabream iridovirus (RSIV), displaying a similar overall trend. However, the skin was not analyzed in any of those studies, and only one studied the gut, reporting increased levels of a carp *crp* isoform in this tissue in response to poly(I:C) [22].

In parallel, apart from the eventual but remarkable decrease in *crp2* transcripts in HK (−140 ± 79 folds), our results also revealed that SVCV infection downregulated the expression of *crp1* in HK, gills, gut, and spleen (the lowest, −47-fold) at 5 dpi, which could be associated with the fact that CRP1 in zebrafish is the only exclusively intracellular isoform, since it lacks the signal peptide. The low participation of CRP1 in not only viral responses but also viral neutralization has been previoulsy described, suggesting that CRPs should be secreted to be efficient against viral infections [25].

The obtained results on the expression of *crps* also suggest that the liver is the major CRP-producing tissue in zebrafish in response to infection (Figure 2), as also occurs in mammals for CRP/SAP and other APPs when the APR is triggered [1,18,51,52]. Under this situation, the expression of CRP in mammals is principally induced in hepatocytes by the action of the cytokine interleukin-6 (IL-6), whose effect is boosted by interleukin-1 (IL-1) [18,51–53]. A similar mechanism might be occurring in fish, as *crp4* and *crp5* are induced in zebrafish embryos microinjected with the *il6* transgene [25] as is *crp1a* by recombinant cytokines *il1b* in primary HK cells from *Salmo salar* [33]. Furthermore, IL-6 has been found to increase in response to viral infections in both mammals [54–56] and fish [57,58]. In this vein, another parallelism to mammals is the observation of a wide extrahepatic expression of CRPs [18,59]. However, this event appears to be more relevant in fish as is inferred from the high basal (Figure 1) and SVCV-induced (Figure 2) levels of *crp* in the HK, gut, and, surprisingly, skin.

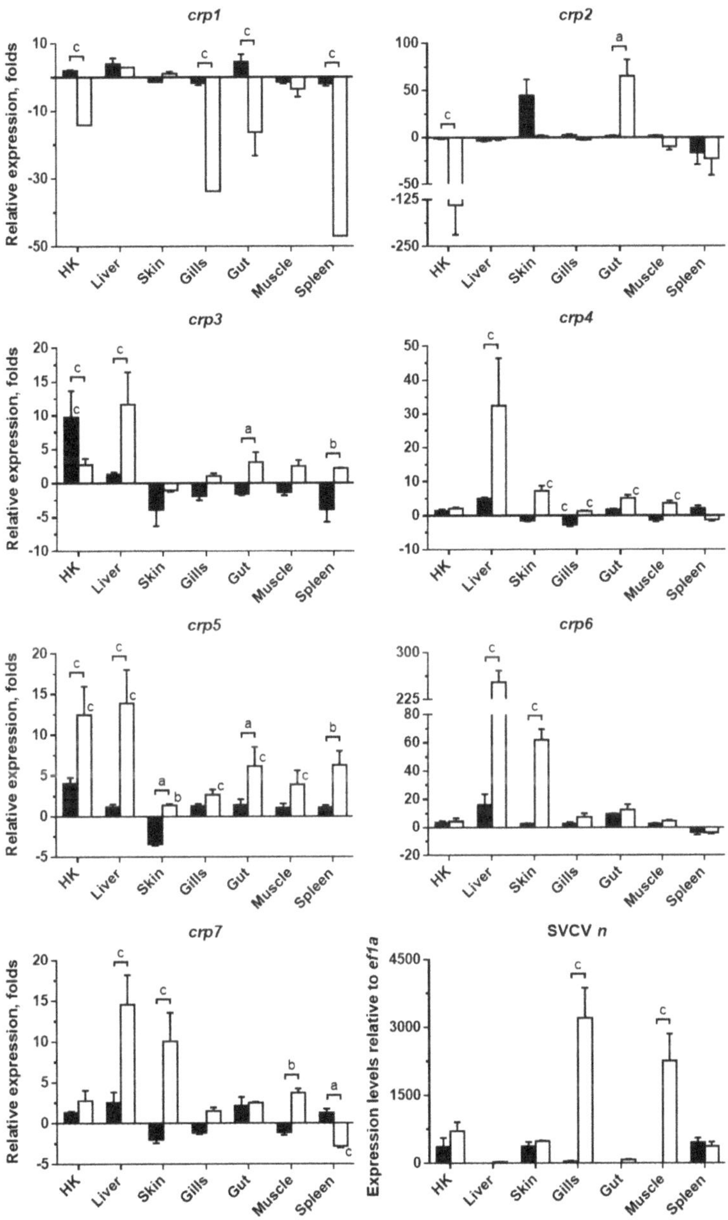

Figure 2. Expression modulation of *crp1–7* in zebrafish tissues in response to spring viremia of carp virus (SVCV) infection. The transcription levels of *crp1–7* and SVCV *n* in tissues from SVCV-infected zebrafish at 2 and 5 dpi (black and white bars, respectively) were quantified by RT-qPCR. *ef1a* mRNA was used as the endogenous control in all cases. *crp1–7* transcription levels were also normalized to the values obtained from the corresponding samples in non-infected fish. Data are represented as the mean fold changes ± SD for *crps* and as the mean relative expression level ± SD for SVCV *n* (four different individuals in all cases). Significant differences were determined by two-way ANOVA and Sidak's multiple comparison test. Statistical differences between the 2- and 5-dpi groups are represented by keys together with a, b, and c letters on top. Statistical differences between the 2- or 5-dpi groups and the non-infected group are represented by a, b, and c letters just on top of the corresponding bars. a, $p \leq 0.05$; b, $p \leq 0.01$; c, $p \leq 0.001$.

3.3. CRPs Are Constitutively Overexpressed in Skin in the Absence of Adaptive Immunity

The modulation of the expression of some zebrafish *crps* in response to SVCV in skin, which had not been described before, attracted our attention. On the one hand, the skin is not considered a major lymphoid [48,60] nor APR-related [18,51] tissue. On the other hand, the skin is one of the preferential entry sites of rhabdovirus in fish [61,62], which explains its early and notable response. In addition, the skin is the largest immunologically active organ in fish and is a type of secondary lymphoid tissue called diffuse mucosa-associated lymphoid tissue (MALT), which is not only endowed with strong innate immune activity, but also with relevant adaptive immune properties [60,63–65].

In order to further characterize this effect in the skin of zebrafish and to explore a potential involvement of the adaptive immune system, a comparative study of the expression changes of *crps* in skin in response to SVCV infection was performed in *rag1* mutants. In this experiment we analyzed the basal and SVCV-induced levels of *crps* in $rag1^{+/+}$ and $rag1^{-/-}$ mutant zebrafish. At this point, it is worth mentioning that wild-type zebrafish (previous experiments) and the $rag1^{+/+}$ zebrafish (following experiments) may differ at some level in the expression of certain genes due to the fact that the former ones include not only homo, but also heterozygous *rag1* individuals ($rag1^{+/+}$ and $rag1^{+/-}$), which may be affected by transcriptional compensatory mechanisms as they are partially deficient in the generation of a mature adaptive immune response [28].

The results (Figure 3) show that basal expression levels of $rag1^{-/-}$ zebrafish were higher for *crp1, 3,* and *5*, and lower for just *crp2*, in comparison to that of $rag1^{+/+}$ zebrafish. After viral challenge, the expression levels of all *crps*, except for *crp2* (which continued to be downregulated), appeared to increase in $rag1^{+/+}$ fish, although only the *crp7* transcription levels were found to be significantly upregulated in this experiment. Regarding $rag1^{+/+}$ and $rag1^{-/-}$ infected groups of zebrafish, expression differences between them were similar to those found at basal conditions, i.e., *crp1, 4, 5,* and *7* expression levels were upregulated in $rag1^{-/-}$ zebrafish, while no differences were detected among their viral loads. When comparing the effect of the SVCV infection on the $rag1^{-/-}$ group, only the expression levels of *crp4* (from 36.25 to 66.90) and *crp7* (13.40 to 46.76) were modulated, suggesting that most *crp* basal expression in this group might be already close to maximum levels.

The modulation of the transcription levels of *mx*, an interferon (IFN) stimulated gene (ISG) often used as a marker of the activation of the type I IFN response, was also analyzed in these samples, resulting in similar results as those previously observed for *crp* genes (Supplementary Figure S1). Thus, the determination of the expression levels of *mxa-g* zebrafish isoforms confirmed that the IFN response to the viral infection was triggered in skin from experimental fish, and that the basal expression of most *mx* isoforms in $rag1^{-/-}$ fish had also reached similar levels to those found after SVCV infection. This observation is in agreement with previous evidence indicating that basal immunological status is elevated in $rag1^{-/-}$ mutants and extends to skin, as well as the parallel modulation of multigene families associated to the antiviral response [66]. In the case of *crp* genes, such compensatory effects might be due not only to their recently described antiviral activity [23–25,40] but also their evolutionary-conserved antibody-like activities [15,67–69]. In addition, the upregulation of this type of genes in $rag1^{-/-}$ mutants has been associated with epigenetic reprogramming [70], similar to the underlying mechanisms regulating innate trained immunity [71,72], and apparently also occurring in the *crp* genes in zebrafish [73].

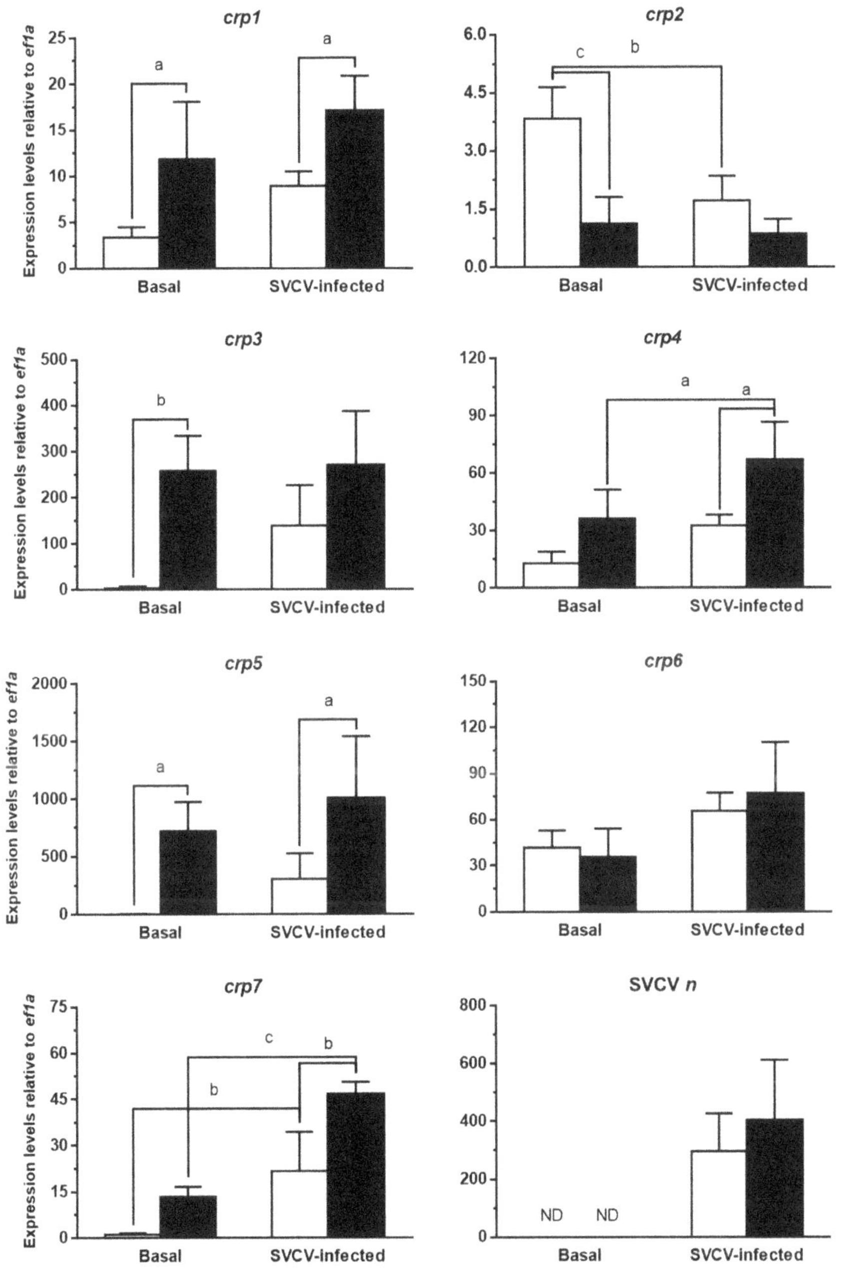

Figure 3. Expression modulation of *crp1–7* in skin from *rag* mutant zebrafish in response to SVCV infection. The transcription levels of *crp1–7* and SVCV *n* were quantified by RT-qPCR in the skin of $rag^{+/+}$ (white bars) and $rag^{-/-}$ (black bars) mutant zebrafish at 2 dpi with SVCV. *ef1a* mRNA was used as the endogenous control to normalize data, which are represented as the mean relative expression level ($\times 10^3$ for *crps*) $\pm$ SD of four different individuals. Significant differences were determined by two-way ANOVAs and Sidak's multiple comparison test. Statistical differences between the experimental groups are represented by keys together with a, b, and c letters on top. a, $p \leq 0.05$; b, $p \leq 0.01$; c, $p \leq 0.001$; ND, not detected.

4. Conclusions

The present study aimed at contributing to the current knowledge on fish short pentraxins by describing extensively the tissue distribution of the transcript expression of all seven zebrafish *crp* genes and their modulation in each tissue, especially skin, in response to infection with SVCV. Thus, all *crps* were found to be constitutively expressed in all tested tissues (i.e., HK, liver, skin, gills, gut, muscle, and spleen). In general terms, *crp4* and *crp5* presented the highest levels of expression, and gills, muscle, and skin were the major sites of expression. In skin, relatively high values were found for *crp4* and *crp5*. After SVCV infection, all *crps* were mostly upregulated, except for *crp1*, which was mainly downregulated. Predominantly, the highest increases occurred at 5 dpi and in the liver. Significant upregulations were also found in skin for *crp5–7*. Additional experiments were completed to further characterize the reactivity of *crps* to SVCV in skin; they included the use of *rag1* mutants to additionally explore the response of *crps* levels at a both the mucosal and SVCV entry sites in the absence of adaptive immunity. In comparison to $rag1^{+/+}$ control zebrafish, $rag1^{-/-}$ mutants showed elevated basal levels of most *crps* but unaltered *crp6* and *7* levels and the downregulated *crp2*. Furthermore, SVCV infection increased just moderately the expression of most *crp* genes, except for *crp2* and *crp6*, which remained unaltered. The analysis of the transcript expression of all *mx* isoforms on these samples indicates a generalized elevation in skin of the basal status of the innate immune system in fish without adaptive immunity. Altogether, the results obtained confirm the reactivity of *crps* to viral infections and suggest the skin, and by extension the skin mucus, as a promising sampling source for biomarker testing purposes.

Supplementary Materials: The following are available online at https://www.mdpi.com/2079-7737/10/2/78/s1, Table S1: qPCR primer sequences for zebrafish genes and SVCV *n*, Figure S1: Expression modulation of *mxa-g* in skin from rag mutant zebrafish in response to SVCV infection.

Author Contributions: Conceptualization, A.F. (Alberto Falco); methodology, M.B.-P., M.A., and A.F. (Alberto Falco); software, M.B.-P., M.A., and A.F. (Alberto Falco); validation, M.B.-P. and A.F. (Alberto Falco); formal analysis, A.F. (Antonio Figueras) and A.F. (Alberto Falco); investigation, M.B.-P., J.C., B.N., A.F. (Antonio Figueras), and A.F. (Alberto Falco); resources, J.C., B.N., A.F. (Antonio Figueras), and A.F. (Alberto Falco); data curation, A.F. (Antonio Figueras) and A.F. (Alberto Falco); writing—original draft preparation, A.F. (Alberto Falco); writing—review and editing, M.B.-P., M.A., J.C., B.N. A.F. (Antonio Figueras), and A.F. (Alberto Falco); visualization, M.B.-P. and A.F. (Antonio Figueras); supervision, A.F. (Antonio Figueras); project administration, B.N., A.F. (Antonio Figueras), and A.F. (Alberto Falco); funding acquisition, J.C. B.N., A.F. (Antonio Figueras), and A.F. (Alberto Falco). All authors have read and agreed to the published version of the manuscript.

Funding: This research was funded by FEDER/Spanish Ministry of Science, Innovation and Universities —State Agency of Research, grant number RTI2018-101969-J-I00 and BIO2017-82851-C3-1R; Xunta de Galicia (GAIN), grant number IN607B 2019/01, and Generalitat Valenciana and Fondo Social Europeo (FSE) 2014–2020, grant number ACIF/2016/207.

Institutional Review Board Statement: All experiments with zebrafish complied with the Spanish Law for Animal Experi-mentation (Royal Decree-Law, 53/2013) and the European Union Council Directive 2010/63/UE. Animal trial procedures were approved by the local government ethics com-mittee on animal experimentation (Dirección General de Agricultura, Ganadería y Pesca, Generalitat Valenciana), the Project Evaluation Board of Miguel Hernández University (permit no. UMH.IBM.JFG.01.14), and the CSIC National Committee on Bioethics (permit no. ES360570202001/16/FUN01/PAT.05/tipoE/BNG).

Informed Consent Statement: Not applicable.

Data Availability Statement: The data presented in this study are available on request from the corresponding author.

Acknowledgments: We thank Victoriano Mulero (University of Murcia) for kindly providing the zebrafish lines.

Conflicts of Interest: The authors declare no conflict of interest.

References

1. Pepys, M.B.; Hirschfield, G.M. C-reactive protein: A critical update. *J. Clin. Investig.* **2003**, *111*, 1805–1812. [CrossRef] [PubMed]
2. Garlanda, C.; Bottazzi, B.; Bastone, A.; Mantovani, A. Pentraxins at the crossroads between innate immunity, inflammation, matrix deposition, and female fertility. *Annu. Rev. Immunol.* **2005**, *23*, 337–366. [CrossRef] [PubMed]
3. Armstrong, P.B. Comparative biology of the pentraxin protein family: Evolutionarily conserved component of innate immune system. *Int. Rev. Cell Mol. Biol.* **2015**, *316*, 1–47. [PubMed]
4. Pathak, A.; Agrawal, A. Evolution of c-reactive protein. *Front. Immunol.* **2019**, *10*. [CrossRef]
5. Shrive, A.K.; Cheetham, G.M.; Holden, D.; Myles, D.A.; Turnell, W.G.; Volanakis, J.E.; Pepys, M.B.; Bloomer, A.C.; Greenhough, T.J. Three dimensional structure of human c-reactive protein. *Nat. Struct. Biol.* **1996**, *3*, 346–354. [CrossRef]
6. Nguyen, N.Y.; Suzuki, A.; Boykins, R.A.; Liu, T.Y. The amino acid sequence of limulus c-reactive protein. Evidence of polymorphism. *J. Biol. Chem.* **1986**, *261*, 10456–10465. [CrossRef]
7. Tharia, H.A.; Shrive, A.K.; Mills, J.D.; Arme, C.; Williams, G.T.; Greenhough, T.J. Complete cdna sequence of sap-like pentraxin from limulus polyphemus: Implications for pentraxin evolution. *J. Mol. Biol.* **2002**, *316*, 583–597. [CrossRef]
8. Introna, M.; Breviario, F.; d'Aniello, E.M.; Golay, J.; Dejana, E.; Mantovani, A. Il-1 inducible genes in human umbilical vein endothelial cells. *Eur. Heart J.* **1993**, *14* (Suppl. K), 78–81.
9. Pepys, M.B.; Baltz, M.L. Acute phase proteins with special reference to c-reactive protein and related proteins (pentaxins) and serum amyloid a protein. *Adv. Immunol.* **1983**, *34*, 141–212.
10. Amatayakul-Chantler, S.; Dwek, R.A.; Tennent, G.A.; Pepys, M.B.; Rademacher, T.W. Molecular characterization of limulus polyphemus c-reactive protein. Ii. Asparagine-linked oligosaccharides. *Eur. J. Biochem.* **1993**, *214*, 99–110. [CrossRef]
11. Pepys, M.B.; Baltz, M.; Gomer, K.; Davies, A.J.; Doenhoff, M. Serum amyloid p-component is an acute-phase reactant in the mouse. *Nature* **1979**, *278*, 259–261. [CrossRef] [PubMed]
12. Schwalbe, R.A.; Dahlback, B.; Coe, J.E.; Nelsestuen, G.L. Pentraxin family of proteins interact specifically with phosphorylcholine and/or phosphorylethanolamine. *Biochemistry* **1992**, *31*, 4907–4915. [CrossRef] [PubMed]
13. Coe, J.E.; Ross, M.J. Hamster female protein, a sex-limited pentraxin, is a constituent of syrian hamster amyloid. *J. Clin. Investig.* **1985**, *76*, 66–74. [CrossRef] [PubMed]
14. Srinivasan, N.; White, H.E.; Emsley, J.; Wood, S.P.; Pepys, M.B.; Blundell, T.L. Comparative analyses of pentraxins: Implications for protomer assembly and ligand binding. *Structure* **1994**, *2*, 1017–1027. [CrossRef]
15. Bello-Perez, M.; Falco, A.; Medina, R.; Encinar, J.A.; Novoa, B.; Perez, L.; Estepa, A.; Coll, J. Structure and functionalities of the human c-reactive protein compared to the zebrafish multigene family of c-reactive-like proteins. *Dev. Comp. Immunol.* **2017**, *69*, 33–40. [CrossRef]
16. Shrive, A.K.; Metcalfe, A.M.; Cartwright, J.R.; Greenhough, T.J. C-reactive protein and sap-like pentraxin are both present in limulus polyphemus haemolymph: Crystal structure of limulus sap. *J. Mol. Biol.* **1999**, *290*, 997–1008. [CrossRef]
17. Chen, R.; Qi, J.; Yuan, H.; Wu, Y.; Hu, W.; Xia, C. Crystal structures for short-chain pentraxin from zebrafish demonstrate a cyclic trimer with new recognition and effector faces. *J. Struct. Biol.* **2015**, *189*, 259–268. [CrossRef]
18. Cray, C. Acute phase proteins in animals. *Prog. Mol. Biol. Transl. Sci.* **2012**, *105*, 113–150.
19. Ansar, W.; Ghosh, S. C-reactive protein and the biology of disease. *Immunol. Res.* **2013**, *56*, 131–142. [CrossRef]
20. Rubio, N.; Sharp, P.M.; Rits, M.; Zahedi, K.; Whitehead, A.S. Structure, expression, and evolution of guinea pig serum amyloid p component and c-reactive protein. *J. Biochem.* **1993**, *113*, 277–284. [CrossRef]
21. MacCarthy, E.M.; Burns, I.; Irnazarow, I.; Polwart, A.; Greenhough, T.J.; Shrive, A.K.; Hoole, D. Serum crp-like protein profile in common carp cyprinus carpio challenged with aeromonas hydrophila and escherichia coli lipopolysaccharide. *Dev. Comp. Immunol.* **2008**, *32*, 1281–1289. [CrossRef] [PubMed]
22. Pionnier, N.; Falco, A.; Miest, J.J.; Shrive, A.K.; Hoole, D. Feeding common carp cyprinus carpio with beta-glucan supplemented diet stimulates c-reactive protein and complement immune acute phase responses following pamps injection. *Fish Shellfish Immunol.* **2014**, *39*, 285–295. [CrossRef] [PubMed]
23. Bello-Perez, M.; Pereiro, P.; Coll, J.; Novoa, B.; Perez, L.; Falco, A. Zebrafish c-reactive protein isoforms inhibit svcv replication by blocking autophagy through interactions with cell membrane cholesterol. *Sci. Rep.* **2020**, *10*, 1–18. [CrossRef]
24. Bello-Perez, M.; Falco, A.; Novoa, B.; Perez, L.; Coll, J. Hydroxycholesterol binds and enhances the anti-viral activities of zebrafish monomeric c-reactive protein isoforms. *PLoS ONE* **2019**, *14*, e0201509. [CrossRef]
25. Bello-Perez, M.; Falco, A.; Medina-Gali, R.; Pereiro, P.; Encinar, J.A.; Novoa, B.; Perez, L.; Coll, J. Neutralization of viral infectivity by zebrafish c-reactive protein isoforms. *Mol. Immunol.* **2017**, *91*, 145–155. [CrossRef] [PubMed]
26. Westerfield, M. *The Zebrafish Book: A Guide for the Laboratory Use of Zebrafish (Brachydanio rerio)*; University of Oregon Press: Eugene, OR, USA, 1995.
27. MacRae, C.A.; Peterson, R.T. Zebrafish as tools for drug discovery. *Nat. Rev. Drug Discov.* **2015**, *14*, 721–731. [CrossRef] [PubMed]
28. Valenzuela-Muñoz, V.; Pereiro, P.; Álvarez-Rodríguez, M.; Gallardo-Escárate, C.; Figueras, A.; Novoa, B. Comparative modulation of lncrnas in wild-type and rag1-heterozygous mutant zebrafish exposed to immune challenge with spring viraemia of carp virus (svcv). *Sci. Rep.* **2019**, *9*, 1–13. [CrossRef]
29. Petrie-Hanson, L.; Hohn, C.; Hanson, L. Characterization of rag 1 mutant zebrafish leukocytes. *BMC Immunol.* **2009**, *10*, 1–8. [CrossRef]

30. Hohn, C.; Petrie-Hanson, L. Rag1$^{-/-}$ mutant zebrafish demonstrate specific protection following bacterial re-exposure. *PLoS ONE* **2012**, *7*, e44451. [CrossRef]
31. Espin-Palazon, R.; Martinez-Lopez, A.; Roca, F.J.; Lopez-Munoz, A.; Tyrkalska, S.D.; Candel, S.; Garcia-Moreno, D.; Falco, A.; Meseguer, J.; Estepa, A. Tnfα impairs rhabdoviral clearance by inhibiting the host autophagic antiviral response. *PLoS Pathog.* **2016**, *12*, e1005699. [CrossRef]
32. Livak, K.J.; Schmittgen, T.D. Analysis of relative gene expression data using real-time quantitative pcr and the 2(-delta delta c(t)) method. *Methods* **2001**, *25*, 402–408. [CrossRef]
33. Lee, P.; Bird, S.; Zou, J.; Martin, S. Phylogeny and expression analysis of c-reactive protein (crp) and serum amyloid-p (sap) like genes reveal two distinct groups in fish. *Fish Shellfish Immunol.* **2017**, *65*, 42–51. [CrossRef] [PubMed]
34. Kovacevic, N.; Hagen, M.O.; Xie, J.; Belosevic, M. The analysis of the acute phase response during the course of trypanosoma carassii infection in the goldfish (*Carassius auratus* L.). *Dev. Comp. Immunol.* **2015**, *53*, 112–122. [CrossRef]
35. Falco, A.; Cartwright, J.R.; Wiegertjes, G.F.; Hoole, D. Molecular characterization and expression analysis of two new c-reactive protein genes from common carp (*Cyprinus carpio*). *Dev. Comp. Immunol.* **2012**, *37*, 127–138. [CrossRef] [PubMed]
36. Choi, K.-M.; Shim, S.H.; An, C.M.; Nam, B.-H.; Jeong, J.-M.; Kim, J.-W.; Park, C.-I. Functional characterisation and expression analysis of recombinant serum amyloid p isoform 1 (rbsap1) from rock bream (*Oplegnathus fasciatus*). *Fish Shellfish Immunol.* **2015**, *45*, 277–285. [CrossRef] [PubMed]
37. Shi, Y.-H.; Chen, K.; Ma, W.-J.; Chen, J. Ayu c-reactive protein/serum amyloid p agglutinates bacteria and inhibits complement-mediated opsonophagocytosis by monocytes/macrophages. *Fish Shellfish Immunol.* **2018**, *76*, 58–67. [CrossRef]
38. Elvitigala, D.A.S.; Wan, Q.; Kim, H.C.; Lee, J. Identification of a c-reactive protein like homologue from black rockfish (*Sebastes schlegelii*) evidencing its potent anti-microbial properties at molecular level. *Dev. Comp. Immunol.* **2015**, *53*, 169–178. [CrossRef]
39. Li, M.-f.; Chen, C.; Li, J.; Sun, L. The c-reactive protein of tongue sole cynoglossus semilaevis is an acute phase protein that interacts with bacterial pathogens and stimulates the antibacterial activity of peripheral blood leukocytes. *Fish Shellfish Immunol.* **2013**, *34*, 623–631. [CrossRef]
40. Wang, T.; Sun, L. Cssap, a teleost serum amyloid p component, interacts with bacteria, promotes phagocytosis, and enhances host resistance against bacterial and viral infection. *Dev. Comp. Immunol.* **2016**, *55*, 12–20. [CrossRef]
41. Pietrzak, E.; Mazurkiewicz, J.; Slawinska, A. Innate immune responses of skin mucosa in common carp (*Cyprinus carpio*) fed a diet supplemented with galactooligosaccharides. *Animals* **2020**, *10*, 438. [CrossRef]
42. Magnadóttir, B.; Hayes, P.; Gísladóttir, B.; Bragason, B.Þ.; Hristova, M.; Nicholas, A.P.; Guðmundsdóttir, S.; Lange, S. Pentraxins crp-i and crp-ii are post-translationally deiminated and differ in tissue specificity in cod (*Gadus morhua* L.) ontogeny. *Dev. Comp. Immunol.* **2018**, *87*, 1–11. [CrossRef] [PubMed]
43. Tsutsui, S.; Yamaguchi, M.; Hirasawa, A.; Nakamura, O.; Watanabe, T. Common skate (raja kenojei) secretes pentraxin into the cutaneous secretion: The first skin mucus lectin in cartilaginous fish. *J. Biochem.* **2009**, *146*, 295–306. [CrossRef] [PubMed]
44. Patel, D.M.; Brinchmann, M.F. Skin mucus proteins of lumpsucker (*Cyclopterus lumpus*). *Biochem. Biophys. Rep.* **2017**, *9*, 217–225. [CrossRef] [PubMed]
45. Ramos, F.; Smith, A.C. The c-reactive protein (crp) test for the detection of early disease in fishes. *Aquaculture* **1978**, *14*, 261–266. [CrossRef]
46. Shao, L.; Xiao, Y.; He, Z.; Gao, L. An n-targeting real-time pcr strategy for the accurate detection of spring viremia of carp virus. *J. Virol. Methods* **2016**, *229*, 27–34. [CrossRef] [PubMed]
47. Wang, Y.; Kuang, M.; Lu, Y.; Lin, L.; Liu, X. Characterization and biological function analysis of the trim47 gene from common carp (*Cyprinus carpio*). *Gene* **2017**, *627*, 188–193. [CrossRef] [PubMed]
48. Uribe, C.; Folch, H.; Enríquez, R.; Moran, G. Innate and adaptive immunity in teleost fish: A review. *Vet. Med.* **2011**, *56*, 486–503. [CrossRef]
49. Encinas, P.; Garcia-Valtanen, P.; Chinchilla, B.; Gomez-Casado, E.; Estepa, A.; Coll, J. Identification of multipath genes differentially expressed in pathway-targeted microarrays in zebrafish infected and surviving spring viremia carp virus (svcv) suggest preventive drug candidates. *PLoS ONE* **2013**, *8*, e73553. [CrossRef]
50. Pionnier, N.; Adamek, M.; Miest, J.J.; Harris, S.J.; Matras, M.; Rakus, K.Ł.; Irnazarow, I.; Hoole, D. C-reactive protein and complement as acute phase reactants in common carp cyprinus carpio during cyhv-3 infection. *Dis. Aquat. Org.* **2014**, *109*, 187–199. [CrossRef]
51. Gruys, E.; Toussaint, M.; Niewold, T.; Koopmans, S. Acute phase reaction and acute phase proteins. *J. Zhejiang Univ. Sci. B* **2005**, *6*, 1045. [CrossRef]
52. Mantovani, A.; Garlanda, C.; Doni, A.; Bottazzi, B. Pentraxins in innate immunity: From c-reactive protein to the long pentraxin ptx3. *J. Clin. Immunol.* **2008**, *28*, 1–13. [CrossRef] [PubMed]
53. Kushner, I.; Jiang, S.-L.; Zhang, D.; Lozanski, G.; Samols, D. Do post-transcriptional mechanisms participate in induction of c-reactive protein and serum amyloid a by il-6 and il-1? *Ann. N. Y. Acad. Sci.* **1995**, *762*, 102–107. [CrossRef] [PubMed]
54. Paludan, S.R. Requirements for the induction of interleukin-6 by herpes simplex virus-infected leukocytes. *J. Virol.* **2001**, *75*, 8008–8015. [CrossRef] [PubMed]
55. Wang, J.; Wang, Q.; Han, T.; Li, Y.-K.; Zhu, S.-L.; Ao, F.; Feng, J.; Jing, M.-Z.; Wang, L.; Ye, L.-B. Soluble interleukin-6 receptor is elevated during influenza a virus infection and mediates the il-6 and il-32 inflammatory cytokine burst. *Cell. Mol. Immunol.* **2015**, *12*, 633–644. [CrossRef] [PubMed]

56. Xia, C.; Liu, Y.; Chen, Z.; Zheng, M. Involvement of interleukin 6 in hepatitis b viral infection. *Cell. Physiol. Biochem.* **2015**, *37*, 677–686. [CrossRef] [PubMed]
57. Øvergård, A.-C.; Nerland, A.H.; Fiksdal, I.U.; Patel, S. Atlantic halibut experimentally infected with nodavirus shows increased levels of t-cell marker and ifnγ transcripts. *Dev. Comp. Immunol.* **2012**, *37*, 139–150. [CrossRef] [PubMed]
58. Wei, X.; Li, X.Z.; Zheng, X.; Jia, P.; Wang, J.; Yang, X.; Yu, L.; Shi, X.; Tong, G.; Liu, H. Toll-like receptors and interferon associated immune factors responses to spring viraemia of carp virus infection in common carp (*Cyprinus carpio*). *Fish Shellfish Immunol.* **2016**, *55*, 568–576. [CrossRef]
59. Black, S.; Kushner, I.; Samols, D. C-reactive protein. *J. Biol. Chem.* **2004**, *279*, 48487–48490. [CrossRef]
60. Lazado, C.C.; Caipang, C.M.A. Mucosal immunity and probiotics in fish. *Fish Shellfish Immunol.* **2014**, *39*, 78–89. [CrossRef]
61. Harmache, A.; LeBerre, M.; Droineau, S.; Giovannini, M.; Brémont, M. Bioluminescence imaging of live infected salmonids reveals that the fin bases are the major portal of entry for novirhabdovirus. *J. Virol.* **2006**, *80*, 3655–3659. [CrossRef]
62. Bearzotti, M.; Delmas, B.; Lamoureux, A.; Loustau, A.-M.; Chilmonczyk, S.; Bremont, M. Fish rhabdovirus cell entry is mediated by fibronectin. *J. Virol.* **1999**, *73*, 7703–7709. [CrossRef] [PubMed]
63. Esteban, M.Á.; Cerezuela, R. Fish mucosal immunity: Skin. In *Mucosal Health in Aquaculture*; Elsevier: Amsterdam, The Netherlands, 2015; pp. 67–92.
64. Salinas, I. The mucosal immune system of teleost fish. *Biology* **2015**, *4*, 525–539. [CrossRef] [PubMed]
65. Xu, Z.; Takizawa, F.; Casadei, E.; Shibasaki, Y.; Ding, Y.; Sauters, T.J.; Yu, Y.; Salinas, I.; Sunyer, J.O. Specialization of mucosal immunoglobulins in pathogen control and microbiota homeostasis occurred early in vertebrate evolution. *Sci. Immunol.* **2020**, *5*. [CrossRef] [PubMed]
66. García-Valtanen, P.; Martínez-López, A.; López-Muñoz, A.; Bello-Perez, M.; Medina-Gali, R.M.; Ortega-Villaizán, M.d.M.; Varela, M.; Figueras, A.; Mulero, V.; Novoa, B. Zebra fish lacking adaptive immunity acquire an antiviral alert state characterized by upregulated gene expression of apoptosis, multigene families, and interferon-related genes. *Front. Immunol.* **2017**, *8*, 121. [PubMed]
67. Agrawal, A.; Singh, P.P.; Bottazzi, B.; Garlanda, C.; Mantovani, A. Pattern recognition by pentraxins. In *Target Pattern Recognition in Innate Immunity*; Springer: Berlin/Heidelberg, Germany, 2009; pp. 98–116.
68. Chorny, A.; Casas-Recasens, S.; Sintes, J.; Shan, M.; Polentarutti, N.; García-Escudero, R.; Walland, A.C.; Yeiser, J.R.; Cassis, L.; Carrillo, J. The soluble pattern recognition receptor ptx3 links humoral innate and adaptive immune responses by helping marginal zone b cells. *J. Exp. Med.* **2016**, *213*, 2167–2185. [CrossRef]
69. Bottazzi, B.; Doni, A.; Garlanda, C.; Mantovani, A. An integrated view of humoral innate immunity: Pentraxins as a paradigm. *Annu. Rev. Immunol.* **2009**, *28*, 157–183. [CrossRef]
70. Novoa, B.; Pereiro, P.; López-Muñoz, A.; Varela, M.; Forn-Cuní, G.; Anchelin, M.; Dios, S.; Romero, A.; Martinez-López, A.; Medina-Gali, R.M. Rag1 immunodeficiency-induced early aging and senescence in zebrafish are dependent on chronic inflammation and oxidative stress. *Aging Cell* **2019**, *18*, e13020. [CrossRef]
71. Netea, M.G.; Joosten, L.A. Master and commander: Epigenetic regulation of macrophages. *Cell Res.* **2016**, *26*, 145–146. [CrossRef]
72. Quintin, J.; Saeed, S.; Martens, J.H.; Giamarellos-Bourboulis, E.J.; Ifrim, D.C.; Logie, C.; Jacobs, L.; Jansen, T.; Kullberg, B.-J.; Wijmenga, C. Candida albicans infection affords protection against reinfection via functional reprogramming of monocytes. *Cell Host Microbe* **2012**, *12*, 223–232. [CrossRef]
73. Medina-Gali, R.; Bello-Perez, M.; Martinez-Lopez, A.; Falco, A.; Ortega-Villaizan, M.M.; Encinar, J.A.; Novoa, B.; Coll, J.; Perez, L. Chromatin immunoprecipitation and high throughput sequencing of svcv-infected zebrafish reveals novel epigenetic histone methylation patterns involved in antiviral immune response. *Fish. Shellfish Immunol.* **2018**, *82*, 514–521. [CrossRef]

biology

MDPI

Article

Potential Involvement of lncRNAs in the Modulation of the Transcriptome Response to Nodavirus Challenge in European Sea Bass (*Dicentrarchus labrax* L.)

Patricia Pereiro [1,2], Raquel Lama [1], Rebeca Moreira [1], Valentina Valenzuela-Muñoz [2], Cristian Gallardo-Escárate [2], Beatriz Novoa [1] and Antonio Figueras [1,*]

1 Instituto de Investigaciones Marinas (IIM), Consejo Superior de Investigaciones Científicas (CSIC), C/Eduardo Cabello 6, 36208 Vigo, Spain; patriciapereiro@iim.csic.es (P.P.); raquell@iim.csic.es (R.L.); rebecamoreira@iim.csic.es (R.M.); beatriznovoa@iim.csic.es (B.N.)

2 Laboratory of Biotechnology and Aquatic Genomics, Interdisciplinary Center for Aquaculture Research (INCAR), University of Concepción, Concepción 4030000, Chile; valevalenzuela@gmail.com (V.V.-M.); crisgallardo@oceanografia.udec.cl (C.G.-E.)

* Correspondence: antoniofigueras@iim.csic.es

Received: 8 June 2020; Accepted: 13 July 2020; Published: 15 July 2020

Abstract: Long noncoding RNAs (lncRNAs) are being increasingly recognised as key modulators of various biological mechanisms, including the immune response. Although investigations in teleosts are still lagging behind those conducted in mammals, current research indicates that lncRNAs play a pivotal role in the response of fish to a variety of pathogens. During the last several years, interest in lncRNAs has increased considerably, and a small but notable number of publications have reported the modulation of the lncRNA profile in some fish species after pathogen challenge. This study was the first to identify lncRNAs in the commercial species European sea bass. A total of 12,158 potential lncRNAs were detected in the head kidney and brain. We found that some lncRNAs were not common for both tissues, and these lncRNAs were located near coding genes that are primarily involved in tissue-specific processes, reflecting a degree of cellular specialisation in the synthesis of lncRNAs. Moreover, lncRNA modulation was analysed in both tissues at 24 and 72 h after infection with nodavirus. Enrichment analysis of the neighbouring coding genes of the modulated lncRNAs revealed many terms related to the immune response and viral infectivity but also related to the stress response. An integrated analysis of the lncRNAs and coding genes showed a strong correlation between the expression of the lncRNAs and their flanking coding genes. Our study represents the first systematic identification of lncRNAs in European sea bass and provides evidence regarding the involvement of these lncRNAs in the response to nodavirus.

Keywords: RNA-Seq; lncRNAs; *Dicentrarchus labrax*; viral infection; nodavirus; immune response

1. Introduction

European sea bass (*Dicentrarchus labrax* L.) is a very valuable fish species, especially for Mediterranean countries. This species was the first non-salmonid species to be cultured in Europe, and since the 1990s, sea bass aquaculture has grown exponentially and is currently one of the main cultured fish species in Europe [1]. However, different diseases cause important economic losses and represent a major limiting factor for production. These effects are observed for nervous necrosis virus (NNV), or nodavirus, a member of the family Nodaviridae, genus *Betanodavirus*, which can affect numerous aquatic animals, including a wide variety of marine and freshwater fish species [2]. This icosahedral, nonenveloped, positive-sense single-stranded RNA virus is the causative agent of viral encephalopathy and retinopathy (VER), which is characterised by damage to its target tissues, the nervous system (e.g., brain, retina and spinal cord) [3]. Among the four nodavirus genotypes, European sea bass seems to be primarily affected by red-spotted grouper nervous necrosis virus (RGNNV), especially during larval and juvenile stages [2,3].

In recent years, some publications have reported the involvement of numerous immune factors in the defence against nodavirus in different tissues from *D. labrax* [4–12]. The development of next-generation sequencing (NGS) technologies also enabled us to analyse the complete transcriptome response after NNV infection both in vitro [13,14] and, more recently, in vivo [15]. Nevertheless, the potential role of noncoding RNAs (ncRNAs) in the modulation of the response to NNV infection in *D. labrax* has not been determined.

The non-coding regions of the genome remained largely unexplored until the last decade. Nevertheless, when it was discovered that the genome is transcribed into many non-coding RNAs (ncRNAs), in addition to the well-known transfer RNAs (tRNAs) or ribosomal RNAs (rRNAs), numerous investigations were conducted in a variety of species in an attempt to identify them and to describe their functions and expression profiles. The long noncoding RNAs (lncRNAs) represent a subset of ncRNAs with a length over 200 nucleotides and transcribed in the same way as mRNA [16]. However, lncRNAs are among least well-understood ncRNAs. This is probably due to the variety of regulatory mechanisms that they could affect [16], but also to their rapid evolution, and consequently, to the absence of recognisable homologs for most of the lncRNAs even in evolutionarily close species [17,18] and to the divergent subcellular location and function of certain conserved lncRNAs [19]. LncRNAs regulate the expression of adjacent genes (*cis*-acting regulation) or genes located at other genomic locations, even on different chromosomes (*trans*-acting) [16]. The gene expression promotion or repression mediated by lncRNAs can be conducted through different mechanisms, including chromatin remodelling, promoter inactivation, transcription interference, activation and/or transport of transcription factors and epigenetic regulation [16]. This wide variety of potential regulatory mechanisms makes it difficult to establish concrete functions for specific lncRNAs. Nevertheless, lncRNAs have been demonstrated to be involved in immune system regulation in mammals [20,21].

In fish, although functional studies remain highly limited, several publications reported the modulation of lncRNAs after different stimuli and/or conditions with a special focus on the immune system [19]. Because lncRNAs can alter the expression of their adjacent genes, the functionality of those lncRNAs identified in fish is usually predicted based on the function of their neighbouring protein-coding genes [22].

In this work, we identified the lncRNA repertoire in the head kidney and brain from European sea bass. We selected these tissues because the head kidney is the main immune and hematopoietic organ in fish and the brain is a target tissue for NNV. Our results showed the existence of tissue-specific lncRNAs, with a specialised role in the maintenance of the basic functions of each tissue, as well as a broad modulation of the lncRNAs after NNV challenge. The functionality of these infection-induced lncRNAs was predicted based on the analysis of their nearby coding genes, revealing a variety of processes in which they could be involved. Because samples were taken at 24 and 72 h post-infection (hpi), we also observed a highly variable lncRNA profile depending on the sampling point, which could be indicative of the fine-tuned gene regulation mediated by these lncRNAs.

2. Materials and Methods

2.1. Fish and Virus

Juvenile *D. labrax* (average weight 70 g) were obtained from the Estación de Ciencias Mariñas de Toralla (ECIMAT, Universidad de Vigo, Vigo, Galicia, Spain) facilities. Prior to experiments, fish were acclimatised to the laboratory conditions for 2 weeks and maintained in 500 litres fibreglass tanks with a re-circulating saline-water system (total salinity approximately 35 g/L) with a light-dark cycle of 12:12 h at 20–22 °C and fed daily with a commercial diet. Animals were euthanised via MS-222 (Sigma-Aldrich, St. Louis, MO, USA) overdose (500 mg/L). All the experimental procedures were reviewed and approved by the CSIC National Committee on Bioethics under approval number ES360570202001/20/FUN.01/INM06/BNG01.

The nodavirus red-spotted grouper nervous necrosis virus (RGNNV) (strain 475-9/99) was propagated at 25 °C in the SSN-1 cell line (ECACC 96082808) cultured in Leibovitz's L-15 medium (Gibco, Carlsbad, CA, USA) supplemented with 2 mM glutamine (Gibco), 2% foetal bovine serum (FBS) (Gibco) and 1% penicillin/streptomycin solution (Invitrogen, Waltham, MA, USA). The virus stock was titrated into 96-well plates according to established protocols [23], and aliquots were stored at −80 °C until use.

2.2. Fish Infection and Sampling

A total of 18 sea bass were intramuscularly injected with 100 µL of a nodavirus suspension (10^6 $TCID_{50}/_{mL}$), whereas the same number of fish served as the control and were inoculated with the same volume of medium (L-15 + glutamine + 2% FBS + penicillin/streptomycin). Head kidney and brain samples were taken from nine fish at 24 and 72 h post-challenge. The same quantity of tissue from three animals was pooled, obtaining three biological replicates (three fish/replicate) per tissue at each sampling point.

For this challenge experiment, mortality data and viral replication analysis in head kidney and brain from infected fish were previously provided [15].

2.3. RNA Isolation and High-Throughput Transcriptome Sequencing

Total RNA from the different samples was extracted using the Maxwell 16 LEV simplyRNA Tissue kit (Promega, Madison, WI, USA) with the automated Maxwell 16 Instrument in accordance with instructions provided by the manufacturer. The quantity and purity of the total RNA was measured in a NanoDrop ND-1000 (NanoDrop Technologies, Inc., Wilmington, DE, USA), and RNA integrity was analysed in the Agilent 2100 Bioanalyzer (Agilent Technologies Inc., Santa Clara, CA, USA) according to the manufacturer's instructions. All the samples showed an RIN over 8.0; therefore, they were used for Illumina library preparation. Double-stranded cDNA libraries were constructed using the TruSeq RNA Sample Preparation Kit v2 (Illumina, San Diego, CA, USA), and sequencing was performed using Illumina HiSeq™ 4000 technology at Macrogen Inc. Korea (Seoul, Republic of Korea). The read sequences were deposited in the Sequence Read Archive (SRA) (http://www.ncbi.nlm.nih.gov/sra) under the accession number PRJNA589774.

2.4. Sequence Assembly and LncRNA Mining

CLC Genomics Workbench, v. 11.0.2 (CLC Bio, Aarhus, Denmark) was used to filter, assemble and perform the RNA-Seq and statistical analyses. Raw reads were trimmed to remove adaptor sequences and low-quality reads (quality score limit 0.05). A reference transcriptome was constructed by de novo assembly using all the samples with an overlap criterion of 70% and a similarity of 0.9 to exclude paralogous sequence variants. The settings used were a mismatch cost = 2, deletion cost = 3, insert cost = 3 and minimum contig length = 200 base pairs. From the de novo assembly, the selection of the lncRNAs was conducted following the pipeline previously described [24] but with some modifications. Briefly, the contigs were annotated using Blastx (e-value < 1×10^{-5}) against the proteins for all the bony fish species included in the NCBI GenBank and UniProt/Swiss-Prot databases. All the annotated contigs were deleted from the assembly, as well as all the contigs with an average coverage <50. The potential open reading frames (ORFs) were predicted for the remaining contigs, and those with a potential ORF >200 bp were discarded. After this step, the Coding Potential Assessment Tool (CPAT) [25] was used to discard sequences with coding potential. The contigs that passed all the filters were considered putative lncRNAs and retained for further analyses.

2.5. Differential Expression Analysis

RNA-Seq analyses of the putative lncRNAs were conducted with the following settings: mismatches = 2, length fraction = 0.8, similarity fraction = 0.8. The expression values were set as transcripts per million (TPM). To identify and quantify the directions of variability in the data, a principal component analysis (PCA) plot was constructed by using the original expression values. Finally, Baggerley's statistical analysis test was used to compare gene expression levels and to identify differentially expressed lncRNAs. Transcripts with absolute fold change (FC) values >2 and false discovery rate (FDR) <0.05 were retained for further analyses. Those significantly modulated lncRNAs were selected, and the average TMP value of the three biological replicates is represented in heat maps. To this end, the distance metric was calculated by Pearson's method, and lncRNAs were hierarchically clustered by average linkage. Venn diagrams were constructed with the Venny 2.1 tool (http://bioinfogp.cnb.csic.es/tools/venny/).

2.6. Genome Mapping and Identification of LncRNA Neighbouring Coding Genes

To position the putative lncRNAs on the genome, the *D. labrax* genome [26], composed of 25 scaffolds, was uploaded into the CLC Genomics Workbench. These scaffolds were constructed based on linkage/radiation hybrid groups (LG1A - LGx), in which approximately 83% of the contigs were located [26]. The remaining unanchored scaffolds were concatenated into a virtual chromosome ('UN') [26].

LncRNAs were mapped using the following parameters: length fraction = 0.8, similarity fraction = 0.8, mismatch cost = 2, insertion cost = 3 and deletion cost = 3. The correlation between the chromosome length and the number of mapped lncRNAs per chromosome was calculated with IBM SPSS Statistics V25.0 by using Pearson's correlation coefficient (r). The coding genes flanking up to 10 kb upstream and downstream from the mapped and differentially expressed lncRNAs for each comparison were identified and selected for Gene Ontology (GO).

2.7. GO Enrichment Analyses

For the significantly differentially expressed lncRNAs in each comparison (Brain 24 h infected vs. Brain 24 h control; Brain 72 h infected vs. Brain 72 h control; HK 24 h infected vs. HK 24 h control; HK 72 h infected vs. HK 72 h control), the lncRNAs-neighbouring coding-genes were extracted for GO enrichment analyses using Blast2GO software [27]. For the GO enrichments, Fisher's exact test was run with default settings, a p-value cut-off of 0.01 was applied, and the enriched list was reduced to the most specific GO terms. The most enriched biological processes among the protein coding genes proximal to lncRNAs were represented. When the number of significantly enriched GO terms exceeded 30, only the 30 most significant terms were represented.

2.8. Correlation Analyses Between LncRNAs and Coding Genes

Correlations in the expression of lncRNAs and their neighbouring genes were calculated by using Pearson's correlation coefficient and Spearman's rank correlation coefficient with IBM SPSS Statistics V25.0. Correlations were considered to be significant when p-values were <0.01. To illustrate the expression correlations, heat maps were illustrated with the free software Heatmapper [28] and the average TPM values of the different experimental conditions and sampling points were normalised by row. The protein–protein interaction networks were constructed with String 11.0 (https://string-db.org) [29].

2.9. Quantitative PCR (qPCR) Validation of LncRNA Expression

cDNA synthesis of the samples was conducted with the NZY First-Strand cDNA Synthesis kit (NZYTech, Lisbon, Portugal) using 0.5 µg of total RNA. A total of 12 lncRNAs were used to validate the RNA-Seq results. Specific qPCR primers for the selected lncRNAs were designed using Primer 3 software [30], and their amplification efficiency was calculated with the threshold cycle (CT) slope method [31]. Primer sequences are listed in Supplementary Table S1. Individual qPCR reactions were carried out in a 25-μL reaction volume containing 12.5 μL of SYBR GREEN PCR Master Mix (Applied Biosystems, Foster City, CA, USA), 10.5 μL of ultrapure water, 0.5 μL of each specific primer (10 μM) and 1 μL of two-fold diluted cDNA template in MicroAmp optical 96-well reaction plates (Applied Biosystems). Reactions were conducted using technical triplicates in a 7300 Real-Time PCR System thermocycler (Applied Biosystems). qPCR conditions consisted of an initial denaturation step (95 °C, 10 min) followed by 40 cycles of a denaturation step (95 °C, 15 s) and one hybridization-elongation step (60 °C, 1 min). The relative expression level of the lncRNAs was normalised following the Pfaffl method [31] and using *18S ribosomal RNA* (*18S*) as a reference gene.

3. Results

3.1. Assembly, Annotation and LncRNA Mining

A summary of the reads, assembly data, contig annotation and lncRNA information is included in Table 1. Approximately 680 million reads were obtained from the different samples of European sea bass with an average of 28 million per sample, and over 99% of raw reads met the quality control standards. From the de novo assembly, a total of 347,317 contigs were obtained with an average length of 723 bp. A total of 69% of the contigs (240,274) were successfully annotated against the bony fish sequences. From the non-annotated contigs (107,043), a total of 12,158 passed all the filters to be considered potential lncRNAs (Supplementary Table S2). The length of these selected contigs ranged from 200 to 6829 bp with an average length of 667 bp (Figure 1a). The GC content (%) of the predicted lncRNAs ranged from 11 to 80% with an average value of 38% (Figure 1b).

Table 1. Summary of the Illumina sequencing, assembly, annotation and lncRNA identification.

READS	
Total reads	679,746,504
Mean reads per sample	28,322,771
ASSEMBLY	
Contigs	347,317
Minimum length	200 bp
Maximum length	26,142 bp
Average length	723 bp
N50	1088 bp
Total assembled reads	215,126,479
ANNOTATION	
Annotated contigs	240,274 (69%)
Non-annotated contigs	107,043
LncRNAs	
Potential lncRNAs	12,158
Minimum length	200
Maximum length	6829
Average length	667
lncRNAs mapped on genome	11,667 (95.96%)

From the 12,158 putative lncRNAs, 95.96% (11,667) were successfully mapped to the *D. labrax* genome (Figure 1c). The scaffold with a higher number of lncRNAs was the virtual chromosome 'UN' (2004 lncRNAs). This finding is most likely due to the greater length of this virtual chromosome compared to the true ones. Indeed, a strong positive correlation (Pearson's r = 0.99) existed between the chromosome length and the number of predicted lncRNAs per chromosome (Figure 1d), reflecting that there is no enrichment in lncRNA abundance in any particular chromosome.

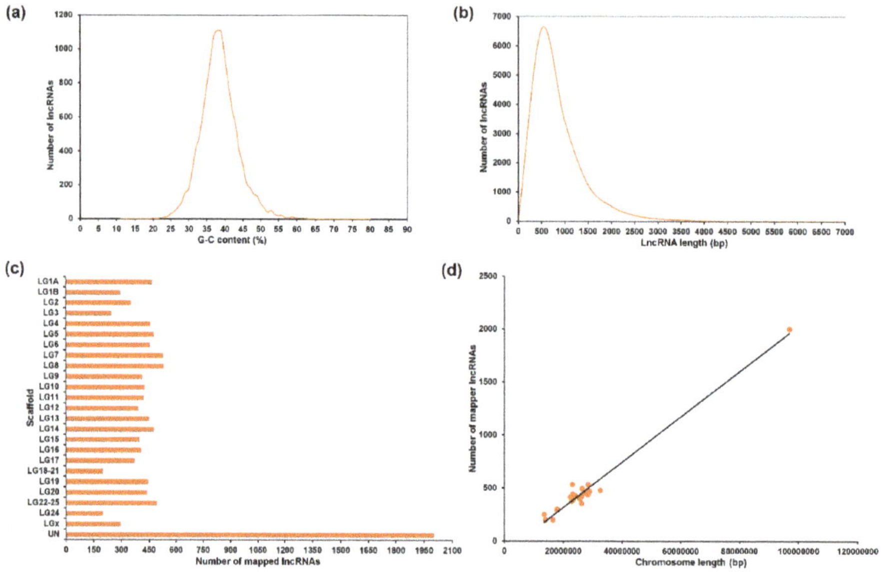

Figure 1. Features of predicted lncRNAs in *D. labrax*. (**a**) Guanine-cytosine (GC) content and (**b**) length distribution of the 12158 predicted lncRNAs. (**c**) LncRNA abundance and localisation per chromosome. (**d**) Correlation between chromosome length and lncRNA abundance.

3.2. Tissue Distribution of the LncRNAs

To identify lncRNAs expressed in head kidney but not in brain and vice versa, only those lncRNAs with a TPM value of 0 in all the samples from the same tissue (12 samples) were considered absent in the corresponding tissue. A Venn diagram of the lncRNAs detected in the head kidney and brain was constructed (Figure 2a). While 30 lncRNAs did not show a TPM value in any of the samples both in head kidney and brain, most of the lncRNAs were detected in both tissues (10823). A total of 971 lncRNAs were expressed in the brain but not in the head kidney, whereas 334 lncRNAs were only found in the head kidney.

As expected, GO enrichment analysis of the lncRNAs expressed only in one of the tissues revealed biological processes directly related to specific aspects of the functionality of each organ (Figure 2b,c). In the head kidney, numerous immune terms were enriched, which were mainly related to the production of cytokines, Toll-like receptor signalling, inflammation, leukocyte activation and proliferation, complement pathway and phagocytosis (Figure 2b). On the other hand, neighbouring genes of those lncRNAs expressed only in the brain showed enrichment in terms directly related to the nervous system (Figure 2c).

To illustrate the relevance of the tissue and of the infection in the lncRNA profile, a PCA score plot was constructed (Figure S1). Interestingly, all the head kidney samples clustered separately from the brain samples, reflecting the higher weight of the tissue compared to the nodavirus infection.

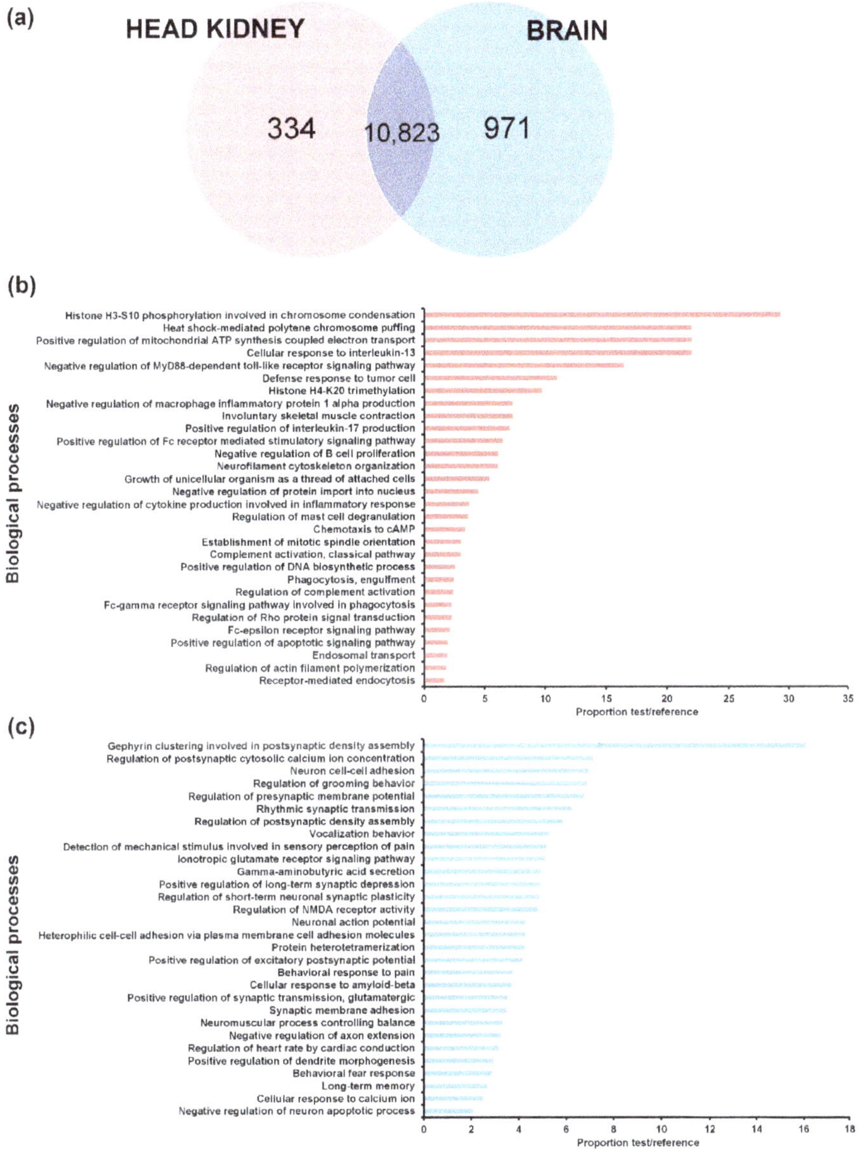

Figure 2. LncRNAs identified per tissue. (**a**) Venn diagram reflecting the number of lncRNAs with a transcripts per million (TPM) value in at least one of the samples from each tissue. Most of the predicted lncRNAs were detected in both the head kidney and brain. In contrast, a total of 30 predicted lncRNAs obtained a TPM value of 0 in all the samples. (**b**,**c**) The neighbouring coding genes of the lncRNAs expressed in the head kidney but not in the brain (**b**) and vice versa (**c**) were analysed by GO enrichment analyses (biological processes). Only the 30 most significant terms were represented.

3.3. *Differential Expression of LncRNAs after Nodavirus Challenge*

The differential expression analysis (FC > 2, FDR < 0.05) for each tissue and sampling point is provided in Table S3. A total of 204 and 93 lncRNAs were significantly modulated in the head kidney at 24 and 72 h after nodavirus infection, respectively (Figure 3a). In the brain, 931 and 342 lncRNAs were affected at these sampling points (Figure 3a). In both tissues, the number of upregulated lncRNAs was higher than that of downregulated lncRNAs at 24 h post-challenge, but these differences were substantially reduced after 72 h.

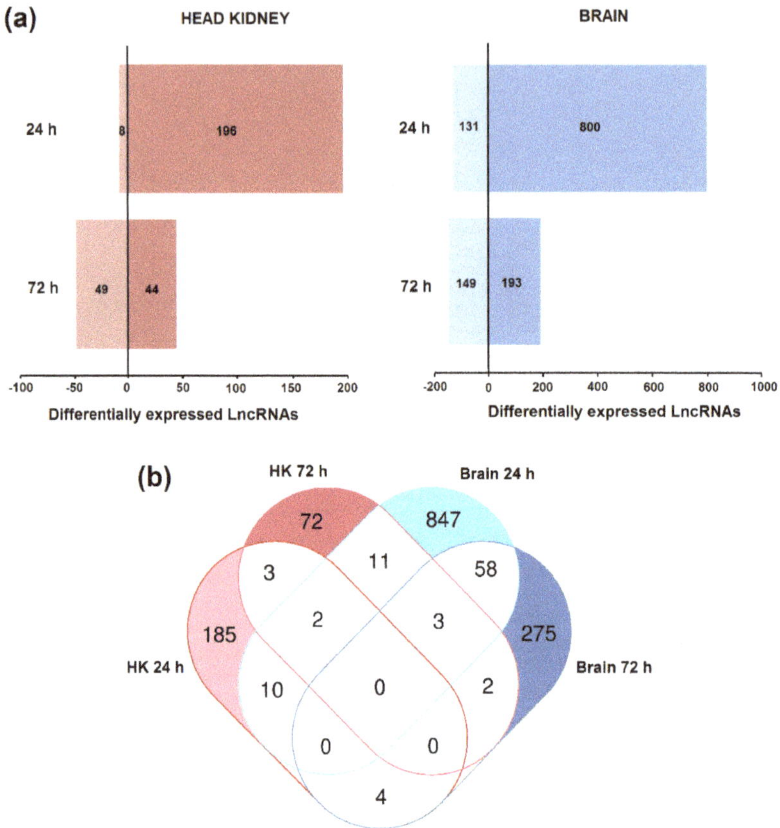

Figure 3. Temporal expression of the predicted lncRNAs after nodavirus infection in head kidney and brain. (**a**) Number of lncRNAs up- and downregulated in each tissue at 24 and 72 hpi with nervous necrosis virus (NNV). (**b**) Venn diagram representing the shared and unshared lncRNAs modulated after NNV challenge in both tissues at the different sampling points.

A Venn diagram was constructed to illustrate the shared and unshared lncRNAs modulated in both tissues at different days post-challenge (Figure 3b). Most of the differentially expressed lncRNAs were only modulated in one of the tissues and their expression pattern also completely changed in the same tissue depending on the day after infection (Figure 3b).

Indeed, by analysing the lncRNAs affected by the infection per tissue, we observed that in the head kidney, only 5 lncRNAs were commonly modulated at both 24 and 72 h post-challenge, representing 1.7% of the total lncRNAs affected by the infection in this tissue (Figure 4a). In the brain, the number of shared lncRNAs between both sampling points was 61, representing 5% of the total (Figure 4b). This almost complete switch in the lncRNA profile over time is also reflected in the corresponding heat maps (Figure 4a,b). Interestingly, in both tissues, the pattern of the analysed lncRNAs showed a notably different profile between the control at 24 h and at 72 h (Figure 4a,b). This highlights the importance of including the corresponding controls for each sampling point, especially in the case of a very stressful manipulation for the challenge (anaesthesia and intramuscular injection). However, the absence of a time 0 control did not allow to determine the basal lncRNA expression profile in the analysed tissues, which could be also an interesting question. In general terms, for both HK and the brain, four main clusters of lncRNAs were observed. For HK, cluster 1 included lncRNAs mainly downregulated at 72 hpi, cluster 2 grouped lncRNAs induced at 24 h by nodavirus, cluster 3 included lncRNAs induced at both 24 and 72 h after infection and cluster 4 included lncRNAs downregulated at both sampling points (Figure 4a). In the brain, cluster 1 was mainly composed of lncRNAs overexpressed at 72 h post-challenge, cluster 2 included those lncRNAs induced both at 24 and 72 h, cluster 3 included lncRNAs highly expressed in controls at 72 h and finally, cluster 4 included the lncRNAs highly represented in the control at 24 h (Figure 4b).

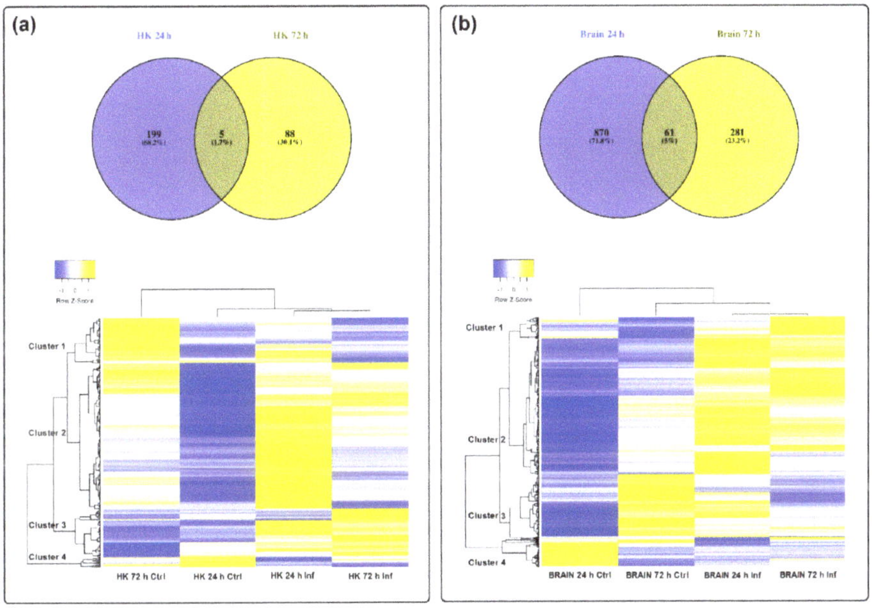

Figure 4. Modulation of lncRNAs in (**a**) head kidney and (**b**) brain after nodavirus challenge. Venn diagrams represent the number of shared and unshared lncRNAs significantly modulated at each sampling point (FC > 2, FDR < 0.05). Heat maps of the lncRNAs significantly affected by the infection in each tissue and hierarchical clustering of the different samples constructed with the TPM values.

3.4. GO Enrichment of the LncRNA Neighbouring Coding Genes

The coding genes flanking the differentially expressed lncRNAs were extracted (Supplementary Table S4) for GO enrichment analysis. The 30 most significantly enriched biological processes are represented in Figures 5 and 6 for head kidney and brain samples, respectively.

For head kidney samples (Figure 5), numerous biological process terms directly involved in immunity were found to be enriched at 24 h post-challenge ('negative regulation of mast cell activation', 'positive regulation of immunoglobulin production', 'positive regulation of B cell proliferation', 'antigen processing and presentation of exogenous peptide antigen via MHC class II' and 'neutrophil mediated immunity'). Nevertheless, this immune representation seemed to be diluted at 72 h, although the immune-related term 'leukotriene production involved in inflammatory response' also appeared to be enriched. In this tissue, some biological process terms suggesting DNA damage and cell cycle arrest were observed both at 24 h ('signal transduction involved in G2 DNA damage checkpoint,' 'signal transduction involved in mitotic DNA damage checkpoint') and 72 h post-challenge ('DNA damage induced protein phosphorylation,' 'cell cycle arrest'). At this sampling point, the term 'positive regulation of endoplasmic reticulum stress-induced intrinsic apoptotic signalling pathway' was also significantly represented.

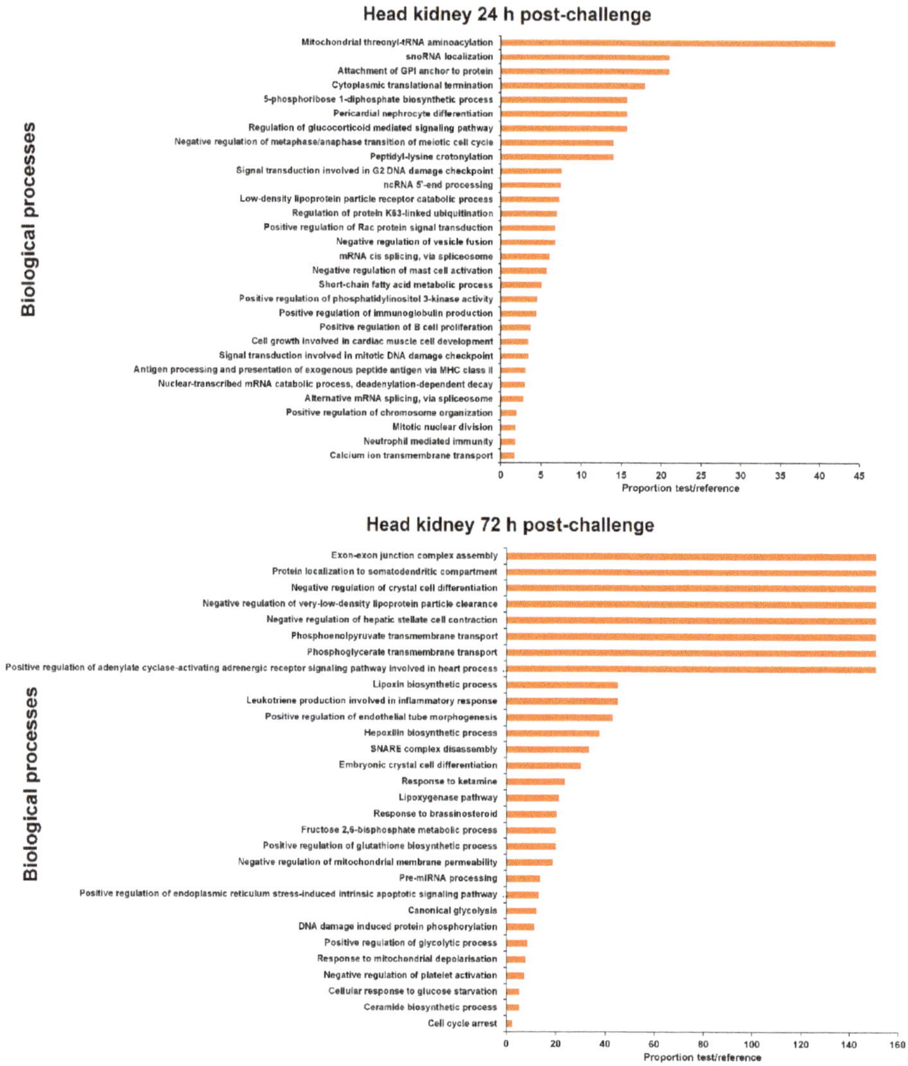

Figure 5. GO enrichment analysis (biological processes) of the neighbouring coding genes of the differentially modulated lncRNAs (FC > 2, FDR < 0.05) in head kidney after viral challenge. Only the 30 most significant terms were represented.

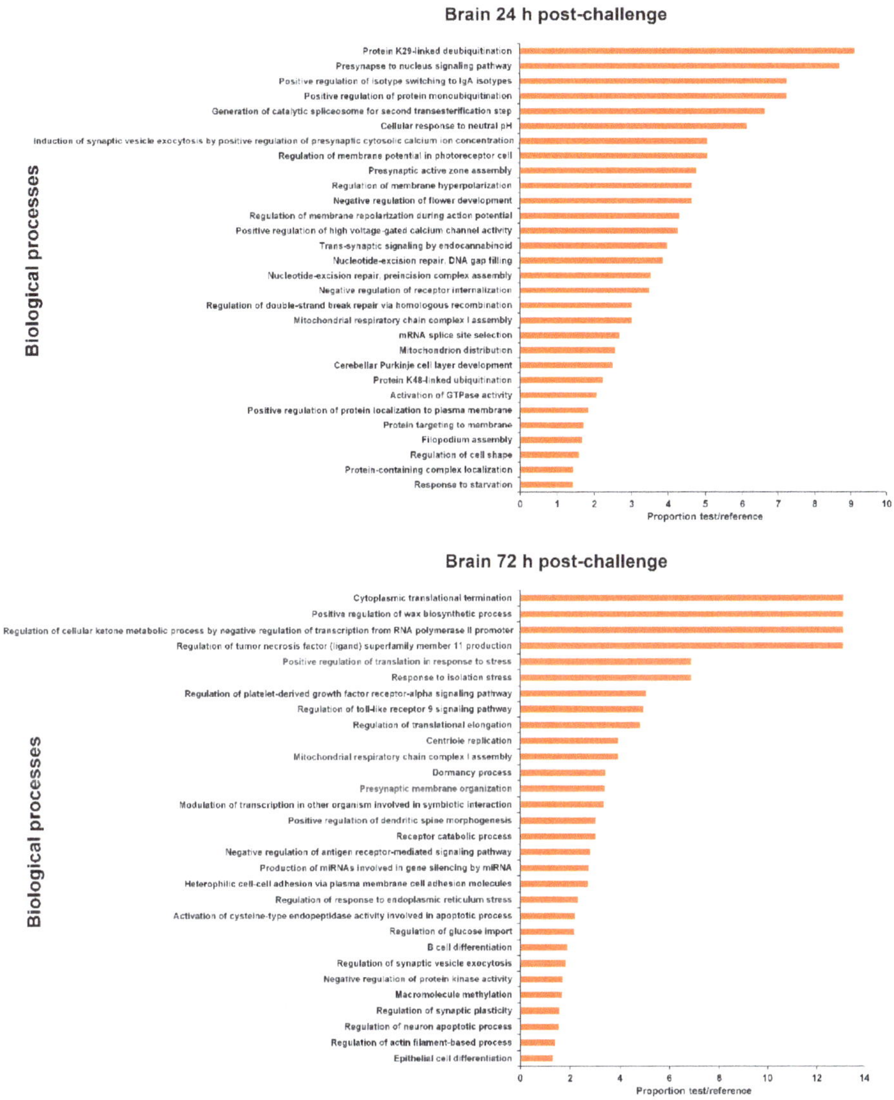

Figure 6. GO enrichment analysis (biological processes) of the neighbouring coding genes of the differentially modulated lncRNAs (FC > 2, FDR < 0.05) in the brain after viral challenge. Only the 30 most significant terms were represented.

In brain samples, the enriched terms were almost completely different from those observed in head kidney (Figure 6). In this case, although a biological process immune term was also found at 24 h ('positive regulation of isotype switching to IgA isotypes'), the representation of immune processes was lower compared to head kidney. As also occurred in the head kidney, some DNA damage terms were also observed ('nucleotide-excision repair, DNA gap filling', 'nucleotide-excision repair, pre-incision complex assembly', 'regulation of double-strand break repair via homologous recombination'), as well as numerous de-ubiquitination-related processes. Moreover, two biological processes related to calcium homeostasis were also represented ('induction of synaptic vesicle exocytosis by positive regulation of presynaptic cytosolic calcium ion concentration' and 'positive regulation of high voltage-gated calcium channel activity'). After 72 h, the representation of immune terms increased in this tissue, with the biological process enriched terms 'regulation of toll-like receptor 9 signalling pathway', 'negative regulation of antigen receptor-mediated signalling pathway' and 'B cell differentiation'. Two stress-related biological processes were represented ('positive regulation of translation in response to stress' and 'response to isolation stress'), and as was observed in head kidney, an endoplasmic reticulum (ER) stress term ('regulation of response to endoplasmic reticulum stress').

The representation of immune biological processes in both tissues could be directly related to the NNV replication level, since the detection of the virus was higher in the head kidney at 24 hpi compared to the brain, but it remained practically unaltered after 72 h [15]. On the other hand, the NNV detection enormously increased in the brain at 72 hpi [15], which could explain the higher representation of immune-related terms at this sampling point in the brain.

3.5. Expression Correlation of LncRNAs and Coding Genes

To compare the magnitude of the modulation after nodavirus infection, we constructed scatter dot plots with the fold-change values of the significantly differentially expressed genes (DEGs) and the lncRNAs in the different samples (FC > 2, FDR > 0.05) (Figure S2). For the DEGs, only those that were successfully annotated were included [15]. In general terms, the fold-change variations of the DEGs were considerably more pronounced compared to those observed for the lncRNAs (Figure S2).

To correlate the expression of the lncRNAs and their flanking coding genes, we randomly selected four DEGs after nodavirus infection [15] with at least one adjacent lncRNA significantly modulated after the viral challenge. The *complement component C3* (*c3*), *NK-tumour recognition protein* (*nktr*), *cerebellin 1* (*cbln1*) and *beta-1,4-galactosyltransferase 1* (*b4galt1*), which were overexpressed in the brain after infection, were represented together with all the predicted lncRNAs flanking and/or overlapping those genes. The TPM values were used to construct the heat maps (Figure 7a) and to calculate the correlation values (Figure 7b). Only one potential lncRNA (Lnc_contig0102476) was found near the *c3* gene, which was significantly upregulated (FC = 3.66) in the brain at 24 hpi. The expression of *c3* and the corresponding lncRNA were strongly correlated, since both were overexpressed in the brain after viral challenge (Figure 7a,b). For *nktr*, four lncRNAs were predicted to be located in the vicinity of the gene (Lnc_contig0004499, Lnc_contig0006737, Lnc_contig0002850, Lnc_contig0029684), and three of them were significantly upregulated in the brain at 24 and/or 72 h after infection. In this case, all the lncRNAs showed the same tendency as that observed for the *nktr* gene (Figure 7a,b). In the case of *cbln1*, four neighbouring lncRNAs were also identified (Lnc_contig0006952, Lnc_contig0221449, Lnc_contig0041997, Lnc_contig0114726), and whereas three of them showed the same pattern as cbln1, the expression of Lnc_contig0221449 was inversely correlated with gene expression (Figure 7a,b). Finally, the lncRNA predicted to be positioned near *b4galt1* (Lnc_contig0036634) was significantly overexpressed in the brain both at 24 and 72 h after NNV infection, and a notably high correlation with gene expression was observed (Figure 7a,b).

(a)

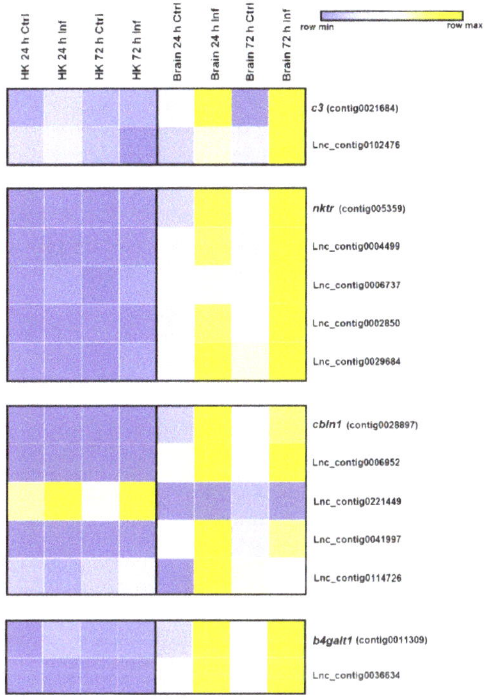

(b)

Gene	lncRNA	Pearson	Spearman
c3 (contig0021684)	Lnc_contig0102476	0.89 *	0.55^{ns}
nktr (contig0005359)	Lnc_contig0004499	0.97 *	0.95 *
	Lnc_contig0006737	0.93*	0.95 *
	Lnc_contig0002850	0.99 *	0.98 *
	Lnc_contig0029684	0.99 *	0.93 *
cbln1 (contig0028897)	Lnc_contig0006952	0.94 *	0.94*
	Lnc_contig0221449	-0.76 *	-0.85 *
	Lnc_contig0041997	0.97 *	0.94 *
	Lnc_contig0114726	0.83 *	0.56^{ns}
b4galt1 (contig0011309)	Lnc_contig0036634	0.98 *	0.91 *

Figure 7. Correlation between differentially modulated coding genes after NNV infection and their flanking lncRNAs. (**a**) Heat map representing the TPM values of four genes and their neighbouring lncRNAs across the different head kidney and brain samples. Expression levels are represented as row-normalised values on a blue–yellow colour scale. (**b**) Correlation values (Pearson's correlation coefficient and Spearman's rank correlation coefficient) between the lncRNAs and their nearby coding genes. * represents significant correlation and ns non-significant correlation.

3.6. LncRNAs Could Mediate Calcium Homeostasis

In a previous analysis of the coding transcriptome, we found a strong regulation of genes involved in calcium homeostasis in the brain, and the concentration of this cation seems to be highly altered during NNV infection and is crucial for infectivity [15]. In this work, the biological process terms 'induction of synaptic vesicle exocytosis by positive regulation of presynaptic cytosolic calcium ion concentration' and 'positive regulation of high voltage-gated calcium channel activity' were also enriched in the brain at 24 hpi (Figure 6), as well as other calcium-related terms not included in the 30 most significantly enriched GO terms but significantly enriched at 24 h and/or at 72 h post-challenge. To illustrate the connections among the different lncRNA neighbouring coding genes involved in calcium homeostasis, a protein–protein network interaction was constructed only for the neighbouring genes of those lncRNAs significantly modulated after viral infection (Figure 8a). Most of the proteins encoded by these genes were interconnected, and the main function of these genes is to directly mediate calcium transport across the cellular, mitochondrial and ER membranes. Among the molecules represented, *sarcoplasmic endoplasmic reticulum calcium atpase 1-like* (*atp2a1* or *serca1*) was the most modulated gene in the coding-transcriptome analysis [15]. On the other hand, the *regulator of g-protein signalling 6-like* (*rgs6*) was slightly modulated [15]. To exemplify the potential involvement of lncRNAs in the modulation of calcium homeostasis-related genes, these two genes were selected, and their expression was correlated to that observed for their neighbouring lncRNAs (Figure 8b). Positive or negative correlations between each coding gene and the corresponding lncRNAs were observed, and these correlations were significant for the *atp21a* gene and the lncRNA Lnc_contig0024843 and for the *rgs6* gene and Lnc_contig0052951. The three lncRNAs were included in the validation of the RNA-Seq results by qPCR.

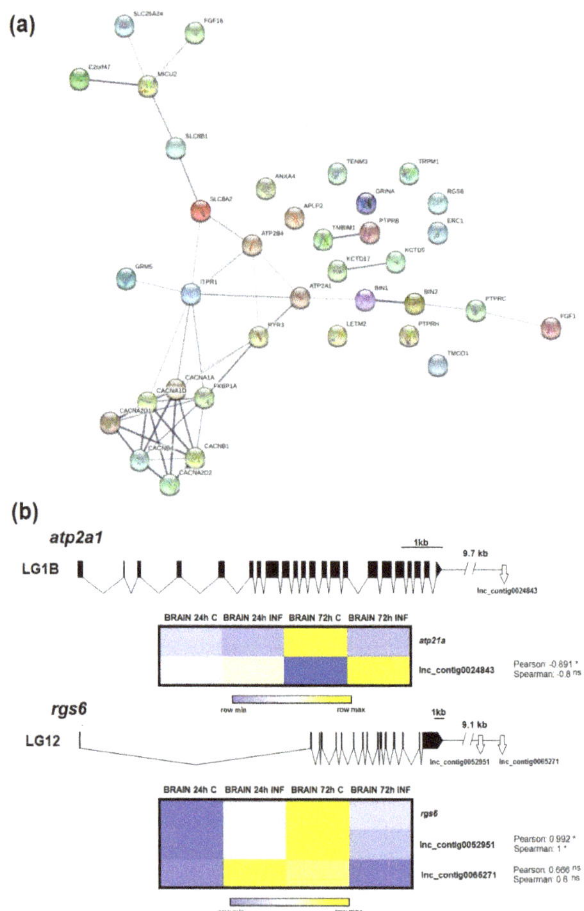

Figure 8. Calcium homeostasis-related genes and their relationship with lncRNAs. (**a**) Protein–protein interaction network of the lncRNAs neighbouring coding genes involved in calcium homeostasis. Numerous lncRNAs significantly modulated in the brain after nodavirus infection were flanked by genes directly related to calcium homeostasis. (**b**) Gene representation of two genes mediating cellular calcium concentration after NNV challenge and genomic location of their neighbouring lncRNAs. Heap maps represent the TPM values of the different contigs. Expression levels are represented as row-normalised values on a blue–yellow colour scale. * represents significant correlation and ns non-significant correlation.

3.7. qPCR Validation of LncRNA Expression

RNA-Seq results were validated by qPCR by analysing the expression of 12 putative lncRNAs significantly modulated in the RNA-Seq results (10 for brain and 2 for head kidney). Although in general terms the magnitude of the modulation was higher in the RNA-Seq data, the qPCR fold-changes followed the same tendency (Figure S3). Indeed, a positive correlation was obtained between the RNA-Seq and qPCR fold-change values by using Pearson's correlation coefficient ($r = 0.84$).

4. Discussion

In mammals, lncRNAs have been linked to a variety of immune processes, including inflammation [32,33], T cell development, differentiation and migration [34], B cell maturation [35], interferon response [36], dendritic cell differentiation [37] and granulocyte maturation [38]. The lncRNA profile has been shown to be modulated after different viral infections in several species [39–42]. Moreover, for several lncRNAs, their specific function in viral infection has been elucidated [43].

Massive analysis of zebrafish (*Danio rerio*) lncRNAs enabled the identification of numerous conserved zebrafish lncRNAs with a putative orthologue in mammals [44], which could be indicative of functional conservation. This analysis of the lncRNA repertoire in zebrafish has derived in the creation of a refined database [45], which could help to facilitate the functional analyses in this model species. However, to the best of our knowledge, only one publication has reported the specific immune function of a lncRNA in fish. In that work, the authors found that the PU.1 gene, which is involved in the expression of adaptive immune genes, is negatively regulated by its antisense lncRNA [46]. In the case of zebrafish, the existence of numerous mutant lines could facilitate the identification of lncRNAs related to specific functions. This finding was observed for the heterozygous $rag1^{+/-}$ zebrafish, which is partially deficient in the Rag1 protein, an endonuclease involved in the assembly of immunoglobulins and T cell receptor (TCR) genes [47]. When wild-type and $rag1^{+/-}$ zebrafish were infected with the spring viremia of carp virus (SVCV), those animals partially deficient in Rag1 showed a modulation of lncRNAs surrounded by genes involved in adaptive immunity, which was not observed in wild-type fish [48]. A plausible explanation is that those animals deficient in adaptive mechanisms potentiated this response to combat the infection compared to the full competent animals [48].

In the past several years, some publications have reported the identification and modulation of lncRNAs in aquacultured fish against a variety of pathogens, especially in salmonids [21,49–52]. However, to date, no lncRNA analyses have been conducted in European sea bass. We first investigated the modulation of the coding transcripts in head kidney and brain from *D. labrax* infected intramuscularly with nodavirus. In that work, we observed a strong modulation of the stress axis during infection with the virus but a slight regulation of immune-related genes [15].

From the Illumina sequencing of the *D. labrax* transcriptome and de novo assembly [15], we selected those contigs that passed all the filters to be considered potential lncRNAs. We obtained a contig list of 12,158 lncRNAs, and 95.96% of them mapped to the genome. Analysis of the lncRNA abundance per chromosome revealed a homogeneous distribution with a strong positive correlation between the number of putative lncRNAs and chromosome length.

RNA-Seq analysis of these contigs in the different samples showed that some lncRNAs were tissue-specific. This tissue specificity was previously reported for different vertebrate species [53,54], including zebrafish [55] and other teleost species [49]. Most lncRNAs influence the expression of their nearby genes, acting as local regulators; therefore, lncRNA expression is often correlated with the expression of their adjacent coding genes [16]. These tissue-specific lncRNAs were flanked by protein-coding genes involved in biological processes closely linked to the functionality of that tissue. Because the head kidney is a major haematopoietic and immune tissue in fish [56], numerous immune-related GO terms were only enriched in this organ. In the brain, almost all the enriched GO terms were directly related to the maintenance of neuronal functions and homeostasis, neurotransmitters and behaviour.

Nodavirus infection significantly modulated the expression of different lncRNAs in both the head kidney and brain. The brain showed a higher number of modulated lncRNAs (931 and 342 at 24 and 72 h, respectively) compared to the head kidney (204 and 93), probably because of the neurotropic nature of the virus. As was also reported for *S. salar* after infectious salmon anaemia virus (ISAV) infection [49] or *Caligus rogercresseyi* infestation [52], lncRNAs were expressed in a temporal-specific manner with a very low percentage of shared lncRNAs between 24 h and 72 h post-challenge in both tissues. Therefore, the lncRNA transcriptome profile changes as the disease progresses with high spatiotemporal specificity.

GO enrichment analyses of the neighbouring coding genes of lncRNAs modulated after NNV challenge in the head kidney revealed a large number of immune-related terms, especially after 24 h. These terms were mainly related to the activation and proliferation of immune cells, antigen presentation, immunoglobulin production and inflammation. However, the analysis of the transcriptome revealed a practically absent immune response in this tissue, and the modulated genes were mainly related to cortisol synthesis and reactive oxygen species (ROS) production [15]. It is worth mentioning that different DNA damage-, cell cycle- and ER stress-related terms were also enriched in this tissue. It is known that oxidative stress caused by ROS production can induce DNA damage and ER stress, which could lead to cell cycle arrest [57–59]. Nevertheless, viruses can manipulate DNA repair pathways and cell cycle control mechanisms to facilitate their own replication [60,61]. On the other hand, ER stress is also intimately related to virus activity. Viruses hijack the host translation machinery to massively produce viral proteins, affecting ER homeostasis and causing ER stress [62]. Moreover, ER stress also induces the production of ROS [63], and ROS are an important component of the immune defence because they can kill pathogens directly by causing oxidative damage or modulating different immune mechanisms [64]. It has been previously described that NNV can induce ROS in European sea bass [15] and oxidative stress-mediated cell death in fish cells [65]. Because ROS are also harmful to the host, the activity of antioxidant molecules, such as glutathione, is also needed to control the damage. The term 'positive regulation of glutathione biosynthetic process' was also enriched among those genes flanking the lncRNAs enriched in head kidney at 72 hpi. A previous analysis of the coding genes differentially modulated in sea bass head kidney after NNV infection revealed enrichment in the oxidation-reduction process [15], which could indicate that these are mechanisms controlled by lncRNAs.

In the brain, we found that those genes located at less than 10 kb from differentially modulated lncRNAs after NNV challenge were also enriched in immune terms both at 24 and 72 hpi. At 24 h, only one immune term was enriched in this tissue ('positive regulation of isotype switching to IgA isotypes'), which was related to antibody production. It has been described in teleosts that host antibody production is an important response to nodavirus infection even at early stages of infection [11,66–68], and the expression level of these immunoglobulins could be related to a higher resistance to the virus [69]. Indeed, antibody production is one of the main immune mechanisms protecting the brain against neurotropic viruses [70]. However, in the transcriptome analyses, the production of antibodies seemed to be downregulated at this early sampling point [15]. According to this finding, although more investigations are needed, inhibition of antibody production could be regulated by lncRNAs.

We previously observed that the main response induced by NNV in European sea bass at these early sampling points is the activation of the hypothalamic-pituitary-interrenal (HPI) axis, which is the stress response [15]. According to this study, some stress-related terms were also enriched in the brain for the neighbouring genes of the lncRNAs modulated at 72 hpi. The activation of the stress axis observed in these same samples could lead to calcium influx into the neurons, generating excitotoxicity, and as a consequence, neuronal damage [71,72]. Indeed, calcium cellular homeostasis was found to be highly altered in the brain during NNV infection [15]. For the neighbouring genes of the lncRNAs significantly modulated in this tissue, different calcium-related biological process terms were also enriched, and as occurred in head kidney, we also found lncRNAs flanked by genes related to the DNA damage response, which could be a consequence of excitotoxicity. Although calcium homeostasis modulations in the brain could be a direct consequence of the stress response activated after NNV challenge, viruses can also perturb it and utilise calcium and calcium-related proteins to benefit their own replication [73]. In fact, nodaviruses require the incorporation of calcium ions into the viral capsid for a correct assembly process and integrity [74,75]. We observed that RGNNV, the genotype infecting European sea bass, needs calcium to replicate correctly [15].

Although the number of annotated DEGs in the different tissues at 24 h and 72 hpi was lower compared to the number of differentially modulated lncRNAs, in general terms, the magnitude of these modulations was higher for the DEGs. This observation could suggest that small variations in lncRNA levels can have a high impact on mRNA expression. Moreover, several lncRNAs can simultaneously affect the expression of the same coding gene, and consequently, their effect can be additive. To link the coding-lncRNA response, we randomly selected four genes that were significantly modulated in the brain after viral challenge and flanked by at least one lncRNA that was also significantly affected by the infection. We observed a strong correlation between the expression of the coding genes and all the closely positioned lncRNAs. Because lncRNAs can activate or repress gene expression [16], negative transcriptional correlations can also be observed. Because of the high modulation of calcium homeostasis-related coding genes and lncRNAs closely located to these genes, we suggest that the expression of these genes is likely affected by lncRNAs.

5. Conclusions

Taking all these observations into account, we can conclude that there exists a high parallelism between the protein coding genes modulated by NNV challenge and the genes located near lncRNAs affected by infection. Therefore, lncRNAs seem to strongly contribute to the response against nodavirus. Further functional studies between significantly correlated lncRNAs and coding genes will help to elucidate the mechanism of the interactions between lncRNAs and immune genes induced after NNV infection in European sea bass.

Supplementary Materials: The following are available online at http://www.mdpi.com/2079-7737/9/7/165/s1. Table S1: Primer pairs used to validate the lncRNA differential expression analysis, Table S2: LncRNA contig list, Table S3: LncRNA differential expression analyses in head kidney and brain at 24 and 72 h post-challenge with NNV, Table S4: Predicted neighbouring coding genes of the potential sea bass lncRNAs, Figure S1: Principal component analysis (PCA) of the lncRNAs in the different samples, Figure S2: Magnitude of transcriptome modulation in the head kidney and brain after NNV infection, Figure S3: Validation of differentially expressed lncRNAs through qPCR.

Author Contributions: Conceptualisation, P.P., B.N. and A.F.; Methodology, P.P., R.L., R.M. and A.F.; Software, P.P., R.L., R.M., V.V.-M., C.G.-E. and A.F.; Validation, R.L.; Formal analysis, P.P. and A.F.; Investigation, P.P. and R.L.; Data curation, P.P.; Writing—original draft preparation, P.P.; Writing—review and editing, P.P., R.L., B.N. and A.F.; Supervision, B.N. and A.F.; Funding acquisition, B.N. All authors have read and agreed to the published version of the manuscript.

Funding: This work was financially supported by the European Union through the funding programme Horizon 2020 (Performfish -727610) and Ministerio de Economía, Industria y Competitividad of Spain (BIO2017-82851-C3-2-R). Patricia Pereiro and Raquel Lama wish to thank the Axencia Galega de Innovación (GAIN, Xunta de Galicia) for their postdoctoral (IN606B-2018/010) and predoctoral contracts (IN606A-2017/011), respectively. Our laboratory is funded by EU Feder Programa Interreg España-Portugal 0474_BLUEBIOLAB and IN607B 2019/01 from Consellería de Economía, Emprego e Industria (GAIN), Xunta de Galicia.

Acknowledgments: We thank the IIM-CSIC aquarium staff for their technical assistance.

Conflicts of Interest: The authors declare no conflict of interest.

References

1. FAO. Cultured Aquatic species information programme. *Dicentrarchus labrax*. In *National Aquaculture Sector Overview*; FAO: Rome, Italy, 2005.
2. Doan, Q.K.; Vandeputte, M.; Chatain, B.; Morin, T.; Allal, F. Viral encephalopathy and retinopathy in aquaculture: A review. *J. Fish Dis.* **2017**, *40*, 717–742. [CrossRef]

3. Munday, B.L.; Kwang, J.; Moody, N. Betanodavirus infections of teleost fish: A review. *J. Fish Dis.* **2002**, *25*, 127–142. [CrossRef]
4. Poisa-Beiro, L.; Dios, S.; Montes, A.; Aranguren, R.; Figueras, A.; Novoa, B. Nodavirus increases the expression of Mx and inflammatory cytokines in fish brain. *Mol. Immunol.* **2008**, *45*, 218–225. [CrossRef]
5. Poisa-Beiro, L.; Dios, S.; Ahmed, H.; Vasta, G.R.; Martínez-López, A.; Estepa, A.; Alonso-Gutiérrez, J.; Figueras, A.; Novoa, B. Nodavirus infection of sea bass (*Dicentrarchus labrax*) induces up-regulation of galectin-1 expression with potential anti-inflammatory activity. *J. Immunol.* **2009**, *183*, 6600–6611. [CrossRef]
6. Sarropoulou, E.; Sepulcre, P.; Poisa-Beiro, L.; Mulero, V.; Meseguer, J.; Figueras, A.; Novoa, B.; Terzoglou, V.; Reinhardt, R.; Magoulas, A.; et al. Profiling of infection specific mRNA transcripts of the European seabass *Dicentrarchus labrax*. *BMC Gen.* **2009**, *10*, 157. [CrossRef]
7. Scapigliati, G.; Buonocore, F.; Randelli, E.; Casani, D.; Meloni, S.; Zarletti, G.; Tiberi, M.; Pietretti, D.; Boschi, I.; Manchado, M.; et al. Cellular and molecular immune responses of the sea bass (*Dicentrarchus labrax*) experimentally infected with betanodavirus. *Fish Shellfish Immunol.* **2010**, *28*, 303–311. [CrossRef]
8. Chaves-Pozo, E.; Guardiola, F.A.; Meseguer, J.; Esteban, M.A.; Cuesta, A. Nodavirus infection induces a great innate cell-mediated cytotoxic activity in resistant, gilthead seabream, and susceptible, European sea bass, teleost fish. *Fish Shellfish Immunol.* **2012**, *33*, 1159–1166. [CrossRef]
9. Novel, P.; Fernandez-Trujillo, M.A.; Gallardo-Galvez, J.B.; Cano, I.; Manchado, M.; Buonocore, F.; Randelli, E.; Scapigliati, G.; Alvarez, M.C.; Bejar, J. Two Mx genes identified in European sea bass (*Dicentrarchus labrax*) respond differently to VNNV infection. *Vet. Immunol. Immunopathol.* **2013**, *153*, 240–248. [CrossRef]
10. Buonocore, F.; Randelli, E.; Tranfa, P.; Scapigliati, G. A CD83-like molecule in sea bass (*Dicentrarchus labrax*): Molecular characterization and modulation by viral and bacterial infection. *Fish Shellfish Immunol.* **2012**, *32*, 1179–1184. [CrossRef]
11. Buonocore, F.; Stocchi, V.; Nunez-Ortiz, N.; Randelli, E.; Gerdol, M.; Pallavicini, A.; Facchiano, A.; Bernini, C.; Guerra, L.; Scapigliati, G.; et al. Immunoglobulin T from sea bass (*Dicentrarchus labrax* L.): Molecular characterization, tissue localization and expression after nodavirus infection. *BMC Mol. Biol.* **2017**, *18*, 8. [CrossRef]
12. Valero, Y.; Morcillo, P.; Meseguer, J.; Buonocore, F.; Esteban, M.A.; Chaves-Pozo, E.; Cuesta, A. Characterization of the interferon pathway in the teleost fish gonad against the vertically transmitted nervous necrosis virus. *J. Gen. Virol.* **2015**, *96*, 2176–2187. [CrossRef] [PubMed]
13. Chaves-Pozo, E.; Valero, Y.; Esteve-Codina, A.; Gómez-Garrido, J.; Dabad, M.; Alioto, T.; Meseguer, J.; Esteban, M.A.; Cuesta, A. Innate cell-mediated cytotoxic activity of European sea bass leucocytes against Nodavirus-infected cells: A functional and RNA-seq study. *Sci. Rep.* **2017**, *7*, 15396. [CrossRef] [PubMed]
14. Chaves-Pozo, E.; Bandín, I.; Olveira, J.G.; Esteve-Codina, A.; Gómez-Garrido, J.; Dabad, M.; Alioto, T.; Esteban, M.A.; Cuesta, A. European sea bass brain DLB-1 cell line is susceptible to nodavirus: A transcriptomic study. *Fish Shellfish Immunol.* **2019**, *86*, 14–24. [CrossRef]
15. Lama, R.; Pereiro, P.; Valenzuela-Muñoz, V.; Gallardo-Escárate, C.; Tort, L.; Figueras, A.; Novoa, B. RNA-Seq analysis of European sea bass (*Dicentrarchus labrax* L.) infected with nodavirus reveals powerful modulation of the stress response. *Vet. Res.* **2020**, *51*, 64. [CrossRef] [PubMed]
16. Ponting, C.P.; Oliver, P.L.; Reik, W. Evolution and functions of long noncoding RNAs. *Cell* **2009**, *136*, 629–641. [CrossRef] [PubMed]
17. Hezroni, H.; Koppstein, D.; Schwartz, M.G.; Avrutin, A.; Bartel, D.P.; Ulitsky, I. Principles of long noncoding RNA evolution derived from direct comparison of transcriptomes in 17 species. *Cell Rep.* **2015**, *11*, 1110–1122. [CrossRef]
18. Perry, R.B.T.; Ulitsky, I. The functions of long noncoding RNAs in development and stem cells. *Development* **2016**, *143*, 3882–3894. [CrossRef]
19. Guo, C.J.; Ma, X.K.; Xing, Y.H.; Zheng, C.C.; Xu, Y.F.; Shan, L.; Zhang, J.; Wang, S.; Wang, Y.; Carmichael, G.G.; et al. Distinct processing of lncRNAs contributes to non-conserved functions in stem cells. *Cell* **2020**, *181*, 621–636. [CrossRef]
20. Aune, T.M.; Spurlock, C.F. Long non-coding RNAs in innate and adaptive immunity. *Virus Res.* **2016**, *212*, 146–160. [CrossRef] [PubMed]
21. Elling, R.; Chan, J.; Fitzgerald, K.A. Emerging role of long noncoding RNAs as regulators of innate immune cell development and inflammatory gene expression. *Eur. J. Immunol.* **2016**, *46*, 504–512. [CrossRef]

22. Wang, M.; Jiang, S.; Wu, W.; Yu, F.; Chang, W.; Li, P.; Wang, K. Non-coding RNAs function as immune regulators in teleost fish. *Front. Immunol.* **2018**, *9*, 2801. [CrossRef]
23. Reed, L.J.; Muench, H. A simple method of estimating fifty percent endpoints. *Am. J. Hyg.* **1938**, *27*, 493–497.
24. Tarifeño-Saldivia, E.; Valenzuela-Miranda, D.; Gallardo-Escárate, C. In the shadow: The emerging role of long non-coding RNAs in the immune response of Atlantic salmon. *Dev. Comp. Immunol.* **2017**, *73*, 193–205. [CrossRef]
25. Wang, L.; Park, H.J.; Dasari, S.; Wang, S.; Kocher, J.P.; Li, W. CPAT: Coding-potential assessment tool using an alignment-free logistic regression model. *Nucleic Acids Res.* **2013**, *41*, e74. [CrossRef] [PubMed]
26. Tine, M.; Kuhl, H.; Gagnaire, P.A.; Louro, B.; Desmarais, E.; Martins, R.S.T.; Hecht, J.; Knaust, F.; Belkhir, K.; Klages, S.; et al. European sea bass genome and its variation provide insights into adaptation to euryhalinity and speciation. *Nat. Commun.* **2014**, *5*, 5770. [CrossRef] [PubMed]
27. Conesa, A.; Götz, S.; García-Gómez, J.M.; Terol, J.; Talón, M.; Robles, M. Blast2GO: A universal tool for annotation, visualization and analysis in functional genomics research. *Bioinformatics* **2005**, *21*, 3674–3676. [CrossRef]
28. Babicki, S.; Arndt, D.; Marcu, A.; Liang, Y.; Grant, J.R.; Maciejewski, A.; Wishart, D.S. Heatmapper: Web-enabled heat mapping for all. *Nucleic Acids Res.* **2016**, *44*, W147–W153. [CrossRef]
29. Szklarczyk, D.; Gable, A.L.; Lyon, D.; Junge, A.; Wyder, S.; Huerta-Cepas, J.; Simonovic, M.; Doncheva, N.T.; Morris, J.H.; Bork, P.; et al. STRING v11: Protein-protein association networks with increased coverage, supporting functional discovery in genome-wide experimental datasets. *Nucleic Acids Res.* **2019**, *47*, D607–D613. [CrossRef]
30. Rozen, S.; Skaletsky, H. Primer3 on the WWW for general users and for biologist programmers. *Methods Mol. Biol.* **2000**, *132*, 365–386.
31. Pfaffl, M.W. A new mathematical model for relative quantification in real-time RT-PCR. *Nucleic Acids Res.* **2001**, *29*, 2002–2007. [CrossRef]
32. Carpenter, S.; Aiello, D.; Atianand, M.K.; Ricci, E.P.; Gandhi, P.; Hall, L.L.; Byron, M.; Monks, B.; Henry-Bezy, M.; O'Neill, L.A.J.; et al. A long noncoding RNA mediates both activation and repression of immune response genes. *Science* **2013**, *341*, 789–792. [CrossRef]
33. Rapicavoli, N.A.; Qu, K.; Zhang, J.; Mikhail, M.; Laberge, R.M.; Chang, H.Y. A mammalian pseudogene lncRNA at the interface of inflammation and anti-inflammatory therapeutics. *eLife* **2013**, *2*, e00762. [CrossRef] [PubMed]
34. Hu, G.; Tang, Q.; Sharma, S.; Yu, F.; Escobar, T.M.; Muljo, S.A.; Zhu, J.; Zhao, K. Expression and regulation of intergenic long noncoding RNAs during T cell development and differentiation. *Nat. Immunol.* **2013**, *14*, 1190–1198. [CrossRef]
35. Agirre, X.; Meydan, C.; Jiang, Y.; Garate, L.; Doane, A.S.; Li, Z.; Verma, A.; Paiva, B.; Martín-Subero, J.I.; Elemento, O.; et al. Long non-coding RNAs discriminate the stages and gene regulatory states of human humoral immune response. *Nat. Commun.* **2019**, *10*, 821. [CrossRef] [PubMed]
36. Kambara, H.; Niazi, F.; Kostadinova, L.; Moonka, D.K.; Siegel, C.T.; Post, A.B.; Carnero, E.; Barriocanal, M.; Fortes, P.; Anthony, D.D.; et al. Negative regulation of the interferon response by an interferon-induced long non-coding RNA. *Nucleic Acids Res.* **2014**, *42*, 10668–10680. [CrossRef] [PubMed]
37. Wang, P.; Xue, Y.; Han, Y.; Lin, L.; Wu, C.; Xu, S.; Jiang, Z.; Xu, J.; Liu, Q.; Cao, X. The STAT3-binding long noncoding RNA lnc-DC controls human dendritic cell differentiation. *Science* **2014**, *344*, 310–313. [CrossRef]
38. Zhang, X.; Lian, Z.; Padden, C.; Gerstein, M.B.; Rozowsky, J.; Snyder, M.; Gingeras, T.R.; Kapranov, P.; Weissman, S.M.; Newburger, P.E. A myelopoiesis-associated regulatory intergenic noncoding RNA transcript within the human HOXA cluster. *Blood* **2009**, *113*, 2526–2534. [CrossRef]
39. Peng, X.; Gralinski, L.; Armour, C.D.; Ferris, M.T.; Thomas, M.J.; Proll, S.; Bradel-Tretheway, B.G.; Korth, M.J.; Castle, J.C.; Biery, M.C.; et al. Unique signatures of long noncoding RNA expression in response to virus infection and altered innate immune signaling. *MBio* **2010**, *1*, e00206–e00210. [CrossRef]
40. Zhang, Q.; Lai, M.M.; Lou, Y.Y.; Guo, B.H.; Wang, H.Y.; Zheng, X.Q. Transcriptome altered by latent human cytomegalovirus infection on THP-1 cells using RNA-seq. *Gene* **2016**, *594*, 144–150. [CrossRef]
41. Zhao, P.; Liu, S.; Zhong, Z.; Jiang, T.; Weng, R.; Xie, M.; Yang, S.; Xia, X. Analysis of expression profiles of long noncoding RNAs and mRNAs in brains of mice infected by rabies virus by RNA sequencing. *Sci. Rep.* **2018**, *8*, 11858. [CrossRef]

42. Hudson, W.H.; Prokhnevska, N.; Gensheimer, J.; Akondy, R.; McGuire, D.J.; Ahmed, R.; Kissick, H.T. Expression of novel long noncoding RNAs defines virus-specific effector and memory CD8+ T cells. *Nat. Commun.* **2019**, *10*, 196. [CrossRef]
43. Fortes, P.; Morris, K.V. Long noncoding RNAs in viral infections. *Virus Res.* **2016**, *212*, 1–11. [CrossRef] [PubMed]
44. Chen, W.; Zhang, X.; Li, J.; Huang, S.; Xiang, S.; Hu, X.; Liu, C. Comprehensive analysis of coding-lncRNA gene co-expression network uncovers conserved functional lncRNAs in zebrafish. *BMC Gen.* **2018**, *19*, 112. [CrossRef] [PubMed]
45. Hu, X.; Chen, W.; Li, J.; Huang, S.; Xu, X.; Zhang, X.; Xiang, S.; Liu, C. ZFLNC: A comprehensive and well-annotated database for zebrafish lncRNA. *Database* **2018**, *2018*, bay114. [CrossRef] [PubMed]
46. Wei, N.; Pang, W.; Wang, Y.; Xiong, Y.; Xu, R.; Wu, W.; Zhao, C.; Yang, G. Knockdown of PU.1 mRNA and AS lncRNA regulates expression of immune-related genes in zebrafish *Danio rerio*. *Dev. Comp. Immunol.* **2014**, *44*, 315–319. [CrossRef] [PubMed]
47. Bassing, C.H.; Swat, W.; Alt, F.W. The mechanism and regulation of chromosomal V(D)J recombination. *Cell* **2002**, *109*, S45–S55. [CrossRef]
48. Valenzuela-Muñoz, V.; Pereiro, P.; Álvarez-Rodríguez, M.; Gallardo-Escárate, C.; Figueras, A.; Novoa, B. Comparative modulation of lncRNAs in wild-type and rag1-heterozygous mutant zebrafish exposed to immune challenge with spring viraemia of carp virus (SVCV). *Sci. Rep.* **2019**, *9*, 14174. [CrossRef]
49. Boltaña, S.; Valenzuela-Miranda, D.; Aguilar, A.; Mackenzie, S.; Gallardo-Escárate, C. Long noncoding RNAs (lncRNAs) dynamics evidence immunomodulation during ISAV-Infected Atlantic salmon (*Salmo salar*). *Sci. Rep.* **2016**, *6*, 22698. [CrossRef]
50. Valenzuela-Miranda, D.; Gallardo-Escárate, C. Novel insights into the response of Atlantic salmon (*Salmo salar*) to *Piscirickettsia salmonis*: Interplay of coding genes and lncRNAs during bacterial infection. *Fish Shellfish Immunol.* **2016**, *59*, 427–438. [CrossRef]
51. Paneru, B.; Al-Tobasei, R.; Palti, Y.; Wiens, G.D.; Salem, M. Differential expression of long non-coding RNAs in three genetic lines of rainbow trout in response to infection with *Flavobacterium psychrophilum*. *Sci. Rep.* **2016**, *6*, 36032. [CrossRef]
52. Valenzuela-Muñoz, V.; Valenzuela-Miranda, D.; Gallardo-Escárate, C. Comparative analysis of long non-coding RNAs in Atlantic and Coho salmon reveals divergent transcriptome responses associated with immunity and tissue repair during sea lice infestation. *Dev. Comp. Immunol.* **2018**, *87*, 36–50. [CrossRef] [PubMed]
53. Kern, C.; Wang, Y.; Chitwood, J.; Korf, I.; Delany, M.; Cheng, H.; Medrano, J.F.; Van Eenennaam, A.L.; Ernst, C.; Ross, P.; et al. Genome-wide identification of tissue-specific long non-coding RNA in three farm animal species. *BMC Genom.* **2018**, *19*, 684. [CrossRef] [PubMed]
54. Ward, M.; McEwan, C.; Mills, J.D.; Janitz, M. Conservation and tissue-specific transcription patterns of long noncoding RNAs. *J. Hum. Transcr.* **2015**, *1*, 2–9. [CrossRef] [PubMed]
55. Kaushik, K.; Leonard, V.E.; Kv, S.; Lalwani, M.K.; Jalali, S.; Patowary, A.; Joshi, A.; Scaria, V.; Sivasubbu, S. Dynamic expression of long non-coding RNAs (lncRNAs) in adult zebrafish. *PLoS ONE* **2013**, *8*, e83616. [CrossRef]
56. Zapata, A.; Diez, B.; Cejalvo, T.; Gutierrez-De Frias, C.; Cortes, A. Ontogeny of the immune system of fish. *Fish Shellfish Immunol.* **2006**, *20*, 126–136. [CrossRef]
57. Lin, S.S.; Huang, H.P.; Yang, J.S.; Wu, J.Y.; Hsia, T.C.; Lin, C.C.; Lin, C.W.; Kuo, C.L.; Gibson-Wood, W.; Chung, J.G. DNA damage and endoplasmic reticulum stress mediated curcumin-induced cell cycle arrest and apoptosis in human lung carcinoma A-549 cells through the activation caspases cascade- and mitochondrial-dependent pathway. *Cancer Lett.* **2018**, *272*, 77–90. [CrossRef]
58. Yang, J.S.; Chen, G.W.; Hsia, T.C.; Ho, H.C.; Ho, C.C.; Lin, M.W.; Lin, S.S.; Yeh, R.D.; Ip, S.W.; Lu, H.F.; et al. Diallyl disulfide induces apoptosis in human colon cancer cell line (COLO 205) through the induction of reactive oxygen species, endoplasmic reticulum stress, caspases cascade and mitochondrial-dependent pathways. *Food Chem. Toxicol.* **2009**, *47*, 171–179. [CrossRef]
59. Zeng, H.; Nanayakkara, G.K.; Shao, Y.; Fu, H.; Sun, Y.; Cueto, R.; Yang, W.Y.; Yang, Q.; Sheng, H.; Wu, N.; et al. DNA checkpoint and repair factors are nuclear sensors for intracellular organelle stresses-inflammations and cancers can have high genomic risks. *Front. Physiol.* **2018**, *9*, 516. [CrossRef]

60. Luftig, M.A. Viruses and the DNA damage response: Activation and antagonism. *Annu. Rev. Virol.* **2014**, *1*, 605–625. [CrossRef]
61. Bagga, S.; Bouchard, M.J. Cell cycle regulation during viral infection. *Methods Mol. Biol.* **2014**, *1170*, 165–227.
62. Li, S.; Kong, L.; Yu, X. The expanding roles of endoplasmic reticulum stress in virus replication and pathogenesis. *Crit. Rev. Microbiol.* **2015**, *41*, 150–164. [CrossRef]
63. Zeeshan, H.M.; Lee, G.H.; Kim, H.R.; Chae, H.J. Endoplasmic reticulum stress and associated ROS. *Int. J. Mol. Sci.* **2016**, *17*, 327. [CrossRef]
64. Paiva, C.N.; Bozza, M.T. Are reactive oxygen species always detrimental to pathogens? *Antioxid. Redox Signal.* **2014**, *20*, 1000–1037. [CrossRef]
65. Chang, C.W.; Su, Y.C.; Her, G.M.; Ken, C.F.; Hong, J.R. Betanodavirus induces oxidative stress-mediated cell death that prevented by anti-oxidants and Zfcatalase in fish cells. *PLoS ONE* **2011**, *6*, e25853. [CrossRef] [PubMed]
66. Grove, S.; Johansen, R.; Reitan, L.J.; Press, C.M.; Dannevig, B.H. Quantitative investigation of antigen and immune response in nervous and lymphoid tissues of Atlantic halibut (*Hippoglossus hippoglossus*) challenged with nodavirus. *Fish Shellfish Immunol.* **2016**, *21*, 525–539. [CrossRef]
67. López-Muñoz, A.; Sepulcre, M.P.; García-Moreno, D.; Fuentes, I.; Béjar, J.; Manchado, M.; Álvarez, M.C.; Meseguer, J.; Mulero, V. Viral nervous necrosis virus persistently replicates in the central nervous system of asymptomatic gilthead seabream and promotes a transient inflammatory response followed by the infiltration of IgM+ B lymphocytes. *Dev. Comp. Immunol.* **2012**, *37*, 429–437. [CrossRef]
68. Piazzon, M.C.; Galindo-Villegas, J.; Pereiro, P.; Estensoro, I.; Calduch-Giner, J.A.; Gómez-Casado, E.; Novoa, B.; Mulero, V.; Sitjà-Bobadilla, A.; Pérez-Sánchez, J. Differential modulation of IgT and IgM upon parasitic, bacterial, viral, and dietary challenges in a perciform fish. *Front. Immunol.* **2016**, *7*, 637. [CrossRef]
69. Wu, M.S.; Chen, C.W.; Liu, Y.C.; Huang, H.H.; Lin, C.H.; Tzeng, C.S.; Chang, C.Y. Transcriptional analysis of orange-spotted grouper reacting to experimental grouper iridovirus infection. *Dev. Comp. Immunol.* **2012**, *37*, 233–242. [CrossRef] [PubMed]
70. Iwasaki, A. Immune regulation of antibody access to neuronal tissues. *Trends Mol. Med.* **2017**, *23*, 227–245. [CrossRef] [PubMed]
71. Arundine, M.; Tymianski, M. Molecular mechanisms of calcium-dependent neurodegeneration in excitotoxicity. *Cell Calcium* **2003**, *34*, 325–337. [CrossRef]
72. Dong, X.X.; Wang, Y.; Qin, Z.H. Molecular mechanisms of excitotoxicity and their relevance to pathogenesis of neurodegenerative diseases. *Acta Pharmacol. Sin.* **2009**, *30*, 379–387. [CrossRef] [PubMed]
73. Zhou, Y.; Frey, T.K.; Yang, J.J. Viral calciomics: Interplays between Ca2+ and virus. *Cell Calcium* **2009**, *46*, 1–17. [CrossRef] [PubMed]
74. Wu, Y.M.; Hsu, C.H.; Wang, C.H.; Liu, W.; Chang, W.H.; Lin, C.S. Role of the DxxDxD motif in the assembly and stability of betanodavirus particles. *Arch. Virol.* **2008**, *153*, 1633–1642. [CrossRef]
75. Chen, N.C.; Yoshimura, M.; Guan, H.H.; Wang, T.Y.; Misumi, Y.; Lin, C.C.; Chuankhayan, P.; Nakagawa, A.; Chan, S.I.; Tsukihara, T.; et al. Crystal structures of a piscine Betanodavirus: Mechanisms of capsid assembly and viral infection. *PLoS Pathog.* **2015**, *11*, e1005203. [CrossRef] [PubMed]

biology

MDPI

Article

Diversity of Rainbow Trout Blood B Cells Revealed by Single Cell RNA Sequencing

Pedro Perdiguero, Esther Morel and Carolina Tafalla *

Fish Immunology and Pathology Group, Animal Health Research Center (CISA-INIA), Valdeolmos, 28130 Madrid, Spain; perdiguero.pedro@inia.es (P.P.); morel.esther@inia.es (E.M.)
* Correspondence: tafalla@inia.es; Tel.: +34-916202300

Simple Summary: Although evolutionarily jawed fish constitute the first group of animals in which a complete adaptive immune system based on immunoglobulins (Igs) is present, many structural immune differences between fish and mammals predict important functional and phenotypical differences between B cells in these two animal groups. However, to date, very few tools are available to study B cell heterogeneity and functionality in fish. Hence, thus far, antibodies targeting the different Igs have been almost exclusively applied as tools to investigate B cell functionality in fish. In the current study, we used the newly developed 10× Genomics single cell RNA sequencing technology and used it to analyze the transcriptional pattern of single B cells from peripheral blood. The results obtained provide us with a transcriptional profile at single cell level of what seem to correspond to different B cell subsets or B cells in different stages of maturation or differentiation. The information provided will not only help us understand the biology of teleost B cells, but also provides us with a repertoire of potential markers that could be used in the future to differentiate trout B cell subsets.

Abstract: Single-cell sequencing technologies capable of providing us with immune information from dozens to thousands of individual cells simultaneously have revolutionized the field of immunology these past years. However, to date, most of these novel technologies have not been broadly applied to non-model organisms such as teleost fish. In this study, we used the 10× Genomics single cell RNA sequencing technology and used it to analyze for the first time in teleost fish the transcriptional pattern of single B cells from peripheral blood. The analysis of the data obtained in rainbow trout revealed ten distinct cell clusters that seem to be associated with different subsets and/or maturation/differentiation stages of circulating B cells. The potential characteristics and functions of these different B cell subpopulations are discussed on the basis of their transcriptomic profile. The results obtained provide us with valuable information to understand the biology of teleost B cells and offer us a repertoire of potential markers that could be used in the future to differentiate trout B cell subsets.

Keywords: teleost; B cells; single cell transcriptomics; immunoglobulins; immune markers; transcription factors; long non-coding RNAs

Citation: Perdiguero, P.; Morel, E.; Tafalla, C. Diversity of Rainbow Trout Blood B Cells Revealed by Single Cell RNA Sequencing. *Biology* **2021**, *10*, 511. https://doi.org/10.3390/biology10060511

Academic Editor: Patricia Pereiro

Received: 28 April 2021
Accepted: 6 June 2021
Published: 9 June 2021

1. Introduction

To date, most studies have agreed on the fact that adaptive immunity appeared during the early stages of vertebrate evolution, most likely in the disappeared placoderms [1]. Accordingly, the genes that define the adaptive immune system such as immunoglobulins (Igs), T cell receptors (TCR), major histocompatibility complex I (MHC I), MHC II, recombination activating gene 1 (RAG1), and RAG2 are present in gnathostomes (jawed vertebrates) including cartilaginous fish such as sharks (the most ancient jawed fish) and teleost fish. While the basic components of adaptive immunity are present, it must be taken into account that the adaptive branch of the immune system continued to evolve in tetrapods, reaching further degrees of specialization and sophistication in mammals. Consequently, there are important differences between the mammalian and teleost adaptive immune system that significantly condition the phenotype and functionality of B cells and how they respond to an antigen encounter. For instance, given the lack of bone marrow, the head kidney is the main hematopoietic organ in teleosts, and the main site for B cell differentiation. Similarly, fish do not have lymph nodes, being the spleen the main secondary immune organ. Within the spleen, the organization of lymphocytes is very primitive, with scattered B and T cells and no clearly defined regions as those found in mammals [2,3]. Thus, no cognate germinal centers (GCs) have ever been identified in the teleost spleen. GCs, formed in mammals during the immune response, promote the close collaboration between proliferating antigen-specific B cells, T follicular helper cells, and specialized follicular dendritic cells (DCs). In this environment, B cells divide in response to antigens and acquire the capacity to differentiate into antibody-secreting cells (ASCs), reaching a terminal state of plasma cells or memory B cells, both having the capacity to secrete high affinity antibodies. It is in these sites, that B cells undergo class switch recombination (CSR) and replace the heavy chain of IgM for IgG, IgA or IgE, antibodies with higher affinity and different effector functions. In the absence of GCs or specialized Igs, whether teleost B cells differentiate to plasma cells or memory B cells equivalent to those found in mammals is still a matter of debate.

Fish also have a more limited Ig array. Fish genomes encode only three classes of Igs, namely IgM, IgD, and the fish specific IgT/Z [4,5]. Thus, in contrast to mammals, no CSR has ever been reported in fish. IgM and IgD are co-expressed on the surface of naïve B cells. In this case, alternative splicing between the recombined variable region and constant regions of IgM and IgD render cells that co-express IgM and IgD of the same specificity. Upon activation, naïve B cells lose surface IgD to become IgM^+IgD^- B cells with a plasmablast profile [6,7]. Moreover, through some still unknown mechanisms, certain cells lose surface IgM and become IgD^+IgM^- B cells. These cells have been identified in catfish blood [8] as well as in rainbow trout gills [9] and gut [10] and have yet unknown functions. Finally, fish B cells expressing IgT cells constitute an independent B cell linage in which IgM and IgD are not expressed [11]. Most teleost species express more than one IgT, three in the case of salmonids [12], although whether individual cells express one or more IgT is still not clear.

To date, antibodies targeting the different Igs have been almost exclusively applied as tools to investigate B cell functionality in fish. In some cases, antibodies recognizing key transcription factors involved in B cell development in mammals (Pax5, Blimp1) have been combined with antibodies recognizing species-specific immunoglobulins (IgM, IgT, and IgD), allowing the identification of B cells in different maturation/differentiation stages by flow cytometry [13,14]. Other studies have focused on identifying in fish orthologue genes of cluster of differentiation (CD) molecules previously characterized and assigned in mammals to specific B cell subsets. However, it should be taken into account that many of these orthologues to key mammalian markers are absent in the teleost genomes or in other cases show a high divergence compared to mammalian proteins, questioning whether they have a similar function. Despite these efforts, the notorious lack of fish-specific markers that differentiate B cell subsets, maturation stages, or differentiated cells have considerably hindered our understanding of teleost B cell functionality.

The recent development of sophisticated tools that permit the analysis of transcriptomes at single cell resolution is allowing a fast progression of our current knowledge regarding the phenotype and function of different leukocyte subsets. Thus, for example, single cell RNA sequencing has been used in Atlantic cod (*Gadus morhua*) spleen cells and peripheral blood leukocytes, identifying the major cell subsets including cytotoxic T cells, B cells, erythrocytes, thrombocytes, neutrophils, and macrophages [15]. Focusing the analysis on one specific immune cell subset, Niu et al. defined different subsets of non-specific cytotoxic cells in Nile tilapia (*Oreochromis niloticus*) [16]. In the present work, we have taken advantage of the recent single cell genomic tools developed to perform an in depth analysis of teleost B cell transcriptomes at single cell resolution. For this study, we used rainbow trout (*Oncorhynchus mykiss*) as a model species and blood as a B cell source given that this is where the higher percentage of B cells are found in homeostasis [17]. To avoid cross-linking of the B cell receptor (BCR) and the subsequent cell activation in response to anti-IgM, we sorted lymphoid (small cells with low complexity) MHC II$^+$ cells, which exclusively corresponded to B cells. Our results evidence the presence of different circulating B cell subpopulations with interesting differences at transcriptional level that seem to reflect different stages of maturation or diverse B cell subsets. These results provide valuable information to elucidate B cell functionality in teleosts. The identification of specific markers for each of these subpopulations will also be of great help to generate in the future novel antibodies that can be used to differentiate teleost B cell subsets.

2. Materials and Methods

2.1. Experimental Fish

Female rainbow trout (*Oncorhynchus mykiss*) of approximately 50–70 g (15–20 cm) were obtained from *Piscifactoria Cifuentes* (Guadalajara, Spain). Fish were maintained at the Animal Health Research Centre (CISA-INIA) laboratory at 16 °C with a re-circulating water system and 12:12 h light:dark photoperiod. The fish were fed a commercial diet twice a day (T4 Royal Optima, Skretting, Spain). The fish were acclimatized to laboratory conditions for at least 2 weeks prior to any experimental procedure. During this period, no clinical signs were ever observed. The procedures described comply with the Guidelines of the European Union Council (2010/63/EU) for the use of laboratory animals and were previously approved by the Ethics committee from the *Instituto Nacional de Investigación y Tecnología Agraria y Alimentaria* (INIA; Code CEEA PROEX002/17).

2.2. Peripheral Blood Leukocyte Isolation and Sorting

Three individual rainbow trout were killed by benzocaine (Sigma) overdose. Blood was extracted with a heparinized needle from the caudal vein and diluted 10 times with Leibovitz medium (L-15, Thermo Fisher Scientific, Walthan, MA, USA) supplemented with 100 IU/mL penicillin and 100 μg/mL streptomycin (P/S, Thermo Fisher Scientific), 2% fetal calf serum (FCS, Thermo Fisher Scientific), and 10 IU/mL heparin (Sigma, St. Louis, MO, USA). Peripheral blood leukocytes (PBLs) were obtained by density gradient centrifugation (500× *g* for 30 min at 4 °C) of diluted blood on 51% continuous Percoll (GE Healthcare, North Richland Hills, TX, USA). The interface cells were collected, washed twice in L-15 containing antibiotics and 5% FCS and adjusted to 2×10^6 cells/mL.

To avoid the activation of B cells by BCR cross-linking and to be able to study different B cell subsets independently of their pattern of surface Ig expression, we decided to sort MHC II$^+$ cells with small size and low complexity (within what has been previously designated as the lymphoid gate), as it was predicted that these cells would correspond mainly to B cells. For this, PBLs were washed in FACS staining buffer (phenol red-free L-15 medium supplemented with P/S and 2% FCS) and incubated with a monoclonal antibody specific for rainbow trout MHC II β chain (mAb mouse IgG1 coupled to allophycocyanin, 2 μg/mL) previously characterized [18]. After 30 min of incubation at 4 °C, the cells were washed with FACS staining buffer and YO-PRO dye (0.05 μM) added to the cell suspension for dead cell exclusion. Lymphoid (small, low complexity) MHC IIβ$^+$ YO-PRO$^-$ (live) cells were then isolated in a FACSAria™ III sorter (BD Biosciences, San Jose, CA, USA) equipped with BD FACSDiva™ software (BD Biosciences). The purity of the sorted population (above 98%) was confirmed in a FACS Celesta flow cytometer (BD Biosciences).

2.3. Library Construction and Sequencing

Isolated MHC II$^+$ cells gently pipetted and diluted to a concentration of 700 cells/μL were used for cell isolation on a 10× Genomics Chromium Controller instrument [19]. All steps, including PBL isolation, sorting, and Chromium™ Single Cell isolation were carried out in the same morning to avoid cell death. A total of 2500 cells per donor were loaded into the chips of the Chromium™ Single Cell 5′ Gel Beads Kit (10× Genomics) and subjected to the Chromium Controller instrument to generate single cell Gel Bead-In Emulsions (GEMs) following the manufacturer's instructions. Next, GEMs were subjected to library construction using the Chromium™ Single Cell 5′ Library Kit v1 (10× Genomics). As a first step, reverse transcription was performed, resulting in cDNA tagged with a cell-specific barcode and a unique molecular index (UMI) per transcript. Fragments were then size selected using SPRIselect magnetic beads (Beckman Coulter). Next, Illumina sequencing adapters were ligated to the size-selected fragments and cleaned up using SPRIselect magnetic beads (Beckman Coulter, Brea, CA, USA). Finally, sample indexes were incorporated and amplified, followed by a double-sided size selection using SPRIselect magnetic beads (Beckman Coulter). The quality of the final library was assessed using an Agilent 2100 Bioanalyzer (Agilent technologies, Amstelveen, The Netherlands). The samples were then sequenced using a NextSeq instrument (Illumina) with 150 PE chemistry.

2.4. Alignment and Initial Processing of the Sequencing Data

The Cell Ranger software (10× Genomics, v3.1) was used to process the sequenced libraries. A rainbow trout reference transcriptome was constructed from the RefSeq *Oncorhynchus mykiss* genome v1.0 using the "Cell Ranger mkref" tool. The sequences encoding the constant regions of rainbow trout Igs were added to the fasta file prior to the construction of the reference transcriptome. The complementary DNA reads from each donor were mapped against this reference using the "Cell Ranger count" tool. Through this system, filtered UMI expression matrices from each donor were generated. As a result, the raw expression data was obtained containing transcriptomes for single MHC II$^+$ cells from the three donor fish.

In accordance with published pipelines and quality control standards, abnormal cells in all datasets were uniformly filtered out based on their gene expression distribution using the Seurat package (version 3.1) [20,21]. Cells with at least 200 detected genes, and only those genes that appeared at least in three cells were included in the initial matrix from each fish. A cell was considered to be abnormal if any of the following criteria were met: (i) detected gene number >2500; (ii) detected count number >15,000 and (iii) >25% of reads mapping for mitochondrial genes or no reads mapping for mitochondrial genes.

2.5. Data Integration and Cell Clustering

The SCTransform method from the Seurat software was applied in order to normalize the three filtered single-cell datasets from different fish. The percentages of mitochondrial and ribosomal proteins previously calculated for each cell were included as variables to regress the data. Then, filtered and normalized datasets were integrated using the PrepSCTIntegration tool to avoid a batch effect, enabling a systematic comparison between the three fish. The merged data was subjected to dimensionality reduction using principal component analysis (PCA) followed by uniform manifold approximation and projection (UMAP) using the first 30 principal components (PCs). This setup was also used to define nearest neighbors among cells with the KNN method using the FindNeighbors function. To group the cells in different subsets according to expression levels, the FindCluster tool was applied using the Louvain algorithm with the resolution set as 0.5, which allowed the correct definition of clearly separated clusters.

2.6. Marker Identification and Functional Analysis

The identification of genes showing differential expression associated to a specific cluster was performed using the FindAllMarkers tool from the Seurat software, considering a significant association for those genes showing an adjusted $p < 0.001$ and $\log FC \geq 0.25$. The information relative to the gene description contained in the *O. mykiss* genome v1.0 was taken into account for gene name association. In order to obtain an actualized functional annotation, the nucleotide sequences from the genes identified as markers were compared with proteins from a set of model species (*Homo sapiens*, *Mus musculus*, *Danio rerio*, *Macata mulata*, *Drosophila melanogaster*, and *Xenopus tropicalis*) using the Blastx software applying as threshold a minimum E value of 10^{-5}. The blast results were subjected to the Blast2GO software for GO term mapping. Sequences were also compared against domain databases using the InterProScan tool implemented in Blast2GO. GO term annotations were inferred for rainbow trout transcripts. Single enrichment analysis was performed by comparing the functions associated to genes from each cluster taking into account differences with an adjusted $p < 0.05$.

3. Results

3.1. Single-Cell Transcriptomic Analysis of Rainbow Trout MHC II⁺ Lymphoid Cells from Peripheral Blood

To acquire a transcriptomic profile of rainbow trout peripheral blood B cells at single-cell resolution, leucocyte isolation following sorting of lymphoid (small size and low complexity) MHC II^+ cells was carried out using blood obtained from three independent unstimulated fish. Cell viability was checked after sorting and confirmed to be higher than 90% in all samples. Single cell cDNA libraries were sequenced using Illumina HiSeq 150PE obtaining a total number of 73,934,335, 83,626,235 and 76,465,953 raw reads per fish. After quality control and mapping using the Cell Ranger software, transcriptomes of 1078, 2178, and 1488 single cells from each fish were acquired, detecting a median of 762, 694, and 1027 genes per cell, respectively.

Single cell sequencing data from each fish was then analyzed to determine the distribution of genes, read counts, percentage of reads mapping mitochondrial genes as well as percentage of reads mapping ribosomal proteins (Figure 1a). The distribution of genes and reads was found to be similar in all fish, with approximately 2500 genes per cell or 15,000 counts per cell (Figure 1a). Some cells, mainly identified in Fish 3, showed a higher number of genes or reads. These cells were considered abnormal cells or potential doublets and were filtered out for successive analysis. Regarding the percentage of mitochondrial reads, a wide distribution was observed in all fish reaching values of ~50% (Figure 1a). Commonly, a high percentage of mitochondrial reads is associated with abnormal cells or cells that have been damaged during isolation. For this reason, cells showing a percentage of mitochondrial reads higher than 25% were considered abnormal and excluded from subsequent analysis. Similarly, a reduced number of cells with no mitochondrial reads were filtered out at this step. After filtering a total of 843, 1814, and 1327 cells were retained for successive analysis.

Once the data was normalized and integrated, the cells from each of the three fish appeared to be equally distributed along the cell projection (Figure 1b). After cell clustering, a global view was generated to illustrate the potential subpopulations of rainbow trout peripheral blood MHC II^+ B cells. It should be noted that 99.28% of the cells analyzed transcribed some type of B cell marker, either genes encoding Ig heavy or light chains or isoforms of CD79, previously identified as a B cell specific marker in Atlantic salmon (*Salmo salar*) [22], confirming that almost all MHC II^+ lymphoid cells in peripheral blood correspond to B cells. The UMAP reduction generates a clear cell clustering highlighting 10 distinct cell populations based on their gene expression profiles (Figure 1c). All the clusters identified were shared by the three fish analyzed (Figure 1b). According to the global expression analysis, the clusters located at the central region of the cell projection (Clusters 0, 1, and 6) seem to contain cells showing a lower number of reads (Figure 1d) as well as a lower number of expressed genes per cell (Figure 1e). In contrast, the clusters located in the periphery of the cell projection show cells with higher number of reads and genes expressed, especially noticeable for clusters 2, 4, 5, and 7 (Figure 1d,e).

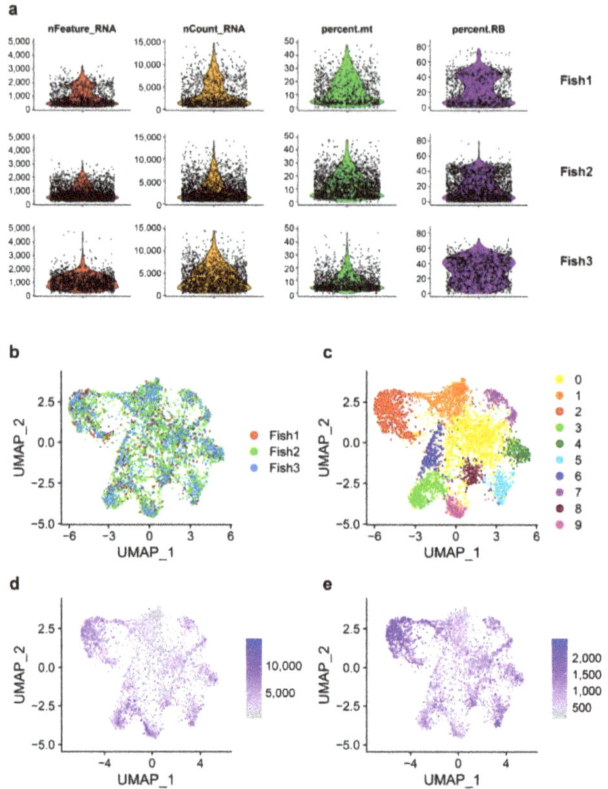

Figure 1. Single cell sequencing quality check and cell projection analysis. (**a**) Quality check of single cell sequencing results. The figure shows the total number of genes (nFeature_RNA), the total number of reads (nCount_RNA), the percentage of reads mapping mitochondrial genes (percent.mt) and the percentage of reads mapping ribosomal proteins (percent.RB) per cell in each individual fish. UMAP visualization of cells integrating data from the three individual fish (**b**); defining clusters of cells with similar transcriptional identities (**c**); showing the total number of reads identified for each cell (**d**) or the total number of genes identified for each cell (**e**).

3.2. Characterization of Marker Transcripts Associated with Identified Cell Clusters

Following the identification of cell clusters of B cells according to their transcriptional profile, a new analysis was performed to identify those genes that were most specific to a given cluster versus the rest. These specific marker transcripts identified for each cluster were retained when the adjusted p value was lower than 0.001 (Table S1). These transcripts that showed significant differences in transcription levels among different B cell subsets, included protein-coding genes as well as genes encoding ribosomal proteins, long non-coding RNAs (lncRNAs), and also mitochondrial genes (Figure 2a,b). Specifically, the gene encoding the 60S ribosomal protein L18a (LOC110516869) was the gene that most significantly represented cluster 0 (Figure 2c). Genes encoding the heat shock protein 70b (hsp70b), the ferritin heavy subunit (frih), an early growth response protein 1 (LOC110488587), and the heat shock protein HSP 90-alpha-like protein (LOC110529844) were the genes that most significantly differentiated clusters 1, 2, 7, and 8 respectively (Figure 2c). Interestingly, clusters 3, 4, 5, and 9 had one lncRNA as their most significant marker transcript (Figure 2c). Finally, cells included in cluster 6 showed a remarkable expression of the mitochondrial 12S ribosomal gene (Figure 2c).

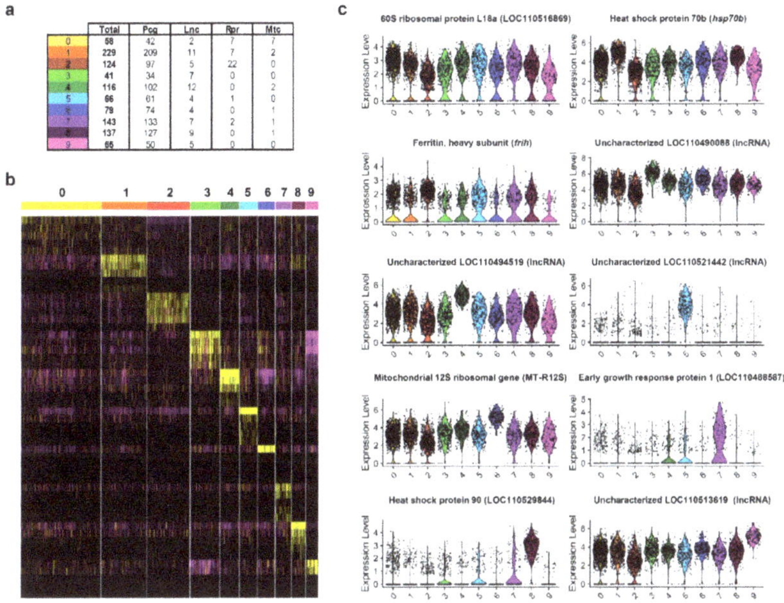

Figure 2. Identification of transcripts specific for each rainbow trout B cell subset. (**a**) Table including the number of transcripts identified for each cell cluster showing significant differences in expression levels (adjusted to $p < 0.001$) differentiating among protein coding genes (Pcg), long non-coding RNAs (Lnc), genes coding for ribosomal proteins (Rpr) and mitochondrial genes (Mtc). (**b**) Heatmap showing expression levels from the top five most significant subset-specific transcripts for each B cell subset (**c**) Violin plots showing normalized expression levels in all B cell subsets for the most significant marker transcript in each B cell subset.

3.2.1. Description of Molecular Markers Associated with Different B Cell Subsets

Marker transcripts for each B cell subset as well as the potential functions associated to each of these genes were further analyzed using the GO term single enrichment analysis. Several of these genes for which significant differences in expression levels were identified among the different clusters included genes encoding important enzymes, genes related to immune functions, as well as several transcription factors involved in important physiological processes (Table S1). Interesting functionalities were also associated with each B cell subset through this analysis (Table S2). The most relevant findings are summarized below.

Cluster 0

Cluster 0 is the largest cluster in number of cells, with 1098 cells in total (taking into account all three fish), and it is located at the center of the cell projection (Figure 3a). According to the transcription of the constant regions of Ig genes around 50% of cells within this cluster are IgM$^+$IgD$^+$ B cells (Figure 3b). A significant enrichment of GO terms "response to chemokines" (GO:1990868) or "cellular response to increased oxygen levels" (GO:0036295) was observed in this cluster within the biological process category (Figure 3c). A gene encoding an interferon-induced protein with tetratricopeptide repeats 5-like (LOC110491392) or a transmembrane protein 42-like (LOC110516550) were found within the genes showing the highest differences in expression levels for this cluster (Table S1). Some interesting immune genes were also found transcribed at significantly higher levels in this cluster. These included genes coding for the constant region of the immunoglobulin mu heavy chain and a C-X-C chemokine receptor type 4-like (LOC110520024 and LOC110501543), which also showed higher expression in Cluster 8 (Figure 4).

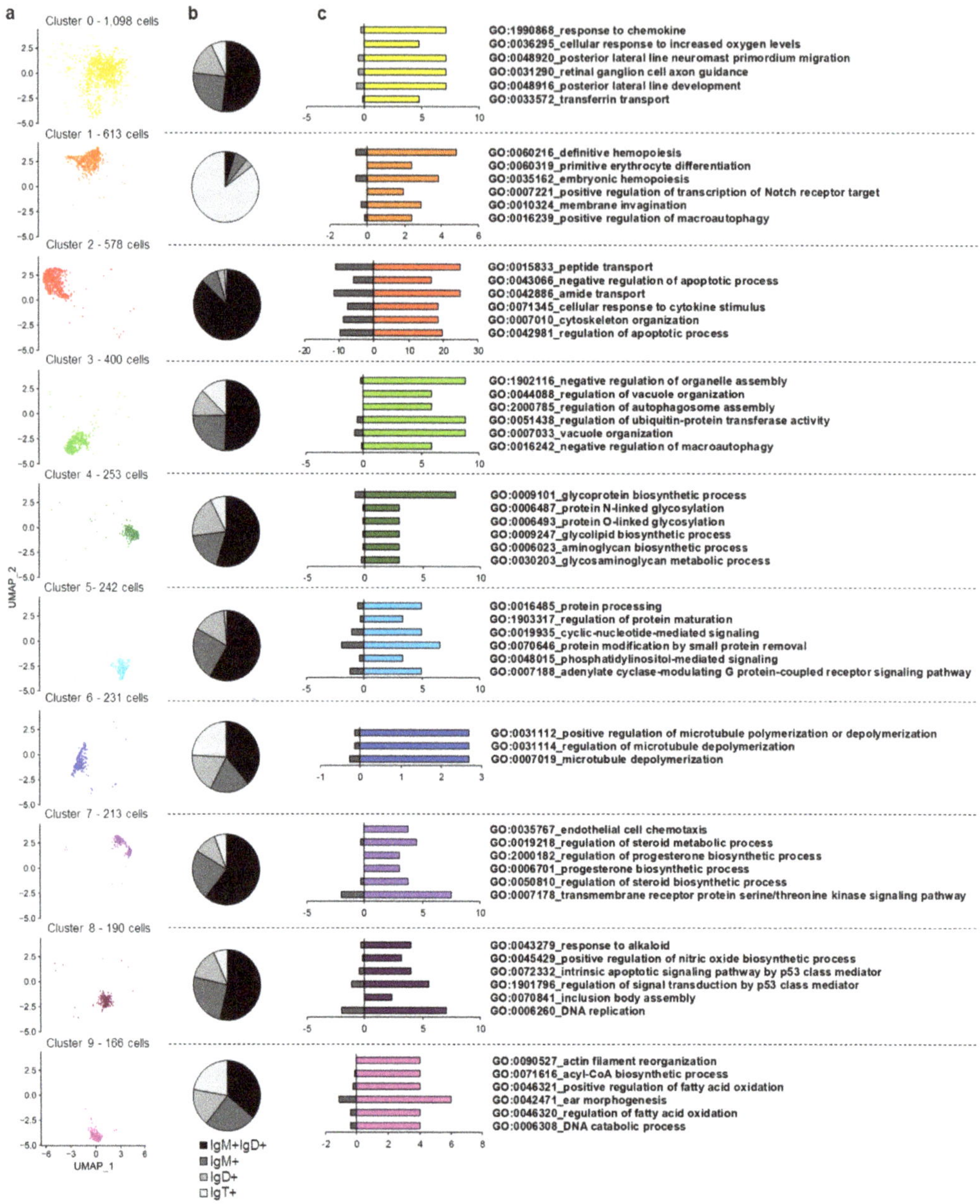

Figure 3. Cluster definition including immunoglobulin profile of cells and functional analysis of marker transcripts. (**a**) UMAP visualization of cells in each cluster. (**b**) Percentage of cells within each cluster assigned to common heavy chain Ig subsets (IgM^+IgD^+, IgM^+, IgD^+ and IgT^+) according to their transcriptional analysis. (**c**) Single enrichment analysis along the different cell clusters. Graphs show the percentage of genes within the most significant GO terms at level six within the biological process category for each specific cluster (colored bars) versus the percentage of genes from this category found in the rest of clusters (grey bars).

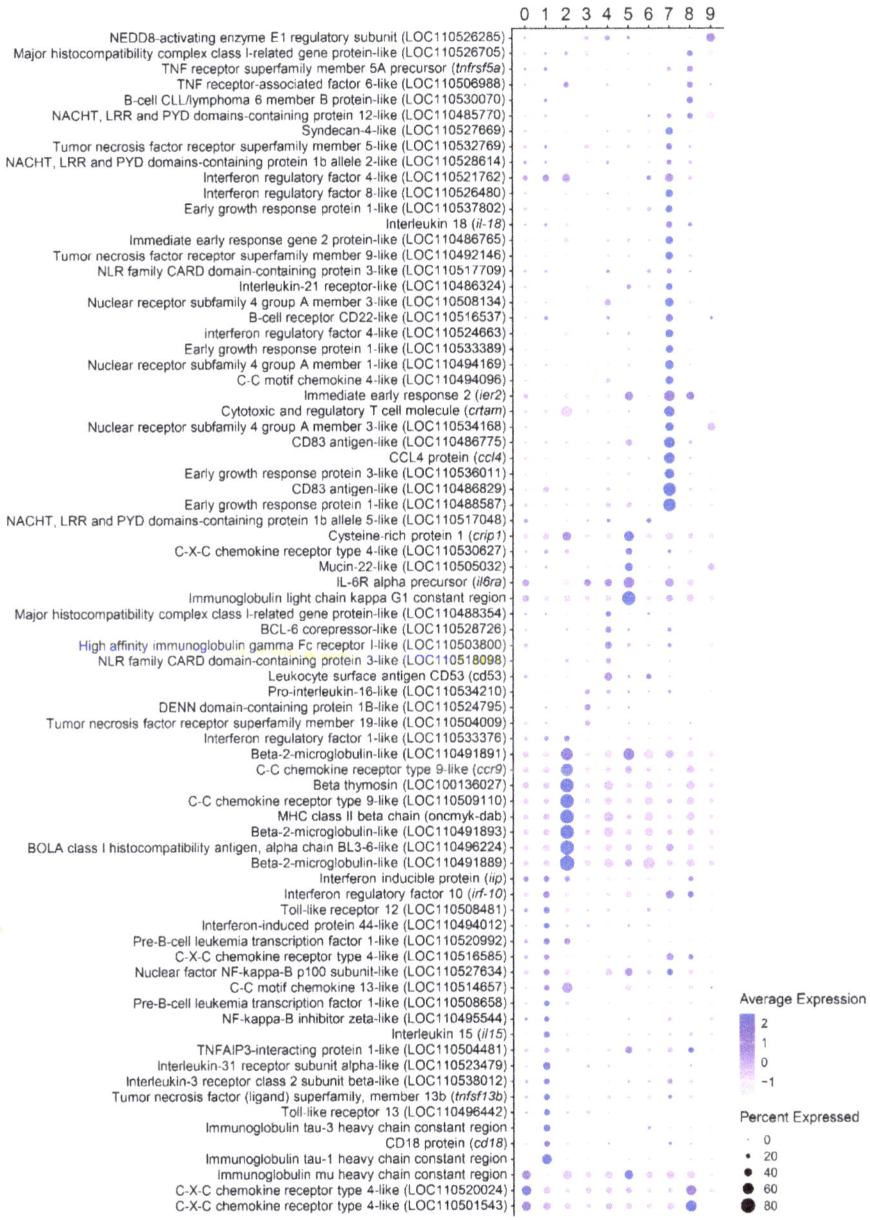

Figure 4. Dot plot analysis showing the marker transcripts identified for each cell cluster related with immune function. For each gene, the average normalized expression (dot color) together with the percentage of cells expressing the gene (dot size) are shown for each cluster.

Cluster 1

Cluster 1 in formed by 613 cells and is located in the upper region of cell projection (Figure 3a). This B cell subset is mainly represented by IgT^{+} B cells (Figure 3b). Intriguingly, this cluster is characterized by a group of markers related with "definitive hemopoiesis" (GO:0060216), "primitive erythrocyte differentiation" (GO:0060319) and "positive regulation of transcription of Notch receptor target" (GO:0007221) (Figure 3c). Regarding gene expression, this cluster is mainly represented by cells expressing immunoglobulin tau-1 and tau-3 heavy chains. Several genes coding for homologues of heat shock protein 70 also appeared among the most significant markers associated with this cluster (*hsp70b*, LOC110486254, LOC110485160, and LOC110486256) (Supplementary Table S1). Other immune genes transcribed at significantly higher levels by several cells within this cluster include genes coding for CD18 (*cd18*), interleukin 15 (*il15*), interleukin 31 receptor subunit alpha-like (LOC110523479), interleukin 3 receptor class 2 subunit beta-like (LOC110538012), tumor necrosis factor superfamily member 13b (*tnfsf13b*), as well as a gene encoding a CXC chemokine receptor type 4-like (LOC110516585) (Figure 4). Some receptors related to innate immune responses were also highly transcribed in cells from this cluster, for example, toll-like receptor 12 (LOC110508481) or 13 (LOC110496442) (Table S1). Among genes that encode transcription factors, these cells are characterized by the transcription of two genes encoding GATA-binding factors 2-like proteins (LOC110494514 and LOC110527950), involved in primitive hematopoiesis in mammals. As mentioned above, some genes related with regulation of the Notch pathway were also up-regulated in this cell cluster. These included two genes encoding pre-B-cell leukemia transcription factors 1-like proteins (LOC110508658 and LOC110520992), and a gene encoding a signal transducer and activator of transcription 1-alpha/beta-like protein (LOC110520020) (Figure 5). Analyzing the GO terms within the molecular function category associated with this cluster, we found "signaling receptor binding" (GO:0005102) and "calcium ion binding" (GO:0005509) (Table S2). In correlation, a representative group of genes highly expressed in cells from this cluster are related with calcium regulation, for example genes encoding ictacalcin (*s10i*), an ictacalcin-like protein (LOC110520890), a C-type lectin domain containing 14A (*clec14a*), and a E3 ubiquitin-protein ligase CBL-like protein (LOC110533571) (Table S1).

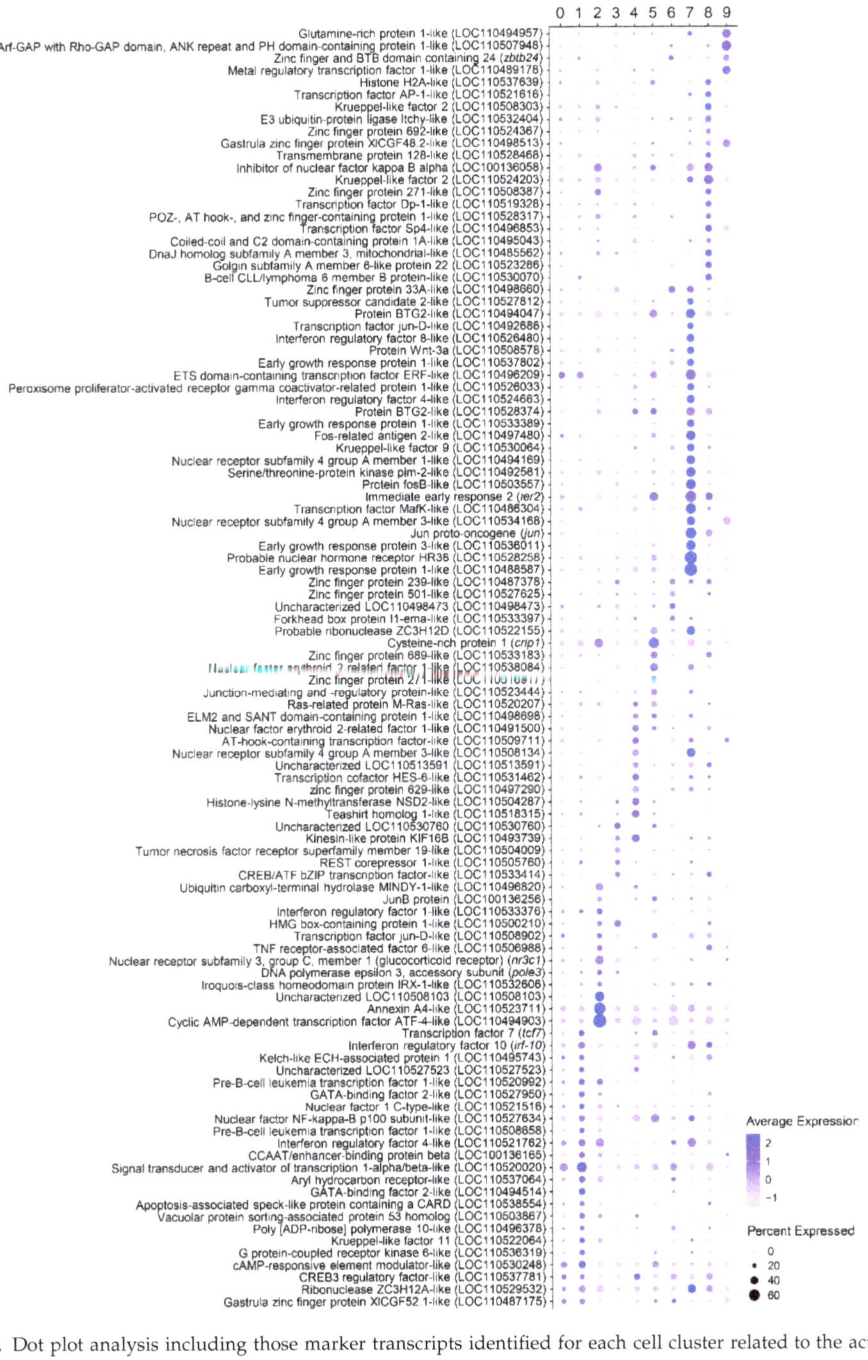

Figure 5. Dot plot analysis including those marker transcripts identified for each cell cluster related to the activity of transcription factors. For each gene, the average normalized expression (dot color) together with the percentage of cells expressing the gene (dot size) are shown for each cluster.

Cluster 2

Cluster 2, located in the left part of the cell projection, is formed by a total of 578 cells (Figure 3a). The cluster is mostly composed of $IgM^{+}IgD^{+}$ B cells, which constitute approximately 87% of these cells according to the levels of transcription of Ig constant regions (Figure 3b). Cluster 2 seems to be the most transcriptionally active cell cluster according to number of reads and genes transcribed per cell (Figure 1e). Genes identified as marker genes specific for this cluster are related to several functions including "peptide transport" (GO:0015833), "negative regulation of apoptotic process" (GO:0043066), "cellular response to cytokine stimulus" (GO:0071345), or "positive regulation of protein binding" (GO:0032092) (Figure 3c, Table S2), among others functionalities (Table S2). Among the genes most significantly expressed by this cluster we found some interesting immune genes, such as two genes encoding CC chemokine receptor type 9 (*ccr9* and LOC110509110), different components of the MHC II complex, including the MHC class II beta chain (*oncmyk-dab*) and the BOLA class I histocompatibility antigen, an alpha chain BL3-6-like (LOC110496224), and several homologues of beta-2-microglobulins (LOC110491889, LOC110491891 and LOC110491893) (Figure 4). In addition, among genes associated with the activity of transcription factors, we found a strong transcription of the factor BTF3 (LOC110523362), annexin A4-like (LOC110523711) and the cyclic AMP-dependent transcription factor ATF-4 (LOC110494903) (Figure 5). Finally, beta thymosin (LOC100136027), a gene associated with plasma cell differentiation, was also identified among the most significant marker transcripts for this cluster (Table S1).

Cluster 3

A total of 400 cells were identified in cluster 3, located at the bottom left region of the cell projection (Figure 3a). According to the levels of transcription of Ig constant regions, this cluster included $IgM^{+}IgD^{+}$ cells, but also IgT^{+}, IgD^{+} or IgM^{+} cells (Figure 3b). The numbers of marker transcripts identified for this cluster were lower than for other clusters but were enriched in functions such as "negative regulation of organelle assembly" (GO:1902116), "regulation of ubiquitin-protein transferase activity" (GO:0051438) and "regulation of gene expression, epigenetic" (GO:0040029) (Figure 3c, Table S2). A gene coding for the PTEN induced putative kinase 1 (*pink1*) was transcribed at significantly higher levels in cells from this cluster, despite the fact that this gene was already highly transcribed in all the cells analyzed. A gene encoding a lysosomal acid phosphatase-like protein (LOC110506315) was also preferentially transcribed in this cluster (Table S1). Concerning immune genes, some cells from this cluster highly expressed a gene coding for a tumor necrosis factor receptor superfamily member 19-like protein (LOC110504009) and a DENN domain-containing 1B-like protein (LOC110524795) (Figure 4). Additionally, a few genes related to transcription factor activity were identified with higher expression in this cluster, such as genes coding for a CREB/ATF bZIP transcription factor-like protein (LOC110533414) and a REST corepressor 1-like protein (LOC110505760) (Figure 5).

Cluster 4

A group of 253 cells localized at the right region of cell projection was designated as cluster 4 (Figure 3a). According to the levels of transcription of Ig constant regions, although the majority of cells from this cluster were $IgM^{+}IgD^{+}$ cells, IgT^{+}, IgD^{+}, or IgM^{+} cells were also included (Figure 3b). Interestingly, marker transcripts for this cluster indicate a significant enrichment in functions related to "glycoprotein biosynthetic process" (GO:0009101) as well as "protein N-" or "O-linked glycosylation" (GO:0006487 and GO:0006493, respectively) (Figure 3c). Among the genes related to these functions that were transcribed at significantly higher levels in this cluster, we found genes coding for several enzymes, such as for example beta-1,3-galactosyltransferase 6-like protein (LOC110527712), beta-1,4-glucuronyltransferase 1 (*b4gat1*), probable C-mannosyltransferase DPY19L3 (LOC110506384), and UDP-GlcNAc:betaGal beta-1,3-N-acetylglucosaminyltransferase 9-like protein (LOC110494808) (Table S1).

Cluster 5

A group of 242 cells localized at the bottom right section of the cell projection was annotated as cluster 5 (Figure 3a). The cells from this cluster were mostly IgM$^+$IgD$^+$ cells but cells transcribing only IgT or IgM were also identified, according to the levels of transcription of Ig constant regions (Figure 3b). The genes identified in this cluster are enriched in functions such as "protein processing" (GO:0016485) or "regulation of inflammatory response" (GO:0050727). (Figure 3c, Table S2). Interestingly, the cells included in this cluster exhibited higher transcription levels of the constant region of immunoglobulin light chain kappa G1. Other genes related with immune functions identified with higher transcription levels in cells from this cluster, included, for instance, genes coding for the IL-6R alpha precursor (*il6ra*), CXC chemokine receptor type 4-like (LOC110530627), mucin-22-like (LOC110505032), or cysteine-rich protein 1 (*crip1*) (Figure 4). Within transcription factors, different zinc finger proteins (585-LOC110515698, 271-LOC110516917, and 689-LOC110533183) were identified with higher expression than other clusters as well as a nuclear factor erythroid 2-related factor 1-like (LOC110538084) (Figure 5).

Cluster 6

A total of 231 cells were included in Cluster 6, situated at the left part of cell projection and near cluster 0 (Figure 3a). Cells from this cluster were identified as a mixture of IgM$^+$IgD$^+$ cells and cells exclusively expressing IgT, IgM or IgD (Figure 3b). Transcripts in this cluster were enriched in a lower number of functionalities than those of other clusters, identifying genes related to "melanocyte differentiation" (GO:0030318) as well as "positive regulation of microtubule polymerization or depolymerization" (GO:0031112) (Figure 3c, Table S2). Cluster 6 is mainly characterized by a high expression of the ribosomal 12S gene mitochondrial gene (Figure 2c). Interestingly, no nuclear genes are clearly associated with cells included in this cluster.

Cluster 7

Cluster 7 is represented by 213 cells located at the upper and right part of the cell projection (Figure 3a). This cluster is mostly composed of IgM$^+$IgD$^+$ B cells, which constitute approximately 63% of these cells according to the levels of transcription of Ig constant regions (Figure 3b). Functions related with "regulation of steroid metabolic process" (GO:0019218) or "transmembrane receptor protein serine/threonine kinase signaling pathway" (GO:0007178) were found within the most significant functions from the biological process category (Figure 3c). This cluster contains several genes associated to important immune system functions, such as "regulation of cell population proliferation" (GO:0042127), "positive regulation of cytokine production" (GO:0001819), "lymphocyte activation" (GO:0046649), and "cell migration" (GO:0016477) (Table S2). Also, significant enrichments in other GO terms from the molecular function category, such as "signaling receptor activator activity" (GO:0030546) or "DNA-binding transcription activator activity" (GO:0001216) were also detected (Table S2). The analysis of cluster 7 highlighted interesting markers related to immune functions. Thus, genes encoding two CD83 antigen-like proteins (LOC110486829 and LOC110486775), CCL4 protein (*ccl4*), three early growth response protein 1-like (LOC110488587, LOC110533389, and LOC110537802) and other early growth response protein 3-like (LOC110536011), the immediate early response 2 protein (*ier2*), as well as several nuclear receptor subfamily 4 genes (LOC110534168, LOC110508134, and LOC110494169) were identified within the most significant markers for this cluster (Figure 4). Additional immune related genes were expressed in several cells from this cluster, for instance, a C-C motif chemokine 4-like protein (LOC110494096), two genes encoding interferon regulatory factor 4-like proteins (LOC110524663 and LOC110521762), an interferon regulatory factor 8-like protein (LOC110526480), a tumor necrosis factor receptor superfamily member 9-like protein (LOC110492146) and a cytotoxic and regulatory T cell molecule (*crtam*) (Figure 4).

Cluster 8

A total of 190 cells were grouped in cluster 8, located at the center of the cell projection near cluster 0 (Figure 3a). According to the levels of transcription of Ig constant regions, although the majority of cells from this cluster were IgM$^+$IgD$^+$ cells, IgT$^+$, IgD$^+$, or IgM$^+$ cells were also present in this cluster (Figure 3b). Genes with higher transcription levels in cluster 8 than in other clusters appeared associated to "cellular response to stress" (GO:0033554) and "programmed cell death" (GO:0012501) (Table S2). Within the molecular function category, a significant enrichment in genes identified in the GO terms "heat shock protein binding" (GO:0031072) and "ubiquitin-like protein ligase binding" (GO:0044389) were also detected (Table S2). These enrichments were reflected in the genes that were identified within the most significant marker transcripts for this cluster, which included genes encoding several heat shock proteins as well as several ubiquitins. Among the heat shock proteins, a gene coding for a HSP90-alpha-like protein (LOC110529844) was the most significant marker showing high transcription levels in all cells included in this cluster (Figure 2c). In addition, a group of genes coding for homologues of the HSP70 kDa protein (LOC110485160, LOC110533353 and LOC110486254) also showed remarkable transcription levels in cells from this cluster. Within ubiquitins, genes encoding polyubiquitin-B (LOC110533627), polyubiquitin (LOC110533628), ubiquitin B (*ubb*), or E3 ubiquitin-protein ligase RNF220-like (LOC110508723) also showed significantly higher mRNA levels in cells from this cluster when compared to cells from other clusters (Table S1). Some genes related to immune functions were also identified as marker transcripts for cluster 8, such as genes coding for a B-cell CLL/lymphoma 6 member B-like protein (LOC110530070), a TNF receptor superfamily member 5A precursor (*tnfrsf5a*), and a TNF receptor-associated factor 6-like protein (LOC110506988) (Figure 4).

Cluster 9

A total of 166 cells were grouped in cluster 9, located at the bottom of the cell projection (Figure 3a). Cells in this cluster corresponded to either IgM$^+$IgD$^+$, IgT$^+$, IgM$^+$ or IgD$^+$ B cells according to their transcriptional Ig profile (Figure 3b). The marker transcripts identified in this subpopulation were mainly associated with functions related to "lipid storage" and "post-translational protein modification" (GO:0043687) (Table S2). This cluster was mainly characterized by the transcription of a gene encoding a glycine-rich RNA-binding -like protein (LOC110508373) (Table S1). Among immune genes, cells from this cluster exhibited higher transcription levels of a cullin-associated NEDD8-dissociated protein 1 (LOC110513018) (Figure 4). Some transcription factors such as the zinc finger and BTB domain containing 24 protein (*zbtb24*) or the metal regulatory transcription factor 1-like protein (LOC110489178) were also transcribed at significantly higher levels by cells in this cluster (Figure 5).

3.2.2. Long Non-Coding RNAs Associated to B Cell Subsets

Several lncRNAs showed differential transcription levels along the different B cell clusters identified among trout peripheral blood B cells (Figure 6a). Remarkably, some of these lncRNAs were found to be widely and strongly expressed all B cells (Figure 6b).

The lncRNAs LOC110490088, LOC110494519, LOC110521442, and LOC110513619 were identified as the most significant marker transcripts for clusters 3, 4, 5, and 9, respectively (Figure 2c). LOC110504686, LOC110520552, and LOC110495721 were also expressed at slightly higher levels in cells from clusters 0, 1 and 2, respectively (Figure 6a,b). In clusters 3 and 4, in addition to the LOC110490088 and LOC110494519 lncRNAs previously mentioned, several other lncRNAs showed strong expression levels. Thus, LOC110499522, LOC110512049, LOC110510226, and LOC110535625 showed higher transcription levels in cells from cluster 3 whereas LOC110494523, LOC110514167, LOC110514392, and LOC110538866 showed higher transcription levels in cluster 4 (Figure 6a,b).

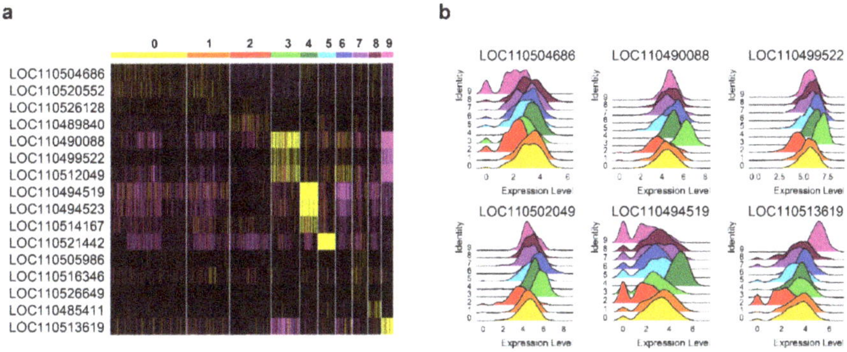

Figure 6. Long non-coding RNAs (lncRNAs) identified as markers for different cell clusters. (**a**) Heatmap showing the most significant lncRNAs with differential expression between clusters. (**b**) Ridge plot showing normalized expression in each cluster of lncRNAs identified as markers for specific subsets.

4. Discussion

Immune studies in mammalian models, especially those undertaken in human and mouse, have identified specific surface markers and/or transcription factors that have allowed the classification of B cells into different subsets and/or different stages of maturation/differentiation. Interestingly, remarkable differences have been observed between human and mouse markers throughout the differentiation process, often sharing just two or three markers between the two species within a specific B cell state [23,24]. In addition, several key markers that define B cell subsets in mammals such as CD19, CD20, or CD24 have no orthologues within the fish genomes, evidencing important differences in B cell biology between mammals and fish. Hence, the analysis of rainbow trout genes encoding homologues to different CD previously described along different states of human and mouse B cells evidenced no expression or residual expression of most of these markers in rainbow trout B cells, with the exception of the different genes encoding CXCR4-like molecules. Surprisingly, CXCR4 was the only marker gene identified in rainbow trout B cells, which can also be considered a marker gene for specific B cell subsets in humans. Other genes identified in our studies as marker genes, have not been identified as such in mammals. Instead, many of these genes correspond to proteins identified in a study recently performed by Peñaranda and collaborators in which a surface proteome of salmon (*Salmo salar*) IgM^{+} B cell was performed [22]. These observations evidence the great differences between markers of B cell subsets between fish and mammals, highlighting the need for high throughput studies that perform an unbiased search for these proteins. Of course, studies such as the one performed here, are not only useful to render a panel of molecules that could be used as markers for specific B cell populations, but also provide further insights on the functionality of teleost B cells which are further discussed.

4.1. Cluster 2 May Represent a Subpopulation of Mature B Cells with High Antigen Presenting Capacities

Cluster 2 was the subpopulation expressing the highest number of genes per cell. This cluster was clearly enriched in IgM^{+}IgD^{+} B cells (~86%). In addition, this cluster transcribed several components of the MHC II complex and associated microglobulin genes as the most significant marker genes. A significant enrichment in genes involved in peptide or amide transport, also seems to indicate that this B cell subpopulation is associated with a high antigen presenting capacity. Two genes encoding orthologues of mammalian CCR9 are also interesting markers for this subpopulation. In mammals, CCR9 has been shown to regulate naïve B cell migration [25] and localization of plasma cells in mucosal tissues [26]. In addition, high CCR9 expression has also been revealed in mammalian memory B cells [27]. To date, the function and ligands of fish CCR9 are still unknown.

Our results also identified a group of transcription factors among the most significant markers of this subpopulation. These include BTF3 as well as the cyclic AMP-dependent transcription factor ATF4. Studies in mammals suggest these transcription factors play different roles associated with the endoplasmic reticulum biology. Thus, BTF3 transcription factor in association with nascent polypeptide-associated complex subunit alpha (LOC110492373), also identified as a marker for this cell cluster, form the nascent polypeptide-associated complex (NAC) which prevents the inappropriate targeting of non-secretory polypeptides to the endoplasmic reticulum. ATF4, on the other hand, has been related with the activation of genes during endoplasmic reticulum stress, a process that regulates activation, B cell differentiation to plasma cell, and cytokine expression mediated by unfolded protein response (UPR) [28].

4.2. *Cluster 7 May Represent a Subpopulation of Activated B Cells with Innate Properties*

Teleost B cells retain several innate functions such as a strong phagocytic capacity, ability to produce pro-inflammatory cytokines or antimicrobial peptides in response to a non-specific stimulation [29–31]. However, whether all teleost B cells or only a specific cell subset has these capacities is currently unknown. In the current study, we found that Cluster 7 had among its marker genes, some that could be catalogued as innate genes, commonly found in DCs or macrophages. Hence, two genes encoding orthologues of CD83 were identified in this subpopulation as marker genes. Interestingly, CD83 has been previously associated to a population of small mononuclear blood cells from Atlantic salmon which also expressed high MHC-II levels, showed a high phagocytic capacity and the ability to differentiate into DC-like cells [32]. However, in Atlantic salmon, no B cell markers were identified in this subpopulation, whereas in our study, the transcription of different immunoglobulins and CD79 unequivocally indicates that these cells correspond to a B cell subset. Other important markers that link this B cell subpopulation with innate responses are the genes annotated in the rainbow trout genome as CCL4 and *tnfrsf9* (CD137), both known to play important roles in monocyte activation. Interestingly, a significant increase in the levels of transcription of both of these genes had also been reported in rainbow trout IgM$^+$ B cells after stimulation with either T-dependent or T-independent antigens [33]. Rainbow trout CCL4 was originally identified as a highly inducible chemokine produced by differentiated rainbow trout macrophages after LPS induction [34]. However, it is important to mention that after this first description, posterior studies focused at establishing phylogenic relationships between mammalian and fish CC chemokines renamed this gene as CK5B [35] and established a close relation with mammalian CCL5 [35,36]. In mammals, CCL5 is known to attract T cells, specifically CD4$^+$ Th1 cells [37]. In rainbow trout, a previous study established that upon viral infection, CpG or poly I:C stimulation, IgM$^+$ B cells increased CK5B transcription. This was interpreted at the time as a mechanism of B cells to attract and obtain co-stimulatory signals from T helper cells [38]. Therefore, it might be possible that this is the case for this B cell subset identified.

Cells from cluster 7 also showed a differential expression of several transcription factors such as *irf4* and *irf8* associated in mammals with different functions throughout the B cell differentiation process. For instance, the combined loss of IRF4 and IRF8 in mammals resulted in a blockage of germline κ and λ gene transcription as well as DNA recombination at the pre-B cell stage [39]. In addition, IRF4 is a common marker for activated B cells known to promote Ig CSR as well as plasma cell differentiation by positive regulation of the *Aicda* and *Blimp1* genes, respectively [40]. Other transcription factors identified as markers for this cluster also point these cells having experienced some kind of activation by either an antigen or another type of stimuli. For instance, an increased transcription of different Egr-1 homologues had also been observed in mature mammalian B cells which undergo proliferation as a consequence of B cell receptor (BCR) cross-linking [41]. The participation of this transcription factor in the differentiation program of B cells into plasma cells has also been evidenced [42]. Egr-3, on the other hand, has been identified as an activator of suppressor of cytokine signaling-1 (SOCS1) and SOCS3, inhibitors of STAT1 and STAT3, which regulate B cell function in adaptive immune responses and homeostasis by promoting antigen receptor signaling [43]. Fra-2, also identified as a key transcription factor in this cluster, is involved in the proliferation and differentiation of B cells acting as an upstream regulator of Foxo1 and Irf4 expression [44]. Thus, all the marker genes identified in the complex regulatory network identified among the markers of cluster 7 seem to indicate that these CD83$^+$ cells characterized by the expression of genes commonly related with macrophages experienced activation.

4.3. Cells in Cluster 1 Correspond to IgT+ B Cells

Cluster 1 corresponds to approximately 85% of cells with a confirmed IgT transcriptional profile. As previously established, these cells constitute an independent B cell linage with no IgM or IgD [29]. Interestingly, the functionalities and marker genes that were associated with this cluster seemed to indicate an initial stage of maturation. For instance, a significantly higher expression of two genes encoding GATA-binding factors 2-like were found in this cell subset when compared to others, and these factors have been commonly related with primitive hematopoiesis in mammals [45]. In mammals, this transcription factor is expressed in hematopoietic progenitor cells (HPCs) and is considered a key component of the transcriptional program required for early hematopoietic development [45,46]. Similarly, the identification as markers of two genes encoding the pre-B cell leukemia transcription factor 1 (*xbp1*), also support that these cells are at an initial stage of maturation, as this is one of the earliest-acting transcription factors that regulates de novo B-lineage lymphopoiesis, with critical functions during the intermediate stage between hematopoietic stem cell development and B cell commitment [47]. As IgT$^+$ cells are known to preferentially colonize mucosal surfaces [48], it might be possible that these cells are still in an immature stage in peripheral blood thus completing their maturation in mucosal tissues.

Cluster 1 also included several HSPs among its most significant markers. In mammals, the interaction of HSPs with antigen presenting cells (APCs) such as macrophages or DCs, through receptors involved in the innate immune response, such as CD40 or Toll-like receptors 2 and 4 (TLR2/4), leads to several non-antigen specific reactions that promote the stimulation of the innate immune system [49,50]. Thus, the strong overrepresentation of these genes in this cluster may indicate that HSPs play a specific role in the maturation and/or differentiation of IgT$^+$ B cells. Remarkably, several innate receptors such as TLR12 and TLR13 were also identified as markers for this subpopulation, being these the only TLRs identified with significant differential expression between cell clusters. Although no homologues have been identified for these genes in humans, both present homologues in mice. Murine TLR12 and TLR13 are known to activate innate immune responses via MYD88 and TRAF6, leading to NF-kappa-B activation, cytokine secretion and an inflammatory response [51,52]. Thus, the expression of these TLRs in IgT$^+$ B cells commonly related with mucosal responses [48], suggests that these cells are traveling from central immune organs to mucosal peripheral tissues where they might be activated through the action of these TLRs.

Several cells within this cluster showed a significant higher expression of two interleukin receptors, *il3r2* and *il31r*. Additionally, some cells from this cluster produced different cytokines that might have effects on other immune cells such as *il15* or *tnfsf13b* (BAFF). Studies in mice have determined that IL-15 together with recombinant CD40 ligand is able to potently induce polyclonal IgM, IgG1, and IgA secretion [53] and together with CpG oligonucleotides to induce the proliferation of class switched human memory B cells [54]. In the case of BAFF, this cytokine is a key factor for B cell survival and maturation, illustrated by the absence of mature B cells in BAFF-deficient mice [55]. Previous work performed in rainbow trout showed the regulation of splenic B cell functionality by BAFF [7] and differential effects on peritoneal IgM$^+$ B cell subpopulations [6]. Furthermore, these studies revealed the production of BAFF by a subset of splenic IgM$^+$ B cells pointing to an auto-regulatory mechanism of B cell survival in this organ [6]. In this case, the expression of these molecules in peripheral blood IgT^+ B cells may indicate the capacity of these cells to regulate other B cell subsets or themselves in an autocrine fashion.

4.4. CXCR4-Like 4 Genes as Key Markers for Fish B Cells

Mammalian CXCR4 has been suggested as a marker for an initial common lymphoid progenitor in mice [56] and for differentiated plasmablasts or plasma cells both in mice and humans [57,58]. Several studies have analyzed the evolution and function of *cxcr4* genes in teleost fish showing the presence of either two paralog genes designated as *cxcr4a and cxcr4b* [59–61] or even four gene copies in cyprinids [62]. Our results identified four different genes annotated as rainbow trout CXCR4-like, which all seemed expressed in B cells. Interestingly, all of them were identified as potential markers for different B cell subpopulations. On one hand, two genes encoding CXCR4 (LOC110520024 and LOC110501543), which seem to be homologues to the molecule designated as CXCR4b in previous studies, were identified as shared markers for cluster 0 and cluster 8. On the other hand, the other two genes encoding homologues of the molecule previously designated as CXCR4a (LOC110516585 and LOC110530627) showed differential expression in cluster 1 and cluster 5, respectively. A previous study in fish pointed to the duplication of CXCR4 genes in teleosts evolving to a gene subfunctionalization that may stabilize the immune system controlling teleost hematopoietic stem/progenitor cell (HSPC) homeostasis [61]. In addition, these authors suggested that HSPCs expressing CXCR4a preferentially bind LPS, whereas those expressing CXCR4b preferentially bind SDF-1 reflecting different pathways of immune cell differentiation [61]. Our analysis focused in B cells seems to indicate that CXCR4 relates to an advanced state of maturation, but with different rainbow trout CXCR4 molecules carrying out a differential regulation of specific B cell subpopulations.

4.5. Long Non-Coding RNAs Seem to Regulate Different B Cell Transitional Stages in Fish

LncRNAs carry out an epigenetic regulation through chromatin modification of differentiating enhancer-associated lncRNAs (eRNAs), acting in cis, and promoter-associated lncRNAs (pRNAs), acting in trans [63]. The potential regulatory function of lncRNAs during B cell differentiation and development has recently emerged [64]. In humans, a set of over 3000 lncRNAs were detected analyzing bone marrow and thymic progenitors spanning the earliest stages of B and T lymphoid specification [65]. The expression patterns observed along these lncRNAs identified were shown to be highly stage-specific and more lineage-specific than protein-coding genes [65]. More recently, RNAseq or microarray studies using isolated B cell subpopulations highlighted equivalent expression patterns between coding genes and lncRNAs throughout B cell differentiation [66–68]. Similarly, in mouse, a repertoire of 4516 lncRNAs were identified with expression along 11 mouse B cell subpopulations, again suggesting that lncRNAs displayed a greater cell-type restriction than coding genes, showing very restricted spatiotemporal roles during B cell development [67]. Interestingly, the expression of lncRNA loci is controlled by transcription factors specific of B cell lineage, such as for example PAX5, known to regulate several lncRNAs in both pro-B and mature B cells [67].

Our results identified 60 transcripts, annotated in the *O. mykiss* genome as lncRNAs, which showed significant differences in expression levels between cell clusters. Thus, four clusters had a lncRNA as its most significant marker, in agreement with the cell-type restriction previously described in mammals.

Interestingly seven lncRNAs were identified as markers for cluster 3 in correlation with a significant enrichment in genes related with "regulation of gene expression, epigenetic". In addition, several functions showing significant enrichment in this cluster were related to the modification or assembly of different cellular organelles. For instance, the *pink1* gene was identified among the most significant markers encoding a protein in this cluster. In mammals, PINK1 modulates mitochondrial trafficking and is involved in the clearance of damaged mitochondria via selective mitophagy [69]. Recent studies in mammals have evidenced important links between mitophagy and immunity [70–72]. Mitochondrial stress in response to pathogens and danger signals induces mitophagy through the PINK1/Parkin pathway [73]. Interestingly, an important role has been attributed to PINK1 in integrating extracellular signals with the metabolic state, thus determining T cell fate [74]. Although there are no evidence in mammals that connect PINK1 with B cell fate, our results suggest a potential connection in fish defining a specific subpopulation of peripheral blood B cells from rainbow trout.

Cluster 4 also has a high number of lncRNAs among their most significant markers. In this cluster, several enzymes related to different mechanisms of glycosylation were also identified as markers. Both N- and O- glycosylations have emerged as important translational modifications of key immune proteins including Igs [75–77]. These modifications have also been observed in the variable regions of these Igs, affecting both antibody specificity and stability [78,79]. In human blood B cells, IgG1 Fc-glycosylation with all-trans retinoic acid has been described upon stimulation [80]. Hence, our results might indicate that B cells in Cluster 4 have already encountered some kind of stimuli that has triggered a glycosylation process.

In what concerns lncRNAs, it should be mentioned that cells from cluster 5, which seemed associated with CXCR4a expression, were also highly represented by the expression of the lncRNA LOC110521442. Finally, cluster 9 is also highly represented by the lncRNA LOC110513619. Interestingly, a significant enrichment in functions related with fatty acid oxidation was observed in this cluster reflecting a specific metabolic state of this B cell subpopulation. Recent results highlighted fatty acid oxidation as a predominant energy source for *ex vivo bona fide* GC B cell growth [81]. Although no GCs have been described in fish, our results suggest that fatty acid metabolism may represent an important function in specific B cell subpopulations, opening a new field to further explore B cell immunometabolism in fish.

Altogether, our results suggest for the first time that fish lncRNAs also carry out key roles in B cell functionality, showing a very restricted spatiotemporal expression. Although the specific functions carried out by these regulators remain to be further analyzed, the correlation with B cell subpopulations associated to important functions suggests a role of lncRNAs in maintaining these different B cell states.

5. Conclusions

In summary, we have taken advantage of the recently developed 10× Genomics single cell sequencing techniques to analyze the transcriptome of 3984 MHC II$^+$ lymphoid cells from rainbow trout peripheral blood. These cells, obtained from three independent fish, mostly corresponded to B cells as established by Ig transcription levels or expression of the B cell marker CD79. This methodology allowed us to establish 10 different B cell clusters, characterized by a specific gene expression profile. These fish were obtained from a fish farm as adults, thus throughout their development they might have encountered different stimuli, microorganisms or even pathogenic agents. For this reason, the B cell population analyzed will surely include cells in different stages of activation. However, whether the different cell clusters identified correspond to cells in different stages of activation/differentiation or in fact are diverse naïve B cell subpopulations that carry out different immune functions should be further explored. In any case, our studies highlight the diversity of the B cell response in teleost and provide further insight on key genes that mediate important processes throughout the B cell maturation/differentiation process in these species, also providing us with a panel of potential markers that might be used in the future to differentiate and further explore the functionality of these B cell subsets.

Supplementary Materials: The following are available online at https://www.mdpi.com/article/10.3390/biology10060511/s1, Table S1. Results from marker identification analysis. Rainbow trout genes showing significant differential expression (adjusted $p < 0.001$) between cell clusters (identified with different colors). Table S2. Results from GO term single enrichment analysis. GO terms significantly enriched (adjusted $p < 0.05$) within markers from each cell cluster for each Gene Ontology category (Biological process, Molecular function and Cellular component). Different clusters are represented by different colors.

Author Contributions: P.P.: Investigation, Formal analysis, Writing—original draft preparation. E.M.: Investigation, Resources. C.T.: Conceptualization, Supervision, Funding acquisition, Writing—reviewing and editing. All authors have read and agreed to the published version of the manuscript.

Funding: This work was supported by the European Research Council (ERC Consolidator Grant 2016 725061 TEMUBLYM).

Institutional Review Board Statement: Procedures described comply with the Guidelines of the European Union Council (2010/63/EU) for the use of laboratory animals and were previously approved by the Ethics committee from the *Instituto Nacional de Investigación y Tecnología Agraria y Alimentaria* (INIA; Code CEEA PROEX002/17).

Informed Consent Statement: Not applicable.

Data Availability Statement: The data discussed in this publication have been deposited in NCBI's Gene Expression Omnibus [82] and are accessible through GEO Series accession number GSE158102 (https://www.ncbi.nlm.nih.gov/geo/query/acc.cgi?acc=GSE158102; Public since 10 May 2021).

Acknowledgments: Lucía González is greatly acknowledged for technical assistance. We also thank Patricia Díaz-Rosales for critically reviewing the manuscript.

Conflicts of Interest: The authors declare no conflict of interest.

References

1. Cooper, M.D.; Alder, M.N. The evolution of adaptive immune systems. *Cell* **2006**, *124*, 815–822. [CrossRef] [PubMed]
2. Press, C.M.; Evensen, Ø. The morphology of the immune system in teleost fishes. *Fish Shellfish Immunol.* **1999**, *9*, 309–318. [CrossRef]
3. Zwollo, P.; Cole, S.; Bromage, E.; Kaattari, S. B cell heterogeneity in the teleost kidney: Evidence for a maturation gradient from anterior to posterior kidney. *J. Immunol.* **2005**, *174*, 6608–6616. [CrossRef] [PubMed]
4. Hansen, J.D.; Landis, E.D.; Phillips, R.B. Discovery of a unique Ig heavy-chain isotype (IgT) in rainbow trout: Implications for a distinctive B cell developmental pathway in teleost fish. *Proc. Natl. Acad. Sci. USA* **2005**, *102*, 6919–6924. [CrossRef] [PubMed]
5. Parra, D.; Takizawa, F.; Sunyer, J.O. Evolution of B cell immunity. *Annu. Rev. Anim. Biosci.* **2013**, *1*, 65–97. [CrossRef]
6. Granja, A.G.; Tafalla, C. Different IgM^+ B cell subpopulations residing within the peritoneal cavity of vaccinated rainbow trout are differently regulated by BAFF. *Fish Shellfish Immunol.* **2019**, *85*, 9–17. [CrossRef]
7. Tafalla, C.; González, L.; Castro, R.; Granja, A.G. B cell-activating factor regulates different aspects of B cell functionality and is produced by a subset of splenic B cells in teleost fish. *Front. Immunol.* **2017**, *8*, 295. [CrossRef]
8. Edholm, E.S.; Bengten, E.; Stafford, J.L.; Sahoo, M.; Taylor, E.B.; Miller, N.W.; Wilson, M. Identification of two IgD^+ B cell populations in channel catfish, *Ictalurus punctatus*. *J. Immunol.* **2010**, *185*, 4082–4094. [CrossRef]
9. Castro, R.; Bromage, E.; Abos, B.; Pignatelli, J.; Gonzalez Granja, A.; Luque, A.; Tafalla, C. CCR7 is mainly expressed in teleost gills, where it defines an IgD^+IgM^- B lymphocyte subset. *J. Immunol.* **2014**, *192*, 1257–1266. [CrossRef]
10. Perdiguero, P.; Martin-Martin, A.; Benedicenti, O.; Diaz-Rosales, P.; Morel, E.; Munoz-Atienza, E.; Garcia-Flores, M.; Simon, R.; Soleto, I.; Cerutti, A.; et al. Teleost IgD^+IgM^- B cells mount clonally expanded and mildly mutated intestinal IgD responses in the absence of lymphoid follicles. *Cell Rep.* **2019**, *29*, 4223–4235. [CrossRef]
11. Zhang, Y.A.; Salinas, I.; Oriol Sunyer, J. Recent findings on the structure and function of teleost IgT. *Fish Shellfish Immunol.* **2011**, *31*, 627–634. [CrossRef]
12. Zhang, N.; Zhang, X.-J.; Chen, D.-D.; Sunyer, J.O.; Zhang, Y.-A. Molecular characterization and expression analysis of three subclasses of IgT in rainbow trout (*Oncorhynchus mykiss*). *Dev. Comp. Immunol.* **2017**, *70*, 94–105. [CrossRef] [PubMed]
13. Zwollo, P. Dissecting teleost B cell differentiation using transcription factors. *Dev. Comp. Immunol.* **2011**, *35*, 898–905. [CrossRef] [PubMed]
14. Zwollo, P.; Haines, A.; Rosato, P.; Gumulak-Smith, J. Molecular and cellular analysis of B-cell populations in the rainbow trout using Pax5 and immunoglobulin markers. *Dev. Comp. Immunol.* **2008**, *32*, 1482–1496. [CrossRef] [PubMed]
15. Guslund, N.C.; Solbakken, M.H.; Brieuc, M.S.; Jentoft, S.; Jakobsen, K.S.; Qiao, S.-W. Single-cell transcriptome profiling of immune cell repertoire of the Atlantic cod which naturally lacks the major histocompatibility class II system. *Front. Immunol.* **2020**, *11*, 2602. [CrossRef] [PubMed]
16. Niu, J.; Huang, Y.; Liu, X.; Zhang, Z.; Tang, J.; Wang, B.; Lu, Y.; Cai, J.; Jian, J. Single-cell RNA-seq reveals different subsets of non-specific cytotoxic cells in teleost. *Genomics* **2020**, *112*, 5170–5179. [CrossRef] [PubMed]
17. Abós, B.; Wang, T.; Castro, R.; Granja, A.G.; Leal, E.; Havixbeck, J.; Luque, A.; Barreda, D.R.; Secombes, C.J.; Tafalla, C. Distinct differentiation programs triggered by IL-6 and LPS in teleost IgM^+ B cells in the absence of germinal centers. *Sci. Rep.* **2016**, *6*, 1–16.

18. Granja, A.G.; Leal, E.; Pignatelli, J.; Castro, R.; Abos, B.; Kato, G.; Fischer, U.; Tafalla, C. Identification of teleost skin CD8α+ dendritic-like cells, representing a potential common ancestor for mammalian cross-presenting dendritic cells. *J. Immunol.* **2015**, *195*, 1825–1837. [CrossRef]
19. Zheng, G.X.; Terry, J.M.; Belgrader, P.; Ryvkin, P.; Bent, Z.W.; Wilson, R.; Ziraldo, S.B.; Wheeler, T.D.; McDermott, G.P.; Zhu, J.; et al. Massively parallel digital transcriptional profiling of single cells. *Nat. Commun.* **2017**, *8*, 14049. [CrossRef]
20. Satija, R.; Farrell, J.A.; Gennert, D.; Schier, A.F.; Regev, A. Spatial reconstruction of single-cell gene expression data. *Nat. Biotechnol.* **2015**, *33*, 495–502. [CrossRef]
21. Stuart, T.; Butler, A.; Hoffman, P.; Hafemeister, C.; Papalexi, E.; Mauck III, W.M.; Hao, Y.; Stoeckius, M.; Smibert, P.; Satija, R. Comprehensive integration of single-cell data. *Cell* **2019**, *177*, 1888–1902.e1821. [CrossRef]
22. Peñaranda, M.; Michelle, D.; Jensen, I.; Tollersrud, L.G.; Bruun, J.-A.; Jørgensen, J.B. Profiling the atlantic salmon IgM+ B cell surface proteome: Novel information on teleost fish B cell protein repertoire and identification of potential B cell markers. *Front. Immunol.* **2019**, *10*, 37. [CrossRef]
23. Mestas, J.; Hughes, C.C. Of mice and not men: Differences between mouse and human immunology. *J. Immunol.* **2004**, *172*, 2731–2738. [CrossRef]
24. Zhang, X.; Lan, Y.; Xu, J.; Quan, F.; Zhao, E.; Deng, C.; Luo, T.; Xu, L.; Liao, G.; Yan, M. CellMarker: A manually curated resource of cell markers in human and mouse. *Nucleic Acids Res.* **2019**, *47*, D721–D728. [CrossRef] [PubMed]
25. Wurbel, M.-A.; Malissen, M.; Guy-Grand, D.; Meffre, E.; Nussenzweig, M.C.; Richelme, M.; Carrier, A.; Malissen, B. Mice lacking the CCR9 CC-chemokine receptor show a mild impairment of early T-and B-cell development and a reduction in T-cell receptor γδ+ gut intraepithelial lymphocytes. *Blood* **2001**, *98*, 2626–2632. [CrossRef]
26. Pabst, O.; Ohl, L.; Wendland, M.; Wurbel, M.-A.; Kremmer, E.; Malissen, B.; Förster, R. Chemokine receptor CCR9 contributes to the localization of plasma cells to the small intestine. *J. Exp. Med.* **2004**, *199*, 411–416. [CrossRef]
27. Demberg, T.; Mohanram, V.; Venzon, D.; Robert-Guroff, M. Phenotypes and distribution of mucosal memory B-cell populations in the SIV/SHIV rhesus macaque model. *Clin. Immunol.* **2014**, *153*, 264–276. [CrossRef]
28. So, J.-S. Roles of endoplasmic reticulum stress in immune responses. *Mol. Cells* **2018**, *41*, 705.
29. Li, J.; Barreda, D.R.; Zhang, Y.-A.; Boshra, H.; Gelman, A.E.; LaPatra, S.; Tort, L.; Sunyer, J.O. B lymphocytes from early vertebrates have potent phagocytic and microbicidal abilities. *Nat. Immunol.* **2006**, *7*, 1116–1124. [CrossRef]
30. Sunyer, J.O. Evolutionary and functional relationships of B cells from fish and mammals: Insights into their novel roles in phagocytosis and presentation of particulate antigen. *Infect. Disord. Drug Targets* **2012**, *12*, 200–212. [CrossRef]
31. Sunyer, J.O. Fishing for mammalian paradigms in the teleost immune system. *Nat. Immunol.* **2013**, *14*, 320. [CrossRef]
32. Haugland, G.T.; Jordal, A.-E.O.; Wergeland, H.I. Characterization of small, mononuclear blood cells from salmon having high phagocytic capacity and ability to differentiate into dendritic like cells. *PLoS ONE* **2012**, *7*, e49260. [CrossRef]
33. Granja, A.G.; Perdiguero, P.; Martín-Martín, A.; Díaz-Rosales, P.; Soleto, I.; Tafalla, C. Rainbow trout IgM+ B cells preferentially respond to thymus-independent antigens but are activated by CD40L. *Front. Immunol.* **2019**, *10*, 2902. [CrossRef]
34. Mackenzie, S.; Liarte, C.; Iliev, D.; Planas, J.; Tort, L.; Goetz, F. Characterization of a highly inducible novel CC chemokine from differentiated rainbow trout (*Oncorhynchus mykiss*) macrophages. *Immunogenetics* **2004**, *56*, 611–615. [CrossRef]
35. Laing, K.J.; Secombes, C.J. Chemokines. *Dev. Comp. Immunol.* **2004**, *28*, 443–460. [CrossRef]
36. Peatman, E.; Liu, Z. Evolution of CC chemokines in teleost fish: A case study in gene duplication and implications for immune diversity. *Immunogenetics* **2007**, *59*, 613–623. [CrossRef]
37. Siveke, J.T.; Hamann, A. Cutting edge: T helper 1 and T helper 2 cells respond differentially to chemokines. *J. Immunol.* **1998**, *160*, 550–554. [PubMed]
38. Abós, B.; Castro, R.; Granja, A.G.; Havixbeck, J.J.; Barreda, D.R.; Tafalla, C. Early activation of teleost B cells in response to rhabdovirus infection. *J. Virol.* **2015**, *89*, 1768–1780. [CrossRef]
39. Ma, S.; Turetsky, A.; Trinh, L.; Lu, R. IFN regulatory factor 4 and 8 promote Ig light chain κ locus activation in pre-B cell development. *J. Immunol.* **2006**, *177*, 7898–7904. [CrossRef]
40. Sciammas, R.; Shaffer, A.; Schatz, J.H.; Zhao, H.; Staudt, L.M.; Singh, H. Graded expression of interferon regulatory factor-4 coordinates isotype switching with plasma cell differentiation. *Immunity* **2006**, *25*, 225–236. [CrossRef]
41. Gururajan, M.; Simmons, A.; Dasu, T.; Spear, B.T.; Calulot, C.; Robertson, D.A.; Wiest, D.L.; Monroe, J.G.; Bondada, S. Early growth response genes regulate B cell development, proliferation, and immune response. *J. Immunol.* **2008**, *181*, 4590–4602. [CrossRef]
42. Oh, Y.-K.; Jang, E.; Paik, D.-J.; Youn, J. Early growth response-1 plays a non-redundant role in the differentiation of B cells into plasma cells. *Immune Netw.* **2015**, *15*, 161–166. [CrossRef] [PubMed]
43. Li, S.; Miao, T.; Sebastian, M.; Bhullar, P.; Ghaffari, E.; Liu, M.; Symonds, A.L.; Wang, P. The transcription factors Egr2 and Egr3 are essential for the control of inflammation and antigen-induced proliferation of B and T cells. *Immunity* **2012**, *37*, 685–696. [CrossRef]
44. Ubieta, K.; Garcia, M.; Grötsch, B.; Uebe, S.; Weber, G.F.; Stein, M.; Ekici, A.; Schett, G.; Mielenz, D.; Bozec, A. Fra-2 regulates B cell development by enhancing IRF4 and Foxo1 transcription. *J. Exp. Med.* **2017**, *214*, 2059–2071. [CrossRef]
45. Labbaye, C.; Valtieri, M.; Barberi, T.; Meccia, E.; Masella, B.; Pelosi, E.; Condorelli, G.; Testa, U.; Peschle, C. Differential expression and functional role of GATA-2, NF-E2, and GATA-1 in normal adult hematopoiesis. *J. Clin. Investig.* **1995**, *95*, 2346–2358. [CrossRef] [PubMed]

46. Bresnick, E.H.; Katsumura, K.R.; Lee, H.-Y.; Johnson, K.D.; Perkins, A.S. Master regulatory GATA transcription factors: Mechanistic principles and emerging links to hematologic malignancies. *Nucleic Acids Res.* **2012**, *40*, 5819–5831. [CrossRef]
47. Sanyal, M.; Tung, J.W.; Karsunky, H.; Zeng, H.; Selleri, L.; Weissman, I.L.; Herzenberg, L.A.; Cleary, M.L. B-cell development fails in the absence of the Pbx1 proto-oncogene. *Blood* **2007**, *109*, 4191–4199. [CrossRef]
48. Zhang, Y.-A.; Salinas, I.; Li, J.; Parra, D.; Bjork, S.; Xu, Z.; LaPatra, S.E.; Bartholomew, J.; Sunyer, J.O. IgT, a primitive immunoglobulin class specialized in mucosal immunity. *Nat. Immunol.* **2010**, *11*, 827–835. [CrossRef]
49. Asea, A.; Rehli, M.; Kabingu, E.; Boch, J.A.; Baré, O.; Auron, P.E.; Stevenson, M.A.; Calderwood, S.K. Novel signal transduction pathway utilized by extracellular HSP70 role of Toll-like receptor (TLR) 2 and TLR4. *J. Biol. Chem.* **2002**, *277*, 15028–15034. [CrossRef]
50. Becker, T.; Hartl, F.-U.; Wieland, F. CD40, an extracellular receptor for binding and uptake of Hsp70–peptide complexes. *J. Cell Biol.* **2002**, *158*, 1277–1285. [CrossRef]
51. Oldenburg, M.; Krüger, A.; Ferstl, R.; Kaufmann, A.; Nees, G.; Sigmund, A.; Bathke, B.; Lauterbach, H.; Suter, M.; Dreher, S. TLR13 recognizes bacterial 23S rRNA devoid of erythromycin resistance–forming modification. *Science* **2012**, *337*, 1111–1115. [CrossRef]
52. Zhang, D.; Zhang, G.; Hayden, M.S.; Greenblatt, M.B.; Bussey, C.; Flavell, R.A.; Ghosh, S. A toll-like receptor that prevents infection by uropathogenic bacteria. *Science* **2004**, *303*, 1522–1526. [CrossRef]
53. Armitage, R.J.; Macduff, B.M.; Eisenman, J.; Paxton, R.; Grabstein, K.H. IL-15 has stimulatory activity for the induction of B cell proliferation and differentiation. *J. Immunol.* **1995**, *154*, 483–490.
54. Bernasconi, N.L.; Traggiai, E.; Lanzavecchia, A. Maintenance of serological memory by polyclonal activation of human memory B cells. *Science* **2002**, *298*, 2199–2202. [CrossRef]
55. Mackay, F.; Browning, J.L. BAFF: A fundamental survival factor for B cells. *Nat. Rev. Immunol.* **2002**, *2*, 465–475. [CrossRef]
56. Ma, Q.; Jones, D.; Springer, T.A. The chemokine receptor CXCR4 is required for the retention of B lineage and granulocytic precursors within the bone marrow microenvironment. *Immunity* **1999**, *10*, 463–471. [CrossRef]
57. Muehlinghaus, G.; Cigliano, L.; Huehn, S.; Peddinghaus, A.; Leyendeckers, H.; Hauser, A.E.; Hiepe, F.; Radbruch, A.; Arce, S.; Manz, R.A. Regulation of CXCR3 and CXCR4 expression during terminal differentiation of memory B cells into plasma cells. *Blood* **2005**, *105*, 3965–3971. [CrossRef] [PubMed]
58. Nie, Y.; Waite, J.; Brewer, F.; Sunshine, M.-J.; Littman, D.R.; Zou, Y.-R. The role of CXCR4 in maintaining peripheral B cell compartments and humoral immunity. *J. Exp. Med.* **2004**, *200*, 1145–1156. [CrossRef] [PubMed]
59. Chong, S.W.; Emelyanov, A.; Gong, Z.; Korzh, V. Expression pattern of two zebrafish genes, *cxcr4a* and *cxcr4b*. *Mech. Dev.* **2001**, *109*, 347–354. [CrossRef]
60. Lu, W.-J.; Zhou, L.; Gao, F.-X.; Sun, Z.-H.; Li, Z.; Liu, X.-C.; Li, S.-S.; Wang, Y.; Gui, J.-F. Divergent expression patterns and function of two *cxcr4* paralogs in hermaphroditic *Epinephelus coioides*. *Int. J. Mol. Sci.* **2018**, *19*, 2943. [CrossRef]
61. Lu, X.-J.; Zhu, K.; Shen, H.-X.; Nie, L.; Chen, J. CXCR4s in teleosts: Two paralogous chemokine receptors and their roles in hematopoietic stem/progenitor cell homeostasis. *J. Immunol.* **2020**, *204*, 1225–1241. [CrossRef]
62. Lu, W.-J.; Zhou, L.; Gao, F.-X.; Zhou, Y.-L.; Li, Z.; Zhang, X.-J.; Wang, Y.; Gui, J.-F. Dynamic and differential expression of duplicated *cxcr4/cxcl12* genes facilitates antiviral response in hexaploid gibel carp. *Front. Immunol.* **2020**, *11*, 2176. [CrossRef]
63. Marques, A.C.; Hughes, J.; Graham, B.; Kowalczyk, M.S.; Higgs, D.R.; Ponting, C.P. Chromatin signatures at transcriptional start sites separate two equally populated yet distinct classes of intergenic long noncoding RNAs. *Genome Biol.* **2013**, *14*, R131. [CrossRef]
64. Winkle, M.; Kluiver, J.; Diepstra, A.; van den Berg, A. Emerging roles for long noncoding RNAs in B-cell development and malignancy. *Crit. Rev. Oncol./Hematol.* **2017**, *120*, 77–85. [CrossRef] [PubMed]
65. Casero, D.; Sandoval, S.; Seet, C.S.; Scholes, J.; Zhu, Y.; Ha, V.L.; Luong, A.; Parekh, C.; Crooks, G.M. Long non-coding RNA profiling of human lymphoid progenitor cells reveals transcriptional divergence of B cell and T cell lineages. *Nat. Immunol.* **2015**, *16*, 1282. [CrossRef]
66. Agirre, X.; Meydan, C.; Jiang, Y.; Garate, L.; Doane, A.S.; Li, Z.; Verma, A.; Paiva, B.; Martín-Subero, J.I.; Elemento, O. Long non-coding RNAs discriminate the stages and gene regulatory states of human humoral immune response. *Nat. Commun.* **2019**, *10*, 1–16. [CrossRef]
67. Brazão, T.F.; Johnson, J.S.; Müller, J.; Heger, A.; Ponting, C.P.; Tybulewicz, V.L. Long noncoding RNAs in B-cell development and activation. *Blood* **2016**, *128*, e10–e19. [CrossRef] [PubMed]
68. Petri, A.; Dybkær, K.; Bøgsted, M.; Thrue, C.A.; Hagedorn, P.H.; Schmitz, A.; Bødker, J.S.; Johnsen, H.E.; Kauppinen, S. Long noncoding RNA expression during human B-cell development. *PLoS ONE* **2015**, *10*, e0138236. [CrossRef]
69. Vives-Bauza, C.; Zhou, C.; Huang, Y.; Cui, M.; De Vries, R.L.; Kim, J.; May, J.; Tocilescu, M.A.; Liu, W.; Ko, H.S. PINK1-dependent recruitment of Parkin to mitochondria in mitophagy. *Proc. Natl. Acad. Sci. USA* **2010**, *107*, 378–383. [CrossRef]
70. Xu, Y.; Shen, J.; Ran, Z. Emerging views of mitophagy in immunity and autoimmune diseases. *Autophagy* **2020**, *16*, 3–17. [CrossRef] [PubMed]
71. Gkikas, I.; Palikaras, K.; Tavernarakis, N. The role of mitophagy in innate immunity. *Front. Immunol.* **2018**, *9*, 1283. [CrossRef] [PubMed]
72. Cho, D.-H.; Kim, J.K.; Jo, E.-K. Mitophagy and Innate Immunity in Infection. *Mol. Cells* **2020**, *43*, 10.

73. Zhang, X.; Yuan, D.; Sun, Q.; Xu, L.; Lee, E.; Lewis, A.J.; Zuckerbraun, B.S.; Rosengart, M.R. Calcium/calmodulin-dependent protein kinase regulates the PINK1/Parkin and DJ-1 pathways of mitophagy during sepsis. *FASEB J.* **2017**, *31*, 4382–4395. [CrossRef]
74. Ellis, G.I.; Zhi, L.; Akundi, R.; Büeler, H.; Marti, F. Mitochondrial and cytosolic roles of PINK 1 shape induced regulatory T-cell development and function. *Eur. J. Immunol.* **2013**, *43*, 3355–3360. [CrossRef]
75. De Haan, N.; Falck, D.; Wuhrer, M. Monitoring of immunoglobulin N-and O-glycosylation in health and disease. *Glycobiology* **2020**, *30*, 226–240. [CrossRef]
76. Collin, M. Antibody glycosylation as an immunological key in health and disease. *Glycobiology* **2020**, *30*, 200. [CrossRef] [PubMed]
77. Arnold, J.N.; Wormald, M.R.; Sim, R.B.; Rudd, P.M.; Dwek, R.A. The impact of glycosylation on the biological function and structure of human immunoglobulins. *Annu. Rev. Immunol.* **2007**, *25*, 21–50. [CrossRef] [PubMed]
78. Van de Bovenkamp, F.S.; Derksen, N.I.; Ooijevaar-de Heer, P.; van Schie, K.A.; Kruithof, S.; Berkowska, M.A.; van der Schoot, C.E.; Ijspeert, H.; van der Burg, M.; Gils, A. Adaptive antibody diversification through N-linked glycosylation of the immunoglobulin variable region. *Proc. Natl. Acad. Sci. USA* **2019**, *115*, 1901–1906. [CrossRef]
79. Koers, J.; Derksen, N.I.; Ooijevaar-de Heer, P.; Nota, B.; van de Bovenkamp, F.S.; Vidarsson, G.; Rispens, T. Biased N-glycosylation site distribution and acquisition across the antibody V region during B cell maturation. *J. Immunol.* **2019**, *202*, 2220–2228. [CrossRef]
80. Wang, J.; Balog, C.I.; Stavenhagen, K.; Koeleman, C.A.; Scherer, H.U.; Selman, M.H.; Deelder, A.M.; Huizinga, T.W.; Toes, R.E.; Wuhrer, M. Fc-glycosylation of IgG1 is modulated by B-cell stimuli. *Mol. Cell. Proteomics* **2011**, *10*. [CrossRef]
81. Weisel, F.J.; Mullett, S.J.; Elsner, R.A.; Menk, A.V.; Trivedi, N.; Luo, W.; Wikenheiser, D.; Hawse, W.F.; Chikina, M.; Smita, S. Germinal center B cells selectively oxidize fatty acids for energy while conducting minimal glycolysis. *Nat. Immunol.* **2020**, *21*, 331–342. [CrossRef] [PubMed]
82. Edgar, R.; Domrachev, M.; Lash, A.E. Gene Expression Omnibus: NCBI gene expression and hybridization array data repository. *Nucleic Acids Res.* **2002**, *30*, 207–210. [CrossRef] [PubMed]

biology

Article

Comparative Transcriptome Analysis Reveals the Molecular Immunopathogenesis of Chinese Soft-Shelled Turtle (*Trionyx sinensis*) Infected with *Aeromonas hydrophila*

Zhao Lv [1], Yazhou Hu [1], Jin Tan [1], Xiaoqing Wang [1], Xiaoyan Liu [1] and Cong Zeng [2,*]

[1] Hunan Engineering Technology Research Center of Featured Aquatic Resources Utilization, Hunan Agricultural University, Changsha 410128, China; lvzhao_0320@hunau.edu.cn (Z.L.); huyazhou@hunau.edu.cn (Y.H.); tan775481248@163.com (J.T.); wangxiao@hunau.net (X.W.); liuxy86@163.com (X.L.)

[2] School of Oceanography, Shanghai Jiao Tong University, Shanghai 200240, China

* Correspondence: congzeng@sjtu.edu.cn

Simple Summary: The Chinese soft-shelled turtle (*Trionyx sinensis*) is an important cultured reptile in East Asia. Hemorrhagic sepsis caused by *Aeromonas hydrophila* infection is the dominant disease in the aquaculture of Chinese soft-shelled turtles, while the molecular pathology is far from clear due to the lag of research on turtle immunology. It has been reported in mammals and fish that the dysfunction of immune responses to pathogen infections causes host tissue hemorrhagic sepsis. In this study, two groups of turtles with different susceptibility to *A. hydrophila* infection are found. A comparative transcriptome strategy is adopted to examine the gene expression profiles in liver and spleen for these two phenotypes of turtles post *A. hydrophila* infection, for the first time revealing the full picture of immune mechanisms against *A. hydrophila*, which provides new insight into the molecular pathology during *A. hydrophila* infection in *T. sinensis*. The findings will promote further investigations on pathogenic mechanisms of hemorrhagic sepsis caused by *A. hydrophila* infection in *T. sinensis*, and also will benefit their culture industry.

Abstract: Although hemorrhagic sepsis caused by *Aeromonas hydrophila* infection is the dominant disease in the aquaculture of Chinese soft-shelled turtle, information on its molecular pathology is seriously limited. In this study, ninety turtles intraperitoneally injected with *A. hydrophila* exhibited two different phenotypes based on the pathological symptoms, referred to as active and inactive turtles. Comparative transcriptomes of liver and spleen from these two groups at 6, 24, and 72 h post-injection (hpi) were further analyzed. The results showed that cytokine–cytokine receptor interaction, PRRs mediated signaling pathway, apoptosis, and phagocytosis enriched in active and inactive turtles were significantly different. Pro-inflammatory cytokines, the TLR signaling pathway, NLR signaling pathway, and RLR signaling pathway mediating cytokine expression, and apoptosis-related genes, were significantly up-regulated in inactive turtles at the early stage (6 hpi). The significant up-regulation of phagocytosis-related genes occurred at 24 hpi in inactive turtles and relatively lagged behind those in active turtles. The anti-inflammatory cytokine, IL10, was significantly up-regulated during the tested periods (6, 24, and 72 hpi) in active turtles. These findings offer valuable information for the understanding of molecular immunopathogenesis after *A. hydrophila* infection, and facilitate further investigations on strategies against hemorrhagic sepsis in Chinese soft-shelled turtle *T. sinensis*.

Keywords: Chinese soft-shelled turtle; *Aeromonas hydrophila*; hemorrhagic sepsis; molecular immunopathogenesis

Citation: Lv, Z.; Hu, Y.; Tan, J.; Wang, X.; Liu, X.; Zeng, C. Comparative Transcriptome Analysis Reveals the Molecular Immunopathogenesis of Chinese Soft-Shelled Turtle (*Trionyx sinensis*) Infected with *Aeromonas hydrophila*. *Biology* **2021**, *10*, 1218. https://doi.org/10.3390/biology10111218

Academic Editor: Patricia Pereiro

Received: 12 October 2021
Accepted: 19 November 2021
Published: 22 November 2021

1. Introduction

Chinese soft-shelled turtle (*Trionyx sinensis*) is an important reptile in East Asia and has been taken as a food resource for a long time in these areas, especially in China and Japan. In ancient Chinese medicine descriptions, consumption of turtles may bring about positive effects for human health, including strengthening immunity, anti-aging, and curing cardiocerebrovascular diseases [1]. The excellent nutriment and medical values mean the turtles have become one of the most critical freshwater aquaculture reptiles, and the annual production is over 325,000 tons in China. However, infectious diseases caused by bacteria and viruses have resulted in severe losses to the aquaculture of turtles [2,3]. It is noteworthy that there are more than 15 kinds of diseases caused by *Aeromonas hydrophila* infection, such as red neck disease, septicemia, furunculosis, etc., accounting for about 60% of the total disease cases in turtles [4,5]. The histopathology of Chinese soft-shelled turtles infected by *A. hydrophila* has been previously described. In general, the pathogenic processes undergo the adhesion of bacteria adhesion factors, and the destruction of host liver, respiratory, and digestive organs by virulence factors such as exotoxin and extracellular enzymes, which finally leads to serious tissue hemorrhagic sepsis and the death of turtles [6,7]. Nevertheless, the molecular pathology of hemorrhagic sepsis caused by *A. hydrophila* infection is far from being elucidated in Chinese soft-shelled turtles.

It has been reported in mammals and fish that the dysfunction of host immune responses to pathogen infections causes tissue hemorrhagic sepsis [8–10]. The immune system generally serves as the security guard for hosts, which holds an immune network composed of many immune cells [11]. The cells of the innate immune system consisting of neutrophils, monocytes, macrophages, dendritic cells, and natural killer cells recognize pathogens, produce cytokines, and engulf pathogens through phagocytosis, which are the first line of host defense against pathogens [12]. These innate immune cells rely on pattern recognition receptors (PRRs) including toll-like receptors (TLRs), nucleotide-binding domain (NOD)-like receptors (NLRs), and retinoic acid-inducible gene I-like receptors (RLRs) to recognize various microbial invaders and produce cytokines that further activate the innate as well as adaptive immune cells [13]. Under normal immune response, immune cells moderately produce and release cytokines, including interleukins (ILs), chemokines, and tumor necrosis factors (TNFs), which can contact more immune cells to participate in the battle between the host and foreign pathogens [14]. When immune cells produce excessive cytokines, especially pro-inflammatory cytokines such as IL1, IL6, IL18, and TNFs ("cytokine storm"), the host immune system is overactivated and attacks self-tissues or cells, causing systemic inflammation, tissue hemorrhagic sepsis, and even death [15,16].

Although the research of turtle immunology lags behind that of mammals and fish, the unique evolutionary status as secondary aquatic reptiles has recently aroused wide concern on the distinct immune response mechanism against pathogens in turtles [17–20]. Fifteen candidate TLR family genes have been identified in *T. sinensis* [21]. After *A. hydrophila* infection, TLR2 and TLR4 are significantly up-regulated in the spleen, indicating the immune response of TLR signaling pathway during bacterial infection in *T. sinensis* [21]. Zhou et al. identified an IL8 homolog from *T. sinensis* and confirmed that IL8 mRNA expression shows significant up-regulation in various tissues, including liver, spleen, kidney, heart, intestine, and blood, after *A. hydrophila* infection [22]. Zhang et al. also investigated the mRNA expression changes of pro-inflammatory cytokines such as IL1β, TNFα, IL6, IL8, and IL12 in *T. sinensis* during acute cold stress, revealing that acute cold stress increases the expression of pro-inflammatory cytokines in the spleen and intestine to withstand *A. hydrophila* infection [23]. These above studies may help us understand the mRNA expression profiles and function of several immune molecules in *T. sinensis*. However, the evidence on the immune response mechanism in *T. sinensis* so far is too scattered and limited, which largely hinders the studying of pathogen–host immunity interaction and the molecular pathology during *A. hydrophila* infection in Chinese soft-shelled turtles.

RNA-sequencing (RNA-Seq) using next-generation sequencing is one of the most useful methods to survey the character of a transcriptome because it offers the whole data on gene expression. To date, transcriptome profiling using next-generation sequencing technologies has provided new insights into pathogen–host immunity interaction in many aquaculture fish species, such as Nile tilapia (*Oreochromis niloticus*) [24], Gold fish (*Carassius auratus* L.) [25], Ussuri catfish (*Pseudobagrus ussuriensis*) [26], Yellow catfish (*Pelteobagrus fulvidraco*) [27], and Atlantic salmon (*Salmo salar*) [28]. More and more studies have used this high-throughput sequencing approach to identify the expression differences of immune molecules between the resistant and susceptible fish to pathogen infection. For example, Moraleda et al. analyzed the comparative transcriptomes of resistant/susceptible salmons with different immune responses to *Piscirickettsia salmonis* infection and revealed that the gene networks involved in the apoptotic, cytoskeletal reorganization, bacterial invasion, and intracellular trafficking processes are tightly associated with disease resistance/susceptibility [29]. We are originally inspired by the previous research results in mammals and fish that the dysfunction of host immune responses to pathogen infections causes tissue hemorrhagic sepsis. In the present study, therefore, a comparative transcriptome strategy is adopted to reveal the immune gene expression profiles in two phenotypes of Chinese soft-shelled turtles with different susceptibility to *A. hydrophila* infection at different time periods (6, 24, and 72 hpi), and to clarify which immune process abnormalities may be related to the occurrence of hemorrhagic sepsis. The results depict the full picture of immune mechanisms in response to *A. hydrophila* infection, suggesting that the dysfunction of cytokine–cytokine receptor interaction, PRRs mediated signaling pathways, phagocytosis, and apoptosis may cause hemorrhagic sepsis during *A. hydrophila* infection in Chinese soft-shelled turtles. This study, for the first time, reveals the host immunity–pathogen infection interaction and molecular immunopathogenesis during *A. hydrophila* infection by comparative transcriptomes, which may contribute to the development of novel management strategies for disease control and prevention in Chinese soft-shelled turtles.

2. Materials and Methods

2.1. Experimental Animals and Bacteria Strain

Experimental turtles (*T. sinensis*) with an average weight of 16.45 ± 1.28 g were obtained from a breeding farm (111°97′ E, 28°90′ N) in Changde City, Hunan Province, China. All the turtles were acclimated in 50 L aquarium with aerated freshwater in a constant temperature laboratory room at 30 °C for two weeks before processing and were fed with a commercial diet (Kesheng Feed Stock Co., Ltd., Hangzhou, Zhejiang, China) twice a day. The animal experiments were according to the rules of the Animal Care and Use Committee of Hunan Agricultural University (Changsha, China; Approval Code: 201903297; Approval Date: 11 October 2019).

The bacteria *A. hydrophila*, isolated from the spleen of clinically diseased turtles, was provided by Professor Zhipeng Gao from the College of Animal Science and Technology, Hunan Agricultural University, China. *A. hydrophila* was cultured in Luria Bertani (LB) medium at 30 °C with the shaking at 200 rpm. After 18 h culturing, the bacteria were harvested by centrifugation, and suspended with sterile phosphate buffer saline (PBS). The bacteria concentration was adjusted to 1.39×10^9 CFU/mL with PBS, which was previously proved as the medium lethal concentration for turtles during *A. hydrophila* infection.

2.2. Experimental Treatments, Pathological Observation, and Sampling

For the *A. hydrophila* challenge, 90 turtles were intraperitoneally injected with bacteria suspension (100 μL per turtle, designated as the treatment group). After bacteria injection, the turtles in the treatment group could be divided into two subgroups based on their pathological symptoms and behavior activities. One subgroup of turtles showed weaker ability to feed and move, with significant hemorrhagic symptoms on the body surface and swelling and congestion on viscera after anatomy, which was designated as the inactive subgroup. The other subgroup of turtles exhibited no obvious pathological symptoms, which was designated as the active subgroup. Ten turtles, injected with 100 μL of PBS, were set up as the control group.

To investigate the expressional profiles of immune-related genes during *A. hydrophila* infection in active and inactive turtles, important immune-related tissues including spleen and liver from the active subgroup turtles ($N = 3$) and inactive subgroup turtles ($N = 3$) were sampled post-injection of *A. hydrophila* at 6, 24, 72 h, respectively. Spleen and liver tissues from the control group turtles ($N = 3$) with PBS injection at 0 h were sampled as the control for the gene expression comparison with those from active subgroup turtles or from inactive subgroup turtles. Each sample was taken three times, and the sampled tissues were stored in liquid nitrogen before RNA extraction.

2.3. RNA Extraction, Library Preparation, and RNA Sequencing (RNA-Seq)

Total RNA from each sampled tissue was isolated by using RNeasy mini kit (QIAGEN, Germantown, MD, USA) and treated with RNase-free DNase I (QIAGEN, Germantown, MD, USA) at 37 °C for 1 h to remove residual genomic DNA. RNA quality and concentration were determined using Agilent Bioanalyser 2100 (Agilent Technologies, Santa Clara, CA, USA) and NanoDrop-1000 (NanoDrop Technologies, Wilmington, DE, USA), respectively. Five micrograms of RNA were used to construct RNA-seq library according to the instruction of Illumina mRNA-Seq Prep Kit (Illumina, San Diego, CA, USA), and the libraries were sequenced by paired-end sequencing on the Illumina HiSeq 2500 sequencing platform (Illumina, San Diego, CA, USA). The quality of RNA-Seq raw reads were assessed with FastQC (version 0.10.1; http://www.bioinformatics.bbsrc.ac.uk/projects/fastqc/ accessed on 11 March 2019), and were cleaned by removing adapter sequences, poly-N sequences, and low-quality sequences. The clean reads were then aligned to the published *T. sinensis* genome using HISAT2 [30]. In total, 291.16 GB clean data were produced, and the average of 20.4~36.3 GB clean reads were obtained from samples. About 80% of the clean reads were mapped to the reference genome by using StringTie software (Center for Computational Biology, Johns Hopkins University, Baltimore, MD, USA) [31], and the Q30 was >92.86%. The general information of RNA-Seq data is listed in Table S1.

2.4. Identification of the Differentially Expressed Genes (DEGs)

The relative transcript abundances in tissues (liver and spleen) from active and inactive subgroup turtles at different time periods of *A. hydrophila* infection compared to the control turtles were, respectively, estimated by using StringTie software with expectation maximization method, based on fragments per kilobase of exon per million fragments mapped (FPKM) [31]. Differential expression analysis was performed by using the DESeq R package with default parameters [32]. Benjamini and Hochberg's approach was used to control the false discovery rate (FDR) by resulting *P*-values adjustment [33]. Genes with an adjusted *P*-value (or q-value) < 0.05 found by DEGSeq were assigned as the differentially expressed.

2.5. KEGG Pathway Enrichment Analysis

Significantly enriched signal transduction pathways represented by DEGs were determined using KEGG pathway enrichment analysis, compared with the whole genome background [34]. The statistical enrichment of DEGs in KEGG pathways was tested using the software KOBAS, with a P-value < 0.05 [34]. The significantly enriched KEGG pathways are listed in Tables S2–S5.

2.6. Gene Expression Validation Using Quantitative PCR (qPCR)

To validate RNA-seq data and gene expression profiles, nine DEGs were randomly selected to perform qPCR. Specific primers were designed using Primer 5 based on the coding sequences of identified genes from the turtle genome; the sequences of primers are listed in Table S6. Primer specificity was ascertained using the following steps: PCR amplification, sequencing of PCR products, and BLAST analysis in the NCBI database.

The RNA samples used for qPCR amplifications were the same as those used to construct the RNA-Seq library mentioned above. The qPCR was performed using the LightCycler®480 Real-Time PCR System (Roche, Basel, Switzerland) with SYBR Green I Master. The reaction mixture (10 μL) comprised 2.5 μL cDNA, 0.5 μL (10 nM) forward primer and 0.5 μL (10 nM) reverse primer, 1.5 μL PCR grade water, and 5 μL Master Mix. Each reaction was performed in triplicate under the following conditions: 95 °C for 10 min, 40 cycles of 95 °C for 15 s, 55 °C for 15 s, and 72 °C for 30 s. The relative expression level of each gene was calculated according to the $2^{-\triangle\triangle Ct}$ method [35] and normalized to the endogenous control genes of β-actin and GAPDH.

3. Results

3.1. Symptom Description of the Turtles Challenged with A. hydrophila

Ninety Chinese soft-shelled turtles were injected with *A. hydrophila* (1.39×10^9 CFU/mL), and the pathological symptoms of turtles were observed. There were no obvious pathological symptoms on the body surface of turtles, and only six turtles showed behavior abnormality with slow-moving action at 6 hpi (post-injection of 6 h). At 24 hpi, the food intake and movement of turtles was overall reduced, and seven turtles died (accounting for about 8%), with the pathological symptom of abdominal congestion. At 72 hpi, 34 turtles (accounting for about 38%) died. The pathological symptoms on the body surface of diseased turtles were obvious, with white spots near the axillae, swelling, and congestion in chest and abdomen (Figure 1A,B). After anatomizing the diseased turtles, pathological symptoms of turtle viscera were easily observed (Figure 1C). The liver and spleen were swelling and congestive, with the tendency to decay, and the color of liver was yellow (Figure 1C). The intestines were filled with no food debris, and the color was white (Figure 1C). There were 13 turtles (accounting for about 14%) without any obvious pathological symptoms at 72 hpi (Figure 1D). During the whole experiment within 72 h, 10 turtles in the control group showed no obvious pathological symptoms and were aggressive and active in feeding and moving (data not shown).

According to pathological symptoms, the Chinese soft-shelled turtles challenged with *A. hydrophila* could be divided into two subgroups, including the active and inactive turtles. The liver and spleen were important immune organs in turtles, and both carried obviously pathological symptoms after *A. hydrophila* infection. Therefore, liver and spleen samples from active and inactive subgroup turtles were taken at 6, 24, and 72 hpi, respectively, and transcriptome sequencing was performed to reveal the differences of gene expression profiles between active and inactive turtles infected by *A. hydrophila*.

Pearson's correlation coefficients were first used to test for biologically repeated correlations between samples (Figure 2A). The generated cluster dendrogram was used to observe the overall correlation of the transcriptomes from the AL group (liver in active turtles), the IL group (liver in inactive turtles), the AS group (spleen in active turtles), and the IS group (spleen in inactive turtles) at different time periods (6, 24, and 72 hpi) (Figure 2A). Three biological replicates of liver and spleen samples from each time period and the transcriptome data both exhibited good correlation (Figure 2A). The similarity test between the three biological replicates required the use of a principal component analysis (PCA) (Figure 2B). Using the first principal component (PC1) and second principal component (PC2), a dimensionality reduction analysis was used to analyze the similarity between each replicate (Figure 2B). Figure 2B showed that biological replicates of samples overall exhibited good similarity. The generated box plot presented the dispersion degree of the gene expression level in a single liver or spleen sample, and intuitively revealed the whole gene expression level difference among all samples (Figure 2C). The results showed that the gene expression level in the spleen was overall higher than that in the liver (Figure 2C).

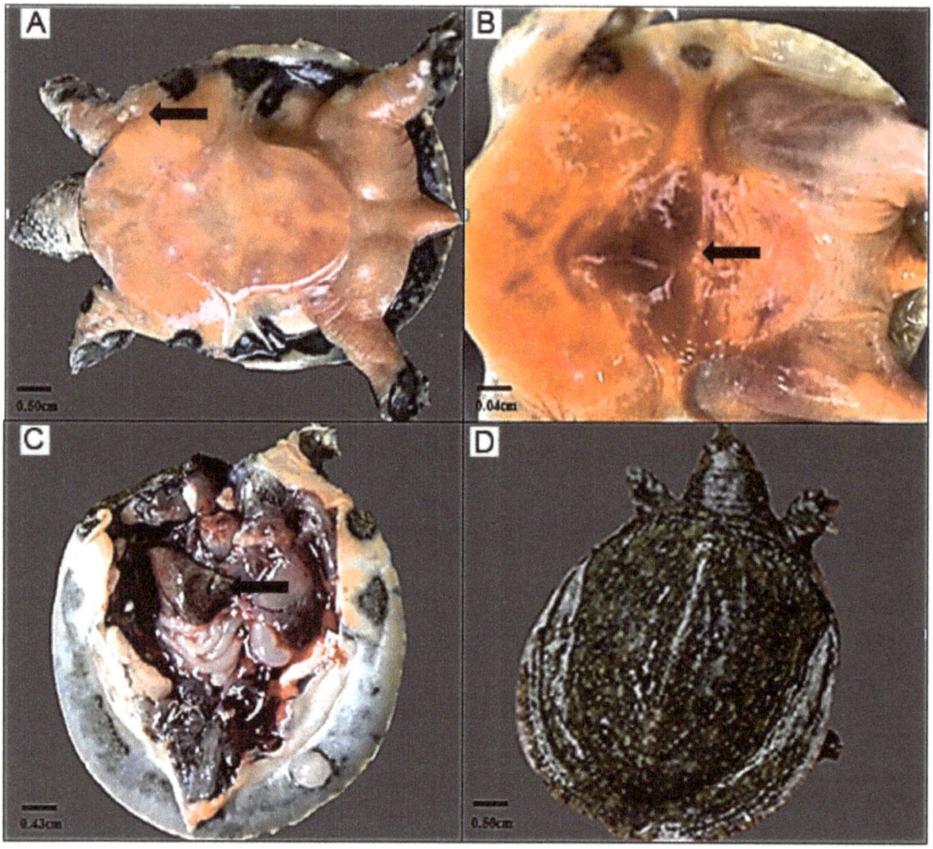

Figure 1. Pathological symptoms of turtles infected by *A. hydrophila*. The pathological symptoms on the body surface including (**A**) white spots near the axillae; (**B**) abdominal congestion in inactive subgroup turtles. (**C**) The pathological symptoms of viscera including liver, spleen, and intestines after anatomy of inactive turtles. (**D**) The active subgroup turtles showed no obvious pathological symptoms and were aggressive and active in feeding and moving. Black arrows indicate the location of pathological symptoms.

Totals of 4092 and 5793 DEGs were obtained in the liver and spleen transcriptomes, respectively (Figure 2D). In the AL group, 321, 401, and 378 genes were significantly up-regulated, and 671, 874, and 151 genes were significantly down-regulated at 6, 24, and 72 hpi, respectively (Figure 2D). In the IL group, the numbers of significantly up-regulated genes were 298, 138, and 79; the numbers of significantly down-regulated genes were 268, 381, and 132 at 6, 24, and 72 hpi, respectively (Figure 2D). In the AS group, 219, 176, and 65 genes were significantly up-regulated, and 576, 437, and 206 genes were significantly down-regulated at 6, 24, and 72 hpi, respectively (Figure 2D). In the IS group, the numbers of significantly up-regulated genes were 1193, 203, and 117; the numbers of significantly down-regulated genes were 1996, 401, and 204 at 6, 24, and 72 hpi, respectively (Figure 2D). The results revealed that the number of DEGs in the spleen transcriptomes was more than that in the liver transcriptomes, and the molecular response peaked at 24 hpi in liver, while it peaked at 6 hpi in spleen (Figure 2D).

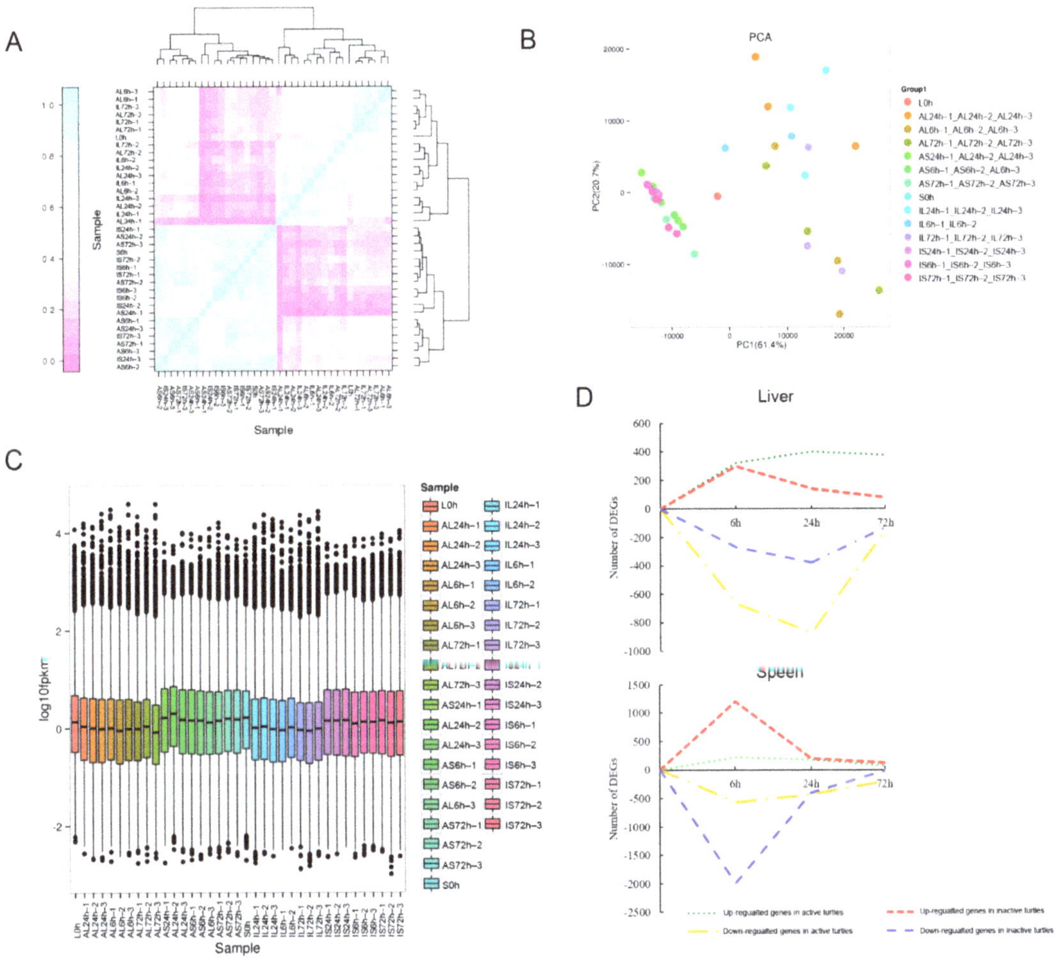

Figure 2. Transcriptional relationship between samples and the overall gene expression profiles. (**A**) Heatmap of correlation value (R square) of 37 libraries from liver or spleen samples. (**B**) Principal component analysis based on all of the expressed genes, showing 14 distinct groups of samples. (**C**) The dispersion degree of the gene expression level in a single liver or spleen sample. (**D**) The significantly up-regulated and down-regulated DEGs identified in livers or spleens from active and inactive turtles at 6, 24, and 72 hpi compared to the control.

3.2. Functional Classification of DEGs in Turtle Liver Transcriptomes by KEGG

In order to investigate the different molecular response mechanisms against *A. hydrophila* infection in livers from active and inactive turtles, the functional classification of DEGs in AL and IL group transcriptomes at different time periods (6, 24, and 72 hpi) were analyzed by KEGG enrichment analysis, and the results are summarized as follows.

3.2.1. Sequential Changes of KEGG Enrichment in AL Group Turtles

In AL group, the up-regulated DEGs were mainly enriched in immune-related pathways including "cytokine–cytokine receptor interaction", phagocytosis-associated processes including "phagosome", "protein processing in endoplasmic reticulum", and "protein export", and pathogen infection-related pathways, while the down-regulated DEGs were intensively involved in 17 metabolism pathways and three cell adhesion-related processes at 6 hpi (Figure 3).

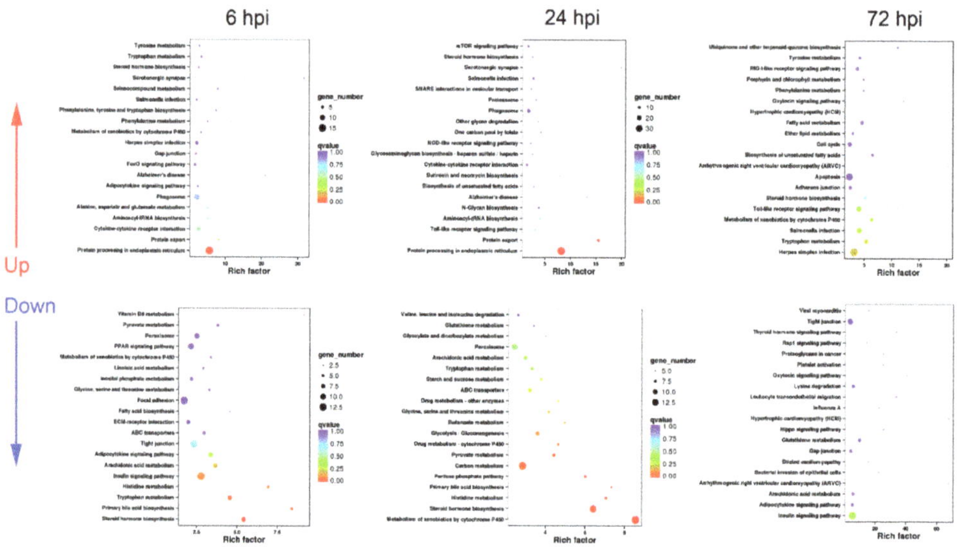

Figure 3. KEGG enrichment analysis in AL group turtles at different time periods. The top 20 KEGG pathways are presented here in the form of scatterplots to show the up-regulated and down-regulated DEGs enriched in livers from active subgroup turtles at 6, 12, and 72 hpi. The enrichment factor is the ratio between the DEG number and the number of all genes in a certain gene enrichment term. The sizes of the dots on these plots denote the number of DEGs, while colors correspond to the q value range.

At 24 hpi, the majority of up-regulated DEGs were also annotated into immune-related pathways including "cytokine–cytokine receptor interaction", "phagosome", "protein processing in endoplasmic reticulum", "proteasome", and "protein export" (Figure 3). In addition, cytokine expression-mediating pathways including "toll-like receptor signaling pathway" and "NOD-like receptor signaling pathway" were significantly up-regulated (Figure 3). Similar to the response at 6 hpi, the down-regulated DEGs were mainly involved in a series of metabolism and cell adhesion pathways (Figure 3).

At 72 hpi, the up-regulated DEGs could be functionally classified into "toll-like receptor signaling pathway" and "RIG-I-like receptor signaling pathway", pathogen infection-related pathways, "apoptosis", and several metabolism-related pathways (Figure 3), while the down-regulated DEGs mainly participated in important metabolism-related pathways such as "insulin signaling pathway" and "adipocytokine signaling pathway" (Figure 3).

3.2.2. Sequential Changes of KEGG Enrichment in IL Group Turtles

Unlike the KEGG enrichment in AL group at 6 hpi, the up-regulated DEGs were mainly enriched in cytokine expression-mediating pathways including "toll-like receptor signaling pathway", "NOD-like receptor signaling pathway", and "RIG-I-like receptor signaling pathway" and "apoptosis", besides "cytokine–cytokine receptor interaction", pathogen infection-related pathways; while phagocytosis-associated processes were not listed in the top 20 up-regulated KEGG pathways in IL group at 6 hpi (Figure 4). The down-regulated DEGs were functionally annotated into cell adhesion-related pathways such as "ECM–receptor interaction" and "focal adhesion", and energetic metabolism pathways (Figure 4).

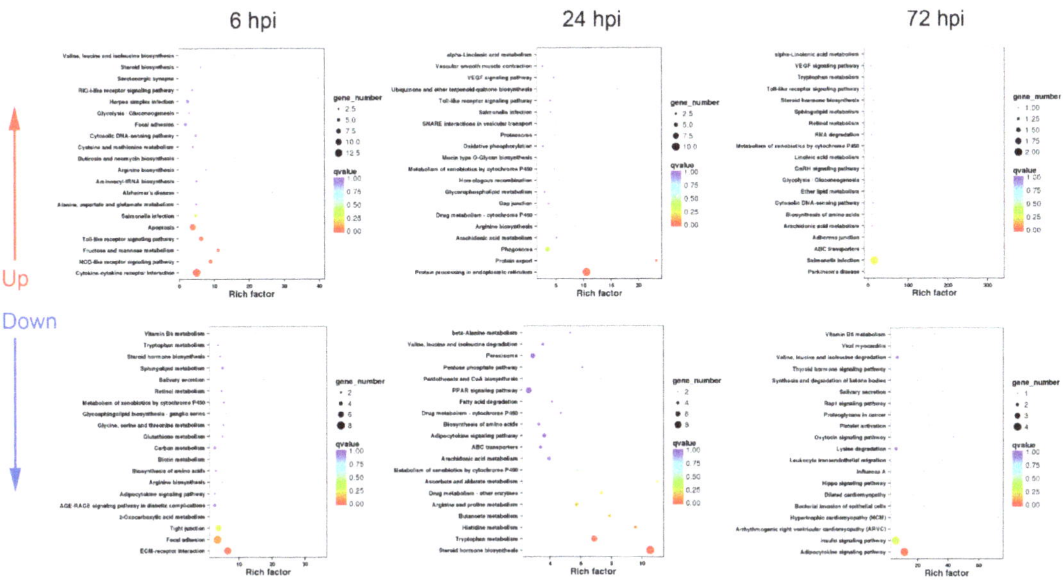

Figure 4. KEGG enrichment analysis in IL group turtles at different time periods. The top 20 KEGG pathways are presented here in the form of scatterplots to show the up-regulated and down-regulated DEGs enriched in livers from inactive subgroup turtles at 6, 12, and 72 hpi. The enrichment factor is the ratio between the DEG number and the number of all genes in a certain gene enrichment term. The sizes of the dots on these plots denote the number of DEGs, while colors correspond to the q value range.

At 24 hpi, the term of "cytokine–cytokine receptor interaction" was not listed in the top 20 up-regulated KEGG pathways, while the up-regulated DEGs were mainly involved in phagocytosis-associated processes such as "phagosome", "protein processing in endoplasmic reticulum", "proteasome", and "protein export" (Figure 4). Additionally, "toll-like receptor signaling pathway" was significantly up-regulated (Figure 4). For the down-regulated DEGs, most of them were associated with hormone synthesis and amino acid metabolism (Figure 4).

At 72 hpi, immune-related terms including "toll-like receptor signaling pathway" and "cytosolic DNA-sensing pathway" were listed in the top 20 up-regulated KEGG pathways (Figure 4). The down-regulated DEGs were mainly enriched in energetic metabolism pathways such as "insulin signaling pathway" and "adipocytokine signaling pathway" (Figure 4).

3.2.3. Expression Difference Analysis of Cytokine, Phagocytosis, and Apoptosis-Related Genes between AL and IL Group Turtles

The KEGG pathways related to immune processes including cytokine–cytokine receptor interaction, phagocytosis, and apoptosis enriched in AL and IL group turtles challenged with *A. hydrophila* were quite different (Tables 1–3). Therefore, the fold changes of differentially expressed cytokine, phagocytosis, and apoptosis-related genes were further analyzed.

Table 1. Fold changes of differentially expressed cytokine and cytokine receptor, and cytokine expression mediating pathway genes in the AL group and the IL group compared to the control.

Categories/ Gene Name	Description	Log$_2$ (Fold Changes)					
		AL Group			IL Group		
		6 Hpi	24 Hpi	72 Hpi	6 Hpi	24 Hpi	72 Hpi
	Interleukins and interleukin receptors						
IL1β	interleukin 1 beta		2.18		4.82		
IL8	interleukin 8				3.55		
IL10	interleukin 10	6.08	4.41	2.31			
IL1R1	interleukin 1 receptor type I		1.66				
IL1R2	interleukin 1 receptor type II	5.88	3.16		7.06		
IL5RA	interleukin 5 receptor alpha		−2.57			−2.40	−3.01
IL18R1	interleukin 18 receptor 1		−5.30				
	Chemokines and chemokine receptors						
CCL5	C-C motif chemokine 5				4.71		
CCL20	C-C motif chemokine 20	9.34	7.63		12.18		
CX3CL1	C-X3-C motif chemokine 1		−4.91		3.28		
CCR5	C-C chemokine receptor type 5	3.02					
	TNF family members and TNF receptors						
TNFSF10	tumor necrosis factor ligand superfamily member 10	−2.56	−2.08				
TNFSF15	tumor necrosis factor ligand superfamily member 15						−2.40
SF6B	tumor necrosis factor receptor superfamily member 6B						−2.57
SF12A	tumor necrosis factor receptor superfamily member 12A	4.49			4.60		
	Toll-like receptor (TLR) signaling pathway						
TLR2	toll-like receptor 2	4.56	2.80				
TLR5	toll-like receptor 5		3.61	1.51			2.89
TRIF	toll-like receptor adapter molecule 1	−3.15	−2.87	−1.99		−3.40	−3.42
TAB1	TAK1-binding protein 1		1.77				
PI3K	phosphoinositide-3-kinase		2.48		2.96		
MAP2K1	mitogen-activated protein kinase kinase 1		2.13				
MAP2K6	mitogen-activated protein kinase kinase 6	−2.04	-2.63			−2.59	
AP-1	proto-oncogene protein c-fos			1.43			
STAT1	signal transducer and activator of transcription 1			1.49			
	RIG-I-like receptor (RLR) signaling pathway						
RIG-I	retinoic acid inducible gene I			2.38			
LGP2	laboratory of genetics and physiology 2			2.21			
MDA5	melanoma differentiation-associated gene 5	2.43					
TRAF3	TNF receptor-associated factor 3				5.07	3.54	
IRF7	interferon regulatory factor 7		−1.83		2.86		
DDX3X	ATP-dependent RNA helicase DDX3X		−1.86	−1.69		−2.08	−2.02

Table 1. *Cont.*

Categories/ Gene Name	Description	Log$_2$ (Fold Changes)					
		AL Group			IL Group		
		6 Hpi	24 Hpi	72 Hpi	6 Hpi	24 Hpi	72 Hpi
		NOD-like receptor (NLR) signaling pathway					
ASC	apoptosis-associated speck-like protein containing a CARD	1.99					
RIPK2	receptor-interacting serine/threonine-protein kinase 2				2.47		
cIAP	baculoviral IAP repeat-containing protein 2/3				2.50		
TNFAIP3	tumor necrosis factor alpha-induced protein 3			1.56	3.75		

Note: This table and the following tables only show the DEGs with the values of log$_2$ (fold change) with FDR < 0.05.

Table 2. Fold changes of differentially expressed phagocytosis-related genes in the AL group and the IL group compared to the control.

Categories/ Gene Name	Description	Log$_2$ (Fold Changes)					
		AL Group			IL Group		
		6 Hpi	24 Hpi	72 Hpi	6 Hpi	24 Hpi	72 Hpi
		Internalization and formation of the phagosomes					
TLR2	toll-like receptor 2	4.56	2.80				
MR	mannose receptor		1.87	1.07		−2.18	
iC3b	the fragment of complement component 3	4.38		2.79			
Collectin	C-type lectin	4.81	3.69	1.47	−2.63		
F-actin	actin beta/gamma 1	5.57	5.31	4.39	5.09	5.68	4.64
		Early phagosome					
Rab5	ras-related protein Rab-5B	2.28					
vATPase	V-type H$^+$-transporting ATPase	2.33	2.05		3.85	−2.08	
CALR	calreticulin		4.05				
		Mature phagosome					
TUBA	tubulin alpha	2.26					
TUBB	tubulin beta	4.63	3.25	1.42		2.72	
vATPase	V-type H$^+$-transporting ATPase	2.33	2.05		3.85	−2.08	
		Phagolysosome					
sec61	protein transport protein SEC61 subunit beta	2.72	3.26			2.95	
vATPase	V-type H$^+$-transporting ATPase	2.33	2.05		3.85	−2.08	
		Activation of NADPH oxidase					
p40phox	neutrophil cytosolic factor 4	3.50					
p47phox	neutrophil cytosolic factor 1		2.79			2.42	
gp91	NADPH oxidase 1			1.59			
		Antigen presentation					
MHC II	MHC class II antigen		−2.21				
sec22	vesicle transport protein SEC22		4.21			3.54	

Table 3. Fold changes of up-regulated apoptosis-related genes in the AL group and the IL group compared to the control.

Gene Name	Description	Log$_2$ (Fold Changes)					
		AL Group			IL Group		
		6 Hpi	24 Hpi	72 Hpi	6 Hpi	24 Hpi	72 Hpi
p53	tumor protein p53	1.94					
IP3R	inositol 1,4,5-triphosphate receptor type 3	1.35	1.56		1.92		
Perforin	perforin 1		3.12	2.57			
PI3K	phosphoinositide-3-kinase		2.48		2.96		
MERK2	mitogen-activated protein kinase kinase 2		2.13				
PERK	protein kinase RNA (PKR)-like ER kinase		1.85	2.30			
Cathepsin	cathepsin B				5.11	4.27	
NOXA	phorbol-12-myristate-13-acetate-induced protein 1			1.62			
AP1	proto-oncogene protein c-fos			1.43			
GZMB	granzyme B				3.87		
IL3R	cytokine receptor common subunit beta				3.43		
A1	hematopoietic Bcl-2-related protein A1				4.59		

The results showed that the sharpest response of cytokines occurred at 24 hpi in the AL group, with the up-regulation of two cytokine and two cytokine receptor genes (Table 1). Among three cytokine expression-mediating pathways, only the response of toll-like receptor signaling pathway was relatively intense, with five up-regulated genes, also at 24 hpi in the AL group (Table 1), while in IL group, the sharpest response of cytokines occurred at 6 hpi, with the up-regulation of five cytokine and two cytokine receptor genes, and few cytokines were differentially expressed at 24 or 72 hpi (Table 1). Moreover, the up-regulation of differentially expressed cytokine genes at 6 hpi in IL group was overall higher than those in AL group at any time periods (Table 1). Nevertheless, IL10, an anti-inflammatory cytokine, was not differentially expressed at any time periods in the IL group, while it was up-regulated at all the time periods in the AL group (Table 1). In addition, the up-regulations of both NOD-like receptor signaling pathway and RIG-I-like receptor signaling pathway genes were intense in the IL group at 6 hpi (Table 1).

Notably, the sharpest response of phagocytosis-related gene occurred at 6 hpi in the AL group, with the up-regulation of 10 DEGs, and lasted until 24 hpi (Table 2). While in the IL group, at 24 hpi, the phagocytosis activity seemed to just start, with five up-regulated phagocytosis-related DEGs (Table 2). In addition, the most intense response of apoptosis-related genes occurred at 6 hpi in the IL group, with the up-regulation of six DEGs, while in the AL group, the apoptosis seemed to just start at 24 hpi with the up-regulation of five DEGs (Table 3).

3.3. Functional Classification of DEGs in Turtle Spleen Transcriptomes by KEGG

The functional classification of DEGs in AS and IS transcriptomes at different time periods were also analyzed by KEGG enrichment to further reveal the different immune response mechanisms against *A. hydrophila* infection in spleens between active and inactive turtles. The results are summarized as follows.

3.3.1. Sequential Changes of KEGG Enrichment in AS Group Turtles

In the AS group, the up-regulated DEGs were functionally associated with immune processes including "cytokine–cytokine receptor interaction", "chemokine signaling pathway", "phagosome", "apoptosis", "leukocyte transendothelial migration", "toll-like receptor signaling pathway", "MAPK signaling pathway", and pathogen infection-related pathways at 6 hpi (Figure 5), while the down-regulated DEGs were involved in cell adhesion and metabolism pathways at 6 hpi (Figure 5).

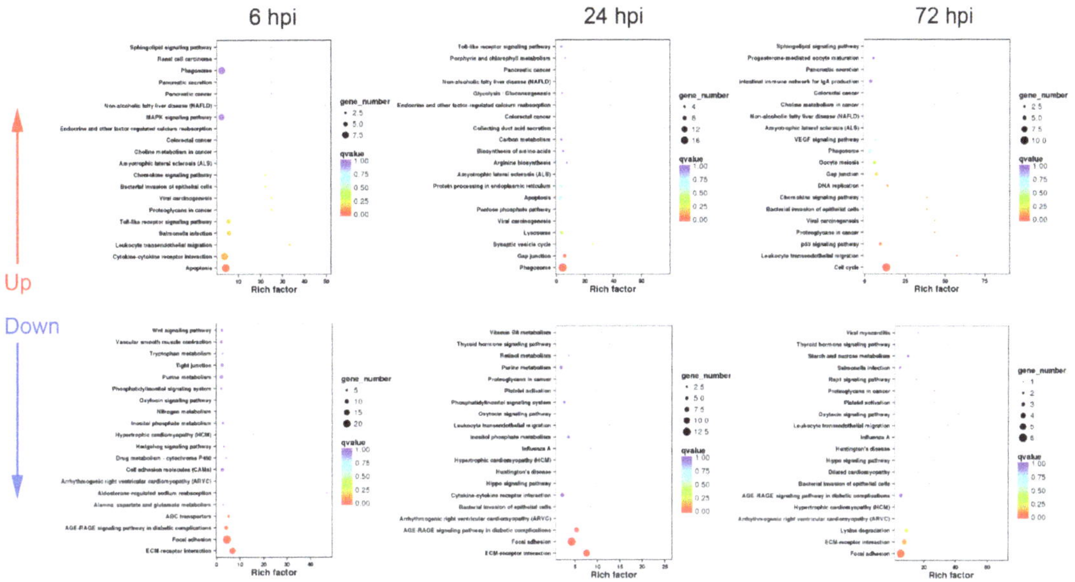

Figure 5. KEGG enrichment analysis in AS group turtles at different time periods. The top 20 KEGG pathways are presented here in the form of scatterplots to show the up-regulated and down-regulated DEGs enriched in spleens from active subgroup turtles at 6, 12, and 72 hpi. The enrichment factor is the ratio between the DEG number and the number of all genes in a certain gene enrichment term. The sizes of the dots on these plots denote the number of DEGs, while colors correspond to the q value range.

At 24 hpi, the up-regulated DEGs were mainly annotated into phagocytosis-related pathways such as "phagosome", "protein processing in endoplasmic reticulum", "synaptic vesicle cycle", and "lysosome". In addition, "toll-like receptor signaling pathway" was significantly up-regulated (Figure 5). Unlike the response at 6 hpi, the down-regulated DEGs at 24 hpi could be enriched in pathways including "cytokine–cytokine receptor interaction" and "bacterial invasion of epithelial cells", besides a series of metabolism and cell adhesion pathways (Figure 5).

At 72 hpi, the up-regulated DEGs were functionally classified into apoptosis-related pathways, including "p53 signaling pathway" and "cell cycle", as well as immune defense processes such as "phagosome", "chemokine signaling pathway", and "leukocyte transendothelial migration" (Figure 5), while the down-regulated DEGs mainly participated in important metabolism and cell adhesion pathways (Figure 5).

3.3.2. Sequential Changes of KEGG Enrichment in IS Group Turtles

The strongest immune response of spleen occurred at 6 hpi in the IS group, with 1193 up-regulated and 1996 down-regulated DEGs. Most up-regulated DEGs were enriched in immune-related pathways including "cytokine–cytokine receptor interaction", "toll-like receptor signaling pathway", "NOD-like receptor signaling pathway", and "RIG-I-like receptor signaling pathway", as well as apoptosis-associated processes (Figure 6), while the down-regulated DEGs were functionally annotated into cell adhesion-related pathways such as "ECM–receptor interaction" and "focal adhesion", and metabolism pathways (Figure 6).

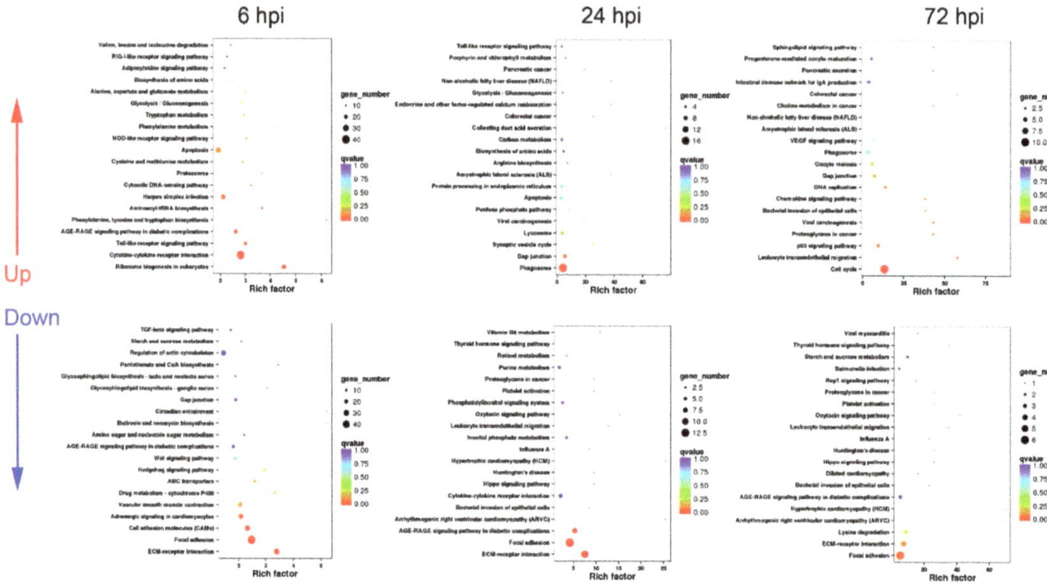

Figure 6. KEGG enrichment analysis in IS group turtles at different time periods. The top 20 KEGG pathways are presented here in the form of scatterplots to show the up-regulated and down-regulated DEGs enriched in spleens from inactive subgroup turtles at 6, 12, and 72 hpi. The enrichment factor is the ratio between the DEG number and the number of all genes in a certain gene enrichment term. The sizes of the dots on these plots denote the number of DEGs, while colors correspond to the q value range.

At 24 hpi, the up-regulated DEGs could be related to phagocytosis such as "phagosome" and "lysosome" (Figure 6). Additionally, "toll-like receptor signaling pathway" was listed in the top 20 up-regulated KEGG pathways (Figure 6). For the down-regulated DEGs, most of them were associated with cell adhesion, hormone synthesis, and amino acid metabolism (Figure 6).

Similar to the response at 72 hpi in the AS group, in the IS group, the up-regulated DEGs mainly participated in apoptosis-related pathways including "p53 signaling pathway" and "cell cycle", as well as "phagosome" and "chemokine signaling pathway" at 72 hpi (Figure 6), while the down-regulated DEGs were enriched in a series of hormone synthesis, amino acid metabolism, and cell adhesion pathways (Figure 6).

3.3.3. Expression Difference Analysis of Cytokine, Phagocytosis, and Apoptosis-Related Genes between AS and IS Group Turtles

The fold changes of differentially expressed cytokine, phagocytosis, and apoptosis-related genes were also analyzed in AS and IS group turtles (Tables 4–6). The results showed that in both AS and IS groups, the sharpest response of cytokines occurred at 6 hpi, while the number of differentially expressed cytokine and cytokine receptor genes in the IS group were overwhelmingly more than that in the AL group (30 up-regulated genes in the IS group vs. seven up-regulated genes in the AS group) (Table 4). Noteworthily, the overall expression of cytokines dramatically decreased at 24 hpi and 72 hpi in the IS group, while gradually decreased in the AS group, and the anti-inflammatory cytokine IL10 was up-regulated at all the time periods in the AS group (Table 4). Consistent with the response of cytokines, the expression of toll-like receptor signaling pathway genes were also relatively intense at 6 hpi, but only with three up-regulated genes in the AS group (Table 4), while in the IS group at 6 hpi, the expressions of three cytokine expression-mediating pathway genes, including toll-like receptor signaling pathway, NOD-like receptor signaling pathway, and RIG-I-like receptor signaling pathway, were overall up-regulated, with 18 up-regulated genes (Table 4).

Table 4. Fold changes of differentially expressed cytokine and cytokine receptor, and cytokine expression mediating pathway genes in the AS group and the IS group compared to the control.

Categories/ Gene Name	Description	Log$_2$ (Fold Changes)					
		AS Group			IS Group		
		6 Hpi	24 Hpi	72 Hpi	6 Hpi	24 Hpi	72 Hpi
		Interleukins and interleukin receptors					
IL1β	interleukin 1 beta				2.46		
IL6	interleukin 6				5.20		
IL7	interleukin 7				−2.79		
IL8	interleukin 8	3.27	2.43		2.41		
IL10	interleukin 10	6.82	5.69	1.21	7.30		
IL1R2	interleukin 1 receptor type II	4.13			3.81		
IL1RAP	interleukin 1 receptor accessory protein				2.24		
IL3RB	cytokine receptor common subunit beta				3.56		
IL4R	interleukin 4 receptor				1.18		
IL5RA	interleukin 5 receptor alpha	−1.95	−3.81			−2.85	
IL5RB	interleukin 5 receptor alpha				3.56		
IL8RB	interleukin 8 receptor	0.07			2.32		
IL12RB1	interleukin 12 receptor beta-1				2.60		
IL12RB2	interleukin 12 receptor beta-1				3.20		
IL15RA	interleukin 15 receptor alpha				3.68		
IL21R	interleukin 21 receptor				1.98		
IL22RA2	interleukin 22 receptor alpha 2	5.21			4.77	4.39	
		Chemokines and chemokine receptors					
CCL20	C-C motif chemokine 20	6.33			6.72		2.56
CXCL10	C-X-C motif chemokine 10				5.36		
CXCL11	C-X-C motif chemokine 11				4.32		
CXCL12	C-X-C motif chemokine 12				−2.00		
CXCL13	C-X-C motif chemokine 13	3.47			2.41		
CXCL14	C-X-C motif chemokine 14				1.81		2.07
CX3CL1	C-X3-C motif chemokine 1		−3.70		3.90		
CCR5	C-C chemokine receptor type 5				1.26		
CXCR4	C-X-C chemokine receptor type 4					−1.75	
XCR1	XC chemokine receptor 1				−4.81		
		TNF family members and TNF receptors					
TNFSF8	tumor necrosis factor ligand superfamily member 8				3.06		
TNFSF10	tumor necrosis factor ligand superfamily member 10		−2.86		−4.23	−3.17	
TNFSF12	tumor necrosis factor ligand superfamily member 12				−1.99		
TNFSF15	tumor necrosis factor ligand superfamily member 15				3.09		

Table 4. *Cont.*

Categories/ Gene Name	Description	Log2 (Fold Changes)					
		AS Group			IS Group		
		6 Hpi	24 Hpi	72 Hpi	6 Hpi	24 Hpi	72 Hpi
TNFSF18	tumor necrosis factor ligand superfamily member 18				4.56		
EDA	ectodysplasin-A				−2.93		
SF6B	tumor necrosis factor receptor superfamily member 6B				7.29		
SF9	tumor necrosis factor receptor superfamily member 9				3.26		
SF12A	tumor necrosis factor receptor superfamily member 12A		6.32		7.32	7.37	
SF13B	tumor necrosis factor receptor superfamily member 13B			2.33			3.08
SF19L	tumor necrosis factor receptor superfamily member 19-like				5.98		
FAS	tumor necrosis factor receptor superfamily member 6				2.34		
NGFR	tumor necrosis factor receptor superfamily member 16				−3.24		
EDAR	ectodysplasin-A receptor				−4.72		
Toll-like receptor (TLR) signaling pathway							
TLR4	toll-like receptor 4				2.30	1.71	
TLR5	toll-like receptor 5	2.77	1.82		3.00		
MyD88	myeloid differentiation factor 88				1.08		
TRIF	toll-like receptor adapter molecule 1				1.96		
Rac	Ras-related C3 botulinum toxin substrate 1	2.37	2.52	1.97	1.84	2.51	2.04
PI3K	phosphoinositide-3-kinase				−4.06		
AKT	RAC serine/threonine-protein kinase				1.26		
MAP2K1	mitogen-activated protein kinase kinase 1	2.88	2.82	2.50	2.85	2.94	2.57
MAP2K6	mitogen-activated protein kinase kinase 6				−2.37		
AP-1	proto-oncogene protein c-fos		1.60		1.33		
STAT1	signal transducer and activator of transcription 1				2.03		
IKBKE	inhibitor of nuclear factor kappa-B kinase subunit epsilon				1.92		

Table 4. *Cont.*

Categories/ Gene Name	Description	Log2 (Fold Changes)					
		AS Group			IS Group		
		6 Hpi	24 Hpi	72 Hpi	6 Hpi	24 Hpi	72 Hpi
	RIG-I-like receptor (RLR) signaling pathway						
RIG-I	retinoic acid inducible gene I				2.32		
LGP2	laboratory of genetics and physiology 2				1.90		
MDA5	melanoma differentiation-associated gene 5				1.54		
TRAF2	TNF receptor-associated factor 2				2.19		
MITA	Mediator of IFN regulatory transcription factor 3 activation				1.29		
	NOD-like receptor (NLR) signaling pathway						
NLRP12	NACHT, LRR and PYD domains-containing protein 12		−2.65		−2.05	−3.11	
ASC	apoptosis-associated speck-like protein containing a CARD				2.27		
RIPK2	receptor-interacting serine/threonine-protein kinase 2				1.46		
CARD8	caspase recruitment domain-containing protein 8				1.49		

Table 5. Fold changes of differentially expressed phagocytosis-related genes in the AS group and the IS group compared to the control.

Categories/ Gene Name	Description	Log2 (Fold Changes)					
		AS Group			IS Group		
		6 Hpi	24 Hpi	72 Hpi	6 Hpi	24 Hpi	72 Hpi
	Internalization and formation of the phagosome						
TLR4	toll-like receptor 4				2.30	1.71	
MR	mannose receptor	−2.71	−1.43		−3.02		
CR1	complement receptor 1		1.69			2.15	
αvβ5	integrin αvβ5				−2.45		
iC3b	the fragment of complement component 3				−3.36		
Collectin	C-type lectin	−4.00	−1.66		−4.85		
F-actin	actin beta/gamma 1	3.60		−3.11	−3.66	−3.71	−2.46
	Early phagosome						
Rab5	ras-related protein Rab-5B	2.47		1.84		1.82	
vATPase	V-type H^+-transporting ATPase	2.21			−1.01	1.59	
CALR	calreticulin					2.10	
	Mature phagosome						
TUBA	tubulin alpha				1.14	1.58	1.63
TUBB	tubulin beta	3.63	1.60		−1.21	2.28	2.34
vATPase	V-type H^+-transporting ATPase	2.21			−1.01	1.59	
Dynein	dynein heavy chain 2				−1.82		

Table 5. *Cont.*

Categories/ Gene Name	Description	Log$_2$ (Fold Changes)					
		AS Group			IS Group		
		6 Hpi	24 Hpi	72 Hpi	6 Hpi	24 Hpi	72 Hpi
		Phagolysosome					
sec61	protein transport protein SEC61 subunit beta				1.36		
vATPase	V-type H^+-transporting ATPase	2.21			−1.01	1.59	
NOS	nitric-oxide synthase				−3.07		
TAP	ATP-binding cassette subfamily B member 3				2.17		
		Activation of NADPH oxidase					
gp91	NADPH oxidase 1		1.62		1.57	1.40	
p40phox	neutrophil cytosolic factor 4				1.27		
p47phox	neutrophil cytosolic factor 1	2.75	2.32		2.53	2.04	1.52
P67phox	neutrophil cytosolic factor 2					1.40	
Rac	Ras-related C3 botulinum toxin substrate 1	2.37	2.52	1.97	1.84	2.51	2.04
		Antigen presentation					
MHC II	MHC class II antigen	3.14	−1.87		−2.54		

Table 6. Fold changes of up-regulated apoptosis-related genes in the AS group and the IS group compared to the control.

Gene Name	Description	Log$_2$ (Fold Changes)					
		AS Group			IS Group		
		6 Hpi	24 Hpi	72 Hpi	6 Hpi	24 Hpi	72 Hpi
p53	tumor protein p53				1.66		
IP3R	inositol 1,4,5-triphosphate receptor type 3				1.17		
Perforin	perforin 1	3.24			5.58		
MERK2	mitogen-activated protein kinase kinase 2	2.88	2.82	2.50	2.85	2.95	2.57
Cathepsin	cathepsin B		6.19	3.24	1.57	2.81	
AP1	proto-oncogene protein c-fos		1.60		1.33		
GZMB	granzyme B	1.22		1.57	5.59		5.34
IL-3R	cytokine receptor common subunit beta				2.56		
TRAF12	TNF receptor-associated factor 2	2.91			2.19		
Fas	tumor necrosis factor receptor superfamily member 6				2.34		
TrkA	neurotrophic tyrosine kinase receptor type 1				6.49		
NIK	mitogen-activated protein kinase kinase kinase 14				1.24		
FLIP	CASP8 and FADD-like apoptosis regulator				1.26		
eiF2α	translation initiation factor 2 subunit 1				1.25		
Calpain	calpain-1				1.49		
ARTS	septin 4				2.45		
AIF	apoptosis-inducing factor 1				1.40		
Gadd45	growth arrest and DNA-damage-inducible protein				2.32		
ASK1	mitogen-activated protein kinase kinase kinase 5					5.27	
CytC	cytochrome c					1.90	1.80

The intense response of phagocytosis started at 6 hpi in the AS group, with the up-regulation of seven DEGs, and lasted until 24 hpi (six up-regulated DEGs) (Table 5), while in the IS group at 6 hpi, the phagocytosis activity seemed to be inhibited, with 10 down-regulated DEGs, and just started at 24 hpi, with 11 up-regulated phagocytosis-related DEGs (Table 5). Nevertheless, in both the AS and IS groups, the sharpest response of apoptosis-related gene expression occurred at 6 hpi, while the number of up-regulated apoptosis-related genes in the IS group (18 DEGs) was significantly more than that in the AL group (four DEGs) (Table 6).

3.4. Validation of DEGs by qPCR

A total of nine DEGs were randomly selected to perform qPCR to validate RNA-Seq data and gene expression profiles (Figure 7). PCR products with expected sizes were successfully amplified with all the nine specific primer pairs, indicating their availabilities for DEG validation (data not shown). The different amplification efficiencies of the nine DEGs between the experimental and control groups were transformed by $\log_2$ (fold change) to compare with the results of RNA-Seq. The results showed that the expression patterns of these genes determined by qPCR were similar to those acquired through RNA-Seq (Figure 7), which confirmed the reliability of the RNA-Seq data. Therefore, the immune-related genes isolated in this study could be useful references for future studies on the molecular mechanisms of Chinese soft-shelled turtles during *A. hydrophila* infection.

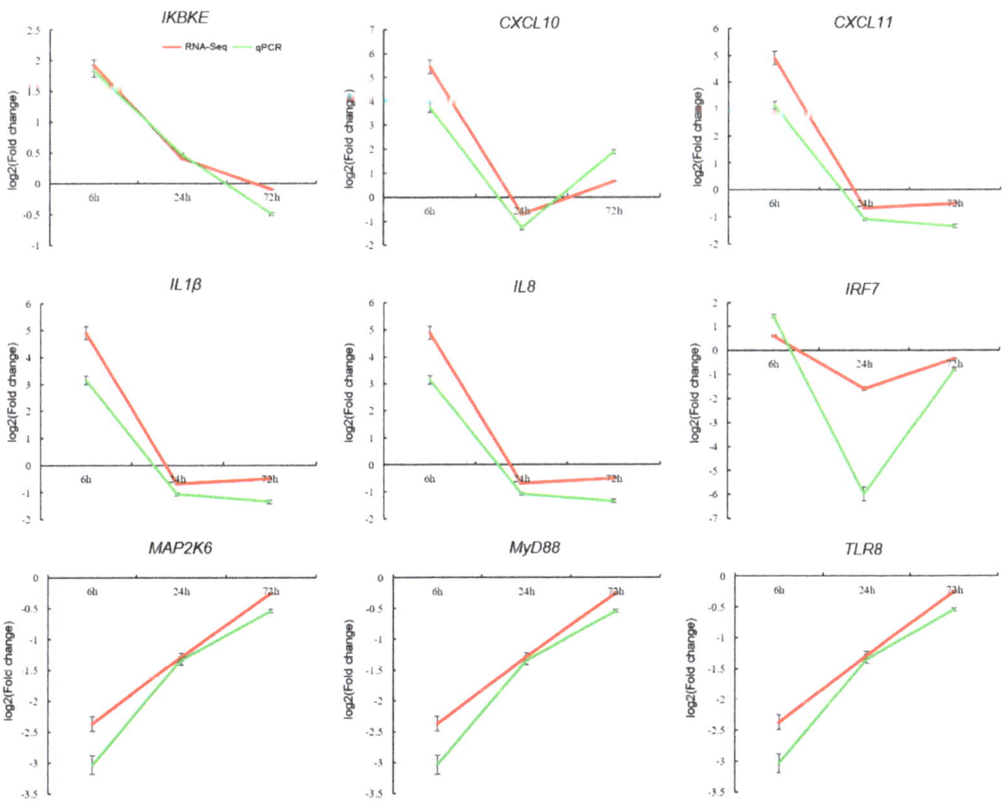

Figure 7. Validation of RNA-Seq results by qPCR. Nine DEGs are randomly selected, and the expressions of genes at different time periods are examined relative to the endogenous control genes (β-actin and GAPDH). The relative expression values are transformed into the $\log_2$ (fold change) form. The results are shown as the mean ± SEM of liver and spleen tissues derived from 3 individual turtles.

4. Discussion

For the nutriment and medical values, Chinese soft-shelled turtle *T. sinensis* has been developed into the largest cultured turtle species in East Asia, especially in China and Japan. Serious infectious diseases caused by pathogens including bacteria and viruses is threatening the aquaculture of turtles [3]; in particular, the hemorrhagic sepsis caused by *A. hydrophila*, with more than 15 kinds of diseases, is the most common and troublesome in turtle disease cases [36,37]. Previous studies have reported in mammals and fish that abnormal immune responses to pathogenic infections, such as excessive activation of immune cells and dysfunction of immune responses, can lead the immune system to attack self-uninfected cells, causing systemic inflammation, tissue hemorrhagic sepsis, and even death [38,39]. However, the research on the immune response mechanisms is limited, and the molecular pathology of turtles infected by *A. hydrophila* remains unclear, which greatly hinders the strategy innovations for disease prevention and control in Chinese soft-shelled turtles.

The susceptible and resistant individuals in the natural population have offered excellent materials to study the molecular pathology or molecular basis of resistance for pathogenic diseases in many animal species [40,41]. In livestock and poultry animals, for example, comparative transcriptomes were analyzed to reveal the molecular mechanism differences in response to *Mycoplasma hyopneumoniae* infection in two pig breeds [42]. These two breeds share DEGs that are involved in immune relevant pathways, including cytokine–cytokine receptor interaction pathway, PI3K-Akt signaling pathway, and chemokine signaling pathway [42]. The study demonstrates that more chemokines and interleukins are specifically and significantly up-regulated, which can enhance the immune responses and reduce the susceptibility to *M. hyopneumoniae* infection in resistant pig breed [42]. When cytokine gene expressions are compared between chicken line 6.3 (Marek's disease-resistant chicken) and line 7.2 (Marek's disease-susceptible chicken) in a transcriptome analysis, among the identified 53 cytokines and 96 cytokine receptors, 15 cytokines and 29 cytokine receptors highly expressed in line 6.3 were detected [43]. In aquaculture fish species, critical cytokines including, IL8 and TNFα, are significantly up-regulated in resistant channel catfish (*Ictalurus punctatus*), while susceptible catfish show high expression levels of IL17 in response to *Flavobacterium columnare* infection [44]. The gene networks involved in the apoptotic process are also associated with disease resistance/susceptibility to *Piscirickettsia salmonis* in Atlantic salmon [29]. This evidence indicates that the expression of immune-relevant genes interrelates with disease resistance/susceptibility to pathogenic diseases in animal hosts. In the present study, two phenotypes of Chinese soft-shelled turtles were found after *A. hydrophila* infection. One group of turtles were active in feeding and moving, with no obvious pathological symptoms, which were considered as the resistant turtles, while the other group of turtles showed obviously pathological symptoms, with swelling and congestion in liver and spleen after *A. hydrophila* infection and the reduction of food intake and movement, which are regarded as susceptible turtles. Comparative liver and spleen transcriptomes from these two groups of turtles at different time periods (6, 24, and 72 hpi) were further analyzed to reveal the molecular basis of resistance/susceptibility for turtles infected by *A. hydrophila*. The results indicate that the expression of cytokine, apoptosis-, and phagocytosis-related genes in both liver and spleen of the inactive turtles is significantly distinct from those in the active turtles analyzed by KEGG pathway enrichment. Therefore, we infer that these gene expression differences may be related to the molecular pathology or resistant basis to *A. hydrophila* infection in Chinese soft-shelled turtles.

Cytokines are a class of low-molecular-weight-secreted proteins that can transduce signals between cells and exert immune regulation and effector functions [45]. They play important roles in the immune system by regulating the intensity and duration of immune responses [46]. During pathogen infection, cytokines produced by immune cells trigger an inflammatory response, which is essential for the early elimination of pathogens [47]. However, the lasting or excessive production and release of cytokines may initiate the "cytokine storm", which often leads to various diseases, including hemorrhagic septicemia and even the failure of key organs or death for animal hosts [48]. It has been reported in many aquaculture fish species that abnormal expression of cytokines is linked to serious hemorrhagic septicemia. For example, hemorrhagic septicemia of mandarin fish (*Siniperca chuatsi*) is mainly caused by *A. hydrophila* infection [49]. Histopathological analysis reveals that inflammation, vacuolization, and extensive necrosis exist in the gill, liver, spleen, and head kidney of the diseased mandarin fish [49]. The mRNA expression levels of pro-inflammatory cytokines including TNFα, CCL3, and IL8 are sharply up-regulated in spleen and head kidney of mandarin fish post-*A. hydrophila* infection [49]. Coincidentally, Chinese perch infected with *A. hydrophila* also shows significantly high mRNA expression levels of pro-inflammatory cytokines, such as TNFα, and IL1β, compared to healthy fish [50]. In tambaqui (*Colossoma macropomum*), IL1β and complement component 4 are intensely up-regulated post-*A. hydrophila* infection [51]. With transcriptome analysis, the KEGG pathways associated with disease and immune responses, such as the cytokine–cytokine receptor interaction, complement and coagulation cascades, and inflammatory bowel disease, are also enriched in *Leiocassis longirostris* with *A. hydrophila* infection [52]. This evidence has suggested the involvement of pro-inflammatory cytokines in pathogenesis of hemorrhagic septicemia caused by *A. hydrophila* infection in fish. In the present study, the up-regulated number of cytokine and cytokine receptor genes are far more, and their up-regulations are more intense in inactive turtles than those in the active turtles. Especially in the spleen of inactive turtles, the significant up-regulations of 12 pro-inflammatory cytokines, including IL1β, IL6, IL8, CCL20, CXCL10, CXCL11, CXCL13, CXCL14, CX3CL1, TNFSF8, TNFSF15, TNFSF18, and 17 cytokine receptors, were identified at 6 hpi. The excessive expression of pro-inflammatory cytokines and their receptors have been confirmed to bring about uncontrolled inflammation, and lead to the pathological changes in tissues or key organs, and systemic hemorrhagic sepsis [48]. Since the high expression of pro-inflammatory cytokines, it is reasonable that the spleen and liver of inactive turtles exhibits the symptoms of hemorrhagic sepsis after *A. hydrophila* infection. In addition, it is worth noting that in both liver and spleen of the active turtles, IL10 is significantly up-regulated at all the tested time periods (6, 24, and 72 hpi). IL10 is well known as an important anti-inflammatory cytokine, which can prevent excessive tissue damage caused by bacterial and viral infections as well as pro-inflammatory responses [53]. Especially in the late phase of pathogen infection, IL10 serves the role in controlling the development of inflammatory diseases [54,55]. These above results collectively hint that the excessive expression of a large number of pro-inflammatory cytokines ("cytokine storm") triggers an imbalanced immune response, which should partly explain the molecular pathology of hemorrhagic sepsis in the liver and spleen of inactive turtles, while the lasting up-regulation of IL10 may be critical for maintaining the immune homeostasis in the active turtles during *A. hydrophila* infection.

In fact, the expression of cytokines is delicately induced and regulated, where signal pathways mediated by PRRs such as TLRs, NLRs, and RLRs undertake the indispensable roles [56]. The innate immune cells utilize PRRs to recognize the invading microorganisms, and trigger downstream immune-related signal cascades [56]. Although the downstream signaling pathways mediated by TLRs, NLRs, and RLRs are different, they all induce the production of specific cytokines. For example, the activation of TLR signaling pathway usually induces the expression of pro-inflammatory cytokines including TNFs, interleukins such as IL1β, IL6, IL8, and IL12, chemokines including CCL3 and CCL5, and interferon genes [57]. It is well known that the activation of RLR signaling pathway initiates interferon production to resist virus, and it can also induce the expression of TNFs, IL8, IL12, and other pro-inflammatory cytokines depending on NF-κB phosphorylation [58], while the activation of NLR signaling pathway is only associated with the induction of pro-inflammatory cytokines such as TNFs, IL1β, IL6, IL8, IL12, CCL3, and CCL5 [59]. In the present study, we found that in both the active and inactive turtles, the TLR-, NLR-, and RLR-mediated signaling pathways exhibited different degrees of activation along with the up-regulation of cytokines after *A. hydrophila* challenge. The difference is that in liver and spleen of the active turtles, only the activation of the TLR signaling pathway is relatively intense when the cytokine expression is peaking at 24 hpi, while in liver and spleen of the inactive turtles, all the TLR, NLR, and RLR signaling pathways are significantly activated when the cytokine expression is peaking at 6 hpi. It has been reported in mammals and several fish species that immoderate activation of the PRRs-mediated signaling pathways causes excessive expression of pro-inflammatory cytokines, leading to the dysfunction of immune regulation and inflammatory disease [56,60]. Similarly, extensive activation of PRRs-mediated signaling pathways may be the key reason for the excessive expression of pro-inflammatory cytokines, which can explain the molecular pathology of hemorrhagic sepsis in inactive turtles after *A. hydrophila* infection.

Apoptosis is defined as programmed cell death, involved in many physiological processes including homeostasis maintenance, and developments of tissue and organ [61]. Conceptually, cell death appears to protect against most acute bacterial pathogens that infect hosts and, in many cases, even more successfully restricts nonpathogenic or opportunistic bacteria [62]. Therefore, apoptosis is considered as an intrinsic immune defense mechanism in response to microbial infections, and the apoptosis of infected cells is an effective way to eliminate pathogenic niches and prevent their further spreading [63]. It has been reported that bacterial infection sensed by PRRs induces NF-κB-dependent inflammatory cytokines, including those of the TNFs and ILs, which further promote inflammatory signaling through death receptors and induce apoptosis [64]. In Japanese flounder, cell apoptosis, along with the up-regulation of NLRP3, ASC, caspase-1, IL1β, and IL18, in the macrophages has been observed after *Edwardsiella tarda* infection [65]. In addition, the apoptosis of erythrocytes can be induced by *A. hydrophila* infection in grass carp, along with the up-regulation of CCL4, CCL11, CCL20, IL4, and IL12 [66]. These studies indicate that the inflammation caused by bacterial infection is often accompanied by cell apoptosis in hosts. In the present study, we found that in active turtles, only several apoptosis-related genes were significantly up-regulated at 6 hpi in both liver and spleen, and their expressions gradually decreased at 24 and 48 hpi, while in inactive turtles, the up-regulation of a large number of apoptosis-related genes, as well as inflammatory cytokines, including TNFs, ILs, and chemokines, were observed at 6 hpi, and the up-regulation of apoptosis-related genes can last to 72 hpi in both liver and spleen. These results confirm that the up-regulation of apoptosis-related genes in the inactive turtles is more intense and lasting than that in the active turtles. Since exuberant cell apoptosis often accompanies tissue damage and causes the pathological changes in parenchymal organs [67], excessive apoptosis-related gene expression may also be involved in the molecular pathology of hemorrhagic sepsis in the liver and spleen of turtles after *A. hydrophila* infection.

Phagocytosis constitutes an important immune response of immunocytes as the first line of defense to recognize and engulf foreign particles or self-apoptotic cells, followed by the digestion and clearance [68]. It is a primitive conserved innate immune defense mechanism for all metazoans, including vertebrates and invertebrates [69]. Macrophages, neutrophils, and dendritic cells are professional phagocytes that are able to phagocytose large foreign particles (with the diameter of >0.5 μm) such as bacteria [70]. Effective phagocytosis requires two components: particle internalization and phagosome maturation [71]. After the bacteria are recognized by the phagocytes, they undergo endocytosis to form the phagosome in the phagocytes [71]. The nascent or early phagosome has no killing activity, and they must transform into the mature phagosome to obtain the bactericidal properties [71]. The mature phagosome further fuses with the lysosome to form the phagolysosome where there are various bactericidal substances such as reactive oxygen species, and hydrolytic enzymes, including protease, polysaccharase, nuclease, and lipase, that can kill and digest the invading bacteria [71]. After digestion, most of the bacterial residues are discharged outside the host phagocytes, and part of the bacterial degradation products are presented onto the surface of antigen-presenting cells by MHC molecules, which promotes adaptive immune responses [72]. In the present study, we observed the expression differences of phagocytosis-related genes that are involved in the processes including internalization and formation of the phagosomes, early/mature phagosome, phagolysosome, activation of NADPH oxidase, and antigen presentation between the active and inactive turtles in response to *A. hydrophila* infection. The results indicate that in active turtles, most of the phagocytosis-related genes are significantly up-regulated at 6 hpi and the up-regulation could last until 24 and 72 hpi in both liver and spleen, while in the inactive turtles, only several phagocytosis-related genes are significantly up-regulated in the liver, and even the majority of phagocytosis-related genes are significantly down-regulated in the spleen at 6 hpi; up to 24 hpi, most of the phagocytosis-related genes are significantly up-regulated in the liver and spleen. At 72 hpi, only a small part of the phagocytosis-related genes is up-regulated. Overall, the activity of phagocytosis in the inactive turtles starts later than that in the active turtles. These results suggest that the lag of phagocytosis can lead to the inability to clear bacteria, which also may be one of the important reasons for the persistent inflammation caused by *A. hydrophila* proliferation in the inactive turtles.

5. Conclusions

In summary, the molecule immune responses of turtles infected by *A. hydrophila* was analyzed, for the first time, by comparative transcriptomes from two group of turtles with different susceptibility to *A. hydrophila* infection. The gene expression profiles indicate that the dysfunction of immune responses, including excessive activation of pro-inflammatory cytokines, PRRs-mediated signaling pathway, and apoptosis, and insufficient phagocytosis activity may contribute to the molecular pathology of hemorrhagic sepsis in liver and spleen of turtles during *A. hydrophila* infection (Figure 8). Although there was a lack of further functional verification for the suspected genes, the data of comparative transcriptomes in this study will provide useful information for future studies on the molecular immunopathogenesis after *A. hydrophila* infection or genetic improvements against hemorrhagic sepsis in Chinese soft-shelled turtles.

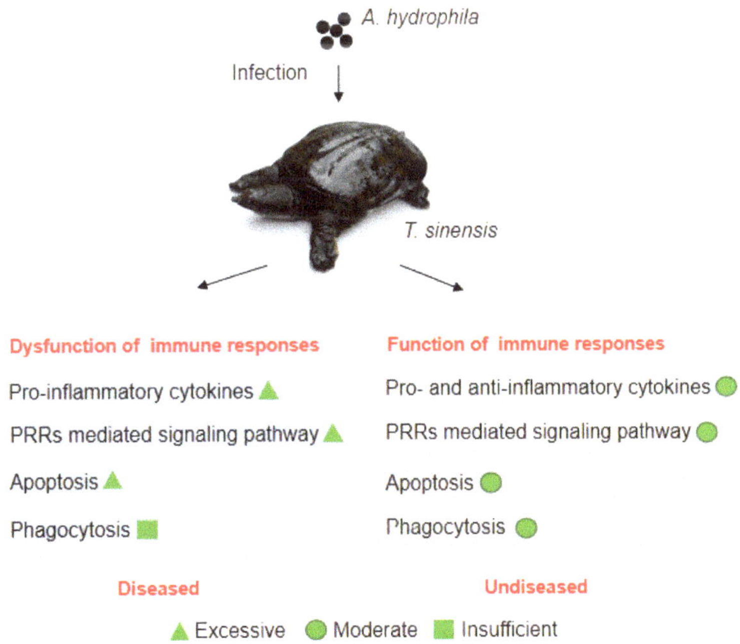

Figure 8. The suspected molecular immunopathogenesis of hemorrhagic sepsis caused by *A. hydrophila* infection in Chinese soft-shelled turtles.

Supplementary Materials: The following are available online at https://www.mdpi.com/article/10.3390/biology10111218/s1. Supplementary Table S1: Details of RNA-Seq data. L, liver; S, spleen; AL group, liver in active turtles; IL group, liver in inactive turtles; AS group, spleen in active turtles; IS group, spleen in inactive turtles. One library was constructed for liver and spleen tissue from the control group, while three libraries were constructed for liver and spleen tissue from AL, IL, AS, and IS groups at 6, 24, and 72 hpi (hours post-injection), except that only two libraries from IL group at 6 hpi because one library was failed in quality. In total, 37 libraries were constructed and sequenced. Supplementary Table S2: The list of up-regulated KEGG pathways in livers from active and inactive turtles compared to that in control group. Supplementary Table S3: The list of down-regulated KEGG pathways in livers from active and inactive turtles compared to that in control group. Supplementary Table S4: The list of up-regulated KEGG pathways in spleens from active and inactive turtles compared to that in control group. Supplementary Table S5: The list of down-regulated KEGG pathways in spleens from active and inactive turtles compared to that in control group. Supplementary Table S6: The sequence information of primers for qPCR in this study.

Author Contributions: Conceptualization, C.Z., J.T. and Z.L.; methodology, J.T., X.L. and Z.L.; validation, J.T.; formal analysis, J.T., Y.H. and Z.L.; investigation, J.T. and Z.L.; data curation, Z.L.; writing—original draft preparation, J.T. and Z.L.; writing—review and editing, Z.L.; visualization, J.T. and Z.L.; supervision, C.Z.; funding acquisition, C.Z., Z.L., Y.H. and X.W. All authors have read and agreed to the published version of the manuscript.

Funding: The work is supported by grants from the Natural Science Foundation of Hunan Province (S2017JJQNJJ1393), Postdoctoral Research Foundation of China (2019M662783), Key projects of Hunan Provincial Education Department (No.19A222) and National Key Research and Development Project: Blue granary (2018YFD0900200).

Institutional Review Board Statement: Animal experiments in this study are approved by the Animal Care and Use Committee of Hunan Agricultural University (Approval Code: 201903297; Approval Date: 11 October 2019).

Informed Consent Statement: Not applicable.

Data Availability Statement: The fastq files containing the raw reads obtained from each Illumina library in this study could be obtained from the Sequence Read Archive (SRA) under BioProject PRJNA781380. All the data are also available from the corresponding author upon reasonable request.

Conflicts of Interest: The authors declare no conflict of interest. The funding sponsors had no role in the design of the study; in the collection, analyses, or interpretation of data; in the writing of the manuscript, and in the decision to publish the results.

Abbreviations

Akt, related to the A and C kinase; CCL, C-C motif chemokine; CFU, colony forming unit; CXCL, C-X-C motif chemokine; ECM, extracellular matrix; GAPDH, glyceraldehyde-3-phosphate dehydrogenase; IL, interleukin; KEGG, Kyoto Encyclopedia of Genes and Genomes; MHC, major histocompatibility complex; NADPH, nicotinamide adenine dinucleotide phosphate coenzyme II; NCBI, National Center for Biotechnology Information; NF-κB, nuclear factor kappa-B; NLR, nucleotide-binding domain-like receptor; NOD, nucleotide-binding domain; PCR, polymerase chain reaction; PI3K, phosphoinositide-3-kinase; PRR, pattern recognition receptor; RIG-I, retinoic acid-inducible gene I; RLR, retinoic acid-inducible gene I-like receptor; TLR, toll-like receptor; TNF, tumor necrosis factor; TNFSF, tumor necrosis factor ligand superfamily member. This list only addresses the abbreviations in contexts. The abbreviations in Tables 1–6 have been explained separately.

References

1. Jia, Y.; Yang, Z.; Hao, Y.; Gao, Y. Effects of animal-plant protein ratio in extruded and expanded diets on nitrogen and energy budgets of juvenile Chinese soft-shelled turtle (*Pelodiscus sinensis* Wiegmann). *Aquac. Res.* **2005**, *36*, 61–68. [CrossRef]
2. Chen, J.; Zhu, N.; Kong, L.; Bei, Y.; Zheng, T.; Ding, X.; He, Z. First reported fatal Bacillus thuringiensis infections in Chinese soft-shelled turtles (*Trionyx sinensis*). *Aquaculture* **2014**, *428–429*, 16–20. [CrossRef]
3. Chen, Z.-X.; Zheng, J.-C.; Jiang, Y.-L. A new iridovirus isolated from soft-shelled turtle. *Virus Res.* **1999**, *63*, 147–151. [CrossRef]

4. Chung, T.; Yi, S.; Kim, B.; Kim, W.; Shin, G. Identification and antibiotic resistance profiling of bacterial isolates from septicaemic soft-shelled turtles (*Pelodiscus sinensis*). *Vet. Med.* **2017**, *62*, 169–177. [CrossRef]
5. Chen, J.; Zhu, N.; Kong, L.; Bei, Y.; Zheng, T.; Ding, X.; He, Z. First case of soft shell disease in Chinese soft-shelled turtle (*Trionyx sinens*) associated with Aeromonas sobria—A. veronii complex. *Aquaculture* **2013**, *406*, 62–67. [CrossRef]
6. Cahill, M.M. Virulence factors in motile *Aeromonas* species. *J. Appl. Bacteriol.* **1990**, *69*, 1–16. [CrossRef]
7. Pasquale, V.; Baloda, S.B.; Dumontet, S.; Krovacek, K. An Outbreak of Aeromonas hydrophila Infection in Turtles (*Pseudemis scripta*). *Appl. Environ. Microbiol.* **1994**, *60*, 1678–1680. [CrossRef] [PubMed]
8. Das, A.; Sahoo, P.; Mohanty, B.; Jena, J. Pathophysiology of experimental Aeromonas hydrophila infection in *Puntius sarana*: Early changes in blood and aspects of the innate immune-related gene expression in survivors. *Veter. Immunol. Immunopathol.* **2011**, *142*, 207–218. [CrossRef]
9. Mares, C.A.; Ojeda, S.S.; Morris, E.G.; Li, Q.; Teale, J.M. Initial delay in the immune response to *Francisella tularensis* is followed by Hypercytokinemia Characteristic of severe sepsis and correlating with upregulation and release of damage-associated molecular patterns. *Infect. Immun.* **2008**, *76*, 3001–3010. [CrossRef]
10. Coperchini, F.; Chiovato, L.; Rotondi, M. Interleukin-6, CXCL10 and infiltrating macrophages in COVID-19-related cytokine storm: Not one for all but all for one! *Front. Immunol.* **2021**, *12*, 668507. [CrossRef]
11. Sattler, S. The role of the immune system beyond the fight against infection. *Adv. Exp. Med. Biol.* **2017**, *1003*, 3–14. [CrossRef]
12. Schultz, M.; van der Poll, M.J.S.A.T. Modulation of innate immune responses in the treatment of sepsis and pneumonia. *Curr. Drug Target.* **2004**, *3*, 11–17. [CrossRef] [PubMed]
13. Qiu, Y.; Tu, G.-W.; Ju, M.-J.; Yang, C.; Luo, Z. The immune system regulation in Sepsis: From innate to adaptive. *Curr. Protein Pept. Sci.* **2019**, *20*, 799–816. [CrossRef] [PubMed]
14. Hotchkiss, R.S.; Monneret, G.; Payen, D. Sepsis-induced immunosuppression: From cellular dysfunctions to immunotherapy. *Nat. Rev. Immunol.* **2013**, *13*, 862–874. [CrossRef]
15. de Castro, I.F.; Guzmán-Fulgencio, M.; García-Álvarez, M.; Resino, S. First evidence of a pro-inflammatory response to severe infection with influenza virus H1N1. *Crit. Care* **2010**, *14*, 115. [CrossRef] [PubMed]
16. Yuan, S.; Jiang, S.-C.; Zhang, Z.-W.; Fu, Y.-F.; Hu, J.; Li, Z.-L. Quantification of cytokine storms during virus infections. *Front. Immunol.* **2021**, *12*, 659419. [CrossRef]
17. Flajnik, M.F. A cold-blooded view of adaptive immunity. *Nat. Rev. Immunol.* **2018**, *18*, 438–453. [CrossRef]
18. Liang, Q.; Li, W.; Guo, N.; Tong, C.; Zhou, Y.; Fang, W.; Li, X. Identification and functional analysis of Interleukin-1β in the Chinese soft-shelled turtle *Pelodiscus sinensis*. *Genes* **2016**, *7*, 18. [CrossRef]
19. Zhang, Z.; Tian, M.; Song, R.; Xing, X.; Fan, Y.; Wang, L.; Niu, C.; Dalmo, R.A. A New IL6 Isoform in Chinese soft-shelled turtle (*Pelodiscus sinesis*) discovered: Its regulation during cold stress and infection. *Biology* **2020**, *9*, 111. [CrossRef]
20. Zhou, X.; Feng, H.; Guo, Q.; Dai, H. Identification and characterization of the first reptilian CD9, and its expression analysis in response to bacterial infection. *Dev. Comp. Immunol.* **2010**, *34*, 150–157. [CrossRef]
21. Liu, T.; Han, Y.; Chen, S.; Zhao, H. Genome-wide identification of Toll-like receptors in the Chinese soft-shelled *turtle Pelodiscus sinensis* and expression analysis responding to *Aeromonas hydrophila* infection. *Fish. Shellfish. Immunol.* **2019**, *87*, 478–489. [CrossRef]
22. Zhou, X.; Guo, Q.; Dai, H. Molecular characterization and expression profiles in response to bacterial infection of Chinese soft-shelled turtle interleukin-8 (IL-8), the first reptilian chemokine gene. *Dev. Comp. Immunol.* **2009**, *33*, 838–847. [CrossRef]
23. Zhang, Z.; Chen, B.; Yuan, L.; Niu, C. Acute cold stress improved the transcription of pro-inflammatory cytokines of Chinese soft-shelled turtle against *Aeromonas hydrophila*. *Dev. Comp. Immunol.* **2015**, *49*, 127–137. [CrossRef] [PubMed]
24. Elbahnaswy, S.; Elshopakey, G.E. Differential gene expression and immune response of Nile tilapia (*Oreochromis niloticus*) challenged intraperitoneally with *Photobacterium damselae* and *Aeromonas hydrophila* demonstrating immunosuppression. *Aquaculture* **2020**, *526*, 735364. [CrossRef]
25. Li, C.; Jiang, J.; Xie, J.; Yang, W.; Wang, Y. Transcriptome profiling and differential expression analysis of the immune-related genes during the acute phase of infection with *Mycobacterium marinum* in the goldfish (*Carassius auratus* L.). *Aquaculture* **2020**, *533*, 736198. [CrossRef]
26. Zhu, C.; Pan, Z.; Chang, G.; Wu, N.; Wang, H.; Ding, H.; Qiang, X.; Zhang, L.; Qiang, J.; Zhang, J.; et al. Transcriptomic insights into immune responses to ulcerative syndrome in *Pseudobagrus ussuriensis*. *Aquaculture* **2021**, *537*, 736504. [CrossRef]
27. Zhou, X.; Zhang, G.-R.; Ji, W.; Shi, Z.-C.; Ma, X.-F.; Luo, Z.-L.; Wei, K.-J. The dynamic immune response of yellow catfish (*Pelteobagrus fulvidraco*) infected with *Edwardsiella ictaluri* presenting the inflammation process. *Front. Immunol.* **2021**, *12*, 625928. [CrossRef] [PubMed]
28. Eslamloo, K.; Caballero-Solares, A.; Inkpen, S.M.; Emam, M.; Kumar, S.; Bouniot, C.; Avendaño-Herrera, R.; Jakob, E.; Rise, M.L. Transcriptomic profiling of the adaptive and innate immune responses of atlantic salmon to *Renibacterium salmoninarum* infection. *Front. Immunol.* **2020**, *11*, 567838. [CrossRef] [PubMed]
29. Moraleda, C.P.; Robledo, D.; Gutiérrez, A.P.; Del-Pozo, J.; Yáñez, J.M.; Houston, R.D. Investigating mechanisms underlying genetic resistance to Salmon Rickettsial Syndrome in Atlantic salmon using RNA sequencing. *BMC Genom.* **2021**, *22*, 156. [CrossRef]
30. Kim, D.; Paggi, J.M.; Park, C.; Bennett, C.; Salzberg, S.L. Graph-based genome alignment and genotyping with HISAT2 and HISAT-genotype. *Nat. Biotechnol.* **2019**, *37*, 907–915. [CrossRef] [PubMed]

31. Pertea, M.; Pertea, G.M.; Antonescu, C.M.; Chang, T.-C.; Mendell, J.T.; Salzberg, S.L. StringTie enables improved reconstruction of a transcriptome from RNA-seq reads. *Nat. Biotechnol.* **2015**, *33*, 290–295. [CrossRef] [PubMed]
32. Wang, L.; Feng, Z.; Wang, X.; Wang, X.; Zhang, X. DEGseq: An R package for identifying differentially expressed genes from RNA-seq data. *Bioinformatics* **2010**, *26*, 136–138. [CrossRef]
33. Camacho, C.; Coulouris, G.; Avagyan, V.; Ma, N.; Papadopoulos, J.; Bealer, K.; Madden, T.L. BLAST+: Architecture and applications. *BMC Bioinform.* **2009**, *10*, 421. [CrossRef] [PubMed]
34. Mao, X.; Tao, C.; Olyarchuk, J.; Wei, L. Automated genome annotation and pathway identification using the KEGG Orthology (KO) as a controlled vocabulary. *Bioinformatics* **2005**, *21*, 3787–3793. [CrossRef]
35. Rao, X.; Huang, X.; Zhou, Z.; Xin, L. An improvement of the 2(-delta delta CT) method for quantitative real-time polymerase chain reaction data analysis. *Biostat. Bioinforma. Biomath.* **2013**, *3*, 71–85. [CrossRef] [PubMed]
36. Yang, Z.; Pan, H.; Sun, H. The immune response and protective efficacy of oral alginate microparticle *Aeromonas sobria* vaccine in soft-shelled turtles (*Trionyx sinensis*). *Veter. Immunol. Immunopathol.* **2007**, *119*, 299–302. [CrossRef]
37. Musgrave, K.E.; Diehl, K.; Mans, C. Aeromonas Hydrophila Keratitis in Freshwater Turtles. *J. Exot. Pet. Med.* **2016**, *25*, 26–29. [CrossRef]
38. Zhang, B.; Li, C.; Wang, X.; Liu, C.; Zhou, H.; Mai, K.; He, G. Administration of commensal *Shewanella* sp. MR-7 ameliorates lipopolysaccharide-induced intestine dysfunction in turbot (*Scophthalmus maximus* L.). *Fish. Shellfish. Immunol.* **2020**, *102*, 460–468. [CrossRef]
39. Pradel, B.; Robert-Hebmann, V.; Espert, L. Regulation of Innate Immune Responses by Autophagy: A Goldmine for Viruses. *Front. Immunol.* **2020**, *11*, 578038. [CrossRef]
40. Yang, M.; Wang, Q.; Wang, S.; Wang, Y.; Zeng, Q.; Qin, Q. Transcriptomics analysis reveals candidate genes and pathways for susceptibility or resistance to Singapore grouper iridovirus in orange-spotted grouper (*Epinephelus coioides*). *Dev. Comp. Immunol.* **2019**, *90*, 70–79. [CrossRef]
41. Zhu, M.; Nan, Y.; Zhai, M.; Wang, M.; Shao, Y.; Blair, H.T.; Morris, S.T.; Kenyon, P.R.; Zhao, Z.; Zhang, H. Comparative profiling of the resistance of different genotypes of mannose-binding lectin to *Mycoplasma pneumoniae* infection in Chinese Merino sheep based on high-throughput sequencing technology. *Veter. Immunol. Immunopathol.* **2021**, *233*, 110183. [CrossRef] [PubMed]
42. Ni, L.G.; Song, C.Y.; Wu, X.S.; Zhao, X.T.; Wang, X.Y.; Li, B.C.; Gan, Y. RNA-seq transcriptome profiling of porcine lung from two pig breeds in response to *Mycoplasma hyopneumoniae* infection. *PeerJ* **2019**, *7*, e7900. [CrossRef] [PubMed]
43. Truong, A.D.; Hong, Y.H.; Lillehoj, H.S. High-throughput sequencing reveals differing immune responses in the intestinal mucosa of two inbred lines afflicted with necrotic enteritis. *Veter. Immunol. Immunopathol.* **2015**, *166*, 116–124. [CrossRef] [PubMed]
44. Peatman, E.; Li, C.; Peterson, B.C.; Straus, D.L.; Farmer, B.D.; Beck, B.H. Basal polarization of the mucosal compartment in Flavobacterium columnare susceptible and resistant channel catfish (*Ictalurus punctatus*). *Mol. Immunol.* **2013**, *56*, 317–327. [CrossRef] [PubMed]
45. Feghali, C.; Wright, T. Cytokines in acute and chronic inflammation. *Front. Biosci.* **1997**, *2*, d12–d26. [CrossRef]
46. Shachar, I.; Karin, N. The dual roles of inflammatory cytokines and chemokines in the regulation of autoimmune diseases and their clinical implications. *J. Leukoc. Biol.* **2013**, *93*, 51–61. [CrossRef] [PubMed]
47. Murtaugh, M.P.; Baarsch, M.J.; Zhou, Y.; Scamurra, R.W.; Lin, G. Inflammatory cytokines in animal health and disease. *Veter. Immunol. Immunopathol.* **1996**, *54*, 45–55. [CrossRef]
48. D'Elia, R.V.; Harrison, K.; Oyston, P.C.; Lukaszewski, R.A.; Clark, G.C. Targeting the "Cytokine Storm" for therapeutic benefit. *Clin. Vaccine Immunol.* **2013**, *20*, 319–327. [CrossRef]
49. Chen, N.; Jiang, J.; Gao, X.; Li, X.; Zhang, Y.; Liu, X.; Yang, H.; Bing, X.; Zhang, X. Histopathological analysis and the immune related gene expression profiles of mandarin fish (*Siniperca chuatsi)* infected with *Aeromonas hydrophila. Fish. Shellfish. Immunol.* **2018**, *83*, 410–415. [CrossRef]
50. Liu, X.; Sun, W.; Zhang, Y.; Zhou, Y.; Xu, J.; Gao, X.; Zhang, S.; Zhang, X. Impact of *Aeromonas hydrophila* and infectious spleen and kidney necrosis virus infections on susceptibility and host immune response in Chinese perch (*Siniperca chuatsi*). *Fish. Shellfish. Immunol.* **2020**, *105*, 117–125. [CrossRef]
51. Gallani, S.U.; Valladão, G.M.R.; Alves, L.D.O.; Kotzent, S.; Hashimoto, D.T.; Wiegertjes, G.; Kirsten, K.; Kreutz, L.C.; Pilarski, F. Patterns of the innate immune response in tambaqui *Colossoma macropomum*: Modulation of gene expression in haemorrhagic septicaemia caused by *Aeromonas hydrophila. Microb. Pathog.* **2020**, *150*, 104638. [CrossRef]
52. Zhou, J.; Zhao, H.; Zhang, L.; Ye, X.; Wang, Z.; Li, Q.; Ke, H.; Zhao, G.; Du, J.; Yang, S.; et al. Effects of bacterial haemorrhagic septicemia on the immune response of *Leiocassis longirostris* by RNA-Seq and microRNA-Seq. *Comp. Biochem. Physiol. Part. D Genom. Proteom.* **2020**, *34*, 100659. [CrossRef]
53. Couper, K.N.; Blount, D.G.; Riley, E.M. IL-10: The master regulator of immunity to infection. *J. Immunol.* **2008**, *180*, 5771–5777. [CrossRef] [PubMed]
54. González, L.A.; Melo-González, F.; Sebastián, V.P.; Vallejos, O.P.; Noguera, L.P.; Suazo, I.D.; Schultz, B.M.; Manosalva, A.H.; Peñaloza, H.F.; Soto, J.A.; et al. Characterization of the anti-inflammatory capacity of IL-10-producing neutrophils in response to *Streptococcus pneumoniae* infection. *Front. Immunol.* **2021**, *12*, 578038. [CrossRef] [PubMed]
55. Saraiva, M.; O'Garra, A. The regulation of IL-10 production by immune cells. *Nat. Rev. Immunol.* **2010**, *10*, 170–181. [CrossRef] [PubMed]

56. Cao, X. Self-regulation and cross-regulation of pattern-recognition receptor signalling in health and disease. *Nat. Rev. Immunol.* **2016**, *16*, 35–50. [CrossRef]
57. Fitzgerald, K.A.; Kagan, J.C. Toll-like receptors and the control of immunity. *Cell* **2020**, *180*, 1044–1066. [CrossRef]
58. Stok, J.E.; Quiroz, M.E.V.; Van Der Veen, A.G. Self RNA Sensing by RIG-I–like Receptors in viral infection and sterile inflammation. *J. Immunol.* **2020**, *205*, 883–891. [CrossRef]
59. Platnich, J.M.; Muruve, D.A. NOD-like receptors and inflammasomes: A review of their canonical and non-canonical signaling pathways. *Arch. Biochem. Biophys.* **2019**, *670*, 4–14. [CrossRef]
60. Lu, Y.; Li, X.; Liu, S.; Zhang, Y.; Zhang, D. Toll-like receptors and inflammatory bowel disease. *Front. Immunol.* **2018**, *9*, 72. [CrossRef]
61. Place, D.E.; Lee, S.; Kanneganti, T.-D. PANoptosis in microbial infection. *Curr. Opin. Microbiol.* **2021**, *59*, 42–49. [CrossRef]
62. Jorgensen, I.; Rayamajhi, M.; Miao, E.A. Programmed cell death as a defence against infection. *Nat. Rev. Immunol.* **2017**, *17*, 151–164. [CrossRef] [PubMed]
63. Lacey, C.; Miao, E. Programmed cell death in the evolutionary race against bacterial virulence factors. *Cold Spring Harb. Perspect. Biol.* **2019**, *12*, a036459. [CrossRef] [PubMed]
64. Amarante-Mendes, G.P.; Adjemian, S.; Branco, L.M.; Zanetti, L.; Weinlich, R.; Bortoluci, K.R. Pattern recognition receptors and the host cell death molecular machinery. *Front. Immunol.* **2018**, *9*, 2379. [CrossRef] [PubMed]
65. Wang, X.; Kong, X.; Liu, X.; Wang, X.; Wang, Z.; Liu, J.; Zhang, Q.; Yu, H. Edwardsiella tarda triggers the pyroptosis of the macrophage of Japanese flounder (*Paralichthys olivaceus*). *Aquaculture* **2020**, *533*, 736153. [CrossRef]
66. Lu, Z.; Yang, M.; Zhang, K.; Zhan, F.; Li, F.; Shi, F.; Li, Y.; Zhao, L.; Li, J.; Lin, L.; et al. *Aeromonas hydrophila* infection activates death receptor apoptosis pathway in the red blood cells of grass carp (*Ctenopharyngodon idellus*). *Aquaculture* **2021**, *532*, 735956. [CrossRef]
67. Wang, Z.; Gu, Z.; Hou, Q.; Chen, W.; Mu, D.; Zhang, Y.; Liu, Q.; Liu, Z.; Yang, D. Zebrafish GSDMEb cleavage-gated pyroptosis drives septic acute kidney injury in vivo. *J. Immunol.* **2020**, *204*, 1929–1942. [CrossRef]
68. Greenberg, S.; Grinstein, S. Phagocytosis and innate immunity. *Curr. Opin. Immunol.* **2002**, *14*, 136–145. [CrossRef]
69. Stuart, L.M.; Ezekowitz, R.A. Phagocytosis and comparative innate immunity: Learning on the fly. *Nat. Rev. Immunol.* **2008**, *8*, 131–141. [CrossRef]
70. Browne, N.; Heelan, M.; Kavanagh, K. An analysis of the structural and functional similarities of insect hemocytes and mammalian phagocytes. *Virulence* **2013**, *4*, 597–603. [CrossRef]
71. Flannagan, R.S.; Cosío, G.; Grinstein, S. Antimicrobial mechanisms of phagocytes and bacterial evasion strategies. *Nat. Rev. Microbiol.* **2009**, *7*, 355–366. [CrossRef] [PubMed]
72. Underhill, D.M.; Goodridge, H.S. Information processing during phagocytosis. *Nat. Rev. Immunol.* **2012**, *12*, 492–502. [CrossRef] [PubMed]

biology

MDPI

Article

Metatranscriptomic Analysis Reveals an Imbalance of Hepatopancreatic Flora of Chinese Mitten Crab *Eriocheir sinensis* with Hepatopancreatic Necrosis Disease

Zeen Shen [1,†], Dhiraj Kumar [1,2,†], Xunmeng Liu [3], Bingyu Yan [1], Ping Fang [3], Yuchao Gu [1], Manyun Li [1], Meiping Xie [1], Rui Yuan [3], Yongjie Feng [1], Xiaolong Hu [1,4], Guangli Cao [1,4], Renyu Xue [1,4], Hui Chen [3], Xiaohan Liu [3] and Chengliang Gong [1,4,*]

1 School of Biology and Basic Medical Science, Soochow University, Suzhou 215123, China; Zeenshen0719@163.com (Z.S.); drkumarindia@163.com (D.K.); Yanbingyu0224@163.com (B.Y.); g1102851085@163.com (Y.G.); limy933@163.com (M.L.); xiemiep@163.com (M.X.); yjfeng@suda.edu.cn (Y.F.); xlhu2013@suda.edu.cn (X.H.); guanglicao@163.com (G.C.); xuery@suda.edu.cn (R.X.)

2 School of Studies in Zoology, Jiwaji University, Gwalior 474011, India

3 Jiangsu Center for Control and Prevention of Aquatic Animal Infectious Disease, Nanjing 210036, China; lxmlxmeng@163.com (X.L.); jsscykzx@163.com (P.F.); yr8624@163.com (R.Y.); chenhuijsbf@163.com (H.C.); xiaohanliu1975@163.com (X.L.)

4 Agricultural Biotechnology Research Institute, Agricultural Biotechnology and Ecological Research Institute, Soochow University, Suzhou 215123, China

* Correspondence: gongcl@suda.edu.cn

† These authors contribute equally to this paper.

Citation: Shen, Z.; Kumar, D.; Liu, X.; Yan, B.; Fang, P.; Gu, Y.; Li, M.; Xie, M.; Yuan, R.; Feng, Y.; et al. Metatranscriptomic Analysis Reveals an Imbalance of Hepatopancreatic Flora of Chinese Mitten Crab *Eriocheir sinensis* with Hepatopancreatic Necrosis Disease. *Biology* **2021**, *10*, 462. https://doi.org/10.3390/biology10060462

Academic Editor: Patricia Pereiro

Received: 16 April 2021
Accepted: 20 May 2021
Published: 23 May 2021

Simple Summary: The cause of Chinese mitten crab Eriocheir sinensis hepatopancreas necrosis disease (HPND) remains a mystery. In this study, metatranscriptomics sequencing was conducted to characterize the changes in the structure and gene expression of hepatopancreatic flora of crabs with and without typical symptoms of HPND; an imbalance of hepatopancreatic flora can be found in the crab with HPND, and the detected microbial taxa decreased, whereas the prevalence of Spiroplasma eriocheiris significantly increased in the hepatopancreatic flora of crabs with typical symptoms of HPND, and the relative abundances of the virus and microsporidia in crabs with HPND were very low and did not increase with disease progression. The differentially-expressed genes (DEGs) in hepatopancreatic flora between crabs with and without HPND were enriched ribosome, retinol metabolism, and biosynthesis of unsaturated fatty acid KEGG pathways. These results suggested that an imbalance of hepatopancreatic flora was associated with crab HPND, and the enriched pathways of DEGs were associated with the pathological mechanism of HPND.

Abstract: Hepatopancreas necrosis disease (HPND) of the Chinese mitten crab Eriocheir sinensis causes huge economic loss in China. However, the pathogenic factors and pathogenesis are still a matter of dissension. To search for potential pathogens, the hepatopancreatic flora of diseased crabs with mild symptoms, diseased crabs with severe symptoms, and crabs without visible symptoms were investigated using metatranscriptomics sequencing. The prevalence of Absidia glauca and Candidatus Synechococcus spongiarum decreased, whereas the prevalence of Spiroplasma eriocheiris increased in the hepatopancreatic flora of crabs with HPND. Homologous sequences of 34 viral species and 4 Microsporidian species were found in the crab hepatopancreas without any significant differences between crabs with and without HPND. Moreover, DEGs in the hepatopancreatic flora between crabs with severe symptoms and without visible symptoms were enriched in the ribosome, retinol metabolism, metabolism of xenobiotics by cytochrome P450, drug metabolism—cytochrome P450, biosynthesis of unsaturated fatty acids, and other glycan degradation. Moreover, the relative abundance of functions of DEDs in the hepatopancreatic flora changed with the pathogenesis process. These results suggested that imbalance of hepatopancreatic flora was associated with crab HPND. The identified DEGs were perhaps involved in the pathological mechanism of HPND; nonetheless, HPND did not occur due to virus or microsporidia infection.

Keywords: *Eriocheir sinensis*; hepatopancreas necrosis disease; metatranscriptomics sequencing; hepatopancreatic flora

1. Introduction

The Chinese mitten crab (*Eriocheir sinensis*) is one of the important crustaceans with great economic value [1,2]. Crab hepatopancreatic necrosis disease (HPND) had a high mortality rate of about 40–50%, which directly affects the crab farmers [1]. The typical clinical symptoms of crabs with HPND are hepatopancreas degeneration and atrophy, muscle atrophy, and a change in the color of the hepatopancreas from golden to pale yellow and white. The gastrointestinal tract is collapsed, and clear dropsy is observed in the inner cavity of the crab as the disease progresses [2]. Diseased crabs can survive for a long time, but have no commercial value because of the low content of lipids and proteins and the low growth rate.

Crab HPND was suggested to be caused by infection of microsporidian *Hepatospora eriocheir* [1]; however, *H. eriocheir* was not detected in all crabs with HPND [3], and the artificially-infected crabs with *H. eriocheir* had no typical HPND symptoms [4]. *Vibrio* was considered to be a causative agent of acute HPND of cultured shrimp, but significant changes in the relative abundance of hepatopancreatic flora were not found in crabs with and without HPND [5–8]. *Vibrio* was isolated from the hemolymph of crabs with HPND. However, typical symptoms of HPND were not generated in animal regression tests performed using *Vibrio*; therefore, *Vibrio* is not thought to be associated with HPND of crab [5–7]. Moreover, HPND of crab was not generated by injecting bacteria-free supernatants from the hepatopancreas of crabs with HPND into healthy crabs, suggesting HPND was not caused by viral infections [2,8].

A previous study found that symptoms of crab HPND could be caused by breeding crabs in water with a pH of 9.5–10.0 [2], and crabs exposed to low concentrations of insecticides caused clinical symptoms of HPND [9]. Epidemiological investigations suggest that HPND in mitten crabs may result from high a pH in surrounding waters, large aquatic plants and an abundance of cyanobacteria, or hypoxia and pesticide residues [10–13]. The association between 55 variables and HPND was assessed by a cross-sectional study method, and 11 risk factors were found to have the greatest impact on HPND prevalence, including "Recent pH in the pond", "Frequency of the abamectin use", "Frequency of switching aerator on in the farm", "Frequency of disinfectant use", "Amount of edible animal ingredient", "Abundance of Cyanobacteria in the pond", and "Frequency of clearing the ponds" [14].

Increasing evidence indicated that Omics provides new clues in the understanding of the etiology and pathogenesis. A metatranscriptomic survey revealed changes in the hepatopancreatic flora of the crab with HPND, but there were no statistically significant difference in viral and microsporidia communities in the hepatopancreas of crab with or without HPND [15]. Metabolomics was used to screen potential causative agents of crab HPND, fatty acid metabolic abnormalities were found in the hepatopancreas of crab with HPND, and high concentrations of propamocarb (a widely used pesticide in vegetables) were detected in the hepatopancreas of crab with HPND, suggesting that pesticide could likely be associated with HPND [12]. The transcriptomic analysis of hepatopancreatic crab with HPND and without HPND showed that the metabolism of xenobiotics by cytochrome P450, drug metabolism-cytochrome P450, chemical carcinogenesis, and material metabolism were the top five significantly enriched pathways for DEGs. The material metabolic abnormalities and drug effects from the external environment were suggested to be associated with crab HPND [16]. Our results obtained from the transcriptomic analysis indicated that crab HPND may be the result of autophagy and apoptosis, the hepatopancreas of crabs with HPND turn from golden yellow/light yellow to almost white was associated with retinol metabolism dysregulation [17].

Although the causes have been discussed from epidemiology, pathogenic microbiology and molecular pathology, the cause of crab HPND is still under debate, and pathological mechanisms of HPND are unknown. Moreover, in aquatic animal diseases, the animal with the same disease may have different symptoms, and the same symptoms may be caused by different diseases. In this study, we hypothesized that the imbalance of hepatopancreatic flora was associated with crab HPND, and the pathological changes of hepatopancreas were involved in the differentially expressed genes (DEGs) in hepatopancreatic flora between crab with and without HPND; therefore, metatranscriptomic sequencing was conducted to characterize the changes in the structure and gene expression of the hepatopancreatic flora of crabs with and without typical symptoms of HPND. Consequently, an imbalance of hepatopancreatic flora was found in the crab with HPND. The relative abundances of virus and microsporidia in crabs with HPND were very low and did not increase with disease progression. The DEGs in hepatopancreatic flora between crab with and without HPND were enriched ribosome, retinol metabolism, and biosynthesis of unsaturated fatty acid KEGG pathways. These results suggested that an imbalance of hepatopancreatic flora was associated with crab HPND and the enriched pathways were associated with the pathological mechanism of HPND.

2. Materials and Methods

2.1. Crabs

Chinese mitten crab *Eriocheir sinensis* with HPND cannot be artificially generated in the laboratory, because the etiology and pathogenesis of HPND are unknown [2,8]. Therefore, two crabs (body weight: 100–150 g) were characterized by degeneration and atrophy of hepatopancreas, where the color of hepatopancrea changes from golden yellow to white: two crabs (body weight: 100–150 g) with mild signs of HPND, where their hepatopancreas were yellow and did not degenerate significantly, and two crabs (body weight: 100–150 g) without visible signs of HPND were collected from Anfeng town of Xinghua city, Jiangsu province, China in 2017.

2.2. cDNA Library Preparation and Metatranscriptomic Sequencing

The crabs for metatranscriptomic sequencing were sampled in the same pond at the same time from Anfeng town. Total RNA was isolated from hepatopancreases (1 g) of 2 crabs without visible signs of HPND (Figure 1A) (healthy crabs), light-yellow hepatopancreases (1 g) from two crabs with mild signs (Figure 1B) (diseased crabs with mild signs) and milky white hepatopancreases (1 g) from two crabs with severe signs (Figure 1D) (diseased crabs with severe signs) using RNeasyR Plus Mini Kits (Qiagen, Valencia, CA, USA) according to the manufacturer's protocol. After genomic DNA was removed by treatment with RNase free DNase (Qiagen, Valencia, CA, USA), the quality and quantity of total RNA were estimated with a NanoDrop 2000 Spectrophotometer (Thermo Scientific, Wilmington, USA) and an Agilent 2100 Bioanalyzer (Agilent Technologies, Palo Alto, CA, USA). RNA integrity number (RIN) was evaluated by electrophoresis on 1% agarose gel. Ribosomal RNA was washed out using the Ribo-Zero^{TM} Magnetic Kit (Epicenter, Charlotte, NC, USA). The cDNA libraries were constructed using TruSeq^{TM} RNA Sample Prep Kits (Illumina, San Diego, CA, USA). The quality of the cDNA libraries was assessed by the Agilent 2100 Bioanalyzer and the complete library was sequenced by Allwegene Technology Co., Ltd. (Nanjing, China) on a HiSeq 2500 Sequencer (Illumina, San Diego, CA, USA) using Mid Output Kits, and 150-bp paired-end reads were obtained for each run. All sequencing data were deposited in the NCBI BioSample database under accessions: SRX6579474 for healthy crabs, SRX6579475 for diseased crabs with mild signs and SRX6579476 for diseased crabs with severe signs.

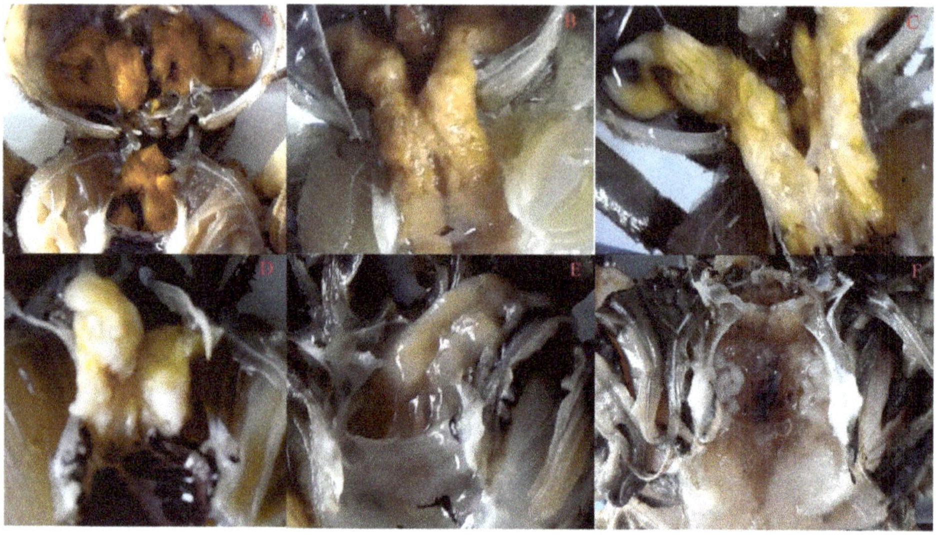

Figure 1. The common clinical signs of crabs with HPND. The body weights were about 100–150 g. (**A**) hepatopancreas of healthy crab, the hepatopancreas golden and plump; (**B**) hepatopancreas of crab with mild signs, the hepatopancreas are yellow and do not degenerate significantly; (**C**) the hepatopancreas turms pale yellow and begins to degeneration; (**D**) the hepatopancreas turns white and gradually atrophies; (**E**) the hepatopancreas erodes; (**F**) the hepatopancreas disappears.

2.3. *Meta-Transcriptomic Data Analysis*

2.3.1. Data Preprocessing

The FastQC toolkit (http://www.bioinformatics.babraham.ac.uk/projects/fastqc/, accessed on 20 June 2019) was used to evaluate the quality of raw reads as Phred score. After eliminating adaptors with SeqPrep software (https://github.com/jstjohn/SeqPrep, accessed on 20 June 2019), low-quality bases (Phred score < 20) were trimmed and reads shorter than 50 bp were discarded using Sickle software (https://github.com/najoshi/sickle, accessed on 21 June 2019). High-quality reads were obtained by discarding rRNA reads after alignment to SILVA SSU (16S/18S) and SILVA LSU (23S/28S) databases with SortMeRNA software (http://bioinfo.lifl.fr/RNA/sortmerna/, accessed on 21 June 2019). The generated high-quality reads were then used for de novo assembly.

2.3.2. De Novo Assembly and ORFs Prediction

High-quality reads were taken in the de novo assemblies with Trinity software (http://trinityrnaseq.github.io/, version trinityrnaseq-r2013-02-25, accessed on 22 June 2019) using default parameters following the previous report [18]. Trans Gene Scan software (http://sourceforge.net/projects/transgenescan/, accessed on 22 June 2019) was applied for the ORF (open reading frame) prediction and CD-HIT software (http://www.bioinformatics.org/cd-hit/, accessed on 22 June 2019) was used to construct nonredundant gene catalogues with identity 95% and coverage 90%.

2.3.3. Gene Expression Level

Gene expression levels were assessed using fragments per kilobase of exon per million fragments mapped (FPKM) values obtained using RNA-Seq by expectation maximization (RSEM) software (http://deweylab.biostat.wisc.edu/rsem/, accessed on 5 July 2019). The empirical analysis of digital gene expression data in R (edger) software (http://www.bioconductor.org/packages/release/bioc/html/edgeR.html, accessed on 10 July 2019) was used to identify differentially expressed genes (DEGs) with false discovery rate (FDR) <0.05 and $|\log 2FC| > 1$ [19,20].

2.3.4. Species Information and Taxonomic Abundance

All genes were aligned against the integrated NCBI NR database with an expectation value of 1e-5 (BLAST Version 2.2.28+, USA, http://blast.ncbi.nlm.nih.gov/Blast.cgi, accessed on 11 July 2019). Species taxonomic information was obtained from the respective taxonomy annotations in NR databases. Species abundance was assessed by calculating FPKM values for the gene in each species and taxonomic abundance was calculated at different taxonomic levels. The taxonomic abundance in each specimen was calculated at the class, order, family, genera, and species levels. Biological replicates were not set; therefore, differences in abundance profiles at each level between two groups were identified with method = "blind", sharingMode = "fit-only" by DEGseq soft [21].

2.4. *Functional Annotations*

To understand the functions of the predicted gene, gene ontology (GO) annotations were obtained from Blast2go (https://www.blast2go.com/, accessed on 12 July 2019) with default parameters based on SWISSPROT (https://web.expasy.org/docs/swiss-prot_guideline.html, accessed on 12 July 2019) annotations. GO terms were classified using Web Gene Ontology (WEGO) annotation software (http://wego.genomics.org.cn/, accessed on 12 July 2019), GO terms with a p-value ≤ 0.05 were designated as a significantly enriched term for DEGs. The Kyoto Encyclopedia of Genes and Genomes (KEGG) (http://www.genome.jp/kegg/, accessed on 12 July 2019) was used for the systematic analysis of gene functions [22,23]. Pathways with a p-value ≤ 0.05 were designated as significantly enriched pathways for DEGs. Evolutionary genealogy of genes: Non-supervised Orthologous Groups (eggNOG) [24] (http://eggnog5.embl.de/download/eggnog_5.0/, accessed on 13 July 2019) and Carbohydrate-Active enzymes (CAZy) Database (http://www.cazy.org/, accesed on 14 July 2019) were used to annotate gene functions [25] Cluster analysis was based on the Bray–Curtis distance to assess the similarities between samples (www.microbiomeanalyst.ca, accessed on 16 July 2019) [26].

2.5. *PCR Detection and Sanger Sequencing*

To validate the sequences determined by metatranscriptomic sequencing, several primer pairs were designed and synthesized based on the sequences obtained from metatranscriptomic sequencing (Table S1). The extracted total RNAs (5 μg), respectively, from hepatopancreases of crabs, sampled from healthy crabs, diseased crabs with mild signs, and diseased crabs with severe signs, were used for RT-PCR. PCR products were recovered and cloned into a pMD-18-T (Takara, Dalian, China) for Sanger sequencing and obtained sequences were compared with corresponding sequences determined by metatranscriptomic sequencing.

3. Results

3.1. *Sign of Crabs with HPND*

Crab HPND may be explored in the terms of the health of farmed crabs, pathogens, pesticides, feed, and ecology. However, the main cause is still unknown. Crab samples with HPND were collected in the different endemic areas. The hepatopancreas of healthy crabs is plumpy and golden in color (Figure 1A). The diseased crabs showed a slow response and poor mobility. The common clinical signs are hepatopancreas degeneration and atrophy, and a change in the color of the hepatopancreas (Figure 1). In the early onset, the hepatopancreas is pale gold (Figure 1B); then hepatopancreas begins to degeneration (Figure 1C). Further, with the development of the disease course, the hepatopancreas turns white and gradually atrophies and erodes (Figure 1D,E). The hepatopancreas of crabs with severe signs of HPND eventually disappear in the later stage of the onset (Figure 1F).

3.2. *Hepatopancreatic Flora of Crabs with HPND*

To find potential pathogens of HPND crab, metatranscriptomic sequencing was conducted to assess changes in the hepatopancreatic flora of the collected crabs with

HPND from Anfeng Town of Xinhua city in 2017. After removing low-quality sequencing, 32,807,952 clean reads were obtained from healthy crabs, 29,532,670 from diseased crabs with mild signs, and 27,582,606 from diseased crabs with severe signs (Table S2). A total of 40,966 transcripts were assembled using clean data and 35,230 unigenes were obtained. Microbial taxonomic information was obtained from the taxonomy annotation NR database using alignment analysis. Species abundance was estimated based on FPKM value (Table 1). A notable change in the number of detected taxa in the hepatopancreatic flora between healthy crabs and diseased crabs with mild signs was not observed; however, taxa were increased in diseased crabs with severe signs.

Table 1. Changes in the relative abundance of microbes at different taxonomic levels in the hepatopancreatic flora of crabs with HPND.

Samples	Kingdoms (Number/Percentages)	Phyla	Classes	Orders	Families	Genera	Species
Healthy crabs	4/30.37%	33/4.48%	64/3.69%	110/3.03%	172/2.87%	197/2.78%	205/2.12%
Diseased crabs with mild signs	4/35.43%	33/3.78%	65/2.92%	110/2.47%	169/2.40%	194/2.37%	205/1.85%
Diseased crabs with severe signs	4/30.86%	33/3.41%	61/2.64%	103/2.19%	162/1.98%	180/1.96%	189/1.50%

The values in the bracket represented the percentage of the microbe which was assigned to a taxon.

The top 10 taxa in relative abundance at different taxonomic levels are in Figure 1. As a whole, the relative abundance of the bacteria belonging to *Proteobacteria* and *Mucoromycota* phyla decreased; moreover, a relative abundance of the bacteria belonging to phylum *Basidiomycota* increased with disease progression, and the relative abundance of phylum *Tenericutes* increased by 106 times in diseased crabs with severe signs (Figure 2A).

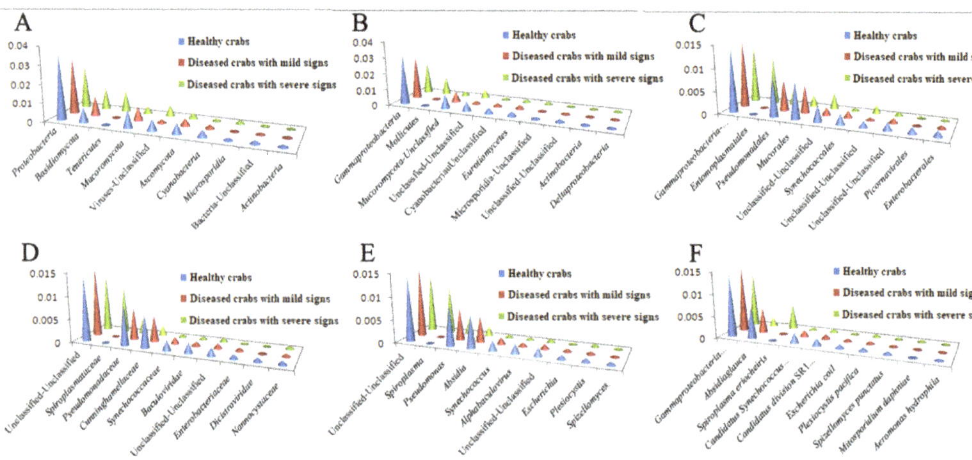

Figure 2. Top 10 taxon in relative abundance at different taxonomic levels. (**A**), (**B**), (**C**), (**D**), (**E**), and (**F**) represented Phyla, Classes, Orders, Families, Genera, and Species, respectively.

At the class level, in the healthy crabs, diseased crabs with mild signs, and diseased crabs with severe signs, the most predominant classes were *GammaProteobacteria* (3.0875%, 2.7123%, and 2.0248%), followed by *Mucoromycota*-Unclassified (0.8729%) and Unclassified-Unclassified (0.6064%) in healthy crabs, *Mucoromycota*-Unclassified (0.6520%) and *Eurotiomycetes* (0.2555%) in diseased crabs with mild signs, and *Mollicutes* (1.1576%) and *Unclassified-Unclassified* (0.5846%) in diseased crabs with severe signs. Compared to healthy crabs, the relative abundance of *Mollicutes* in diseased crabs with severe signs increased by 106 times (Figure 2B).

In the case of order, the abundance of *GammaProteobacteria*-Unclassified was maximum (1.4327%) in healthy crabs, followed by *Pseudomonadales* (0.8613%) and *Mucorales* (0.8316%). The relative abundances of these three orders was, respectively, 1.4975%, 0.6911%, and 0.6200% in diseased crabs with mild signs, and 1.2718%, 0.3552%, and 0.2702% in diseased crabs with severe signs. The relative abundance of *Entomoplasmatales* in diseased crabs with severe signs increased 987 times compared to healthy crabs (Figure 2C).

In family, the top 3 families for relative abundance in both healthy crabs and diseased crabs with mild signs were Unclassified-Unclassified (1.4327% for healthy crabs, 1.4975% for diseased crabs with mild signs), *Pseudomonadaceae* (0.8580%, 0.6906%), and *Cunninghamellaceae* (0.7395%, 0.5681%). In diseased crabs with severe signs, the most predominant families were Unclassified-Unclassified (1.2718%), followed by *Spiroplasmataceae* (1.0452%) and *Pseudomonadaceae* (0.3521%). The relative abundance of *Spiroplasmataceae* in diseased crabs with severe signs increased by 977 times compared to healthy crabs (Figure 2D).

In the case of genus, the top 3 genera for relative abundance in both healthy crabs and diseased crabs with mild signs were Unclassified-Unclassified (1.4327% for healthy crabs, 1.4975% for diseased crabs with mild signs), *Pseudomonas* (0.8580%, 0.6906%), and *Absidia* (0.7395%, 0.5681%). In XFAF-2, the top 3 genera were Unclassified-Unclassified (1.2718%), *Spiroplasma* (1.0452%), and *Pseudomonas* (0.3521%) (Figure 2E).

At the species level, the top 3 predominant species in both healthy crabs and diseased crabs with mild signs were *Gamma Proteobacteria bacterium* 2W06 (1.4186% for healthy crabs, 1.4900% for diseased crabs with mild signs), *Absidia glauca* (0.7395%, 0.5681%), and *Candidatus Synechococcus spongiarum* (0.2311%, 0.1457%). In XFAF-2, the top 3 predominant species were *Gamma Proteobacteria bacterium* 2W06 (1.2654%), *Spiroplasma eriocheiris* (0.5885%), and *Absidia glauca* (0.2116%) (Figure 1F). The prevalence of bacteria belonging to *A. glauca* and *C. Synechococcus spongiarum* species decreased in crabs with HPND, whereas the prevalence of *S. eriocheiris* species in diseased crabs with severe signs increased by 697 times compared to healthy crabs (Figure 2F).

At different taxonomic levels, 127 families, 141 genera, and 165 species were shared by healthy crabs, diseased crabs with mild signs and diseased crabs with severe signs. A total of 6 families in healthy crabs, 5 in diseased crabs with mild signs, and 4 in diseased crabs with severe signs; 9 genera in healthy crabs, 7 in diseased crabs with mild signs, and 4 in diseased crabs with severe signs; and 8 species in healthy crabs, 11 in diseased crabs with mild signs, and 8 in diseased crabs with severe signs (Figure S1) were specifically recorded. Although differences in the composition of hepatopancreatic flora in crabs without and with HPND were found, microbial taxa specifically detected in crabs with HPND were not dominant, their relative abundances were very low and did not increase with disease progression.

3.3. HPND Is Associated with a Change in the Construction of Hepatopancreatic Flora

Cluster analysis based on the top 10 taxa by relative abundance was performed using Bray–Curtis distance matrixes to assess the similarity of samples in the hepatopancreatic flora at different taxonomic levels. Diseased crabs with mild signs were similar to healthy crabs, while diseased crabs with severe signs were different from healthy crabs (data not shown). To further assess differences in hepatopancreatic flora among healthy crabs, diseased crabs with mild signs, and diseased crabs with severe signs, the top 35 genera in relative abundance were used to construct heatmaps. The hepatopancreatic flora of diseased crabs with mild signs and diseased crabs with severe signs was different from healthy crabs (Figure 3).

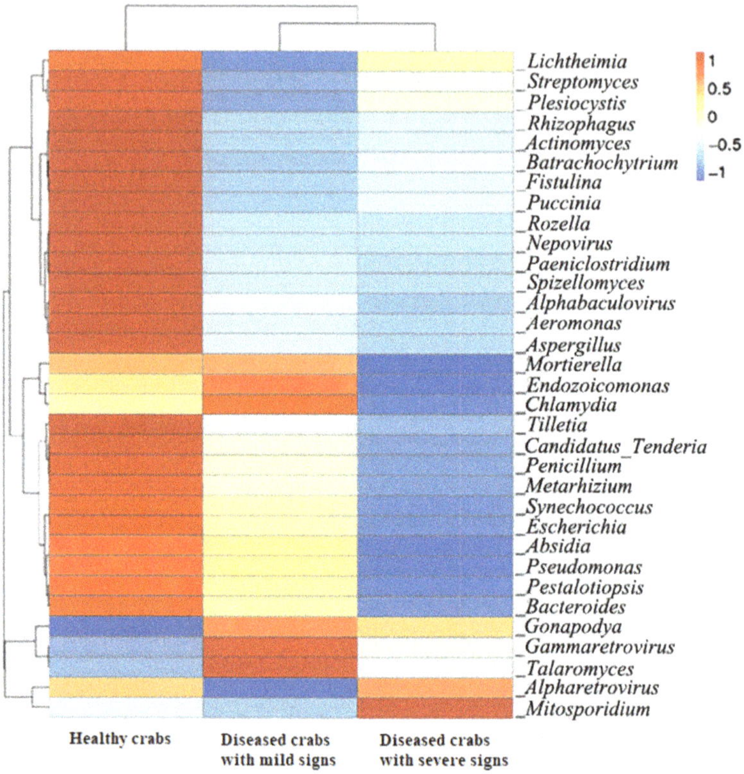

Figure 3. Clustering of relative abundance at the genus level. The horizontal axis represented sample information; vertical axis represented species information; the cluster tree on the left side was taxon clustering tree; the color intensity in the square grid represented the bacterial relative abundance, which is named as z-value and generated by the relative abundance of the genus in each line after normalization treatment.

3.4. Viral Infection Is Not Involved in HPND

The metatranscriptomic sequencing was used to find the potential pathogens of crab HPND. Homologous sequences of 27 viral genera were found in crab hepatopancreases, among the recorded genera 24 were found in healthy crabs and diseased crabs with mild signs, and 21 in diseased crabs with severe signs. The homologous sequences of specific viral genera in crabs with HPND were not found. The top 4 sequences in relative abundance were homologous to *Alphabaculovirus*, Unclassified genera, *Nepovirus*, and *Alpharetrovirus*; however, their relative abundances in diseased crabs with mild signs and diseased crabs with severe signs were lower compared to healthy crabs (Figure S2). At the species level, homologous sequences of 34 viral species were found in crab hepatopancreases. Out of 34 species, 31 were in healthy crabs, 28 in diseased crabs with mild signs, and 25 in diseased crabs with severe signs. The top 3 sequences in relative abundance were the homologous sequences of the cherry leaf roll virus, avian leukosis virus, and *Penaeus monodon* nudivirus in healthy crabs; avian leukosis virus, *Heliothis virescens* ascovirus 3a, and Tanapox virus in diseased crabs with mild signs; and avian leukosis virus, *P. monodon* nudivirus, and reticuloendotheliosis virus in diseased crabs with severe signs. The relative abundances of these viral homologous sequences in crabs with HPND were low and did not increase with disease progression (Table 2), therefore crabs with HPND did not involve in viral infection.

Table 2. Relative abundances of the detected viruses in the samples.

Homologous Sequences of Detected Virus	Relative Abundance (%)			Fold Change	
	Healthy Crabs	Diseased Crabs with Mild Signs	Diseased Crabs with Severe Signs	Diseased Crabs with Mild Signs/Healthy Crabs	Diseased Crabs with Severe Signs/Healthy Crabs
Cherry leaf roll virus	0.0246	0.0006	0.0002	0.0243	0.0081
Avian leukosis virus	0.0163	0.0134	0.0156	0.8221	0.9571
Penaeus monodon nudivirus	0.0161	0	0.0114	0	0.7081
Heliothis virescens ascovirus 3a	0.0065	0.0038	0.0025	0.5846	0.3846
Tanapox virus	0.0048	0.0033	0.002	0.6875	0.4167
Swinepox virus	0.0034	0.0014	0.0013	0.4117	0.3823
Reticuloendotheliosis virus	0.0026	0.0021	0.0033	0.8077	1.2692
Tipula oleracea nudivirus	0.002	0.0026	0.0017	1.3	0.85
Cotesia sesamiae bracovirus	0.0015	0.0006	0.0004	0.4	0.2667
Chelonus inanitus bracovirus	0.0008	0.0007	0.0005	0.875	0.625
Lymphocystis disease virus Sa	0.0008	0.0001	0.0001	0.125	0.125
Metopaulias depressus WSSV-like virus	0.0008	0	0.0003	0	0.375
Pigeonpox virus	0.0007	0.0002	0.0005	0.2857	0.7143
Marine RNA virus SF-1	0.0007	0.0002	0	0.2857	0
Cyprinid herpesvirus 3	0.0005	0.0003	0.0004	0.6	0.8
Saimiriine herpesvirus 4	0.0005	0.0002	0	0.4	0
Oryctes rhinoceros nudivirus	0.0004	0.0001	0.0004	0.25	1
Avian musculoaponeurotic fibrosarcoma virus AS42	0.0004	0.0001	0.0003	0.25	0.75
Murine leukemia virus	0.0003	0.0004	0.0005	1.3333	1.6667
Abelson murine leukemia virus	0.0003	0.0004	0.0004	1.3333	1.3333
Deerpox virus W-848-83	0.0003	0.0002	0	0.6667	0
Glypta fumiferanae ichnovirus	0.0003	0.0001	0.0001	0.3333	0.3333
Antarctic picorna-like virus 1	0.0003	0.0001	0.0001	0.3333	0.3333
Eriocheir sinensis reovirus	0.0003	0	0.0001	0	0.3333
UR2 sarcoma virus	0.0003	0	0	0	0
Aureococcus anophagefferens virus	0.0002	0.0001	0.0001	0.5	0.5
Cotesia congregata bracovirus	0.0002	0.0001	0	0.5	0
Bovine papular stomatitis virus	0.0001	0.0002	0.0008	2	8
Infectious spleen and kidney necrosis virus	0.0001	0	0.0002	0	2
Hepelivirus	0	0.0011	0	0	0
Canarypox virus	0	0.0004	0	0	0
Yaba monkey tumor virus	0	0	0.0002	0	0
Avian sarcoma virus	NA				
Alphapapillomavirus 7	NA				

NA, homologous sequences could be detected, but the abundance was very low. Relative abundance of viruses in healthy crabs was used as the control in the calculation of fold change.

3.5. Microsporidia Infection Is Not Involved in HPND

Homologous sequences of the phylum Microsporidia that included *Mitosporidium*, *Nosema* and *Anncaliia* genera were detected in crab hepatopancreas; their relative abundance in crabs with HPND did not show any significant differences between groups (Table 3). As *H. eriocheir* was supposed to be a pathogen of crab HPND. Results of metatranscriptomic sequencing did not record *H. eriocheir*.

Table 3. Relative abundances of the detected microsporidium in the samples.

Detected Homologous Sequences of Microsporidian	Relative Abundance (%)		
	Healthy Crabs	Diseased Crabs with Mild Signs	Diseased Crabs with Severe Signs
Mitosporidium daphniae	8.39×10^{-5}	3.91×10^{-5}	2.79×10^{-4}
Nosema bombycis	8.07×10^{-6}	1.08×10^{-5}	4.73×10^{-6}
Nosema apis	7.99×10^{-6}	6.73×10^{-6}	1.44×10^{-6}
Anncaliia algerae	2.13×10^{-6}	1.47×10^{-6}	1.64×10^{-6}

3.6. The Number of DEGs in Hepatopancreatic Flora Increased with HPND Progression

To explore the association of gene expression profiles of hepatopancreatic flora with HPND, DEGs with FDR < 0.05 and ∣log2FC∣ > 1 were identified using empirical analysis of digital gene expression data without biological replicates in edgeR software. Linking diseased crabs with mild signs and healthy crabs, data showed 489 DEGs (up: 283, down: 206); diseased crabs with severe signs and healthy crabs showed 968 (up: 310, down: 658), and diseased crabs with severe signs and diseased crabs with mild signs disclosed 673 (up: 207, down: 466) (Figure 4). Venn diagrams showed 45 DEGs were shared by diseased crabs with mild signs vs. healthy crabs, diseased crabs with severe signs vs. healthy crabs, and diseased crabs with severe signs vs. diseased crabs with mild signs. In common, were 6 upregulated genes and 12 downregulated genes (Figure 5). DEGs heat maps showed differences in gene expression profiles in hepatopancreatic flora among healthy crabs, diseased crabs with mild signs and diseased crabs with severe signs. Among all, healthy crabs and diseased crabs with mild signs were grouped, and diseased crabs with severe signs were in another group, and the number and expression level of DEGs changed as disease progressed (Figure S3).

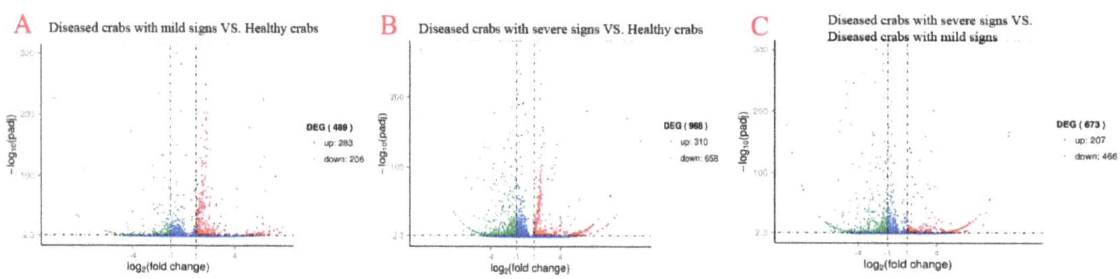

Figure 4. Volcano plot for gene expression among different samples. (**A**) Diseased crabs with mild signs VS. Healthy crabs; (**B**) Diseased crabs with severe signs VS. Healthy crabs; (**C**) Diseased crabs with severe signs VS. Diseased crabs with mild signs.

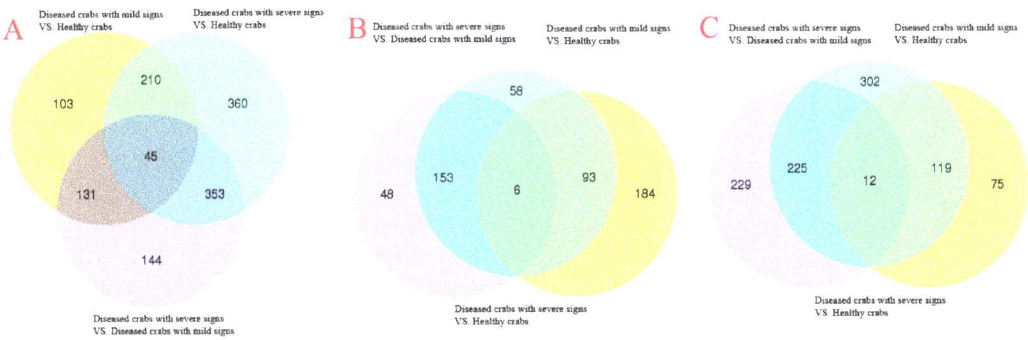

Figure 5. Venn diagram of DEGs identified by metatranscriptomic sequencing. (**A**) Venn diagrams of all DEGs among Diseased crabs with mild signs and Healthy crabs, Diseased crabs with severe signs and Healthy crabs, and Diseased crabs with severe signs and Diseased crabs with mild signs; (**B**) Venn diagrams of up-regulated DEGs among Diseased crabs with mild signs and Healthy crabs, Diseased crabs with severe signs and Healthy crabs, and Diseased crabs with severe signs and Diseased crabs with mild signs; (**C**) Venn diagram of down-regulated DEGs among Diseased crabs with mild signs and Healthy crabs, Diseased crabs with severe signs and Healthy crabs, and Diseased crabs with severe signs and Diseased crabs with mild signs.

3.7. The Function of Differentially Expressed Genes (DEGs) Associated with HPND Pathological Mechanism

To understand the pathological mechanism of HPND, GO enrichment of DEGs was performed. DEGs for diseased crabs with mild signs vs. healthy crabs were enriched to 30 GO terms such as metabolic, organic substance metabolic, and primary metabolic processes in the biological process (BP); intracellular, organelle, and intracellular organelle in the cellular component (CC); and structural molecule activity and structural constituent of the ribosome in molecular function (MF) (Figure 6A). DEGs for diseased crabs with severe signs vs. healthy crabs were enriched to 28 GO terms: metabolic process, biosynthetic process, organic substance biosynthetic process in BP; cytoplasm, the cytoplasmic part in CC; and oxidoreductase activity and structural molecule activity in MF (Figure 6B); DEGs for diseased crabs with severe signs vs. diseased crabs with mild signs were enriched to 17 GO terms: metabolic process, single-organism metabolic process and oxidation-reduction process in BP, and catalytic activity, oxidoreductase activity, and hydrolase activity in MF (Figure 6C). Three enriched GO terms (hydrolase activity-hydrolyzing O-glycosyl compounds, carbohydrate metabolic process, and hydrolase activity-acting on glycosyl bonds) were shared by diseased crabs with mild signs and healthy crabs, diseased crabs with severe signs and healthy crabs, and diseased crabs with severe signs and diseased crabs with mild signs. These results indicated that carbohydrates in the hepatopancreas of crab were used by hepatopancreatic flora.

KEGG enrichment was used to annotate DEG functions. The top 20 enriched KEGG pathways are in Figure S4: ribosome and another glycan degradation for diseased crabs with mild signs vs. healthy crabs (Figure S4A); ribosome, retinol metabolism, metabolism of xenobiotics by cytochrome P450, drug metabolism—cytochrome P450, and other glycan degradation for diseased crabs with severe signs vs. healthy crabs (Figure S4B); and lysosome, and sphingolipid metabolism for diseased crabs with severe signs vs. healthy crabs (Figure S4C). Seven enriched KEGG pathways (retinol metabolism, lysosome, other glycan degradation, metabolism of xenobiotics by cytochrome P450, drug metabolism cytochrome P450, ribosome, and steroid hormone biosynthesis) were shared by diseased crabs with mild signs and healthy crabs, and diseased crabs with severe signs and healthy crabs. Of these, the expression of all 68 enriched genes to ribosomes in diseased crabs with mild signs vs. healthy crabs and 71 in diseased crabs with severe signs vs. healthy crabs was upregulated (Figure S5), indicating that nutrients in hepatopancreas were used to synthesize microbial proteins. The expression of all 14 enriched genes to retinol metabolism in diseased crabs with severe signs vs. healthy crabs was downregulated (Figure 7), implying that the levels of retinal, retinoate, rhodopsin, and beta-carotene, which were associated with the color of hepatopancreas, decreased in the diseased crabs with severe signs. Moreover, the expression of all 7 enriched genes to the biosynthesis of unsaturated fatty acids in diseased crabs with severe signs vs. healthy crabs was downregulated (Figure S6).

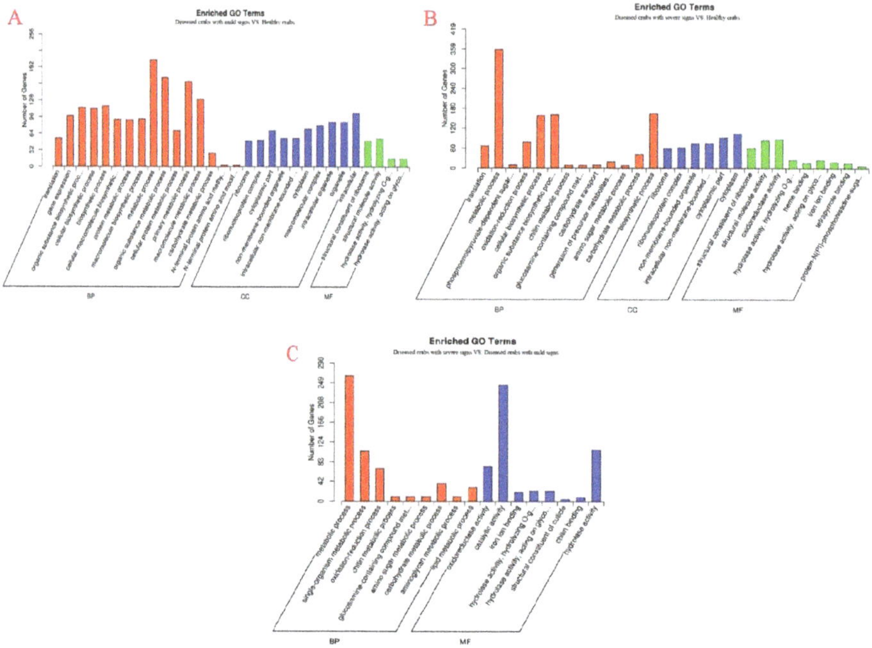

Figure 6. GO enrichment analysis of DEGs identified in the hepatopancreatic flora. (**A**) diseased crabs with mild signs vs. healthy crabs; (**B**) diseased crabs with severe signs vs. healthy crabs; (**C**) diseased crabs with severe signs vs. diseased crabs with mild signs. BP, CC, and MF represented biological processes, cellular components and molecular functions, respectively.

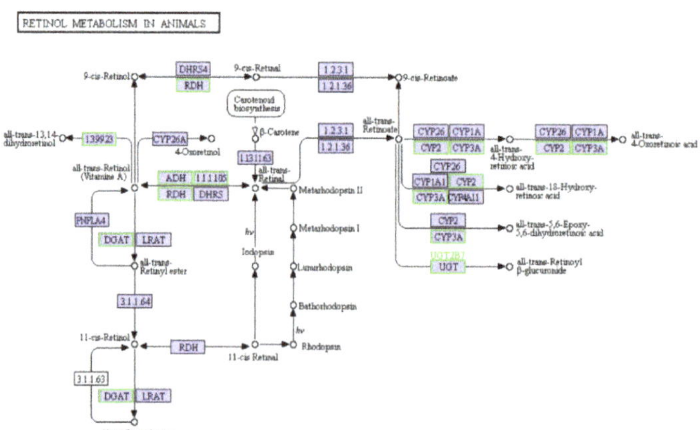

Figure 7. 14 enriched genes to retinol metabolism in diseased crabs with severe signs vs. healthy crabs were downregulated. The green boxes represent the downregulated genes.

3.8. Relative Abundances of Functions Changed with the Pathogenesis

Changes in functional abundance of genes in the hepatopancreatic flora were indicated in Figure 8. Compared to healthy crabs, the relative abundance of genetic information processing increased by 1.3798 times for diseased crabs with mild signs and 1.5341 times for diseased crabs with severe signs. The relative abundance of other functions except for human diseases decreased with disease progression in KEGG annotations (Figure 8A). For eggNOG, the relative abundance of translation-ribosomal structure and biogenesis increased. The relative abundance of general function prediction only and posttranslational modification-protein turnover-chaperones decreased with pathogenesis (Figure 8B). In the CAZy, the relative abundance of xenobiotics and carbohydrate esterases increased, and the relative abundance of glycoside hydrolases decreased in diseased crabs with mild signs and diseased crabs with severe signs compared to healthy crabs (Figure 8C). Clustering analysis using relative abundances of functions showed that diseased crabs with mild signs and diseased crabs with severe signs were together and the relative abundances of functions changed with pathogenesis (Figure S7).

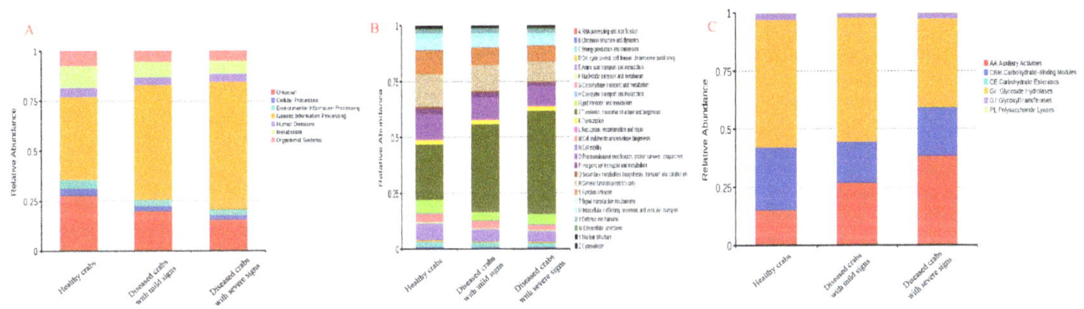

Figure 8. Functional abundances of different samples at eggNOG, KEGG, and CaZy levels. (**A**) KEGG; (**B**) eggNOG; (**C**) CaZy.

3.9. PCR Detection and Sanger Sequencing

To validate the sequences detected by metatranscriptomic analysis, the homologous sequences of genes from *C. sesamiae* bracovirus, *P. monodon* nudivirus, *E. sinensis* reovirus, *M. depressus* WSSV-like virus, and *S. eriocheiris*, respectively, were amplified from the cDNA of diseased crabs with severe signs by PCR. The identified sequences of PCR products were identical with the corresponding sequences determined by metatranscriptomic sequencing. Moreover, *P. monodon* nudivirus, *E. sinensis* reovirus, and the *M. depressus* WSSV-like virus were only found in healthy crabs and diseased crabs with severe signs by PCR processes. *C. sesamiae* bracovirus and *S. eriocheiris* were found in all three samples as detected by metatranscriptomic sequencing.

4. Discussion

The aetiology of HPND was explored in the pathogens [1,4], pathological changes [2,8,17], ecotype [9–11,14], hepatopancreatic flora [15], the transcriptome of hepatopancreas [1,6,17], metabolite profiling of the hepatopancreas [12], and offspring seed and epidemiological surveys [8,13]; however, the aetiology of HPND crab remains unidentified.

The etiology of aquatic animal diseases can be roughly divided into biological and abiotic factors. *H. eriocheir* belonging to microsporidian was considered as the pathogen of crab HPND, because *H. eriocheir* was found in the crabs with HPND [1]. However, *H. eriocheir* was not found in some crabs with HPND [2,17,27]. The typical HPND symptoms could not be generated by artificial infection with *H. eriocheir* [4], Moreover, the sequences of *H. eriocheir* were not found in the hepatopancreatic flora by a meta-transcriptomic survey [15]. In the present study, the *Mitosporidium*, *Nosema*, and *Anncaliia* genera belonging to phylum *Microsporidia* were recorded in hepatopancreatic flora of the crabs with HPND by the meta-transcriptomics data. The significant differences in relative abundances between crabs with and without HPND were not observed. Moreover, sequences of *H. eriocheir* were not identified by metatranscriptomic sequencing. Therefore, it can be concluded that crab HPND was not caused by the microsporidian infection.

To date, studies on *E. sinensis* disease caused by viral infection is limited. White spot syndrome virus [28], Reovirus [29–33], and Roni-like virus [34] were noticed in the diseased *E. sinensis*. However, signs of HPND were not found in the infected crabs with these viruses. Eleven families of viruses were identified in the hepatopancreata of crab by meta-transcriptome analysis; however, the expression levels of these viruses were extremely low in the crabs with and without HPND [14]. In our previous study, virus-like particles were not found in the hepatopancreatic cells of crabs with HPND by electron microscope observation [2,17]. Moreover, HPND was not generated by inoculation with bacteria-free supernatants from the hepatopancreas of the crabs with HPND [2]. In this study, homologous sequences of 34 viral species were identified in the hepatopancreases of crab from the analysis of the meta-transcriptome data; however, the relative abundances of these viral homologous sequences in crabs with HPND were very low and did not increase with disease progression. These results suggested that crab HPND was not caused by viral infection. However, the identified sequences by metatranscriptomic sequencing are derived from transcripts of microbial genes, suggesting that the detected viruses, especially DNA viruses, infect crab. Our findings provide clues for molecular epidemiological investigations of crab diseases.

The *Vibrio* is considered as the causal agent of acute HPND of cultured shrimp [5–8]. Bacteria belonging to *Vibrio* spp. could be 10~23.33% in crabs with HPND; however, HPND-like signs were not generated in animal regression tests performed with *Vibrio* spp. [13]. In our findings, hepatopancreatic flora in crabs with HPND was not dominated by the genera Vibrio. Therefore, *Vibrio* spp. was not a pathogen of HPND.

Microsymbionts in the hepatopancreas of isopods were involved in digestion, nutrition and absorption, reproduction, and immunity [35,36]. The imbalance of symbionts in the hepatopancreas is associated with some illnesses [37–39]. Symbiotic bacteria in crabs belong to the phyla Bacteroidetes, Proteobacteria, Firmicutes, and Tenericutes [40–42]. The genus *Candidatus hepatoplasma* in phylum Tenericutes is beneficial to its isopod host under low-nutrient conditions [43]. The genus *Spiroplasma* belonging to phylum Tenericutes is deemed a pathogen of Chinese mitten crab tremor disease [39,44], and *Acholeplasma* spp. are linked to clearwater disease of the mud crab *Scylla serrata* [38]. A previous report indicated that the relative abundance of phylum Tenericutes increases in crabs with HPND [15]. Similar results were found in our study. The largest increase in relative abundances was *S. eriocheiris* belonging to Tenericutes. However, typical tremor symptoms were not observed in crabs with HPND. The possibility that an increase in the abundance of *S. eriocheiris* is associated with HPND should be further studied.

Some diseases are not caused by a specific microbial infection, but by an imbalance in microbial flora. Host disease can be induced by symbiotic microbes after this balance is disrupted [43,45]. Significant differences in the hepatopancreatic flora were observed between crabs with and without HPND. The microbial diversity reduced and the microbial amount increased in the hepatopancreatic flora of crabs with HPND. Specifically, the prevalence of the Tenericutes phylum increase, the prevalence of the proteobacteria and Bacteroidetes phyla decreased in crabs with HPND [15]. Similar results were found in this study. Therefore, we strongly suggested imbalance in the hepatopancreatic flora is associated with HPND. However, determining if a change in hepatopancreatic flora is the cause or result of HPND is difficult. Nonetheless, a large number of nutrients are consumed by bacteria, which may result in hepatopancreas atrophy. Apoptosis and autophagy were found in the hepatopancreases of crabs with HPND [2,17]. It has been known that both apoptosis and autophagy can be induced by nutrition deficiency [46]. Therefore, apoptosis and autophagy of hepatopancreases of crabs with HPND may associate with nutrient deficiency caused by the multiplication of symbiotic bacteria.

To explore the association of gene expression profiles of hepatopancreatic flora with HPND, DEGs in hepatopancreatic flora between crabs with and without HPND were identified by analyzing meta-transcriptomic sequencing data. The results indicated that the number of DEGs increased with HPND progression. Therefore, these DEGs could be associated with the pathological mechanism of HPND. The expression of all 71 genes enriched in ribosomes was upregulated in diseased crabs with severe signs vs. healthy crabs, which revealed that microbial protein synthesis was enhanced and host protein synthesis was obstructed in the crab hepatopancreas. This may be one of the causes of hepatopancreas degeneration of crabs with HPND. The hepatopancreas color is believed to be associated with pigments including β-carotene [17]. β-carotene, a red-orange pigment, can be transformed into retinol and retinene [47]. In this study, we found that the expression of all 14 enriched genes to retinol metabolism in diseased crabs with severe signs vs. healthy crabs was down-regulated. The color of the hepatopancreases of crabs with HPND turns from golden/light yellow to almost white during pathogenesis [1]. Therefore, we suggested that the color change of the hepatopancreases of crabs with HPND was associated with the down-regulation of genes related to retinol metabolism in the hepatopancreatic flora. Downregulated genes in the hepatopancreatic flora of crab with HPND were enriched to the biosynthesis of unsaturated fatty acids, suggesting that liposoluble carotenoid accumulation was reduced.

Crab HPND was suggested to be the result of both autophagy and apoptosis induced by some unknown abiotic factors including pesticides, toxins from environmental bacteria and algae, chemical fertilizer, antibiotics in food, and farming water [2,17]. This speculation is supported by epidemiological investigations that crab HPND may result from hypoxia and pesticide residues, high pH in surrounding waters, massive aquatic plants and abundance of cyanobacteria, or in water environments with empty-illumination, without aquatic plants and reduced feeding, or hypoxia and pesticide residues [2,4,9,11,13]. The association between 55 variables and HPND was assessed. Recent pH in the pond, frequency of the abamectin and disinfectant used, frequency of switching aerator on the farm, and abundance of Cyanobacteria in the pond etc. were found to have the greatest impact on HPND prevalence. Bacterial symbionts change with the ecological environment in which their hosts live [35,36]. Therefore, the relationships among environment, hepatopancreatic flora and crab HPND need to be explored.

Several studies have indicated that increased occurrence rates of HPND with increased pesticide applications [9,10,12]. A high concentration of propamocarb was detected in the hepatopancreas of crabs with HPND [11]. Moreover, crabs exposed to a low concentration of insecticides caused clinical symptoms of HPND [9]. Interestingly, DEGs in the hepatopancreas between crab with HPND and without HPND were enriched to the metabolism of xenobiotics by cytochrome P450, drug metabolism-cytochrome P450, and material metabolism [16]. Our previous transcriptomic analysis indicated that the down-regulated DEGs in the hepatopancreas of crabs with HPND were enriched in chlorocyclohexane and chlorobenzene degradation KEGG pathways [16]. In this study, we found that DEGs in the hepatopancreatic flora between crab with HPND and without HPND were also enriched in the metabolism of xenobiotics by cytochrome P450 and drug metabolism-cytochrome P450 KEGG pathways. A comparison of the relative abundance of functions for DEGs in the hepatopancreatic flora among samples indicated an increase in the relative abundance of xenobiotics in the diseased crabs. These results strongly implied that unknown toxic substances including pesticide residues in the aquaculture environment are associated with HPND.

5. Conclusions

The imbalance of hepatopancreatic flora is associated with crab HPND. However, HPND does not correlate with virus or microsporidia infection. Changes in hepatopancreatic flora and the enriched pathways of DEGs may associate with the pathological mechanism of HPND.

Supplementary Materials: The following are available online at https://www.mdpi.com/article/10.3390/biology10060462/s1, Figure S1: Venn diagram of species identified by metatranscriptomic sequencing in different samples at different taxonomic levels, Figure S2: Relative abundance of viral genera among different samples, Figure S3: Heatmap of all DEGs among diseased crabs with mild signs and healthy crabs, diseased crabs with severe signs and healthy crabs, and diseased crabs with severe signs and diseased crabs with mild signs, Figure S4: Top 20 enriched KEGG pathways of DEGs identified in the hepatopancreatic flora, Figure S5: 71 enriched genes to ribosomes in diseased crabs with severe signs vs. healthy crabs, Figure S6: Enriched genes to the biosynthesis of unsaturated fatty acids in diseased crabs with severe signs vs. healthy crabs was down-regulated. Figure S7: Heatmap of KEGG, eggNOG and CaZy between samples, Table S1: Primers used in this study, Table S2: Data quality of metatranscriptomic sequencing.

Author Contributions: Conceptualization, C.G.; methodology, C.G. and X.L. (Xiaohan Liu); software, X.H.; validation, C.G.; formal analysis, Y.G., X.H., G.C. and R.X.; investigation, Z.S., X.L. (Xunmeng Liu), B.Y., M.L. and M.X.; resources, Y.G., P.F., R.Y., Y.F. and H.C.; data curation, X.H.; writing original draft preparation, Z.S., D.K; writing—review and editing, D.K.; supervision, C.G.; project administration, X.L. (Xiaohan Liu). All authors have read and agreed to the published version of the manuscript.

Funding: This research was funded by the Triple-New Project of Aquaculture of Jiangsu Province, China (D2017-3), the Key Research and Development Program of Jiangsu Province (Modern Agriculture) (BE2016322), and the Priority Academic Program of Development of Jiangsu Higher Education. UGC, BSR, New Delhi, India sincerely acknowledged.

Institutional Review Board Statement: Not applicable.

Informed Consent Statement: Not applicable.

Data Availability Statement: Not applicable.

Conflicts of Interest: The authors declare no conflict of interest.

References

1. Ding, Z.; Meng, Q.; Liu, H.; Yuan, S.; Zhang, F.; Sun, M.; Zhao, Y.; Shen, M.; Zhou, G.; Pan, J.; et al. First case of hepatopancreatic necrosis disease in pond-reared Chinese mitten crab, *Eriocheir sinensis*, associated with microsporidian. *J. Fish Dis.* **2016**, *39*, 1043–1051. [CrossRef] [PubMed]
2. Pan, Z.H.; Song, X.H.; Hu, X.L.; Xue, R.Y.; Cao, G.L.; Zar, M.S.; Kumar, D.; Feng, Y.; Wei, Y.; Zhang, W.Y.; et al. Pathological Changes and Risk Factors of Hepatopancreas Necrosis Disease of Mitten Crab, *Eriocheir sinensis*. *Fish. Aqua. J.* **2017**, *8*, 220. [CrossRef]
3. Cui, L.B.; Tang, S.L.; Qi, R.R.; Lei, Y.; Li, Y.B.; Wang, J. Pathological study on "Shuibiezi" disease of crab *Eriocheir sinensis*. *J. Yantai Univ.* **2017**, *30*, 313–316.
4. Luo, D.; Zhang, J.Y.; Zhao, Z.M.; Chen, H. Relevant research in "Shuibiezi" disease of crab *Eriocheir sinensis* and microsporidian *H. eriocheir*. *Sci. Fish Farming* **2018**, *4*, 61–62.
5. Soto-Rodriguez, S.A.; Gomez-Gil, B.; Lozano-Olvera, R.; Betancourt-Lozano, M.; Morales-Covarrubias, M.S. Field and experimental evidence of Vibrio parahaemolyticus as the causative agent of acute hepatopancreatic necrosis disease of cultured shrimp (*Litopenaeus vannamei*) in Northwestern Mexico. *Appl. Environ. Microbiol.* **2015**, *81*, 1689–1699. [CrossRef]
6. Li, P.; Kinch, L.N.; Ray, A.; Dalia, A.B.; Cong, Q.; Nunan, L.M.; Camilli, A.; Grishin, N.V.; Salomon, D.; Orth, K. Acute Hepatopancreatic Necrosis Disease-Causing Vibrio parahaemolyticus Strains Maintain an Antibacterial Type VI Secretion System with Versatile Effector Repertoires. *Appl. Environ. Microbiol.* **2017**, *83*. [CrossRef] [PubMed]
7. Devadas, S.; Banerjee, S.; Yusoff, F.M.; Bhassu, S.; Shariff, M. Experimental methodologies and diagnostic procedures for acute hepatopancreatic necrosis disease (AHPND). *Aquaculture* **2019**, *499*, 389–400. [CrossRef]
8. Yang, Z.Y.; Zhang, Y.L.; Hu, K.; Liu, L.S.; Cai, H.G.; Zhang, F.X.; Yang, X.L. Etiological and histoparhological study on hepatopancreatic necrosis syndrome in *Eriocheir sinensis*. *Acta Hydrobiol. Sin.* **2018**, *42*, 17–25.
9. Shen, G.; Shui, Y.; Zhang, X.; Song, K.; Wang, Y.; Xu, Z.; Shen, H. Hepatopancreatic necrosis disease (HPND) in Chinese mitten crab *Eriocheir sinensis* tightly linked to low concentration of two insecticides. *Aquac. Res.* **2020**, *52*, 2294–2304. [CrossRef]
10. Ye, J.S.; Yuan, S.; Zhao, Z.M.; Liu, M.J.; Han, X.P.; Dai, J.H. Investigation and analysis of the syndrome of hepatopancreas of *Eriocheir sinensis* in Jiangsu province. *Feed Ind.* **2017**, *38*, 61–64.
11. Gu, X.L.; Jiang, G.M.; Wei, B.; Niu, J.; Song, X.H. Investigation and analysis on the correlation of pesticides and hypoxia with "Shuibiezi" disease of *Eriocheir sinensis*. *Sci. Fish Farming* **2017**, *4*, 59–61.

12. Gao, T.; Xu, Y.; Wang, K.; Deng, Y.; Yang, Y.; Lu, Q.; Pan, J.; Xu, Z. Comparative LC-MS based non-targeted metabolite profiling of the Chinese mitten crab *Eriocheir sinensis* suffering from hepatopancreatic necrosis disease (HPND). *Aquaculture* **2018**, *491*, 338–345. [CrossRef]
13. Zhu, J.M.; Wang, Z.; Cai, C.F.; Tang, X.S.; Shen, J.M.; Wu, D.F. Inducement and prevention technology of "Shuibiezi" disease of Chinese mitten crab *Eriocheir sinensis*. *Sci. Fish Farming* **2016**, *6*, 13–15.
14. Huang, X.; Feng, Y.; Xiong, G.; Zhong, L.; Liu, S.; Fang, P.; Chen, H.; Wang, K.; Geng, Y.; Ouyang, P.; et al. A cross-sectional study of Bayesian belief network modeling: Risk factors for hepatopancreatic necrosis syndrome of the Chinese mitten crab (*Eriocheir sinensis*) in China. *Aquaculture* **2020**, *524*, 735293. [CrossRef]
15. Shen, H.S.; Zang, Y.N.; Song, K.; Ma, Y.C.; Dai, T.H.; Serwadda, A. A Meta-Transcriptomics Survey Reveals Changes in the Microbiota of the Chinese Mitten Crab *Eriocheir sinensis* Infected with Hepatopancreatic Necrosis Disease. *Front. Microbiol.* **2017**, *8*, 732. [CrossRef] [PubMed]
16. Yang, Z.; Hu, K.; Hou, Y.; Wang, Y.; Yao, Y.; Lei, X.; Yan, B.; Jiang, Q.; Xiong, C.; Xu, L.; et al. Transcriptome analysis of hepatopancreas of *Eriocheir sinensis* with hepatopancreatic necrosis disease (HPND). *PLoS ONE* **2020**, *15*, e0228623. [CrossRef]
17. Yan, B.; Liu, X.; Zhou, Y.; Zhang, M.; Fang, P.; Jiang, M.; Yuan, R.; Hu, X.; Cao, G.; Xue, R.; et al. Transcriptomic analysis reveals that hepatopancreatic necrosis disease in *Eriocheir sinensis* (Chinese mitten crabs) may be the result of autophagy and apoptosis. *Aquaculture* **2020**, *515*, 734579. [CrossRef]
18. Grabherr, M.G.; Mauceli, E.; Ma, L.J. Genome sequencing and assembly. *Methods Mol. Biol.* **2011**, *722*, 1–9. [PubMed]
19. Robinson, M.D.; McCarthy, D.J.; Smyth, G.K. edgeR: A Bioconductor package for differential expression analysis of digital gene expression data. *Bioinformatics* **2010**, *26*, 139–140. [CrossRef] [PubMed]
20. Reiner, A.; Yekutieli, D.; Benjamini, Y. Identifying differentially expressed genes using false discovery rate controlling procedures. *Bioinformatics* **2003**, *19*, 368–375. [CrossRef] [PubMed]
21. Wang, L.; Feng, Z.; Wang, X.; Wang, X.; Zhang, X. DEGseq: An R package for identifying differentially expressed genes from RNA-seq data. *Bioinformatics* **2010**, *26*, 136–138. [CrossRef] [PubMed]
22. Kanehisa, M.; Goto, S.; Hattori, M.; Aoki-Kinoshita, K.F.; Itoh, M.; Kawashima, S.; Katayama, T.; Araki, M.; Hirakawa, M. From genomics to chemical genomics: New developments in KEGG. *Nucleic Acids Res.* **2006**, *34*, D354–D357. [CrossRef]
23. Kanehisa, M.; Goto, S.; Sato, Y.; Kawashima, M.; Furumichi, M.; Tanabe, M. Data, information, knowledge and principle: Back to metabolism in KEGG. *Nucleic Acids Res.* **2001**, *12*, D199–D205. [CrossRef]
24. Powell, S.; Forslund, K.; Szklarczyk, D.; Trachana, K.; Roth, A.; Huerta-Cepas, J.; Gabaldón, T.; Rattei, T.; Creevey, C.; Kuhn, M.; et al. eggNOG v4. 0: Nested orthology inference across 3686 organisms. *Nucleic Acids Res.* **2014**, *42*, D231–D239. [CrossRef] [PubMed]
25. Cantarel, B.L.; Coutinho, P.M.; Rancurel, C.; Bernard, T.; Lombard, V.; Henrissat, B. The Carbohydrate-Active EnZymes database (CAZy): An expert resource for Glycogenomics. *Nucleic Acids Res.* **2009**, *37*, 233–238. [CrossRef]
26. Morlon, H.; Chuyong, G.; Condit, R.; Hubbell, S.; Kenfack, D.; Thomas, D.; Valencia, R.; Green, J.L. A general framework for the distance–decay of similarity in ecological communities. *Ecol. Lett.* **2008**, *11*, 904–917. [CrossRef]
27. Yang, Z.Y.; Zeng, L.G.; Wang, Y.L.; Yao, Y.; Hou, Y.J.; Lei, X.Q.; Xu, L.Q.; Xiong, C.X.; Yang, X.L. Ultramicroscopical histopathology and pathophysiology of "Shuibiezi" disease of *Eriocheir sinensis*. *Acta Agric. Zhejiangensis* **2018**, *30*, 1137–1148.
28. Ding, Z.; Yao, Y.; Zhang, F.; Wan, J.; Sun, M.; Liu, H.; Zhou, G.; Tang, J.; Pan, J.; Xue, H.; et al. The first detection of white spot syndrome virus in naturally infected cultured Chinese mitten crabs, *Eriocheir sinensis* in China. *J. Virol. Methods* **2015**, *220*, 49–54. [CrossRef] [PubMed]
29. Gong, C.L.; Xue, R.Y.; Cao, G.L.; Wei, Y.H.; Zhu, Y.X.; Chen, H.; Wu, X.F. Study on reovirus-like virus of *Eriocheir sinensis*. *Virol. Sin.* **2000**, *15*, 395–399.
30. Zhang, S.; Shi, Z.; Zhang, J.; Bonami, J.R. Purification and characterization of a new reovirus from the Chinese mitten crab, *Eriocheir sinensis*. *J. Fish Dis.* **2004**, *27*, 687–692. [CrossRef]
31. Zhang, S.; Zhang, J.; Huang, C.; Jean-Robert, B.; Shi, Z. Preliminary studies on two strains of reovirus from crab *Eriocheir sinensis*. *Virol. Sin.* **2002**, *17*, 263–265.
32. Ma, Y.; Dai, T.; Serwadda, A.; Shen, H. Detecting a novel *Eriocheir sinensis* reovirus by reverse transcription loop-mediated isothermal amplification assay. *Lett. Appl. Microbiol.* **2016**, *63*, 363–368. [CrossRef] [PubMed]
33. Shen, H.; Ma, Y.; Hu, Y. Near-Full-Length Genome Sequence of a Novel Reovirus from the Chinese Mitten Crab, *Eriocheir sinensis*. *Genome Announc.* **2015**, 3. [CrossRef]
34. Zhang, S.; Bonami, J.R. A roni-like virus associated with mortalities of the freshwater crab, *Eriocheir sinensis* Milne Edwards, cultured in China, exhibiting 'sighs disease' and black gill syndrome. *J. Fish Dis.* **2007**, *30*, 181–186. [CrossRef]
35. Cheung, M.K.; Yip, H.Y.; Nong, W.; Law, P.T.; Chu, K.H.; Kwan, H.S.; Hui, J.H. Rapid change of microbiota diversity in the Gut but not the Hepatopancreas during gonadal development of the new shrimp model Neocaridina denticulata. *Mar. Biotechnol.* **2015**, *17*, 811–819. [CrossRef] [PubMed]
36. Bouchon, D.; Zimmer, M.; Dittmer, J. The Terrestrial Isopod Microbiome: An All-in-One Toolbox for Animal-Microbe Interactions of Ecological Relevance. *Front. Microbiol.* **2016**, *7*, 1472. [CrossRef]
37. Olmos, J.; Ochoa, L.; Paniagua-Michel, J.; Contreras, R. Functional Feed Assessment on Litopenaeus vannamei Using 100% Fish Meal Replacement by Soybean Meal, High Levels of Complex Carbohydrates and Bacillus Probiotic Strains. *Mar. Drugs* **2011**, *9*, 1119–1132. [CrossRef] [PubMed]

38. Chen, J.G.; Lou, D.; Yang, J.F. Isolation and Identification of *Acholeplasma* sp. from the Mud Crab, Scylla serrata. *Evid. Based Complement. Altern. Med.* **2011**, *2011*, 209406. [CrossRef]
39. Wang, W.; Wen, B.; Gasparich, G.E.; Zhu, N.; Rong, L.; Chen, J.; Xu, Z. A spiroplasma associated with tremor disease in the Chinese mitten crab (*Eriocheir sinensis*). *Microbiology* **2004**, *150*, 3035–3040. [CrossRef] [PubMed]
40. Givens, C.E.; Burnett, K.G.; Burnett, L.E.; Hollibaugh, M.B.T. Microbial communities of the carapace, gut, and hemolymph of the Atlantic blue crab, *Callinectes sapidus*. *Mar. Biol.* **2013**, *160*, 2841–2851. [CrossRef]
41. Chen, X.; Di, P.; Wang, H.; Li, B.; Pan, Y.; Yan, S.; Wang, Y. Bacterial community associated with the intestinal tract of Chinese mitten crab (*Eriocheir sinensis*) farmed in Lake Tai, China. *PLoS ONE* **2015**, *10*, e0123990. [CrossRef]
42. Zhang, M.; Sun, Y.; Chen, L.; Cai, C.; Qiao, F.; Du, Z.; Li, E. Symbiotic Bacteria in Gills and Guts of Chinese Mitten Crab (*Eriocheir sinensis*) Differ from the Free-Living Bacteria in Water. *PLoS ONE* **2016**, *11*, e0148135. [CrossRef] [PubMed]
43. Fraune, S.; Zimmer, M. Host-specificity of environmentally transmitted Mycoplasma-like isopod symbionts. *Environ. Microbiol.* **2008**, *10*, 2497–2504. [CrossRef]
44. Wang, W.; Rong, L.; Gu, W.; Du, K.; Chen, J. Study on experimental infections of Spiroplasma from the Chinese mitten crab in crayfish, mice and embryonated chickens. *Res. Microbiol.* **2003**, *154*, 677–680. [CrossRef]
45. Jorth, P.; Turner, K.H.; Gumus, P.; Nizam, N.; Buduneli, N.; Whiteley, M. Metatranscriptomics of the human oral microbiome during health and disease. *MBio* **2014**, *5*, e01012–e01014. [CrossRef]
46. He, Z.; Pu, L.; Yuan, C.; Jia, M.; Wang, J. Nutrition deficiency promotes apoptosis of cartilage endplate stem cells in a caspase-independent manner partially through upregulating BNIP3. *Acta Biochim. Biophys. Sin.* **2017**, *49*, 25–32. [CrossRef] [PubMed]
47. Biesalski, H.K.; Chichili, G.R.; Frank, J.; von Lintig, J.; Nohr, D. Conversion of betacarotene to retinal pigment. *Vitam. Horm.* **2007**, *75*, 117–130. [PubMed]

biology 2022, 11, 364

Article

Acute Inflammation Induces Neuroendocrine and Opioid Receptor Genes Responses in the Seabass *Dicentrarchus labrax* Brain

Rita Azeredo [1,*], Marina Machado [1], Patricia Pereiro [1,2], Andre Barany [3], Juan Miguel Mancera [3] and Benjamín Costas [1,4,*]

1 Centro Interdisciplinar de Investigação Marinha e Ambiental (CIIMAR), Universidade do Porto, 4450-208 Matosinhos, Portugal; mcasimiro@ciimar.up.pt (M.M.); patriciapereiro@iim.csic.es (P.P.)
2 Instituto de Investigaciones Marinas (IIM-CSIC), 36208 Vigo, Spain
3 Department of Biology, Faculty of Marine and Environmental Sciences, Instituto Universitario de Investigación Marina (INMAR), Campus de Excelencia Internacional del Mar (CEI·MAR), University of Cadiz, 11519 Puerto Real, Spain; andre.barany@uca.es (A.B.); juanmiguel.mancera@uca.es (J.M.M.)
4 Instituto de Ciências Biomédicas Abel Salazar (ICBAS-UP), Universidade do Porto, 4050-313 Porto, Portugal
* Correspondence: mleme@ciimar.up.pt (R.A.); bcostas@ciimar.up.pt (B.C.)

Simple Summary: It is generally accepted (in mammals and in teleost fish, too) that stressful conditions affect the performance of an immune response. What is still far from being known is at what extend does an immune process affects the neuroendocrine system. Vaccination for instance, is nowadays a common practice in aquaculture and little is known about its physiological implications other than immunization. Here is a first approach to the study of the European seabass' brain gene expression patterns in response to a peripheral inflammatory process. Genes related to the stress response were focused, along with those related to the opioid system. Increased expression of certain genes suggests the activation of a stress response triggered by inflammatory signals. Additionally, contrasting expression patterns of the same gene (increased vs decreased) in the different brain regions (as well as the time needed for changes to happen) point at different functions. These results clearly show the reactivity of different brain responses to an immune response, highlighting the importance of further studies on downstream implications (behavior, feeding, welfare, reproduction).

Citation: Azeredo, R.; Machado, M.; Pereiro, P.; Barany, A.; Mancera, J.M.; Costas, B. Acute Inflammation Induces Neuroendocrine and Opioid Receptor Genes Responses in the Seabass *Dicentrarchus labrax* Brain. *Biology* **2022**, *11*, 364. https://doi.org/10.3390/biology11030364

Academic Editors: Geert F. Wiegertjes and Brian Dixon

Received: 19 January 2022
Accepted: 21 February 2022
Published: 24 February 2022

Abstract: In fish, as observed in mammals, any stressful event affects the immune system to a larger or shorter extent. The neuroendocrine-immune axis is a bi-directional network of mobile compounds and their receptors that are shared between both systems (neuroendocrine and immune) and that regulate their respective responses. However, how and to what extent immunity modulates the neuroendocrine system is not yet fully elucidated. This study was carried out to understand better central gene expression response patterns in a high-valued farmed fish species to an acute peripheral inflammation, focusing on genes related to the hypothalamus-pituitary-interrenal axis and the opioid system. European seabass, *Dicentrarchus labrax*, were intra-peritoneally injected with either Freund's Incomplete Adjuvant to induce a local inflammatory response or Hanks Balances Salt Solution to serve as the control. An undisturbed group was also included to take into account the effects due to handling procedures. To evaluate the outcomes of an acute immune response, fish were sampled at 4, 24, 48, and 72 h post-injection. The brain was sampled and dissected for isolation of different regions: telencephalon, optic tectum, hypothalamus, and pituitary gland. The expression of several genes related to the neuroendocrine response was measured by real-time PCR. Data were statistically analyzed by ANOVA and discriminant analyses to obtain these genes' responsiveness for the different brain regions. Serotonergic receptors were upregulated in the telencephalon, whereas the optic tectum inhibited these transcription genes. The hypothalamus showed a somewhat delayed response in which serotonin and glucocorticoid receptors were concerned. Still, the hypothalamic corticotropin-releasing hormone played an important role in differentiating fish undergoing an inflammatory response from those not under such conditions. Opioid receptors gene expression increased in both the hypothalamus and the telencephalon, while in the optic tectum, most were downregulated. However, no changes in the pituitary gland were observed. The different brain regions under immune stimulation demonstrated clear, distinct responses regarding gene transcription rates as well as the time period needed for the effect to occur. Further, more integrative studies are required to associate functions to the evaluated genes more safely and better understand the triggering mechanisms.

Keywords: genomics; stress response; HPI-axis; neuroendocrine-immune interaction

1. Introduction

Inadequate aquaculture rearing conditions (husbandry, transportation, crowding densities, water parameters, etc.) often result in stressful environments that compromise fish growth and welfare [1–3]. Although these are the most obvious and relevant outcomes for fish farmers, other physiological responses, such as fish immune defenses, can be similarly compromised. Indeed, the extension of stress effects on fish health has mainly been investigated. It is now generally acknowledged that chronic stressful rearing conditions jeopardize fish immune response since it downregulates several immune defense mechanisms [4]. A fish farm ultimately decreases fish immune resistance upon a disease outbreak, leading to high mortalities.

Neuroendocrine and immune responses are tightly connected in what is known as the neuroendocrine-immune axis that comprises both the brain and the head-kidney. Teleost fish head-kidney presents endocrine and immune tissues as well as a complex paracrine signaling network, acting as an intermediary organ between neuroendocrine and immune systems. Neuroendocrine-immune processes are bidirectional, and so not only does stress (internal or external) modulate the immune response, but immunological processes are also able to trigger the hypothalamus-pituitary-interrenal (HPI) axis [5].

The regulators and effectors of these modulatory mechanisms belong to several molecular classes—from neuropeptides, opioids, and neurotransmitters to interleukins and chemokines. Moreover, several of these players are shared by both systems (immune and neuroendocrine). A fair amount of studies has been devoted to evaluating stress-induced effects on fish immunity. Most of these studies used cortisol as the primary stress marker as well as plasma glucose and lactate as indicators of secondary stress responses [6]. Upstream, hypothalamic corticotropin-releasing hormone (*crh*) and corticotropin-releasing hormone binding protein (*crhbp*), as well as pituitary proopiomelanocortin (*pomc*) gene expression are also good primary stress markers [7]. Serotonin, a tryptophan metabolite that mostly acts as a neurotransmitter at central levels, also presents a role in the stress responses modulating corticotropin-releasing hormone (CRF) and adrenocorticotropin hormone (ACTH) secretion. Often, when stress-inducing factors are persistent, this neuroendocrine response becomes chronic and suppresses immune mediators' function, both by a metabolic reorganization that reduced influx of energy and by a direct inhibitory effect of cortisol [4,8,9].

Although the bidirectional aspect of this neuroendocrine-immune axis is well-acknowledged, the effects that an immunological process have on the HPI axis are far more disregarded than those inflicted by stress on immunity. In fish, such mechanisms have been reviewed by Verburg-van Kemenade and co-workers [5] and in more detail by Engelsma and colleagues [10], where the role of cytokines is given particular emphasis, as well as the pathways through which an immune process communicates with the central nervous system. Nonetheless, the neuronal arm of the HPI axis has far more branches connecting to the central backbone of the CRH and ACTH [7]. In this way, the serotonergic system, for instance, not only operates mood and behaviour but it also regulates the stress response (and ultimately cortisol release) using a considerable network of serotonin receptors [11–13]. Specifically, these receptors present a widespread brain distribution; however, it is not homogenous throughout different brain regions both in terms of abundancy and of identity, suggesting distinct functions and reactivity [14].

Likewise, and though comparatively even less explored, opioid receptors display a ubiquitous central distribution that is linked to their diverse plethora of functions. Opioids have long been associated to mood, behaviour, and nociception in fish [15]. However, they are also involved in regulatory mechanisms of both the immune and the stress responses [16,17]. At least in what carp (*Cyprinus carpio*) is concerned, opioids effects in leucocytes have been shown to be evolutionarily conserved [18]. Nevertheless, their exact roles and responsiveness to immune stimulation in the brain are still not known.

In an attempt to further understand the extent of peripheral immune signaling impact on central neuroendocrine responses, the present study intends to unveil and characterize the central neuronal gene expression profile, focusing on neuroendocrine and opioid receptors, in response to an acute peripheral inflammation in a marine fish species, the European seabass (*Dicentrarchus labrax*).

2. Material and Methods

2.1. Fish and Experimental Design

European seabass juveniles (n = 72, 87.3 g ± 16.5 body mass) were acquired from a certificated hatchery (MARESA, Spain) and maintained at the facilities of *Servicios Centrales de Investigación en Cultivos Marinos* (SCI-CM, CASEM, University of Cadiz, Puerto Real, Cádiz, Spain; Facilities for Breeding, Supplying and Users of Experimental Animals; Spanish Operational Code REGA ES11028000312). The fish were acclimated for 2 weeks in a flow-through 2 m^3-tank. They were then transferred to a flow-through seawater system composed of sixteen 80 L-fiber glass tanks and fed a commercial diet for 30 days. The fish were maintained under natural photoperiod (June–July 2017, 36°31′45″ N, 6°11′31″ W), temperature (18–19 °C), and salinity (39 g L^{-1}). Supplemental aeration was provided to maintain dissolved oxygen at 6.8 ± 0.4 mg L^{-1}. Ammonia (<0.1 mg L^{-1}), nitrite (<0.2 mg L^{-1}), and nitrate (<50 mg L^{-1}) were determined once weekly. Fish were fed twice per day (9:00 a.m. and 1:00 p.m.) at a rate of 2% of their body weight over the 30-day feeding trial. At the end of this period, eight fish were netted and euthanized by anaesthetic overdose (1 mL of 2-phenoxyethanol l^{-1} seawater; Merck, Darmstadt, Germany). Brains were collected and dissected into telencephalon, optic tectum, hypothalamus, and pituitary gland. Samples were kept in RNAlater (Sigma) at 4 °C for 24 h and finally stored at −20 °C until further processing. This group of fish, sampled before any intervention, was subsequently designated as undisturbed fish (0 h). The remaining fish were anesthetized (0.5 mL of 2-phenoxyethanol l^{-1} seawater) and intraperitoneally (i.p.) injected with 100 µL of either Freund's Incomplete Adjuvant (FIA) to induce inflammation [19], or Hanks Balanced Salt Solution (HBSS) to serve as a sham group (CTRL), and reallocated in duplicate tanks of the original system for each experimental condition (n = 8 per condition). The fish were then sampled at 4, 24, 48, and 72 h, as previously described, following i.p. injection (n = 4 per tank, n = 8 per time and experimental group). They were fasted for 24 h before sampling and i.p. injection as well as during the experimental time.

All the experimental procedures complied with the University of Cádiz (Spain) guidelines and the European Union Council (2010/63/EU) to use animals in research. The experimental procedures were previously approved by the Spanish Government's Ethics and Animal Welfare Committee (RD53/2013) and endorsed by the Regional Government (Junta de Andalucía reference number 28-04-15-241). All animal protocols were performed under Group-D licenses accredited by FELASA (Federation of European Laboratory Animal Science Associations).

2.2. Gene Expression

Total RNA isolation was conducted with the NZY Total RNA Isolation kit (NZYTech, Lisbon, Portugal) following the manufacturer's specifications. RNA was quantified using the DS-11 Spectrophotometer (DeNovix), and first-strand cDNA was synthesized with NZY First-Strand cDNA Synthesis Kit (NZYTech, Lisbon, Portugal). Quantitative PCR assays were performed with CFX384 Touch Real-Time PCR Detection System, using 4.4 µL of diluted cDNA mixed with 5 μL of NZYSpeedy qPCR Green Master Mix® and 0.3 μL (10 μM) of each specific primer in a final volume of 10 μL. The cDNA amplification was carried out with specific primers for genes that have been selected for their involvement in the neuroendocrine response. Primers were designed with NCBI Primer Blast Tool and IDT OligoAnalyzer Tool™, respecting known qPCR restrictions (amplicon size, Tm difference between primers, GC content, and self-dimer or crossdimer formation). Part of the template sequences were obtained from available data in NCBI, while others were identified after searching the databases dicLab v1.0c seabass genome [20] and designed as previously described. The efficiency of primer pairs was analysed in serial, 2-fold dilutions of cDNA by calculating the slope of the regression line of the cycle thresholds (Ct) vs. the relative concentration of cDNA. Accession number, efficiency values, annealing temperature, product length, and primers sequences are presented in Table 1. Melting curve analysis was also performed to verify that no primer dimers were amplified. The standard cycling conditions were 95 °C initial denaturation for 10 min, followed by 40 cycles of two steps (95 °C denaturation for 15 s followed by primer annealing temperature for 1 min), 95 °C for 1 min followed by 35 s at the annealing temperature, and finally, 95 °C for 15 s. All reactions were carried out as technical duplicates. The expression of the target genes was normalized using the geometric mean of European seabass ribosome 40s subunit (*40s*) and elongation factor 1α (*ef1α*) expression levels and calculated according to the Pfaffl method [21].

2.3. Statistical Analysis

Gene expression values are presented as mean ± standard deviation (mean ± SD). Data were analysed for normality and homogeneity of variance, and, when necessary, outliers were removed, and data were log-transformed before being treated statistically. Possible i.p. injection effects were detected by One-way ANOVA, while inflammation- and sampling time-induced effects were identified using a two-way ANOVA. When statistical significance was detected, ANOVA analyses were followed by Tukey post-hoc test to identify differences within experimental treatments. These statistical analyses were performed using the computer package Statistica 13 for Windows. The level of significance used was $p \leq 0.05$ for all statistical tests. In an attempt to discriminate and characterize brain regions of fish under inflammatory conditions, a multivariate canonical discriminant analysis was performed on each brain region dataset. Thereby, numerous combinations of the original variables (discriminant functions) were evaluated. Each discriminant function explains part of the total variance of the dataset and is loaded by variables contributing the most to that variation. Wilk's λ test assessed discriminatory effectiveness, and the distance between group centroids was measured by squared Mahalanobis distance. To attest whether these distances were statistically significant, Fisher's F statistic was performed. Discriminant analyses were carried out using the data analysis tool XLSTAT for Microsoft Office Excel, and a significance level of 95% ($p \leq 0.05$) was used.

Table 1. Forward and reverse primers for real-time PCR.

Acronym	GenBank	Eff [2]	AT [3]	Product Length [4]	Forward Primer Sequence	Reverse Primer Sequence
40s	HE978789.1	108.8	60	79	TGATTGTGACAGACCCTCGTG	CACAGAGCAATGGTGGGGAT
ef1α	AJ866727.1	92.8	57	144	AACTTCAACGCCCAGGTCAT	CTTCTTGCCAGAACGACGGT
gr1	AY619996.1	114.19	60	100	AAATCTGCCTGGTGTGTTCC	TGCCCTTTCACTGCTCTCTT
gr2	AY549305.1	109.4	55	142	CTTCTACAGCACCAGCACCA	TCTCCTGTTTGACCACACCA
crh	JF274994.1	110.21	60	200	AACCCAAAACTCCCAGCAG	TGTTCCCAACTTTCCCTTGT
crhbp	MG832822.1	105.47	60	199	TGTCATCTCCCAGTCACCAG	GCCATTTCCTCCAAGCAAC
pomc	AY691808.1	101.98	60	158	TCTTCCTCCTCCTCTCCACA	CGCCTTCTCATCTCTTCAGG
htr1aβ	DLAgn_00119560 [1]	102.0	60	176	GGAGCGTAAAACGGTGAAAA	TGGGGTTGAGGAGAGAGTTG
htr2a	DLAgn_00222310 [1]	103.8	60	18	CCTCTGACCTCTGTCCCATC	ACTGAAATCGTCCACACTGC
htr2b	DLAgn_00148380 [1]	109.4	60	165	ATTGCCCTCGTCACTGTTCT	GCTGTGTTGGATTGGCTTCT
htr2c	DLAgn_00037670 [1]	118.0	60	195	CATCCGCAACCCCATAGAG	ACGAAGGAGCCAATCAGCAT
tph1	DLAgn_00154580 [1]	107.0	62	114	CGCATAGACTTCACAACAGAGG	CAGCAGAGGGAGGTTCTTCA
ogfr1	DLAgn_00128530 [1]	96.8	60	185	GTTGGGAATGGAGATGGAAA	GCTTCAGATTTTGGCTCAGG
ogfr2	DLAgn_00089660 [1]	96.6	60	146	CTTGCCTTCCTGTCTCCAGT	CTTGTCTCGGTTTCCTTTGG
kor1	DLAgn_00007470 [1]	89.7	60	249	TCTGGTGCTTGTGGTAGTCG	TGGCAGTCTCTGTGTCCTTG
kor2	DLAgn_00077520 [1]	82.0	60	163	CTCGTCAGTGTCCCCGAAAC	CCCCCTTCAGTTTGGCCGAGAG
nopr	DLAgn_00125610 [1]	97.5	60	106	CTCCTTTCTCATCCCTGTGG	GTTGCGGTCCTTTTCCTTG
muor	DLAgn_00015310 [1]	99.8	60	240	GTCACCAGCACCCTACCATT	CGAGGAGAGAATCCAGTTGC
dor2	DLAgn_00062690 [1]	108.1	60	81	CGCTTCTCGGTCTCCATAACT	GGTCTCATTACTACTTGAAG

[1] Sequences obtained from databases dicLab v1.0c seabass genome. [2] Efficiency of PCR reactions were calculated from serial dilutions of tissue RT reactions in the validation procedure. [3] Annealing temperature (°C). [4] Amplicon (nt).

3. Results

For clarity, the results are presented in two main subsections: (i) the first one gathers genes more directly related to the HPI-axis response (results from Section 3.1), and (ii) the second one looks separately at the opioid receptors response. Notwithstanding their involvement in the same neuroendocrine pathways, they are relatively poorly studied for their role during inflammation (results from Section 3.2). Moreover, within each subsection and given the amount of data collected, relevance will be granted to (i) the i.p. injection effect and (ii) inflammation-induced changes. Intraperitoneal injection per se (regardless of content nature) was considered to modulate neuroendocrine gene expression patterns when both groups simultaneously behaved significantly differently from undisturbed fish (0 h). Inflammation was thought to affect gene expression whenever (i) there were significant differences between CTRL and FIA groups or (ii) whenever there was a difference between 0 h and FIA fish, without CTRL being different from 0 h. Due the high amount of results obtained, the complete set of gene expression results is provided as a Supplementary File.

3.1. HPI-Axis Response

3.1.1. Telencephalon

Serotonin receptor 2A (*htr2b*) expression was upregulated in FIA-injected group, regardless of sampling point (Supplementary File, Table S1). Telencephalic expression of *htr2b* and serotonin receptor 2C (*htr2c*) was higher in FIA-injected fish at 4 h than in 0 h group (Figure 1A,B respectively). FIA-injected fish enhanced serotonin receptor 1Aβ (*htr1aβ*) expression levels with respect to CTRL at 48 h (Supplementary File, Table S1). Glucocorticoid receptor 2 (*gr2*) was downregulated at 72 h in both injected groups (Supplementary File, Table S1). No significant differences between FIA, CTRL, and 0 h groups were observed regarding glucocorticoid receptor 1 (*gr1*) and tryptophan hydroxylase 1 (*tph1*) gene expression, although both genes were downregulated over time, irrespective of treatment.

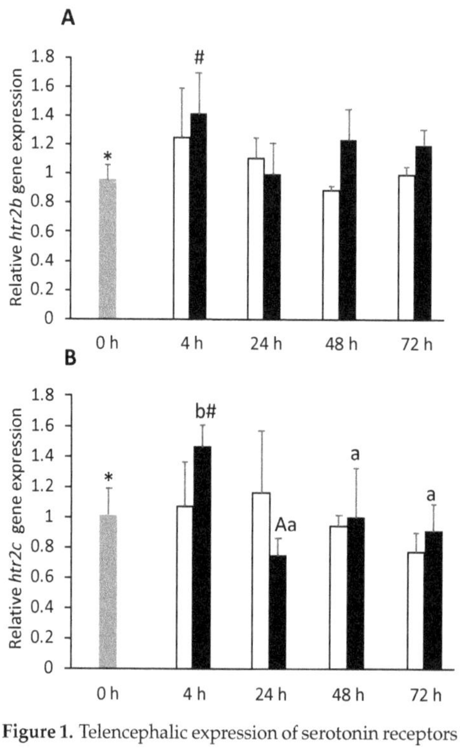

Figure 1. Telencephalic expression of serotonin receptors 2B (*htr2b*, **A**) and 2C (*htr2c*, **B**) in undisturbed

European seabass (0 h, ▣) or i.p.-injected with a sham solution (CTRL, □) or Freund's Incomplete Adjuvant (FIA, ■) and sampled at 4, 24, 48, 72 h post-injection (means ± SD, n = 8). One-way ANOVA was performed to identify differences between i.p.-injected fish and the undisturbed group, followed by a Tukey post-hoc test. Different symbols (*, #) stand for significant differences between i.p.-injected groups and the undisturbed group (0 h). Two-way ANOVA was performed to identify significant differences within the i.p.-injected fish, followed by a Tukey post-hoc test. Capital letters stand for significant differences between stimuli. Lower-case letters indicate significant differences between sampling times ($p \leq 0.05$).

When evaluating linear functions of HPI-related variables in the telencephalon and their contributions to differences between 0 h and FIA groups (FIA4, FIA24, FIA48, FIA72), the overall discriminant analysis performance was very reasonable (Wilks $\lambda = 0.19$, $p = 0.04$). It resulted in four discriminant functions, with the first two accounting for 79.2% of the total variability (Figure 2A). The first discriminant function (F1, 44%) was negatively loaded by *gr1*, *gr2*, *htr2a*, and *htr2c* (i.e., lower gene expression) (Figure 2A, correlations of −0.62, −0.64, −0.62, −0.78, respectively) whereas the second function (F2, 35.25%) was positively loaded by *htr2b* (i.e., higher gene expression) (Figure 2A, correlation of 0.86). The analysis of Mahalanobis distances between groups' multivariate means demonstrated that FIA4, FIA48, and FIA72 differed from 0 h, and that FIA4 differed from FIA24 and FIA72 ($p < 0.05$, Figure 2B).

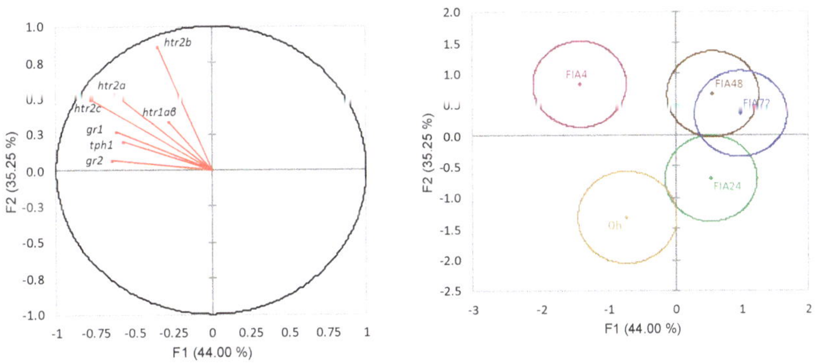

(A) **(B)**

Figure 2. Canonical discriminant analysis of European seabass telencephalic expression of HPI-axis-related genes. (**A**) Correlation variables/factors (factor loads) for two main discriminant functions (F1 and F2); *gr1*, glucocorticoid receptor 1; *gr2*, glucocorticoid 2; *htr1aβ*, serotonin receptor 1Aβ; *htr2a*, serotonin receptor 2A; *htr2b*, serotonin receptor 2B; *htr2c*, serotonin receptor 2C; *tph1*, tryptophan hydroxylase 1. (**B**) Canonical discriminant scores of each group. Group centroids are marked by a small diamond, whereas circles indicate data distribution per group.

3.1.2. Optic Tectum

An extended gene expression suppression was observed in the FIA group compared to the CTRL group, regardless of the sampling point. Similar to *htr2c* (Figure 3A), *gr1*, *htr1aβ*, and *htr2a* were all downregulated (Supplementary File, Table S2). In the same way, *gr2* gene expression decreased in FIA-injected fish at 4 h compared to both 0 h and CTRL groups (Figure 3B). No significant differences were observed regarding both *htr2b* and *tph1*.

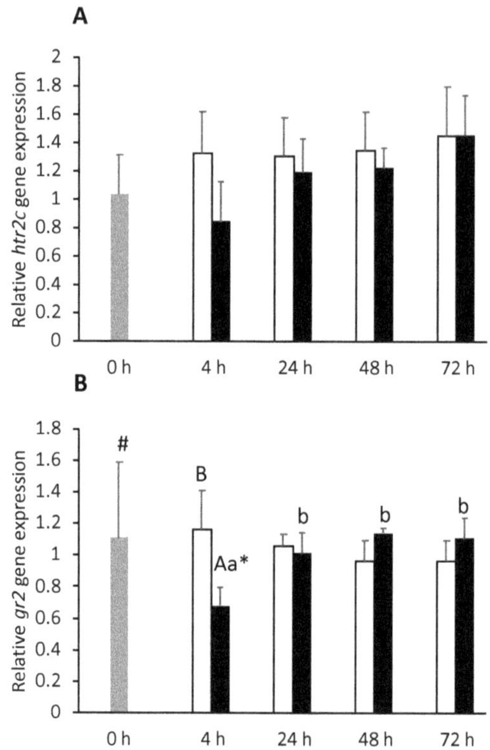

Figure 3. Optic tectum expression of serotonin receptor 2C (*htr2c*, **A**) and glucocorticoid receptor 2 (*gr2*, **B**) in undisturbed European seabass (0 h, ▩) or i.p.-injected with a sham solution (CTRL, ☐) or Freund's Incomplete Adjuvant (FIA, ■) and sampled at 4, 24, 48, 72 h post-injection (means ± SD, n = 8). Different symbols (* and #) stand for significant differences between i.p.-injected groups and the undisturbed group (0 h). Capital letters stand for significant differences between stimuli. Lower-case letters indicate significant differences between sampling times. Further details in legend of Figure 1.

The discriminant analysis to neuroendocrine-related genes was also statistically significant (Wilks $\lambda = 0.27$, $p = 0.04$), with the first two discriminant functions explaining 93.3% of the data total variability (Figure 4). The first discriminant function (F1, 85.5%) was positively loaded by both *gr2* and *htr2c* (i.e., higher expression) (Figure 4A, correlations of 0.62 and 0.68, respectively), whereas the second function (F2, 7.8%) was positively loaded by *gr1* (i.e., higher expression) (Figure 4A, correlation of 0.70). The analysis of Mahalanobis distances between groups' multivariate means demonstrated that FIA4 was significantly different from FIA24, FIA48, and FIA72 ($p < 0.05$, Figure 4B).

3.1.3. Hypothalamus

Gene expression of *gr1* was upregulated by i.p. injection, being higher at 48 h in the FIA group, compared to the CTRL group (Figure 5A). A similar feature was observed for *gr2* expression (data not shown). In addition, corticotropin-releasing hormone (*crh*) expression was upregulated by i.p. injection whereas *tph1* transcription significantly decreased (Supplementary File, Table S3). Expression levels of *htr1aβ* were higher at 48 h in the FIA group compared to the CTRL group (Figure 5B). No significant differences between the FIA and CTRL groups were detected regarding corticotropin-releasing hormone-binding protein (*crhbp*) and *tph1* gene expression (Supplementary File, Table S3).

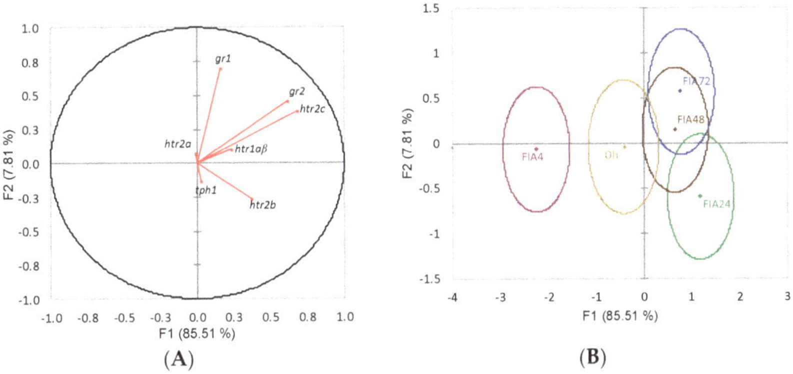

Figure 4. Canonical discriminant analysis of European seabass optic tectum expression of HPI-axis-related genes. (**A**) Correlation variables/factors (factor loads) for two main discriminant functions (F1 and F2); *gr1*, glucocorticoid receptor 1; *gr2*, glucocorticoid 2; *htr1aβ*, serotonin receptor 1Aβ; *htr2a*, serotonin receptor 2A; *htr2b*, serotonin receptor 2B; *htr2c*, serotonin receptor 2C; *tph1*, tryptophan hydroxylase 1. (**B**) Canonical discriminant scores of each group. Group centroids are marked by a small diamond, whereas circles indicate data distribution per group.

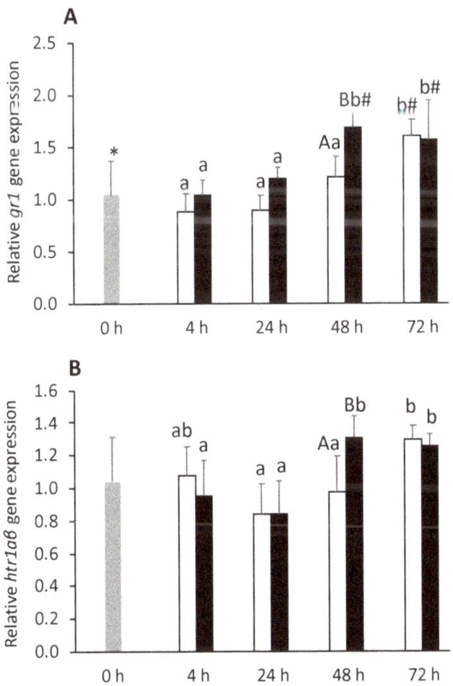

Figure 5. Hypothalamic expression of glucocorticoid receptor 1 (*gr1*, **A**) and serotonin receptor 1A*β* (*htr1aβ*, **B**) in undisturbed European seabass (0 h, ▩) or i.p.-injected with a sham solution (CTRL, □) or Freund's Incomplete Adjuvant (FIA, ■) and sampled at 4, 24, 48, 72 h post-injection (means ± SD, n = 8). Different symbols (* and #) stand for significant differences between i.p.-injected groups and the undisturbed group (0 h). Capital letters stand for significant differences between stimuli. Lower-case letters indicate significant differences between sampling times. Further details in legend of Figure 1.

Neuroendocrine variables discriminant analysis (Wilks $\lambda = 0.12$, $p = 0.002$) produced four discriminant functions from which the first two accounted for 77.4% of the total dataset variability (Figure 6). The first function (F1, 53.6%) was positively loaded by *gr1*, *gr2*, *crh*, *crhbp*, and *htr1aβ* (i.e., higher expression) (Figure 6A, correlations of 0.70, 0.68, 0.66, 0.69, and 0.60, respectively) while no significant loadings were attributed to the second function (F2, 23.8%).The analysis of Mahalanobis distances between groups' multivariate means demonstrated that the t0h group was significantly different from FIA24 and FIA72, FIA4 was different from FIA72, and FIA24 differed from FIA48 and FIA72 ($p < 0.05$, Figure 6B).

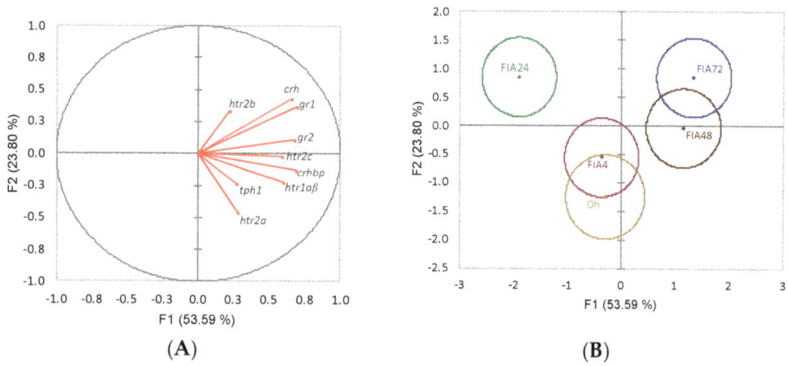

Figure 6. Canonical discriminant analysis of European seabass hypothalamic expression of HPI-axis-related genes. (**A**) Correlation variables/factors (factor loads) for two main discriminant functions (F1 and F2); *gr1*, glucocorticoid receptor 1; *gr2*, glucocorticoid 2; *htr1aβ*, serotonin receptor 1Aβ; *htr2a*, serotonin receptor 2A; *htr2b*, serotonin receptor 2B; *htr2c*, serotonin receptor 2C; *tph1*, tryptophan hydroxylase 1; *crh*, corticotropin-releasing hormone; *crhbp*, crh-binding protein. (**B**) Canonical discriminant scores of each group. Groups centroids are marked by a small diamond, whereas circles indicate data distribution per group.

3.1.4. Pituitary Gland

Serotonin receptors *htr1aβ* and *htr2a* were downregulated in the pituitary gland of all injected fish, at all time-points (Figure 7A,B, respectively). Expression level of *htr2c* was downregulated in FIA-injected fish sampled at 24 h compared to the 0 h group and was also lower in this group respect to CTRL regardless of sampling time (Supplementary File, Table S4).

The discriminant analysis to neuroendocrine variables had an overall satisfactory performance (Wilks $\lambda = 0.43$, $p = 0.028$), and the first two discriminant functions accounted for 88.4% of total data variability (Figure 8). The first function (F1, 58%) was negatively loaded by *htr1aβ* and *htr2a* (i.e., lower expression) (Figure 8A, correlations of -0.80 and -0.95, respectively) whereas the second discriminant function (F2, 30.4%) was positively loaded by *gr1* (i.e., higher expression) (Figure 8A, correlation of 0.86). The analysis of Mahalanobis distances showed that group 0 h was significantly different from all FIA groups, FIA4 was different from FIA24, FIA48 and FIA72, and FIA24 also different from FIA48 ($p \leq 0.05$, Figure 8B).

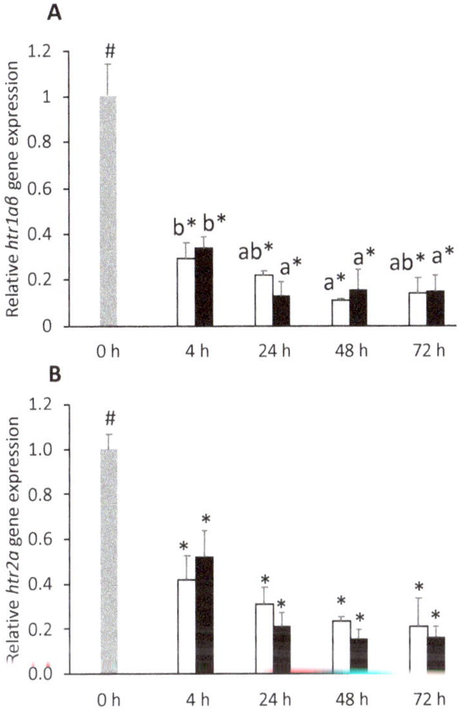

Figure 7. Pituitary gland expression of serotonin receptor 1Aβ (*htr1aβ*, **A**) and serotonin receptor 2A (*htr2a*, **B**) in undisturbed European seabass (0 h, ▩) or i.p.-injected with a sham solution (CTRL, □) or Freund's Incomplete Adjuvant (FIA, ■) and sampled at 4, 24, 48, 72 h post-injection (means ± SD, n = 8). Different symbols (* and #) stand for significant differences between i.p.-injected groups and the undisturbed group (0 h). Capital letters stand for significant differences between stimuli. Lower-case letters indicate significant differences between sampling times. Further details in legend of Figure 1.

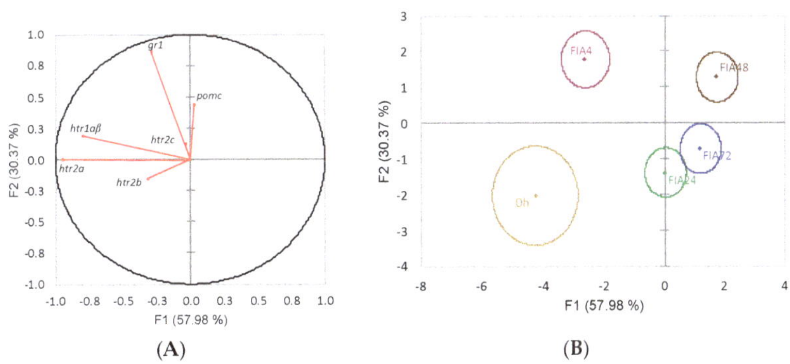

Figure 8. Canonical discriminant analysis of European seabass pituitary gland expression of HPI-axis-related genes. (**A**) Correlation variables/factors (factor loads) for two main discriminant functions (F1 and F2); *gr1*, glucocorticoid receptor 1; *htr1aβ*, serotonin receptor 1Aβ; *htr2a*, serotonin receptor 2A; *htr2b*, serotonin receptor 2B; *htr2c*, serotonin receptor 2C; *pomc*, proopiomelanocortin. (**B**) Canonical discriminant scores of each group. Group centroids are marked by a small diamond, whereas circles indicate data distribution per group.

3.2. Opioid Receptors Response

3.2.1. Telencephalon

Intra-peritoneal injection suppressed nociception receptor (*nopr*) gene expression in the telencephalon of CTRL and FIA groups from 24 h post-injection until the end of the experiment (Supplementary File, Table S1). The inflammatory condition enhanced opioid growth factor receptor 1 (*ogfr1*) gene expression at 4 h respect to the CTRL group (Figure 9A), but expression levels significantly decreased at 4 h post-injection to values similar to those of the CTRL group. The mu opioid receptor (*muor*) expression was similarly upregulated in FIA-injected fish, in which expression was higher than that of the CTRL group at both 4 and 48 h post-injection (Figure 9B). At 4 h, *muor* expression was also higher in FIA than 0 h. Finally, the delta opioid receptor 2 (*dor2*) increased also significantly in fish under inflammation respect to CTRL fish, regardless of sampling time (Supplementary File, Table S1).

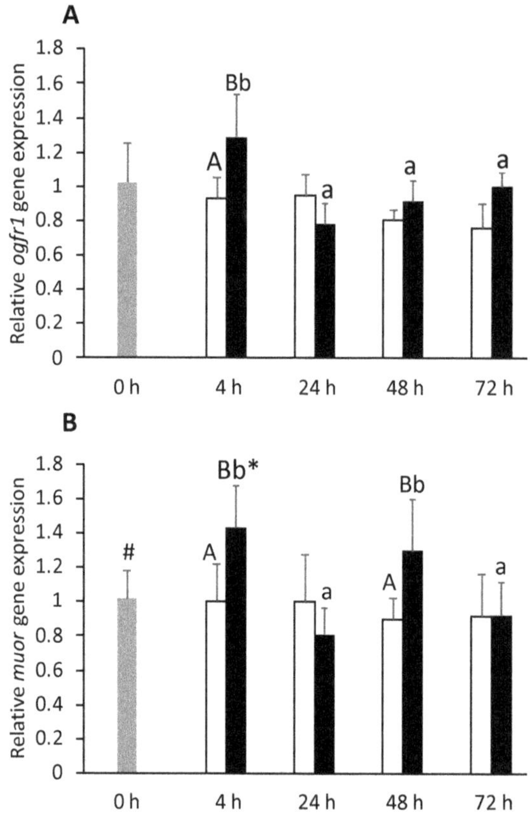

Figure 9. Telencephalic expression of opioid growth factor 1 (*ogfr1*, **A**) and mu opioid receptors (*muor*, **B**) in undisturbed European seabass (t0h, ▩) or i.p.-injected with a sham solution (CTRL, □) or Freund's Incomplete Adjuvant (FIA, ■) and sampled at 4, 24, 48, 72 h post-injection (means ± SD, n = 8). Different symbols (* and #) stand for significant differences between i.p.-injected groups and the undisturbed group (0 h). Capital letters stand for significant differences between stimuli. Lower-case letters indicate significant differences between sampling times. Further details in legend of Figure 1.

The discriminant analysis for opioid receptors' gene expression of 0 h and FIA groups (Wilks $\lambda = 0.2$, $p = 0.004$) resulted in four linear functions from which the first two accounted for 89.8% of the data total variability (Figure 10). The first discriminant function (F1, 53.5%) was negatively loaded by kappa opioid receptor 2 (*kor2*) and nociceptin opioid receptor (*nopr*) (i.e., lower gene expression) (Figure 10A, correlations of −0.61 and −0.93, respectively) while the second function (F2, 36.4%) was positively loaded by *muor* (i.e., higher gene expression) (Figure 10A, correlation of 0.85%). The analysis of Mahalanobis distances between group's multivariate means showed that t0h was different from FIA24, FIA48, and FIA72. The FIA4 group was different from both FIA24 and FIA72. FIA24 was also different from FIA48 ($p < 0.05$, Figure 10B).

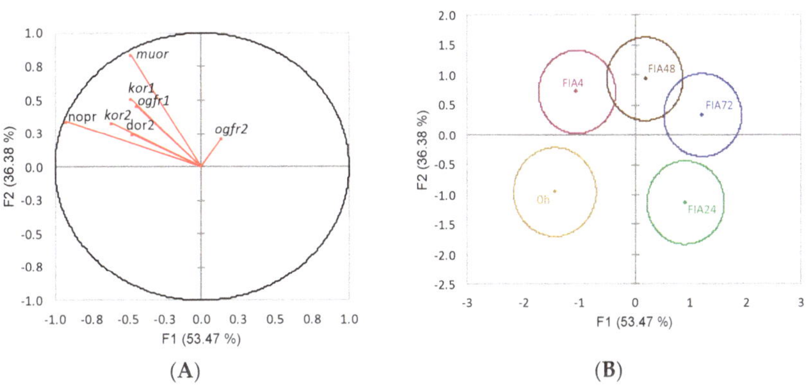

Figure 10. Canonical discriminant analysis of European seabass telencephalic expression of opioid receptor genes. (**A**) Correlation variables/factors (factor loads) for two main discriminant functions (F1 and F2); *muor*, mu opioid receptor; *kor1*, kappa opioid receptor 1; *kor2*, kappa opioid receptor 2; *dor2*, delta opioid receptor 2; *ogfr1*, opioid growth factor receptor 1, *ogfr2*, opioid growth factor receptor 2; *nopr*, nociceptin opioid receptor. (**B**) Canonical discriminant scores of each group. Group centroids are marked by a small diamond, whereas circles indicate data distribution per group.

3.2.2. Optic Tectum

In the optic tectum, inflammation seemed to carry out a transversal inhibitory effect where *ogfr1* (at 4 h post-injection, Supplementary File, Table S2) and opioid growth factor receptor 2 (*ogfr2*) (at 24 and 48 h post-injection, Figure 11A) expression levels decreased in FIA-injected fish compared to CTRL fish. In addition, *kor2*, *muor*, and *dor2* were also down-regulated in FIA fish, compared to CTRL fish, regardless of sampling point (Figure 11B). Moreover, *nopr* expression was lower in the FIA group than in CTRL group at both 4 and 24 h post-injection (Supplementary File, Table S2).

The performance of the discriminant analysis to opioid receptors gene expression showed no significant differences amongst data variability (Wilks $\lambda = 0.4$, $p = 0.13$).

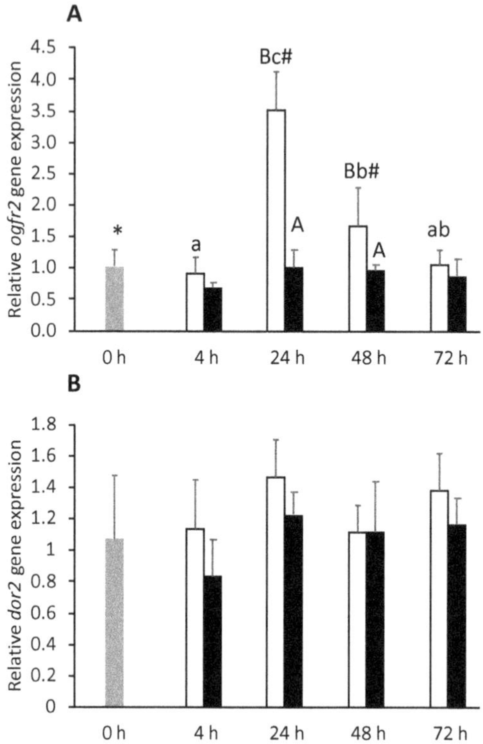

Figure 11. Optic tectum expression of opioid growth factor receptor 2 (*ogfr2*, **A**) and delta opioid receptor 2 (*dor2*, **B**) in undisturbed European seabass (0 h, ▩) or i.p.-injected with a sham solution (CTRL, □) or Freund's Incomplete Adjuvant (FIA, ■) and sampled at 4, 24, 48, 72 h post-injection (means ± SD, n = 8). Different symbols (* and #) stand for significant differences between i.p.-injected groups and the undisturbed group (0 h). Capital letters stand for significant differences between stimuli. Lower-case letters indicate significant differences between sampling times. Further details in legend of Figure 1.

3.2.3. Hypothalamus

Expression levels of both *ogfr1* and *kor2* were upregulated by inflammation whereas *muor* was downregulated. For the three affected transcripts, hypothalamic reaction to i.p. injection was earlier in the FIA group compared to CTRL group responses. Fish undergoing an inflammatory response increased expression levels of *ogfr1*, *kor2* (Figure 12A), and *dor2* (Figure 12B) in FIA fish at 48 h compared to their CTRL counterparts. Differently, it downregulated *dor2* transcription at 24 h respect to 0 h fish (Figure 12B).

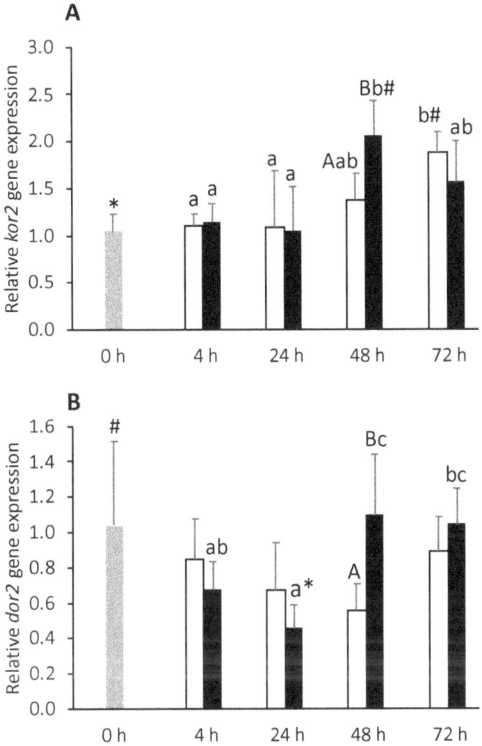

Figure 12. Hypothalamic expression of kappa opioid receptor 2 and delta opioid receptor 2 (*kor2*, **A** and *dor2*, **B**) in undisturbed European seabass (0 h, ▩) or i.p.-injected with a sham solution (CTRL, □) or Freund's Incomplete Adjuvant (FIA, ■) and sampled at 4, 24, 48, 72 h post-injection (means ± SD, n = 8). Different symbols (* and #) stand for significant differences between i.p.-injected groups and the undisturbed group (0 h). Capital letters stand for significant differences between stimuli. Lower-case letters indicate significant differences between sampling times. Further details in legend of Figure 1.

The discriminant analysis (Wilks $\lambda = 0.09$, $p < 0.0001$) produced four discriminant functions from which the first two accounted for 83.5% of the total dataset variability (Figure 13). The first function (F1, 63.1%) was positively loaded by *ogfr2* (i.e., higher expression) (Figure 13A, correlation of 0.75) while the second function (F2, 20.4%) was negatively loaded by kappa opioid receptor 1 (*kor1*), *nopr*, and *dor2* (i.e., lower gene expression) (Figure 13A, correlations of −0.62, −0.76, and −0.93, respectively). The Mahalanobis distances between groups' multivariate means showed that t0h and FIA4 groups were significantly different from FIA24, FIA48, and FIA72, but not different from one another. Furthermore, FIA24 differed from FIA48 and FIA72, while FIA48 was also different from FIA72 ($p < 0.05$, Figure 13B).

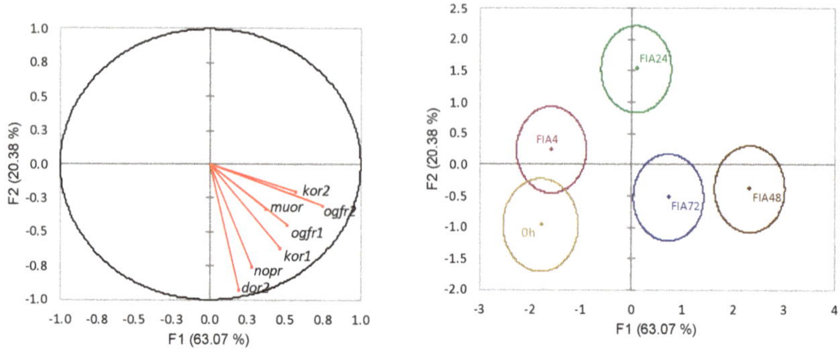

Figure 13. Canonical discriminant analysis of European seabass hypothalamic expression of opioid receptor genes. (**A**) Correlation variables/factors (factor loads) for two main discriminant functions (F1 and F2); *muor*, mu opioid receptor; *kor1*, kappa opioid receptor 1; *kor2*, kappa opioid receptor 2; *dor2*, delta opioid receptor 2; *ogfr1*, opioid growth factor receptor 1, *ogfr2*, opioid growth factor receptor 2; *nopr*, nociceptin opioid receptor. (**B**) Canonical discriminant scores of each group. Group centroids are marked by a small diamond, whereas circles indicate data distribution per group.

3.2.4. Pituitary Gland

In the pituitary gland, *dor2* gene expression was downregulated at 24 h in both injected groups (Figure 14). Otherwise, no effects of inflammation were detected in this tissue opioid receptors gene expression.

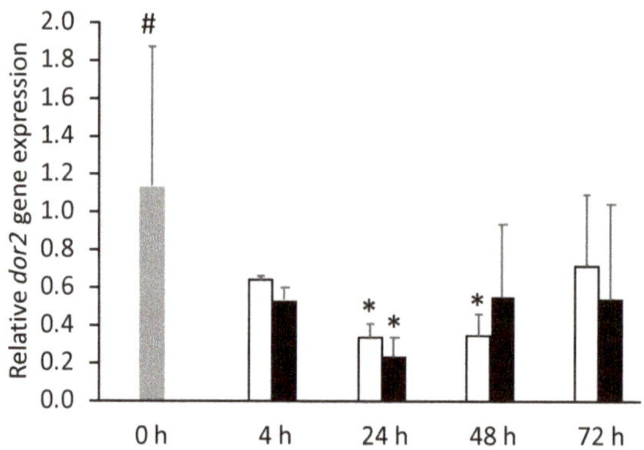

Figure 14. Pituitary gland expression of delta opioid 1 (*dor2*) in undisturbed European seabass (0 h, ▩) or i.p.-injected with a sham solution (CTRL, ☐) or Freund's Incomplete Adjuvant (FIA, ■) and sampled at 4, 24, 48, 72 h post-injection (means ± SD, n = 8). Different symbols (* and #) stand for significant differences between i.p.-injected groups and the undisturbed group (0 h). Further details in legend of Figure 1.

The first two linear functions resulting from data discriminant analysis (Wilks $\lambda = 0.04$, $p < 0.03$) explained 96.3% of the data variability (Figure 15). Both functions were negatively loaded by *dor2* (i.e., lower gene expression) (Figure 15A, correlations values of −0.61 and −0.75, respectively). The analysis of Mahalanobis distances between groups' multivariate means showed that the FIA4 group was significantly different from and FIA24 and from FIA72 ($p \leq 0.05$, Figure 15B).

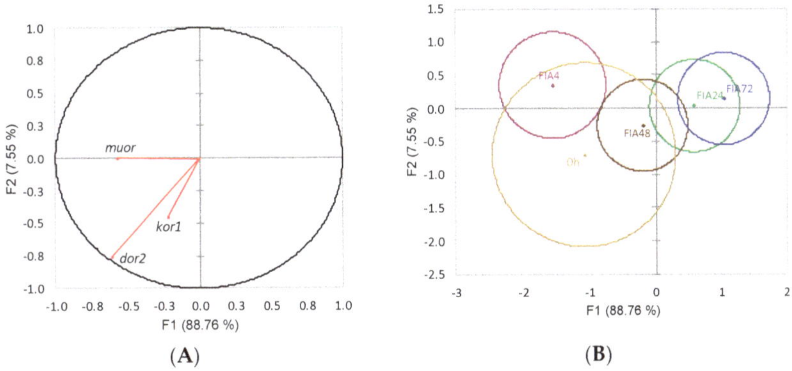

Figure 15. Canonical discriminant analysis of European seabass pituitary gland expression of opioid receptor genes. (**A**) Correlation variables/factors (factor loads) for two main discriminant functions (F1 and F2); *muor*, mu opioid receptor; *kor1*, kappa opioid receptor 1; *dor2*, delta opioid receptor 2. (**B**) Canonical discriminant scores of each group. Groups centroid are marked by a small diamond, whereas circles indicate data distribution per group.

4. Discussion

Fish handling is amongst the top stress-inducing procedures in aquaculture. Chasing and intra-peritoneally injecting fish, which involves air exposure, is a procedure often carried out during vaccination. It represents an acute stress that activates the HPI-axis and initiates a neuroendocrine response, irrespective of potential long-term physiological consequences [22,23]. In this study, while being aware of the acute stress (i.p. injection and/or air exposure) implications on brain gene expression, there was no intention to evaluate such effects. However, innate immune mechanisms are also triggered right after the immune challenge (FIA i.p. injection). Hence, our choice of sampling at such an early time point (4 h) implies that it is likely to witness some lingering stress effects (induced artifact due to injection procedure). Thus, they might mask those inflicted by the development of the inflammatory response at this early stage.

4.1. HPI-Axis Response

A stress-induced effect was observed in both experimental groups when *crh*, encoding for CRH—one of the first compounds to be released by the hypothalamic tissue upon neuroendocrine stimulation [7]—was found upregulated at 72 h post-injection. However, our sampling scheme most certainly missed a much earlier induction, previously demonstrated by Skrzynska and co-workers [22]. One would then expect rising plasma ACTH levels due to CRH-induced pituitary secretion of this hormone [7]. Despite the fact that we did not measure it directly, expression of *pomc* (ACTH genetic precursor) was evaluated in the pituitary gland but acute stress did not change its transcriptional rate. Transcription of *pomc* is regulated by several factors, including CRH (activation) and GR (impairment) [24]. GR1 was down-regulated in the pituitary gland at 72 h, which would increase the expectations of observing *pomc* upregulation. Nonetheless, the absence of *pomc* transcriptional changes does not imply that there was no induction of ACTH response, since an expression enhancement could have occurred at an earlier time. Indeed, Liu and co-workers have observed the upregulation of *pomcb* in the pituitary gland of gilthead seabream, one hour following air exposure, with gene expression returning to basal levels at 6 and 24 h post stress [23]. Yet, transcription of *pomca* was not significantly altered by the acute stress, pointing at potentially different isoform functions. In the present study, the coding sequence selected for this gene does not specify to a particular isoform, but it shows high similarity degree to other teleost species' *pomca* nucleotide sequences.

There are contrasting reports of stress and cortisol modulatory effects in which GR transcriptomic regulation is concerned [25–28]. In the head-kidney of European seabass leucocytes, both *gr1* and *gr2* were upregulated upon in vitro cortisol treatment [28]. The present study shows that these genes' behavior towards acute stress was the opposite. Specifically, *gr1* was upregulated in the hypothalamus, whereas the telencephalic *gr2* transcription was downregulated. There is a great gap in the knowledge of brain glucocorticoid receptors' distribution and dynamics in teleost fish, with most studies focusing on other tissues [28–30]. This central localization, where neuroendocrine pathways are initiated, indicates their involvement in mediating corticosteroid feedback mechanisms through transcriptomic regulation. The hypothalamic stress-responsive *gr1* enhancement expression appears to be a regulatory mechanism triggered upon neuroendocrine stimulation. Accordingly, it was concomitant to *crh* higher expression levels in the same tissue.

In parallel to the stimulation at the central nervous system of the neuroendocrine response, acute stress also leads to a rise in monoaminergic activity. Nevertheless, these changes are known to occur almost immediately after the acute stress (chasing/handling/injection), with levels typically returning to basal values within 4–8 h post-stress [31]. At 4 h post-injection, two of the evaluated serotonin receptors (*htr1aβ* and *htr2a*) were markedly downregulated in the pituitary gland. Moreover, in the hypothalamus, i.p. injection inhibited the gene expression of the serotonin synthesizing enzyme, *tph1*. Despite this being in opposition to what was observed by Gesto and co-workers [31], a similar inhibition of *htr1aα* expression was observed in the telencephalon of rainbow trout (*Oncorhynchus mykiss*) 4 h after being subjected to acute stress (confinement, [32]). Gene expression data from this study do not allow to discriminate auto- from heteroreceptors. Nevertheless, and regardless of the extension and direction of this modulatory effect (which was not within the scope of this study), stress-induced downregulation of these receptors' gene expression, together with a decreased availability of serotonin synthesizing enzyme, suggests that the neuroendocrine response at some level regulates serotonergic activity. In support of this hypothesis, Medeiros and McDonald [33] showed in Gulf toad-fish that *htr1a* was downregulated by cortisol treatment, indicating that serotonin receptors were under negative feedback control of this hormone.

Intraperitoneal injection with a phlogistic agent elicits a local inflammatory response (FIA-injected fish). Overall some of these genes' behaviour was inverted while others were unaltered. What was remarkably clear was a regionalization of the brain response, i.e., establishing marked response patterns (stimulation vs. inhibition) according to different isolated regions. In what the telencephalon is concerned, i.p., injection with FIA induced the expression of three serotonin receptors, two of them (*htr2b* and *htr2c*) at 4 h post-injection. Note that in mammals, this part of the brain includes the so-called limbic system consisting of structures that support several functions such as emotion, behaviour, and long-term memory. It is vastly innervated by serotonergic neurons coming from the raphe nuclei, and it interacts with the mammalian HPI-axis homolog, the hypothalamus-pituitary-adrenal axis [34]. In teleosts, an evident connection between telencephalic serotonin and HPI-axis has been demonstrated too [34–37]. Yet, little is known about the effect of peripheral immune signals on serotonergic pathways. The fact that there was an early induction of these serotonin receptors following the inflammatory insult might be related to circulating immune mediators such as cytokines, which are intensively produced at the onset of an immune response and have also been associated to the activation of mammalian central neuroendocrine pathways, including serotonergic pathways [10,38].

In line with the stimulatory pattern observed in the telencephalic region, the development of inflammatory response was accompanied by upregulation of both glucocorticoid receptors (*gr1* and *gr2*) and *htr1aα* genes in the hypothalamus at 48 h post-injection. This was a response unexpectedly delayed in time considering these are mechanisms generally known to be mounted soon after the initial trigger [39], even when the trigger is an immune mediator (which begins to be synthesized at the onset of the inflammatory response [40]). Interestingly, this seemed to be a critical time point for the hypothalamic response, given several genes were differently expressed between control and FIA-injected fish at this time (including opioid-related genes). The ANOVA statistical approach did not retrieve significant alterations on the more obvious *crh* or *crhbp* genes. However, the discriminant analysis of data from the hypothalamus of fish undergoing inflammation attributed a significant role to these two genes which, together with both glucocorticoids (*gr1* and *gr2*) and one serotonin receptor (*htr1aβ*), showed time-dependent increased expression. Altogether, it is important to be aware of the effects of an acute stress, such as a peritoneal injection and/or air exposure, particularly at this first stage of the HPI-axis physiology. The absence of a more rapid and stronger hypothalamic response might be explained by a masking effect of the intraperitoneal puncture.

In contrast to the observed in telencephalon and hypothalamus, the optic tectum reacted to immune signaling with a general inhibitory behaviour. Although this region is the primary visual center in the brain of the teleost [41], several studies (including non-mammalian) have demonstrated its involvement in other physiological mechanisms, such as the stress response [25,36,42]. In our experiment, both glucocorticoid receptors (*gr1* and *gr2*) and three serotonergic receptors (*htr1aβ*, *htr2a*, and *htr2c*) were less expressed in the optic tectum of fish undergoing an inflammatory response than control fish (injected with HBSS). Lower expression levels of these receptors suggest a general shutdown of glucocorticoid-mediated regulatory pathways as well as an impaired serotonergic activity. However, whether these are direct consequences of inflammatory signals and to what extent these changes affect other tectal functions is far from being understood. On matters of the teleostean brain function, information is scarce at best. However, these divergent responses amongst different brain regions indicate different functions of these tissues in order to face the same stimulus.

4.2. Opioid Receptors

Opioid neuropeptides are mainly synthesized at the central nervous system, but they are ubiquitously produced in the organism. Met-enkephalin, β-endorphin and dynorphin are produced from different gene precursors and bind to their specific receptors, except for met-enkephalin that may bind to delta opioid receptor, mu opioid receptor and the opioid growth factor receptor [43]. Similarly, opioid receptors are expressed in the teleost brain and other organs such as the head-kidney [17]. Recent findings further support their involvement in the immune response, demonstrating a direct effect of opioids on carp phagocyte immune function and immune-related gene expression [18,44]. On the other hand, opioid receptors expression has also been shown to be modulated by in vitro immune stimulation [17]. In mammals, opioids are mostly known to mediate nociception and mood and have long been studied regarding their clinical use as pain killers and their addiction properties [45]. However, there is quite a significant gap in fish regarding their central nervous system function since nociception is still a critically debated issue amongst fish biologists [15,46]. Even less effort has been put into understanding their main dynamics in response to peripheral immune stimuli.

As reviewed by Wei and Loh [47], *muor* transcriptional regulation was found to be mediated—among others—by endocrine factors, such as cytokines and interferon-γ, which demonstrates how sensitive these receptors are to immune stimulation in mammals. However, such knowledge has resulted from in vitro studies with immune-related cell lines, indicating that it is important to consider these receptors' localization when evaluating their transcriptional dynamics. In the brain, opioid receptors transcription is therefore likely to be differently regulated, even within different regions. In the present study, FIA-induced inflammation evoked a more noticeable hypothalamic *ogfr1*, *kor2*, and *dor2* gene expression enhancement than induced by the i.p. injection itself but only 48 h after the injection onset of inflammation. In addition, in the telencephalon, *muor* and *ogfr1* expressions were upregulated in FIA-injected fish, but this altered pattern was observed earlier at 4 h post-injection. Therefore, the two regions' reactivity to an immune challenge is differently characterized in receptor types and in the time it takes for their transcription rate to change, suggesting a differential role of telencephalon and hypothalamus after an inflammatory change. interestingly telencephalic serotonin receptors were, as aforementioned, similarly upregulated in parallel to the unfolding inflammatory process, suggesting that both receptor families (i.e., serotonin and opioids receptors) might be involved in the same neurologic response to immune signals. Indeed, Tao and Auberbach [48] have shown that opioid infusion into the rat brain induced a rise in extracellular serotonin, an effect blocked by a selective μ-receptor antagonist, demonstrating the existence of a close relationship between both systems, at least in mammals. However, further studies are necessary in order to demonstrate a similar relationship in teleost.

Tectum opioid receptors behaved differently from the telencephalon and the hypothalamus and certainly to the pituitary gland, seemingly unaffected by inflammation. Indeed, and concerning what was observed with other neuroendocrine-related genes, most of the responding opioid receptor genes in the optic tectum were downregulated by inflammatory signaling, including those upregulated elsewhere. Far more than just the main visual center, the optic tectum is believed to gather other sensory modalities (electroreception, infra-red sensitivity, mechanoreception, etc.), then convey the acquired and processed information to motor neurons. Thereby, it is also involved in reactive behavior [49]. The responsiveness of opioid receptors in this particular region suggests that opioid peptides might also modulate sensory function. Additionally, in an inflammatory setting, these potential modulatory pathways seem to be altered, and so might be the resulting sensory and behavior functions.

5. Conclusions

Understanding how the fish brain processes peripheral inflammatory signals and whether such signaling molecules might trigger possible reactions is of great importance in fish health and welfare. The current approach yielded new data on the effect of a peritoneal inflammation on brain neuroendocrine response. Changes in various gene transcriptional rates in different brain regions—and in spite of the i.p. injection-induced stress—demonstrate the role of immune signals as transcriptional factors. It also shows their potential to regulate several brain-originated physiological responses such as the HPI-axis, behavior, nociception, and sensory functions. Finally, opioid receptors and other genes more directly involved with the neuroendocrine response are differently expressed in the evaluated brain regions, possibly pointing out the existence of site-specific functions.

Supplementary Materials: The following supporting information can be downloaded at: https://www.mdpi.com/article/10.3390/biology11030364/s1, Table S1. Telencephalic expression of neuroendocrine genes in the European seabass i.p. injected with Freud's Incomplete Adjuvant (FIA) or Hank's Balanced Salt Solution (CTRL) and sampled at 4, 24, 48 or 72 h post injection; Table S2. Optic tectum expression of neuroendocrine genes in the European seabass i.p. injected with Freud's Incomplete Adjuvant (FIA) or Hank's Balanced Salt Solution (CTRL) and sampled at 4, 24, 48 or 72 h post injection; Table S3. Hypothalamic expression of neuroendocrine genes in the European seabass i.p. injected with Freud's Incomplete Adjuvant (FIA) or Hank's Balanced Salt Solution (CTRL) and sampled at 4, 24, 48 or 72 h post injection; Table S4. Pituitary gland expression of neuroendocrine genes in the European seabass i.p. injected with Freud's Incomplete Adjuvant (FIA) or Hank's Balanced Salt Solution (CTRL) and sampled at 4, 24, 48 or 72 h post injection.

Author Contributions: Conceptualization, B.C. and J.M.M.; in vivo experiment, A.B. and B.C.; in silico search for the target genes coding sequences and primers design, R.A. and P.P.; RNA extraction and cDNA synthesis, M.M.; gene expression analysis and statistical analysis, R.A.; writing—original draft preparation, R.A.; writing—review and editing, R.A., P.P., M.M., A.B., J.M.M. and B.C. All authors have read and agreed to the published version of the manuscript.

Funding: This work was supported by the project INFLAMMAA (PTDC/CVT-CVT/32349/2017), financed by Portugal and the European Union through FEDER, COMPETE 2020 and CRESC Algarve 2020, in the framework of Portugal 2020, and by national funds through Fundação para a Ciência e a Tecnologia (FCT, Portugal). National funds also supported this research through FCT within the scope of UIDB/04423/2020 and UIDP/04423/2020. FCT supported BC through 2020.00290.CEECIND. Patricia Pereiro wishes to thank the Axencia Galega de Innovación (GAIN, Xunta de Galicia) for her postdoctoral contract (IN606B-2018/010). This study was also partially supported by grant PID2020-117557RB-C22 (funded by MCIN/AEI/10.13039/501100011033 and by the European Union) to JMM.

Institutional Review Board Statement: The study was conducted according to the guidelines of the Declaration of Helsinki, and approved by the Institutional Review Board (or Ethics Committee) of the University of Cádiz (Spain) guidelines and the European Union Council (2010/63/EU) to use animals in research. The experimental procedures were previously approved by the Spanish Government's Ethics and Animal Welfare Committee (RD53/2013) and endorsed by the Regional Government (Junta de Andalucía reference number 28-04-15-241).

Informed Consent Statement: Not applicable.

Data Availability Statement: All data is provided in the main text or Supplementary File.

Conflicts of Interest: The authors declare no competing interest.

References

1. Ashley, P.J. Fish welfare: Current issues in aquaculture. *Appl. Anim. Behav. Sci.* **2007**, *104*, 199–235. [CrossRef]
2. Sneddon, L.; Wolfenden, D.; Thomson, J. Stress management and welfare. In *Fish Physiology: Biology of Stress in Fish*; Schreck, C.B., Tort, L., Farrell, A.P., Brauner, C.J., Eds.; Elsevier: Amsterdam, The Netherlands, 2016; Volume 35, pp. 463–539.
3. Sadoul, B.; Vijayan, M.M. Stress and Growth. In *Fish Physiology: Biology of Stress in Fish*; Schreck, C.B., Tort, L., Farrell, A.P., Brauner, C.J., Eds.; Elsevier: Amsterdam, The Netherlands, 2016; pp. 167–205.
4. Tort, L. Stress and immune modulation in fish. *Dev. Comp. Immunol.* **2011**, *35*, 1366–1375. [CrossRef]
5. Verburg-Van Kemenade, B.M.L.; Stolte, E.H.; Metz, J.R.; Chadzinska, M. Neuroendocrine–immune interactions in teleost fish. In *Fish Neuroendocrinology*, 1st ed.; Bernier, N.J., Kraak, G.V.D., Farrell, A.P., Colin, J.B., Eds.; Fish Physiology; Academic Press: Burlington, ON, Canada, 2009; Volume 28, pp. 313–364.
6. Costas, B.; Aragao, C.; Dias, J.; Afonso, A.; Conceicao, L.E.C. Interactive effects of a high-quality protein diet and high stocking density on the stress response and some innate immune parameters of Senegalese sole *Solea senegalensis*. *Fish Physiol. Biochem.* **2013**, *39*, 1141–1151. [CrossRef] [PubMed]
7. Gorissen, M.; Flik, G. The Endocrinology of the Stress Response in Fish. In *Fish Physiology: Biology of Stress in Fish*; Schreck, C.B., Tort, L., Farrell, A.P., Brauner, C.J., Eds.; Elsevier: Amsterdam, The Netherlands, 2016; pp. 75–111.
8. Azeredo, R.; Machado, M.; Martos-Sitcha, J.A.; Martinez-Rodriguez, G.; Moura, J.; Peres, H.; Oliva-Teles, A.; Afonso, A.; Mancera, J.M.; Costas, B. Dietary tryptophan induces opposite health-related responses in the Senegalese sole (*Solea senegalensis*) reared at low or high stocking densities with implications in disease resistance. *Front. Physiol.* **2019**, *10*, 508. [CrossRef] [PubMed]
9. Verburg-Van Kemenade, B.M.L.; Ribeiro, C.M.S.; Chadzinska, M. Neuroendocrine-immune interaction in fish: Differential regulation of phagocyte activity by neuroendocrine factors. *Gen. Comp. Endocrinol.* **2011**, *172*, 31–38. [CrossRef] [PubMed]
10. Engelsma, M.Y.; Huising, M.O.; van Muiswinkel, W.B.; Flik, G.; Kwang, J.; Savelkoul, H.F.J.; Verburg-van Kemenade, B.M.L. Neuroendocrine–immune interactions in fish: A role for interleukin-1. *Vet. Immunol. Immunopathol.* **2002**, *87*, 467–479. [CrossRef]
11. Hoglund, E.; Balm, P.H.M.; Winberg, S. Stimulatory and inhibitory effects of 5-HT1A receptors on adrenocorticotropic hormone and cortisol secretion in a teleost fish, the Arctic charr (*Salvelinus alpinus*). *Neurosci. Lett.* **2002**, *324*, 193–196. [CrossRef]
12. Winberg, S.; Nilsson, A.; Hylland, P.; Soderstom, V.; Nilsson, G.E. Serotonin as a regulator of hypothalamic-pituitary-interrenal activity in teleost fish. *Neurosci. Lett.* **1997**, *230*, 113–116. [CrossRef]
13. Khan, N.A.; Troutaud, D.; Moulinoux, J.P.; Deschaux, P. Characterization of serotonin receptors in fish brain: Polyamines inhibit the binding process. *Neurosci. Res. Commun.* **1996**, *18*, 97–105. [CrossRef]
14. Timothy, M.; Forlano, P.M. Serotonin distribution in the brain of the plainfin midshipman: Substrates for vocal-acoustic modulation and a reevaluation of the serotonergic system in teleost fishes. *J. Comp. Neurol.* **2020**, *528*, 3451–3478. [CrossRef]
15. Chervova, L.S.; Lapshin, D.N. Nociceptive thresholds in fish and its modulation by opioids. *Eur. J. Neurosci.* **2000**, *12*, 50.
16. Eisenstein, T.K. The Role of Opioid Receptors in Immune System Function. *Front. Immunol.* **2019**, *10*, 2904. [CrossRef] [PubMed]
17. Chadzinska, M.; Hermsen, T.; Savelkou, H.F.J.; van Kemenade, B.M.L.V. Cloning of opioid receptors in common carp (*Cyprinus carpio* L.) and their involvement in regulation of stress and immune response. *Brain Behav. Immun.* **2009**, *23*, 257–266. [CrossRef] [PubMed]
18. Verburg-van Kemenade, B.M.L.; Savelkoul, H.F.J.; Chadzinska, M. Function of the Opioid System during Inflammation in Carp. *Trends Comp. Endocrinol. Neurobiol.* **2009**, *1163*, 528–532. [CrossRef] [PubMed]

19. Tafalla, C.; Bogwald, J.; Dalmo, R.A. Adjuvants and immunostimulants in fish vaccines: Current knowledge and future perspectives. *Fish Shellfish Immunol.* **2013**, *35*, 1740–1750. [CrossRef] [PubMed]
20. Kuhl, H.; Beck, A.; Wozniak, G.; Canario, A.V.M.; Volckaert, F.A.M.; Reinhardt, R. The European sea bass *Dicentrarchus labrax* genome puzzle: Comparative BAC-mapping and low coverage shotgun sequencing. *BMC Genom.* **2010**, *11*, 68. [CrossRef]
21. Pfaffl, M.W. A new mathematical model for relative quantification in real-time RT-PCR. *Nucleic Acids Res.* **2001**, *29*, e45. [CrossRef]
22. Skrzynska, A.K.; Maiorano, E.; Bastaroli, M.; Naderi, F.; Miguez, J.M.; Martinez-Rodriguez, G.; Mancera, J.M.; Martos-Sitcha, J.A. Impact of Air Exposure on Vasotocinergic and Isotocinergic Systems in Gilthead Sea Bream (*Sparus aurata*): New Insights on Fish Stress Response. *Front. Physiol.* **2018**, *9*, 96. [CrossRef]
23. Liu, X.H.; Khansari, A.R.; Teles, M.; Martinez-Rodriguez, G.; Zhang, Y.G.; Mancera, J.M.; Reyes-Lopez, F.E.; Tort, L. Brain and Pituitary Response to Vaccination in Gilthead Seabream (*Sparus aurata* L.). *Front. Physiol.* **2019**, *10*, 717. [CrossRef]
24. Drouin, J. 60 YEARS OF POMC Transcriptional and epigenetic regulation of POMC gene expression. *J. Mol. Endocrinol.* **2016**, *56*, T99–T112. [CrossRef]
25. Yao, M.; Hu, F.; Denver, R.J. Distribution and corticosteroid regulation of glucocorticoid receptor in the brain of *Xenopus laevis*. *J. Comp. Neurol.* **2008**, *508*, 967–982. [CrossRef] [PubMed]
26. Gądek-Michalska, A.; Spyrka, J.; Rachwalska, P.; Tadeusz, J.; Bugajski, J. Influence of chronic stress on brain corticosteroid receptors and HPA axis activity. *Pharmacol. Rep.* **2013**, *65*, 1163–1175. [CrossRef]
27. Terova, G.; Gornati, R.; Rimoldi, S.; Bernardini, G.; Saroglia, M. Quantification of a glucocorticoid receptor in sea bass (*Dicentrarchus labrax*, L.) reared at high stocking density. *Gene* **2005**, *357*, 144–151. [CrossRef] [PubMed]
28. Vazzana, M.; Vizzini, A.; Sanfratello, M.A.; Celi, M.; Salerno, G.; Parrinello, N. Differential expression of two glucocorticoid receptors in seabass (teleost fish) head kidney after exogeneous cortisol inoculation. *Comp. Biochem. Physiol. A-Mol. Integr. Physiol.* **2010**, *157*, 49–54. [CrossRef]
29. Vazzana, M.; Vizzini, A.; Salerno, G.; Di Bella, M.L.; Celi, M.; Parrinello, N. Expression of a glucocorticoid receptor (D1GR1) in several tissues of the teleost fish *Dicentrarchus labrax*. *Tissue Cell* **2008**, *40*, 89–94. [CrossRef]
30. Vizzini, A.; Vazzana, M.; Cammarata, M.; Parrinello, N. Peritoneal cavity phagocytes from the teleost sea bass express a glucocorticold receptor (cloned and sequenced) involved in genomic modulation of the in vitro chemiluminescence response to zymosan. *Gen. Comp. Endocr.* **2007**, *150*, 114–123. [CrossRef]
31. Gesto, M.; Lopez-Patino, M.A.; Hernandez, J.; Soengas, J.L.; Miguez, J.M. The response of brain serotonergic and dopaminergic systems to an acute stressor in rainbow trout: A time course study. *J. Exp. Biol.* **2013**, *216*, 4435–4442. [CrossRef]
32. Moltesen, M.; Laursen, D.C.; Thornqvist, P.O.; Andersson, M.A.; Winberg, S.; Hoglund, E. Effects of acute and chronic stress on telencephalic neurochemistry and gene expression in rainbow trout (*Oncorhynchus mykiss*). *J. Exp. Biol.* **2016**, *219*, 3907–3914. [CrossRef]
33. Medeiros, L.R.; McDonald, M.D. Cortisol-mediated downregulation of the serotonin 1A receptor subtype in the Gulf toadfish, *Opsanus beta*. *Comp. Biochem. Physiol. A-Mol. Integr. Physiol.* **2013**, *164*, 612–621. [CrossRef]
34. Lillesaar, C. The serotonergic system in fish. *J. Chem. Neuroanat.* **2011**, *41*, 294–308. [CrossRef]
35. Hoglund, E.; Balm, P.H.M.; Winberg, S. Skin darkening, a potential social signal in subordinate Arctic charr (*Salvelinus alpinus*): The regulatory role of brain monoamines and pro-opiomelanocortin-derived peptides. *J. Exp. Biol.* **2000**, *203*, 1711–1721. [CrossRef]
36. Overli, O.; Winberg, S.; Pottinger, T.G. Behavioral and neuroendocrine correlates of selection for stress responsiveness in rainbow trout—A review. *Integr. Comp. Biol.* **2005**, *45*, 463–474. [CrossRef]
37. Silva, P.I.M.; Martins, C.I.M.; Khan, U.W.; Gjoen, H.M.; Overli, O.; Hoglund, E. Stress and fear responses in the teleost pallium. *Physiol. Behav.* **2015**, *141*, 17–22. [CrossRef] [PubMed]
38. Hassanain, M.; Zalcman, S.; Bhatt, S.; Siegel, A. Interleukin-1 beta in the hypothalamus potentiates feline defensive rage: Role of serotonin-2 receptors. *Neuroscience* **2003**, *120*, 227–233. [CrossRef]
39. Arends, R.J.; Mancera, J.M.; Munoz, J.L.; Bonga, S.E.W.; Flik, G. The stress response of the gilthead sea bream (*Sparus aurata* L.) to air exposure and confinement. *J. Endocrinol.* **1999**, *163*, 149–157. [CrossRef] [PubMed]
40. Machado, M.; Azeredo, R.; Fontinha, F.; Fernandez-Boo, S.; Conceicao, L.E.C.; Dias, J.; Costas, B. Dietary methionine improves the European seabass (*Dicentrarchus labrax*) immune status, inflammatory response, and disease resistance. *Front. Immunol.* **2018**, *9*, 2672. [CrossRef]
41. Northmore, D.P.M. Holding visual attention for 400 million years: A model of tectum and torus longitudinalis in teleost fishes. *Vis. Res.* **2017**, *131*, 44–56. [CrossRef]
42. Lepage, O.; Vilchez, I.M.; Pottinger, T.G.; Winberg, S. Time-course of the effect of dietary L-tryptophan on plasma cortisol levels in rainbow trout *Oncorhynchus mykiss*. *J. Exp. Biol.* **2003**, *206*, 3589–3599. [CrossRef]
43. Przewlocki, R.; Przewlocka, B. Opioids in neuropathic pain. *Curr. Pharm. Des.* **2005**, *11*, 3013–3025. [CrossRef]
44. Chadzinska, M.; Savelkoul, H.F.J.; Verburg-van Kemenadea, B.M.L. Morphine affects the inflammatory response in carp by impairment of leukocyte migration. *Dev. Comp. Immunol.* **2009**, *33*, 88–96. [CrossRef]
45. Budd, K. Pain management: Is opioid immuno suppression a clinical problem? *Biomed. Pharmacother.* **2006**, *60*, 310–317. [CrossRef] [PubMed]
46. Sneddon, L.U. Evolution of nociception in vertebrates: Comparative analysis of lower vertebrates. *Brain Res. Rev.* **2004**, *46*, 123–130. [CrossRef] [PubMed]

47. Wei, L.-N.; Loh, H.H. Transcriptional and epigenetic regulation of opioid receptor genes: Present and future. *Annu. Rev. Pharmacol.* **2011**, *51*, 75–97. [CrossRef] [PubMed]
48. Tao, R.; Auerbach, S.B. Opioid receptor subtypes differentially modulate serotonin efflux in the rat central nervous system. *J. Pharmacol. Exp. Ther.* **2002**, *303*, 549–556. [CrossRef] [PubMed]
49. Northmore, D.P.M. The Optic Tectum. In *Encyclopedia of Fish Physiology: From Genome to Environment*; Farrell, A.P., Ed.; Elsevier: Amsterdam, The Netherlands, 2011; Volume 1, pp. 131–142.

biology 22, 11, 764

Article

Chronic Inflammation Modulates Opioid Receptor Gene Expression and Triggers Respiratory Burst in a Teleost Model

Diogo Peixoto [1,2,3,*], Marina Machado [2], Rita Azeredo [2] and Benjamín Costas [2,*]

1 ICBAS—Instituto de Ciências Biomédicas Abel Salazar, Universidade do Porto, 4050-313 Porto, Portugal
2 CIIMAR—Centro Interdisciplinar de Investigação Marinha e Ambiental, 4450-208 Matosinhos, Portugal; mcasimiro@ciimar.up.pt (M.M.); mleme@ciimar.up.pt (R.A.)
3 Departamento de Biología, Facultad de Ciencias del Mar y Ambientales, Instituto Universitario de Investigación Marina (INMAR), Campus de Excelencia Internacional del Mar (CEIMAR), Universidad de Cádiz, 11519 Puerto Real, Spain
* Correspondence: dpeixoto@ciimar.up.pt (D.P.); bcostas@ciimar.up.pt (B.C.)

Simple Summary: The natural presence of opportunistic pathogens in aquatic rearing systems in alignment with favorable conditions (compromised fish immune status and/or inappropriate rearing conditions) might result in serious acute disease episodes that can develop into chronic immune responses. The present study characterizes molecular, cellular and humoral markers of chronic inflammation in a fish species with high commercial value. The intense recruitment of immune cells to the inflammatory focus 21 days after triggering an immune response illustrates a clear chronic character. The cellular response was also noticed with circulating leukocyte numbers rising in the blood of the inflamed fish. Furthermore, the cellular-mediated respiratory burst peaked at 21 days post-injection, suggesting that phagocytes were still actively fighting the inflammatory agent. Regarding the molecular analysis, certain genes appear to be good markers of a chronic inflammation response due to their importance in pathways with high relevance in chronic inflammation settings. The present study can serve as a baseline to assess long-term immune-related responses in future studies.

Abstract: Stress-inducing husbandry and rearing conditions, bacterial infections or parasitic diseases may all lead to chronic inflammation. The immune response will then channel energy away from growth, reproduction and other important physiological processes, to fuel immune-related metabolic responses. The present study aims to unravel the mechanisms and contribute with new information on the molecular, cellular and humoral parameters of European seabass (*Dicentrarchus labrax*) undergoing chronic inflammation that can be used as health indicators for application in fish health management. European seabass individuals were intra-peritoneally injected with either Freund's Incomplete Adjuvant (FIA) to induce inflammation or Hanks Balanced Salt Solution (HBSS) to serve as sham. Fish were sampled at 24 h, 7, 14 and 21 days post-injection and blood, plasma and head-kidney were collected. The results found were clear indicators of an inflamed peritoneal cavity and an ongoing systemic immune response that persisted for at least 21 days. Locally, inflammation was characterized by an intense recruitment of immune cells that was still evident 21 days after injection, thus illustrating the chronic character of the immune response. Cellular response was also noticed peripherally with leukocyte numbers rising in the blood of FIA-injected fish. Furthermore, the cellular-mediated respiratory burst peaked at 21 days post-FIA injection, suggesting that phagocytes were still actively fighting the phlogistic agent. Regarding the head-kidney molecular analysis, *cxcr4* and *il34* appear to be good markers of a chronic inflammation response due to their importance for pathways with high relevance in chronic inflammation settings. In addition, opioid receptor *nopr* seems to be a good marker of a chronic inflammation response due to its role in detecting noxious stimuli. The present study can serve as a baseline to assess long-term immune-related responses in future studies. For that, more research is nonetheless required to select more responsive and specific molecular markers.

Citation: Peixoto, D.; Machado, M.; Azeredo, R.; Costas, B. Chronic Inflammation Modulates Opioid Receptor Gene Expression and Triggers Respiratory Burst in a Teleost Model. *Biology* **2022**, *11*, 764. https://doi.org/10.3390/biology11050764

Academic Editor: Alberto Teodorico Correia

Received: 30 March 2022
Accepted: 15 May 2022
Published: 17 May 2022

Keywords: European seabass; chronic inflammation; opioid receptors; immune status

1. Introduction

Increasing consumer concerns about the quality, safety, freshness and health value aquaculture products have been pushing farmers into becoming increasingly aware of fish welfare and on how intimately connected it is with final product quality [1,2]. Fish health is compromised by several external and internal factors that often occur in a persistent manner throughout the production cycle. The natural presence of opportunistic pathogens in rearing systems in alignment with favorable conditions (compromised fish immune status and/or inappropriate rearing conditions) might result in serious acute disease episodes, or even develop into a chronic immune response. Chronic immune responses might also arise with the misuse of alternative ingredients in aquafeeds, given the high amount of antinutritional factors and/or due to an inadequate feed formulation/regime [3]. On top of that, these situations might be exacerbated by rearing conditions such as high animal densities. It is now generally accepted that stressful rearing conditions are described as affecting flesh quality (lower muscle pH and faster meat quality deterioration), while decreasing growth rates and rendering fish more prone to pathologies [1,4–10].

When antigens are detected by the host immune recognition complexes, an array of soluble and cellular defense mechanisms is activated, sustained by both the head-kidney and blood, to fuel a local inflammatory response. Inflammation is a protective reaction of the host in response to injury, infection, or simply the presence of a foreign body, that consists of complex series of homeostatic mechanisms affecting the immune, nervous and circulatory systems [11–13]. Inflammatory processes are orchestrated by pro- and anti-inflammatory mediators working in different timeframes and aimed at eliminating the threat and achieving a new homeostasis without or with limited self-damage [14]. Despite its protective character, it involves an increase in the oxidative stress with increased production of reactive oxygen and nitrogen species that are not only deleterious for the invading microorganism but also for the host. Even more so, an unrestrained inflammatory response is often linked to an extended exhaustion of resources, tissue damage and fibrosis [15–17].

Fish opioid receptors have been shown to be expressed in the hypothalamus, pituitary gland and head-kidney with a huge range of physiological roles, including nociception and cognition, as well as affective, endocrine, cardiovascular, gastrointestinal, immune, metabolic and respiratory functions, stress and autonomic regulation [10,18]. A vice versa modulation occurs between the immune and neuroendocrine systems where they both share a network of compounds and their receptors that are involved in the regulation of both responses. As active compounds of this network, opioids and their receptors participate in regulatory mechanisms of the cellular immune response, as demonstrated by the immunomodulatory role of Mu opioid (*mu*) on immune cell migration [17] and by regulatory functions on neuroendocrine and immune responses in common carp, *Cyprinus carpio* L. [19].

A prolonged inflammation involves longer time periods in which more energy and resources are allocated to cope with the demands of the response [9]. At this point, it may develop into a systemic situation typically associated with leukocytosis, activation of complement cascades, lowering plasma iron levels and amino acid mobilization from muscle to liver for intense protein synthesis and secretion [14,20]. Chronic immune responses are, therefore, highly energy and nutrient demanding, thus deviating resources away from other key physiological processes such as growth and reproduction.

Despite a reasonable amount of studies on fish immune responses, there is no solid chronic inflammation model to be applied in teleost fish studies and a battery of biological markers that can more accurately characterize this particular immune context. The present study aims to gather information on the molecular, cellular and humoral parameters of

European seabass (*Dicentrarchus labrax*) undergoing chronic inflammation that can be used as health indicators for application in fish health management.

2. Materials and Methods

2.1. Experimental Design

European seabass were acquired from a certificated hatchery (MARESA, Ayamonte, Spain) and maintained in quarantine for 2 weeks at the Centro Interdisciplinar de Investigação Marinha e Ambiental (CIIMAR; University of Porto, Portugal) fish holding facilities under culture conditions described below. Groups of 24 individuals (≈300 g) were randomly distributed into 2 tanks (3000 L) of a recirculating seawater system in which O^2 saturation (7.38 ± 0.01 mg L^{-1}), salinity (35), temperature 20 ± 0.5 °C and photoperiod (10:14 h dark:light) were kept unchanged throughout the experiment. Ammonium and nitrite levels were kept below 0.025 and 0.3 mg L^{-1}, respectively. At the beginning of the trial, 100 µL of Freund's Incomplete Adjuvant (FIA; Sigma-Aldrich, St. Louis, MI, USA) were intra-peritoneally (i.p.) injected in half of the fish (FIA group) to induce inflammation, while the other half was injected with 100 µL of a saline solution (Hanks' Balanced Salt solution; HBSS, Sigma-Aldrich) to serve as sham (CTRL group). The trial lasted for 21 days and during this period the individuals were daily hand-fed a commercial diet. At 24 h, 7, 14 and 21 days after i.p. injection, 6 fish from each tank were euthanized by an overdose of anesthetic (2-phenoxyethanol) and weighed. Blood was sampled from the caudal vein using heparinized syringes and fresh samples were used to assess the hematological profile and extracellular respiratory burst. The remaining blood was centrifuged at 10,000× *g* for 10 min at 4 °C, and plasma was collected and stored at −80 °C for evaluating innate humoral immune response parameters. Peritoneal exudates were collected afterwards as described below, and the head-kidney was sampled and stored at −80 °C until further processing for gene expression studies (Figure 1). No mortality was observed during the trial. Experiment trials were directed by trained scientists (following FELASA category C recommendations) and conducted according to the guidelines on the protection of animals used for scientific purposes from the European directive 2010/63/UE.

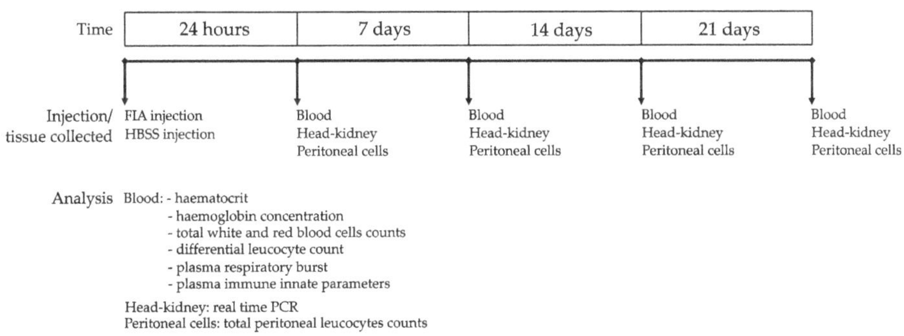

Figure 1. Experimental design.

2.2. Peritoneal Exudates

Peritoneal cells were collected according to the procedure described by [21,22]. Quickly, 5 mL of cold HBSS supplemented with 30 units heparin mL^{-1} was injected into the peritoneal cavity. The peritoneal area was then slightly massaged and the exudate containing suspended cells was withdrawn. Finally, total peritoneal leukocytes counts were performed using a hemocytometer.

2.3. Hematological Parameters

The hematological profiling was performed according to Machado et al. [23] and comprised the analysis of hematocrit, hemoglobin concentration, total white (WBC) and red (RBC) blood cells counts as well as differential leukocyte counts. Immediately after blood collection, blood smears were performed and air dried. Fixation and staining protocols (Wright's stain; Hemacolor) were performed according to Afonso et al. [24] including peroxidase granule staining step in order to facilitate neutrophil detection. Slides were examined under optical microscope (×1000) and at least 200 leukocytes per slide were counted and classified as thrombocytes, lymphocytes, monocytes and neutrophils. Absolute concentration of each leukocyte type was calculated based on total WBC counts ($\times 10^4$ mL^{-1}).

2.4. Respiratory Burst

Respiratory burst in peripheral leukocytes was evaluated according to Nikoskelainen et al. [25] with some modifications [21]. Briefly, 4 μL of blood was added to 96 μL of HBSS in a 96-well flat bottom (white plate). Then, 100 μL of freshly prepared luminol solution (2 mM luminol in 0.2 M borate buffer pH 9.0, with 2 μg mL^{-1} phorbol 12-myristate 13-acetate) was added to each well. Luminol-amplified chemiluminescence was measured every 3 min for 2 h in a luminescence reader for generation of kinetic curves. The integral luminescence in relative light units was calculated.

2.5. Plasma Innate Immune Parameters

The following parameters were previously validated in plasma samples of European seabass. All analyses were conducted in triplicate and absorbance was read on a microplate reader.

2.5.1. Antiprotease Activity

Antiprotease activity was determined as described by Ellis [26]. Briefly, 10 μL of plasma was incubated with 10 μL of trypsin solution (5 mg mL^{-1} in sodium bicarbonate ($NaHCO_3$), 5 mg mL^{-1}, pH 8.3) for 10 min at 22 °C in polystyrene microtubes. Then, 100 μL of phosphate buffered saline (NaH_2PO_4, 13.9 mg mL^{-1}, pH 7.0) and 125 μL of azocasein (20 mg mL^{-1} in $NaHCO_3$, 5 mg mL^{-1}, pH 8.3) were added and the mixture was incubated for 1 h at 22 °C. Finally, 250 μL of trichloroacetic acid (TCA) was added to each microtube and incubated for 30 min at 22 °C. The mixture was centrifuged at 10,000× *g* for 5 min at room temperature. Afterwards, 100 μL of the supernatant was transferred to a 96-well plate that contained 100 μL of NaOH (40 mg mL^{-1}) per well. The absorbance was read at 450 nm. Phosphate buffer in place of plasma and trypsin served as blank, whereas a reference sample was prepared using phosphate buffer in place of plasma samples. The percentage inhibition of trypsin activity compared to the reference sample was calculated.

2.5.2. Protease Activity

Protease activity was quantified using the azocasein hydrolysis assay according to Azeredo [14], using 10 μL of plasma sample and adding 100 μL of NaH_2PO_4 (13.9 mg mL^{-1}, pH 7.0) and 125 μL azocasein (20 mg mL^{-1} in $NaHCO_3$, pH 8.3). After incubation for 24 h at 22 °C in polystyrene microtubes, 250 μL of 10% TCA was added to each mixture, followed by centrifugation (10,000× *g* for 5 min). In a microplate, 100 μL of 1 N NaOH was added to 100 μL of the supernatant and the absorbance was read at 450 nm. The percentage of protease activity compared to the reference sample (trypsin solution, 5 mg mL^{-1} in $NaHCO_3$, pH 8.3) was calculated.

2.5.3. Lysozyme Concentration

Lysozyme concentration was measured using a turbidimetric assay as described by Ellis [27]. Briefly, a solution of *Micrococcus lysodeikticus* (0.5 mg mL^{-1}, 0.05 M sodium phosphate buffer, pH 6.2) was prepared and added (250 μL) to 10 μL of each plasma sample in a microplate to give a final volume of 260 μL. The reaction was carried out at 25 °C and the absorbance was measured at 450 nm after 0.5 and 5 min. Lyophilized hen's egg white lysozyme was diluted in sodium phosphate buffer (0.05 M, pH 6.2) and used to develop a standard curve from which the amount of lysozyme in each sample was calculated.

2.5.4. Peroxidase Activity

Total peroxidase activity in plasma was measured according to Quade [28]. Briefly, 15 μL of each sample was diluted in 135 μL of HBSS without Ca^{+2} and Mg^{+2} in flat-bottomed 96-well plates. Then, 50 μL of 20 mM 3,3′,5,5′-tetramethylbenzidine hydrochloride (TMB) and 50 μL of 5 mM hydrogen peroxide (H_2O_2) were added. The color-change reaction was stopped after 2 min by adding 50 μL of 2 M sulfuric acid (H_2SO_4) and the OD was read at 450 nm. HBSS instead of plasma was used to serve as blank. One unit of peroxidase activity (units mL^{-1} sample) was defined by the quantity of peroxidase that produces an absorbance change of 1 OD.

2.5.5. Immunoglobulin M

Plasma immunoglobulin M (IgMp) levels were analyzed by an enzyme-linked immunosorbent assay (ELISA) [29,30]. Plate wells were coated with plasma proteins, washed 3 times with T-TBS (1× Tris-buffered saline and 0.1% Tween 20, pH 7.3), blocked for 2 h at room temperature with blocking buffer (5% low fat milk in T-TBS) and rinsed again with T-TBS. Plates were then incubated for 1 h with 100 μL per well of anti-European seabass IgM monoclonal antibody (1:100 in blocking buffer), washed and incubated with the secondary antibody anti-mouse IgG-HRP (1:1000 in blocking buffer). Afterwards, 100 μL of TMB was added and the color-change reaction was stopped after 5 min by adding 100 μL of 2 M H_2SO_4. The OD was read at 450 nm. Negative controls consisted of samples without plasma or without primary antibody, from which OD values were subtracted for each sample.

2.6. Gene Expression Analysis

Head-kidney samples were used for total RNA isolation using the NZY Total RNA Isolation kit (NZYTech) following manufacturer's instructions and first-strand cDNA was synthesized with NZY First-Strand cDNA Synthesis Kit (NZYTech). DNA amplification was carried out with specific primers for each gene that had been selected for its involvement in immune responses and oxidative stress. Sequences encoding European seabass *il34, ptx, tgfβ, cxcr4, il1β, casp1, il10, inf-γ, IgM, mmp9, muor, kor1, nopr, dor2* and *ogfr1* were identified after carrying out a search in the databases and primers were designed as described in [31]. Accession number, efficiency values, annealing temperature, product length and primers sequences are presented in Table 1. Real-time quantitative PCR was carried out in a CFX384 Touch Real-Time PCR Detection System (Biorad), using 4.4 μL of diluted cDNA mixed with 5 μL of iTaq Universal SYBR Green Supermix (BioRad) and 0.3 μL (10 μM) of each specific primer in a final volume of 10 μL. The standard cycling conditions were 95 °C initial denaturation for 10 min, followed by 40 cycles of two steps (95 °C denaturation for 15 s followed by primer annealing temperature for 1 min), 95 °C for 1 min followed by 35 s at the annealing temperature, and finally, 95 °C for 15 s. All reactions were carried out as technical duplicates. Melting curve analysis was also performed to verify that no primer dimers were amplified. The expression of the target genes was normalized using the expression of European seabass elongation factor 1-α (*ef1α*), 40s ribosomal protein (*40s*) and ribosomal protein L13 mRNA (*i13α*).

Table 1. Specifications of real-time qPCR assays including forward (F) and reverse (R) primers, GenBank ID (NCBI), efficiencies (Eff) of qPCR reactions and annealing temperature (Ta).

Gene	Acronym	GenBank ID	Eff (%) [1]	Ta (°C)	Primer Sequence (5′–3′)
Elongation factor 1-α	*ef1α*	AJ866727.1	96.45	57	F: AACTTCAACGCCCAGGTCAT R: CTTCTTGCCAGAACGACGGT
40s ribosomal protein	*40s*	HE978789.1	92.96	55	F: TGATTGTGACAGACCCTCGTG R: CACAGAGCAATGGTGGGGAT

Table 1. *Cont.*

Gene	Acronym	GenBank ID	Eff (%) [1]	Ta (°C)	Primer Sequence (5′–3′)
ribosomal protein L13 mRNA	*L13α*	DQ836931.1	127.31	55	F: AGTCCGGTGTCCCACTATCA R: GAGTTCCTCCAGGGTGAAGC
Interleukin 34	*il34*	DLAgn_00164750 [2]	120.94	60	F: GGAAATACGCTTCAGGGATG R: GGCACTCTGTCGGGTTCTT
C-Reactive protein	*ptx*	EU660933.1	114.76	60	F: TGAAGTTTTTGCTGCTGGTG R: GGTTTCTTGTGGGAAGGTGA
Chemokine CXC receptor 4	*cxcr4*	FN687464.1	93.43	60	F: ACCAGACCTTGTGTTTGCCA R: ATGAAGCCCACCAGGATGTG
Interleukin 1 β	*il1β*	AJ269472.1	96.7	57	F: AGCGACATGGTGCGATTTCT R: CTCCTCTGCTGTGCTGATGT
Interleukin 10	*il10*	AM268529.1	116	55	F: ACCCCGTTCGCTTGCCA R: CATCTGGTGACATCACTC
Transforming growth factor-β	*tgf β*	AM421619.1	105.56	55	F: ACCTACATCTGGAACGCTGA R: TGTTGCCTGCCCACATAGTAG
Caspase 1	*casp1*	DQ198377.1	124	55	F: GTGTTTCAGATGCGGGAGGA R: ATTTAAGTTAACTCACCGGGGG
Interferon γ	*ifn-γ*	FQ310507.3	118.38	55	F: GTACAGACAGGCGTCCAAAGCATCA R: CAAACAGGGCAGCCGTCTCATCAA
Inmunoglobulin M	*IgM*	FN908858	91.64	60	F: AGGACAGGACTGCTGCTGTT R: CACCTGCTGTCTGCTGTTGT
Matrix-metalloproteinase 9	*mmp9*	FN908863.1	98.44	57	F: TGTGCCACCACAGACAACTT R: TTCCATCTCCACGTCCCTCA
μ opioid receptor	*muor*	DLAgn_00015310 [2]	99.81	60	F: GTCACCAGCACCCTACCATT R: CGAGGAGAGAATCCAGTTGC
κ-type opioid receptor-like 1	*kor1*	DLAgn_00007470 [2]	89.71	60	F: TCTGGTGCTTGTGGTAGTCG R: TGGCAGTCTCTGTGTCCTTG
Nociceptin opioid receptor	*nopr*	DLAgn_00125610 [2]	97.57	60	F: CTCCTTTCTCATCCCTGTGG R: GTTGCGGTCCTTTTCCTTG
δ-opioid receptor	*dor2*	DLAgn_00062690 [2]	108.06	60	F: CGCTTCTCGGTCTCCATAACT R: GGTCTCATTACTACTTGAAG
Opioid growth factor receptor-like protein 1-like	*ogfr1*	DLAgn_00128530 [2]	96.79	60	F: GTTGGGAATGGAGATGGAAA R: GCTTCAGATTTTGGCTCAGG

[1] Efficiency of qPCR reactions (represented in percentage) were calculated from serial dilutions of tissue RT reactions in the validation procedure. [2] Sequences obtained from databases dicLab v1.0c seabass genome.

2.7. Data Analysis

All results are expressed as mean ± standard deviation (SD). Shapiro–Wilk test was used for normality of variances, as well as Pearson skewness coefficient. When normality was not observed, a non-parametric test with Kruskal–Wallis pairwise comparisons was used to compare significant differences in all the parameters. Differences were tested by a two-way ANOVA, with time and treatment (FIA or CTRL) as factors, followed by a *post-hoc* Tukey HSD test, used to identify significant differences amongst groups. Statistical analyses were carried out using IBM SPSS v27.0 Statistics with a significance level of 95% ($p \leq 0.05$). A principal component analysis (PCA) using Addinsoft XLSTAT 2021 system software and a discriminant analysis (DA) using IBM SPSS v27.0 Statistics were applied to assess the consensus among the variables that differed significantly among dietary treatments.

3. Results

3.1. Hematological Response

The complete set of results is available in the Supplementary File (Table S1). The total white blood cell population increased at 21 days post-injection regardless of treatment, while no significant differences were observed in the total red blood cell population

(Figure 2A,B). The hematocrit decreased from 7 to 14 days post-injection regardless of treatment but levels were observed to increase back to values similar to those found in the first sampling time (Figure 2C). Peripheral neutrophils were higher in FIA-injected fish than in the CTRL group in the first two sampling points and concentration decreased from 24 h to 14 days post-injection in inflamed fish (Figure 2D). In contrast, in the CTRL group, neutrophils were higher at 21 days compared to all other sampling points. Monocyte concentration was higher in FIA-injected fish compared to the levels in the CTRL group at 7 and 21 days post-injection and it significantly increased over time in inflamed fish (Figure 2E). Monocytes were also observed to increase in time in FIA-injected fish, peaking at 21 days post-injection. FIA-injected fish had higher lymphocyte numbers at 24 h post-injection compared to their CTRL counterparts, but a significant increase was observed in both groups at 21 days, relative to the other sampling points (Figure 2F). Peripheral thrombocyte concentration was observed to increase at 21 days post-injection relative to the other sampling points regardless of treatment, while CTRL fish had a higher concentration than FIA-injected fish, regardless of time of sampling (Figure 2G).

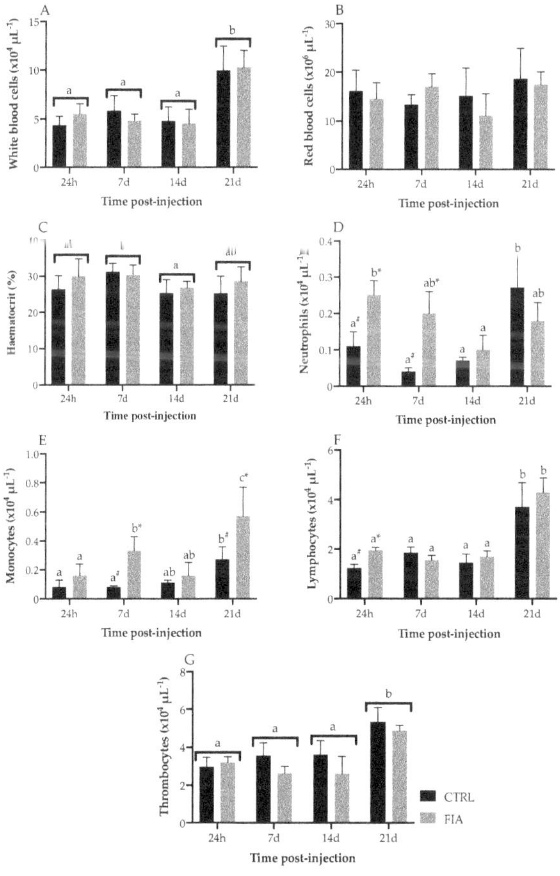

Figure 2. Absolute values of total white and red, peripheral leukocytes (neutrophils, monocytes, lymphocytes and thrombocytes) of European seabass sampled at 24 h, 7, 14 and 21 days after i.p. injection. (**A**)—white blood cells, (**B**)—red blood cells, (**C**)—hematocrit, (**D**)—neutrophils, (**E**)—monocytes, (**F**)—lymphocytes, (**G**)—thrombocytes. Values are presented as means ± SD (n = 6). Different low case letters stand for statistically significant differences attributed to sampling time. Symbols stand for significant differences between stimuli. (Two-way ANOVA; Tukey *post-hoc* test; $p \leq 0.05$).

3.2. *Total Peritoneal Cell Counts and Respiratory Burst of Circulating Leukocytes*

Regarding total peritoneal cells, an increase was observed in FIA-injected fish sampled at 7 days after i.p. injection compared to those sampled at 24 h. Leukocyte numbers remained high in this group until the last sampling point. Moreover, total peritoneal cell numbers in FIA-injected fish were significantly higher than those of CTRL fish at 7, 14 and 21 days post-injection (Figure 3).

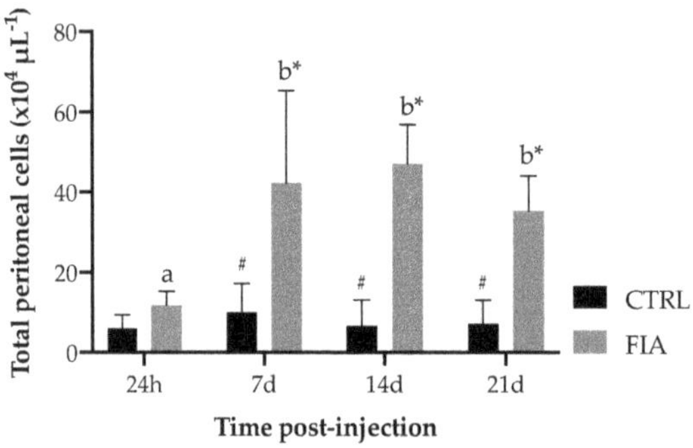

Figure 3. Absolute values of total peritoneal leukocytes of European seabass sampled at 24 h, 7, 14 and 21 days after i.p. injection. Values are presented as means ± SD (n = 6). Different low case letters stand for statistically significant differences attributed to sampling time. Symbols stand for significant differences between stimuli. (Two-way ANOVA; Tukey *post-hoc* test; $p \leq 0.05$).

Respiratory burst decreased from 24 h to 7 days post-injection in FIA-injected fish, and remained low until it increased again from 14 to 21 days following FIA injection (Figure 4).

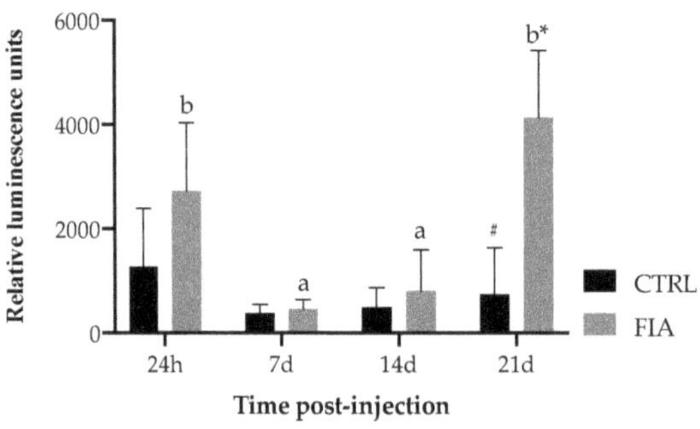

Figure 4. Relative luminescence units of respiratory burst of European seabass sampled at 24 h, 7, 14 and 21 days after i.p. injection. Values are presented as means ± SD (n = 6). Different low case letters stand for statistically significant differences attributed to sampling time. Symbols stand for significant differences between stimuli. (Two-way ANOVA; Tukey *post-hoc* test; $p \leq 0.05$).

3.3. *Innate Humoral Parameters*

The complete set of results is available in the Supplementary File (Table S2).

Total proteases and lysozyme concentrations were enhanced in FIA-injected fish compared to the CTRL group, regardless of sampling time (Figure 5A,B, respectively). Moreover, total protease activity increased from 24 h to 7 days post-injection regardless of treatment and remained so until the end of the experiment (Figure 5A). Differently, the lysozyme concentration increased at 7 days post-injection regardless of treatment but decreased at 21 days back to levels similar to those observed in the first sampling time (Figure 5B). Plasma IgM remained stable until it peaked at 21 days post-injection, irrespective of treatment (Figure 5C). No significant differences were observed regarding antiprotease and peroxidase activities (Supplementary File, Table S2).

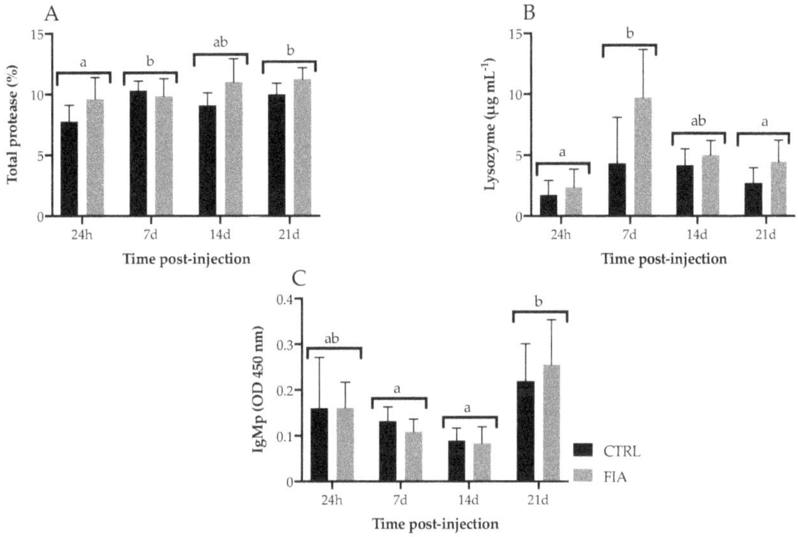

Figure 5. Plasma innate immune response parameters of European seabass sampled at 24 h, 7, 14 and 21 days after i.p. injection. (A)—total protease content, (B)—lysozyme and (C)—immunoglobulins M. Values are presented as means ± SD (n = 6). If interaction was significant, Tukey *post-hoc* test was used to identify differences among treatments. Different low case letters stand for statistically significant differences between sampling points. (Two-way ANOVA; Tukey *post-hoc* test; $p \leq 0.05$).

3.4. Head-Kidney Gene Expression

The complete set of results is available in the Supplementary File (Table S3).

Regarding head-kidney gene expression, *il1β* and *casp1* were both upregulated in FIA-injected fish, irrespective of sampling time (Figure 6A,B, respectively). Moreover *il1β*, *casp1*, *tgfβ* and *il10* relative expressions decreased from 24 h to 7 days regardless of treatment (Figure 6A, B, C and D, respectively). However, while the *tgfβ* and *il1β* decrease was followed by an upregulation at 14 days, *casp1* and *il10* expression remained at lower levels until the end of the experiment. Differently, *cxcr4* relative expression was upregulated at 14 days post-injection, and levels were kept high until the last sampling point, regardless of treatment (Figure 6E). *mmp9* relative expression in FIA-injected and CTRL fish dropped from 24 h to 7 days post-injection and was still lower at 21 days (Figure 6F). Although not statistically significant, *il34* increased in time until 14 days in FIA-injected fish (Figure 6G). Opioid receptors did not significantly differ between FIA and HBSS groups but mu and kappa opioid receptors were downregulated in fish sampled at 7-, 14- and 21-days post-injection compared to levels measured at 24 h (*muor* and *kor1*, Figure 7A and B, respectively). Likewise, the nociception opioid receptor (*nopr*) expression was downregulated from 24 h to 7 days post-injection and levels were still comparably lower at 21 days, irrespective of treatment (Figure 7C). However, *nopr* expression was increased from 7 days to 14 days, regardless of treatment. Pentraxin (*ptx*), immunoglobulin M (*igm*), interferon-γ (*inf-γ*), delta opioid receptor 2 (*dor2*) and opioid growth factor receptor 1 (*ogfr1*) were not significantly modulated by sampling time nor treatment (Supplementary File, Table S3).

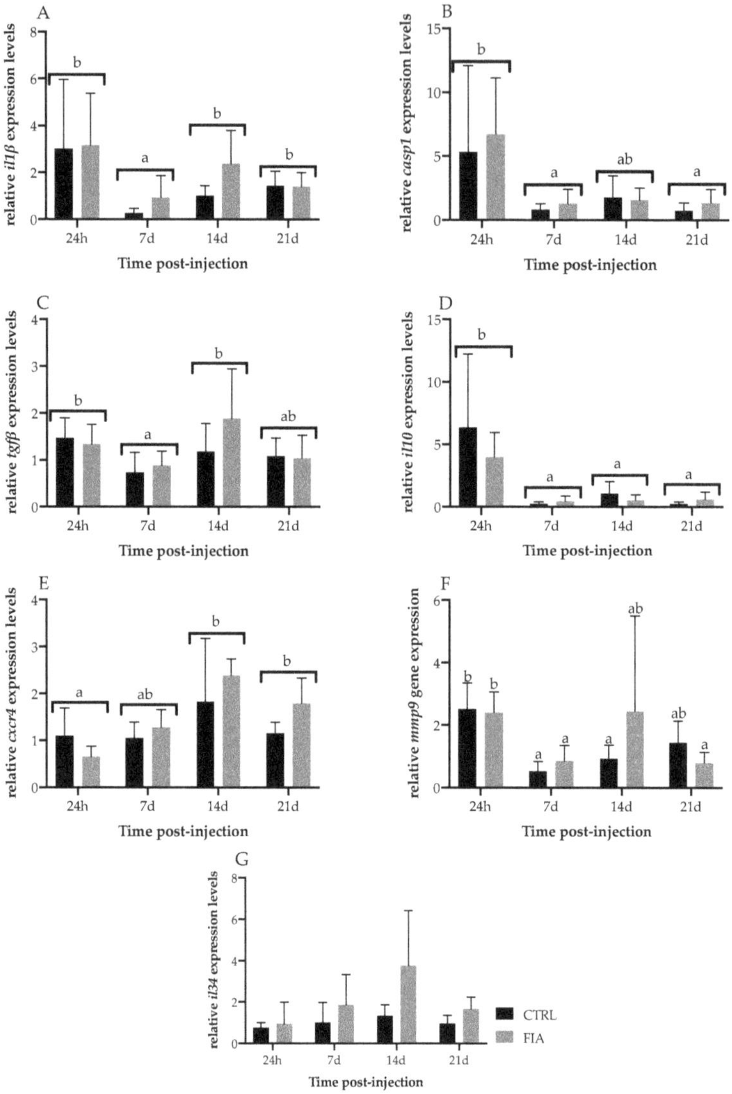

Figure 6. Relative expression of immune-related genes in the head-kidney of European seabass sampled at 24 h, 7, 14 and 21 days after i.p. injection. (**A**)—*il1β*, (**B**)—*casp1*, (**C**)—*tgfβ*, (**D**)—*il10*, (**E**)—*cxcr4*, (**F**)—*mmp9* and (**G**)—*il34*. Values are presented as means ± SD (n = 6). If interaction was significant, Tukey *post-hoc* test was used to identify differences among treatments. Different low case letters stand for statistically significant differences between sampling points. (Two-way ANOVA; Tukey *post-hoc* test; $p \leq 0.05$).

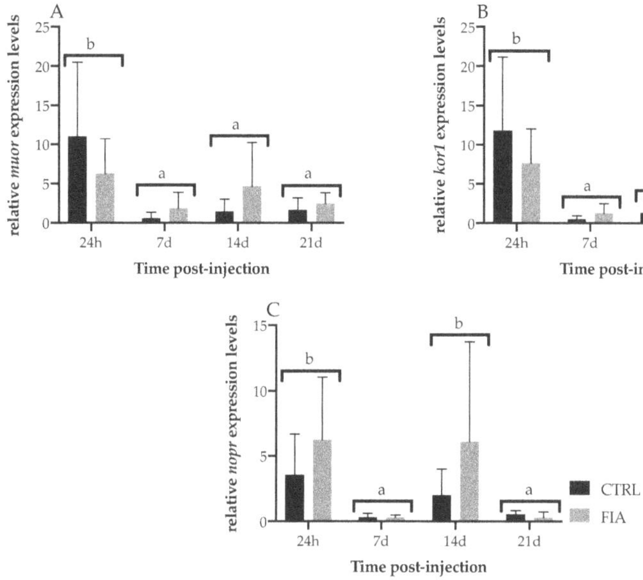

Figure 7. Relative expression of opioid receptor-related genes in the head-kidney of European seabass sampled at 24 h, 7, 14 and 21 days after i.p. injection. (**A**)—*muor*, (**B**)—*kor1* and (**C**)—*nopr*. Values are presented as means ± SD ($n = 6$). If interaction was significant, Tukey *post-hoc* test was used to identify differences among treatments. Different low case letters stand for statistically significant differences between sampling points. (Two-way ANOVA; Tukey *post-hoc* test; $p \leq 0.05$).

The principal component analysis (PCA) methodology attempted to identify underlying variables or factors that might explain the pattern of correlations within a set of observed variables. Figure 8 shows the two first dimensions of the PCA consensus, which together account for 69.65% of the variability of the experimental data (F1 51.04%; F2 18.61%), being the overall performance of the analysis medium (Kaiser–Meyer–Olkin index = 0.701; Bartlett's sphericity test $p < 0.001$). PCA indicated that the fish sampled at 24 h post-injection had different contributions to the model in F1, where FIA-injected fish contributed with 41.5% whereas HBSS-injected fish (CTRL) contributed with 20.6% (see Supplementary File, Tables S4 and S5). Moreover, increased gene expression seemed to be strongly discriminating the group of fish sampled at 24 h post-injection from those sampled at 7, 14 and 21 days post-injection. On the other hand, the humoral parameters were enhanced in these later-sampled groups, except for FIA-injected fish sampled at 21 days which were strongly loaded by *cxcr4* and *il34* gene expression. The discriminant analysis (M Box test $p < 0.05$; Wilk's Lambda value < 0.2) confirmed a clear separation between FIA- and HBSS-injected fish sampled at 24 h and the other sampling points represented in the biplot in Figure 8, whereas no significant differences among both FIA- and HBSS-injected fish at 7, 14 and 21 days were observed (F1; Figure 8).

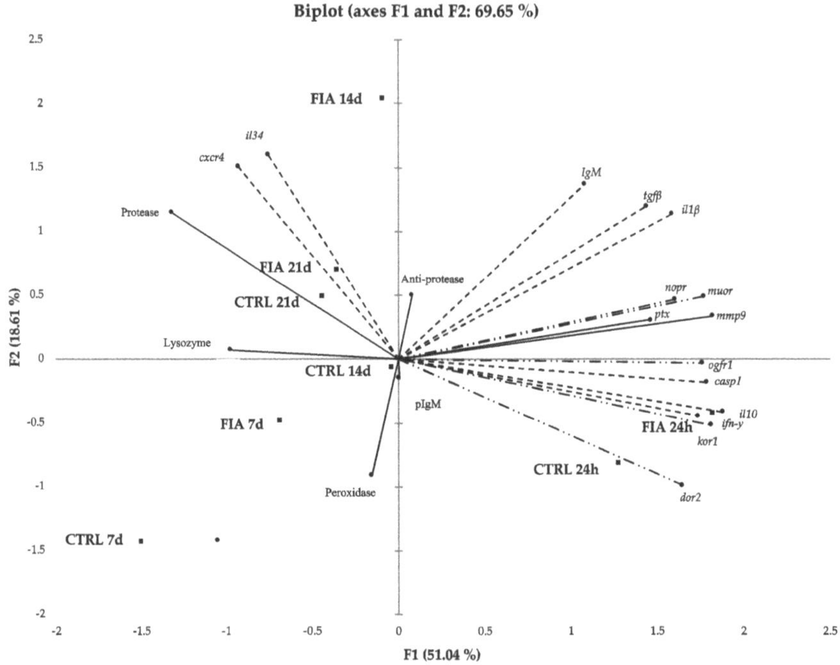

Figure 8. Principal component analysis biplot of the mean scores and loadings for plasma immune and gene expression variables during the inflammation period in European seabass. (—) Plasma immune response, (- - -) gene expression and (—··—) opioid receptors variables at 24 h, 7, 14 and 21 days after FIA- and HBSS-injection. IgMp, plasma immunoglobulins M; *cxcr4*, chemokine CXC receptor 4; *il34*, interleukin 34; *ptx*, C-reactive protein; *igm*, immunoglobulin M; *il1β*, interleukin 1 β; *tgfβ*, transforming growth factor-β; *mmp9*, matrix-metalloproteinase 9; *casp1*, caspase 1; *il10*, interleukin 10; *ifn-γ*, interferon γ; *kor1*, κ-type opioid receptor-like 1; *dor2*, δ-opioid receptor; *nopr*, nociceptin opioid receptor; *muor*, μ opioid receptor and *ogfr1*, opioid growth factor receptor-like protein 1-like.

4. Discussion

Although there are a reasonable number of studies regarding fish immune responses against different stimuli, few are clear about the mechanisms involved in chronic inflammatory insults. Thus, this study aimed to gather information on the molecular, cellular and humoral parameters of European seabass experiencing chronic inflammation which could be used as health indicators for application in fish health management.

Phagocytes are differentiated myeloid cells (mostly neutrophils and macrophages) with a crucial importance in fish innate immune responses, in particular during the inflammatory response, as central regulators of both proinflammatory and homeostatic anti-inflammatory processes [32]. Vasodilatation and cell migration are two classical aspects of the acute phase of inflammation [13] and neutrophils are known to be the first leukocytes recruited to the inflammation site, and to launch an innate immune response [33–35]. Accordingly, in this study, peripheral neutrophils were indeed higher in those fish undergoing an inflammatory response and sampled at earlier timepoints (24 h or even 7 days post-injection). At later sampling points, however, neutrophils were no longer standing out in this group, in comparison to the CTRL group. Differently, a clear prolonged cellular response denoting a chronic immune response was displayed by monocytes. Higher peripheral monocyte concentration was observed in FIA-injected fish at 7 days and, as the response progressed in time, it was once higher 21 days after the insult. The absence of differences between FIA and CTRL fish in peripheral monocyte numbers at 14 days post-injection could be related to a slowdown of cell recruitment, since total peritoneal leukocyte concentration in FIA treated fish was already much higher at this timepoint than in CTRL fish. A clear emphasized cell response was observed at the inflammatory focus of FIA-injected fish, as attested by a conspicuous accumulation of leukocytes in the peritoneal cavity 7 days following injection. The numbers were as high at 21 days, suggesting that the inflammatory response was still on course. Similar phagocyte responses were observed in European seabass and Senegalese sole (*Solea senegalensis*) upon i.p. injection of live or inactivated *Photobacterium damselae piscicida*, respectively, with a reduction in peripheral lymphocyte and monocyte numbers, in contrast to an increase in the inflamed peritoneal cavity [23,31,36]. In the present study, the sustained cell recruitment after so many days post-injection (i.e., 21 days) highlights that leukocyte concentration is a good peripheral marker of an activated immune response following chronic immune stimulation in the European seabass.

Fish activated thrombocytes, with regard to their hemostatic and immunological activity, are equivalent to mammalian platelets, taking their part in the process of inflammation following the hemostatic process and thrombus formation [37]. To a certain extent, thrombocytes are also able to perform phagocytosis and have the ability to neutralize and degrade microorganisms accelerating monocytes and dendritic cell activation and antigen presentation to T-cells [37–39]. Circulating thrombocyte populations proved to be lower in FIA-injected fish than in the CTRL group. This result is in accordance with previous studies stating that thrombocytes accumulate locally at the damage site, releasing and expressing proteins and substances to deal with the inflammatory insult [37,40,41].

In immune cells performing phagocytosis, this phenomenon is correlated with an increase in oxygen consumption, known as respiratory burst, that is immediately converted into reactive oxygen species [42–45]. The initial drop observed from 24 h to 7 days post-injection in fish injected with FIA emphasizes the importance of this cytotoxic mechanism in the acute phase of the immune response. Interestingly, this cellular-mediated process peaked at 21 days post-FIA injection suggesting that phagocytes were still actively fighting the phlogistic agent. Since monocytes were also abundant at this time point in the same group, it is likely that these cells were the main players. Respiratory burst is a particularly interesting parameter to be enhanced at such a late time-point, as it is a key discriminant feature of active phagocytes [43], very much present at the onset of inflammation.

Being amongst the most significant innate immune mechanisms, lysozyme activity has an important role against Gram-positive and -negative bacteria, as well as in activating the complement system and phagocytes [46]. Proteases, besides other regulatory mechanisms [47], are also directly involved with the immune response as they cleave the pathogens' proteins, activate and enhance the production of other immune components such as complement or immunoglobins [48]. Results from the present study point out a clear difference between FIA and CTRL fish, with the fish undergoing an inflammatory response having higher plasma protease activity and lysozyme concentration than those of the CTRL group, regardless of the stage of the response. Yet, as no interactive effects have been observed between inflammation and sampling time in any of the evaluated humoral parameters, their specific behavior throughout the immune response was not distinguishable in the present experiment, despite their importance throughout an activated immune response.

As the inflammatory response continues, there is an intensification of transcriptional activation of several immune-related molecular markers, such as inflammatory cytokines and chemokines that activate and regulate the immune response, antimicrobial peptides, acute phase proteins, and immune stimuli-dependent enzymes, among others [49]. In this study, a principal component analysis (PCA) showed that at 24 h post-infection in both FIA- and HBSS-injected fish, the expression of most genes analyzed in the head-kidney was increased compared to the other sampling points. Moreover, in spite of being not statistically significant, this general molecular upregulation was more prominent in FIA-injected fish than in those from the CTRL group. IL1β is a potent proinflammatory cytokine which induces gene expression of other cytokines such as tumor necrosis factor α, interleukins 1α, 6 and 8, prostaglandin-endoperoxide synthase 2 and monocyte chemoattractant protein 1 [50–52]. For full functionality, IL1β has to be cleaved by the cysteine protease caspase 1, which thereby has a regulatory role in the inflammatory response [50–52]. Similarly, to that reported in previous studies [53], the highest expression levels of both *il1β* and *casp1* were observed in fish sampled at 24 h, and levels were higher in the FIA-injected compared to the CTRL fish, regardless of sampling point. However, after a first decrease to much lower levels from 24 h to 7 days, head-kidney *il1β* was upregulated again at 14 days, pointing to the ongoing chronic inflammatory response that is not only focused at the inflammation site, but is also triggering a systemic response. The presence of similar gene expression patterns (despite at lower scales) in the CTRL group might be an effect of the injection per se and the associated stress due to handling/air exposure procedures which are known to induce neuroendocrine and immune responses [10].

The anti-inflammatory cytokine TGFβ is a potent negative regulator of hematopoiesis, modulating the proliferation, differentiation and function of several cell types [54]. TGFβ counteracts other earlier cytokine actions (such as IL1β) as the immune response develops, controlling the inflammatory process in a later stage [14,55]. Similarly, IL10 is also a regulatory cytokine described to peak during the late phase of an inflammatory response, which inhibits excessive activation of the immune response and initiate processes of wound healing, tissue remodeling and recovery [10,56,57]. Both genes did not seem to be modulated by the inflammatory stimulus, although *il10* expression significantly dropped from 24 h to 7 days post-injection, regardless of the nature of the injection. Despite these being among the most significant anti-inflammatory mediators, those anti-inflammatory cytokines do not exclusively regulate immune responses. In line with this, the multivariate analysis showed that *cxcr4* and *il34* were particularly important in FIA-injected fish at 14 and 21 days, contrasting with most of the molecular markers discussed previously. Indeed, relatively high levels of *il34* expression across different tissues suggest a regulatory and homeostatic role of *il34* during the immune response [58,59]. It has been reported that IL34 plays a role in macrophage biology and in autoimmune and chronic inflammatory diseases [60]. *cxcr4* is an exclusive receptor of CXC chemokine ligand 12 (*cxcl12*), showing higher expression in immune tissues of teleost fishes [61,62] and functioning to maintain active neutrophils at the inflammatory site [63]. In accordance with results from the present study, van der Sar et al. [63] reported that the *cxcl12/cxcr4* signaling axis could modulate the inflammation resolution in zebrafish larvae.

As a compound that, amongst other functions, is involved in leukocyte migration [64], and based on previous studies [53], *mmp9* mRNA expression level was found higher at 24 h post-injection than at subsequent sampling points [65]. Unfortunately, and possibly due to a stress-induced masking effect, no differences were observed between FIA and CTRL groups. However, although not statistically significant, the expression levels of *mmp9* in FIA-injected fish at 14 days post-injection seemed to be higher than those of the CTRL group, suggesting that at this later stage, and in this particular context, chronic inflammation is also marked by a second wave of peripheral cell migration to the inflammation site.

Amongst the broad spectrum of physiological roles that opioids and their receptors seem to play, this study intended to evaluate their ability to regulate the fish immune response. The clear inhibitory effects of morphine—an opiate that binds MUOR—have been observed on the gene expression of head-kidney proinflammatory cytokines and their receptors during the innate immune response of carp, *Cyprinus carpio* [19]. Opioids are also known to be involved in neuroendocrine mechanisms, in which the head-kidney plays a part, too. This multidisciplinary profile of opioids' physiological functions could help explain the absence of differences between CTRL and FIA fish at all sampling points, if an acute stress induced by the i.p. injection was in place. Nonetheless, irrespective of the role played by these opioids and their receptors, the present results clearly point out their importance during the earlier sampling points, as *muor*, *kor1* and *nopr* gene expression was strikingly lower at 7 days compared to 24 h post-injection, and was still lower at 21 days. Accordingly, carp head-kidney opioid receptors were observed to be upregulated in the first sampling points after i.p. injection with zymosan (6 and 24 h post-injection) but were gradually downregulated [19].

An exception was observed in *nopr*, the expression of which was upregulated at 14 days, only to recede to lower levels at 21 days. Specialized nociceptors, such as NOPR, detect noxious stimuli which by definition can or do cause tissue damage [66] and are usually associated to the inflammatory response and to later phases of this response [67]. Nociception in fish is linked to opioids and their receptors [66]. From the opioid receptors analyzed in the present study, it can be hypothesized that the upregulation of *nopr* mRNA levels can be related to ongoing chronic inflammation at 14 days post-injection, as already perceived by other parameters mentioned above.

Considering the present results, a timeline diagram of hypothetical mechanisms involved during a chronic inflammation on the peritoneal cavity and their potential markers is illustrated on Figure 9.

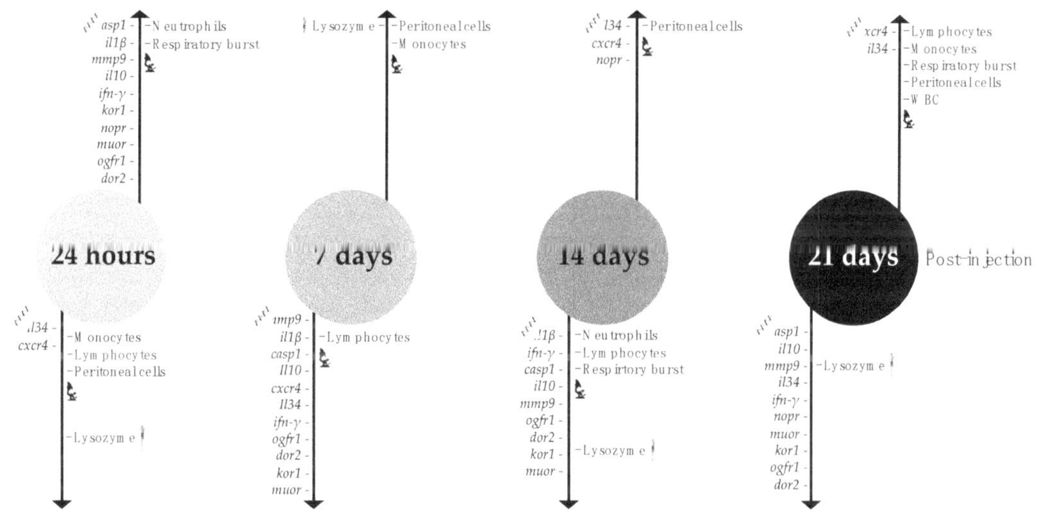

Figure 9. Timeline diagram of hypothetical mechanisms involved during a chronic inflammation on the peritoneal cavity and their potential markers. The diagram shows a qualitative assessment of the relative abundance of each parameter in inflamed fish over time. WBC—white blood cells, IgMp, plasma immunoglobulins M; *casp1*, caspase 1; *il1β*, interleukin 1 β; *mmp9*, matrix-metalloproteinase 9; *il10*, interleukin 10; *ifn-γ*, interferon γ; *cxcr4*, chemokine CXC receptor 4; *il34*, interleukin 34; *kor1*, κ-type opioid receptor-like 1; *nopr*, nociceptin opioid receptor; *muor*, μ opioid receptor; *ogfr1*, opioid growth factor receptor-like protein 1-like and *dor2*, δ-opioid receptor.

5. Conclusions

In conclusion, and having in mind the possible presence of a stress-induced effect (i.p. air exposure and injection) that seemed to have masked part of the evaluated immune parameters, the results from the present study illustrate an orchestrated response to a peripheral and local immune stimulation. Immune defenses such as peripheral neutrophil populations, plasma proteases and lysozyme levels, as well as head-kidney mRNA expression of *il1β* and *casp1* genes were observed to change with inflammation but not in a particularly chronic way.

Locally, inflammation was characterized by an intense recruitment of immune cells that was still evident 21 days after injection thus illustrating the chronic character of the immune response. Cellular response was also noticed peripherally with leukocyte numbers rising in the blood of FIA-injected fish, accompanied by an intensification of respiratory burst, suggesting that these cells were still actively fighting the phlogistic agent after 3 weeks (Figure 10A). Regarding the head-kidney molecular markers, *cxcr4* and *il34*, whose pathways have high relevance in chronic inflammation settings, stood out as relevant markers of chronic inflammatory responses (Figure 10B). The present study offers new data on the dynamics of a chronic inflammation context, and might serve as a baseline for the evaluation of immune responses in the European seabass.

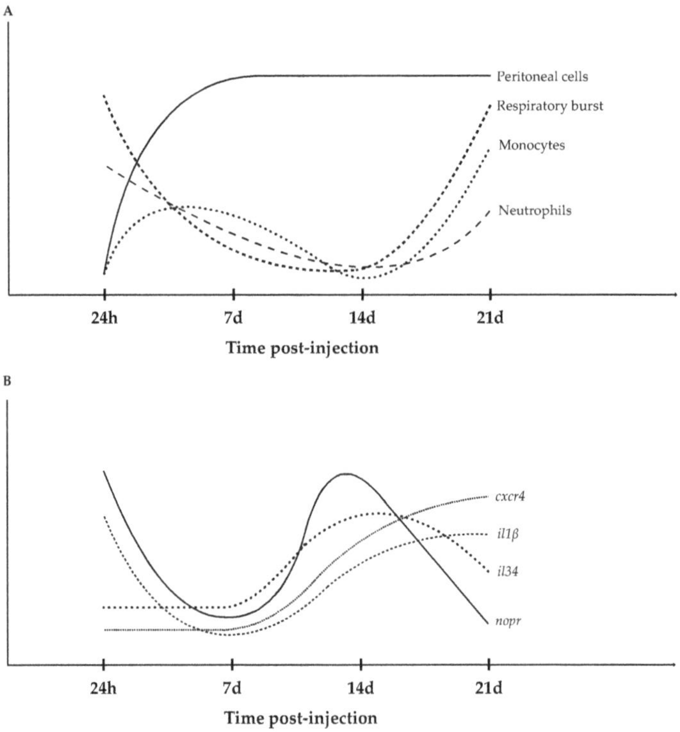

Figure 10. Schematic representation of main hematological (**A**) and gene expression (**B**) results from the present study. *cxcr4*, chemokine CXC receptor 4; *il34*, interleukin 34; *il1β*, interleukin 1 β and *nopr*, nociceptin opioid receptor.

Supplementary Materials: The following supporting information can be downloaded at: https://www.mdpi.com/article/10.3390/biology11050764/s1, Table S1: Absolute values of total white, red and peritoneal cells, peripheral leukocytes (neutrophils, monocytes, lymphocytes and thrombocytes) and respiratory burst of European seabass 24 h, 7, 14 and 21 days after inflammation; Table S2: Plasma innate immune response parameters of European seabass 24 h, 7, 14 and 21 days after inflammation; Table S3: Quantitative expression of immune-related gene and opioid receptors of head-kidney of European seabass 24 h, 7, 14 and 21 days after inflammation; Table S4: Principal component analysis (PCA) eigenvalues; Table S5: Principal component analysis (PCA) eigenvectors.

Author Contributions: Conceptualization, B.C. and M.M.; methodology, B.C., R.A. and M.M.; validation, B.C., R.A. and M.M.; investigation, D.P., M.M. and R.A.; data curation, D.P. and M.M.; writing—original draft preparation, D.P.; writing—review and editing, M.M., R.A. and B.C.; supervision, B.C. and R.A.; funding acquisition, B.C. All authors have read and agreed to the published version of the manuscript.

Funding: This research was funded by INFLAMMAA (reference PTDC/CVT-CVT/32349/2017), financed by Portugal and the European Union through FEDER, COMPETE 2020 and CRESC Algarve

2020, in the framework of Portugal 2020, and national funds through Fundação para a Ciência e a Tecnologia (FCT, Portugal). D.P. and B.C. were supported by FCT, Portugal (UI/BD/150900/2021 and IF/00197/2015, respectively).

Institutional Review Board Statement: The animal study protocol was approved by the CIIMAR Animal Welfare Committee and DGAV (ORBEA-CIIMAR_26_2018) and was carried out under license number 0421/000/000/2020 in a registered facility (N16091.UDER). Experiments were directed by trained scientists (following FELASA category C recommendations) and were conducted according to the guidelines on the protection of animals used for scientific purposes from the European directive 2010/63/UE.

Informed Consent Statement: Not applicable.

Data Availability Statement: All data is provided in the main text or Supplementary Files.

Conflicts of Interest: The authors declare no conflict of interest.

References

1. Matos, E.; Gonçalves, A.; Nunes, M.L.; Dinis, M.T.; Dias, J. Effect of harvesting stress and slaughter conditions on selected flesh quality criteria of gilthead seabream (*Sparus aurata*). *Aquaculture* **2010**, *305*, 66–72. [CrossRef]
2. Poli, B.M.; Parisi, G.; Scappini, F.; Zampacavallo, G. Fish welfare and quality as affected by pre-slaughter and slaughter management. *Aquac. Int.* **2005**, *13*, 29–49. [CrossRef]
3. Costas, B.; Aragão, C.; Dias, J.; Afonso, A.; Conceição, L.E.C. Interactive effects of a high-quality protein diet and high stocking density on the stress response and some innate immune parameters of Senegalese sole *Solea senegalensis*. *Fish Physiol. Biochem.* **2013**, *39*, 1141–1151. [CrossRef] [PubMed]
4. Acerete, L.; Reig, L.; Alvarez, D.; Flos, R.; Tort, L. Comparison of two stunning/slaughtering methods on stress response and quality indicators of European sea bass (*Dicentrarchus labrax*). *Aquaculture* **2009**, *287*, 139–144. [CrossRef]
5. Engelsma, M.; Hougee, S.; Nap, D.; Hofenk, M.; Rombout, J.; van Muiswinkel, W.; Verburg-van Kemenade, B. Multiple acute temperature stress affects leucocyte populations and antibody responses in common carp, *Cyprinus carpio* L. *Fish Shellfish Immunol.* **2003**, *15*, 397–410. [CrossRef]
6. Ribas, L.; Flos, R.; Reig, L.; MacKenzie, S.; Barton, B.; Tort, L. Comparison of methods for anaesthetizing Senegal sole (*Solea senegalensis*) before slaughter: Stress responses and final product quality. *Aquaculture* **2007**, *269*, 250–258. [CrossRef]
7. Sadoul, B.; Vijayan, M.M. Stress and Growth. In *Biology of Stress in Fish Fish Physiology*; Schreck, C., Tort, L., Farrell, A., Brauner, C., Eds.; Fish Physiology; Academic Press: Cambridge, MA, USA, 2016; Volume 35.
8. Terova, G.; Gornati, R.; Rimoldi, S.; Bernardini, G.; Saroglia, M. Quantification of a glucocorticoid receptor in sea bass (*Dicentrarchus labrax*, L.) reared at high stocking density. *Gene* **2005**, *357*, 144–151. [CrossRef] [PubMed]
9. Tort, L. Stress and immune modulation in fish. *Dev. Comp. Immunol.* **2011**, *35*, 1366–1375. [CrossRef]
10. Verburg-van Kemenade, B.M.L.; Stolte, E.H.; Metz, J.R.; Chadzinska, M. Neuroendocrine–Immune Interactions in Teleost Fish. In *Fish Neuroendocrinology*; Bernier, N., Van Der Kraak, G., Farrel, A., Brauner, C., Eds.; Fish Physiology; Academic Press: Cambridge, MA, USA, 2009; Volume 28, pp. 313–364.
11. Dezfuli, B.; Lui, A.; Boldrini, P.; Pironi, F.; Giari, L. The inflammatory response of fish to helminth parasites. *Parasite* **2008**, *15*, 426–433. [CrossRef]
12. Sharkey, K. Substance P and Calcitonin Gene-Related Peptide (CGRP) in Gastrointestinal Inflammation. *Ann. N. Y. Acad. Sci.* **1992**, *664*, 425–442. [CrossRef]
13. Suzuki, Y.; Iida, T. Fish granulocytes in the process of inflammation. *Annu. Rev. Fish Dis.* **1992**, *2*, 149–160. [CrossRef]
14. Azeredo, R. Amino Acids as Novel Nutraceutics to Modulate Immune Mechanisms and Increase Disease Resistance in Fish. Ph.D. Thesis, Faculty of Sciences of University of Porto, Porto, Portugal, 2017.
15. Havixbeck, J.J.; Rieger, A.M.; Wong, M.E.; Hodgkinson, J.W.; Barreda, D.R. Neutrophil contributions to the induction and regulation of the acute inflammatory response in teleost fish. *J. Leukoc. Biol.* **2016**, *99*, 241–252. [CrossRef] [PubMed]
16. Finn, J.P.; Nielson, N.O. The inflammatory response of rainbow trout. *J. Fish Biol.* **1971**, *3*, 463–478. [CrossRef]
17. Chadzinska, M.; Savelkoul, H.F.; Kemenade, B.L.V.-V. Morphine affects the inflammatory response in carp by impairment of leukocyte migration. *Dev. Comp. Immunol.* **2009**, *33*, 88–96. [CrossRef] [PubMed]
18. Herrero-Turrión, M. Opioids and Opioid Receptors in Fishes. In *Reference Module in Life Sciences*; Elsevier: Amsterdam, The Netherlands, 2017. [CrossRef]
19. Chadzinska, M.; Hermsen, T.; Savelkoul, H.F.; Verburg-van Kemenade, B.M. Cloning of opioid receptors in common carp (*Cyprinus carpio* L.) and their involvement in regulation of stress and immune response. *Brain Behav. Immun.* **2009**, *23*, 257–266. [CrossRef]
20. Bayne, C.J.; Gerwick, L. The acute phase response and innate immunity of fish. *Dev. Comp. Immunol.* **2001**, *25*, 725–743. [CrossRef]
21. Azeredo, R.; Machado, M.; Afonso, A.; Fierro-Castro, C.; Reyes-López, F.E.; Tort, L.; Gesto, M.; Conde-Sieira, M.; Míguez, J.M.; Soengas, J.; et al. Neuroendocrine and Immune Responses Undertake Different Fates following Tryptophan or Methionine Dietary Treatment: Tales from a Teleost Model. *Front. Immunol.* **2017**, *8*, 1226. [CrossRef]

22. Afonso, A.; Ellis, A.E.; Silva, M.T. The leucocyte population of the unstimulated peritoneal cavity of rainbow trout (*Oncorhynchus mykiss*). *Fish Shellfish Immunol.* **1997**, *7*, 335–348. [CrossRef]
23. Machado, M.; Azeredo, R.; Díaz-Rosales, P.; Afonso, A.; Peres, H.; Oliva-Teles, A.; Costas, B. Dietary tryptophan and methionine as modulators of European seabass (*Dicentrarchus labrax*) immune status and inflammatory response. *Fish Shellfish Immunol.* **2015**, *42*, 353–362. [CrossRef]
24. Afonso, A.; Lousada, S.; Silva, J.; Ellis, A.E.; Silva, M.T. Neutrophil and macrophage responses to inflammation in the peritoneal cavity of rainbow trout *Oncorhynchus mykiss*. A light and electron microscopic cytochemical study. *Dis. Aquat. Org.* **1998**, *34*, 27–37. [CrossRef]
25. Nikoskelainen, S.; Ouwehand, A.; Bylund, G.; Salminen, S.; Lilius, E.-M. Immune enhancement in rainbow trout (*Oncorhynchus mykiss*) by potential probiotic bacteria (*Lactobacillus rhamnosus*). *Fish Shellfish Immunol.* **2003**, *15*, 443–452. [CrossRef]
26. Ellis, A. Serum antiproteases in fish. In *Techiniques in Fish Immunology*; Stolen, F.T., Andrerson, D.P., Roberson, B.S., Van Muiswinkel, W.B., Eds.; SOS, Warehouse: Fair Haven, NJ, USA, 1990.
27. Ellis, A. Lysozyme assays. In *Techiniques in Fish Immunology*; Stolen, F.T., Andrerson, D.P., Roberson, B.S., Van Muiswinkel, W.B., Eds.; SOS, Warehouse: Fair Haven, NJ, USA, 1990.
28. Quade, M.J.; Roth, J.A. A rapid, direct assay to measure degranulation of bovine neutrophil primary granules. *Vet. Immunol. Immunopathol.* **1997**, *58*, 239–248. [CrossRef]
29. Cuesta, A.; Meseguer, J.; Esteban, M. Total serum immunoglobulin M levels are affected by immunomodulators in seabream (*Sparus aurata* L.). *Vet. Immunol. Immunopathol.* **2004**, *101*, 203–210. [CrossRef] [PubMed]
30. Azeredo, R.; Machado, M.; Guardiola, F.A.; Cerezuela, R.; Afonso, A.; Peres, H.; Oliva-Teles, A.; Esteban, M.; Costas, B. Local immune response of two mucosal surfaces of the European seabass, *Dicentrarchus labrax*, fed tryptophan- or methionine-supplemented diets. *Fish Shellfish Immunol.* **2017**, *70*, 76–86. [CrossRef]
31. Machado, M.; Azeredo, R.; Domingues, A.; Boo, S.F.; Dias, J.; Conceicao, L.; Costas, B. Dietary tryptophan deficiency and its supplementation compromises inflammatory mechanisms and disease resistance in a teleost fish. *Sci. Rep.* **2019**, *9*, 7689. [CrossRef]
32. Rieger, A.M.; Konowalchuk, J.D.; Grayfer, L.; Katzenback, B.; Havixbeck, J.J.; Kiemele, M.D.; Belosevic, M.; Barreda, D.R. Fish and Mammalian Phagocytes Differentially Regulate Pro-Inflammatory and Homeostatic Responses In Vivo. *PLoS ONE* **2012**, *7*, e47070. [CrossRef]
33. Griffin, B. Random and directed migration of trout (*Salmo gairdneri*) leukocytes: Activation by antibody, complement, and normal serum components. *Dev. Comp. Immunol.* **1984**, *8*, 589–597. [CrossRef]
34. do Vale, A.; Afonso, A.; Silva, M.T. The professional phagocytes of sea bass (*Dicentrarchus labrax* L.): Cytochemical characterisation of neutrophils and macrophages in the normal and inflamed peritoneal cavity. *Fish Shellfish Immunol.* **2002**, *13*, 183–198. [CrossRef]
35. Rossaint, J.; Margraf, A.; Zarbock, A. Role of Platelets in Leukocyte Recruitment and Resolution of Inflammation. *Front. Immunol.* **2018**, *9*, 2712. [CrossRef]
36. Costas, B.; Rêgo, P.C.N.P.; Simões, I.; Marques, J.; Castro-Cunha, M.; Afonso, A. Cellular and humoral immune responses of Senegalese sole, *Solea senegalensis* (Kaup), following challenge with two *Photobacterium damselae* subsp. *piscicida* strains from different geographical origins. *J. Fish Dis.* **2013**, *36*, 543–553. [CrossRef]
37. Ferdous, F.; Scott, T. A comparative examination of thrombocyte/platelet immunity. *Immunol. Lett.* **2015**, *163*, 32–39. [CrossRef] [PubMed]
38. Nagasawa, T.; Nakayasu, C.; Rieger, A.M.; Barreda, D.; Somamoto, T.; Nakao, M. Phagocytosis by Thrombocytes is a Conserved Innate Immune Mechanism in Lower Vertebrates. *Front. Immunol.* **2014**, *5*, 445. [CrossRef] [PubMed]
39. Stosik, M.; Tokarz-Deptuła, B.; Deptuła, W. Characterisation of thrombocytes in Osteichthyes. *J. Vet. Res.* **2019**, *63*, 123–131. [CrossRef] [PubMed]
40. Klinger, M.H.; Jelkmann, W. Review: Role of Blood Platelets in Infection and Inflammation. *J. Interf. Cytokine Res.* **2002**, *22*, 913–922. [CrossRef]
41. Semple, J.W.; Italiano, J.E., Jr.; Freedman, J. Platelets and the immune continuum. *Nat. Rev. Immunol.* **2011**, *11*, 264–274. [CrossRef]
42. Secombes, C.J.; Cross, A.R.; Sharp, G.J.; Garcia, R. Nadph oxidase-like activity in rainbow trout *Oncorhynchus mykiss* (walbaum) macrophages. *Dev. Comp. Immunol.* **1992**, *16*, 405–413. [CrossRef]
43. Plouffe, D.A.; Hanington, P.C.; Walsh, J.G.; Wilson, E.C.; Belosevic, M. Comparison of select innate immune mechanisms of fish and mammals. *Xenotransplantation* **2005**, *12*, 266–277. [CrossRef]
44. Campos-Pérez, J.; Ellis, A.; Secombes, C. Investigation of factors influencing the ability of *Renibacterium salmoninarum* to stimulate rainbow trout macrophage respiratory burst activity. *Fish Shellfish Immunol.* **1997**, *7*, 555–566. [CrossRef]
45. Lamas, J.; Ellis, A.E. Atlantic salmon (*Salmo salar*) neutrophil responses to *Aeromonas salmonicida*. *Fish Shellfish Immunol.* **1994**, *4*, 201–219. [CrossRef]
46. Saurabh, S.; Sahoo, P.K. Lysozyme: An important defence molecule of fish innate immune system. *Aquac. Res.* **2008**, *39*, 223–239. [CrossRef]
47. Murphy, A.E.; Stokesbury, M.J.W.; Easy, R.H. Exploring epidermal mucus protease activity as an indicator of stress in Atlantic sturgeon (*Acipenser oxyrinchus oxyrhinchus*). *J. Fish Biol.* **2020**, *97*, 1354–1362. [CrossRef] [PubMed]
48. Esteban, M.Á. An Overview of the Immunological Defenses in Fish Skin. *ISRN Immunol.* **2012**, *2012*, 853470. [CrossRef]

49. Ahmed, A.U.; Williams, B.R.G.; Hannigan, G.E. Transcriptional Activation of Inflammatory Genes: Mechanistic Insight into Selectivity and Diversity. *Biomolecules* **2015**, *5*, 3087–3111. [CrossRef]
50. Reis, M.I.R.; Vale, A.D.; Pereira, P.J.B.; Azevedo, J.E.; dos Santos, N.M.S. Caspase-1 and IL-1β Processing in a Teleost Fish. *PLoS ONE* **2012**, *7*, e50450. [CrossRef]
51. Weber, A.; Wasiliew, P.; Kracht, M. Interleukin-1 (IL-1) Pathway. *Sci. Signal.* **2010**, *3*, cm1. [CrossRef] [PubMed]
52. Zou, J.; Secombes, C.J. The Function of Fish Cytokines. *Biology* **2016**, *5*, 23. [CrossRef]
53. Machado, M.; Azeredo, R.; Fontinha, F.; Fernández-Boo, S.; Conceicao, L.; Dias, J.; Costas, B. Dietary Methionine Improves the European Seabass (*Dicentrarchus labrax*) Immune Status, Inflammatory Response, and Disease Resistance. *Front. Immunol.* **2018**, *9*, 2672. [CrossRef]
54. Kubiczkova, L.; Sedlarikova, L.; Hajek, R.; Sevcikova, S. TGF-β—An excellent servant but a bad master. *J. Transl. Med.* **2012**, *10*, 183. [CrossRef]
55. Reyes-Cerpa, S.; Maisey, K.; Reyes-Lpez, F.; Toro-Ascuy, D.; Mara, A.; Imarai, M. Fish Cytokines and Immune Response. In *New Advances and Contributions to Fish Biology*; Türker, H., Ed.; IntechOpen: London, UK, 2012. [CrossRef]
56. Pinto, R.D.; Nascimento, D.S.; Reis, M.I.; do Vale, A.D.; dos Santos, N.M. Molecular characterization, 3D modelling and expression analysis of sea bass (*Dicentrarchus labrax* L.) interleukin-10. *Mol. Immunol.* **2007**, *44*, 2056–2065. [CrossRef]
57. Tafalla, C.; Coll, J.; Secombes, C. Expression of genes related to the early immune response in rainbow trout (*Oncorhynchus mykiss*) after viral haemorrhagic septicemia virus (VHSV) infection. *Dev. Comp. Immunol.* **2005**, *29*, 615–626. [CrossRef]
58. Wang, T.; Kono, T.; Monte, M.M.; Kuse, H.; Costa, M.M.; Korenaga, H.; Maehr, T.; Husain, M.; Sakai, M.; Secombes, C.J. Identification of IL-34 in teleost fish: Differential expression of rainbow trout IL-34, MCSF1 and MCSF2, ligands of the MCSF receptor. *Mol. Immunol.* **2013**, *53*, 398–409. [CrossRef] [PubMed]
59. Grayfer, L.; Kerimoglu, B.; Yaparla, A.; Hodgkinson, J.W.; Xie, J.; Belosevic, M. Mechanisms of Fish Macrophage Antimicrobial Immunity. *Front. Immunol.* **2018**, *9*, 1105. [CrossRef] [PubMed]
60. Masteller, E.L.; Wong, B.R. Targeting IL-34 in chronic inflammation. *Drug Discov. Today* **2014**, *19*, 1212–1216. [CrossRef] [PubMed]
61. Liu, X.; Kang, L.; Liu, W.; Lou, B.; Wu, C.; Jiang, L. Molecular characterization and expression analysis of the large yellow croaker (*Larimichthys crocea*) chemokine receptors CXCR2, CXCR3, and CXCR4 after bacterial and poly I:C challenge. *Fish Shellfish Immunol.* **2017**, *70*, 228–239. [CrossRef] [PubMed]
62. Fu, Q.; Yang, Y.; Li, C.; Zeng, Q.; Zhou, T.; Li, N.; Liu, Y.; Liu, S.; Liu, Z. The CC and CXC chemokine receptors in channel catfish (*Ictalurus punctatus*) and their involvement in disease and hypoxia responses. *Dev. Comp. Immunol.* **2017**, *77*, 241–251. [CrossRef] [PubMed]
63. Isles, H.M.; Herman, K.D.; Robertson, A.L.; Loynes, C.A.; Prince, L.R.; Elks, P.M.; Renshaw, S.A. The CXCL12/CXCR4 Signaling Axis Retains Neutrophils at Inflammatory Sites in Zebrafish. *Front. Immunol.* **2019**, *10*, 1784. [CrossRef]
64. van der Sar, A.M.; Spaink, H.P.; Zakrzewska, A.; Bitter, W.; Meijer, A.H. Specificity of the zebrafish host transcriptome response to acute and chronic mycobacterial infection and the role of innate and adaptive immune components. *Mol. Immunol.* **2009**, *46*, 2317–2332. [CrossRef]
65. Machado, M. Mediators of Inflammation: Unravelling the Role of Methionine and Tryptophan during Infection. Ph.D. Thesis, Univerisity of Porto, Porto, Portugal, 2020.
66. Sneddon, L.U. Comparative Physiology of Nociception and Pain. *Physiology* **2018**, *33*, 63–73. [CrossRef]
67. Enyedi, B.; Kala, S.; Nikolich-Zugich, T.; Niethammer, P. Tissue damage detection by osmotic surveillance. *Nat. Cell Biol.* **2013**, *15*, 1123–1130. [CrossRef]

biology

MDPI

Article

High-Temperature Stress Effect on the Red Cusk-Eel (*Geypterus chilensis*) Liver: Transcriptional Modulation and Oxidative Stress Damage

Phillip Dettleff [1,*], Rodrigo Zuloaga [2,3], Marcia Fuentes [2,3], Pamela Gonzalez [2,3], Jorge Aedo [2,3], Juan Manuel Estrada [4], Alfredo Molina [2,3] and Juan Antonio Valdés [2,3,*]

1 Escuela de Medicina Veterinaria, Facultad de Agronomía e Ingeniería Forestal, Facultad de Ciencias Biológicas y Facultad de Medicina, Pontificia Universidad Católica de Chile, Santiago 7820436, Chile
2 Facultad de Ciencias de la Vida, Universidad Andrés Bello, Santiago 8370186, Chile; rodrigo.zuloaga.r@gmail.com (R.Z.); fvmarcia@gmail.com (M.F.); pamela.gonzalez.tr@gmail.com (P.G.); jor.aedo.a@gmail.com (J.A.); amolina@unab.cl (A.M.)
3 Interdisciplinary Center for Aquaculture Research (INCAR), Concepción 4030000, Chile
4 Centro de Investigación Marina Quintay (CIMARQ), Universidad Andrés Bello, Quintay 2340000, Chile; mestrada@unab.cl
* Correspondence: phillip.dettleff@uc.cl (P.D.); jvaldes@unab.cl (J.A.V.); Tel.: +56-223-547-494 (P.D.); +56-227-705-815 (J.A.V.)

Simple Summary: The red cusk-eel (*Genypterus chilensis*) is a native Chilean species important for aquaculture diversification in Chile. The effect of high-temperature stress on the liver, a key organ for fish metabolism, is unknown. In this study we determined for the first time the effects of high-temperature stress on the liver of red cusk-eel. The results showed that high-temperature stress increased hepatic enzyme activity in the plasma of stressed fish. Additionally, this stressor generated oxidative damage in liver, and generated a transcriptional response with 1239 down-regulated and 1339 up-regulated transcripts associated with several processes, including unfolded protein response, heat shock response and oxidative stress, among others. Together, these results indicate that high-temperature stress generates a relevant impact on liver, with should be considered for the aquaculture and fisheries industry of this species under a climate change scenario.

Abstract: Environmental stressors, such as temperature, are relevant factors that could generate a negative effect on several tissues in fish. A key fish species for Chilean aquaculture diversification is the red cusk-eel (*Genypterus chilensis*), a native fish for which knowledge on environmental stressors effects is limited. This study evaluated the effects of high-temperature stress on the liver of red cusk-eel in control (14 °C) and high-temperature (19 °C) groups using multiple approaches: determination of plasmatic hepatic enzymes (ALT, AST, and AP), oxidative damage evaluation (AP sites, lipid peroxidation, and carbonylated proteins), and RNA-seq analysis. High-temperature stress generated a significant increase in hepatic enzyme activity in plasma. In the liver, a transcriptional regulation was observed, with 1239 down-regulated and 1339 up-regulated transcripts. Additionally, high-temperature stress generated oxidative stress in the liver, with oxidative damage and transcriptional modulation of the antioxidant response. Furthermore, an unfolded protein response was observed, with several pathways enriched, as well as a heat shock response, with several heat shock proteins up regulated, suggesting candidate biomarkers (i.e., *serpinh1*) for thermal stress evaluation in this species. The present study shows that high-temperature stress generated a major effect on the liver of red cusk-eel, knowledge to consider for the aquaculture and fisheries of this species.

Keywords: red cusk-eel; thermal stress; RNA-seq; liver transcriptome; oxidative damage; protein folding; hepatic enzymes

Citation: Dettleff, P.; Zuloaga, R.; Fuentes, M.; Gonzalez, P.; Aedo, J.; Estrada, J.M.; Molina, A.; Valdés, J.A. High-Temperature Stress Effect on the Red Cusk-Eel (*Geypterus chilensis*) Liver: Transcriptional Modulation and Oxidative Stress Damage. *Biology* **2022**, *11*, 990. https://doi.org/10.3390/biology11070990

Academic Editor: Patricia Pereiro

Received: 10 May 2022
Accepted: 27 June 2022
Published: 29 June 2022

1. Introduction

Environmental factors are important for the physiology of fish, particularly those associated with water conditions. Among these, temperature, pH, and dissolved oxygen (DO) are crucial for the homeostasis of the fish in marine environments, and changes in these factors could lead to generating a stressful status for the animal [1]. It has been reported that an increase in water temperature could lead to stress and negative effects on marine fish, including salmonids [2,3], catfish [4], Atlantic cod [5], and gilthead seabream [6]. Understanding the effects of water temperature on fish is crucial in the actual scenario of climate change, considering the average rising sea temperature per decade [7]. Additionally, the effect of climate change on sea temperatures is expected to influence relevant phenomena of the Pacific coast, such as El Niño–Southern Oscillation (ENSO), increasing their intensity and frequency [8], which is relevant for marine species of the South Pacific coast, considering the increase in water temperatures associated with this phenomena. Therefore, it is important to understand the effect of an increase in temperature and how it affects the stress response in marine fish with aquaculture potential.

Chile is a relevant country in seafood production, with an important fishery and aquaculture industry. Additionally, the Chilean aquaculture sector is recognized for its important salmon and mussel industries [9]. However, in the last decade, an important effort has been made by the public private association to diversify Chilean aquaculture with native fish of commercial value. One of these species is the red cusk-eel (*Genypterus chilensis*), part of the *Genypterus* genus, endemic to the South Pacific coast, an economically relevant fish for fisheries and, recently, part of the Chilean aquaculture diversification program [10,11], with recent elucidation of the complete production cycle [12]. The red cusk-eel is a demersal fish with a carnivorous diet; it lives on rocky bottoms [13] and is characterized reproductively as a multiple spawner [11]. The high value of its flesh makes this species an attractive product; however, tons of fisheries have presented variable levels, with a decreasing tendency in the last decade [14]. In this sense, it is important to understand how environmental factors could affect this species and the mechanisms involved.

Stress in fish can be characterized as an adaptative response to danger, which generates physiological changes to prepare the fish to respond and survive threats. This stress response in fish is mediated by the hypothalamic–pituitary–interrenal (HPI) axis though several key chemical mediators, including corticotropin releasing factor, adrenocorticotropic hormone, α-melanocyte-stimulating hormone, adrenaline, and cortisol, a key hormone that increases in plasma under stress [15]. If the stressor is maintained for prolongated periods, a chronic stress status is generated in the fish, leading to several negative physiological effects, including decreased growth, reproductive problems, behavior modifications, and immune response [16,17]. Moreover, at a cellular level, the stress can lead to an increase in reactive oxygen species (ROS), which could lead to an oxidative stress status, as previously observed for several teleost fish species [18–22], an effect also observed for red cusk-eel in response to several stressors [23–25]. Red cusk-eel has shown low tolerance to intensive farming stressors, with limited information related to the stress response capacity in this species [13]. However, the specific stress response varies according to species, as well as the effect on each tissue. Our previous studies on *G. chilensis* showed a variable tissue response under handling stress, with altered metabolic status in the liver, a modulation of the immune response in the head kidney, and an induction of atrophy in skeletal muscle through coding and noncoding regulation [26]. Additionally, it has been observed that thermal stress could induce muscle atrophy in this species [24], as well as oxidative damage in eggs, with a minor effect on the ovary [23]. One of the most important organs for fish metabolism is the liver, which is directly involved in stress response by metabolizing and liberating stored energy to respond to stress [17]. Nevertheless, the liver response to thermal stress in *Genypterus* species has not been previously studied, nor has the impact of this stressor on the oxidative status of this tissue. It is important to consider that sea temperatures will increase through sea heat waves due to the effect of ENSO under a climate change scenario, which will affect the Chilean coast associated with the geographic range of *G. chilensis*. Therefore, the objective of this study was to evaluate the effect of high-temperature stress on the liver of *G. chilensis* in terms of the transcriptomic and oxidative stress status to determine the negative impact of this type of stressor on liver.

2. Materials and Methods

2.1. Ethics Statement

All procedures with the red cusk-eel individuals and all scientific activities adhered to animal welfare procedures and were approved by the bioethical committees of the Universidad Andres Bello (007/2018) and the National Commission for Scientific and Technological Research (CONICYT) of the Chilean government.

2.2. Fish Sampling and Experimental Design

In this study, we used reproductively immature red cusk-eel juveniles (*G. chilensis*) of 12 months of age (average weight of 665 ± 52.7 g; average length of 60 ± 4.8 cm), collected from the Centro de Investigación Marina de Quintay (CIMARQ), maintained under natural photoperiod conditions (L:D 12:12), and controlled temperature (14 ± 1 °C), and fed daily with commercial pellet food. Fish were separated into control and stress groups, with the stress group subjected to a standardized thermal stress protocol proven to generate stress in red cusk-eel [24]. Briefly, this protocol consists of increasing the temperature over 24 h at a rate of 1 °C in 5 h. This protocol maintains the thermal stress temperature (19 ± 1 °C) for 5 days. This high temperature protocol was selected considering heat waves observed in the summer season on the Chilean coast in recent years [27]. The control group was maintained at the control temperature (14 ± 1 °C) during the assay. At the end of the experiment, six individuals per group (two tanks per group, with three animals sampled per tank, total of N = 12 sampled fish) were netted and sampled. Blood samples were collected via caudal puncture using heparinized tubes, immediately centrifuged at 5000× *g* for 10 min for serum obtention and stored at −80 °C until analysis. After blood sampling, fish were euthanized (overdose of anesthetic 3-aminobenzoic acid ethyl ester, 300 mg/L). Fish livers were collected and stored for RNA extraction in RNAsave solution (Biological Industries, Cromwell, CT, USA) or immediately frozen in liquid nitrogen and stored at −80 °C until analysis for oxidative damage evaluation.

2.3. AST, ALT and AP Evaluation

The plasmatic activities of aspartate aminotransferase (AST), alanine aminotransferase (ALT), and alkaline phosphatase (AP) were determined using commercially available kits from Valtek (Santiago, Chile) following the manufacturer's instruction. Briefly, these kits determined the enzymatic activity (IU/L) via the generation of colorimetric products from glutamate (colorimetric product: 450 nm), pyruvate (535 nm), and p-nitrophenol (405 nm) for AST, ALT and AP activity, respectively.

2.4. Oxidative Stress Assays in Liver

To determine the oxidative damage in the liver of red cusk-eel in response to high-temperature stress, DNA oxidative damage, protein carbonylation, and lipid peroxidation were evaluated using commercially available kits. The DNA oxidative damage assay was performed with genomic DNA (gDNA) purified from 25 mg of the liver using the DNAzol reagent (Invitrogen, Carlsbad, CA, USA) following the manufacturer's protocol and quantified with the Epoch Spectrophotometer System (BioTek, Winooski, VT, USA). Then, the apurinic/apyrimidinic sites (AP sites) were determined in the gDNA using the OxiSelect Oxidative DNA Damage Quantification Kit (Cell Biolabs, CA, USA) following the manufacturer's protocol. The protein carbonylation assay was performed using total protein extracted from 100 mg of the liver in 1 mL of lysis buffer containing 50 mM Tris-HCl pH 7.4, 150 mM NaCl, 1 mM EDTA, and 1% NP-40, solubilized at 4 °C after 12,000× *g* centrifugation. Proteins were quantified using the Pierce BCA Protein Assay Kit (Thermo Scientific, Batavia, IL, USA). Then, carbonylated protein content was quantified using the OxiSelect Protein Carbonyl Spectrophotometric Assay (Cell Biolabs, San Diego, CA, USA) following the manufacturer's protocol. The lipid peroxidation determination in the liver was performed using the OxiSelect HNE Adduct Competitive ELISA Kit (Cell Biolabs, San Diego, CA, USA), following the manufacturer's protocol. This kit determines lipid peroxidation through the quantification of hydroxynonenal (HNE) protein adducts in the extracted proteins of the liver.

2.5. Liver RNA Extraction, Library Preparation and Illumina Sequencing

Total RNA was extracted from the liver stored in RNAsave solution (Biological Industries, Cromwell, CT, USA) using the TRIzol® reagent (Invitrogen, Carlsbad, CA, USA) protocol. Total RNA was quantified by fluorometry with the Qubit® RNA quantitation assay (Invitrogen, Carlsbad, CA, USA) and purity was determined according to 260/280 ratio using the Epoch Spectrophotometer System (BioTek, Winooski, VT, USA). The RNA integrity was measured according to RNA Quality Measurement Number (RQN) through a Fragment Analyzer with the Standard Sensitivity RNA Analysis kit (Advanced Analytical Technologies, Fiorenzuola, Italy), selecting samples with RQN $\geq$ 8. The cDNA libraries construction were performed with 1 µg of total RNA per sample using the Illumina® TruSeq RNA Sample Prep Kit v2 (Illumina®, San Diego, CA, USA), following the manufacturer's protocol. The sizes of the mRNA libraries were determined through a Fragment Analyzer using the NGS Fragment Analysis kit (Advanced Analytical Technologies) and quantified by qPCR using the Kapa Library Quantification kit (Roche, Little Falls, NJ, USA). Paired-end sequencing (2 $\times$ 100 bp) was performed on a Hiseq 2500 (Illumina®) platform in Macrogen Inc. (Seul, South Korea).

2.6. Reads Filtering, Differential Expression, and GO Enrichment Analysis

The raw reads obtained from Illumina sequencing were trimmed to remove the remaining Illumina adapter, low-quality sequences, and short sequences (<50 bp), using the CLC Genomics Workbench v.7.0.3 software. The filtered reads were mapped to a *G. chilensis* reference transcriptome previously annotated by our group ([28], NCBI accession number SRS614525) using the CLC Genomics Workbench v.7.0.3 software with the following parameters: mismatch cost = 2, insertion cost = 3, deletion cost = 3, length fraction = 0.8, and a similarity fraction = 0.8. The expression values were used for clustering and heatmap chart generation with R. Differential expression analysis was performed with the R package DESeq2 (version 1.2.10) [29] to determine differentially expressed transcripts in the liver between the control and stress groups. Transcripts presenting an adjusted *p*-value of <0.05 and an absolute log2 fold change of >1 were considered as differentially expressed between the groups. Enrichment analysis was performed on the list of differentially expressed transcripts and GO terms to determine the overrepresented processes in response to high-temperature stress in the liver, considering up and down-expressed transcripts. This analysis was performed using the enrichment analysis tool implemented in the Blast2GO software [30]. Additionally, the Kyoto Encyclopedia of Genes and Genomes (KEGG) metabolic pathway database was used to build the represented pathways through the KEGG Automatic Annotation Server (KAAS) [31] using the KEGG Orthology (KO) identifiers of the differentially expressed list.

2.7. RNA-seq Validation by qPCR

Total RNA previously extracted was DNAse-treated to remove residual gDNA, and 1 µg of RNA was reverse transcribed into cDNA using the QuantiTect® Reverse Transcription kit (Qiagen, Germantown, MD, USA), following the manufacturer's protocol. A total of thirteen differentially expressed transcripts were selected for qPCR validation, corresponding to: *hsp60*, *hsp70*, *gpx7*, *ddit4*, *leptin*, *msh2*, *msh3*, *c1ql1*, *ccl20*, *atg12*, *atg4b*, *casp3* and *c3*. Primers were designed with Primer 3 software v0.4.0 (http://frodo.wi.mit.edu/primer3/, accessed on 6 May 2021) using the reference transcriptome previously described. The primer sequences, amplicon size, Tm, and efficiency are presented in Table 1. The qPCR was performed in a Stratagene MX3000P qPCR system (Stratagene, La Jolla, CA, USA). All qPCR assays were performed in triplicates, using no-template and no-RT controls, in compliance with the MIQE guidelines [32]. The qPCR reaction mixture contained 7.5 µL of 2× Brilliant® II SYBR® Green master mix (Agilent Technologies, CA, USA), 100 ng of cDNA per reaction and 200 nM of each primer in a 15 µL final volume. Thermal cycling conditions were an initial activation of 2 min at 95 °C, followed by 40 cycles of 30 s at 95 °C, 30 s at 62 °C, and 30 s at 72 °C. A melt curve analysis was performed to confirm a single qPCR product and a standard curve using two-fold series dilutions was used to estimate the efficiency of each primer set. The expression of target genes was normalized using the geometric means of two reference genes (*actb* and *taf12*) previously validated for red cusk-eel in the liver [26] and following the methodology described by [33].

Table 1. Primer sequence, amplicon size, Tm and efficiency of validated genes by qPCR.

Abbreviation	Gene Name	Forward Sequence	Reverse Sequence	Amplicon Size (bp)	Tm	Efficiency (%)
taf12	Transcription initiation factor TFIID subunit 12	GATCTGTAACGACGACGAAGAA	CAAATCAGAGGGACGTCATGTA	92	62	101
actb	Beta actin	TGTCCCTGTATGCTTCTGGT	CCCCTCTCAGTCAGAATCTTCAT	172	62	104
hsp60	Heat shock protein 60	GACGGTTCCAATCTCTACATCTC	CGCTCTCCAAACCAGTTACA	86	62	99
hsp70	Heat shock protein 70	AAGATCAGCGACGACGATAAG	CTGGTGCTCATACTCCTCTTTC	105	62	95
ddit4	DNA damage-inducible transcript 4	GGGAATGAGGAGTTTGGTACAT	GAAGGAAGTGGTGGACCTTATT	88	62	96
gpx7	Glutathione peroxidase 7-like	TCTCCTTCCCTCTGTTCAGTAA	GAAATTCCAGTCGGGCTCTT	104	62	99
leptin	Leptin	CGAAGAGACTTCCTGCTTCAC	CTGATGATCTGGGTGGACTTTC	110	62	104
msh2	DNA mismatch repair Msh2-like	GCCCGTTCCCAGATATTTGAT	CGACCGCAATGACTACTACAC	100	62	98
msh3	DNA mismatch repair Msh3-like	CGACTTCTTCAGGGACTTTGG	TGGCTCTCTGAGTGTCTGT	78	62	104
clql1	Complement C1q 2-like	GATGTTTGTGGCGACGTATTTG	GTTGCTTTCTCAGCCTCTGTA	99	62	104
ccl20	C-C motif chemokine 20-like	CAGCCGTGTGTTAGGGAATA	CAGTTGTCTCGTGTCTCTCTATC	123	62	98
atg12	Ubiquitin ATG12-like	GCCCTCACCAGATCAAGAAG	AGAGAGTCAGAGTGGAGTTAGAG	133	62	102
atg4b	Cysteine protease ATG4B-like	ATCTGGGCGATCTGATGAATG	CGGAGGGCAGAAACAAAGA	96	62	102
casp3	Caspase 3	GCTCCAATTCTTTCCCGTATTT	CAGATTTCCTCTACGCCTACTC	123	61	103
c3	Complement C3-like	CTGCTTCTGGTGACCTGTTTA	CTTCGTGTCCTCTCCATCTTTC	99	62	103

2.8. Statistical Analysis

Significant differences between means of the control and stress groups for DNA damage (AP sites), enzymatic activity (AST, ALT, and AP) and differential expression of the qPCR validated genes were determined using a *t*-test with a significance threshold of $p < 0.05$. All statistical analyses were performed using GraphPad Prism, v.5.00 (GraphPad Software, San Diego, CA, USA).

3. Results

3.1. Hepatic Enzyme Activity and Oxidative Stress Response to High-Temperature Stress

We previously reported that thermal stress for 5 days significantly increased plasmatic levels of cortisol and glucose in *G. chilensis* [24]. To understand how thermal stress affects the metabolism in this species, we studied the effect of high temperature on the liver. To evaluate the effect on the hepatic function of this type of stressor, we measured plasmatic markers of liver damage, i.e., ALT, AST, and AP enzymatic activity. High-temperature stress significantly increased the plasmatic activity of ALT (Figure 1A), AST (Figure 1B) and AP (Figure 1C) in the stress group, evidencing altered hepatic function in the liver of stressed fish. To determine the oxidative damage in the liver generated by high-temperature stress, the DNA oxidative damage, protein carbonylation and lipid peroxidation were determined in the liver. High temperatures generated DNA damage, evidenced by the significant increase in apurinic/apyrimidinic sites in the stressed group (Figure 2A). Oxidative damage was also observed in lipid and proteins, determined by a significant increase in protein carbonylation (Figure 2B) and lipid peroxidation (Figure 2C) in response to high temperature in the stress group.

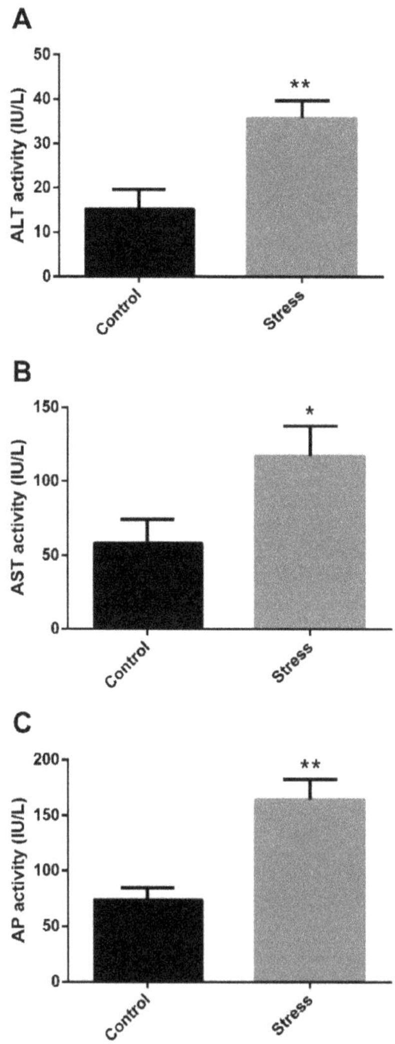

Figure 1. Plasmatic activity of alanine aminotransferase (ALT), aspartate aminotransferase (AST), and alkaline phosphatase (AP) in control and high-temperature stress groups of *G. chilensis*. Levels of ALT (**A**), AST (**B**), and AP (**C**) activity are expressed in IU/L. Bars represent the mean ± SEM. Significant differences between the control and stress groups are indicated by asterisks; * ($p < 0.05$) and ** ($p < 0.01$).

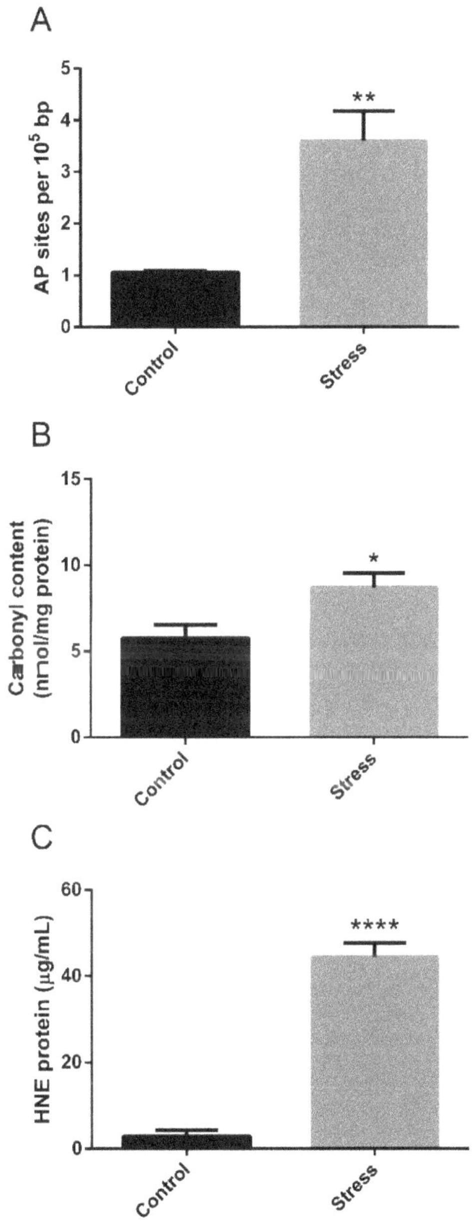

Figure 2. Evaluation of oxidative damage in the liver of *G. chilensis* in control and high-temperature stress groups. Level of DNA damage in terms of AP sites (**A**), protein carbonylation (**B**), and lipid peroxidation in terms of HNE adducts (**C**) in control and stress groups. Bars represent the mean ± SEM. Significant differences between the control and stress groups are indicated by asterisks; * ($p < 0.05$), ** ($p < 0.01$) and **** ($p < 0.0001$).

3.2. Differentially Expressed Transcripts in Hepatic Response to High-Temperature Stress

To understand the complexity of the stress response in the liver of *G. chilensis,* we performed RNA-seq analysis on the liver of each experimental group. The sequencing generated a total of 754,678,455 paired-end reads, with an average of 58,052,189 raw paired-end reads per library. The raw data are available from NCBI under BioProject PRJNA835467 with BioSamples accession number SAMN28102858, SAMN28102859, SAMN28102860, SAMN28102861, SAMN28102862, SAMN28102863, SAMN28102864, SAMN28102865, SAMN28102866, and SAMN28102867. After trimming by quality, adapters, and size, we obtained an average of 58,018,122 high-quality filtered paired-end reads per library (Table 2). These reads were mapped to the *G. chilensis* reference transcriptome (NCBI accession number SRS614525), obtaining an average of 85.2% mapped reads (Table 2). Expression values were used for normalization and differential expression analysis with the R package DESeq2 (version 1.2.10) [29], obtaining a total of 2578 differentially expressed transcripts between control and stressed groups. Of these transcripts, 1239 were down-regulated (Table S1) and 1339 were up-regulated (Table S2) in the stressed group (Figure 3 and Figure S1), evidencing five clusters of transcripts with different patterns of expression for the control and thermal stress groups (Figure S2).

Table 2. Sequencing and mapping statistics of liver libraries from control and thermal stress groups of *G. chilensis*.

Experimental Samples	Number of Reads	Average Length Number of Reads	Number of Reads after Trimming	Average Length after Trimming	Percentage of Mapped Reads
Control 1	51,220,468	101	51,103,248	95.0	85.1
Control 2	43,384,328	101	43,371,547	94.9	86.0
Control 3	50,445,708	101	50,425,183	94.9	85.4
Control 4	70,505,074	101	70,462,858	94.9	84.7
Control 5	65,838,802	101	65,800,592	94.9	85.7
Control 6	57,706,004	101	57,677,646	94.9	85.3
Stress 1	61,733,308	101	61,711,175	94.9	86.8
Stress 2	50,630,806	101	50,612,518	94.9	84.1
Stress 3	58,086,704	101	58,053,699	94.9	84.5
Stress 4	65,071,756	101	65,045,365	94.9	85.5
Stress 5	62,370,664	101	62,343,212	94.8	84.0
Stress 6	59,632,644	101	59,610,421	94.9	85.8
Average	**58,052,189**	**101**	**58,018,122**	**94.9**	**85.2**

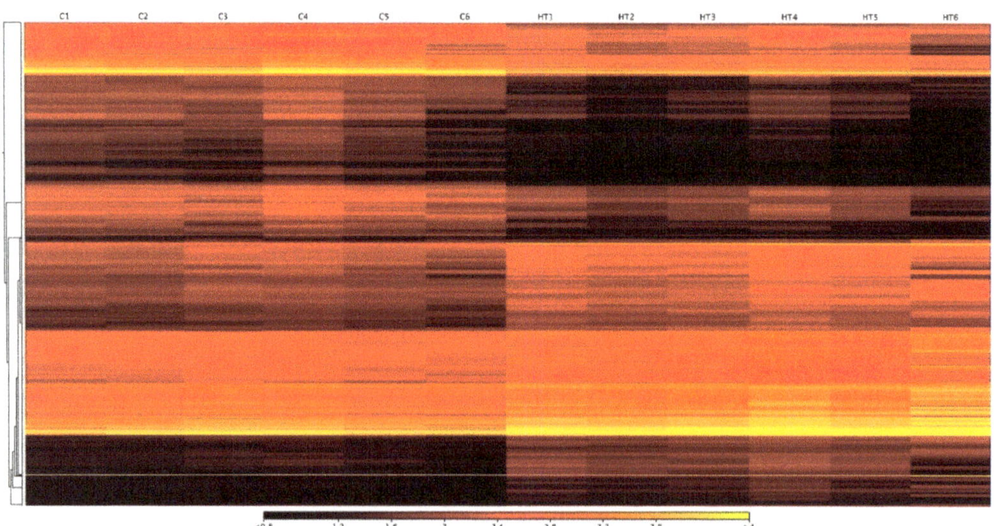

Figure 3. Differentially expressed transcripts in response to high-temperature stress in *G. chilensis.* The heatmap presents the differentially expressed transcripts between control (C) and stress groups (HT). The numbers 1 to 6 indicate the sample library of control and stress fish.

3.3. GO Enrichment and Pathway Analysis in the Liver

The differentially expressed transcripts were used for an enrichment analysis using GO terms to determine the overrepresented processes in response to high-temperature stress in the up- and down-regulated transcripts of the liver. The enrichment analysis showed 38 enriched processes for up-regulated transcripts (Table 3), including 18 biological processes (BP), 16 cellular components (CC), and 4 molecular functions (MF). Several of the enriched terms were related to protein metabolism, including protein folding (GO:0006457), protein transport (GO:0015031), protein localization (GO:0008104), protein retention in ER lumen (GO:0006621), and unfolded protein binding (GO:0051082). Additionally, the BP term response to heat (GO:0009408) was also enriched in the up-regulated transcripts in the high-temperature stress group. No enriched GO terms were identified in the down-regulated transcripts (FDR < 0.05).

Table 3. Enriched GO terms represented in up-regulated transcripts in response to high-temperature stress in the liver of *G. chilensis*.

GO Name	GO Category	GO ID	FDR	N° of Transcripts
Protein folding	BP	GO:0006457	2.96×10^{-11}	28
Establishment of protein localization	BP	GO:0045184	6.62×10^{-3}	30
Protein transport	BP	GO:0015031	1.36×10^{-2}	29
Nitrogen compound transport	BP	GO:0071705	1.65×10^{-2}	33
Protein localization to endoplasmic reticulum	BP	GO:0070972	3.41×10^{-2}	6
Macromolecule localization	BP	GO:0033036	3.41×10^{-2}	36
Transport	BP	GO:0006810	3.41×10^{-2}	96
Cellular protein localization	BP	GO:0034613	3.41×10^{-2}	25
Organic substance transport	BP	GO:0071702	3.41×10^{-2}	38
Protein localization	BP	GO:0008104	3.41×10^{-2}	31
Response to heat	BP	GO:0009408	3.41×10^{-2}	5
Cellular macromolecule localization	BP	GO:0070727	3.41×10^{-2}	25
Establishment of localization	BP	GO:0051234	3.80×10^{-2}	96
Golgi organization	BP	GO:0007030	3.80×10^{-2}	4
Protein retention in ER lumen	BP	GO:0006621	4.09×10^{-2}	3
Maintenance of protein localization in organelle	BP	GO:0072595	4.09×10^{-2}	3
Maintenance of protein localization in endoplasmic reticulum	BP	GO:0035437	4.09×10^{-2}	3
Intracellular transport	BP	GO:0046907	4.09×10^{-2}	28
Endoplasmic reticulum	CC	GO:0005783	3.20×10^{-6}	33
Cytoplasm	CC	GO:0005737	9.42×10^{-6}	133
Nuclear outer membrane-endoplasmic reticulum membrane network	CC	GO:0042175	1.31×10^{-5}	24
Endoplasmic reticulum sub-compartment	CC	GO:0098827	3.68×10^{-5}	23
Endoplasmic reticulum membrane	CC	GO:0005789	3.68×10^{-5}	23
Endomembrane system	CC	GO:0012505	6.46×10^{-5}	50
Organelle sub-compartment	CC	GO:0031984	7.63×10^{-4}	29
Organelle membrane	CC	GO:0031090	2.14×10^{-3}	44
Sarcomere	CC	GO:0030017	3.00×10^{-2}	7
Myofibril	CC	GO:0030016	3.00×10^{-2}	7
Coated membrane	CC	GO:0048475	3.41×10^{-2}	10
Membrane	CC	GO:0016020	3.41×10^{-2}	225
Membrane coat	CC	GO:0030117	3.41×10^{-2}	10
Coated vesicle membrane	CC	GO:0030662	3.41×10^{-2}	7
Cytoplasmic vesicle membrane	CC	GO:0030659	3.41×10^{-2}	8
Contractile fiber	CC	GO:0043292	3.41×10^{-2}	7
Unfolded protein binding	MF	GO:0051082	1.67×10^{-9}	19
Signal sequence binding	MF	GO:0005048	2.70×10^{-3}	5
ER retention sequence binding	MF	GO:0046923	4.09×10^{-2}	3
Phosphofructokinase activity	MF	GO:0008443	4.69×10^{-2}	5

In the KEGG pathway analysis, we identified several pathways represented in the down-regulated and up-regulated transcripts, including apoptosis, cell cycle, MAPK signaling pathway, autophagy, protein processing in endoplasmic reticulum, and ubiquitin-mediated proteolysis. Among these pathways, protein processing in endoplasmic reticulum was one of the most represented on the differentially expressed transcripts (Figure 4), with 31 up-regulated and three down-regulated transcripts represented in this pathway, including several heat shock proteins; heat shock protein 40 (*hsp40*, also known as *dnaJ homolog subfamily B member 1*), heat shock protein 70 (*hsp70*) and heat shock protein 90 (*hsp90*).

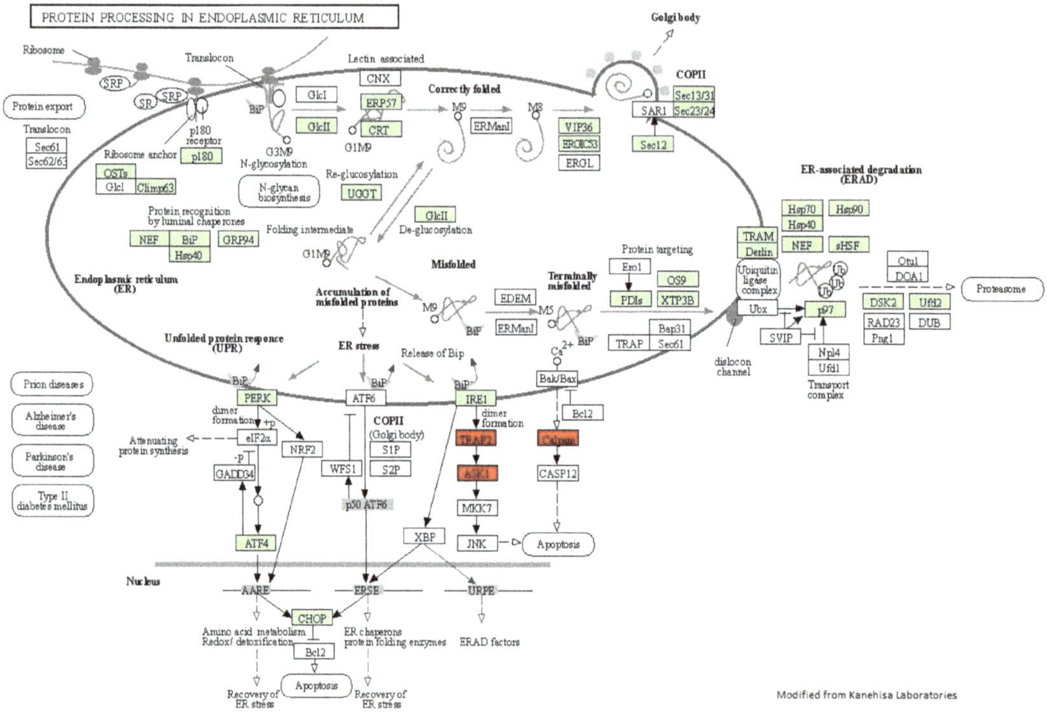

Figure 4. Illustration of transcriptional changes in liver of *G. chilensis* in response to high-temperature stress using modified KEGG pathway map of protein processing in the endoplasmic reticulum. The red rectangles indicate down-regulated transcripts, while green rectangles indicates up-regulated transcripts between the control and high-temperature groups.

3.4. qPCR Validation of Differentially Expressed Genes

The validation of the RNA-seq results was performed by selecting 13 differentially expressed transcripts between the control and stressed groups, including heat shock protein 60 (*hsp60*), hsp70, DNA damage-inducible transcript 4 (*ddit4*), glutathione peroxidase 7-like (*gpx7*), leptin, DNA mismatch repair Msh2-like (*msh2*), DNA mismatch repair Msh3-like (*msh3*), complement C1q 2-like (*clql1*), C-C motif chemokine 20-like (*ccl20*), ubiquitin ATG12-like (*atg12*), cysteine protease ATG4B-like (*atg4b*), caspase 3 (*casp3*) and complement C3-like (*c3*) (Table 1). All of the selected differentially expressed transcripts on RNA-seq were validated by qPCR, confirming the results of the RNA-seq analysis (Figure 5), with a Pearson's correlation coefficient of r = 0.847, confirming the significant differential expression of all tested transcripts by qPCR between the control and high-temperature group (Figure S3).

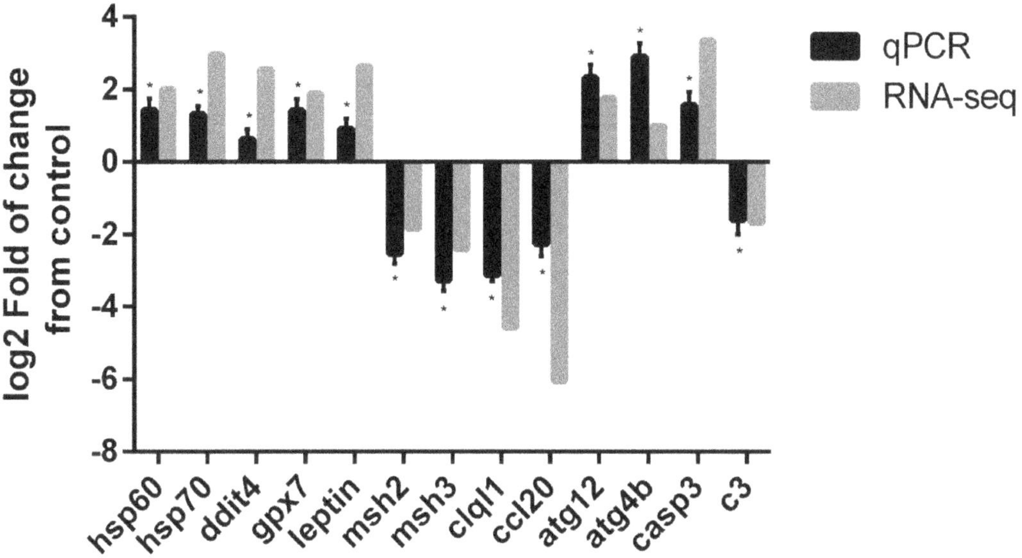

Figure 5. qPCR validation of selected differentially expressed transcripts in the liver of *G. chilensis* in response to high-temperature stress. The transcript expression levels were normalized with the geometric means of *actb* and *taf12*. The differential expression levels according to qPCR (black bars) and RNA-seq (gray bars) for these selected genes are expressed as log_2 fold changes. The log_2FC represents the expression change in the stress group compared with the control group. Results are expressed as the mean ± standard error. Significant differences in the validated qPCR data between control and stress groups are indicated by asterisks in the log2 Fold qPCR bars; (* *p*-value < 0.05).

4. Discussion

Stress in fish is a relevant issue for aquaculture species and native populations of fish [15]. The environmental stress factor associated with water parameters represents a key issue in fish, especially considering the variations in the environment in the short- and long-term associated with global warming, climate change, ENSO, and pollution of the oceans, which could lead to modification of sea temperature, pH, and DO level, as well as increased levels of pollutants, including microplastics and toxic compounds, which could generate stress in teleost fish [7,8,34–36]. These environmental stressors could also affect the red cusk-eel, a relevant species for Chilean fisheries and aquaculture diversification. However, studies aimed at understanding the effect of stressors in *Genypterus* species are limited [25,26,37,38], with no information about the impact of thermal stress on the liver for *Genypterus* species. In this sense, thermal stress has been previously studied in other tissues of *Genypterus* species, including the skeletal muscle, ovary, and post-ovulatory eggs [23,24]. Therefore, to understand the effect of high-temperature stress at the hepatic level, we evaluated the hepatic enzymes, oxidative stress response, and transcriptional regulation in red cusk-eel.

4.1. High-Temperature Effect on Hepatic Enzymes

In a previous study, we determined that high temperature (19 °C) could generate a stress response in red cusk-eel with an increase in the plasmatic level of cortisol and glucose [24]. In the present study, we determined that a high temperature increased the plasmatic activity of ALT, AST, and AP enzymes, evidencing the effect at the hepatic level. This effect was previously observed in other teleost fish under stress conditions, including the plasmatic activity of ALT and AST under high densities and metal pollution in rohu (*Labeo rohita*) [39,40], metal toxicity and pesticides in Nile tilapia (*Oreochromis niloticus*) [41,42] and spotted snakehead (*Channa punctatus*) [43], and elevated ALT, AST and AP under handling stress in red cusk-eel [25]. In terms of high-temperature stress, different results have been observed, with an increase in the plasmatic activity of AST and ALT in pufferfish (*Takifugu obscurus*) [44], which is consistent with our results for red cusk-eel, while no effect for AST and ALT was observed in Turbot (*Scophthalmus maximus*) [45] under high-temperature stress, evidencing that the hepatic response to thermal stress could vary according to fish species. The increased levels on ALT, AST an AP enzymes could be indicative of liver dysfunction, reflecting hepatocyte damage due to thermal stress, which is concordant with the results observed in mammals [46].

4.2. Oxidative Stress under High-Temperature Stress

Oxidative stress corresponds to a disturbance between the production of ROS, which can accumulate in cells, and the antioxidant defenses generated by the cellular systems to detoxify these ROS [47]. OS participate in different normal cellular functions, acting as a second messenger in signal transduction [48], but they can also generate cellular damage, including oxidative damage generating lipid peroxidation, as well as damage to proteins and DNA [49]. Environmental stress, including thermal stress, can lead to oxidative stress status in marine animals [3]. It was previously observed that thermal stress could modulate the oxidative stress status in several tissues of teleost fish [50,51]. One of the relevant organs in which oxidative status could be affected by thermal stress is the liver, as it was previously observed that thermal stress by low temperature could increase antioxidant enzymes in milkfish (*Chanos Chanos*) [52], as well as generate oxidative damage, leading to lipid peroxidation in Pacu (*Piaractus mesopotamicus*) under low-temperature stress [53,54]. High-temperature stress could also generate a relevant impact on the oxidative status of fish, with increased antioxidant enzyme activities, as observed for Senegalese sole (*Solea senegalensis*) [55]. Additionally, oxidative damage was observed in the liver of rohu (*Labeo rohita*), with lipid peroxidation and DNA fragmentation [56]. This is consistent with the findings of our study, evidencing that thermal stress generates an important oxidative effect on the liver of red cusk-eel. This response was also observed at the transcriptional level, with the up-regulation of GPx genes in several tissues of teleost fish under thermal stress, including black porgy (*Acanthopagrus schlegeli*) [57] and pufferfish [58], as well as in red cusk-eel eggs under thermal stress [23], where *gpx1* was increased, an effect not observed for this species in skeletal muscle [24]. Additionally, the effect of high-temperature stress on the liver related to DNA damage was also observed at the transcriptional level, with the up-regulation of genes involved in DNA mismatch repair (*msh2* and *msh3*), concordant with the previously reported effect of thermal stress on zebrafish (*Danio rerio*) [59] and American lobster [60]. However, the thermal stress response associated with oxidative stress in teleost fish could vary according to species and tissues, as observed in sheepshead minnow (*Cyprinodon variegatus*), where a limited effect on antioxidant enzymes and no lipid peroxidation were present [61], in contrast to the variable lipid peroxidation observed in Senegalese sole [55]. We previously observed in red cusk-eel that the impact of high-temperature stress on oxidative damage could vary according to tissue, with lipid peroxidation and DNA damage observed for skeletal muscle [24], but no oxidative damage was observed in the ovaries under high-temperature stress [23], showing that the liver is a sensitive organ under thermal stress in red cusk-eel under an ENSO temperature increase scenario.

4.3. High-Temperature Stress in Hepatic Protein Processing and Folding

Temperature can modulate several cellular processes through gene expression regulation [2]. Here, we used RNA-seq analysis to evaluate, for the first time, the effect of thermal stress on the liver of red cusk-eel, considering that no studies have previously evaluated the temperature effect on the liver in any species of the *Genypterus* genus. We observed a higher hepatic transcriptional response to high-temperature stress compared to other types of stressors, such as handling stress, previously observed for red cusk-eel (4.6 times the differentially expressed transcripts) [25], evidencing that a high temperature could have a higher transcriptional impact on the liver than others stressors in this species. At the cellular level, the function of endoplasmic reticulum is key to protein synthesis, folding, and exporting [62]. However, external processes such as thermal stress could generate alterations in homeostasis, affecting normal protein folding, which leads to endoplasmic reticulum (ER) stress. To alleviate this stress, the unfolded protein response molecular mechanism is activated in cells [63]. This ER stress and subsequent unfolded protein response is concordant with our results in the liver, where we observed enriched processes associated with this type of response in the up-regulated genes in the stress group, including protein folding, protein transport, protein localization, protein retention in ER lumen, and unfolded protein binding. In this sense, we found several genes associated with protein processing and folding pathways, including transcripts involved in: protein export, such as signal recognition particle 14 and 19 kDa (*srp14* and *srp19*) and signal peptidase complex subunit 2-like (*spcs2*) [64]; transcripts associated with protein processing in endoplasmic reticulum (Figure 4), with part of this pathway down-regulated, including TNF receptor-associated factor 2-like (*traf2*), and mainly an up-regulation of this pathway, including ubiquilin-4 (*ubqln4*), cytoskeleton-associated protein 4 (*ckap4*); ribosome-binding protein 1 (*rrbp1*, also known as *p180*); and glucosidase 2 subunit beta (*prkcsh*). Others include transcripts associated with the unfolded protein response, including eukaryotic translation initiation factor 2 alpha kinase 1 (*eik2ak1*, also known as *hri*), 3 (*eik2ak3*, also known as *perk*) and 5 (*eik2ak5*), cyclic AMP-dependent transcription factor ATF-4-like (*atf4*) and DNA damage-inducible transcript 3 -like (*ddit4*, also known as *chop*). The unfolded protein response could be initiated by EIK2AK3 kinase activation of eIF2α, leading to ribosome inhibition and attenuating protein synthesis [65]. Additionally, the ATF4 gene was activated, which would lead to DDIT4 gene regulation and an antioxidant response in cells [66], evidencing that thermal stress in the liver of red cusk-eel modulates the protein processing and generates an unfolded protein response that is not able to control the oxidative stress and damage in this tissue. This is concordant with the results observed in previous studies where unfolded protein response genes were activated under thermal stress in mammals [67,68] and fish, e.g., in the liver of Tambaqui (*Colossoma macropomum*) [69] and gilthead sea bream (*Sparus aurata*) [70].

4.4. Heat Shock Protein as Thermal Stress Biomarkers

The cellular response to thermal stress is a key process to preserve protein integrity; this is known as the heat shock response, which includes heat shock proteins (Hsps) to re-fold the proteins damaged by temperature [71]. This response was observed in the liver of red cusk-eel with the enrichment of the term response to heat (GO:0009408), highlighting several heat shock proteins differentially expressed in response to high-temperature stress, including *hsp40*, *hsp60*, *hsp70*, *hsp90*, and *serpinh1*. The up-regulation of these Hsps was previously observed in other fish under thermal stress, with an increase in the liver of three-spined stickleback (*Gasterosteus aculeatus*) (*hsp60*, *hsp70*, and *hsp90*) [72], Atlantic salmon (*Salmo salar*) (*hsp70*) [73,74], Atlantic cod (*Gadus morhua*) (*hsp70* and *hsp90*) [74], and Wuchang bream (*Megalobrama amblycephala*) (*hsp60*, *hsp70*, and *hsp90*) [75]. The *hsp60* and *hsp70* expression levels were also regulated under thermal stress in the postovulatory eggs and skeletal muscle of red cusk-eel [23,24], showing that these genes could be valuable biomarkers of thermal stress in this species. However, the most up-regulated gene in the liver under high-temperature stress was *serpinh1* (5.7$\log_2$ Fold Change, Table S2). *SerpinH1* (also known as *Hsp47*) is a chaperone involved in the biosynthesis of collagen at the ER [76], with a key role in the restoration of homeostasis during high-temperature stress and oxidative stress [22]. It was shown to be a good biomarker for thermal stress in several salmonid species, including rainbow trout (*Oncorhynchus mykiss*), sockeye salmon (*Oncorhynchus nerka*), and Chinook salmon (*Oncorhynchus tshawytscha*) [3,77], as well as zebrafish [78], evidencing that *sherpinh1* is one of the most suitable transcriptional biomarkers of high-temperature stress in fish. This was also the case for red cusk-eel in this study, representing a useful tool to evaluate thermal stress status in this species under a climate change scenario.

5. Conclusions

The present study evaluated for the first time the effects of high-temperature stress in the liver of *G. chilensis*, using a multiple approach of plasmatic hepatic enzymes, oxidative damage evaluation and RNA-seq analysis. We showed that high-temperature stress under heatwaves ENSO-associated scenario generated a major effect in the liver, affecting hepatic enzymes, generating oxidative damage in this tissue, as well as generating an unfolded protein response at the molecular level in several associated pathways, including a heat shock response, evidencing the affection of red cusk-eel under this type of stressor. This study contributes to knowledge about thermal stress under a climate change scenario, generating candidate biomarkers for thermal stress evaluation in this species, information that should be relevant for the aquaculture and fisheries industry of red cusk-eel.

Supplementary Materials: The following supporting information can be downloaded at: https://www.mdpi.com/article/10.3390/biology11070990/s1, Figure S1: Differentially expressed transcripts in response to high-temperature stress in *G. chilensis* expressed as average transcription per group; Figure S2: Differentially expressed transcripts and clustering groups in response to high-temperature stress in *G. chilensis*; Figure S3: Validation of selected differentially expressed transcripts by qPCR on liver of *G. chilensis* on response to high-temperature stress; Table S1: Down-regulated differentially expressed transcripts between control and stress groups of *G. chilensis*; Table S2: Up-regulated differentially expressed transcripts between control and stress groups of *G. chilensis*.

Author Contributions: Conceptualization and methodology, P.D. and J.A.V.; bioinformatic analysis, P.D.; animal experiment and sampling, P.D., R.Z., M.F., P.G., J.A. and J.M.E.; resources and funding acquisition, P.D., J.M.E., J.A.V. and A.M., writing-draft preparation, P.D. and J.A.V.; writing—review and editing, P.D., J.A.V., R.Z. and J.A. All authors have read and agreed to the published version of the manuscript.

Funding: This study was supported by CONICYT/FONDAP (grant number 15110027) awarded to Juan Antonio Valdés and Alfredo Molina, CONICYT FONDECYT Postdoctorado (grant number 3180283) awarded to Phillip Dettleff, and CONICYT FONDECYT Regular (grant number 1201498).

Institutional Review Board Statement: All procedures with the red cusk-eel individuals in this study were conducted according to the guidelines of the Declaration of Helsinki and approved by the Bioethics Committee of Universidad Andres Bello and the National Commission for Scientific and Technological Research (CONICYT) of the Chilean government (N 007/2018).

Informed Consent Statement: Not applicable.

Data Availability Statement: The raw read sequences obtained from liver sequencing of red cusk-eel were deposited in NCBI under BioProject PRJNA835467, with BioSample accession number SAMN28102858, SAMN28102859, SAMN28102860, SAMN28102861, SAMN28102862, SAMN28102863, SAMN28102864, SAMN28102865, SAMN28102866, and SAMN28102867. The datasets generated and/or analyzed in the present study are available from the corresponding author upon reasonable request.

Acknowledgments: The authors would like to thank the staff at CIMARQ, Universidad Andres Bello, and Italo Barrios for their assistance with experiments and fish management.

Conflicts of Interest: The authors declare no conflict of interest. The sponsors had no role in the design, execution, interpretation, or writing of the study.

References

1. Somero, G.N. The physiology of climate change: How potentials for acclimatization and genetic adaptation will determine 'winners' and 'losers'. *J. Exp. Biol.* **2010**, *213*, 912–920. [CrossRef] [PubMed]
2. Huang, J.; Li, Y.; Liu, Z.; Kang, Y.; Wang, J. Transcriptomic responses to heat stress in rainbow trout *Oncorhynchus mykiss* head kidney. *Fish Shellfish Immunol.* **2018**, *82*, 32–40. [CrossRef] [PubMed]
3. Akbarzadeh, A.; Gunther, O.P.; Houde, A.L.; Li, S.; Ming, T.J.; Jeffries, K.M.; Hinch, S.G.; Miller, K.M. Developing specific molecular biomarkers for thermal stress in salmonids. *BMC Genom.* **2018**, *19*, 749. [CrossRef] [PubMed]

4. Dalvi, R.S.; Pal, A.K.; Tiwari, L.R.; Baruah, K. Influence of acclimation temperature on the induction of heat-shock protein 70 in the catfish *Horabagrus brachysoma* (Gunther). *Fish Physiol. Biochem.* **2012**, *38*, 919–927. [CrossRef]
5. Aursnes, I.A.; Rishovd, A.L.; Karlsen, H.E.; Gjoen, T. Validation of reference genes for quantitative RT-qPCR studies of gene expression in Atlantic cod (*Gadus morhua* L.) during temperature stress. *BMC Res. Notes* **2011**, *4*, 104. [CrossRef] [PubMed]
6. Madeira, D.; Vinagre, C.; Costa, P.M.; Diniz, M.S. Histopathological alterations, physiological limits, and molecular changes of juvenile *Sparus aurata* in response to thermal stress. *Mar. Ecol. Prog. Ser.* **2014**, *505*, 256–266. [CrossRef]
7. Dobretsov, S.; Coutinho, R.; Rittschof, D.; Salta, M.; Ragazzola, F.; Hellio, C. The oceans are changing: Impact of ocean warming and acidification on biofouling communities. *Biofouling* **2019**, *35*, 585–595. [CrossRef]
8. Wang, B.; Luo, X.; Yang, Y.M.; Sun, W.; Cane, M.A.; Cai, W.; Yeh, S.W.; Liu, J. Historical change of El Nino properties sheds light on future changes of extreme El Nino. *Proc. Natl. Acad. Sci. USA* **2019**, *116*, 22512–22517. [CrossRef]
9. Flores-Kossack, C.; Montero, R.; Kollner, B.; Maisey, K. Chilean aquaculture and the new challenges: Pathogens, immune response, vaccination and fish diversification. *Fish Shellfish Immunol.* **2020**, *98*, 52–67. [CrossRef]
10. Vega, R.; Pradenas, M.; Estrada, J.M.; Ramirez, D.; Valdebenito, I.; Mardones, A.; Dantagnan, P.; Alfaro, D.; Encina, F.; Pichara, C. Evaluation and comparison of the efficiency of two incubation systems for *Genypterus chilensis* (Guichenot, 1848) eggs. *Lat. Am. J. Aquat. Res.* **2012**, *40*, 187–200. [CrossRef]
11. Chong, J.; Gonzalez, P. Reproductive cycle and maturity mean size of the red cusk eel, *Genypterus chilensis* (Guichenot, 1881) in the coast off Talcahuano, Chile. *Rev. Biol. Mar. Oceanogr.* **2009**, *44*, 257–262. [CrossRef]
12. Gonzalez, P.; Dettleff, P.; Valenzuela, C.; Estrada, J.M.; Valdes, J.A.; Meneses, C.; Molina, A. Evaluating the genetic structure of wild and commercial red cusk-eel (*Genypterus chilensis*) populations through the development of novel microsatellite markers from a reference transcriptome. *Mol. Biol. Rep.* **2019**, *46*, 5875–5882. [CrossRef] [PubMed]
13. Vega, R.; Estrada, J.M.; Ramirez, D.; Flores, C.; Zamorano, J.; Encina, F.; Mardones, A.; Valdebenito, I.; Dantagnan, P. Growth of cusk eel *Genypterus chilensis* juveniles in culture conditions. *Lat. Am. J. Aquat. Res.* **2015**, *43*, 344–350. [CrossRef]
14. SERNAPESCA. *Anuario Estadístico de Pesca y Acuicultura 2020*; Ministeriode Economía: Santiago, Chile, 2021.
15. Aluru, N.; Vijayan, M.M. Stress transcriptomics in fish: A role for genomic cortisol signaling. *Gen. Comp. Endocrinol.* **2009**, *164*, 142–150. [CrossRef]
16. Ellis, T.; Yildiz, H.Y.; Lopez-Olmeda, J.; Spedicato, M.T.; Tort, L.; Overli, O.; Martins, C.I.M. Cortisol and finfish welfare. *Fish Physiol. Biochem.* **2012**, *38*, 163–188. [CrossRef]
17. Faught, E.; Aluru, N.; Vijayan, M. The molecular stress response. In *Biology of Stress in Fish*; Schreck, C.B., Tort, L., Farrell, A.P., Brauner, C.J., Eds.; Elsevier: Amsterdam, The Netherlands, 2016; Volume 35, pp. 114–149.
18. Logan, C.A.; Somero, G.N. Effects of thermal acclimation on transcriptional responses to acute heat stress in the eurythermal fish *Gillichthys mirabilis* (Cooper). *Am. J. Physiol.-Regul. Integr. Comp. Physiol.* **2011**, *300*, R1373–R1383. [CrossRef]
19. Wu, S.M.; Liu, J.H.; Shu, L.H.; Chen, C.H. Anti-oxidative responses of zebrafish (*Danio rerio*) gill, liver and brain tissues upon acute cold shock. *Comp. Biochem. Phys. A* **2015**, *187*, 202–213. [CrossRef]
20. Yang, C.G.; Jiang, M.; Wen, H.; Tian, J.; Liu, W.; Wu, F.; Gou, G.W. Analysis of differential gene expression under low-temperature stress in Nile tilapia (*Oreochromis niloticus*) using digital gene expression. *Gene* **2015**, *564*, 134–140. [CrossRef]
21. Nakano, T.; Kameda, M.; Shoji, Y.; Hayashi, S.; Yamaguchi, T.; Sato, M. Effect of severe environmental thermal stress on redox state in salmon. *Redox Biol.* **2014**, *2*, 772–776. [CrossRef]
22. Wang, Y.; Liu, Z.; Li, Z.; Shi, H.; Kang, Y.; Wang, J.; Huang, J.; Jiang, L. Effects of heat stress on respiratory burst, oxidative damage and SERPINH1 (HSP47) mRNA expression in rainbow trout *Oncorhynchus mykiss*. *Fish Physiol. Biochem.* **2016**, *42*, 701–710. [CrossRef]
23. Dettleff, P.; Zuloaga, R.; Fuentes, M.; Estrada, J.M.; Molina, A.; Valdes, J.A. Temperature effect on oxidative stress and egg quality-related genes on post-ovulatory eggs and ovary of red cusk-eel (*Genypterus chilensis*). *J. Fish Biol.* **2021**, *98*, 1475–1480. [CrossRef] [PubMed]
24. Dettleff, P.; Zuloaga, R.; Fuentes, M.; González, J.; Aedo, J.; Estrada, J.M.; Molina, A.; Valdés, J.A. Physiological and molecular responses to thermal stress in red cusk-eel (*Genypterus chilensis*) juveniles reveals atrophy and oxidative damage in skeletal muscle. *J. Therm. Biol.* **2020**, *94*, 102750. [CrossRef] [PubMed]
25. Naour, S.; Espinoza, B.M.; Aedo, J.E.; Zuloaga, R.; Maldonado, J.; Bastias-Molina, M.; Silva, H.; Meneses, C.; Gallardo-Escarate, C.; Molina, A.; et al. Transcriptomic analysis of the hepatic response to stress in the red cusk-eel (*Genypterus chilensis*): Insights into lipid metabolism, oxidative stress and liver steatosis. *PLoS ONE* **2017**, *12*, e0176447. [CrossRef] [PubMed]
26. Dettleff, P.; Hormazabal, E.; Aedo, J.; Fuentes, M.; Meneses, C.; Molina, A.; Valdes, J.A. Identification and Evaluation of Long Noncoding RNAs in Response to Handling Stress in Red Cusk-Eel (*Genypterus chilensis*) via RNA-seq. *Mar. Biotechnol.* **2020**, *22*, 94–108. [CrossRef] [PubMed]
27. SHOA. *Servicio Hidrográfico y Oceanográfico de la Armada de Chile 2019*; Servicio Hidrografico y Oceanografico de la Armada: Valparaiso, Chile, 2019.
28. Aedo, J.E.; Maldonado, J.; Estrada, J.M.; Fuentes, E.N.; Silva, H.; Gallardo-Escarate, C.; Molina, A.; Valdes, J.A. Sequencing and de novo assembly of the red cusk-eel (*Genypterus chilensis*) transcriptome. *Mar. Genom.* **2014**, *18*, 105–107. [CrossRef]
29. Love, M.I.; Huber, W.; Anders, S. Moderated estimation of fold change and dispersion for RNA-seq data with DESeq2. *Genome Biol.* **2014**, *15*, 550. [CrossRef]

30. Conesa, A.; Gotz, S.; Garcia-Gomez, J.M.; Terol, J.; Talon, M.; Robles, M. Blast2GO: A universal tool for annotation, visualization and analysis in functional genomics research. *Bioinformatics* **2005**, *21*, 3674–3676. [CrossRef]
31. Moriya, Y.; Itoh, M.; Okuda, S.; Yoshizawa, A.C.; Kanehisa, M. KAAS: An automatic genome annotation and pathway reconstruction server. *Nucleic Acids Res.* **2007**, *35*, W182–W185. [CrossRef]
32. Bustin, S.A.; Benes, V.; Garson, J.A.; Hellemans, J.; Huggett, J.; Kubista, M.; Mueller, R.; Nolan, T.; Pfaffl, M.W.; Shipley, G.L.; et al. The MIQE guidelines: Minimum information for publication of quantitative real-time PCR experiments. *Clin. Chem.* **2009**, *55*, 611–622. [CrossRef]
33. Vandesompele, J.; De Preter, K.; Pattyn, F.; Poppe, B.; Van Roy, N.; De Paepe, A.; Speleman, F. Accurate normalization of real-time quantitative RT-PCR data by geometric averaging of multiple internal control genes. *Genome Biol.* **2002**, *3*, research0034.1. [CrossRef]
34. Doney, S.C.; Fabry, V.J.; Feely, R.A.; Kleypas, J.A. Ocean acidification: The other CO_2 problem. *Annu. Rev. Mar. Sci.* **2009**, *1*, 169–192. [CrossRef] [PubMed]
35. Alimi, O.S.; Farner Budarz, J.; Hernandez, L.M.; Tufenkji, N. Microplastics and Nanoplastics in Aquatic Environments: Aggregation, Deposition and Enhanced Contaminant Transport. *Environ. Sci. Technol.* **2018**, *52*, 1704–1724. [CrossRef] [PubMed]
36. Bhagat, J.; Nishimura, N.; Shimada, Y. Toxicological interactions of microplastics/nanoplastics and environmental contaminants: Current knowledge and future perspectives. *J. Hazard. Mater.* **2021**, *405*, 123913. [CrossRef] [PubMed]
37. Aballai, V.; Aedo, J.E.; Maldonado, J.; Bastias-Molina, M.; Silva, H.; Meneses, C.; Boltana, S.; Reyes, A.; Molina, A.; Valdes, J.A. RNA-seq analysis of the head-kidney transcriptome response to handling-stress in the red cusk-eel (*Genypterus chilensis*). *Comp. Biochem. Phys. D* **2017**, *24*, 111–117. [CrossRef]
38. Aedo, J.E.; Maldonado, J.; Aballai, V.; Estrada, J.M.; Bastias-Molina, M.; Meneses, C.; Gallardo-Escarate, C.; Silva, H.; Molina, A.; Valdes, J.A. mRNA-seq reveals skeletal muscle atrophy in response to handling stress in a marine teleost, the red cusk-eel (*Genypterus chilensis*). *BMC Genom.* **2015**, *16*, 1024. [CrossRef]
39. Pakhira, C.; Nagesh, T.S.; Abraham, T.J.; Dash, G.; Behera, S. Stress responses in rohu, *Labeo rohita* transported at different densities. *Aquac. Rep.* **2015**, *2*, 39–45. [CrossRef]
40. Kumar, N.; Krishnani, K.K.; Singh, N.P. Oxidative and Cellular Metabolic Stress of Fish: An Appealing Tool for Biomonitoring of Metal Contamination in the Kolkata Wetland, a Ramsar Site. *Arch. Environ. Contam. Toxicol.* **2019**, *76*, 469–482. [CrossRef]
41. Veisi, S.; Sarkheil, M.; Johari, S.A.; Safari, O. Dietary supplementation with melatonin: Influence on growth performance, oxidative stress status, and amelioration of silver nanoparticles-induced toxicity in Nile tilapia (*Oreochromis niloticus*). *Trop. Anim. Health Prod.* **2021**, *53*, 314. [CrossRef]
42. Firat, O.; Tutus, R. Comparative Acute Toxicity Assessment of Organophosphate and Avermectin Insecticides on a Freshwater Fish *Oreochromis niloticus*. *Bull. Environ. Contam. Toxicol.* **2020**, *105*, 582–587. [CrossRef]
43. Bharti, S.; Rasool, F. Analysis of the biochemical and histopathological impact of a mild dose of commercial malathion on *Channa punctatus* (Bloch) fish. *Toxicol. Rep.* **2021**, *8*, 443–455. [CrossRef]
44. Cheng, C.H.; Guo, Z.X.; Luo, S.W.; Wang, A.L. Effects of high temperature on biochemical parameters, oxidative stress, DNA damage and apoptosis of pufferfish (*Takifugu obscurus*). *Ecotoxicol. Environ. Saf.* **2018**, *150*, 190–198. [CrossRef] [PubMed]
45. Yang, S.; Zhao, T.; Ma, A.; Huang, Z.; Liu, Z.; Cui, W.; Zhang, J.; Zhu, C.; Guo, X.; Yuan, C. Metabolic responses in *Scophthalmus maximus* kidney subjected to thermal stress. *Fish Shellfish Immunol.* **2020**, *103*, 37–46. [CrossRef] [PubMed]
46. Das, A. Heat stress-induced hepatotoxicity and its prevention by resveratrol in rats. *Toxicol. Mech. Methods* **2011**, *21*, 393–399. [CrossRef] [PubMed]
47. Pizzino, G.; Irrera, N.; Cucinotta, M.; Pallio, G.; Mannino, F.; Arcoraci, V.; Squadrito, F.; Altavilla, D.; Bitto, A. Oxidative Stress: Harms and Benefits for Human Health. *Oxid. Med. Cell. Longev.* **2017**, *2017*, 8416763. [CrossRef]
48. Betteridge, D.J. What is oxidative stress? *Metab. Clin. Exp.* **2000**, *49*, 3–8. [CrossRef]
49. Lesser, M.P. Oxidative stress in marine environments: Biochemistry and physiological ecology. *Annu. Rev. Physiol.* **2006**, *68*, 253–278. [CrossRef]
50. Madeira, D.; Narciso, L.; Cabral, H.N.; Vinagre, C.; Diniz, M.S. Influence of temperature in thermal and oxidative stress responses in estuarine fish. Comparative Biochemistry and Physiology. *Part A Mol. Integr. Physiol.* **2013**, *166*, 237–243. [CrossRef]
51. Vinagre, C.; Madeira, D.; Narciso, L.; Cabral, H.N.; Diniz, M.S. Effect of temperature on oxidative stress in fish: Lipid peroxidation and catalase activity in the muscle of juvenile seabass, *Dicentrarchus labrax*. *Ecol. Indic.* **2012**, *23*, 274–279. [CrossRef]
52. Chang, C.H.; Mayer, M.; Rivera-Ingraham, G.; Blondeau-Bidet, E.; Wu, W.Y.; Lorin-Nebel, C.; Lee, T.H. Effects of temperature and salinity on antioxidant responses in livers of temperate (*Dicentrarchus labrax*) and tropical (*Chanos Chanos*) marine euryhaline fish. *J. Therm. Biol.* **2021**, *99*, 103016. [CrossRef]
53. Ale, A.; Bacchetta, C.; Rossi, A.S.; Scarabotti, P.A.; Cazenave, J. Low temperature stress in a cultured fish (*Piaractus mesopotamicus*) fed with *Pyropia columbina* red seaweed-supplemented diet. *Fish Physiol. Biochem.* **2021**, *47*, 829–839. [CrossRef]
54. Bacchetta, C.; Ale, A.; Rossi, A.S.; Karakachoff, M.; Cazenave, J. Effects of cold stress on juvenile *Piaractus mesopotamicus* and the mitigation by beta-carotene. *J. Therm. Biol.* **2020**, *88*, 102497. [CrossRef] [PubMed]
55. Castro, C.; Perez-Jimenez, A.; Guerreiro, I.; Peres, H.; Castro-Cunha, M.; Oliva-Teles, A. Effects of temperature and dietary protein level on hepatic oxidative status of Senegalese sole juveniles (*Solea senegalensis*). *Comp. Biochem. Physiol. Part A Mol. Integr. Physiol.* **2012**, *163*, 372–378. [CrossRef] [PubMed]

56. Roychowdhury, P.; Aftabuddin, M.; Pati, M.K. Thermal stress-induced oxidative damages in the liver and associated death in fish, *Labeo rohita*. *Fish Physiol. Biochem.* **2021**, *47*, 21–32. [CrossRef] [PubMed]
57. Lopez-Anido, R.N.; Harrington, A.M.; Hamlin, H.J. Coping with stress in a warming Gulf: The postlarval American lobster's cellular stress response under future warming scenarios. *Cell Stress Chaperones* **2021**, *26*, 721–734. [CrossRef]
58. An, K.W.; Kim, N.N.; Shin, H.S.; Kil, G.S.; Choi, C.Y. Profiles of antioxidant gene expression and physiological changes by thermal and hypoosmotic stresses in black porgy (*Acanthopagrus schlegeli*). *Comp. Biochem. Physiol. Part A Mol. Integr. Physiol.* **2010**, *156*, 262–268. [CrossRef]
59. Cheng, C.H.; Yang, F.F.; Liao, S.A.; Miao, Y.T.; Ye, C.X.; Wang, A.L.; Tan, J.W.; Chen, X.Y. High temperature induces apoptosis and oxidative stress in pufferfish (*Takifugu obscurus*) blood cells. *J. Therm. Biol.* **2015**, *53*, 172–179. [CrossRef]
60. Ho, T.N.; Paul, G.V.; Chen, Y.H.; Hsu, T. Heat stress upregulates G-T mismatch binding activities in zebrafish (*Danio rerio*) embryos preexposed and nonexposed to a sublethal level of cadmium (Cd). *Chemosphere* **2019**, *218*, 179–188. [CrossRef]
61. Baker, B.P.; Van Wie, I.; Braun, E.; Jimenez, A.G. Thermal stability vs. variability: Insights in oxidative stress from a eurytolerant fish. *Comp. Biochem. Physiol. Part A Mol. Integr. Physiol.* **2020**, *249*, 110767. [CrossRef]
62. Anelli, T.; Sitia, R. Protein quality control in the early secretory pathway. *EMBO J.* **2008**, *27*, 315–327. [CrossRef]
63. Malini, M.K.; Lekshmy, V.S.; Pal, M.; Chinnusamy, V.; Nagaraj Kumar, M. Unfolded protein response (UPR) mediated under heat stress in plants. *Plant Physiol. Rep.* **2020**, *25*, 569–582. [CrossRef]
64. Faoro, C.; Ataide, S.F. Noncanonical Functions and Cellular Dynamics of the Mammalian Signal Recognition Particle Components. *Front. Mol. Biosci.* **2021**, *8*, 679584. [CrossRef] [PubMed]
65. Harding, H.P.; Zhang, Y.H.; Ron, D. Protein translation and folding are coupled by an endoplasmic-reticulum-resident kinase. *Nature* **1999**, *397*, 271–274. [CrossRef] [PubMed]
66. Vattem, K.M.; Wek, R.C. Reinitiation involving upstream ORFs regulates ATF4 mRNA translation in mammalian cells. *Proc. Natl. Acad. Sci. USA* **2004**, *101*, 11269–11274. [CrossRef] [PubMed]
67. Adachi, M.; Liu, Y.; Fujii, K.; Calderwood, S.K.; Nakai, A.; Imai, K.; Shinomura, Y. Oxidative Stress Impairs the Heat Stress Response and Delays Unfolded Protein Recovery. *PLoS ONE* **2009**, *4*, A57–A66. [CrossRef]
68. Ahmed, K.; Zaidi, S.F.; Mati-ur-Rehman; Rehman, R.; Kondo, T. Hyperthermia and protein homeostasis: Cytoprotection and cell death. *J. Therm. Biol.* **2020**, *91*, 102615. [CrossRef]
69. Fe-Goncalves, L.M.; Araujo, J.D.A.; dos Santos, C.H.D.; Val, A.L.; de Almeida-Val, V.M.F. How will farmed populations of freshwater fish deal with the extreme climate scenario in 2100? Transcriptional responses of *Colossoma macropomum* from two Brazilian climate regions. *J. Therm. Biol.* **2020**, *89*, 102487. [CrossRef]
70. Mininni, A.N.; Milan, M.; Ferraresso, S.; Petochi, T.; Di Marco, P.; Marino, G.; Livi, S.; Romualdi, C.; Bargelloni, L.; Patarnello, T. Liver transcriptome analysis in gilthead sea bream upon exposure to low temperature. *BMC Genom.* **2014**, *15*, 765. [CrossRef]
71. Feder, M.E.; Hofmann, G.E. Heat-shock proteins, molecular chaperones and the stress response: Evolutionary and ecological physiology. *Annu. Rev. Physiol.* **1999**, *61*, 243–282. [CrossRef]
72. Dammark, K.B.; Ferchaud, A.L.; Hansen, M.M.; Sorensen, J.G. Heat tolerance and gene expression responses to heat stress in threespine sticklebacks from ecologically divergent environments. *J. Therm. Biol.* **2018**, *75*, 88–96. [CrossRef]
73. Corey, E.; Linnansaari, T.; Cunjak, R.A.; Currie, S. Physiological effects of environmentally relevant, multi-day thermal stress on wild juvenile Atlantic salmon (*Salmo salar*). *Conserv. Physiol.* **2017**, *5*, cox014. [CrossRef]
74. Hori, T.S.; Gamperl, A.K.; Afonso, L.O.; Johnson, S.C.; Hubert, S.; Kimball, J.; Bowman, S.; Rise, M.L. Heat-shock responsive genes identified and validated in Atlantic cod (*Gadus morhua*) liver, head kidney and skeletal muscle using genomic techniques. *BMC Genom.* **2010**, *11*, 72. [CrossRef] [PubMed]
75. Liu, B.; Xu, P.; Brown, P.B.; Xie, J.; Ge, X.P.; Miao, L.H.; Zhou, Q.L.; Ren, M.C.; Pan, L.K. The effect of hyperthermia on liver histology, oxidative stress and disease resistance of the Wuchang bream, *Megalobrama amblycephala*. *Fish Shellfish Immunol.* **2016**, *52*, 317–324. [CrossRef] [PubMed]
76. Nakai, A.; Satoh, M.; Hirayoshi, K.; Nagata, K. Involvement of the stress protein HSP47 in procollagen processing in the endoplasmic reticulum. *J. Cell Biol.* **1992**, *117*, 903–914. [CrossRef] [PubMed]
77. Pandey, A.; Rajesh, M.; Baral, P.; Sarma, D.; Tripathi, P.H.; Akhtar, M.S.; Ciji, A.; Dubey, M.K.; Pande, V.; Sharma, P.; et al. Concurrent changes in thermal tolerance thresholds and cellular heat stress response reveals novel molecular signatures and markers of high temperature acclimation in rainbow trout. *J. Therm. Biol.* **2021**, *102*, 103124. [CrossRef]
78. Long, Y.; Li, L.; Li, Q.; He, X.; Cui, Z. Transcriptomic characterization of temperature stress responses in larval zebrafish. *PLoS ONE* **2012**, *7*, e37209. [CrossRef] [PubMed]

biology

MDPI

Article

Targeting the Mild-Hypoxia Driving Force for Metabolic and Muscle Transcriptional Reprogramming of Gilthead Sea Bream (*Sparus aurata*) Juveniles

Fernando Naya-Català [1], Juan A. Martos-Sitcha [1,2], Verónica de las Heras [1], Paula Simó-Mirabet [1], Josep À. Calduch-Giner [1] and Jaume Pérez-Sánchez [1,*]

[1] Nutrigenomics and Fish Growth Endocrinology Group, Institute of Aquaculture Torre de la Sal, CSIC, 12595 Ribera de Cabanes, Spain; fernando.naya@iats.csic.es (F.N.-C.); juanantonio.sitcha@uca.es (J.A.M.-S.); veronica.delasher@alum.uca.es (V.d.l.H.); paula.simo@csic.es (P.S.-M.); j.calduch@csic.es (J.À.C.-G.)

[2] Department of Biology, Faculty of Marine and Environmental Sciences, Instituto Universitario de Investigación Marina (INMAR), Campus de Excelencia Internacional del Mar (CEI-MAR), University of Cádiz, 11519 Cádiz, Spain

* Correspondence: jaime.perez.sanchez@csic.es

Simple Summary: Reduced oxygen availability generates a number of adaptive features across all the animal kingdom, and the goal of this study was targeting the mild-hypoxia driving force for metabolic and muscle transcriptional reprogramming of gilthead sea bream juveniles. Attention was focused on blood metabolic and muscle transcriptomic landmarks before and after exhaustive exercise. Our results after mild-hypoxia conditioning highlighted an increased contribution of lipid metabolism to whole energy supply to preserve the aerobic energy production, a better swimming performance regardless of changes in feed intake, as well as reduced protein turnover and improved anaerobic fitness with the restoration of normoxia.

Abstract: On-growing juveniles of gilthead sea bream were acclimated for 45 days to mild-hypoxia (M-HYP, 40–60% O_2 saturation), whereas normoxic fish (85–90% O_2 saturation) constituted two different groups, depending on if they were fed to visual satiety (control fish) or pair-fed to M-HYP fish. Following the hypoxia conditioning period, all fish were maintained in normoxia and continued to be fed until visual satiation for 3 weeks. The time course of hypoxia-induced changes was assessed by changes in blood metabolic landmarks and muscle transcriptomics before and after exhaustive exercise in a swim tunnel respirometer. In M-HYP fish, our results highlighted a higher contribution of aerobic metabolism to whole energy supply, shifting towards a higher anaerobic fitness following normoxia restoration. Despite these changes in substrate preference, M-HYP fish shared a persistent improvement in swimming performance with a higher critical speed at exercise exhaustion. The machinery of muscle contraction and protein synthesis and breakdown was also largely altered by mild-hypoxia conditioning, contributing this metabolic re-adjustment to the positive regulation of locomotion and to the catch-up growth response during the normoxia recovery period. Altogether, these results reinforce the presence of large phenotypic plasticity in gilthead sea bream, and highlights mild-hypoxia as a promising prophylactic measure to prepare these fish for predictable stressful events.

Keywords: hypoxia; hypo-metabolic state; growth; swimming performance; metabolic landmarks; muscle transcriptome; glycolysis; lipid metabolism; protein turnover; gilthead sea bream

Citation: Naya-Català, F.; Martos-Sitcha, J.A.; de las Heras, V.; Simó-Mirabet, P.; Calduch-Giner, J.À.; Pérez-Sánchez, J. Targeting the Mild-Hypoxia Driving Force for Metabolic and Muscle Transcriptional Reprogramming of Gilthead Sea Bream (*Sparus aurata*) Juveniles. *Biology* **2021**, *10*, 416. https://doi.org/10.3390/biology10050416

Academic Editor: M. Gonzalo Claros

Received: 25 February 2021
Accepted: 4 May 2021
Published: 8 May 2021

1. Introduction

Reduced oxygen (O_2) availability generates physiological and anatomical changes that increase ventilation rates, erythropoiesis, and tissue vascularization, with a decrease in muscle oxidative capacity and a switch in substrate preference towards more O_2-efficient fuels [1–3]. These adaptive features occur across all the animal kingdom, contributing to epigenetic mechanisms to depress metabolic rates when individuals are facing predictable seasonal signals (hibernation, cold hardening, or diapause) or unpredictable episodic stresses, such as hypoxia, desiccation, or traumatic surgical situations [4–6]. Besides, epigenetics allows pre-programming of offspring to high-altitude hypoxic environments by imprinting genes at the embryonic or placental interface, resulting in transgenerational and/or intra-generational heritable changes that affect gene expression [7,8].

Hypoxia is also a common stressor in aquatic ecosystems [9–11] and "dead zones" expand rapidly in oceans as climate emergency causes unprecedented O_2 losses [12,13]. This will have a strong negative impact on fisheries and aquaculture production [14], and future selective breeding will need to be directed towards more robust and resilient farmed fish to mitigate the effects of climate change [15,16]. The first sign of a mismatch between O_2 supply and demand is the reduction of appetite, varying the O_2 threshold level for maximal feed intake in Atlantic salmon (*Salmo salar*) and gilthead sea bream (*Sparus aurata*) between 40% and 75% saturation within the range of temperature tolerance [17,18]. This threshold level is decreased at high stocking densities [19–21], probably due to the unbalanced production and scavenging of reactive O_2 species (ROS) [22]. However, acclimation to one stressor can also improve the capacity to cope with another critical co-occurring stressor, and warm acclimation improves the hypoxia tolerance in Atlantic killifish (*Fundulus heteroclitus*) [23]. Meanwhile, cold exposure also facilitates hypoxia adaptation, as the reduction of metabolic rates is likely accompanied by a reduction in mitochondrial O_2 use [24]. These different metabolic strategies to cope with changing temperature and reduced O_2 availability are also evidenced on a seasonal and developmental basis [25]. Thus, European sea bass (*Dicentrarchus labrax*) cope with moderate hypoxia at the expenses of a delayed larval maturation of digestive function [26]. Likewise, early acute hypoxia has transgenerational impairment effects on the reproductive performance of medaka (*Oryzias latipes*) [27]. By contrast, mild-hypoxia exposure during the embryonic development of zebrafish (*Danio rerio*) is protective against severe hypoxia insults later in life [28].

It is important to note that hypoxia acclimation affects endurance training in athletes and other animal models, including fish, usually via increased O_2 uptake capacity and aerobic metabolic capacity [29–31]. In juveniles and fingerlings of farmed gilthead sea bream, successful adaption to severe and moderate hypoxia has been demonstrated to occur by the induction of hypo-metabolic states, increased O_2-mitochondria affinity, and/or aerobic/anaerobic metabolic switches in substrate preference as metabolic fuels [20,32–34]. Thus, the goal of the present study is to underline new insights on the mild-hypoxia driving force for reprograming growth and swimming performance of on-growing juveniles of gilthead sea bream in order to prepare individuals to better respond to predictable stresses. For this purpose, the time course of metabolic responses after mild-hypoxia conditioning and normoxia recovery periods was assessed by changes in blood metabolic landmarks and muscle transcriptomics before and after exhaustive exercise in a swim tunnel respirometer.

2. Materials and Methods

2.1. Ethics Statement

All procedures were approved by the Ethics and Animal Welfare Committees of the Institute of Aquaculture Torre de la Sal (IATS) and CSIC. The study was conducted in the IATS's registered aquaculture infrastructure facility (code ES120330001055), in accordance with the principles published in the European Animal Directive (2010/63/EU) and Spanish laws (Royal Decree RD53/2013) for the protection of animals used in scientific experiments.

2.2. Experimental Setup of Hypoxia Conditioning

Gilthead sea bream juveniles of Atlantic origin (Ferme Marine du Douhet, Bordeaux, France) were reared from early life stages (3–5 g initial body weight) in the indoor experimental facilities of the Institute of Aquaculture Torre de la Sal (IATS, CSIC, Spain) under the natural photoperiod and temperature conditions at our latitude (40°5′ N; 0°10′ E). In June

2017, fish of 21–28 g body weight were randomly distributed in twelve 90 L tanks (n = 20), connected to two separated recirculating aquaculture systems (RAS) with regulation of the water temperature (24–26 °C) and O_2 concentration (Supplementary Figure S1). As shown in Figure 1, fish were allowed to acclimate to experimental tanks for 5 days before any manipulation of O_2 concentration, keeping the unionized ammonia below 0.02 mg/L. After this acclimation period, the O_2 concentration of one RAS (six 90 L tanks) was ramped through 20 h to achieve a mild-hypoxia condition (M-HYP: 3–4 ppm, 40–60% O_2 saturation), according to the values of limiting oxygen saturation (LOS, defined as O_2 levels where the maximal metabolic rates start to decrease with further reduction in dissolved O_2) reported for this fish species [17] at a given temperature. The remaining fish, coupled to a second RAS, were maintained under normoxic conditions (5.5–6 ppm, 85–90% O_2 saturation). These fish constituted two different normoxic groups, depending on if they were fed to visual satiation (N) or pair-fed (N-PF) to the M-HYP group, fed to visual satiation, with a commercial diet (EFICO YM 853 3 mm, BioMar, Palencia, Spain) once daily (12:00 a.m., six days per week).

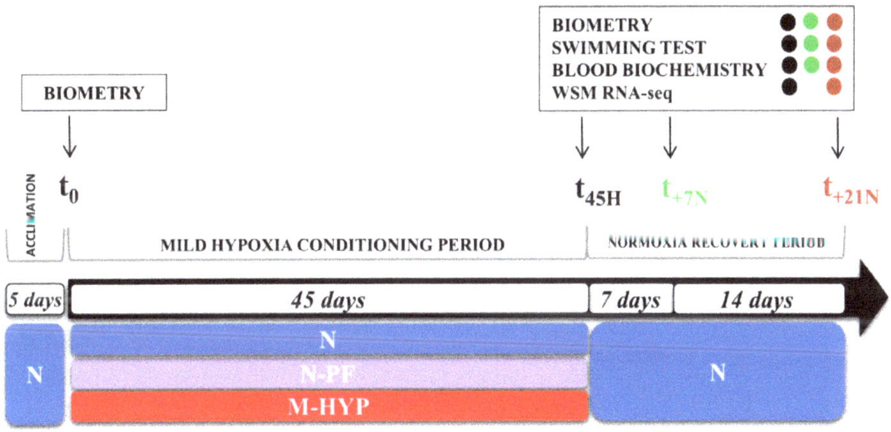

Figure 1. Experimental setup showing the timing and type of data recorded at each sampling point.

After 45 days of mild-hypoxia conditioning (t_{45H}), 12 overnight-fasted fish per experimental condition were randomly selected and anesthetized (between 10:00 and 12:00 a.m.) with 100 mg/L 3-aminobenzoic acid ethyl ester (MS-222, Sigma, Saint Louis, MO, USA). Blood was taken from caudal vessels with heparinized syringes in less than 3 min for all the fish from the same tank. The haematocrit (Ht) and haemoglobin concentration (Hb) were determined in fresh samples. The remaining blood was centrifuged at 3000× *g* for 20 min at 4 °C, and plasma samples were frozen and stored at −20 °C until biochemical and hormonal analyses were performed. Prior to skeletal muscle collection, fish were killed by cervical section and representative portions of the dorsal tissue were excised and immediately snap-frozen in liquid nitrogen and stored at −80 °C until extraction of total RNA and tissue lactate quantification. At this stage, 6–7 additional fish per experimental condition were used for swim tests (see Section 2.3), and blood and skeletal muscle were rapidly taken from exhausted fish for biochemical and transcriptomic analyses. The remaining fish were kept under normoxia and continued to be fed until visual satiation for three additional weeks, which constituted the normoxia restoration period with additional sampling points at Week 1 (t_{+7N}; 6 fish) and Week 3 (t_{+21N}; 7 fish) for swim tests as well as biochemical and transcriptomic analyses. Data on body weight were retrieved for all fish at t_0, t_{45H}, t_{+7N}, and t_{+21N} (see Figure 1).

2.3. *Swim Tunnel Respirometer*

Fish were exercised during hypoxia and normoxia restoration periods in an intermittent-closed swim tunnel respirometer of 10 L water volume (Loligo®Systems, Viborg, Denmark), as reported elsewhere [20,31]. To ensure a high water quality, the water bath was connected to a RAS with water temperature and O_2 concentration set at 26 ± 0.5 °C and 60% saturation (4 ppm), respectively. For the testing procedures, slightly anesthetized fish were transferred into the swim tunnel, after obtaining their biometrical parameters, and recovered and acclimated at a swimming speed of 0.5–1.0 body lengths per second (BL/s). Acclimation was achieved when the O_2 consumption rates (MO_2) reached a constant low plateau, which typically happened after 30–45 min with an MO_2 around 220–240 mgO_2/kg/h [35]. After this acclimation period, the water velocity was increased in 0.5 BL/s steps, and fish were submitted to controlled speeds until exhaustion. Each swimming interval at a given velocity lasted 5 min, consisting of "flush–wait–measurement" cycles (60 s flush interval to exchange the respirometer water = "flush"; 30 s mixing phase in closed mode = "wait"; and a 210 s MO_2 measuring period in closed mode = "measurement"). During the measurement interval, O_2 saturation of the swim tunnel water was recorded every second. MO_2 was automatically calculated by the AutoResp™ software from linear decreases (r^2 = 0.98–1.0) in chamber O_2 saturation during the measurement period at each discrete and specific speed, using the appropriate constants for O_2 solubility in seawater (salinity, temperature, and barometric pressure).

2.4. *Blood Biochemistry*

Haemoglobin (Hb) concentration was determined with a HemoCue B-Haemoglobin Analyser®(AB, Leo Diagnostic, Sweden). The haematocrit (Hc) was measured after centrifugation of blood in heparinized capillary tubes at 13,000× *g* for 10 min in a Sigma 1-14 centrifuge (Sigma). Blood lactate was measured in deproteinized samples (8% perchloric acid) by an enzymatic method based on the use of lactate oxidase and peroxidase (Ref. 1001330; SpinReact S.A., Girona, Spain). The same kit was used to determine muscular lactate concentrations after mincing and homogenization of samples by mechanic disruption in 7.5 volumes ice-cold 0.6 N perchloric acid, neutralized using 1 M KCO_3, and centrifuged at 3000× *g* for 30 min at 4 °C. Plasma glucose was determined by the glucose oxidase method (ThermoFisher Scientific, Waltham, MA, USA) according to the manufacturer's instructions. Plasma triglycerides (TAGs) were determined using lipase/glycerol kinase/glycerol-3-phosphate oxidase reagent. Plasma free fatty acids (FFA) were analysed using a commercial enzymatic method (NEFA-C, Wako Test, Neuss, Germany). Plasma cortisol levels were measured with a commercial Cortisol Enzyme Immunoassay Kit from Arbor Assays™ (NCal™ International Standard Kit, DetectX®, K003; Ann Arbor, MI, USA), following the manufacturer's instructions. Plasma growth hormone (Gh) was determined by a homologous gilthead sea bream radioimmunoassay (RIA) [36]. Plasma insulin-like growth factor-1 (Igf-1) was extracted by acid-ethanol cryoprecipitation, and its concentration was determined by means of a generic fish Igf-1 RIA validated for Mediterranean perciform fish [37].

2.5. *Illumina Sequencing and Sample Quality Assessment*

Total RNA from tissue portions of white skeletal muscle was extracted using the MagMAX™-96 for Microarrays total RNA isolation kit (Life Technologies, Carlsbad, CA, USA). The quality and integrity of the isolated RNA was checked on an Agilent Bioanalyzer 2100 total RNA Nano series II chip (Agilent, Santa Clara, CA, USA) with RIN (RNA Integrity Number) values varying between 8 and 10. Illumina RNA-seq libraries were prepared from 500 ng total RNA using the Illumina TruSeq™ Stranded mRNA LT Sample Prep Kit (Illumina Inc. San Diego, CA, USA) according to the manufacturer's instructions. All RNA-seq libraries were sequenced on an Illumina HiSeq2500 sequencer as a 1 × 75 nucleotides single-end (SE) read format, according to the manufacturer's protocol. Raw sequenced data were deposited in the Sequence Read Archive (SRA) of the National Center for

Biotechnology Information (NCBI) under the Bioproject accession number PRJNA679473 (BioSample accession numbers: SAMN16834555-597). Approximately 882 million SE reads were obtained from the 50 samples sequenced, with an average of ~18 million reads per sample. Quality analysis was performed with FASTQC v0.11.7 (https://www.bioinformatics.babraham.ac.uk/projects/fastqc/ accessed on 27 April 2019), and libraries were filtered with Prinseq [38] for quality > 28 and < 5% of Ns in the sequence. Then, libraries were mapped and annotated using TopHat2 [39] and the gilthead sea bream draft genome as reference [40]. A representative transcriptome per sample was constructed using Cufflinks, with the data quality checked with CummeRbund [41].

2.6. Statistics

Changes in the growth performance and blood parameters through all the experiment were analysed by t-test or one-way ANOVA followed by Student–Newman–Keuls post-test. At t_{45H} and t_{+21N}, the differentially expressed (DE) genes were retrieved with normalized fragment per kilobase per million (FPKM) values using Cuffdiff [41], with false discovery rates (FDR) adjustment with a cut-off of 0.05. To increase the number of DE genes without loss of statistical robustness, supervised partial least-squares discriminant analysis (PLS-DA) and hierarchical clustering of samples were sequentially applied using EZinfo v3.0 (Umetrics, Umea, Sweden) and the R package gplots, respectively. The genes included in this analysis were filtered by ANOVA p-values < 0.05. The final list of genes contributing to group separation was determined by the minimum Variable Importance in the Projection (VIP) values [42,43], driving the right clustering of all individuals in the heatmap analysis. To discard the possibility of over-fitting of the supervised discriminant model, a validation test consisting in 500 random permutations was performed using SIMCA-P+ v11.0 (Umetrics). The heatmap representation was constructed using the average linkage method and Euclidean distance.

Genes above the VIP threshold were analysed for gene ontology (GO) with the R package ShinyGO v0.61 [44], after conversion of the gilthead sea bream annotated sequences to human equivalents. Significantly enriched GO categories were obtained after FDR correction using a cut-off of 0.05. The list of genes associated with enriched GO terms was introduced in the Search Tool for the Retrieval of Interacting Genes (STRING v.11) database [45]. Functional protein–protein association networks were considered statistically significant at FDR p-values < 0.05 and a high confidence score of 0.7. The tools used for the sequencing quality analysis, cleaning, mapping, transcriptome assembly, and differential gene expression are contained in the GPRO suite [46].

3. Results

3.1. Growth Performance during Mild-Hypoxia and Normoxia Restoration

Control fish (N group) grew during the hypoxia conditioning period (45 days), from 24 g to 79 g, at high specific growth rates (SGR = 2.59) for this species and class of size. Feed intake (g dry matter/fish) was reduced by 25% in fish exposed to mild-hypoxia, whereas the feed conversion ratio (FCR = dry feed intake/wet weigh gain) remained within optimum levels (0.98–0.95) in both N and M-HYP fish (Table 1). Normoxic pair-fed fish (N-PF) also grew efficiently (FCR = 0.96) at the same growth rate than M-HYP fish (SGR= 2.24–2.25%). During the first week of the subsequent normoxic and unrestricted feeding period (t_{45H}–t_{+7N}), the growth performance parameters remained similar in all groups, although during the last two weeks (t_{+7N}–t_{+21N}), growth rates of N-PF and M-HYP fish were higher ($p < 0.001$) than in control fish, as denoted by the SGR values (2.26 and 2.19 vs. 1.79, respectively), helping to compensate, at least in part, the initial growth impairment. Furthermore, the feed conversion ratio was improved to some extent ($p = 0.103$), with the achieved FCR varying between 1.21 in N fish to 1.13–1.14 in M-HYP/N-PF fish.

Table 1. Effects of mild-hypoxia conditioning on the growth performance of gilthead sea bream juveniles in a 45-day trial followed by a normoxia recovery period of 21 days. Values are the mean ± SEM of triplicate tanks. The *p*-values are the result of one-way ANOVA. Different superscript letters indicate significant differences between the experimental groups (SNK test, $p < 0.05$).

	N	N-PF	M-HYP	*p*-Value
Mild-hypoxia conditioning (t_0–t_{45H})				
Initial body weight (g)	24.58 ± 0.11	24.1 ± 0.10	24.19 ± 0.03	0.112
Final body weight (g)	78.69 ± 0.79 ^{b}	66.13 ± 1.41 ^{a}	66.06 ± 0.97 ^{a}	<0.001
Feed intake (g DM/fish)	53.36 ± 0.15 ^{b}	40.77 ± 0.22 ^{a}	40.08 ± 0.84 ^{a}	<0.001
Weight gain (%) [1]	220.30 ± 2.03 ^{b}	174.51 ± 4.50 ^{a}	173.22 ± 3.86 ^{a}	<0.001
SGR (%) [2]	2.59 ± 0.01 ^{b}	2.24 ± 0.04 ^{a}	2.23 ± 0.03 ^{a}	<0.001
FCR (%) [3]	0.98 ± 0.009	0.96 ± 0.02	0.95 ± 0.008	0.285
Normoxia recovery period (t_{+7N}–t_{+21N})				
Initial body weight (g)	98.76 ± 1.20 ^{b}	83.50 ± 0.50 ^{a}	82.00 ± 1.14 ^{a}	<0.001
Final body weight (g)	126.5 ± 1.30 ^{b}	114.7 ± 0.33 ^{a}	111.3 ± 1.81 ^{a}	0.001
Feed intake (g DM/fish)	37.62 ± 1.28	36.57 ± 1.50	35.66 ± 0.50	0.329
Weight gain (%) [1]	28.54 ± 0.46 ^{a}	37.22 ± 1.33 ^{b}	35.50 ± 0.85 ^{b}	0.001
SGR (%) [2]	1.79 ± 0.03 ^{a}	2.26 ± 0.07 ^{b}	2.19 ± 0.04 ^{b}	<0.001
FCR (%) [3]	1.21 ± 0.03	1.12 ± 0.02	1.14 ± 0.02	0.103

[1] Weight gain (%) = (100 × body weigh increase)/initial body weight; [2] Specific growth rate = 100 × (ln final body weight − ln initial body weight)/days; [3] Feed conversion ratio = dry feed intake/wet weight gain.

3.2. Blood Patterns at the End of the Mild-Hypoxia Conditioning Period

Data on blood haematology and biochemistry in free-swimming fish at t_{45H} are shown in Table 2. In this fish group, the reduction of feed intake was associated to lowered ($p = 0.011$) Hb concentrations in N-PF fish, but control values were restored with the combined reduction of feed intake and O_2 availability in M-HYP fish. Circulating levels of lactate were lowered in both N-PF and M-HYP fish, with the lowest levels in fish exposed to a low O_2 concentration ($p < 0.001$). In contrast, feed intake and O_2 availability showed an opposite effect on plasma levels of FFAs, achieving the highest concentrations in N-PF fish and the lowest in M-HYP fish ($p = 0.029$). No statistically significant differences were found in the other analysed parameters (Hc, glucose, TAGs, cortisol, Gh, Ifg-1), but the calculated Gh/Igf-1 ratio increased significantly ($p < 0.05$) from 0.13 in N fish to 0.25 in M-HYP fish.

Table 2. Effects of mild-hypoxia conditioning on blood haematology and blood biochemistry of gilthead sea bream juveniles. Values are the mean ± SEM of 6–10 fish (2–3 fish per replicate tank). The *p*-values are the result of one-way ANOVA. Different superscript letters indicate significant differences between the experimental groups (SNK test, $p < 0.05$).

	N	N-PF	M-HYP	*p*-Value
Haemoglobin (g/dL)	8.36 ± 0.38 ^{b}	6.43 ± 0.64 ^{a}	7.88 ± 0.22 ^{b}	0.011
Haematocrit (%)	34.7 ± 1.24	33.7 ± 0.99	31.0 ± 1.41	0.175
Lactate (mg/dL)	14.1 ± 0.15 ^{b}	6.32 ± 0.57 ^{a}	4.18 ± 0.77 ^{a}	<0.001
Glucose (mg/dL)	57.1 ± 5.98	55.7 ± 2.29	56.8 ± 2.35	0.493
Triglycerides (mg/dL)	2.80 ± 0.28	4.02 ± 0.34	3.02 ± 0.46	0.128
Free fatty acids (nmol/μL)	0.426 ± 0.052 ^{ab}	0.595 ± 0.045 ^{b}	0.388 ± 0.045 ^{a}	0.029
Cortisol (ng/mL)	24.1 ± 5.43	29.3 ± 10.56	14.3 ± 4.71	0.270
Growth hormone (ng/mL)	9.19 ± 3.94	12.4 ± 5.30	13.9 ± 4.87	0.752
Insulin-like growth factor-1 (ng/mL)	69.3 ± 5.74	60.5 ± 3.42	55.5 ± 3.94	0.285
Gh/Igf-1	0.13 ± 0.058 ^{a}	0.20 ± 0.081 ^{ab}	0.25 ± 0.041 ^{b}	0.032

3.3. Swim Tests: Critical Swimming and Blood Patterns after Exhaustive Exercise

Results of the swim tests at different times over the experimental period (t_{45H}, t_{+7N} and t_{+21N}) are shown in Figure 2. Overall, MO_2 increased linearly with the increase of water speed until a maximum metabolic rate (MMR) that varied non-significantly between 400 and 357 $mgO_2/kg/h$. Then, fish of all experimental groups showed a sharp decrease in O_2 consumption until being exhausted at their own critical speed (U_{crit}). At t_{45H}, the achieved U_{crit} was higher ($p < 0.01$) in M-HYP fish (7.6 BL/s) than in the other two experimental groups that shared undistinguishable critical swimming (6.8–6.9 BL/s) (Figure 2A). The subsequent swim tests at t_{+7N} and t_{+21N} were only conducted in control fish and M-HYP fish, which highlighted persistent higher U_{crit} values in M-HYP ($p < 0.001$) over the course of all the experimental period (Figure 2B,C). The effect of exhaustive exercise on circulating levels of metabolites and hormones in N-PF (t_{45H}) and M-HYP fish (t_{45H}, t_{+7N}, t_{+21N}) is shown as a percentage of change of N fish (Figure 3). Circulating levels of glucose, lactate, cortisol, Gh, and Igf-1 were lowered in N-PF and/or M-HYP in comparison to N fish at the end of the hypoxia conditioning period. However, this trend was reversed over time, especially in the case of lactate ($p < 0.05$). The opposite pattern was found for circulating FFAs, which shared raised levels in M-HYP fish at the beginning of the normoxia recovery period ($p < 0.05$), with a restoration of values of control fish at the last testing point (t_{+21N}). Raw data on blood parameters are shown in Supplementary Table S1.

3.4. Analysis of RNA-seq Libraries and DE Genes by Stringent FDR

After trimming and quality filtering, around 3% of all skeletal muscle reads were discarded, with the remaining reads ranging between 91 million (6.83 Gb) and 121 million (9.08 Gb) in all experimental groups (see details in Supplementary Table S2). Up to 82% of these pre-processed reads were mapped against the reference genome, which retrieved 33,756 muscle transcripts. At t_{45H}, 151 muscle transcripts (134 unique gene descriptions) were differentially expressed (FDR-adjusted p-value < 0.05) in free-swimming fish (Figure 4A). Among them, 108 genes were differentially regulated when comparisons are made between N-PF and N fish, decreasing these numbers to 72 and 21 transcripts when comparing M-HYP against N-PF fish, and M-HYP against N fish, respectively. After exercise exhaustion at t_{45H}, the number of DE transcripts was 114 (101 unique gene descriptions) for an FDR-adjusted p-value < 0.05 (Figure 4B). Among them, 18 transcripts were differentially regulated when comparing N-PF and N fish, 41 when comparing M-HYP and N-PF groups, and 78 when comparing M-HYP and N fish. The magnitude of change was apparently decreased over time with only 15 DE transcripts (FDR-adjusted p-value $<$ 0.05) when comparisons are made between M-HYP and N fish at t_{+21N} (Figure 4C).

3.5. Discriminant Classifiers and Enriched GO Terms

For a given sampling time, supervised PLS-DA models of the skeletal muscle transcriptome clearly separated along the X-axis the N fish from M-HYP fish (Supplementary Figure S2) in the analysis of free-swimming fish at t_{45H}. Otherwise, the two first components explained more than 85% and 95% of total variance in forced exercise fish after conditioning (t_{45H}) (Supplementary Figure S2A,B) and recovery (t_{+21N}) (Supplementary Figure S2C,D,E,F), respectively. This classifier performance was validated by 500-model permutation tests (Supplementary Figure S3), which was reinforced by the right hierarchical clustering of samples when applying different cut-offs for the VIP values. At t_{45H}, such an approach yielded two main clusters corresponding to N fish and N-PF/M-HYP fish. However, the number of DE transcripts among groups was increased from 222 (219 unique gene descriptions) to 421 (400 unique gene descriptions) by exhaustive exercise, decreasing in parallel the VIP cut-off value from 1.2 to 1 (Figure 5A,B). The VIP cut-off for right clustering remained low at t_{+21N}, but the number of DE transcripts between N and M-HYP fish decreased until 180 (179 unique gene descriptions) (Figure 5C).

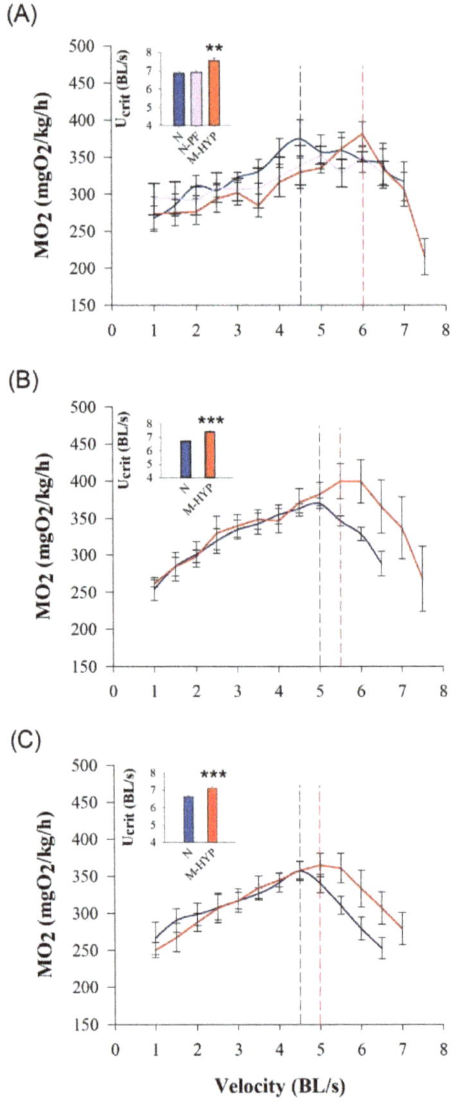

Figure 2. Swim tests conducted at the end of the mild-hypoxia conditioning period (t_{45H}) (**A**), after a 1-week normoxia recovery period (t_{+7N}) (**B**), and after a 3-week normoxia recovery period (t_{+21N}) (**C**). Mild-hypoxia fish (M-HYP) are in red, pair-fed fish (N-PF) are in violet, and normoxic fish (N) are in blue. Values showing oxygen consumption (MO_2), maximum metabolic Rate (MMR), and U_{crit} are the mean ± SEM of 4–6 fish. Asterisks indicate statistically significant differences among groups (one-way ANOVA, SNK test, ** $p < 0.01$; *** $p < 0.001$).

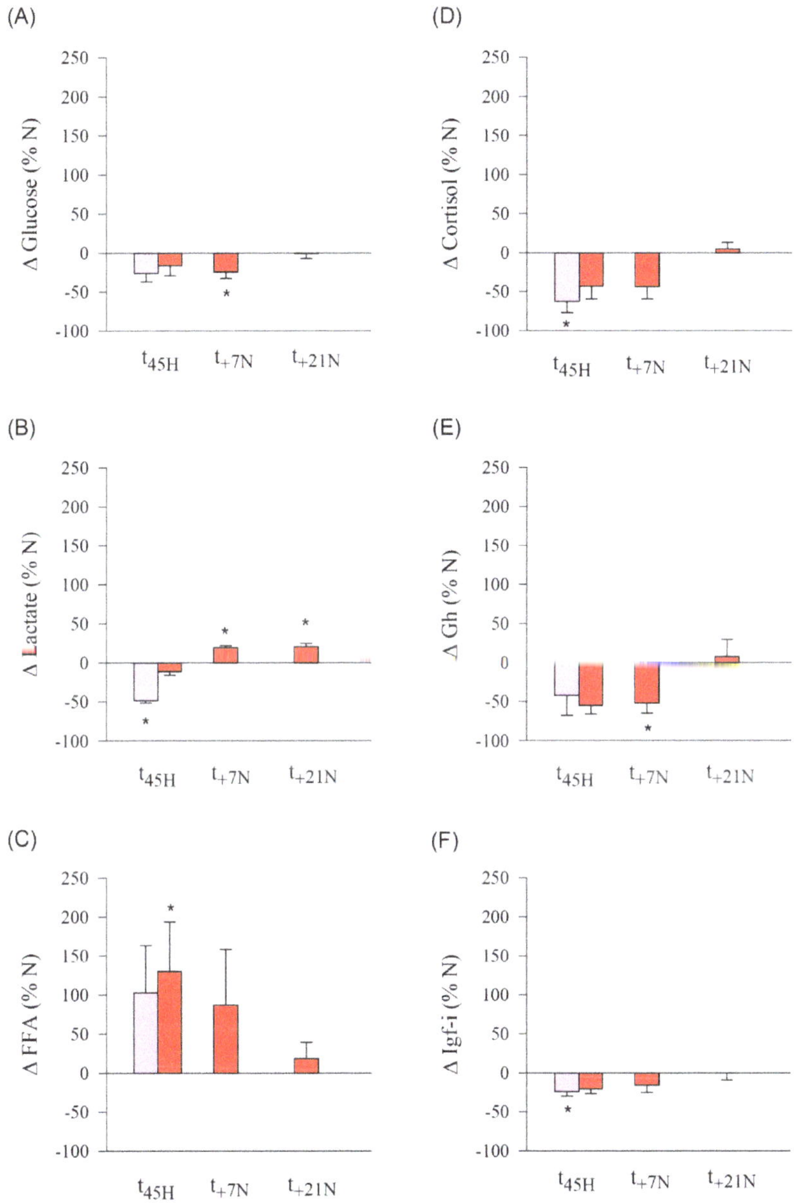

Figure 3. Changes in circulating levels of glucose (**A**), lactate (**B**), free fatty acids (**C**), cortisol (**D**), Gh (**E**), and Igf-I (**F**) in hypoxic (M-HYP, red) and pair-fed fish (N-PF, violet) following exhaustive exercise at the end of the mild-hypoxia conditioning period (t_{45H}) and over the course of the normoxia recovery period (t_{+7N}, t_{+21N}). Values are the mean ± SEM of 4–6 fish. Asterisks indicate statistically significant differences with N fish at each experimental time (SNK test, * $p < 0.05$).

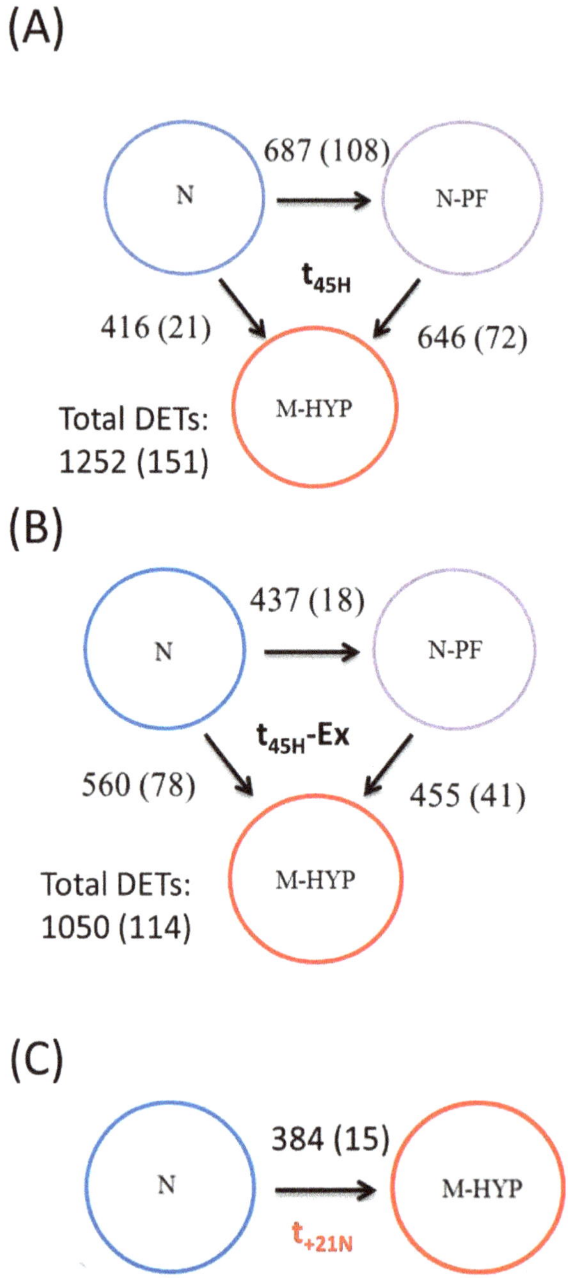

Figure 4. Differential expression analysis results at the end of the mild-hypoxia conditioning period (t_{45H}) (**A**), after exhaustive exercise at t_{45H} (**B**), and after a 3-week normoxia recovery period (t_{+21N}) (**C**). Numbers over or next to the arrows indicate differentially expressed genes (ANOVA, $p < 0.05$) between groups. Numbers between parentheses indicate differentially expressed genes (FDR-adjusted p-value < 0.05) between groups.

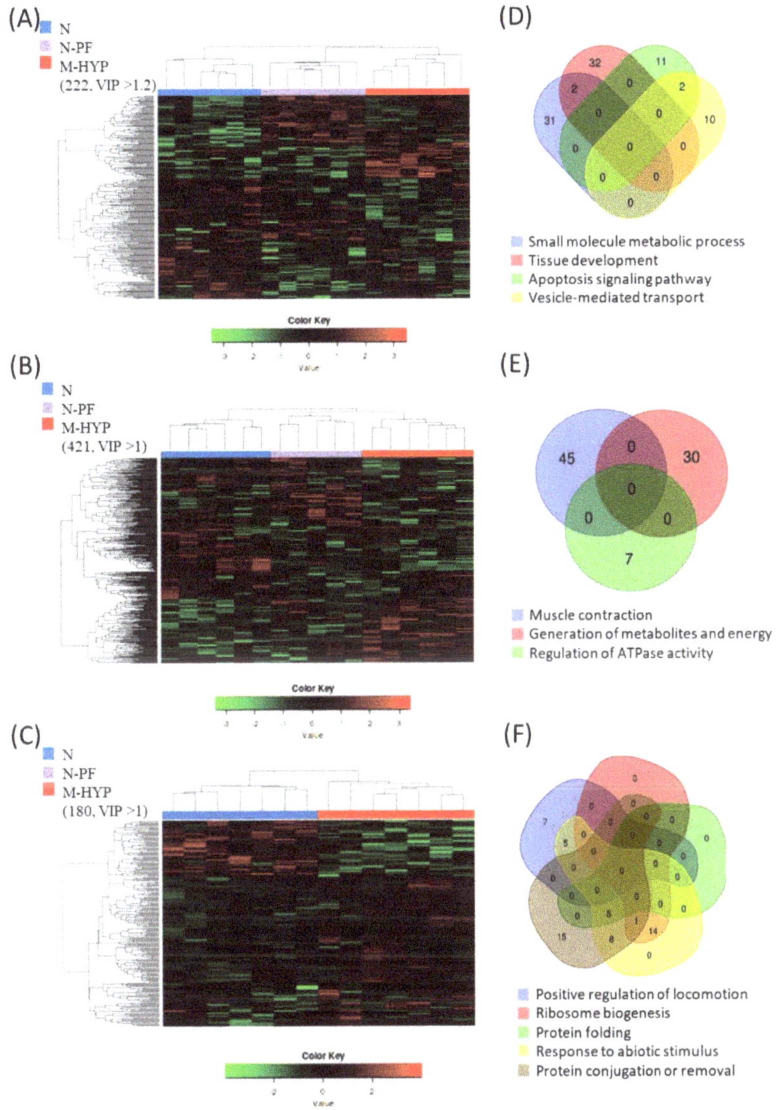

Figure 5. Clustering and gene enrichment analysis. Heatmap showing the abundance distribution (z-score) of the DE genes identified to be driving the separation between groups at the end of the mild-hypoxia conditioning period (t_{45H}) before (**A**) and after exhaustion exercise (**B**), and at the end of the normoxia recovery period (t_{+21N}) after exhaustion exercise (**C**). Venn diagrams show the level of overlapping of enriched GO-BP categories at the end of the mild-hypoxia conditioning period before (**D**) and after exhaustion exercise (**E**), and at the end of the normoxia recovery period after exhaustion exercise (**F**).

After gene clustering, a functional enrichment analysis was performed at each sampling time. The enriched categories in Biological Processes (BP), together with their allocated genes annotation and expression values in each comparison, are shown in Supplementary Table S3. For mild-hypoxia in free-swimming fish (Figure 5D), the enriched processes were (1) small molecule metabolic process, mainly lipid metabolism (GO:0044281;

33 allocated genes); (2) tissue development (GO:0009888; 34 allocated genes); (3) apoptosis signalling pathway (GO:0097190; 13 allocated genes); and (4) vesicle-mediated transport, mainly endocytosis (GO:0016192; 12 allocated genes). After exhaustive exercise at t_{45H} (Figure 5E), the enriched processes were (1) muscle contraction (GO:0006936; 42 allocated genes); (2) generation of precursors of metabolites and energy (GO:0006091; 30 allocated genes); and (3) regulation of ATPase activity (GO:0043462; 7 allocated genes). After normoxia recovery and exhaustive exercise at t_{+21N} (Figure 5F), the enriched processes were (1) positive regulation of locomotion (GO:0040017; 12 allocated genes); (2) ribosome biogenesis (GO:0042254; 15 allocated genes); (3) protein folding (GO:0006457; 5 allocated genes); (4) response to abiotic stimulus (GO:0009628; 33 allocated genes); and (5) protein conjugation or removal (GO:0070647; 29 allocated genes).

3.6. Linking Enriched Processes with Gene Expression Patterns

According to the protein–protein network analysis, a total of 31 interactions corresponding to 32 genes allocated to enriched processes were disclosed in free-swimming fish after the mild-hypoxia conditioning period (Figure 6A). One major link comprised molecules linked to endocytosis (KEGG: ko04144; 7 genes) and apoptotic signalling in response to DNA damage processes (GO:0008630; 6 genes), with ubiquitin-60S ribosomal protein L40 (*uba52*) connecting up to 8 genes involved in both processes. However, there is not a clear gene expression pattern associated with reduced O_2 availability or restricted feed intake (Figure 6D). Conversely, most lipid-related genes of the interaction plot disclosed a clear upregulation in M-HYP fish. The peroxisome proliferator-activated receptor gamma (*ppar*γ) worked as a pivotal molecule connecting up to 4 genes, all of them framed in the PPAR signalling pathway (KEGG: ko03320). This pivotal gene was significantly upregulated in the comparisons of M-HYP vs. N and M-HYP vs. N-PF. A similar expression pattern was found for diacylglycerol O-acyltransferase 2 (*dgat2*), 2-acylglycerol O-acyltransferase 2-A-like (*mogat2*), NADP-dependent malic enzyme (*me1*), farnesyl pyrophosphate synthase (*fdps*), diphosphomevalonate decarboxylase (*mvd*), 3-ketoacyl-CoA thiolase B, peroxisomal (*acaa1*), and alcohol dehydrogenase 5 (*adh5*). The long-chain-fatty-acid–CoA ligase 1 (*acsl1*) and prostaglandin E synthase 3 (*ptges3*) were equally upregulated in both M-HYP and N-PF fish in comparison to N fish. By contrast, in comparison to the controls, the expression of carnitine O-palmitoyltransferase 2 (*cpt2*) was activated in N-PF fish but not in M-HYP fish.

After the swim test following mild-hypoxia conditioning, the number of connections increased up to 146 with 40 genes in the enriched GO terms (Figure 6B), which indicates an increased cohesion pattern between DE genes following stringent exercise. Besides, such physiological response rendered the interaction of 22 genes related to muscle contraction and sliding. In this sampling point, a strong overall downregulation of genes involved in muscular machinery was found in the comparisons M-HYP vs. N and M-HYP vs. N-PF, with no differences in the comparison N-PF vs. N (Figure 6E). Among them, myosin-1, -6 and -7 (*myh1, myh6, myh7*), myosin light chain kinase (*mylk*), troponin C (*tnnc2*), troponin I, slow and fast skeletal muscle (*tnni1, tnni2*), troponin T, cardiac and skeletal muscle isoform (*tnnt2, tnnt3*), and tropomyosin alpha (*tpm3, tpm4*) and beta (*tpm2*) chains were disclosed under this type of response.

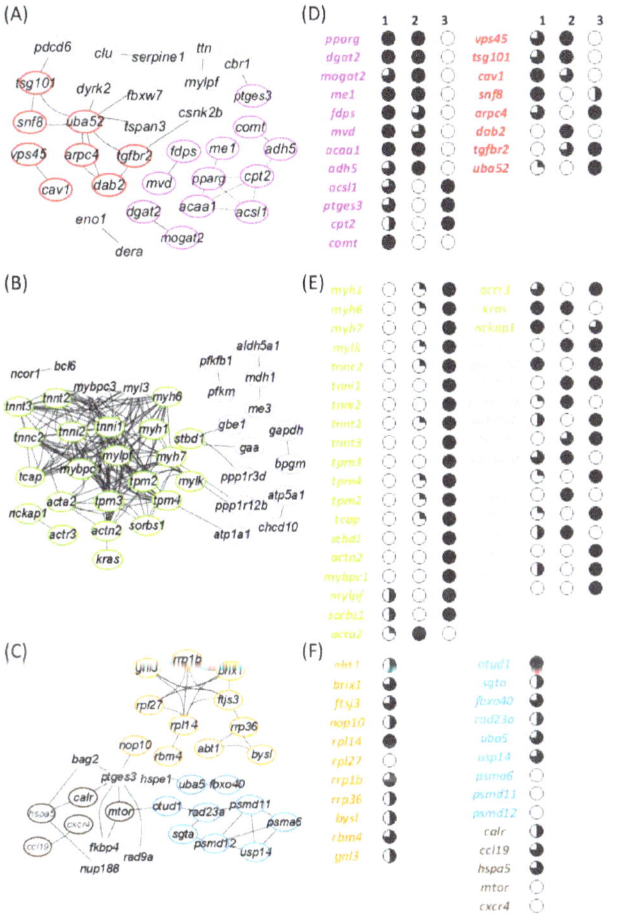

Figure 6. Protein–protein interaction plots and expression patterns of enriched processes at the end of the mild-hypoxia conditioning period (t_{45H}) before (**A**,**D**) and after exhaustive exercise (**B**,**E**), and at the end of the normoxia recovery period (t_{+21N}) after exhaustion exercise (**C**,**F**). Edges between nodes shows significant relations (FDR < 0.05; STRING confidence score > 0.7). Colours represent genes related to the following enriched biological processes: small molecule metabolic process (purple), vesicle-mediated transport (red), muscle contraction (green), generation of precursors of metabolites and energy (violet), ribosome biogenesis (orange), small protein conjugation or removal (blue), and positive regulation of locomotion (brown). Numbers above columns indicate the comparison in the RNA-seq analysis (1: M-HYP vs. N; 2: M-HYP vs. N-PF; 3: N-PF vs. N). Sliced symbols in (**D**) and (**E**) represent the comparison between the gene expression log_2FC values in the respective comparison for each gene. White circles represent the lowest log_2FC values, whereas the black circles represent the highest. Genes in (**C**), forming a category in the interaction plot, were ordered by their log_2FC in comparison 1, and sliced symbols in (**F**) were then applied for all the genes at the same time.

Following the normoxia recovery period (t_{+21N}), targeted genes of the enriched BP were also changing, the number of interactions decreasing to 43 and corresponding to 31 genes (Figure 6C). One major link in this group involved a total of 11 genes related at the same time with ribonucleoprotein complex biogenesis and response to abiotic stimulus (Figure 6F). The generalized response of these genes was their upregulation in M-HYP fish in comparison to N fish. Among others, the activator of basal transcription 1 (*abt1*), the ribosome biogenesis protein BRX1 homolog (*brix1*), and the ribosomal RNA processing proteins 1B and 36 (*rrp1b*, *rrp36*) were found. A second link grouped 9 genes related with the protein modification by small protein conjugation or removal. Here, genes related with deubiquitination processes, as the OTU domain-containing protein 1 (*otud1*), the UV excision repair protein RAD23 homolog A (*rad23a*), the ubiquitin-like modifier-activating enzyme 5 (*uba5*), and the ubiquitin carboxyl-terminal hydrolase 14 (*usp14*), were upregulated in M-HYP fish. Otherwise, genes involved in the proteasome degradation such as the proteasome subunit alpha type-6 (*psma6*), the 26S proteasome non-ATPase regulatory subunit 11 (*psmd11*), and the 26S proteasome non-ATPase regulatory subunit 12 (*psmd12*) were downregulated in M-HYP fish. A third important link was found between 5 genes related to the regulation of locomotion and the response to abiotic stimulus with the upregulation of calreticulin (*calr*), C-C motif chemokine 19 (*ccl19*), and heat shock 70kDa Protein 5 (*hspa5/grp78*), and the downregulation of serine/threonine-protein kinase mTOR (*mtor*) and C-X-C chemokine receptor type 4 (*cxcr4*) in M-HYP fish.

4. Discussion

Episodes of high temperature and hypoxia are increasing in extent and severity in coastal marine ecosystems, and these stressors have the capacity to reinforce each other because increasing temperature decreases O_2 solubility [47]. Hypoxia is, thereby, a major aquaculture stressor around the world [48]. The first sign of O_2 scarcity is the reduction of appetite, and subsequent growth impairments reflect the different temperature and O_2 tolerance ranges of living organisms [18,49–51], as well as their plasticity for prioritizing feed efficiency at the expenses of maximum growth "oxystatic theory" [52,53]. Thus, both in this and a previous study [20], we found that mild-hypoxia acclimation in summer (40–60% saturation) deaccelerated growth of fast-growing juveniles of gilthead sea bream, whereas FCR was not impaired or even improved during mild-hypoxia and normoxia recovery periods, respectively. This is because the best feed conversion and hormonal signatures for fast and efficient growth generally occur before the achievement of maximum growth at the greater ration size [54,55]. This also applies at the cellular level, where the maximum ATP yield per molecule of O_2 (P/O ratio) is increased during food shortage [56,57] or hypometabolic hypoxia [33]. Such adaptive feature was supported herein by lowered plasma levels of lactate, which would reflect a reduced basal metabolism rather than a switch of aerobic to anaerobic metabolism during mild-hypoxia exposure. This was also previously stated [20], but herein the pair-fed experimental design allowed us to disclose that low plasma lactate levels arise from a reduced feed intake that becomes slightly although non-significantly lowered by limited O_2 availability. Conversely, circulating levels of haemoglobin and FFAs evolved differentially when fish faced changes in feed intake and O_2 availability, resulting in reduced erythropoiesis and enhanced lipid mobilization, which can be interpreted as a discriminant feature of hypo-metabolic states triggered by feed restriction during normoxia. Otherwise, circulating levels of cortisol and growth-promoting factors did not differ significantly among groups, though the highest Gh/Igf-1 ratio of M-HYP fish during hypoxia exposure underlines a limited growth potentiality under reduced O_2 and metabolic fuels availability [58].

The changing metabolic phenotype is also indicative of a number of whole-organism traits of ecological and economic importance, such as dominance, aggression, and swimming performance [59]. A high percentage of this genetic variance is expected to be unexplained [60], but recent research in gilthead sea bream associated reduced locomotor activity with more continuous growth [61], less size heterogeneity [34,61], and enhanced phenotypic plasticity of gut microbiota [62]. Likewise, large domesticated strains of Atlantic salmon and rainbow trout (*Oncorhynchus mykiss*) become athletically less robust that wild fish [63–66]. However, the opposite is also true and different exercise protocols are able to increase the growth performance of a wide range of farmed fish, including gilthead sea bream [67–69]. Moreover, critical swimming is a highly predictive marker of fillet yield in both gilthead sea bream and Atlantic salmon [70]. The mechanisms by which hypoxia acclimation or hypoxia priming during early life affect fillet yield remain elusive, though the present study highlighted that M-HYP fish showed a higher U_{crit} than the two other experimental groups. This observation underscored the improved availability of this group of fish to resist fatigue during training endurance, as it has been reported for hypoxia acclimation in athletes and different animal models [29,30]. Interestingly, exercised M-HYP fish would benefit of this metabolic advantage through all the normoxia recovery period (21 days), though the time course of changes in blood landmarks evidenced a dynamic metabolic trade-off with an increased availability of aerobic metabolic fuels (increase of circulating levels of FFA for aerobic ATP production) after hypoxia conditioning, which shifted towards a higher anaerobic fitness (increased lactate production from glucose) at the end of the normoxia recovery period.

At the transcriptional level, up to 222 genes have a discriminant role for classification of all individuals in their respective group (N, N-PF, or M-HYP) at the end of the hypoxia conditioning period. During hypoxia exposure, the number of discriminant genes was also increased up to 421 by exhaustive exercise, which is in line with previous studies in rainbow trout where sustained swimming increased the transcriptional activity of skeletal muscle [71]. Certainly, in our experimental model, the number of interactions and interacting genes in the protein interaction plots increased largely in exercised fish. In any case, cluster analysis grouped together N-PF and M-HYP fish before and after exhaustive exercise, which indicates that most hypoxia-mediated changes in mRNA transcripts were mediated, at least in part, by a reduced feed intake. This assumption was supported by a high representation of discriminant genes (68–70%) that were up- or downregulated in parallel in both N-PF and M-HYP fish (Supplemental Table S3).

At a closer look, the protein interaction plots disclose lipid metabolism (small molecule metabolic process) as a main differentially regulated process in our hypoxia model (Figure 6A,D). This is not surprising given that many aspects of carbohydrates, but also of lipid metabolism, are modified in humans and transgenic animals by HIF (hypoxia inducible factor) during physiological or pathological hypoxia, contributing significantly to the pathogenesis and/or progression of cancer and metabolic disorders [72]. The ultimate mechanisms remain controversial, though experimental evidence supports that TAG synthesis and the extracellular uptake of FAs are promoted under hypoxia by the transcription factor PPARγ in a HIF-dependent manner [73,74]. Lipid accumulation under hypoxia is further supported by the HIF inhibition of FA oxidation via the downregulation of the transcriptional coactivator PGC-1α (proliferator-activated receptor-γ coactivator-1α) [75,76]. Thus, a number of studies have shown that obesity is increased by inhibition of HIF and decreased by HIF activation [77], but other authors claimed that HIF activation induces obesity [78,79]. Indeed, the activation of both lipid catabolism and anabolism was co-occurring in the present study, the hypoxic induction of *pparg* being associated with the upregulation of *acaa1* (peroxisomal β-oxidation enzyme) and *acsl1*, which catalyses the conversion of long-chain fatty acids to their active form acyl-CoAs for both oxidation and biosynthesis processes. This hypoxic activation of lipid anabolism was also supported by the upregulation of key-rate limiting enzymes of FA synthesis (*me1*) and esterification into diacylglycerols (*mogat2*) and TAG (*dgat2*). A similar enhancement of lipid metabolism at brain, liver and muscle has been recently reported in Ya fish *Schizothorax prenanti* as an adaptation to acute hypoxia stress, together with an upregulation of antioxidant genes [80]. This overall activation of lipid metabolism reflects an adaptive trade-off that might serve to facilitate the aerobic energy metabolism during prolonged reduced O_2 availability, limiting at the same time the risk of lipotoxicity (excess of intracellular FAs) through the induction of TAG accumulation [81,82]. Thus, the muscle expression of *pparg* and other lipogenic/anabolic enzymes (*me1, mogat2, dgat2*) was repressed in N-PF, whereas the trend for catabolic enzymes of peroxisomal (*acsl1*) and mitochondrial (*cpt2*) β-oxidation was the upregulation in N-PF fish and to a lower extent in M-HYP fish. In other words, the activation of lipid cell storage would be primarily mediated by O_2 availability, whereas the overall stimulation of lipid catabolic enzymes would be triggered by a deficit in metabolic fuels. In this line, hypoxia induced a pronounced mobilization of stored TAGs in the euryoxic goby *Gillichthys mirabilis* [83]. Otherwise, the activation of the transcription factor SREPB (sterol regulatory element binding protein) signalling pathway was required in the poor survival glioblastoma multiforme to preserve lipid biosynthesis and cell viability under lipid- and O_2-deprived conditions [84], which was herein in agreement with the upregulated expression of *fdps* and *mvd* in both N-PF and M-HYP fish.

The hypoxic induction of *pparg* persisted after exhaustive exercise in M-HYP fish at the t_{45H} sampling point. However, protein interaction plots rendered the over-representation of the two other main processes (muscle contraction and generation of metabolites and energy), with an overall downregulation in comparison to normoxic control fish that was extensive to a lower extent in N-PF fish (Figure 6B,E). This metabolic feature highlighted the more efficient energy metabolism of M-HYP fish, probably with a higher contribution of aerobic metabolism to whole energy supply, as part of the mechanisms of a wide range of animal taxa to cope with high-altitude hypoxia [85–91]. The precise mechanisms remain unsolved in our experimental model, though the downregulated expression of muscle lactate dehydrogenase should contribute to regulate muscle contractions. Indeed, several studies stated that lactate decreases the blood flow to the working muscle, in a process that induces fatigue [92,93]. Marathoners deal with this fatigue with a higher proportion of oxidative fibres, which contract slowly and use aerobic respiration to produce ATP [94,95]. This type of adaptive response might also contribute to drive the muscle transcriptome of M-HYP fish, resulting in a persistent increase in critical swimming speed, due to their improved endurance training during the hypoxia conditioning period. Reinforcing this assumption, mRNA transcripts of myosin, troponin, and tropomyosin were strongly downregulated in M-HYP fish, but remained unaltered in N-PF fish, which suggests that this feature is a good example of a muscle transcriptional response mediated by changes in O_2 availability rather than feed intake.

From our results, it was also conclusive that normoxic M-HYP fish (t_{+21N}) shared a catch-up growth response, in concurrence with a persistent improvement in the swimming performance and changing muscle transcriptome after exercise exhaustion. These hypoxia-driven effects included a clear exercise activation of the anaerobic glycolysis pathway in M-HYP fish in comparison to normoxic control fish (Supplemental Table S3). Certainly, the regulation of glycolysis has been largely studied in fish [96–98], and wide-transcriptomic studies highlighted a fast transition from aerobic oxidation to anaerobic glycolysis when individuals faced restrictive O_2 concentrations [99]. However, anaerobic metabolism is less efficient than aerobic ATP production, and the enhanced exercise activation of glycolysis by hypoxia conditioning was apparently delayed over the course of the normoxia recovery period. Intriguingly, this glycolysis activation was supported by increased levels of blood lactate (Figure 3B) and the upregulated expression of muscle lactate dehydrogenase. However, the protein interaction plot showed the main enrichment of ribosome biogenesis and protein conjugation GO terms with the upregulation of markers of protein synthesis (*abt1, brix1, rrp1b, rrp36*) and deubiquitination (*otud1, rad23a, usp14*) pathways, in combination with the strong downregulation of several proteasome subunits (*psma6, psmd11, psmd12*) (Figure 6C,F). This finding highlighted a reduced muscle protein turnover after exhaustive exercise in normoxic M-HYP, which would also contribute to support the catch-up growth of N-PF and M-HYP fish during the normoxia recovery period. Indeed, a large body of evidence has revealed that faster growing and/or more efficient fish have lower rates of protein turnover (equivalent to protein breakdown in growing individuals) [100,101]. Otherwise, a number of overlapping genes for positive regulation of locomotion and response to stimulus were also induced or repressed by exhaustive exercise in M-HYP fish. Among them, noteworthy is the upregulated expression of *hspa5/grp78* and *calr*, highly conserved chaperone proteins of endoplasmic reticulum (ER) that reduce ER stress and apoptosis through the enhancement of the cellular folding capacity [102,103]. Certainly, micro-RNAs (miRs) have emerged as key gene regulators in many diseases, as reduced miR30a increased the HSPA5 level and attenuated ischemic brain infarction in focal ischemia stroked mice [104]. Likewise, both *hspa5* and *calr* belong to the antioxidant defence system of the epithelial layers [105,106], and their upregulated expression in M-HYP fish would contribute to prevent the disruption of the intracellular redox state and the cellular folding capacity of skeletal muscle under metabolic states, resulting in high O_2 consumption rates and enhanced ROS production.

5. Conclusions

Mild-hypoxia acclimation acts as a driving force with effects on growth and swimming performance through changes in metabolic and muscle transcriptomic landmarks. The first consequence of mild-hypoxia exposure was to prioritize feed efficiency at the expenses of maximum growth. Such adaptive feature would take advantage of the benefits of a reduced feed intake and hypometabolic state, resulting in a higher contribution of aerobic metabolism to whole energy supply that shifted towards a higher anaerobic fitness following normoxia restoration. Despite the changes in substrate preference for metabolic fuels, mild-hypoxia acclimation led to higher U_{crit} at exhaustive exercise before and after normoxia restoration, which would reflect the metabolic imprint of a lower O_2 availability rather than a reduced feed intake in our hypoxia pair-fed model, at least in the short/medium-term (21 days post-recovery). At the transcriptomic level, our results depicted the overall activation of lipid metabolism to facilitate the aerobic energy metabolism during prolonged reduced O_2 availability, with the limitation at the same time of the risk of lipotoxicity through the enhanced TAG accumulation of the excess of intracellular FAs. The machinery of muscle contraction and protein synthesis and breakdown was also largely altered by mild-hypoxia conditioning, and the achieved responses contributed to mitigate fatigue response under exhaustive exercise, and to preserve a catch-up growth during the normoxia recovery period. Altogether, these results reinforce the high phenotypic plasticity of gilthead sea bream, which is supported at the genomic level by a high rate of recent local gene duplications [40] that might favour the acquisition of novel gene functions, and a rapid and efficient adaption of individuals to a changing and challenging environment. In a practical sense, as summarized in Figure 7, mild-hypoxia pre-programming emerges as a promising prophylactic measure to depress metabolic rates and so prepare individuals to respond to predictable stressful events, preserving and even improving FCR and swimming performance.

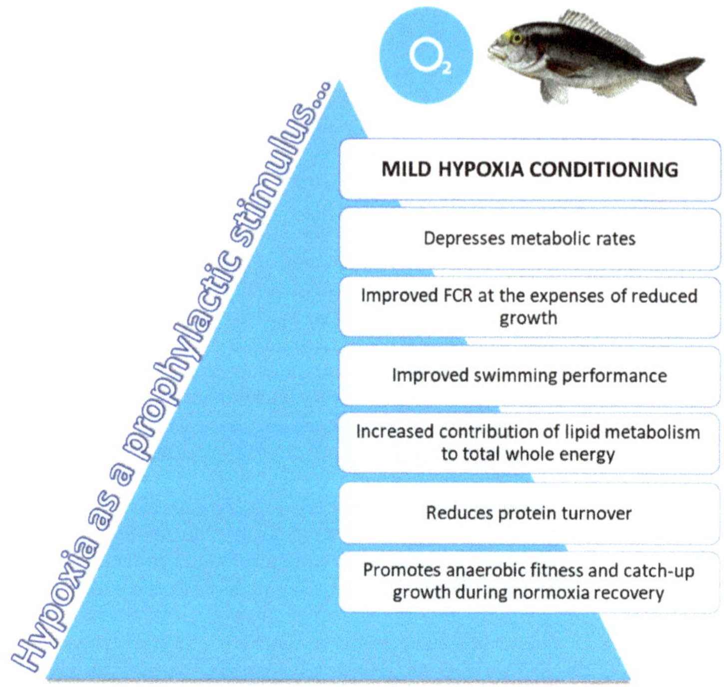

Figure 7. Schematic representation of the proposed model for integrative responses of gilthead sea bream exposed to mild-hypoxia stress conditioning.

Supplementary Materials: The following are available online at https://www.mdpi.com/article/10.3390/biology10050416/s1. Figure S1: Closed-system scheme for the experimental set-up. Figure S2: Graphical representation of the goodness-of-fit and scores of the PLS-DA model in each sampling time along the experiment. (A), (C) and (E) Graphical representation of the goodness-of-fit of the PLS-DA model of t_{45H}, t_{45H} after exercise and t_{+21N} after exercise, respectively. (B), (D), (F) Two-dimensional PLS-DA score plot representing the distribution of the samples between the first two components in the model of t_{45H}, t_{45H} after exercise and t_{+21N} after exercise, respectively. Figure S3: Validation plots of the PLS-DA models consisting in 500 random permutations. (A), (B), (C) Validation plots of the PLS-DA models of t_{45H}, t_{45H} after exercise and t_{+21N} after exercise, respectively. Table S1: Effects of mild-hypoxia conditioning and the subsequent normoxia recovery (1 or 3 weeks thereafter) on plasma and muscle tissue levels of metabolites and hormones after exercise exhaustion of gilthead sea bream juveniles. Values are the mean ± SEM of 6–7 fish (2–3 fish per replicate tank). p-values are the result of one-way ANOVA or t-test analysis where appropriate. Different superscript letters or asterisks (*) indicate significant differences between experimental groups ($p < 0.05$). Table S2: Statistics of the preprocessing and mapping of RNA-seq libraries. t_{45H}, t_{45H}-Ex and t_{+21N} correspond to experimental groups conditioning, Exercise in t_{45H} and Exercise in t_{+21N}, respectively. N, N-PF and M-HYP correspond to the different fish conditions: Normoxia, normoxia pair-fed and mild-hypoxia, respectively. Table S3: Differentially expressed genes and associated enriched pathways.

Author Contributions: J.P.-S. conceived and designed the study; V.d.l.H. and P.S.-M. conducted fish rearing; J.A.M-S. performed blood and swim tests; F.N.-C. and J.À.C.-G. performed transcriptional and bioinformatics analyses; All authors analyzed and interpreted the data; F.N.-C., J.A.M.-S. and J.P.-S. drafted the original manuscript. All authors reviewed, edited and approved the final manuscript.

Funding: This work was financially supported by a grant from the European Commission of the European Union under the Horizon 2020 research infrastructure project AQUAEXCEL^{2020} (652831) to J.P-S. Additional funding was obtained by a Spanish MICINN project (Bream-AquaINTECH,

RTI2018–094128-B-I00). J.A.M.-S. received a Postdoctoral Research Fellowship (*Juan de la Cierva-Formación*, Reference FJCI-2014-20,161).

Institutional Review Board Statement: All procedures were approved by the Ethics and Animal Welfare Committees of Institute of Aquaculture Torre de la Sal (IATS) and CSIC. The study was conducted in the IATS registered aquaculture infrastructure facility (code ES120330001055) in accordance with the principles published in the European Animal Directive (2010/63/EU) and Spanish laws (Royal Decree RD53/2013) for the protection of animals used in scientific experiments.

Informed Consent Statement: Not applicable.

Data Availability Statement: The datasets presented in this study can be found in online repositories. The names of the repository/repositories and accession number(s) can be found below: NCBI (accession: SAMN16834555-597, PRJNA679473).

Acknowledgments: The authors are grateful to Biotechvana S.L. for the allowance to use its computational cluster servers.

Conflicts of Interest: The authors declare no conflict of interest.

References

1. Hoppeler, H.; Vogt, M. Muscle tissue adaptations to hypoxia. *J. Exp. Biol.* **2001**, *204*, 3133–3139. [CrossRef]
2. Murray, A.J. Metabolic adaptation of skeletal muscle to high altitude hypoxia: How new technologies could resolve the controversies. *Genome Med.* **2009**, *1*, 117. [CrossRef] [PubMed]
3. Gamboa, J.L.; Andrade, F.H. Muscle endurance and mitochondrial function after chronic normobaric hypoxia: Contrast of respiratory and limb muscles. *Eur. J. Physiol.* **2012**, *463*, 327–338. [CrossRef]
4. Larson, J.; Drew, K.L.; Folkow, L.P.; Milton, S.L.; Park, T.J. No oxygen? No problem! Intrinsic brain tolerance to hypoxia in vertebrates. *J. Exp. Biol.* **2014**, *217*, 1024–1039. [CrossRef] [PubMed]
5. Storey, K.B. Regulation of hypometabolism: Insights into epigenetic controls. *J. Exp. Biol.* **2015**, *218*, 150–159. [CrossRef]
6. Wu, H.; Chen, W.; Zhao, F.; Zhou, Q.; Reinach, P.S.; Deng, L.; Ma, L.; Luo, S.; Srinivasalu, N.; Pan, M.; et al. Scleral hypoxia is a target for myopia control. *Proc. Natl. Acad. Sci. USA* **2018**, *115*, E7091–E7100. [CrossRef] [PubMed]
7. Brown, C.J.; Rupert, J.L. Hypoxia and environmental epigenetics. *High Alt. Med. Biol.* **2014**, *15*, 323–330. [CrossRef]
8. Julian, C.G. Epigenomis and human adaptation to high altitude. *J. Appl. Physiol.* **2017**, *123*, 1362–1370. [CrossRef] [PubMed]
9. Richards, J.G. Physiological, behavioral and biochemical adaptations of intertidal fishes to hypoxia. *J. Exp. Biol.* **2011**, *214*, 191–199. [CrossRef] [PubMed]
10. Zhu, C.D.; Wang, Z.H.; Yan, B. Strategies for hypoxia adaptation in fish species: A review. *J. Comp. Physiol. B Biochem. Syst. Environ. Physiol.* **2013**, *183*, 1005–1013. [CrossRef]
11. Deutsch, C.; Ferrel, A.; Seibel, B.; Pörtner, H.O.; Huey, R.B. Ecophysiology. Climate change tightens a metabolic constraint on marine habitats. *Science* **2015**, *348*, 1132–1135. [CrossRef]
12. Schmidtko, S.; Stramma, L.; Visbeck, M. Decline in global oceanic oxygen content during the past five decades. *Nature* **2017**, *542*, 335–339. [CrossRef] [PubMed]
13. Oschlies, A.; Brandt, P.; Stramma, L.; Schmidtko, S. Drivers and mechanisms of ocean deoxygenation. *Nat. Geosci.* **2018**, *11*, 467–473. [CrossRef]
14. FAO. *The State of World Fisheries and Aquaculture 2018. Contributing to Food Security and Nutrition for All*; Food and Agriculture Organization of the United Nations: Rome, Italy, 2018.
15. Sae-Lim, P.; Kause, A.; Mulder, H.A.; Olesen, I. Breeding and genetics symposium: Climate change and selective breeding in aquaculture. *J. Anim. Sci.* **2017**, *95*, 1801–1812. [CrossRef]
16. Abdel-Tawwab, M.; Monier, M.N.; Hoseinifar, S.H.; Faggio, C. Fish response to hypoxia stress: Growth, physiological, and immunological biomarkers. *Fish Physiol. Biochem.* **2019**, *45*, 997–1013. [CrossRef]
17. Remen, M.; Nederlof, M.A.J.; Folkedal, O.; Thorsheim, G.; Sitjà-Bobadilla, A.; Pérez-Sánchez, J.; Oppedal, F.; Olsen, R.E. Effect of temperature on the metabolism, behaviour and oxygen requirements of *Sparus aurata*. *Aquacult. Environ. Interact.* **2015**, *7*, 115–123. [CrossRef]
18. Remen, M.; Sievers, M.; Torgersen, T.; Oppedal, F. The oxygen threshold for maximal feed intake of Atlantic salmon post-smolts is highly temperature-dependent. *Aquaculture* **2016**, *464*, 582–592. [CrossRef]
19. Araújo-Luna, R.; Ribeiro, L.; Bergheim, A.; Pousão-Ferreira, P. The impact of different rearing condition on gilthead seabream welfare: Dissolved oxygen levels and stocking densities. *Aquacult. Res.* **2018**, *49*, 3845–3855. [CrossRef]
20. Martos-Sitcha, J.A.; Simó-Mirabet, P.; de Las Heras, V.; Calduch-Giner, J.À.; Pérez-Sánchez, J. Tissue-specific orchestration of gilthead sea bream resilience to hypoxia and high stocking density. *Front. Physiol.* **2019**, *10*, 840. [CrossRef]
21. Parma, L.; Pelusio, N.F.; Gisbert, E.; Esteban, M.A.; D'amico, F.; Soverini, M.; Candela, M.; Dondi, F.; Gatta, P.P.; Bonaldo, A. Effects of rearing density on growth, digestive conditions, welfare indicators and gut bacterial community of gilthead sea bream (*Sparus aurata, L.* 1758) fed different fishmeal and fish oil dietary levels. *Aquaculture* **2020**, *518*, 734854. [CrossRef]

22. Calduch-Giner, J.À.; Davey, G.; Saera-Vila, A.; Houeix, B.; Talbot, A.; Prunet, P.; Cairns, M.T.; Pérez-Sánchez, J. Use of microarray technology to assess the time course of liver stress response after confinement exposure in gilthead sea bream (*Sparus aurata L.*). *BMC Genom.* **2010**, *11*, 193. [CrossRef]
23. McBryan, T.L.; Healy, T.M.; Haakons, K.L.; Schulte, P.M. Warm acclimation improves hypoxia tolerance in *Fundulus heteroclitus*. *J. Exp. Biol.* **2016**, *219*, 474–484. [CrossRef] [PubMed]
24. Healy, T.M.; Chung, D.J.; Crowther, K.G.; Schulte, P.M. Metabolic and regulatory responses involved in cold acclimation in Atlantic killifish, *Fundulus heteroclitus*. *J. Comp. Physiol. B.* **2017**, *187*, 463–475. [CrossRef]
25. Salin, K.; Auer, S.K.; Villasevil, E.M.; Anderson, G.J.; Cairns, A.G.; Mullen, W.; Hartley, R.C.; Metcalfe, N.B. Using the MitoB method to assess levels of reactive oxygen species in ecological studies of oxidative stress. *Sci. Rep.* **2017**, *7*, 41228. [CrossRef]
26. Vanderplancke, G.; Claireaux, G.; Quazuguel, P.; Huelvan, C.; Corporeau, C.; Mazurais, D.; Zambonino-Infante, J.L. Exposure to chronic moderate hypoxia impacts physiological and developmental traits of European sea bass (*Dicentrarchus labrax*) larvae. *Fish Physiol. Biochem.* **2015**, *41*, 233–242. [CrossRef]
27. Wang, S.Y.; Lau, K.; Lai, K.P.; Zhang, J.W.; Tse, A.C.; Li, J.W.; Tong, Y.; Chan, T.F.; Wong, C.K.; Chiu, J.M.; et al. Hypoxia causes transgenerational impairments in reproduction of fish. *Nat. Commun.* **2016**, *7*, 12114. [CrossRef] [PubMed]
28. Manchenkov, T.; Pasillas, M.P.; Haddad, G.G.; Imam, F.B. Novel genes critical for hypoxic preconditioning in zebrafish are regulators of insulin and glucose metabolism. *G3-Genes Genom. Genet.* **2015**, *5*, 1107–1116. [CrossRef] [PubMed]
29. Sinex, J.A.; Chapman, R.F. Hypoxic training methods for improving endurance exercise performance. *J. Sport Health Sci.* **2015**, *4*, 4. [CrossRef]
30. Hawley, J.A.; Lundby, C.; Cotter, J.D.; Burke, L.M. Maximizing cellular adaptation to endurance exercise in skeletal muscle. *Cell Metab.* **2018**, *27*, 962–976. [CrossRef]
31. Fu, S.-J.; Brauner, C.J.; Cao, Z.-D.; Richards, J.G.; Peng, J.-L.; Dhillon, R.; Wang, Y.-X. The effect of acclimation to hypoxia and sustained exercise on subsequent hypoxia tolerance and swimming performance in goldfish (*Carassius auratus*). *J. Exp. Biol.* **2011**, *214*, 2080–2088. [CrossRef]
32. Magnoni, L.J.; Martos-Sitcha, J.A.; Queiroz, A.; Calduch-Giner, J.A.; Magalhàes Gonçalves, J.F.; Rocha, C.M.R.; Abreu, H.T.; Schrama, J.W.; Ozorio, R.O.A.; Pérez-Sánchez, J. Dietary supplementation of heat-treated *Gracilaria* and *Ulva* seaweeds enhanced acute hypoxia tolerance in gilthead seabream (*Sparus aurata*). *Biol. Open* **2017**, *6*, 897–908. [CrossRef] [PubMed]
33. Martos-Sitcha, J.A.; Bermejo-Nogales, A.; Calduch-Giner, J.A.; Pérez-Sánchez, J. Gene expression profiling of whole blood cells supports a more efficient mitochondrial respiration in hypoxia-challenged gilthead sea bream (*Sparus aurata*). *Front. Zool.* **2017**, *14*, 34. [CrossRef] [PubMed]
34. Perera, E.; Rosell-Moll, E.; Naya-Català, F.; Simó-Mirabet, P.; Calduch-Giner, J.; Pérez-Sánchez, J. Effects of genetics and early-life mild hypoxia on size variation in farmed gilthead sea bream (*Sparus aurata*). *Fish Physiol. Biochem.* **2020**. [CrossRef] [PubMed]
35. Martos-Sitcha, J.A.; Simó-Mirabet, P.; Piazzon, M.C.; de las Heras, V.; Calduch-Giner, J.A.; Puyalto, M.; Tinsley, J.; Makol, A.; Sitjà-Bobadilla, A.; Pérez-Sánchez, J. Dietary sodium heptanoate helps to improve feed efficiency, growth hormone status and swimming performance in gilthead sea bream (*Sparus aurata*). *Aquac. Nutr.* **2018**, *24*, 1638–1651. [CrossRef]
36. Martínez-Barberá, J.P.; Pendón, C.; Martí-Palanca, H.; Calduch-Giner, J.À.; Rodríguez, R.B.; Valdivia, M.M.; Pérez-Sánchez, J. The use of recombinant gilthead sea bream (*Sparus aurata*) growth hormone for radioiodination and standard preparation in radioimmunoassay. *Comp. Biochem. Physiol. A Physiol.* **1995**, *110*, 335–340. [CrossRef]
37. Vega-Rubín de Celis, S.; Gómez-Requeni, P.; Pérez-Sánchez, J. Production and characterization of recombinantly derived peptides and antibodies for accurate determinations of somatolactin, growth hormone and insulin-like growth factor-I in European sea bass (*Dicentrarchus labrax*). *Gen. Comp. Endocrinol.* **2004**, *139*, 266–277. [CrossRef] [PubMed]
38. Schmieder, R.; Edwards, R. Quality control and preprocessing of metagenomic datasets. *Bioinformatics* **2011**, *27*, 863–864. [CrossRef] [PubMed]
39. Kim, D.; Pertea, G.; Trapnell, C.; Pimentel, H.; Kelley, R.; Salzberg, S.L. TopHat2: Accurate alignment of transcriptomes in the presence of insertions, deletions and gene fusions. *Genome Biol.* **2013**, *14*, R36. [CrossRef] [PubMed]
40. Pérez-Sánchez, J.; Naya-Català, F.; Soriano, B.; Piazzon, M.C.; Hafez, A.; Gabaldón, T.; Llorens, C.; Sitjà-Bobadilla, A.; Calduch-Giner, J.A. Genome sequencing and transcriptome analysis reveal recent species-specific gene duplications in the plastic gilthead sea bream (*Sparus aurata*). *Front. Mar. Sci.* **2019**, *6*, 760. [CrossRef]
41. Trapnell, C.; Roberts, A.; Goff, L.; Pertea, G.; Kim, D.; Kelley, D.R.; Pimentel, H.; Salzberg, S.L.; Rinn, J.L.; Pachter, L. Differential gene and transcript expression analysis of RNA-seq experiments with TopHat and Cufflinks. *Nat. Protoc.* **2017**, *7*, 562–578. [CrossRef]
42. Li, H.; Ma, M.L.; Luo, S.; Zhang, R.M.; Han, P.; Hu, W. Metabolic responses to ethanol in *Saccharomyces cerevisiae* using a gas chromatography tandem mass spectrometry-based metabolomics approach. *Int. J. Biochem. Cell Biol.* **2012**, *44*, 1087–1096. [CrossRef]
43. Kieffer, D.A.; Piccolo, B.D.; Vaziri, N.D.; Liu, S.; Lau, W.L.; Khazaeli, M.; Nazertehrani, S.; Moore, M.E.; Marco, M.L.; Martin, R.J.; et al. Resistant starch alters gut microbiome and metabolomic profiles concurrent with amelioration of chronic kidney disease in rats. *Am. J. Physiol. Renal Physiol.* **2016**, *310*, F857–F871. [CrossRef]
44. Ge, S.X.; Jung, D.; Yao, R. ShinyGO: A graphical gene-set enrichment tool for animals and plants. *Bioinformatics* **2020**, *36*, 2628–2629. [CrossRef] [PubMed]

45. Szklarczyk, D.; Gable, A.L.; Lyon, D.; Junge, A.; Wyder, S.; Huerta-Cepas, J.; Simonovic, M.; Doncheva, N.T.; Morris, J.H.; Bork, P.; et al. STRING v11: Protein-protein association networks with increased coverage, supporting functional discovery in genome-wide experimental datasets. *Nucleic Acids Res.* **2019**, *47*, D607–D613. [CrossRef] [PubMed]
46. Futami, R.; Muñoz-Pomer, A.; Viu, J.M.; Domínguez-Escribà, L.C.L.; Bernet, G.P.; Sempere, J.M.; Moya, A.; Llorens, C. GPRO: The professional tool for management, functional analysis and annotation of omic sequences and databases. *Biotech. Bioinf.* **2011**, *1*, 1–5.
47. Garcia, H.E.; Gordon, L.I. Oxygen solubility in seawater: Better fitting equations. *Limnol. Oceanogr.* **1992**, *37*, 1307–1312. [CrossRef]
48. Intergovernmental Panel on Climate Change. *Climate Change 2014-Impacts, Adaptation and Vulnerability: Part A: Global and Sectoral Aspects: Working Group II Contribution to the IPCC Fifth Assessment Report*; Cambridge University Press: Cambridge, UK, 2014.
49. Pichavant, K.; Person-Le-Ruyet, J.; Bayon, N.L.; Severe, A.; Roux, A.L.; Boeuf, G. Comparative effects of long-term hypoxia on growth, feeding and oxygen consumption in juvenile turbot and European sea bass. *J. Fish Biol.* **2001**, *59*, 875–883. [CrossRef]
50. Cadiz, L.; Zambonino-Infante, J.L.; Quazuguel, P.; Madec, L.; Le Delliou, H.; Mazurais, D. Metabolic response to hypoxia in European sea bass (*Dicentrarchus labrax*) displays developmental plasticity. *Comp. Biochem. Physiol.* **2018**, *215*, 1–9. [CrossRef]
51. Vikeså, V.; Nankervis, L.; Hevrøy, E.M. High dietary energy level stimulates growth hormone receptor and feed utilization in large Atlantic salmon (*Salmo salar* L.) under hypoxic conditions. *Aquac. Nutr.* **2017**, *23*, 1193–1203. [CrossRef]
52. Dam, A.V.; Pauly, D. Simulation of the effects of oxygen on food consumption and growth of Nile tilapia, *Oreochromis niloticus* (L.). *Aquacult. Res.* **1995**, *26*, 427–440. [CrossRef]
53. Saravanan, S.; Geurden, I.; Figueiredo-Silva, A.C.; Kaushik, S.J.; Haidar, M.N.; Verreth, J.A.; Schrama, J.W. Control of voluntary feed intake in fish: A role for dietary oxygen demand in Nile tilapia (*Oreochromis niloticus*) fed diets with different macronutrient profiles. *Br. J. Nutr.* **2012**, *108*, 1519–1529. [CrossRef]
54. Brett, J.R. Environmental factors and growth. In *Fish Physiology*, 1st ed.; Hoar, W.S., Randall, D.J., Brett, J.R., Eds.; Academic Press: London, UK, 1977; Volume VIII, pp. 599–675.
55. Pérez-Sánchez, J.; Martí-Palanca, H.; Kaushik, S. Ration size and protein intake affect growth hormone (GH) levels, hepatic GH-binding and plasma insulin-like growth factor-I immunoreactivity in a marine teleost, gilthead sea bream (*Sparus aurata*). *J. Nutr.* **1995**, *125*, 546–552. [PubMed]
56. Monternier, P.A.; Marmillot, V.; Rouanet, J.L.; Roussel, D. Mitochondrial phenotypic flexibility enhances energy savings during winter fast in king penguin chicks. *J. Exp. Biol.* **2014**, *217*, 2691–2697. [CrossRef] [PubMed]
57. Salin, K.; Villasevil, E.M.; Auer, S.K.; Anderson, G.J.; Selman, C.; Metcalfe, N.B.; Chinopoulos, C. Simultaneous measurement of mitochondrial respiration and ATP production in tissue homogenates and calculation of effective P/O ratios. *Phys. Rep.* **2016**, *4*, e13007. [CrossRef] [PubMed]
58. Pérez-Sánchez, J.; Simó-Mirabet, P.; Naya-Català, F.; Martos-Sitcha, J.A.; Perera, E.; Bermejo-Nogales, A.; Benedito-Palos, L.; Calduch-Giner, J.À. Somatotropic axis regulation unravels the differential effect of nutritional and environmental factors in growth performance of marine farmed fish. *Front. Endocrinol.* **2018**, *9*, 687. [CrossRef] [PubMed]
59. Metcalfe, N.B.; Van Leeuwen, T.E.; Killen, S.S. Does individual variation in metabolic phenotype predict fish behavior and performance? *J. Fish Biol.* **2016**, *88*, 298–321. [CrossRef] [PubMed]
60. Knap, P.W.; Kause, A. Phenotyping for genetic improvement of feed efficiency in fish: Lessons from pig breeding. *Front. Genet.* **2018**, *9*, 184. [CrossRef] [PubMed]
61. Perera, E.; Simó-Mirabet, P.; Shin, H.S.; Rosell-Moll, E.; Naya-Català, F.; De las Heras, V.; Martos-Sitcha, J.A.; Karalazos, V.; Armero, E.; Arizcun, M.; et al. Selection for growth is associated in gilthead sea bream (*Sparus aurata*) with diet flexibility, changes in growth patterns and higher intestine plasticity. *Aquaculture* **2019**, *507*, 349–360. [CrossRef]
62. Piazzon, M.C.; Naya-Català, F.; Perera, E.; Palenzuela, O.; Sitjà-Bobadilla, A.; Pérez-Sànchez, J. Genetic selection for growth drives differences in intestinal microbiota composition and parasite disease resistance in gilthead sea bream. *Microbiome* **2020**, *8*, 168. [CrossRef]
63. Anttila, K.; Mänttäri, S. Ultrastructural differences and histochemical characteristics in swimming muscles between wild and reared Atlantic salmon. *Acta Physiol.* **2009**, *196*, 249–257. [CrossRef]
64. Zhang, Y. *Estimating Aerobic and Anaerobic Capacities Using the Respiratory Assessment Paradigm: A Validation Using Atlantic Salmon (Salmo salar) and European Sea Bass (Dicentrarchus labrax)*; The University of British Columbia: Vancouver, BC, Canada, 2016.
65. Bellringer, K.L.; Thorgaard, G.H.; Carter, P.A. Domestication is associated with reduced burst swimming performance and increased body size in clonal rainbow trout lines. *Aquaculture* **2014**, *420–421*, 154–159. [CrossRef]
66. Claireaux, G.; McKenzie, D.J.; Genge, A.G.; Chatelier, A.; Aubin, J.; Farrell, A.P. Linking swimming performance, cardiac pumping ability and cardiac anatomy in rainbow trout. *J. Exp. Biol.* **2005**, *208*, 1775–1784. [CrossRef]
67. Blasco, J.; Moya, A.; Millán-Cubillo, A.; Vélez, E.J.; Capilla, E.; Pérez-Sánchez, J.; Gutiérrez, J.; Fernández-Borrás, J. Growth-promoting effects of sustained swimming in fingerlings of gilthead sea bream (*Sparus aurata L.*). *J. Comp. Physiol. B* **2015**, *185*, 859–868. [CrossRef] [PubMed]
68. Vélez, E.J.; Lutfi, E.; Azizi, S.; Perelló, M.; Salmerón, C.; Riera-Codina, M.; Ibarz, A.; Fernández-Borràs, J.; Blasco, J.; Capilla, E.; et al. Understanding fish muscle growth regulation to optimize aquaculture production. *Aquaculture* **2017**, *467*, 28–40. [CrossRef]
69. Palstra, A.P.; Kals, J.; Böhm, T.; Bastiaansen, J.; Komen, H. Swimming performance and oxygen consumption as non-lethal indicators of production traits in atlantic salmon and gilthead seabream. *Front. Physiol.* **2020**, *11*, 759. [CrossRef]

70. Palstra, A.P.; Planas, J.V. *Swimming Physiology of Fish: Towards Using Exercise to Farm a Fit Fish in Sustainable Aquaculture*, 1st ed.; Springer: Berlin, Germany, 2013; pp. 1–429.
71. Palstra, A.P.; Beltran, S.; Burgerhout, E.; Brittijn, S.A.; Magnoni, L.J.; Henkel, C.V.; Jansen, H.J.; van den Thillart, G.E.; Spaink, H.P.; Planas, J.V. Deep RNA sequencing of the skeletal muscle transcriptome in swimming fish. *PLoS ONE* **2013**, *8*, e53171. [CrossRef] [PubMed]
72. Mylonis, I.; Simos, G.; Paraskeva, E. Hypoxia-inducible factors and the regulation of lipid metabolism. *Cells* **2019**, *8*, 214. [CrossRef]
73. Krishnan, J.; Suter, M.; Windak, R.; Krebs, T.; Felley, A.; Montessuit, C.; Tokarska-Schlattner, M.; Aasum, E.; Bogdanova, A.; Perriard, E.; et al. Activation of a HIF1α -PPARγ axis underlies the integration of glycolytic and lipid anabolic pathways in pathologic cardiac hypertrophy. *Cell Metab.* **2009**, *9*, 512–524. [CrossRef]
74. Shi, H.; Zhao, W.; Zhang, C.; Shahzad, K.; Luo, J.; Loor, J.J. Transcriptome-wide analysis reveals the role of PPARγ controlling the lipid metabolism in goat mammary epithelial cells. *PPAR Res.* **2016**, *2016*, 9195680. [CrossRef]
75. Liu, Y.; Ma, Z.; Zhao, C.; Wang, Y.; Wu, G.; Xiao, J.; McClain, C.J.; Li, X.; Feng, W. HIF-1α and HIF-2α are critically involved in hypoxia-induced lipid accumulation in hepatocytes through reducing PGC-1α-mediated fatty acid β-oxidation. *Toxicol. Lett.* **2014**, *226*, 117–123. [CrossRef] [PubMed]
76. Ban, J.J.; Ruthenborg, R.J.; Cho, K.W.; Kim, J.W. Regulation of obesity and insulin resistance by hypoxia-inducible factors. *Hypoxia (Auckl)* **2014**, *2*, 171–183.
77. Gaspar, J.M.; Mendes, N.F.; Corrêa-da-Silva, F.; Lima-Junior, J.C.; Gaspar, R.C.; Ropelle, E.R.; Araujo, E.P.; Carvalho, H.M.; Velloso, L.A. Downregulation of HIF complex in the hypothalamus exacerbates diet-induced obesity. *Brain. Behav. Immun.* **2018**, *73*, 550–561. [CrossRef] [PubMed]
78. Virtue, S.; Vidal-Puig, A. Nothing iffy about HIF in the hypothalamus. *PLoS Biol.* **2011**, *9*, e1001116. [CrossRef]
79. Gaspar, J.M.; Velloso, L.A. Hypoxia inducible factor as a central regulator of metabolism—Implications for the development of obesity. *Front. Neurosci.* **2018**, *12*, 813. [CrossRef] [PubMed]
80. Zhao, L.L.; Sun, J.L.; Liang, J.; Liu, Q.; Luo, J.; Li, Z.Q.; Yan, T.M.; Zhou, J.; Yang, S. Enhancing lipid metabolism and inducing antioxidant and immune responses to adapt to acute hypoxic stress in Schizothorax prenanti. *Aquaculture* **2020**, *519*, 734933. [CrossRef]
81. Listenberger, L.L.; Han, X.; Lewis, S.E.; Cases, S.; Farese, R.V., Jr.; Ory, D.S.; Schaffer, J.E. Triglyceride accumulation protects against fatty acid-induced lipotoxicity. *Proc. Natl. Acad. Sci. USA* **2003**, *100*, 3077–3082. [CrossRef] [PubMed]
82. Mylonis, I.; Sembongi, H.; Befani, C.; Liakos, P.; Siniossoglou, S.; Simos, G. Hypoxia causes triglyceride accumulation by HIF-1-mediated stimulation of lipin 1 expression. *J. Cell Sci.* **2012**, *125*, 3485–3493. [CrossRef] [PubMed]
83. Gracey, A.Y.; Troll, J.V.; Somero, G.N. Hypoxia-induced gene expression profiling in the euryoxic fish *Gillichthys mirabilis*. *Proc. Natl. Acad. Sci. USA* **2001**, *98*, 1993–1998. [CrossRef] [PubMed]
84. Lewis, C.A.; Brault, C.; Peck, B.; Bensaad, K.; Griffiths, B.; Mitter, R.; Chakravarty, P.; East, P.; Dankworth, B.; Alibhai, D.; et al. SREBP maintains lipid biosynthesis and viability of cancer cells under lipid- and oxygen-deprived conditions and defines a gene signature associated with poor survival in glioblastoma multiforme. *Oncogene* **2015**, *34*, 5128–5140. [CrossRef]
85. Wang, M.S.; Li, Y.; Peng, M.S.; Zhong, L.; Wang, Z.J.; Li, Q.Y.; Tu, X.L.; Dong, Y.; Zhu, C.L.; Wang, L. Genomic analyses reveal potential independent adaptation to high altitude in Tibetan chickens. *Mol. Biol. Evol.* **2015**, *32*, 1880–1889. [CrossRef] [PubMed]
86. Ge, R.L.; Simonson, T.S.; Gordeuk, V.; Prchal, J.T.; McClain, D.A. Metabolic aspects of high-altitude adaptation in Tibetans. *Exp. Biol.* **2015**, *100*, 1247–1255. [CrossRef]
87. Horscroft, J.A.; Kotwica, A.O.; Laner, V.; West, J.A.; Hennis, P.J.; Levett, D.; Howard, D.J.; Fernandez, B.O.; Burgess, S.L.; Ament, Z.; et al. Metabolic basis to Sherpa altitude adaptation. *Proc. Natl. Acad. Sci. USA* **2017**, *114*, 6382–6387. [CrossRef]
88. Ge, R.L.; Cai, Q.; Shen, Y.Y.; San, A.; Ma, L.; Zhang, Y.; Yi, X.; Chen, Y.; Yang, L.; Huang, Y. Draft genome sequence of the Tibetan antelope. *Nat. Commun.* **2013**, *1858*, 4. [CrossRef]
89. Qiu, Q.; Zhang, G.J.; Ma, T.; Qian, W.B.; Wang, J.Y.; Ye, Z.Q.; Cao, C.C.; Hu, Q.J.; Kim, J.; Larkin, D.M.; et al. The yak genome and adaptation to life at high altitude. *Nat. Genet.* **2012**, *44*, 946–949. [CrossRef] [PubMed]
90. Zhao, D.; Zhang, Z.; Cease, A.; Harrison, J.; Kang, L. Efficient utilization of aerobic metabolism helps Tibetan locusts conquer hypoxia. *BMC Genom.* **2013**, *14*, 631. [CrossRef] [PubMed]
91. Kang, J.; Ma, X.; He, S. Evidence of high-altitude adaptation in the glyptosternoid fish, *Creteuchiloglanis macropterus* from the Nujiang River obtained through transcriptome analysis. *BMC Evol. Biol.* **2017**, *17*, 229. [CrossRef]
92. Wright, J.R.; McCloskey, D.I.; Fitzpatrick, R.C. Effects of muscle perfusion pressure on fatigue and systemic arterial pressure in human subjects. *J. Appl. Physiol.* **1999**, *86*, 845–851. [CrossRef]
93. Wan, J.J.; Qin, Z.; Wang, P.Y.; Sun, Y.; Liu, X. Muscle fatigue: General understanding and treatment. *Exp. Mol. Med.* **2017**, *49*, e384. [CrossRef]
94. Zierath, J.R.; Hawley, J.A. Skeletal muscle fiber type: Influence on contractile and metabolic properties. *PLoS Biol.* **2004**, *2*, e348. [CrossRef] [PubMed]
95. Trappe, S.; Harber, M.; Creer, A.; Gallagher, P.; Slivka, D.; Minchev, K.; Whitsett, D. Single muscle fiber adaptations with marathon training. *J. Appl. Phisiol.* **2006**, *101*, 721–727. [CrossRef] [PubMed]
96. Nikinmaa, M. Oxygen-dependent cellular functions—Why fishes and their aquatic environment are a prime choice of study. *Comp. Biochem. Physiol. Part A Mol. Integr. Physiol.* **2002**, *133*, 1–16. [CrossRef]

97. Ton, C.; Stamatiou, D.; Liew, C.C. Gene expression profile of zebrafish exposed to hypoxia during development. *Physiol. Genom.* **2003**, *13*, 97–106. [CrossRef]
98. Zhong, X.P.; Wang, D.; Zhang, Y.B.; Gui, J.F. Identification and characterization of hypoxia-induced genes in *Carassius auratus* blastulae embryonic cells using suppression subtractive hybridization. *Comp. Biochem. Phys. B* **2009**, *152*, 161–170. [CrossRef]
99. Qi, D.; Chao, Y.; Wu, R.; Xia, M.; Chen, Q.; Zheng, Z. Transcriptome analysis provides insights into the adaptive responses to hypoxia of a Schizothoracine Fish (*Gymnocypris eckloni*). *Front. Physiol.* **2018**, *9*, 1326. [CrossRef] [PubMed]
100. Houlihan, D.F.; Carter, C.G.; McCarthy, I.D. Protein synthesis in animals. In *Nitrogen Metabolism and Excretion*, 1st ed.; Wright, P.J., Walsh, P.A., Eds.; CRC Press: Boca Ratón, FL, USA, 1995; pp. 1–29.
101. McCarthy, I.D.; Owen, S.F.; Watt, P.W.; Houlihan, D.F. Individuals maintain similar rates of protein synthesis over time on the same plane of nutrition under controlled environmental conditions. *PLoS ONE* **2016**, *11*, e0152239. [CrossRef]
102. Dauer, P.; Sharma, N.S.; Gupta, V.K.; Durden, B.; Hadad, R.; Banerjee, S.; Dudeja, V.; Saluja, A.; Banerjee, S. ER stress sensor, glucose regulatory protein 78 (GRP78) regulates redox status in pancreatic cancer thereby maintaining "stemness". *Cell Death Dis.* **2019**, *10*, 132. [CrossRef] [PubMed]
103. Park, K.W.; Eun Kim, G.; Morales, R.; Moda, F.; Moreno-Gonzalez, I.; Concha-Marambio, L.; Lee, A.S.; Hetz, C.; Soto, C. The endoplasmic reticulum chaperone grp78/bip modulates prion propagation in vitro and in vivo. *Sci. Rep.* **2017**, *7*, 44723. [CrossRef] [PubMed]
104. Wang, P.; Zhang, N.; Liang, J.; Li, J.; Han, S.; Li, J. Micro-RNA-30a regulates ischemia-induced cell death by targeting heat shock protein HSPA5 in primary cultured cortical neurons and mouse brain after stroke. *J. Neurosci. Res.* **2015**, *93*, 1756–1768. [CrossRef] [PubMed]
105. Pérez-Sánchez, J.; Terova, G.; Simó-Mirabet, P.; Rimoldi, S.; Folkedal, O.; Calduch-Giner, J.A.; Olsen, R.E.; Sitjà-Bobadilla, A. Skin mucus of gilthead sea bream (*Sparus aurata* L.). Protein mapping and regulation in chronically stressed fish. *Front. Phisiol.* **2017**, *8*, 34. [CrossRef] [PubMed]
106. Gisbert, E.; Andree, K.B.; Quintela, J.C.; Calduch-Giner, J.A.; Ipharraguerre, I.R.; Pérez-Sánchez, J. Olive oil bioactive compounds increase body weight, and improve gut health and integrity in gilthead sea bream (*Sparus aurata*). *Br. J. Nutr.* **2017**, *117*, 351–363. [CrossRef]

biology MDPI

Article

Role of Transportome in the Gills of Chinese Mitten Crabs in Response to Salinity Change: A Meta-Analysis of RNA-Seq Datasets

Adeel Malik [1,*] and Chang-Bae Kim [2,*]

1 Institute of Intelligence Informatics Technology, Sangmyung University, Seoul 30316, Korea
2 Department of Biotechnology, Sangmyung University, Seoul 30316, Korea
* Correspondence: adeel@procarb.org (A.M.); evodevo@smu.ac.kr (C.-B.K.)

Simple Summary: *Eriocheir sinensis* is a freshwater crab and is considered as one of the most important cost-effective species for freshwater aquaculture. *E. sinensis* can grow in both freshwater and brackish waters. In order to adapt to this changing salinity, *E. sinensis* can regulate the osmotic concentration of its hemolymph. Additionally, studies have shown that gills are one of the most important tissues in osmoregulation. In this work, we performed the first meta-analysis of publicly available RNA-Seq datasets to identify differentially expressed genes in the gills under different salinity conditions. The results highlighted that many different types of transporters show altered expression because of salinity change. Some of these transporters may serve as novel or new biomarkers for osmoregulation. The findings of this work also suggest that cellular processes related to many morphological changes are also affected.

Abstract: Chinese mitten crab (CMC) or *Eriocheir sinensis* is a strong osmoregulator that can keep rigorous cellular homeostasis. CMC can flourish in freshwater, as well as seawater, habitats and represents the most important species for freshwater aquaculture. Salt stress can have direct effects on several stages (e.g., reproduction, molting, growth, etc.) of the CMC life cycle. To get a better overview of the genes involved in the gills of CMC under different salinity conditions, we conducted an RNA-Seq meta-analysis on the transcriptomes of four publicly available datasets. The meta-analysis identified 405 differentially expressed transcripts (DETs), of which 40% were classified into various transporter classes, including accessory factors and primary active transporters as the major transport classes. A network analysis of the DETs revealed that adaptation to salinity is a highly regulated mechanism in which different functional modules play essential roles. To the best of our knowledge, this study is the first to conduct a transcriptome meta-analysis of gills from crab RNA-Seq datasets under salinity. Additionally, this study is also the first to focus on the differential expression of diverse transporters and channels (transportome) in CMC. Our meta-analysis opens new avenues for a better understanding of the osmoregulation mechanism and the selection of potential transporters associated with salinity change.

Keywords: Chinese mitten crab; *Eriocheir sinensis*; transportome; transporters; salinity; osmoregulation; transcriptome; RNA-Seq; meta-analysis; gills

Citation: Malik, A.; Kim, C.-B. Role of Transportome in the Gills of Chinese Mitten Crabs in Response to Salinity Change: A Meta-Analysis of RNA-Seq Datasets. *Biology* **2021**, *10*, 39. https://doi.org/10.3390/biology10010039

Received: 22 December 2020
Accepted: 6 January 2021
Published: 8 January 2021

Publisher's Note: MDPI stays neutral with regard to jurisdictional claims in published maps and institutional affiliations.

1. Introduction

Chinese mitten crab (CMC), also known as *Eriocheir sinensis*, is a freshwater crab native to China and found as an invasive species in Europe and the United States of America (USA) [1]. This crab is one of the most essential cost-effective species for freshwater aquaculture in China, the Eastern Pacific coast, and the Korean Peninsula [2]. CMC is one of the unique crustaceans in a sense that they require two differing environmental conditions to complete their life cycle. As an adult, CMC spends its life in freshwater and later moves to a brackish (saline) water for reproduction [3]. Thus, CMC can regulate the osmotic concentration of its hemolymph so that it can better adjust to the new saline environment [4]. Therefore, these unique features make CMC a model organism among crustaceans to study the mechanism of osmoregulation under varying salinity conditions. All organisms exhibit adaptation to extracellular salinity within clearly defined ranges [5]. However, the mechanism is quite complex, and therefore, to better understand the response of these organisms towards salinity stress requires not only the knowledge of its molecular components but a thorough understanding of the complete biological system. To maintain the osmotic homeostasis during salinity stress, suitable signal transduction pathways are triggered in the aquatic animals [6–8]. However, there is limited knowledge regarding such signal transduction events that regulate osmoregulation in an aquatic organism [7]. The regulation of osmotic pressure and ionic balance in response to a change in salinity is an energy-dependent process in which lots of energy is utilized [9]. The significance of such energy metabolism-related pathways has been highlighted in previous studies [9,10]. One of the key groups of proteins that maintain osmotic pressure at the expense of huge quantities of energy are various ion transporters and ion transport channels [11]. All these membrane transporters and channels are collectively known as the "transportome", which regulates the movement of various ions, nutrients, and drugs across biological membranes [12]. Sodium/potassium-transporting ATPase (Na^+/K^+-ATPase) of the gill epithelium is one of the essential ion transporters for hyperosmotic crustaceans and has been implicated in osmoregulation, as well as ion regulation [13]. Transporters such as Na^+-K^+-2Cl cotransporter-1 (*NKCC1*) that belongs to the chloride–cation cotransporters gene family [14] are involved in the secretion of Cl^- from the body of crustaceans to accomplish salt and water retention [9]. Similarly, V-type ATPase is one of the essential components of sodium uptake in the gills [15] and, therefore, provides a driving force for sodium to enter into gill epithelial cells via an epithelial sodium channel [13].

Some of these transporters have been highlighted in transcriptome-based studies of CMC during salinity stress. Specifically, such RNA-Seq-based studies in the gills of *E. sinensis* have mainly focused on the key transporter genes, such as Na^+/K^+-ATPase, *NKCC1*, and V-type ATPase [3,9,16]. However, in addition to these well-known biomarker genes, there are several other ion transport enzymes that are involved in ion transport [17]. Moreover, the uptake of ions by branchia in osmoregulating crustaceans is accomplished by the regulated activity of an array of transport proteins and transport-related enzymes [18]. Currently, more than 1500 transporter families have been described based on their functional and phylogenetic information [19]. Broadly, all these families are grouped into seven major classes in this transport classification (TC) system and include (i) channels or pores (TC 1), (ii) electrochemical potential-driven transporters (TC 2), (iii) primary active transporters (TC 3), (iv) group translocators (TC 4), (v) transmembrane electron carriers (TC 5), (vi) accessory factors involved in transport (TC 8), and (vii) incompletely characterized transport systems (TC 9). Many of these transporters are multi-subunit channels encoded by several genes [20], of which only a limited number of such subunits are highlighted. Therefore, in order to get a broader understanding of the diverse types of transporters involved in osmotic regulation in the gills of CMC (which lacked in the previous RNA-Seq-based studies of *E. sinensis*), we carried out a meta-analysis of four independent RNA-Seq studies. RNA-Seq has become the method of choice for differential expression analyses; however, due to their high costs, limited number of biological replicates are common in studies, which, in turn, leads to a low detection of differentially expressed genes (DEGs) [21]. Therefore, a meta-analysis of several independent studies focused on a specific biological question offers a useful way to overcome the limited detection power of independent studies. Meta-analysis approaches are a set of statistical techniques that merge multiple and independent studies to get a single common and significant outcome [22]. In a meta-analysis, transcriptome data from multiple individual studies can be analyzed in order to define common molecular signatures, improve reproducibility, or find more reliable biomarkers [23,24]. To the best of our knowledge, this study is the first ever RNA-Seq meta-analysis to investigate the in-depth account of various transporters implicated in salinity stress in *E. sinensis*.

2. Materials and Methods

2.1. Datasets

The National Center for Biotechnology Information (NCBI) BioProject (https://www.ncbi.nlm.nih.gov/bioproject) database was queried to search for RNA-Seq studies in response to salinity changes in *E. sinensis*. Studies that focused on a change in salinity from freshwater to higher salinity using tissue samples from gills were selected only. RNA-Seq reads (fastq files) for the selected samples from each study (Table 1) were retrieved from the NCBI's sequence read archive (SRA: https://trace.ncbi.nlm.nih.gov/Traces/sra/) database by using the fastq-dump program of the SRA Toolkit (http://ncbi.github.io/sra-tools/).

Table 1. Summary of the datasets and the number of trimmed reads retained postfiltering.

Dataset	BioProject Accession	Type *	#Raw Reads	#Trimmed Reads	Reads Retained Postfiltering (%)	Reference
DS1	PRJNA481259		49,648,084	47,331,397	95.33	[3]
DS2	PRJNA488907	PE	17,817,553,2	14,353,948,4	80.42	[16]
DS3	PRJNA80779		64,212,034	60,050,212	93.51	[10]
DS4	PRJNA508867	SE	75,845,511	75,292,415	99.28	[9]

* PE = Paired-end. SE = Single-end.

2.2. De Novo Transcriptome Assembly

Prior to assembly, Trimmomatic [25] within the Trinity v.2.10.0 [26] pipeline was used to filter the reads by removing the adapter sequences, and then, the reads were scanned with a 4-base-wide sliding window, cutting when the average quality per base dropped below 20. Furthermore, the reads were removed if their lengths dropped below 50% of the original read length. De novo transcriptome assembly was performed on the resulting quality filtered reads by using the docker image of Trinity v.2.10.0. Each dataset was assembled separately.

2.3. Assembly Statistics and Completeness

Assembly statistics were calculated by using the TrinityStats.pl script provided within the Trinity pipeline. The percentage of reads mapping back to the assembly was computed with Bowtie2 [27]. BUSCO v.4.0.6 [28] was used to assess the transcriptome completeness of all datasets against the arthropoda_odb10 lineage. To quantify the transcript abundance, we used the ultra-fast alignment-free method kallisto [29] bundled within the docker image of Trinity.

2.4. Functional Annotation of the Transcripts and Enrichment Analysis

Potential coding regions and open reading frames (ORFs) were predicted with the TransDecoder [26] pipeline using the default parameters. The resultant protein sequences were then scanned against the UniProtKB/Swiss-Prot [30] database using BLASTp [31]. In contrast, HMMER v.3.1b2 [32] was used to identify the protein domains against the Pfam [33] database. Additionally, we also scanned the protein sequences from each dataset against the amino acid sequences (Crabdb) previously identified from the draft crab genome [34]. Orthologous groups were identified by using the online functional annotation tool eggNOG-mapper [35]. The classification of differentially expressed transcripts (DETs) into various transporter classes was performed by a BLASTp-based search against the Transport Classification Database (TCDB) [19]. Enrichment of gene ontology (GO) terms and KEGG pathways of the DETs were performed using the ShinyGO v.0.61 [36] and KOBAS v.3.0 [37] web-based tools. Enriched categories with a corrected p-value < 0.05 cutoff were selected only.

2.5. *Differential Expression and Transcriptome Meta-Analysis*

Differential expression (DE) analysis for individual studies was performed at the isoform level using the edgeR [38] package within the Trinity pipeline. The output of the DE analysis from all datasets was further employed for the meta-analysis, which was carried out by using the metaRNASeq [39] package. metaRNASeq uses Fisher's combined probability test [40] for meta-analyses and has been used in several studies [41–43]. As the name suggests, this method combines the *p*-values for each gene from individual studies using the following formula [39,40]:

$$F_g = -2\sum_{s=1}^{S} ln(p_{gs})$$

Here, p_{gs} corresponds to the raw *p*-value calculated for a gene (g) in a differential analysis for the study (s). With independent *p*-values, F_g has a χ^2 distribution with 2S degrees of freedom. Smaller *p*-values result in larger F_g, and therefore, the null hypothesis is rejected [43]. *p*-values were adjusted for the Benjamini–Hochberg false discovery rate (FDR), and a corrected *p*-value < 0.05 was considered as statistically significant [44]. Transcripts (genes) that exhibited an FDR of <0.05 and an average fold change (FC) of ≥ 2 ($log_2FC \leq -1$ or $log_2FC \geq 1$) were considered as DETs in the meta-analysis. Heatmap and correlation analyses were conducted by using the pheatmap (https://cran.r-project.org/package=pheatmap) and Hmisc (https://cran.r-project.org/package=Hmisc) packages, respectively, within the R (https://www.r-project.org/) programming environment.

2.6. *Network Analysis and Community Detection*

The amino acid sequences of the DETs were mapped to the STRING v.11.0 [45] database (STRING-DB) to predict the interactions between these sequences using *Drosophila melanogaster* (fruit fly) as the reference organism. The interaction network of the DETs was visualized and further analyzed with Cytoscape v.3.8.0 [46]. Node degree distribution and community structure were estimated by using the NetworkAnalyzer [47] and GLay [48] plugins of Cytoscape, respectively.

3. Results

3.1. *Datasets and De Novo Transcriptome Assembly*

Based on the search criteria (see Methods), four RNA-Seq datasets were retrieved for this study. Although all these studies used freshwater and high salinity samples, the concentration of salt in higher salinity groups varied. For example, the salinity in two of these studies was 30% [3,10], whereas the other two used 20-ppt and 25-ppt saline water [9,16], respectively. All the selected datasets were previously published, and among them, three were pair ends, and one study represented single-end reads. Table 1 summarizes these datasets and provides statistics on reads from each individual dataset. We de novo assembled each of these RNA-Seq datasets separately. Furthermore, to assess the read composition of our assembly, we generated statistics on all reads that map to our assembled transcripts. We observed that the number of reads that mapped back to the assembly ranged between 94% to 97% (Table S1). These numbers are above 80% and signify a good quality assembly [26]. The average contig length ranged between 630.8 and 889.08 for DS4 and DS3, respectively. Similarly, the N50 value for DS4 was lower (974), whereas DS3 had the largest (1526) N50 value. The completeness of the assembled transcripts with BUSCO showed the percentage of orthologs among the Arthropods represented in our assembled CMC transcriptomes. Specifically, of the 1013 BUSCO groups searched, a high percentage (72–94%) of complete orthologs belonging to the phylum Arthropods were detected for each dataset (Figure S1A). This includes both the complete single copy, as well as complete duplicates (putative paralogs or complete genes with multiple copies). For non-model organisms such as *E. sinensis*, BUSCO scores within the range of 50–95% are expected [28].

3.2. Identification of Coding Regions and Functional Annotation

TransDecoder identified 179,118, 279,120, 52,132, and 147,537 transcripts with the longest open reading frames (ORFs) for four CMC datasets (DS1, DS2, DS3, and DS4), respectively. Each of these ORFs were at least 100 amino acids long, and sequences shorter than 100 residues were excluded. These transcripts were then scanned against the UniProt, Crabdb, and Pfam databases to identify ORFs with homology to known proteins. Based on the BLASTp search, the percentage of sequences that mapped to the UniProt database ranged between 15.71% (DS2) and 27.89% (DS1). These numbers are consistent and slightly better in some cases as compared to the previous studies on this non-model organism [9,10,16]. Interestingly, the number of longest ORF sequences that mapped to Crabdb were also in similar ranges, e.g., between 15.83% (DS2) and 23.39% (DS3) (Figure S1B). These low numbers could be attributed to the fact that, at present, the number of annotated sequences in Crabdb is only around 7500, suggesting that many of the transcripts would have no known match in this database. Similarly, about 12–22% of transcripts with long ORFs had orthologs in the KEGG database. In contrast, the number of orthologs identified by eggNOG-mapper were higher as compared to KEGG (Figure S1B). Specific COG categories for each dataset are provided as a supplementary figure (Figure S2). Among all the databases, transcripts across all datasets showed higher annotations against the Pfam database. The results from the UniProt and Pfam searches were then integrated to further select the coding regions by using TransDecoder. The results indicated that the total number of protein coding transcripts that shared some similarity with the known sequences in the reference databases ranged between 43.39% (DS2) and 48% (DS3) (Figure S1C). This suggests that many of the CMC transcripts could be novel that are not present in these databases. Nevertheless, it may still be possible to identify a higher number of transcripts sharing similarity with known sequences using other large databases such as the NCBI nonredundant (NR) protein database [9,16]. It may be noted that the size of the NR database is huge, and performing similarity searches against such databases is computationally very expensive.

3.3. Meta-Analysis of CMC Gills Transcriptome

Following the DE analysis of each individual dataset, we used the Fisher method and identified 3574 DETs across all four datasets. From this list, we first selected only those transcripts that exhibited an average fold change of $\geq$2 ($\log_2$FC ≤ -1 or $\log_2$FC ≥ 1) and an adjusted *p*-value of <0.05. This resulted in 1748 DETs, of which a large number of transcripts (~1300) were further excluded from the analysis because of their inconsistent or conflicting expression patterns. Such inconsistent transcripts were upregulated in some studies and exhibited downregulation in others. Therefore, only those transcripts that showed consistent expression in the four datasets were considered as DE (Figure 1).

For example, a transcript upregulated in one dataset was also upregulated in the other three datasets and vice versa. No gene that was identified as DE in the individual studies was lost in the meta-analysis. Although the meta-analysis identified four transcripts that were not detected as being DE in individual studies alone (Table S2), their average fold change was slightly lower (FC 1.5) as compared to the 2 FC criteria set in the current study. Hence, 405 DETs were finally retained for further downstream analysis (Figure 2).

Of the 405 DETs, 272 were downregulated, whereas 133 were upregulated between freshwater and high salinity conditions. Figure 3 provides the overview of top 20 up- and downregulated transcripts, and the complete list of 405 transcripts with detailed annotations is provided as additional information in Table S3.

Moreover, several of the 405 DETs exhibited very high correlations (Table 2). From these data, we observed that the topmost downregulated transcript encodes for protein argonaute-2 (*AGO2*), whereas mitochondrial succinate–CoA ligase (GDP-forming) subunit beta (*SUCLG2*), also known as GTP-specific succinyl-CoA synthetase (G-SCS), was encoded by the topmost upregulated transcript.

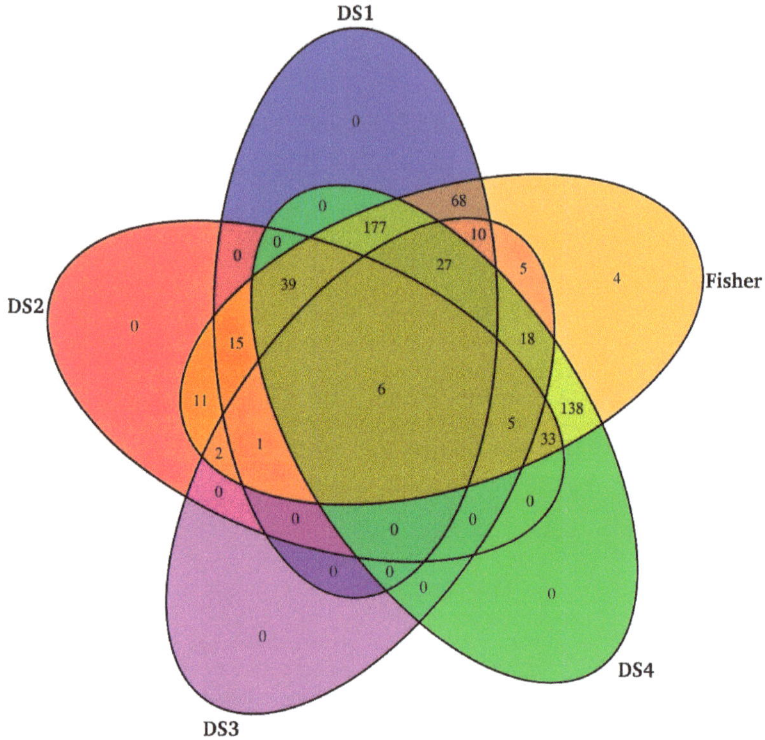

Figure 1. Venn diagram showing the number of differentially expressed transcripts identified from individual studies and meta-analysis based on the Fisher method.

Among the list of 405 DETs, six (including one of the topmost upregulated, *SUCLG2*) were identified as DE across all four datasets, as well as by the Fisher method (Table 3).

3.4. GO and KEGG Enrichment Analyses of DETs

Using the ShinyGO tool, enriched GO categories associated with DETs were identified in terms of the biological process (BP), molecular function (MF), and cellular component (CC). Based on this analysis, the GO term "response to stimulus" was the most enriched biological process, followed by several others related to development, such as anatomical structure development, multicellular organism development, and system development (Figure 4A and Table S4). Besides the enrichment of various developmental processes, an interesting observation was that out of 30 topmost enriched GO biological processes, at least six were related to the nervous system and included nervous system development, neuron differentiation, the generation of neurons, neurogenesis, neuron development, and neuron projection development (Table S4). Reports have suggested that, during osmotic changes, the neuroendocrine system may modulate ion transport in the gill epithelium through neurotransmitter signaling [9]. One of the key enzymes that was identified to be downregulated in high salinity as compared to freshwater conditions in the meta-analysis was the alpha subunit of sodium/potassium-transporting ATPase (Na^+/K^+-ATPase). This enzyme was found to be part of various enriched biological processes, such as the response to stimulus and nervous system development, in the meta-analysis.

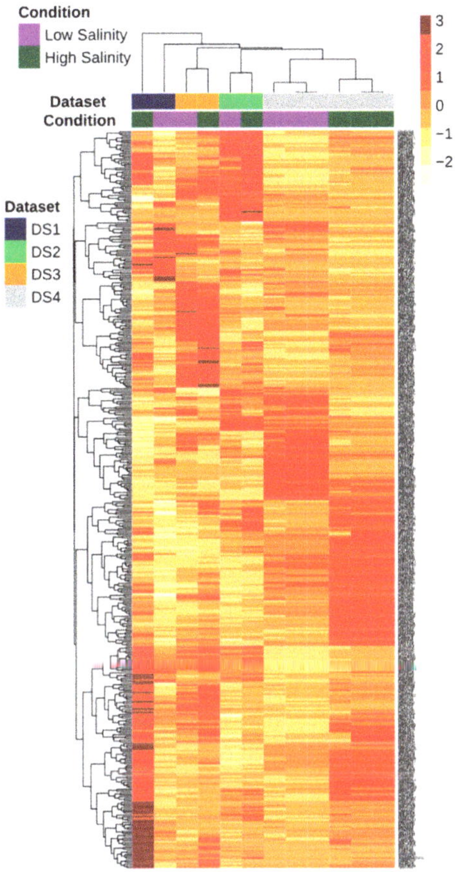

Figure 2. Heatmap showing differentially expressed transcripts identified by meta-analysis in the gills transcriptome of four different Chinese mitten crab (CMC) datasets under salinity. Only transcripts with an average fold change (FC) of $\geq$2.0 and a Benjamini-Hochberg adjusted p-value of <0.05 were considered to be differentially expressed.

We also identified DETs that encode five V-ATPase subunits (subunits a1, C, D1, H, and the proteolipid). The transcripts for all these subunits were upregulated by more than two FC, except in the case of subunit a1 (V-type proton ATPase 116-kDa subunit a1 encoded by the *ATP6V0A1* gene), which was downregulated. Among these, some of them, such as subunit H, were annotated to be involved in developmental processes (BP), whereas others (e.g., subunit D1) were involved in various molecular functions (MF), including nucleoside-triphosphatase activity, ATPase activity, etc. Additionally, the meta-analysis also identified subunits B, E, and G of the V1 cytoplasmic domain and d1 and e2 of the V0 membrane-bound domain of the V-ATPase complex to be differentially expressed. However, their expression patterns across all studies were conflicting.

One of the downregulated transcripts in the meta-analysis was identified to share similarity with integrin alpha-PS2 of *Drosophila* encoded by an inflated (*if*) gene. This transcript was identified in 27 out of 30 enriched GO biological processes, e.g., development, differentiation, morphogenesis, locomotion, and chemotaxis. The remaining three enriched processes in which integrin alpha-PS2 was not found were related to mRNA splicing. In *Drosophila*, such integrins are known to mediate the specific migration of tracheal cells in a given direction [49]. In contrast, this integrin was found in only two (protein binding and protein dimerization activity) enriched MF categories. Most of the enriched GO molecular functions were either related to the binding of various types of molecular compounds (e.g., nucleotide, purine, carbohydrate, small molecule, drug, etc.) or exhibited phosphatase/kinase activities (Figure 4B). The gene that was part of many enriched molecular functions encodes unconventional myosin IC (*Myo1C*), which has been implicated in many roles, including tension sensing and calcium-dependent cargo binding [50]. *Myo1C*, which was identified to be downregulated in the current study, belongs to a superfamily of actin-based motors that have roles in various cellular processes. Another unconventional myosin that was downregulated by more than four-fold was *Myo18a*, which plays an essential role in the organization and structure of the Golgi complex.

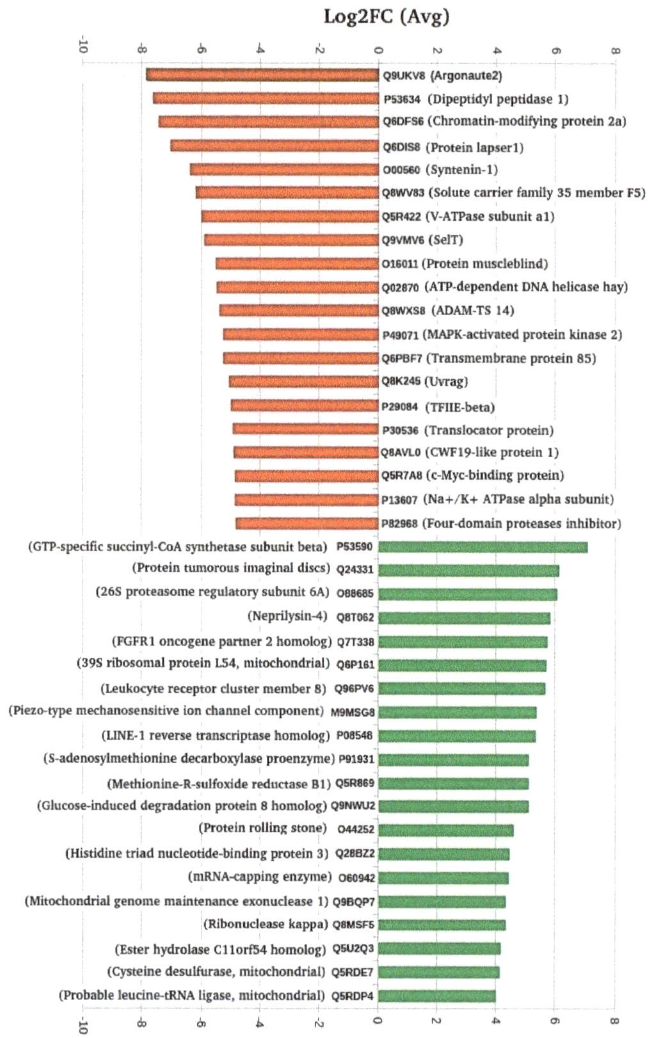

Figure 3. Top 20 up- and downregulated differentially expressed transcripts.

Table 2. Top 20 highly correlated differentially expressed transcripts (DETs). P1 and P2 represent the first and the second proteins in a correlated pair. Direction of the arrow represents whether the representative transcript is downregulated (↓) or upregulated (↑).

P1	P1 Description	P2	P2 Description	Effect	Cor	Corrected *p*-Value
Q9D4P0	ADP-ribosylation factor-like protein 5B	B2D0J5	Venom carboxylesterase-6	↓	1	9.4355×10^{-7}
Q7ZV80	Survival of motor neuron-related-splicing factor 30	B0WTN3	Eukaryotic translation initiation factor 3 subunit M	↓	0.99	1.2953×10^{-6}
Q04164	Putative epidermal cell surface receptor	Q9UQC9	Calcium-activated chloride channel regulator 2	↓	0.99	1.5143×10^{-6}
Q8C0K5	Graves disease carrier protein homolog (GDC) (Mitochondrial solute carrier protein homolog)	Q13336	Urea transporter 1	↑	0.99	3.4321×10^{-6}
Q5REW1	Iodotyrosine deiodinase 1 (IYD-1)	Q5I7G2	Retinoic acid receptor RXR	↑	0.99	3.4321×10^{-6}
Q9Y535	DNA-directed RNA polymerase III subunit RPC8	Q5VWG9	Transcription initiation factor TFIID subunit 3	↓	0.99	3.4321×10^{-6}
Q04164	Putative epidermal cell surface receptor	Q27421	Protein outspread	↓	0.99	3.4321×10^{-6}
Q8WXS8	A disintegrin and metalloproteinase with thrombospondin motifs 14	O16011	Protein muscleblind	↓	0.99	3.4321×10^{-6}
O74503	Upstream activation factor subunit spp27	Q53CF6	Cytochrome c oxidase subunit 7A1, mitochondrial	↑	0.99	3.7031×10^{-6}
Q4R3Y4	Long-chain fatty acid transport protein 4 (FATP-4)	Q7ZV80	Survival of motor neuron-related-splicing factor 30	↓	0.99	3.9073×10^{-6}
Q68HB4	Profilin	P56616	Ubiquitin-conjugating enzyme E2 C	↓	0.99	6.1125×10^{-6}
Q4R3Y4	Long-chain fatty acid transport protein 4(FATP-4)	B0WTN3	Eukaryotic translation initiation factor 3 subunit M	↓	0.99	6.1125×10^{-6}
Q86WZ6	Zinc finger protein 227	Q8C0K5	Graves disease carrier protein homolog	↑	0.99	7.7076×10^{-6}
P53590	Succinate–CoA ligase (GDP-forming) subunit beta, mitochondrial	P21158	C-factor	↑	0.99	7.7076×10^{-6}
Q9D4P0	ADP-ribosylation factor-like protein 5B	P82968	Four-domain proteases inhibitor	↓	0.99	7.7076×10^{-6}
Q04833	Low-density lipoprotein receptor-related protein	P82968	Four-domain proteases inhibitor	↓	0.99	8.8569×10^{-6}
Q9VB68	Serine protease grass	Q5VWG9	Transcription initiation factor TFIID subunit 3	↓	0.99	8.8569×10^{-6}
Q8K0U4	Heat shock 70-kDa protein 12A	Q8N539	Fibrinogen C domain-containing protein 1	↓	0.99	1.2159×10^{-5}
A6QP05	Dehydrogenase/reductase SDR family member 12	Q9CY58	Plasminogen activator inhibitor 1 RNA-binding protein	↑	0.99	1.5922×10^{-5}
P29844	Endoplasmic reticulum chaperone BiP	Q91V92	ATP-citrate synthase	↓	0.99	1.6280×10^{-5}

Table 3. List of six transcripts and their annotations (based on BLASTp) that were commonly identified in the individual datasets and the Fisher method. Direction of the arrow represents whether the representative transcript is downregulated (↓) or upregulated (↑).

UniProt ID	Gene Name	Organism	Protein Name	Gene Ontology (Biological Process)	Effect
P17789	*ttk*	*Drosophila melanogaster* (Fruit fly)	Protein tramtrack, beta isoform	branch fusion, open tracheal system (GO:0035147)	↓
P53590	*SUCLG2*	*Sus scrofa* (Pig)	Succinate–CoA ligase (GDP-forming) subunit beta, mitochondrial	succinyl–CoA metabolic process (GO:0006104)	↑
Q28BZ2	*Hint3*	*Xenopus tropicalis* (Western clawed frog)	Histidine triad nucleotide-binding protein 3	NA	↑
Q5U2Q3	NA	*Rattus norvegicus* (Rat)	Ester hydrolase C11orf54 homolog	NA	↑
Q6DIS8	*Lzts2*	*Xenopus tropicalis* (Western clawed frog)	Leucine zipper putative tumor suppressor 2 homolog	microtubule severing (GO:0051013)	↓
Q8BTN6	*Leng9*	*Mus musculus* (Mouse)	Leukocyte receptor cluster member 9	NA	↓

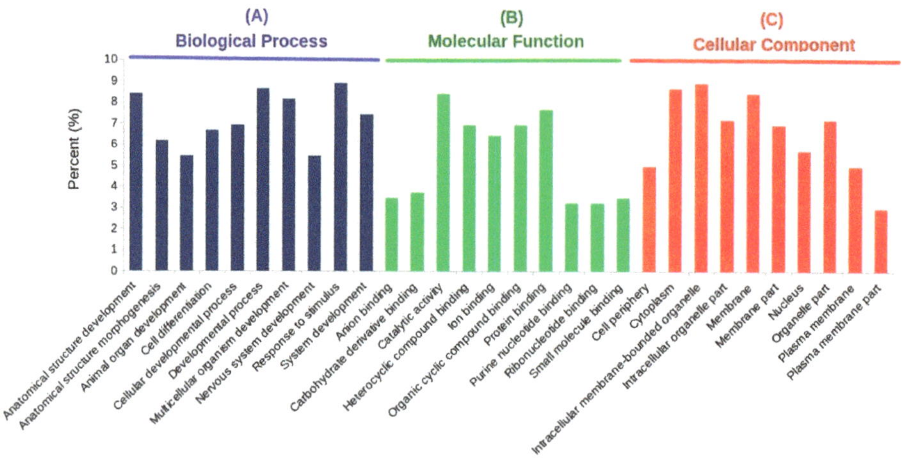

Figure 4. Gene ontology (GO) enrichment of differentially expressed transcripts identified by the meta-analysis. Only the top 10 processes for each GO category in terms of the biological process (BP), molecular function (MF), and cellular component (CC) are shown. A complete list of enriched terms is provided as Table S4.

Among the GO cellular components, the most enriched were related to membranes (e.g., intracellular membrane-bounded organelles, the plasma membrane, the Golgi membrane, etc.); the cytoplasm; cell junction; and nucleus (Figure 4C). The proteins most frequently found in these categories include different subunits of V-type proton ATPase.

The KEGG enrichment analysis identified 12 statistically significant pathways to be most important in the DETs. Most of these enriched pathways were related to metabolism, ECM-receptor interaction, and oxidative phosphorylation (Table 4). A total of 93 DETs were mapped to these pathways, of which 60 were downregulated, while the remaining 33 exhibited upregulation. One of the topmost downregulated transcripts encoded a charged multivesicular body protein 2a (*CHMP2a*), which belongs to the Snf family of small coil-coiled proteins. Such proteins are key components of the endosomal sorting required for transport complex III (*ESCRT-III*), which are essential for the formation of multivesicular bodies (MVBs) and the sorting of cargo proteins to MVBs [51]. V-ATPase subunit a1 was another topmost downregulated transcript identified to play essential roles in three different enriched KEGG pathways—namely, metabolic pathways and phagosome and oxidative phosphorylation. In contrast, the topmost upregulated gene, *SUCLG2*, was also implicated in KEGG metabolic pathways. The regulation of osmotic pressure via different transporters and channels is an energy-dependent process [11]. In a given tissue, the concentration of mitochondrial oxidative phosphorylation complexes is adjusted to the highest energy conversion requirements [52]. The activity of such mitochondrial complexes is regulated by tissue metabolic stress to sustain energy metabolism homeostasis. In addition to metabolic pathways, the roles of other pathways, such as the environmental information processing, signal transduction, and antioxidant pathways, were previously highlighted [3,9,10].

3.5. Interaction Network of DETs

STRING-DB was used to predict the interactions between these DETs using the common fruit fly (*Drosophila melanogaster*) as the source organism. Of these DETs, the amino acid sequences of only 57 (13 up- and 44 downregulated) transcripts were mapped to the fruit fly proteome, whereas an additional 231 proteins were predicted as their interacting partners. Therefore, the interaction network consisted of 288 nodes, each representing a gene, and 905 edges (interactions between nodes) (Figure S3). The node degree of each node in the network was calculated with the NetworkAnalyzer plugin of Cytoscape and ranged between 1 and 31. The topmost two nodes were considered as hubs, because they exhibited a node degree of ≥30. These nodes were identified to encode for a bifunctional glutamate/proline–tRNA ligase (*GluProRS*: node degree = 31) and a DNA-directed RNA polymerase II subunit RPB2 (*RpII140*: node degree = 30). Of these two, *GluProRS* was predicted by STRING-DB, whereas *RpII140* was one of the downregulated transcripts identified in the meta-analysis.

Table 4. Enriched KEGG pathways detected in the DETs identified by the RNA-Seq meta-analysis of Chinese mitten crab (CMC) gills under salinity.

Pathway Name	Pathway ID	No. of Input Genes	Total No. of Genes	Corrected *p*-Value
Metabolic pathways	dme01100	52	1111	0.00001
Apoptosis—fly	dme04214	9	63	0.00034
Phagosome	dme04145	10	89	0.00067
ECM-receptor interaction	dme04512	4	12	0.00316
Oxidative phosphorylation	dme00190	11	144	0.00471
Endocytosis	dme04144	10	122	0.00497
Glycerophospholipid metabolism	dme00564	6	63	0.02182
Pyruvate metabolism	dme00620	5	46	0.02654
Sphingolipid metabolism	dme00600	4	28	0.02680
Spliceosome	dme03040	8	128	0.04231
Glycolysis/Gluconeogenesis	dme00010	5	55	0.04283
Protein processing in endoplasmic reticulum	dme04141	8	133	0.04868

There is a growing interest in the application of biological networks to address important biological phenomena [53]. Network-based studies offer a framework to get a broad overview of data that includes several interacting groups. This potential of a community-based network analysis allows researchers to investigate and compare a diverse set of proteins and their interactions within the context of modules or communities identified in the network [54]. Therefore, we attempted to identify the modular structure of the interaction network and identified that the network could be divided into 12 modules or communities (clusters) (Figure 5). However, only eight modules with 10 or more nodes (genes) were identified by using the GLay implementation of the fast-greedy algorithm [55] within the clusterMaker [56] plugin of Cytoscape. The algorithm finds clusters by repetitiously discarding edges from the network and then looks again for the linked nodes [57]. We performed an enrichment analysis for each of these eight individual clusters to identify enriched GO terms and KEGG pathways associated with these modules. The number of nodes in each of these clusters ranged between 11 (cluster 5) and 60 (cluster 2), respectively. The enrichment analysis of all the clusters suggested that each of these clusters was involved in performing specific functional roles. For example, most of the enriched GO processes in the largest cluster 2 were related to morphogenesis, development, or differentiation (Figure 5 and Table S5). In contrast, the smallest cluster 2 was enriched with processes related to oxidoreductase or sulfurtransferase activities, including others. Many transporters were also found in almost every cluster; however, only clusters 2, 5, and 9 were enriched with different transporter classes (Figure 5).

3.6. Transporters Implicated in Salinity Change

The above network analysis highlighted that almost all the clusters contained transporters; however, only clusters 2, 5, and 9 were enriched ($p < 0.05$) with different transporter classes (Figure 5). Moreover, most of the nodes in the network were proteins predicted as the interacting partners of a limited number of DETs that were mapped to STRING-DB. Therefore, to get a detailed account of the diversity of various types of transporters, we scanned the DETs against the TCDB using the BLASTp program. Although the role of transporters in the gills of CMC in response to salinity was suggested in previous studies, only a limited number and types of such transporters are highlighted. Most of these studies focused on the roles of only well-known markers, such as Na+/K+-ATPase, V-ATPases, and Cu2+ or Ca2+ transport ATPases [3,9]. Moreover, recognizing the fact that, at present, hundreds of transporter families have been described [19], focusing on only few marker genes does not provide the global overview of the role of the transportome in a salinity change. In this work, we identified that about 40% (162/405) of the DETs were classified as belonging to one or the other transporter class. These transporters are further described based on the most abundant classes or families identified in the DETs (Table 5).

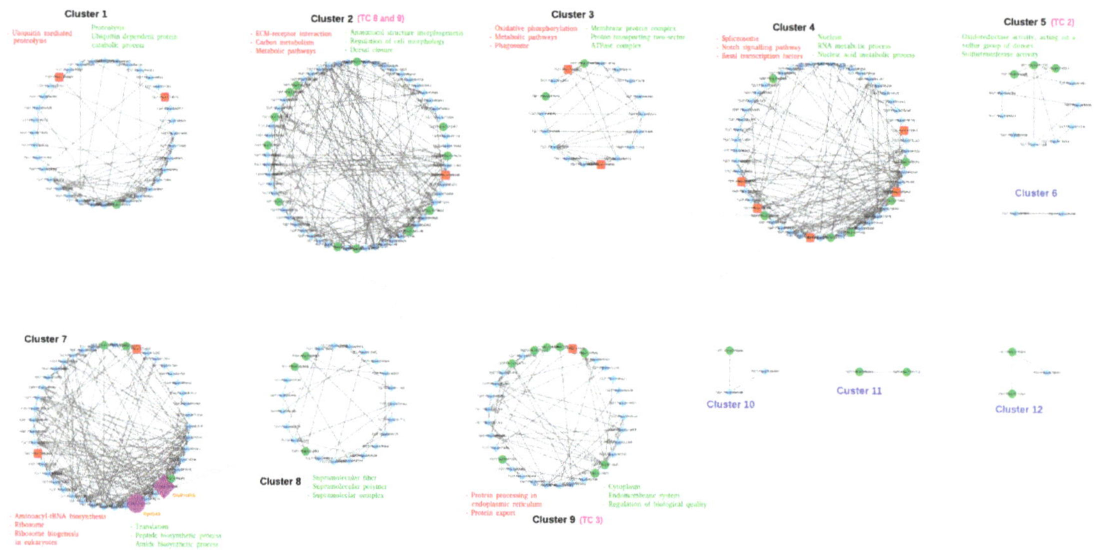

Figure 5. Community analysis of the interaction network. Red, green, and cyan nodes represent up- (square), downregulated (circle), and STRING-predicted (diamonds) genes, respectively. Nodes in purple indicate the hub genes labeled with orange text. The topmost enriched KEGG pathways (red text), GO terms (green text), and transporter class (pink text) are mentioned for each cluster. Clusters with <10 nodes (blue text) were excluded from the analysis.

3.6.1. Accessory Factors Involved in Transport (TC 8)

This transporter class includes proteins that function with or are complexed to other known transport proteins. Of the 162 DE transporters, about 28% (45) represented class 8 transporters (Figure 6A). Almost all these transcripts belonged to subclass 8.A, except two, which were classified as subclass 8.B transporters (Table 5). Most of the transcripts (35/43) belonging to subclass 8.A were downregulated, whereas both of the 8.B transcripts were downregulated. Some of the most abundant TC 8 families are discussed below.

Auxiliary Transport Proteins (TC 8.A)

This subclass represents proteins that aid in transporting across one or more biological membranes but do not themselves take part in direct transport [19]. Such transporters consistently function cooperatively with well-established transport systems. Almost all the transcripts (43/45) assigned to TC 8 belonged to subclass 8.A, which also represented the most abundant transport subclass among all the identified subclasses (Figure 6B). These 43 transcripts were shared between 24 different families (Table S6).

Table 5. List of the most abundant transporter classes, including their representative subclasses and families. Numbers in parenthesis indicate the total number of differentially expressed transcripts assigned to that specific class, subclass, and family. Direction of the arrow represents whether the representative transcript is downregulated (↓) or upregulated (↑).

Class	Class Description	Subclass	Subclass Description	Family	Family Description	Representative Transcript	Effect
TC 8 (45)	Accessory Factors Involved in Transport	TC 8.A (43)	Auxiliary transport proteins	TC 8.A.24 (6	The Ezrin/Radixin/Moesin-binding Phosphoprotein 50 (EBP50) family	Syntenin-1	↓
				TC 8.A.23 (4	The Basigin family	Tyrosine-protein kinase Abl	↓
				TC 8.A.104 (4	The 5′-AMP-activated protein kinase (AMPK) family	Serine/threonine-protein kinase pim-3	↓
		TC 8.B (2)	Ribosomally synthesized protein/peptide toxins/agonists that target channels and carriers	TC 8.B.14 (2	The Sea Anemone Peptide Toxin, Class 1 (BgK) family	Matrix metalloproteinase-24	↓
TC 3 (37)	Primary Active Transporters	TC 3.A (33)	P-P-bond hydrolysis-driven transporters	TC 3.A.2 (6	The H+- or Na+-translocating F-type, V-type and A-type ATPase (F-ATPase) superfamily	V-type proton ATPase subunit a1	↓
				TC 3.A.16 (6	The Endoplasmic Reticular Retrotranslocon (ER-RT) family	26S proteasome regulatory subunit 6A	↑
				TC 3.A.5 (4	The General Secretory Pathway (Sec) family	Putative U5 small nuclear ribonucleoprotein 200-kDa helicase	↓
		TC 3.D (4)	Oxidoreduction-driven transporters	TC 3.D.1 (3	The H+ or Na+-translocating NADH Dehydrogenase (NDH) family	NADH dehydrogenase (ubiquinone) iron-sulfur protein 8, mitochondrial	↑
				TC 3.D.4 (1	The Proton-translocating Cytochrome Oxidase (COX) Superfamily	Cytochrome c oxidase subunit 7A1, mitochondrial	↑
TC 1 (34)	Channels/Pores	TC 1.A (17)	α-Type Channels	TC 1.A.115 (	The Nonselective Cation Channel-2 (NSCC2) family	Dehydrogenase/reductase SDR family member 12	↑
				TC 1.A.17 (2	The Calcium-dependent Chloride Channel (Ca-ClC) family	Transmembrane channel-like protein 7	↓
		TC 1.I (10)	Membrane-bounded Channels	TC 1.I.1 (10	The Nuclear Pore Complex (NPC) family	MAP kinase-activated protein kinase 2	↓
TC 9 (27)	Incompletely Characterized Transport Systems	TC 9.A (12)	Recognized Transporters of Unknown Biochemical Mechanism	TC 9.A.3 (4	The Sorting Nexin27 (SNX27)-Retromer Assembly Apparatus	Ras-related protein Rap-1b	↓
				TC 9.A.63 (	The Retromer-dependent Vacuolar Protein Sorting (R-VPS) family	Cell division control protein 42 homolog	↓
		TC 9.B (15)	Putative uncharacterized transport proteins.	TC 9.B.87 (	The Selenoprotein P Receptor (SelP-Receptor) family	Cubilin	↓
TC 2 (16)	Electrochemical Potential-driven Transporters	TC 2.A (16)	Porters (uniporters, symporters, antiporters)	TC 2.A.1 (4	The Major Facilitator Superfamily (MFS)	Solute carrier family 49 member 4 homolog	↓
				TC 2.A.29 (	The Mitochondrial Carrier (MC) family	Mitochondrial coenzyme A transporter SLC25A42	↑
				TC 2.A.7 (3	The Drug/Metabolite Transporter (DMT) Superfamily	Solute carrier family 35 member F5	↓
TC 4 (2)	Group Translocators	TC 4.C (1)	Acyl CoA ligase-coupled transporters	TC 4.C.1 (1	The Fatty Acid Transporter (FAT) Family	Long-chain fatty acid transport protein 4	↓
		TC 4.D (1)	Polysaccharide Synthase/Exporters	TC 4.D.1 (1	The Putative Vectorial Glycosyl Polymerization (VGP) Family	Beta-1,4-mannosyltransferase egh	↓
TC 5 (1)	Transmembrane Electron Carriers	TC 5.B (1)	Transmembrane 1-Electron Transfer Carriers	TC 5.B.2 (1	The Eukaryotic Cytochrome b561 (Cytb561) Family	Putative ferric-chelate reductase 1 homolog	↓

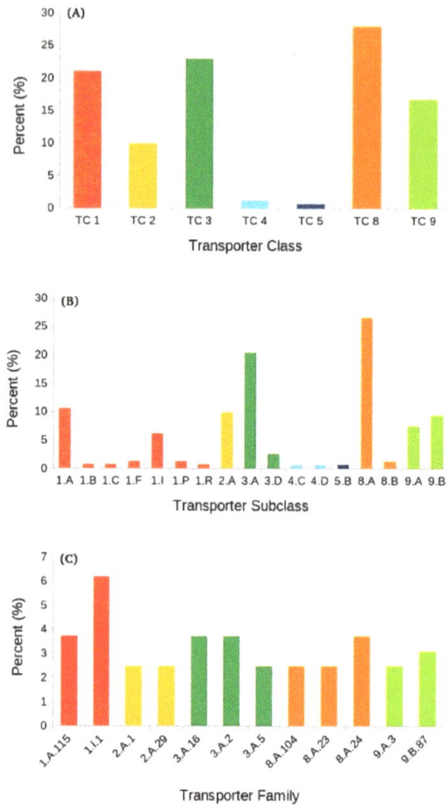

Figure 6. Classification of differentially expressed transcripts into various types of transporters at the (**A**) class, (**B**) subclass, and (**C**) family levels, as described in the Transporter Classification Database (TCDB) system.

Family TC 8.A.24: This was the second-most abundant transport family found in the meta-analysis (Figure 6C). Proteins belonging to this group represent the Ezrin/Radixin/Moesin-binding Phosphoprotein 50 (EBP50) family of transporters. EBP50, also known as Na^+/H^+ exchange regulatory cofactor (NHE-RF), is an adaptor protein that is known to manage a number of cell receptors and channels [58]. Several proteins of this family contain PDZ domains that are found in a wide variety of signaling proteins [59]. Six DETs were assigned to this family, of which five were downregulated, and only one transcript that encodes Salvador homolog 1 (*SAV1*) was upregulated. SAV1 consists of two WW domains, each consisting of about 33 amino acids. The name (WW) refers to two trademark tryptophan (W) residues placed 20–22 amino acids apart. Proteins with WW domains are involved in functions related to signaling or may play structural roles [60]. In addition to SAV1, the other seven upregulated transcripts representing class 8.A transporters were assigned to six other TC families (TC 8.A.28, 8.A.92, 8.A.96, 8.A.112, 8.A.128, and 8.A.131).

Family TC 8.A.23: Transcripts belonging to this group represent the basigin family of proteins that regulate transporters [61]. All the four DETs that were found in this family were downregulated and were annotated as LIM domain kinase 1, contactin, activated Cdc42 kinase-like, and tyrosine-protein kinase Abl.

Family TC 8.A.104: This family comprises the catalytic subunits of AMP-activated protein kinases (*AMPK*), energy sensor protein kinases that are involved in regulating cellular energy metabolism [19]. All the four DET transcripts that fell within this family were downregulated by more than four-fold and represented cAMP-dependent protein kinase catalytic subunit beta, ribosomal protein S6 kinase alpha-5, serine/threonine-protein kinase pim-3, and dual serine/threonine and tyrosine protein kinase. In mammals, as the ATP levels within the cells decrease, *AMPK* triggers energy-producing pathways and suppress the processes that consume energy [62].

Ribosomally Synthesized Protein/Peptide Toxins/Agonists that Target Channels and Carriers (TC 8.B)

In contrast to 8.A transporters, the only two DETs that represented subclass 8.B were downregulated and belonged to the sea anemone peptide toxin, class 1 (BgK) family of transporters (TC 8.B.14). Such toxins are known to block potassium channels [63].

3.6.2. Primary Active Transporters (TC 3)

TC 3 was the second-most abundant transport class after TC 8 and constituted about 23% (37 DETs) of the total transporters identified in the DETs (Figure 6A). These transporters aid in the active transport of solutes against a concentration gradient that is derived by a primary source of energy in the form of chemical, electrical, or solar energy [19]. Although there are five different subclasses of primary active transporters in the TCDB, only subclasses 3.A (89%) and 3.D (11%) were identified (Table 5).

P-P-Bond Hydrolysis-Driven Transporters (TC 3.A)

Of the 33 DETs identified in this group, TC 3.A was the second-most abundant subclass (Figure 6B), representing 14 different families. However, the three major families were identified as the H^+- or Na^+-translocating F-type, V-type, and A-type ATPase (F-ATPase) superfamily (TC 3.A.2), the Endoplasmic Reticular Retrotranslocon (ER-RT or ERAD) family (TC 3.A.16), and the general secretory pathway (Sec) family (TC 3.A.5).

Family TC 3.A.2: Of the six DETs representing the TC 3.A.2 family, five depicted various V-ATPase subunits (subunits a1, C, D1, H, and proteolipid), as described previously (see above), whereas the sixth DET encodes ribonuclease kappa (*RNASEK*).

Family TC 3.A.16: In the case of the six transporters belonging to the ERAD family (TC 3.A.16.), DETs representing 26S proteasome regulatory subunit 6A, cell division cycle protein 48 homolog MJ1156, and putative deoxyribonuclease TATDN1 were upregulated, whereas E3 ubiquitin-protein ligase AMFR, ubiquitin-conjugating enzyme E2 C, and ER degradation-enhancing alpha-mannosidase-like protein 2 were downregulated. Initially, it was thought that the ERAD system was entirely involved in the degradation of misfolded and orphan secretory proteins; however, the system has broad implications in various cellular processes, such as protein folding and transport, the regulation of metabolism, immune response, and ubiquitin-proteasome-dependent degradation [64].

Family TC 3.A.5: In addition to the TC families 3.A.2 and 3.A.16, the third topmost family in which all the four mapped DETs were downregulated by more than two-fold was TC 3.A.5 (the general secretory pathway (Sec) family). These 3.A.5 members were identified as calpain-3, signal recognition particle 54-kDa protein, transport protein Sec31a, and putative u5 small nuclear ribonucleoprotein 200-kDa helicase. Protein complexes belonging to the Sec family are found in both the prokaryotes, as well as eukaryotes. The Sec system essentially consists of two major components: (a) three integral inner membrane proteins, SecYEG, and (b) the ATPase motor protein SecA and represents one of the major ways by which protein export or membrane integration occurs in bacteria [65].

Oxidoreduction-Driven Transporters (TC 3.D)

Only four DETs distributed between two families—namely, the H^+ or Na^+-translocating NADH dehydrogenase (NDH) family (TC 3.D.1) and the proton-translocating cytochrome oxidase (COX) superfamily (3.D.4)—were included in this subclass. All the four transcripts assigned to this subclass were upregulated by $\geq$two-fold.

Family TC 3.D.1: As the name suggests, NADH:ubiquinone oxidoreductases type I (NDH-Is) of bacterial, as well as of eukaryotic, mitochondria and chloroplast origin, coupled with electron transfer to the electrogenic transport of protons or Na^+ [66,67]. Of the four DETs identified in the 3.D subclass, three were found in this family and represent NADH dehydrogenase (ubiquinone) iron-sulfur protein 8 (mitochondrial) (*NDUFS8*), NADH dehydrogenase (ubiquinone) 1 alpha subcomplex subunit 13 (*NDUFA13*), and probable NADH dehydrogenase (ubiquinone) 1 alpha subcomplex subunit 12 (*Y94H6A.8*), respectively. The vertebrate H+-translocating NADH dehydrogenase (NDH) complex consists of 45 subunits [68], and among these, *NDUFA13* is crucial for membrane potential formation and NADH assembly [69].

Family TC 3.D.4: Only one DET was found in this family and represents mitochondrial cytochrome c oxidase subunit 7A1 (*COX7A1*). These multi-subunit enzyme complexes reduce O_2 to water and, consequently, pump four protons across the membrane [19].

3.6.3. Channels/Pores (TC 1)

The third-most abundant class of transporters was channels/pores with about 34/162 (~21%) DE transporters found in this class (Figure 6A and Table 5). Transporters in this group have transmembrane channels mainly consisting of α-helical or β-strand-type spanners [19]. In the TCDB, at least 23 different subclasses of channels are recognized; however, only seven such categories were found in this study (Table S6).

α-Type Channels (TC 1.A)

In general, these transporters consist mostly of α-helical spanners, although they may also contain some β-strands [19]. Of all the TC 1 transporters, 17/34 (50%) were found in this subclass. Furthermore, 11 different types of TC 1.A families were identified in this category.

Family TC 1.A.115: This was the most abundant family among all α-type channel transporters, with six DETs found in this group, and is represented by the pore-forming NADPH-dependent 1-acyldihydroxyacetone phosphate reductase (*AYR1*) family. Transporters like *AYR1* form an NADPH-regulated channel in lipid bilayers, as well as in the outer mitochondrial membranes [70]. Of the six transporters belonging to TC 1.A.115, four were downregulated and represented dehydrogenase/reductase SDR family member 11, D-beta-hydroxybutyrate dehydrogenase (mitochondrial), D-beta-hydroxybutyrate dehydrogenase (mitochondrial), and retinol dehydrogenase. The remaining two—namely, dehydrogenase/reductase SDR family member 12 and C-factor—were upregulated.

Family TC 1.A.17: This category represents transporters from the calcium-dependent chloride channel (Ca-ClC) family. Only two (calcium permeable stress-gated cation channel 1 and transmembrane channel-like protein 7) DETs were mapped to the Ca-ClC family. Both these transporters were downregulated.

In addition to the above-mentioned families, there were nine additional TC 1.A families, each represented by only one representative on the list of transporters found in this work (Table S6). Among these, five were downregulated, representing the families 1.A.79 (SID1 transmembrane family member 1), 1.A.33 (Endoplasmic reticulum chaperone BiP), 1.A.21 (Bcl-2-related ovarian killer protein), 1.A.13 (Calcium-activated chloride channel regulator 2), and 1.A.101 (Peroxisomal membrane protein 11C), respectively. The Bcl-2-related ovarian killer protein was the most downregulated of all these five transporters. The Bcl-2 family is represented by apoptosis regulator Bcl-X and its homologs [71]. Such proteins play essential roles in apoptosis, the cell suicide program necessary for development, tissue homeostasis, and protection against pathogens.

In contrast, the transporters that represented families 1.A.77 (Protein MAK16 homolog B), 1.A.75 (Piezo-type mechanosensitive ion channel component), 1.A.46 (Bestrophin-2), and 1.A.28 (Urea transporter 1) were all upregulated (Table S6). Of these, the family represented by the Piezo-type mechanosensitive ion channel component, a multi-pass, mechanically activated cation channel [72], was the most upregulated.

Membrane-Bound Channels (TC 1.I)

These transporters allow the transport of molecules across cells or organelle membranes [19]. Ten transporters belonging exclusively to the nuclear pore complex (NPC) family (TC 1.I.1) were found in this subclass and were identified as the most abundant transporter family (Figure 6C). Proteins that belong to the NPC family aid in the exchange of molecules between the interior of the nucleus and the cytoplasm [73]. Serine/threonine-protein kinase OSR1 (Oxidative stress-responsive 1 protein) was the only transporter of this family that was upregulated, whereas all other nine transcripts were downregulated. OSR1 is known to bind and phosphorylate PAK1 (Serine/threonine-protein kinase PAK 1) and interacts with chloride channel proteins (*SLC12A6* isoform 2, *SLC12A1*, and *SLC12A2* (*NKCC1*)), thereby initiating the cellular response to environmental stress [74]. In contrast, the downregulated transcripts belonging to this (TC 1.I.1) family comprise a polycomb protein (*EED*), two unconventional myosins (*Myo1C* and *Myo18a*), serine/threonine-protein kinase 16 (*STK16*), activating molecule in BECN1-regulated autophagy protein 1 (*AMBRA1*), mitogen-activated protein kinase kinase kinase kinase 3 (*MAP4K3*), ATP-dependent RNA helicase (*DDX3X*), cyclin-dependent kinase 9 (*CDK9*), and *MAPk-Ak2*). Among these, *MAPk-Ak2* was the most downregulated. *MAPk-Ak2* is a substrate of the p38 MAPK that is responsible for the signaling events affecting several cellular processes such as inflammation, division and differentiation, apoptosis, and motility in response to a variety of extracellular stimuli [75]. p38 MAPK is generally regulated by alterations in environmental osmolarity by dual tyrosine/threonine phosphorylation [76] mediated by dual specificity mitogen-activated protein kinase kinase 3 (MAPKK3 or MEK-3) and dual specificity mitogen-activated protein kinase kinase 6 (MAPKK6 or MEK-6) [77,78].

Miscellaneous Channels/Pore Families

In addition to TC 1.A and 1.I, two proteins each from subclass vesicle fusion pores (TC 1.F) and the non-envelope virus penetration complex (1.P) were seen among the list of DETs (Table S6). Both the TC 1.F transporters were downregulated and represented the synaptosomal vesicle fusion pores family (TC 1.F.1). Family 1.F.1 transporters are the power engines that are known to bring the membranes together [79]. The two transporters were annotated as protein ROP and vacuolar protein sorting-associated protein 45, respectively. On the other hand, one of the DETs from subclass 1.P representing mitochondrial protein tumorous imaginal discs (TID58) was highly upregulated, whereas another mitochondrial chaperone protein dnaJ 1 (*ATJ1*) was downregulated. Currently, there is only one family, the polyoma virus SV40 ER penetration channels (VPEC) (TC 1.P.1), known in this subclass. It is reported that a non-enveloped virus stimulates its own membrane translocation by initiating the release and recruitment of essential transport factors to the endoplasmic reticulum (ER) membrane [80]. The other under-represented transporter families from channels/pores (TC 1) were the Pro-Pro-Glu actinobacterial outer membrane porin (PPE) (TC 1.B.94), aerolysin channel-forming toxin (aerolysin) (TC 1.C.4), and the membrane contact site (MCS) (TC 1.R.1) families, respectively. Each of these three families were represented by a single member. Among these three, only the transcript representing TC 1.B.94 was downregulated (Table S6).

3.6.4. Incompletely Characterized Transport Systems (TC 9)

This class consists of transporters that are unclassified and comprises three main subclasses (TC 9.A, 9.B, and 9.C) in the TCDB classification system. A total of 27/162 DE transcripts (that encode transporters) shared between subclasses 9.A (12) and 9.B (15) were identified from this class of transporters (Table 5). No transporter from subclass TC 9.C was found in the list of DETs identified in this work.

Recognized Transporters of Unknown Biochemical Mechanism (TC 9.A)

This subclass represents proteins that are recognized as transporters, but their mechanism of action is unknown [19]. TC 9.A is divided into at least 77 different families, although only five such families were identified in the meta-analysis.

Family TC 9.A.3: The members of this family represent the sorting nexin 27 (*SNX27*)-retromer assembly apparatus (RetromerAA) of transporters. Four transporters were identified in this family, and all of them were downregulated. The PDZ domain of SNX27 helps in the recycling of internalized transmembrane proteins from endosomes to the plasma membrane by linking PDZ-dependent cargo identification to retromer-mediated transport [19].

Family TC 9.A.63: The second-most abundant category with three members identified in this group represented the retromer-dependent vacuolar protein sorting (R-VPS) family. Both of these components are necessary for maintaining the vacuole membrane organization [81]. All three transporters from this family were also downregulated.

Two DETs were also found in the autophagy-related phagophore-formation transporter (APT) family (TC 9.A.15) (Table S6). One of them was upregulated (serine/threonine-protein kinase tousled-like 2: *TLK2*) by more than four-fold, while the other one—namely, lactoylglutathione lyase (*GLO1*)—was downregulated. Autophagy-related genes are known to play essential roles in the development and stress responses in plants [82]. In addition to these, three more 9.A families with a single member in each were found. These include the mitochondrial cholesterol/porphyrin/5-aminolevulinic acid uptake translocator protein (TSPO) family (TC 9.A.24), Ca2+-dependent phospholipid scramblase (Scramblase) (TC 9.A.36), and small nuclear RNA exporter (snRNA-E) (TC 9.A.60) families, respectively. All the three transcripts representing these families were downregulated; however, the transcript that potentially codes for the mitochondrial translocator protein (TSPO) was the most downregulated.

Putative Transport Proteins (TC 9.B)

This category consists of proteins for which a transport function has been proposed; however, there is no information to back this suggestion [19]. The proteins in this subclass could be assigned to a well-defined class when their role in transport is verified, else they will be removed from the TCDB. Currently, with 408 different families, TC 9.B is the largest subclass in the TC (date accessed: September 22, 2020) classification system. We found 15 DETs distributed in 11 different TC 9.B families.

Family TC 9.B.87: This group represents the selenoprotein P receptor (SelP-Receptor) family of putative transporters. SelP comprises most of the selenium in blood plasma, and it is used by various organs (e.g., kidneys, brain, and testes) as a selenium source for the biosynthesis of selenoprotein [19]. SelP homologs are implicated in diverse functional roles, including the regulation of ion transporters. Five DETs were identified in this family, and all of them were downregulated. Among these, Cubilin, a 460-kDa endocytic receptor, was the most downregulated transcript. Cubilin is known to be involved in several essential functions that include the absorption of vitamin B12 in the intestines, catabolism of apolipoprotein A-I, and, in general, renal protein reabsorption [83].

Additional families in which at least one DET was found are TCs 9.B.105, 9.B.135, 9.B.142, 9.B.198, 9.B.208, 9.B.229, 9.B.265, 9.B.278, 9.B.311, and 9.B.371 (Table S6). All the transcripts representing each individual family were downregulated, except in the case of the transcript (retinoic acid receptor: *RXR*) found in the TC 9.B.208 (vitamin D3 receptor (VDR) family), which was upregulated. RXR is a ligand-dependent transcription factor that may play a role in the retinoic acid response pathway [84].

3.6.5. Electrochemical Potential-Driven Transporters (TC 2)

Additionally known as secondary carrier-type facilitators, TC 2 transporters use a carrier-mediated process to catalyze uniport, antiport, or symport systems [19]. Among all the abundant transporter classes, only 16/162 (9.87%) DE transporters were found in this class. This class comprises four subclasses (TCs 2.A-D); however, only DETs belonging to subclass 2.A were observed (Table 5). TC 2.A is further divided into 133 families, of which only eight were present.

Family TC 2.A.1: Proteins of this ancient, huge, and diverse family represent the major facilitator superfamily (MFS) and comprise millions of sequenced proteins [19]. Members of this superfamily catalyze uniport, symport, and/or antiport. DETs that represented glucose-6-phosphate exchanger (*SLC37A2*) and facilitated trehalose transporter Tret1-2 homolog (*TRET1-2*), putative inorganic phosphate cotransporter (Picot), and solute carrier family 49-member 4 homolog (*SLC49A4*) were identified in this family. Among these, only *SLC37A2* was upregulated, whereas the other three were downregulated.

Family TC 2.A.29: Similar to TC 2.A.1, only four DETs that encode for mitochondrial coenzyme A transporter (*SLC25A42*), mitochondrial ornithine transporter 1 (*SLC25A15*), Graves' disease carrier protein homolog (*SLC25A16*), and mitochondrial glycine transporter (*SLC25A38*) were found in this family. Among these, only *SLC25A38* was downregulated.

Family TC 2.A.7: Only three DETs—namely, adenosine 3′-phospho 5′-phosphosulfate transporter 2 (*SLC35B3*), acid sphingomyelinase-like phosphodiesterase 3b (*SMPDL3B*), and solute carrier family 35 member F5 (*SLC35F5*)—were assigned to this drug/metabolite transporter (DMT) superfamily. All the three transcripts representing this superfamily were downregulated by more than four-fold. In general, most of the DMT superfamily members are nucleotide–sugar transporters that are involved in the endoplasmic reticulum and Golgi of eukaryotic cells [85].

Other categories of subclass TC 2.A included the zinc (Zn^{2+})-iron (Fe^{2+}) permease (ZIP) (2.A.5); neurotransmitter:sodium symporter (NSS) (2.A.22); bile acid:Na+ symporter (BASS) (2.A.28); glycerol uptake (GUP) or membrane-bound acyl transferase (MBOAT) (2.A.50); and the sweet, PQ-loop, saliva;, and MtN3 (Sweet) (2.A.123) families. Only a single representative from each of these families was found in the DETs, and each of these transcripts were downregulated (Table S6).

In addition to the above-discussed transporter families that were over-represented in the DETs identified in the CMC transcriptome by the RNA-Seq meta-analysis, three additional transcripts, each belonging to different families, were also found. Among them, a transcript representing putative ferric-chelate reductase 1 homolog (*CG8399*) was the most downregulated and belonged to the eukaryotic cytochrome b561 (Cytb561) family (TC 5.B.2) of the class transmembrane electron carriers (TC 5) (Table S6). Trans-plasma membrane electron transfer is achieved by the b-type cytochromes from various families and is involved in broad cellular processes that involve two interacting redox couples physically isolated by a phospholipid bilayer—for example, the uptake of iron and redox signaling [86]. The remaining two transcripts belonged to the fatty acid transporter (FAT) (TC 4.C.1) and the putative vectorial glycosyl polymerization (VGP) (TC 4.D.1) families of the class group translocators (TC 4), respectively (Table 5 and Table S6). Members of TC 4 modify the substrate during the transport process, which involves combined chemical and vectorial reactions [19].

Overall, of the 162 DETs that encode various transporters, the majority of them were downregulated, whereas only 41 were upregulated. The analysis suggested that change in salinity triggers the differential expression of almost all the major transport classes (e.g., TC 1, TC 2, TC 3, and TC 8), as well as several unclassified transporters (TC 9). Moreover, transporters belonging to TC 8 were the most diverse and represented 25 different families. In contrast, only three transporters belonging to TC 4 and TC 5 were found. Some of these numbers might change in the future when TC 9 (unclassified) transporters are assigned to well-defined classes or removed from the transporter classification system.

4. Discussion

Studies have shown that gills are the most relevant tissues during osmoregulation in CMC, and to understand the expression profiles and the associated pathways, many studies have applied RNA-Seq techniques under different salinity conditions [3,9,10,16].

However, differences in the experimental conditions and the analytical pipelines used to investigate the same biological question might affect which genes are differentially expressed in each study [39]. To overcome some of these issues, a meta-analysis offers a way to systematically integrate results from multiple individual studies and remove the inconsistencies in these datasets by increasing the sample size and statistical power to detect more robust biomarkers associated with a specific condition [87]. In this work, we performed the first comprehensive meta-analysis of four separate RNA-Seq studies involving the gills of CMC under variable salinity conditions to get an overview of their gene expression profiles and, also, to identify novel or more reliable genes associated with salinity change. The integration of multiple RNA-Seq datasets in this meta-analysis identified several new potential transcripts that may serve as potential biomarkers and highlights that a broad variety of transporters are associated with salinity change.

Our meta-analysis identified 405 DETs between freshwater and high-salinity conditions. Of these, the protein argonaute-2 (*AGO2*) and mitochondrial succinate–CoA ligase (GDP-forming) subunit beta (*SUCLG2*) were encoded by the topmost down- and upregulated transcripts, respectively. Proteins belonging to the argonaute family associate with small RNAs and help in mRNA degradation, translational repression, or both [88]. In contrast, *SUCLG2* catalyzes the only step in the citric acid cycle that provides substrate-level phosphorylation to synthesize GTP from GDP [89]. While ATP is the main source of energy for many biological processes, *SUCLG2* may directly influence the anabolic functions of the citric acid cycle inside the mitochondria by supplying GTP for the GTP-dependent steps of protein synthesis [90]. One of the highly downregulated genes found in this work was tramtrack (*ttk*) (Table 3), which encodes for protein tramtrack (beta isoform) in *Drosophila melanogaster* (fruit fly). *ttk* is a transcription factor that has been implicated in several roles, including cell fate specification, cell proliferation, and cell cycle regulation. Besides these established roles, *ttk* also plays a crucial role in tracheal development [91]. Gills are one of the primary sites of ion transport and the movement of water in aquatic organisms [92]. Therefore, a series of structural and functional modifications are developed within these osmoregulatory systems so that a constant ion and osmotic homeostasis is maintained as the organism moves from low- to high-salinity conditions or vice versa. Studies have also shown that salinity also regulates the number of branchial chloride cells and their shape, as well as the expression level of several transport proteins [93]. These results are consistent with the GO and KEGG pathway enrichment analysis performed in this work, where it was found that several processes related to development and morphogenesis were over-represented (Figure 4 and Table S4).

As discussed above, a large number (162/405) of DETs were found to be transporters belonging to various classes. Transporters such as Na^+/K^+-ATPase and V-type ATPase are well-known biomarkers of crustacean gills [3] and have been highlighted in several studies. Although the function of these ATPases in ion transport is well-established, the role of transporters and enzymes from other classes to complete the process cannot be undermined [94]. This is supported by the fact that, in this work, a large number of differentially expressed TC 8 (accessory factors) transporters were identified. Such transporters are not directly involved in the transport of ions, but they function or form complexes with other known transporters to complete the transport [19]. Several representatives of this class were downregulated. For example, Syntenin-1 (*SDCBP*), also known as syndecan-binding protein 1, was the most downregulated transcript from the family TC 8.A.23. Syndecans are receptors for a diverse type of proteins and play essential roles in many cellular processes, such as development, maintenance, repair, cell adhesion, and migration. Reports have shown that syndecans can also regulate stretch-activated ion channels [95]. Additionally, several members of TC 8 transporters (TC 8.A.23) are indispensable for maintaining voltage-gated potassium channels [96] and regulating Na^+/K^+-ATPase and connexin 43 [97] or are involved in several other cellular processes, such as proliferation, migration, and survival [98]. Therefore, the identification of several accessory factors in response to salinity stress suggests that the significance of these proteins is like that of well-known transporters. These data should trigger more interest towards their functional roles in osmoregulation.

The number of TC 3 (primary active) transporters was slightly lower as compared to the accessory factors but higher than other transport classes. TC 3 transporters help in the active transport of solutes against a concentration gradient that is derived by a primary source of energy [19]. For example, some of these transporters (TC 3.A) carry out the active uptake or removal of solutes by the hydrolysis of a diphosphate bond of inorganic pyrophosphate, ATP, or other nucleoside triphosphates [19]. In contrast, other members of this class represent transport systems that are energized by the exothermic flow of electrons and consists of proteins (TC 3.D) that facilitate the transport of solutes from a reduced to an oxidized substrate [19]. Na^+/K^+-ATPase (TC 3.A.3) and V-type ATPase (TC 3.A.2) are two well-known examples of this class. Studies have shown that Na^+/K^+-ATPase is an essential component for low-salinity adaptation in blue crab (*Callinectes sapidus*) [99]. Similarly, V-ATPases (large, multi-subunit proton pumps) are another very important class of transporters that are involved in transporting hydrogen ions in exchange for energy in the form of ATP [20]. In *Drosophila*, there are 14 subunits of V-ATPase encoded by at least 33 genes [20]. Some of these subunits also have shown splice variants. Previous studies on the gills of mitten crabs have either reported no DE V-ATPase subunits [10] or highlighted only few subunits [3,9,16]. As discussed above, we identified several DE subunits of V-ATPase, of which some exhibited a similar trend in their expression, whereas, for others, an inconsistent expression pattern was observed. In spite of the well-recognized role of V-ATPase, it is worth mentioning that its function depends on *RNASEK*, which closely associates with this enzyme [100]. Decline in the expression of *RNASEK* changes the localization of various V-ATPase subunits and effects the expression levels of some of its subunits. Therefore, to better understand a complete mechanism such as salinity change, it is of importance to identify the interacting partners (or co-expressed genes) of the well-known biomarkers and explore the consequent effects on the changing condition.

TC 1 proteins aid in transport through an energy-independent (facilitated diffusion) process in which the substrate moves across the channel or pore without the coupling of the translocation step to another chemical or vectorial process [101]. Some of these channels (e.g., TC 1.A) are found in all organisms and catalyze the transport of solutes by an energy-independent process by passage through a channel or pore with no indication of a carrier-mediated mechanism [19]. These data suggest that such transporters have widespread and conserved physiological functions [102]. The transport of ions through channels, especially the ingress of calcium ions into the cytosol, serves as an indication for environmental responses in eukaryotic organisms [103]. In fact, in plants, a swift upsurge in the cytosolic-free Ca^{2+} concentration is one of the initial events observed after osmotic stress [104]. Members of the Ca-ClC family are cation channels that are permeable to calcium and gated by physical signals such as osmotic stress [103]. In animals, such channels are necessary for normal electrolyte and fluid secretion, olfactory perception, and neuronal, as well as smooth muscle, excitability [105]. Similarly, mechanosensitive channels are transmembrane proteins with pores that mediate the flow of ions or osmolytes across membranes in response to mechanical stimuli. In animals, they are essential for somatosensory perception, whereas, in plants, in addition to their sensory and regulatory roles, they are indispensable for responses to osmotic shock [106].

In our meta-analysis, we observed that the number of secondary carriers is much lower as compared to other transporter classes (Figure 6). Additionally, only one subclass (TC 2.A) was found in this work and includes transporters that employ a carrier-mediated process to catalyze uniport (a single species is transported), antiport (two or more species are transported in opposite directions), and/or symport (two or more species are transported together in the same direction) [19]. One of the most widely distributed families of this subclass among all living organisms is the major facilitator superfamily (MFS), also known as the uniporter–symporter–antiporter family [107]. Such transporters are single-polypeptide secondary carriers that can transport only small solutes in response to chemiosmotic ion gradients. MFS transporters utilize the ionic (H^+) gradient across the membrane as the source of energy, a mechanism well-conserved among living organisms [108]. Another important family within secondary carriers is the mitochondrial carrier (MC) family (the human SLC25 family), which, in general, prefers the antiport (exchange of a solute for some other) system [19]. Individuals of this family participate in the transport of keto acids, amino acids, nucleotides, inorganic ions, and cofactors across the mitochondrial inner membrane. SLC25 transporters serve as a bridge between metabolic reactions of the cytosol and mitochondrial matrix (or various cell compartments) by catalyzing the transfer of diverse solutes across the membrane. They also participate in numerous crucial metabolic pathways, such as oxidative phosphorylation, the citric acid cycle, Ca^{2+}-cell signaling, and cell death [109]. Although only a limited number of DE transporters belonging to TC 2 were identified, the importance of this transporter class during changes in the environmental conditions such as salinity cannot be under-represented.

Several transporters of unknown classification were also found distributed between two subclasses (TC 9.A and 9.B) of this group. The proteins of TC 9.A include transporters with unknown biochemical mechanisms; however, their role in transport has been established [19]. In contrast, TC 9.B consists of putative transporters with no evidence of a transport-related function. Once their role in transport is established, they will be assigned to a relevant group or removed from the transport classification system if proven otherwise. In our network analysis, many of the TC 9 transporters are grouped together with TC 8 transporters (Figure 5), suggesting that some of these may either be assigned to the TC 8 category or, probably, they might carry out the transport-related functions in coordination.

5. Conclusions

Overall, in this meta-analysis of RNA-Seq data, we successfully identified a number of transcripts whose expressions were apparently altered in the gills of Chinese mitten crabs in response to ambient changes in the salinity conditions. Of these DETs, a large number were found to be transporters from diverse families, again highlighting the significance and greater contribution of the transportome in osmotic regulation. Coordinated responses induced by salinity changes triggered several processes related to cell development, differentiation, motility, and protein synthesis, including others, as some of the most enriched processes. The processes and the DETs identified from our meta-analysis will serve as more robust candidates for understanding the mechanism of osmotic regulation in *E. sinensis* as compared to the genes and pathways reported from individual studies. This study, in addition to being the first meta-analysis of CMC in response to salinity, also represents the first such study that provides a focused overview of various transporters involved in salinity change. The various types of transporters highlighted in this meta-analysis can provide significant clues for experimental scientists to design future experiments to get a better understanding of the adaptation to salinity change and, consequently, the underlying mechanisms of osmotic regulation at varying salinity conditions in Chinese mitten crabs.

Supplementary Materials: The supplementary materials can be found at https://www.mdpi.com/2079-7737/10/1/39/s1: Figure S1: Assembly statistics showing (A) the percentage of orthologs identified from the BUSCO search against the Arthropoda lineage; (B) annotation of transcripts against Pfam, UniProt, Crabdb, and KEGG databases using BLASTp; and (C) total number of transcripts with open reading frames in each species identified using the TransDecoder. For BLASTp, transcripts with the topmost hits at 30% coverage and percent (%) identity at an evalue cutoff of 1e-5 were selected. For Pfam, the numbers represent the unique number of transcripts that are mapped to this database. Figure S2: The distribution of various COG categories assigned to the transcripts originating from each individual study. The COG categories are Function unknown (S); Signal transduction mechanisms (T); Posttranslational modification, protein turnover, and chaperones (O); Translation, ribosomal structure, and biogenesis (J); Transcription (K); RNA processing and modification (A); Intracellular trafficking, secretion, and vesicular transport (U); Carbohydrate transport and metabolism (G); Energy production and conversion (C); Amino acid transport and metabolism (E); Cytoskeleton (Z); Lipid transport and metabolism (I); Inorganic ion transport and metabolism (P);

Replication, recombination, and repair (L); Cell cycle control, cell division, and chromosome partitioning (D); Secondary metabolites biosynthesis, transport, and catabolism (Q); Chromatin structure and dynamics (B); Nucleotide transport and metabolism (F); Coenzyme transport and metabolism (H); Cell wall/membrane/envelope biogenesis (M); Extracellular structures (W); Defense mechanisms (V); Cell Motility (N); and Nuclear structure (Y). Figure S3: Interaction network of differentially expressed transcripts. Red, green, and cyan nodes represent up- (square), downregulated (circle), and STRING-predicted (diamonds) genes, respectively. Nodes in purple indicate the hub genes labeled with orange text. Table S1: Assembly statistics of each dataset. Table S2: List of additional DETs identified by the meta-analysis. (↓) indicates that the representative transcript is downregulated. Table S3: List of DETs (2-Fold change) identified by the RNA-Seq meta-analysis. Annotations for each transcript are based on the BLASTp analysis against the UniProt database. Direction of the arrow represents whether the representative transcript is downregulated (↓) or upregulated (↑). Table S4: List of enriched GO categories in terms of the biological process (BP), molecular function (MF), and cellular component (CC). Table S5: Enrichment analysis of each individual cluster identified in the interaction network. Clusters with <10 nodes were excluded from the enrichment analysis. Table S6: List of DETs that were assigned to a transporter family based on the BLASTp analysis against the TCDB. Direction of the arrow represents whether the representative transcript is downregulated (↓) or upregulated (↑).

Author Contributions: Conceptualization, A.M. and C.-B.K.; formal Analysis, A.M. and C.-B.K.; writing—original draft preparation, A.M. and C.-B.K.; and writing—review and editing, A.M. and C.-B.K. All authors have read and agreed to the published version of the manuscript.

Funding: This research received no external funding.

Institutional Review Board Statement: Not applicable.

Informed Consent Statement: Not applicable.

Data Availability Statement: The data presented in this study are available within the article. If required, any additional data is available on request from the authors.

Conflicts of Interest: The authors declare no conflict of interest.

References

1. Anger, K. Effects of temperature and salinity on the larval development of the Chinese mitten crab Eriocheir sinensis (Decapoda: Grapsidae). *Mar. Ecol. Prog.* **1991**, *72*, 103–110. [CrossRef]
2. Sui, L.; Zhang, F.; Wang, X.; Bossier, P.; Sorgeloos, P.; Hänfling, B. Genetic diversity and population structure of the Chinese mitten crab Eriocheir sinensis in its native range. *Mar. Biol.* **2009**, *156*, 1573–1583. [CrossRef]
3. Zhang, D.; Qi, T.; Liu, J.; Liu, Q.; Jiang, S.; Zhang, H.; Wang, Z.; Ding, G.; Tang, B. Adaptively differential expression analysis in gill of Chinese mitten crabs (*Eriocheir japonica sinensis*) associated with salinity changes. *Int. J. Biol. Macromol.* **2018**, *120*, 2242–2246. [CrossRef] [PubMed]
4. Dittel, A.I.; Epifanio, C.E. Invasion biology of the Chinese mitten crab *Eriochier sinensis*: A brief review. *J. Exp. Mar. Biol. Ecol.* **2009**, *374*, 79–92. [CrossRef]
5. Kültz, D.; Fiol, D.; Valkova, N.; Gomez-Jimenez, S.; Chan, S.Y.; Lee, J. Functional genomics and proteomics of the cellular osmotic stress response in 'non-model' organisms. *J. Exp. Biol.* **2007**, *210*, 1593–1601. [CrossRef] [PubMed]
6. Evans, D.H. Cell signaling and ion transport across the fish gill epithelium. *J. Exp. Zool.* **2002**, *293*, 336–347. [CrossRef]
7. Evans, T.G.; Somero, G.N. A microarray-based transcriptomic time-course of hyper- and hypo-osmotic stress signaling events in the euryhaline fish Gillichthys mirabilis: Osmosensors to effectors. *J. Exp. Biol.* **2008**, *211*, 3636–3649. [CrossRef]
8. Fiol, D.F.; Kültz, D. Osmotic stress sensing and signaling in fishes. *Febs J.* **2007**, *274*, 5790–5798. [CrossRef]
9. Yang, Z.; Zhou, J.; Wei, B.; Cheng, Y.; Zhang, L.; Zhen, X. Comparative transcriptome analysis reveals osmotic-regulated genes in the gill of Chinese mitten crab (*Eriocheir sinensis*). *PLoS ONE* **2019**, *14*, e0210469. [CrossRef]
10. Li, E.; Wang, S.; Li, C.; Wang, X.; Chen, K.; Chen, L. Transcriptome sequencing revealed the genes and pathways involved in salinity stress of Chinese mitten crab, *Eriocheir sinensis*. *Physiol. Genom.* **2014**, *46*, 177–190. [CrossRef]
11. Tseng, Y.C.; Hwang, P.P. Some insights into energy metabolism for osmoregulation in fish. *Comp. Biochem. Physiol. Toxicol. Pharmacol.* **2008**, *148*, 419–429. [CrossRef] [PubMed]
12. Huang, Y.; Anderle, P.; Bussey, K.J.; Barbacioru, C.; Shankavaram, U.; Dai, Z.; Reinhold, W.C.; Papp, A.; Weinstein, J.N.; Sadée, W. Membrane transporters and channels: Role of the transportome in cancer chemosensitivity and chemoresistance. *Cancer Res.* **2004**, *64*, 4294–4301. [CrossRef] [PubMed]
13. Castilho, P.C.; Martins, I.A.; Bianchini, A. Gill Na(+),K(+)-ATPase and osmoregulation in the estuarine crab, Chasmagnathus granulata Dana, 1851 (Decapoda, Grapsidae). *J. Exp. Mar. Biol. Ecol.* **2001**, *256*, 215–227. [CrossRef]

14. Cutler, C.P.; Cramb, G. Molecular physiology of osmoregulation in eels and other teleosts: The role of transporter isoforms and gene duplication. *Comp. Biochem. Physiol. Part A Mol. Integr. Physiol.* **2001**, *130*, 551–564. [CrossRef]
15. Evans, D.H.; Piermarini, P.M.; Potts, W.T.W. Ionic transport in the fish gill epithelium. *J. Exp. Zool.* **1999**, *283*, 641–652. [CrossRef]
16. Chen, X.; Peng, Z.; Hou, X.; Wang, J.; Wang, C. The Molecular Basis of Osmoregulation and Physiological Processes Associated with Salinity Changes in the Chinese Mitten Crab *Eriocheir sinensis*. *J. Shellfish Res.* **2019**, *38*, 643–653. [CrossRef]
17. Charmantier, G.; Charmantier-Daures, M.; Bouaricha, N.; Thuet, P.; Trilles, J.P.; Aiken, D.E. Ontogeny of Osmoregulation and Salinity Tolerance in Two Decapod Crustaceans: *Homarus americanus* and *Penaeus japonicus*. *Biol. Bull.* **1988**, *175*, 102–110. [CrossRef]
18. Henry, R.P.; Lucu, C.; Onken, H.; Weihrauch, D. Multiple functions of the crustacean gill: Osmotic/ionic regulation, acid-base balance, ammonia excretion, and bioaccumulation of toxic metals. *Front. Physiol.* **2012**, *3*, 431. [CrossRef]
19. Saier, M.H., Jr.; Reddy, V.S.; Tsu, B.V.; Ahmed, M.S.; Li, C.; Moreno-Hagelsieb, G. The Transporter Classification Database (TCDB): Recent advances. *Nucleic Acids Res.* **2016**, *44*, D372–D379. [CrossRef]
20. Allan, A.K.; Du, J.; Davies, S.A.; Dow, J.A. Genome-wide survey of V-ATPase genes in Drosophila reveals a conserved renal phenotype for lethal alleles. *Physiol. Genom.* **2005**, *22*, 128–138. [CrossRef]
21. Marot, G.; Jaffr'ezic, F.; Rau, A. metaRNASeq: Differential meta-analysis of RNA-seq data. *dim (param)* **2020**, *1*, 3.
22. Toro-Domínguez, D.; Villatoro-García, J.A.; Martorell-Marugán, J.; Román-Montoya, Y.; Alarcón-Riquelme, M.E.; Carmona-Sáez, P. A survey of gene expression meta-analysis: Methods and applications. *Brief. Bioinform.* **2020**. [CrossRef] [PubMed]
23. Cho, H.; Kim, H.; Na, D.; Kim, S.Y.; Jo, D.; Lee, D. Meta-analysis method for discovering reliable biomarkers by integrating statistical and biological approaches: An application to liver toxicity. *Biochem. Biophys. Res. Commun.* **2016**, *471*, 274–281. [CrossRef] [PubMed]
24. Piras, I.S.; Manchia, M.; Huentelman, M.J.; Pinna, F.; Zai, C.C.; Kennedy, J.L.; Carpiniello, B. Peripheral Biomarkers in Schizophrenia: A Meta-Analysis of Microarray Gene Expression Datasets. *Int. J. Neuropsychopharmacol.* **2019**, *22*, 186–193. [CrossRef] [PubMed]
25. Bolger, A.M.; Lohse, M.; Usadel, B. Trimmomatic: A flexible trimmer for Illumina sequence data. *Bioinformatics* **2014**, *30*, 2114–2120. [CrossRef]
26. Haas, B.J.; Papanicolaou, A.; Yassour, M.; Grabherr, M.; Blood, P.D.; Bowden, J.; Couger, M.B.; Eccles, D.; Li, B.; Lieber, M.; et al. De novo transcript sequence reconstruction from RNA-seq using the Trinity platform for reference generation and analysis. *Nat. Protoc.* **2013**, *8*, 1494–1512. [CrossRef]
27. Langmead, B.; Salzberg, S. Fast gapped-read alignment with Bowtie 2. *Nat. Methods* **2012**, *9*, 357–359. [CrossRef]
28. Seppey, M.; Manni, M.; Zdobnov, E.M. BUSCO: Assessing Genome Assembly and Annotation Completeness. *Methods Mol. Biol.* **2019**, *1962*, 227–245. [CrossRef]
29. Bray, N.L.; Pimentel, H.; Melsted, P.; Pachter, L. Near-optimal probabilistic RNA-seq quantification. *Nat. Biotechnol.* **2016**, *34*, 525–527. [CrossRef]
30. Consortium, T.U. UniProt: A worldwide hub of protein knowledge. *Nucleic Acids Res.* **2019**, *47*, D506–D515. [CrossRef]
31. Camacho, C.; Coulouris, G.; Avagyan, V.; Ma, N.; Papadopoulos, J.; Bealer, K.; Madden, T.L. BLAST+: Architecture and applications. *BMC Bioinform.* **2009**, *10*, 421. [CrossRef] [PubMed]
32. Finn, R.D.; Clements, J.; Arndt, W.; Miller, B.L.; Wheeler, T.J.; Schreiber, F.; Bateman, A.; Eddy, S.R. HMMER web server: 2015 update. *Nucleic Acids Res.* **2015**, *43*, W30–W38. [CrossRef] [PubMed]
33. El-Gebali, S.; Mistry, J.; Bateman, A.; Eddy, S.R.; Luciani, A.; Potter, S.C.; Qureshi, M.; Richardson, L.J.; Salazar, G.A.; Smart, A.; et al. The Pfam protein families database in 2019. *Nucleic Acids Res.* **2019**, *47*, D427–D432. [CrossRef] [PubMed]
34. Song, L.; Bian, C.; Luo, Y.; Wang, L.; You, X.; Li, J.; Qiu, Y.; Ma, X.; Zhu, Z.; Ma, L.; et al. Draft genome of the Chinese mitten crab, Eriocheir sinensis. *GigaScience* **2016**, *5*, 5. [CrossRef]
35. Huerta-Cepas, J.; Forslund, K.; Coelho, L.P.; Szklarczyk, D.; Jensen, L.J.; von Mering, C.; Bork, P. Fast Genome-Wide Functional Annotation through Orthology Assignment by eggNOG-Mapper. *Mol. Biol. Evol.* **2017**, *34*, 2115–2122. [CrossRef]
36. Ge, S.X.; Jung, D.; Yao, R. ShinyGO: A graphical gene-set enrichment tool for animals and plants. *Bioinformatics* **2020**, *36*, 2628–2629. [CrossRef]
37. Xie, C.; Mao, X.; Huang, J.; Ding, Y.; Wu, J.; Dong, S.; Kong, L.; Gao, G.; Li, C.Y.; Wei, L. KOBAS 2.0: A web server for annotation and identification of enriched pathways and diseases. *Nucleic Acids Res.* **2011**, *39*, W316–W322. [CrossRef]
38. Robinson, M.D.; McCarthy, D.J.; Smyth, G.K. edgeR: A Bioconductor package for differential expression analysis of digital gene expression data. *Bioinformatics* **2010**, *26*, 139–140. [CrossRef]
39. Rau, A.; Marot, G.; Jaffrézic, F. Differential meta-analysis of RNA-seq data from multiple studies. *BMC Bioinform.* **2014**, *15*, 91. [CrossRef]
40. Fisher, R.A. *Statistical Methods for Research Workers*; Oliver and Boyd: Edinburgh, UK, 1925.
41. Panahi, B.; Frahadian, M.; Dums, J.T.; Hejazi, M.A. Integration of Cross Species RNA-seq Meta-Analysis and Machine-Learning Models Identifies the Most Important Salt Stress-Responsive Pathways in Microalga Dunaliella. *Front. Genet.* **2019**, *10*, 752. [CrossRef]
42. Chan, M.Y.; Efthymios, M.; Tan, S.H.; Pickering, J.W.; Troughton, R.; Pemberton, C.; Ho, H.H.; Prabath, J.F.; Drum, C.L.; Ling, L.H.; et al. Prioritizing Candidates of Post-Myocardial Infarction Heart Failure Using Plasma Proteomics and Single-Cell Transcriptomics. *Circulation* **2020**, *142*, 1408–1421. [CrossRef] [PubMed]

43. Alimadadi, A.; Munroe, P.B.; Joe, B.; Cheng, X. Meta-Analysis of Dilated Cardiomyopathy Using Cardiac RNA-Seq Transcriptomic Datasets. *Genes* **2020**, *11*, 60. [CrossRef] [PubMed]
44. Benjamini, Y.; Hochberg, Y. Controlling the False Discovery Rate: A Practical and Powerful Approach to Multiple Testing. *J. R. Stat. Soc. Ser. B* **1995**, *57*, 289–300. [CrossRef]
45. Szklarczyk, D.; Gable, A.L.; Lyon, D.; Junge, A.; Wyder, S.; Huerta-Cepas, J.; Simonovic, M.; Doncheva, N.T.; Morris, J.H.; Bork, P.; et al. STRING v11: Protein-protein association networks with increased coverage, supporting functional discovery in genome-wide experimental datasets. *Nucleic Acids Res.* **2019**, *47*, D607–D613. [CrossRef] [PubMed]
46. Shannon, P.; Markiel, A.; Ozier, O.; Baliga, N.S.; Wang, J.T.; Ramage, D.; Amin, N.; Schwikowski, B.; Ideker, T. Cytoscape: A software environment for integrated models of biomolecular interaction networks. *Genome Res.* **2003**, *13*, 2498–2504. [CrossRef] [PubMed]
47. Assenov, Y.; Ramírez, F.; Schelhorn, S.E.; Lengauer, T.; Albrecht, M. Computing topological parameters of biological networks. *Bioinformatics* **2008**, *24*, 282–284. [CrossRef]
48. Su, G.; Kuchinsky, A.; Morris, J.H.; States, D.J.; Meng, F. GLay: Community structure analysis of biological networks. *Bioinformatics* **2010**, *26*, 3135–3137. [CrossRef]
49. Boube, M.; Martin-Bermudo, M.D.; Brown, N.H.; Casanova, J. Specific tracheal migration is mediated by complementary expression of cell surface proteins. *Genes Dev.* **2001**, *15*, 1554–1562. [CrossRef]
50. Hartman, M.A.; Finan, D.; Sivaramakrishnan, S.; Spudich, J.A. Principles of unconventional myosin function and targeting. *Annu. Rev. Cell Dev. Biol.* **2011**, *27*, 133–155. [CrossRef]
51. Peck, J.W.; Bowden, E.T.; Burbelo, P.D. Structure and function of human Vps20 and Snf7 proteins. *Biochem. J.* **2004**, *377*, 693–700. [CrossRef]
52. Phillips, D.; Covian, R.; Aponte, A.M.; Glancy, B.; Taylor, J.F.; Chess, D.; Balaban, R.S. Regulation of oxidative phosphorylation complex activity: Effects of tissue-specific metabolic stress within an allometric series and acute changes in workload. *Am. J. Physiol. Regul. Integr. Comp. Physiol.* **2012**, *302*, R1034–R1048. [CrossRef] [PubMed]
53. Malik, A.; Lee, E.J.; Jan, A.T.; Ahmad, S.; Cho, K.H.; Kim, J.; Choi, I. Network Analysis for the Identification of Differentially Expressed Hub Genes Using Myogenin Knock-down Muscle Satellite Cells. *PLoS ONE* **2015**, *10*, e0133597. [CrossRef] [PubMed]
54. Malik, A.; Lee, J.; Lee, J. Community-based network study of protein-carbohydrate interactions in plant lectins using glycan array data. *PLoS ONE* **2014**, *9*, e95480. [CrossRef] [PubMed]
55. Newman, M.E.; Girvan, M. Finding and evaluating community structure in networks. *Phys. Rev. E* **2004**, *69*, 026113. [CrossRef]
56. Morris, J.H.; Apeltsin, L.; Newman, A.M.; Baumbach, J.; Wittkop, T.; Su, G.; Bader, G.D.; Ferrin, T.E. clusterMaker: A multi-algorithm clustering plugin for Cytoscape. *BMC Bioinform.* **2011**, *12*, 436. [CrossRef]
57. Koh, G.C.; Porras, P.; Aranda, B.; Hermjakob, H.; Orchard, S.E. Analyzing protein-protein interaction networks. *J. Proteome Res.* **2012**, *11*, 2014–2031. [CrossRef]
58. Li, J.; Dai, Z.; Jana, D.; Callaway, D.J.; Bu, Z. Ezrin controls the macromolecular complexes formed between an adapter protein Na^+/H^+ exchanger regulatory factor and the cystic fibrosis transmembrane conductance regulator. *J. Biol. Chem.* **2005**, *280*, 37634–37643. [CrossRef]
59. Ponting, C.P.; Phillips, C.; Davies, K.E.; Blake, D.J. PDZ domains: Targeting signalling molecules to sub-membranous sites. *Bioessays* **1997**, *19*, 469–479. [CrossRef]
60. Macias, M.J.; Wiesner, S.; Sudol, M. WW and SH3 domains, two different scaffolds to recognize proline-rich ligands. *FEBS Lett.* **2002**, *513*, 30–37. [CrossRef]
61. Muramatsu, T. Basigin (CD147), a multifunctional transmembrane glycoprotein with various binding partners. *J. Biochem.* **2016**, *159*, 481–490. [CrossRef]
62. Egan, D.F.; Shackelford, D.B.; Mihaylova, M.M.; Gelino, S.; Kohnz, R.A.; Mair, W.; Vasquez, D.S.; Joshi, A.; Gwinn, D.M.; Taylor, R.; et al. Phosphorylation of ULK1 (hATG1) by AMP-activated protein kinase connects energy sensing to mitophagy. *Science* **2011**, *331*, 456–461. [CrossRef] [PubMed]
63. Cotton, J.; Crest, M.; Bouet, F.; Alessandri, N.; Gola, M.; Forest, E.; Karlsson, E.; Castañeda, O.; Harvey, A.L.; Vita, C.; et al. A potassium-channel toxin from the sea anemone Bunodosoma granulifera, an inhibitor for Kv1 channels. Revision of the amino acid sequence, disulfide-bridge assignment, chemical synthesis, and biological activity. *Eur. J. Biochem.* **1997**, *244*, 192–202. [CrossRef] [PubMed]
64. Meusser, B.; Hirsch, C.; Jarosch, E.; Sommer, T. ERAD: The long road to destruction. *Nat. Cell Biol.* **2005**, *7*, 766–772. [CrossRef] [PubMed]
65. Collinson, I.; Breyton, C.; Duong, F.; Tziatzios, C.; Schubert, D.; Or, E.; Rapoport, T.; Kühlbrandt, W. Projection structure and oligomeric properties of a bacterial core protein translocase. *Embo J.* **2001**, *20*, 2462–2471. [CrossRef]
66. Brandt, U. Energy converting NADH:quinone oxidoreductase (complex I). *Annu. Rev. Biochem.* **2006**, *75*, 69–92. [CrossRef]
67. Gemperli, A.C.; Schaffitzel, C.; Jakob, C.; Steuber, J. Transport of Na(+) and K (+) by an antiporter-related subunit from the Escherichia coli NADH dehydrogenase I produced in Saccharomyces cerevisiae. *Arch. Microbiol.* **2007**, *188*, 509–521. [CrossRef]
68. Cardol, P.; Vanrobaeys, F.; Devreese, B.; Van Beeumen, J.; Matagne, R.F.; Remacle, C. Higher plant-like subunit composition of mitochondrial complex I from Chlamydomonas reinhardtii: 31 conserved components among eukaryotes. *Biochim. Et Biophys. Acta* **2004**, *1658*, 212–224. [CrossRef]

69. Lu, H.; Cao, X. GRIM-19 is essential for maintenance of mitochondrial membrane potential. *Mol. Biol. Cell* **2008**, *19*, 1893–1902. [CrossRef]
70. Krüger, V.; Becker, T.; Becker, L.; Montilla-Martinez, M.; Ellenrieder, L.; Vögtle, F.N.; Meyer, H.E.; Ryan, M.T.; Wiedemann, N.; Warscheid, B.; et al. Identification of new channels by systematic analysis of the mitochondrial outer membrane. *J. Cell Biol.* **2017**, *216*, 3485–3495. [CrossRef]
71. Adams, J.M.; Cory, S. The Bcl-2 protein family: Arbiters of cell survival. *Science* **1998**, *281*, 1322–1326. [CrossRef]
72. Coste, B.; Mathur, J.; Schmidt, M.; Earley, T.J.; Ranade, S.; Petrus, M.J.; Dubin, A.E.; Patapoutian, A. Piezo1 and Piezo2 are essential components of distinct mechanically activated cation channels. *Science* **2010**, *330*, 55–60. [CrossRef] [PubMed]
73. Jahed, Z.; Soheilypour, M.; Peyro, M.; Mofrad, M.R. The LINC and NPC relationship—It's complicated! *J. Cell Sci.* **2016**, *129*, 3219–3229. [CrossRef] [PubMed]
74. Burington, B.; Barlogie, B.; Zhan, F.; Crowley, J.; Shaughnessy, J.D., Jr. Tumor cell gene expression changes following short-term in vivo exposure to single agent chemotherapeutics are related to survival in multiple myeloma. *Clin. Cancer Res. Off. J. Am. Assoc. Cancer Res.* **2008**, *14*, 4821–4829. [CrossRef] [PubMed]
75. Gurgis, F.M.; Ziaziaris, W.; Munoz, L. Mitogen-activated protein kinase-activated protein kinase 2 in neuroinflammation, heat shock protein 27 phosphorylation, and cell cycle: Role and targeting. *Mol. Pharmacol.* **2014**, *85*, 345–356. [CrossRef] [PubMed]
76. Han, J.; Lee, J.D.; Bibbs, L.; Ulevitch, R.J. A MAP kinase targeted by endotoxin and hyperosmolarity in mammalian cells. *Science* **1994**, *265*, 808–811. [CrossRef] [PubMed]
77. Dérijard, B.; Raingeaud, J.; Barrett, T.; Wu, I.H.; Han, J.; Ulevitch, R.J.; Davis, R.J. Independent human MAP-kinase signal transduction pathways defined by MEK and MKK isoforms. *Science* **1995**, *267*, 682–685. [CrossRef]
78. Raingeaud, J.; Whitmarsh, A.J.; Barrett, T.; Dérijard, B.; Davis, R.J. MKK3- and MKK6-regulated gene expression is mediated by the p38 mitogen-activated protein kinase signal transduction pathway. *Mol. Cell. Biol.* **1996**, *16*, 1247–1255. [CrossRef]
79. Rizo, J.; Xu, J. Synaptic vesicle fusion without SNARE transmembrane regions. *Dev. Cell* **2013**, *27*, 124–126. [CrossRef]
80. Inoue, T.; Tsai, B. Regulated Erlin-dependent release of the B12 transmembrane J-protein promotes ER membrane penetration of a non-enveloped virus. *PloS Pathog.* **2017**, *13*, e1006439. [CrossRef]
81. Arlt, H.; Reggiori, F.; Ungermann, C. Retromer and the dynamin Vps1 cooperate in the retrieval of transmembrane proteins from vacuoles. *J. Cell Sci.* **2015**, *128*, 645–655. [CrossRef]
82. Lv, X.; Pu, X.; Qin, G.; Zhu, T.; Lin, H. The roles of autophagy in development and stress responses in Arabidopsis thaliana. *Apoptosis Int. J. Program. Cell Death* **2014**, *19*, 905–921. [CrossRef] [PubMed]
83. Kozyraki, R. Cubilin, a multifunctional epithelial receptor: An overview. *J. Mol. Med.* **2001**, *79*, 161–167. [CrossRef] [PubMed]
84. Bouton, D.; Escriva, H.; de Mendonça, R.L.; Glineur, C.; Bertin, B.; Noël, C.; Robinson-Rechavi, M.; de Groot, A.; Cornette, J.; Laudet, V.; et al. A conserved retinoid X receptor (RXR) from the mollusk Biomphalaria glabrata transactivates transcription in the presence of retinoids. *J. Mol. Endocrinol.* **2005**, *34*, 567–582. [CrossRef]
85. Song, Z. Roles of the nucleotide sugar transporters (SLC35 family) in health and disease. *Mol. Asp. Med.* **2013**, *34*, 590–600. [CrossRef] [PubMed]
86. Picco, C.; Scholz-Starke, J.; Festa, M.; Costa, A.; Sparla, F.; Trost, P.; Carpaneto, A. Direct Recording of Trans-Plasma Membrane Electron Currents Mediated by a Member of the Cytochrome b561 Family of Soybean. *Plant Physiol.* **2015**, *169*, 986–995. [CrossRef]
87. Walker, E.; Hernandez, A.V.; Kattan, M.W. Meta-analysis: Its strengths and limitations. *Cleve Clin. J. Med.* **2008**, *75*, 431–439. [CrossRef]
88. Meister, G.; Landthaler, M.; Patkaniowska, A.; Dorsett, Y.; Teng, G.; Tuschl, T. Human Argonaute2 mediates RNA cleavage targeted by miRNAs and siRNAs. *Mol. Cell* **2004**, *15*, 185–197. [CrossRef]
89. Huang, J.; Fraser, M.E. Structural basis for the binding of succinate to succinyl-CoA synthetase. *Acta Crystallogr. Sect. D Struct. Biol.* **2016**, *72*, 912–921. [CrossRef]
90. Lambeth, D.O.; Tews, K.N.; Adkins, S.; Frohlich, D.; Milavetz, B.I. Expression of two succinyl-CoA synthetases with different nucleotide specificities in mammalian tissues. *J. Biol. Chem.* **2004**, *279*, 36621–36624. [CrossRef]
91. Araújo, S.J.; Cela, C.; Llimargas, M. Tramtrack regulates different morphogenetic events during Drosophila tracheal development. *Development* **2007**, *134*, 3665–3676. [CrossRef]
92. Shirangi, S.A.; Kalbassi, M.R.; Khodabandeh, S.; Jafarian, H.; Lorin-Nebel, C.; Farcy, E.; Lignot, J.H. Salinity effects on osmoregulation and gill morphology in juvenile Persian sturgeon (*Acipenser persicus*). *Fish Physiol. Biochem.* **2016**, *42*, 1741–1754. [CrossRef] [PubMed]
93. McCormick, S.D.; Regish, A.M.; Christensen, A.K. Distinct freshwater and seawater isoforms of Na^+/K^+-ATPase in gill chloride cells of Atlantic salmon. *J. Exp. Biol.* **2009**, *212*, 3994–4001. [CrossRef] [PubMed]
94. Towle, D.W.; Weihrauch, D. Osmoregulation by Gills of Euryhaline Crabs: Molecular Analysis of Transporters1. *Am. Zool.* **2015**, *41*, 770–780. [CrossRef]
95. Mitsou, I.; Multhaupt, H.A.B.; Couchman, J.R. Proteoglycans, ion channels and cell-matrix adhesion. *Biochem. J.* **2017**, *474*, 1965–1979. [CrossRef] [PubMed]
96. Stogmann, E.; Reinthaler, E.; Eltawil, S.; El Etribi, M.A.; Hemeda, M.; El Nahhas, N.; Gaber, A.M.; Fouad, A.; Edris, S.; Benet-Pages, A.; et al. Autosomal recessive cortical myoclonic tremor and epilepsy: Association with a mutation in the potassium channel associated gene CNTN2. *Brain* **2013**, *136*, 1155–1160. [CrossRef]
97. Giepmans, B.N.G. Role of connexin43-interacting proteins at gap junctions. *Adv. Cardiol.* **2006**, *42*, 41–56. [CrossRef]

98. Ojala, V.K.; Knittle, A.M.; Kirjalainen, P.; Merilahti, J.A.M.; Kortesoja, M.; Tvorogov, D.; Vaparanta, K.; Lin, S.; Kast, J.; Pulliainen, A.T.; et al. The guanine nucleotide exchange factor VAV3 participates in ERBB4-mediated cancer cell migration. *J. Biol. Chem.* **2020**, *295*, 11559–11571. [CrossRef]
99. Towle, D.W.; Palmer, G.; Harris, J. Role of gill Na+ + K+ -dependent ATPase in acclimation of blue crabs (*Callinectes sapidus*) to low salinity. *J. Exp. Zool.* **1976**, *196*, 315–321. [CrossRef]
100. Perreira, J.M.; Aker, A.M.; Savidis, G.; Chin, C.R.; McDougall, W.M.; Portmann, J.M.; Meraner, P.; Smith, M.C.; Rahman, M.; Baker, R.E.; et al. RNASEK Is a V-ATPase-Associated Factor Required for Endocytosis and the Replication of Rhinovirus, Influenza A Virus, and Dengue Virus. *Cell Rep.* **2015**, *12*, 850–863. [CrossRef]
101. Busch, W.; Saier, M.H., Jr. The IUBMB-endorsed transporter classification system. *Mol. Biotechnol.* **2004**, *27*, 253–262. [CrossRef]
102. Anwar, T.; Samudrala, G. Bioinformatics Analysis and Functional Prediction of Transmembrane Proteins in Entamoeba histolytica. *Genes* **2018**, *9*, 499. [CrossRef] [PubMed]
103. Hou, C.; Tian, W.; Kleist, T.; He, K.; Garcia, V.; Bai, F.; Hao, Y.; Luan, S.; Li, L. DUF221 proteins are a family of osmosensitive calcium-permeable cation channels conserved across eukaryotes. *Cell Res.* **2014**, *24*, 632–635. [CrossRef] [PubMed]
104. Liu, X.; Wang, J.; Sun, L. Structure of the hyperosmolality-gated calcium-permeable channel OSCA1.2. *Nat. Commun.* **2018**, *9*, 5060. [CrossRef] [PubMed]
105. Pang, C.; Yuan, H.; Ren, S.; Chen, Y.; An, H.; Zhan, Y. TMEM16A/B associated CaCC: Structural and functional insights. *Protein Pept. Lett.* **2014**, *21*, 94–99. [CrossRef] [PubMed]
106. Prole, D.L.; Taylor, C.W. Identification and analysis of putative homologues of mechanosensitive channels in pathogenic protozoa. *PLoS ONE* **2013**, *8*, e66068. [CrossRef]
107. Pao, S.S.; Paulsen, I.T.; Saier, M.H., Jr. Major facilitator superfamily. *Microbiol. Mol. Biol.* **1998**, *62*, 1–34. [CrossRef]
108. Kumar, S.; Lekshmi, M.; Parvathi, A.; Ojha, M.; Wenzel, N.; Varela, M.F. Functional and Structural Roles of the Major Facilitator Superfamily Bacterial Multidrug Efflux Pumps. *Microorganisms* **2020**, *8*, 266. [CrossRef]
109. Palmieri, F. The mitochondrial transporter family SLC25: Identification, properties and physiopathology. *Mol. Asp. Med.* **2013**, *34*, 465–484. [CrossRef]

biology

MDPI

Article

Genomic and Transcriptomic Landscape and Evolutionary Dynamics of Heat Shock Proteins in Spotted Sea Bass (*Lateolabrax maculatus*) under Salinity Change and Alkalinity Stress

Xujian Li [1], Saisai Liu [1], Yapeng Wang [1], Wei Lu [1], Quanqi Zhang [1,2,3] and Jie Cheng [1,2,3,*]

1 Key Laboratory of Marine Genetics and Breeding, Ministry of Education, Ocean University of China, 5 Yushan Road, Qingdao 266003, China; lxj1316@stu.ouc.edu.cn (X.L.); liusaisai@stu.ouc.edu.cn (S.L.); wangyapeng@stu.ouc.edu.cn (Y.W.); lw1981@ouc.edu.cn (W.L.); qzhang@ouc.edu.cn (Q.Z.)
2 Laboratory for Marine Fisheries Science and Food Production Processes, Pilot National Laboratory for Marine Science and Technology, 1 Wenhai Road, Qingdao 266237, China
3 Key Laboratory of Tropical Aquatic Germplasm of Hainan Province, Sanya Oceanographic Institution, Ocean University of China, Sanya 572024, China
* Correspondence: jiecheng@ouc.edu.cn; Tel.: +86-0532-82031986

Simple Summary: Heat shock proteins (Hsps) are ubiquitous and conserved in almost all living organisms and are involved in a wide spectrum of cellular responses against diverse environmental stresses. However, our knowledge about the coordinated Hsp co-chaperon interaction is still limited, especially in aquatic animals facing dynamic water environments. In this study, we provided the systematic analysis of 95 *Hsp* genes (*LmHsps*) in spotted sea bass (*Lateolabrax maculatus*), an important aquaculture species in China, under salinity change and alkalinity stress through in silico analysis. The coordinated expression of *LmHsps* in response to salinity change and alkalinity stress in the gills was determined. Our results confirmed the diverse regulated expression of *Hsps* in *L. maculatus*, and that the responses to alkalinity stress may have arisen through the adaptive recruitment of *LmHsp40-70-90* co-chaperons. Our results provide vital insights into the function and adaptation of aquatic animal Hsps in response to salinity-alkalinity stress.

Abstract: The heat shock protein (Hsp) superfamily has received accumulated attention because it is ubiquitous and conserved in almost all living organisms and is involved in a wide spectrum of cellular responses against diverse environmental stresses. However, our knowledge about the Hsp co-chaperon network is still limited in non-model organisms. In this study, we provided the systematic analysis of 95 *Hsp* genes (*LmHsps*) in the genome of spotted sea bass (*Lateolabrax maculatus*), an important aquaculture species in China that can widely adapt to diverse salinities from fresh to sea water, and moderately adapt to high alkaline water. Through in silico analysis using transcriptome and genome database, we determined the expression profiles of *LmHsps* in response to salinity change and alkalinity stress in *L. maculatus* gills. The results revealed that *LmHsps* were sensitive in response to alkalinity stress, and the *LmHsp40-70-90* members were more actively regulated than other *LmHsps* and may also be coordinately interacted as co-chaperons. This was in accordance with the fact that members of *LmHsp40*, *LmHsp70*, and *LmHsp90* evolved more rapidly in *L. maculatus* than other teleost lineages with positively selected sites detected in their functional domains. Our results revealed the diverse and cooperated regulation of *LmHsps* under alkaline stress, which may have arisen through the functional divergence and adaptive recruitment of the *Hsp40-70-90* co-chaperons and will provide vital insights for the development of *L. maculatus* cultivation in alkaline water.

Keywords: heat shock protein; co-chaperon network; salinity-alkalinity adaptation; molecular evolution; *Lateolabrax maculatus*

Citation: Li, X.; Liu, S.; Wang, Y.; Lu, W.; Zhang, Q.; Cheng, J. Genomic and Transcriptomic Landscape and Evolutionary Dynamics of Heat Shock Proteins in Spotted Sea Bass (*Lateolabrax maculatus*) under Salinity Change and Alkalinity Stress. *Biology* **2022**, *11*, 353. https://doi.org/10.3390/biology11030353

Academic Editor: Patricia Pereiro

Received: 6 February 2022
Accepted: 21 February 2022
Published: 23 February 2022

1. Introduction

Heat shock proteins (Hsps) are highly conserved stress proteins existing broadly from prokaryotes to mammals with chaperone activity [1,2], which play essential roles in maintaining the stability of cell structure and mitigating the effect of protein misfolding and aggregation in almost all aspects of protein metabolism [3]. The Hsp superfamily could be classified into Hsp10 (Hspe, 10 kDa), small Hsp (sHsp/Hspb, 20–30 kDa), Hsp40 (Dnaj, 40 kDa), Hsp60 (Hspd, 60 kDa), Hsp70 (Hspa, 70 kDa), Hsp90 (Hspc, 83~90 kDa), and Hsp100/110 (Hsph, 100–110 kDa) according to their relative molecular mass, structure homology, and function [4]. For example, sHsps can bind a large range of non-native substrate proteins to form a sHsp substrate complex prior to irreversible aggregation, which becomes the target of ATP-dependent Hsp70–Hsp100 disaggregation activity [5]. Hsp40s, also known as J-domain proteins (Dnajs), have a key role in the process of transferring substrates to their Hsp70 partners by stimulating the ATP hydrolysis of Hsp70 [6]. Hsp10/60s are co-chaperones mainly implicated in mitochondrial protein import and macromolecular assembly, which function together to facilitate the correct folding of imported proteins [7]. Hsp70s and Hsp90s are both ATP-dependent molecular chaperones, which perform numerous functions in a wide variety of cellular processes, including the protection of proteins from stresses [8]. In addition, the Hsp chaperons also cooperate with each other to constitute a dynamic and functionally versatile network for diverse cellular functions [9]. For instance, Hsp70 and Hsp40 could interact with a substrate protein and co-chaperones to facilitate the transfer of the substrate to Hsp90 [9,10], while Hsp90 and Hsp70 could also directly collaborate with each other [10]. In contrast, sHsps function independently of ATP so as to prevent the formation of large insoluble protein aggregates [11].

With the increasing impacts of climate change and anthropogenic stresses, the function of the heat-shock system affecting organisms' evolutionary fitness in diverse environments is a growing and important topic [12]. A large number of studies have focused on *Hsp* families in animals living in a highly dynamic aquatic environment. For instance, changes in abiotic factors (temperature, salinity, pH, dissolved oxygen, heavy metals, or other pollutants) [13–15] as well as biotic pathogen and toxic algae infection [16–20] could cause stresses in aquatic animals. Accordingly, *Hsps* are expressed significantly and act as chaperones to help cells maintain normal physiological activities related to growth, reproduction, and physiological homeostasis. In particular, salinity changes in aquatic environments can increase the lysozyme activity, mucosal production, and immune defense in teleosts, and high alkalinity is harmful to most fishes, leading to health damage and even degradation of biomolecules [21]. Recently, although numerous studies have strongly implicated a critical role of *Hsps* in salinity-alkalinity adaptation [22–24], most of them are only focused on responses for a limited number of *Hsps* rather than the full complement of genes that comprise the Hsp co-chaperone network in salinity-alkalinity adaptation. Therefore, with the saline-alkaline water aquaculture becoming a promising way to accommodate the growing need for the aquaculture industry, it is important to investigate the physiological change and molecular response in fishes adapting to saline-alkaline waters, which could also help to understand their adaptation in the aquatic ecosystem.

Spotted sea bass (*Lateolabrax maculatus*) is an economically important fish with a high nutritional value that is widely distributed in the coastal and estuarine areas of China, Japan, and the Korean peninsula [25]. *L. maculatus* has an excellent ability to adapt to a broad variety of salinity environments ranging from fresh to sea water [26]; therefore, its aquaculture is viable in both freshwater ponds and sea water cages in China. Additionally, *L. maculatus* has been proved to be able to survive in high alkaline water (carbonate alkalinity = 10 mmol/L) for a long period of time [27], and the aquaculture of *L. maculatus* in alkaline water has been under development recently. To explore the function of *Hsp* superfamily in *L. maculatus* (*LmHsp*) under salinity change and alkalinity stress, in this study, 95 *LmHsp* genes were identified in *L. maculatus* genome, and their phylogeny, conserved structure, expression profile, possible co-chaperon network under stress, as well as molecular evolution pattern were investigated. Our results provide comprehensive insights for further understanding the environmental adaptation of *Hsp* co-chaperons in *L. maculatus* and will help to establish the alkaline water aquaculture of *L. maculatus*.

2. Materials and Methods

2.1. Genome-Wide Identification and Annotation of Hsp Families in L. maculatus

To identify *LmHsp* genes, the coding sequences (cds) and amino acid (aa) sequences of *L. maculatus* genome (PRJNA408177) were blast searched using available *Hsp* cds and aa sequences from nine fish species, including zebrafish (*Danio rerio*), channel catfish (*Ictalurus punetaus*), stickleback (*Gasterosteus aculeatus*), medaka (*Oryzias latipes*), tilapia (*Oreochromis niloticus*), turbot (*Scophthalmus maximus*), fugu (*Takifugu rubripes*), Asian sea bass (*Lates calcarifer*), spotted gar (*Lepisosteus oculatus*), and four tetrapod species, including human (*Homo sapiens*), mouse (*Mus musculus*), chicken (*Gallus gallus*), and Xenopus (*Xenopus tropicalis*) from Ensemble (http://asia.ensembl.org accessed on 9 November 2021) and NCBI (http://www.ncbi.nlm.nih.gov accessed on 9 November 2021). These sequences were used as query to perform BLASTN and BLASTP (e-value = 1×10^{-5}) against the sequence resources of *L. maculatus*. The conserved domains of candidate Hsp sequences were predicted using SMART [28] and Pfam [29], and the sequences without a corresponding complete Hsp domain were excluded from further analysis. When a gene could not be accurately distinguished, it was named after the zebrafish orthologs with letter "L" (meaning "like").

For the identified LmHsp aa sequences, the Multiple Em for Motif Elicitation (MEME, https://meme-suite.org/meme accessed on 18 November 2021) program was applied to evaluate their conserved motifs, with the parameters of any number of repetitions, optimum width of motifs from 6 to 200 with 4 motifs for sHsp, 5 motifs for Hsp40, 3 motifs for Hsp10/60, 13 motifs for Hsp70, 6 motifs for Hsp90, and 3 motifs for Hsp100. The Gene Structure Display Sever (GSDS 2.0, http://gsds.gao-lab.org accessed on 20 November 2021) was used to visualize the exon-intron of *LmHsp* genes, and the diagrammatic sketches were illustrated using TBtools v1.09 [30].

2.2. Phylogenetic Analysis of Hsp Families

All Hsp aa sequences from *L. maculatus* and other selected vertebrates mentioned before were used to construct phylogenetic trees. Multiple protein sequence alignments were conducted with Muscle [31]. According to the Bayesian information criterion (BIC), the best models selected for phylogeny construction were JTT + F + G4 model for sHsp, VT + G4 model for Hsp40, LG + I + G4 model for Hsp60, JTT + G4 model for Hsp70, LG + G4 model for Hsp90, and JTT + I for Hsp100. The Maximum Likelihood (ML) phylogenetic trees of each Hsp family were generated using IQ-Tree [32], respectively, with a bootstrap of 1000 replicates. Furthermore, iTOL [33] was used to visualize the phylogenetic trees.

2.3. Expression of LmHsp Genes through Transcriptome Analysis

To investigate the expression profiles of *LmHsp* genes in *L. maculatus* tissues, transcriptome data of seven adult tissues were obtained from NCBI (brain-SRR7528887, stomach-SRR7528884, spleen-SRR7528888, liver-SRR7528886, gill-SRR7528883, testis-SRR7528885, and ovary-SRR2937376). Through the Hisat and StringTie pipeline [34], the Fragments Per Kilobase of exon per Million mapped reads (FPKM) of each *LmHsp* gene were obtained. To understand the expression dynamics of *LmHsps* under stresses, the RNA-seq data was also obtained from NCBI as gill tissues under salinity change (PRJNA515986) and alkalinity stress (PRJNA611641), respectively. The former data represented *L. maculatus* specimens (158.23 ± 18.77 g) originated from sea water (13.5–14.5 °C, pH 7.8–8.15, salinity 30 ppt) with salinity acclimation in fresh water (0 ppt, control group), brackish water (15 ppt), and sea water (30 ppt) for 30 days, respectively [35]. The latter data represented *L. maculatus* specimens (body weight: 140.32 ± 2.56 g) acclimated in fresh water (pH 7.8–8.15) for 30 days (control group) and then challenged with alkalinity stress (carbonate alkalinity = 18 mmol/L with $NaHCO_3$ and Na_2CO_3 added to fresh water) at 0 h, 12 h, 24 h, and 72 h [27].

All the experimental groups contained three individuals for biological replicate, which represented a good correlation among groups from principal component analysis (PCA) (Figure S1). These RNA-seq data were processed with the Hisat and StringTie pipeline and the $\log_2$ (fold change) ($\log_2$FC) for *LmHsp* genes between the test and control group was calculated by edgeR [36]. TBtools was employed to draw heatmaps with $\log_2$ (FPKM + 1) and $\log_2$FC values, respectively. In addition, the trend analysis of gene expression was conducted to cluster *LmHsp* genes with similar expression patterns along the test time points under alkalinity stress via online toolkit OmicShare (https://www.omicshare.com/ accessed on 17 Feburary 2022).

2.4. Protein-Protein Interaction and Co-Expression Analysis

Protein-protein interaction (PPI) networks among *Hsp* families were analyzed via the online STRING 11.5 tool (https://string-db.org/ accessed on 10 October 2021) for human and zebrafish, with the parameter minimum required interaction score of 0.7. For *L. maculatus Hsps* in response to alkalinity stress, the expression correlation between each *LmHsp* gene was evaluated using pairwise Pearson's correlation coefficient (PCC) with their FPKM and the correlation coefficient value of 0.7. The networks were visualized using Cytoscape [37].

2.5. Molecular Evolution Analysis

The cds of each *Hsp* family was aligned by Muscle, and MEGA7 [38] was used to generate ML trees to determine their phylogeny. The EasyCodeML v1.21 [39] was used to perform the molecular evolution analysis, with the site model (SM), branch model (BM), and branch-site model (BSM) tests in the PAML package [40]. Firstly, the ratio of non-synonymous to synonymous substitution (dN/dS, ω) and the likelihood ratio tests (LRTs) were employed to evaluate the selective pressures in each *LmHsp* family. Six SMs were used to test the positive selection in each codon: M0 assumes to have the same ω for all codons, M3 assumes the ω of all codons showing a simple discrete division trend; M1a assumes only conservative sites ($0 < \omega < 1$) and neutral sites ($\omega = 1$) for all codons, while M2a is considered to increase the existence of positive sites ($\omega > 1$) for all codons on the basis of M1a; the ω of all codons in M7 are assumed to belong to the matrix (0, 1) with a beta distribution, while M8 adds another type of ω ($\omega > 1$) on the basis of M7. LRTs were used to judge whether the paired models (M0 vs. M3; M1a vs. M2a; M7 vs. M8) are significantly different [40], and to estimate whether there are positive selected sites (PSSs) (M2a vs. M1a and M8 vs. M7) under the premise of significant p value ($p < 0.05$).

Furthermore, BM were conducted to test the selective pressure of each *LmHsp* family. Phylogeny was constructed with all *LmHsp* cds, and each *Hsp* family was marked as the foreground branch (ω_1), respectively, to explore the evolution rate compared with the remaining *LmHsp* families (background branch, ω_0). BSM were conducted to test the selective pressure of single *LmHsp* gene (foreground branch, ω_1) with 10 fish species (background branch, ω_0), including spotted sea bass (*L. maculatus*), zebrafish (*D. rerio*), channel catfish (*I. punetaus*), stickleback (*G. aculeatus*), medaka (*O. latipes*), tilapia (*O. niloticus*), turbot (*S. maximus*), fugu (*T. rubripes*), Asian sea bass (*L. calcarifer*), and spotted gar (*L. oculatus*). With positive selection suggested ($p < 0.05$), the PSSs were further verified with high BEB posterior probabilities (>0.95). The PSSs were then retrieved from the aa sequences within functional Hsp domains from the Pfam database. 3D Structure prediction analysis was performed via Phyre2 online tool (https://www.sbg.bio.ic.ac.uk/phyre2/ accessed on 29 December 2021) [41] and modified by PyMol software (https://pymol.org/2/ accessed on 30 December 2021).

3. Results and Discussion

3.1. Genomic Landscape, Functional Domain, and Phylogeny of Hsp Superfamily in L. maculatus

Through genome-wide screen, 95 *Hsp* genes (*LmHsps*) were identified in the *L. maculatus* genome, including 12 *sHsps* (*LmHspb*), 50 *Hsp40s* (*LmDnaj*), 9 *Hsp10/60s* (*LmHspe*, *LmHspd*,

Lmcct), 17 *Hsp70*s (*LmHspa*), 5 *Hsp90*s (*LmHspc*), and 2 *Hsp100*s (*LmXlp*) (Figure 1 and Table S1). The general features of *LmHsps* were listed in Table S2, with their length ranging from 56 to 2220 amino acids (aa), and exon numbers between 1 and 53. The copy number of each *LmHsp* family was generally conserved among the selected teleost and tetrapod species (Figure 1 and Table S1), some of which (*sHsp*, *Hsp40* and *Hsp90*) were more in teleosts than that in the tetrapod lineages, most likely due to the teleost specific genome duplication (TSGD) event in the teleost lineages [42].

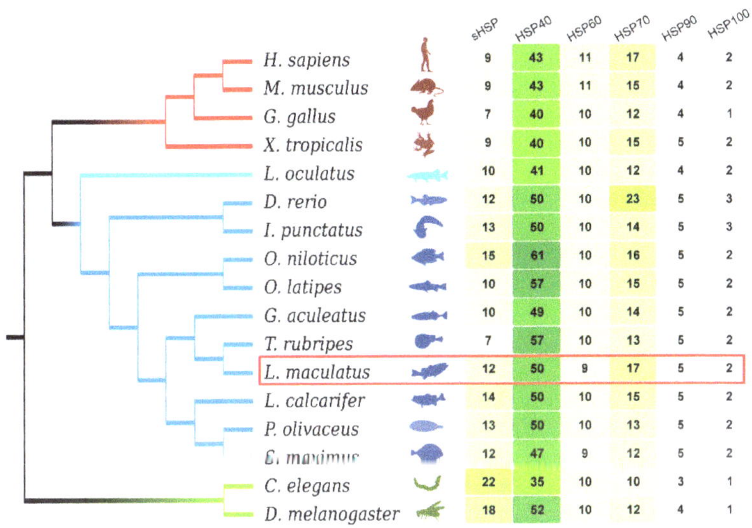

Figure 1. Genome-wide identification of *Hsp* families in 10 teleost species and other metazoan species, with *L. maculatus* in the red frame. Branches with different colors represent different metazoan groups. The gene numbers are illustrated with the heatmap.

Specifically, 12 *sHsp* genes were identified in the *L. maculatus* genome, which were clustered into 8 sub-families, but absent in other 4 sub-families as *Hspb1*, *Hspb2*, *Hspb3*, and *Hspb9* (Figure S2A). The *Hsp30*s was not identified in the tetrapods, but was only present in teleosts with independent duplication. In addition, as the ubiquitous family of chaperones, the structure of sHsps can be defined into three domains [43]: a conserved α-crystallin domain (ACD) with N-terminal region (NTR) and C-terminal region (CTR) on either side [44]. Most *LmsHsps* comprised three conserved motifs (motif 1/2/4, Figure 2A) except *Hsp30L_10022466*, while *Cryab*, *Hspb11*, *Hsp30L_10002822*, and *Hsp30L_10002821* contained one extra motif 3, suggesting strong structure conservation among *LmsHsp* genes (Figure 2A).

A total of 50 *Hsp40* genes (*Dnaj*s) were identified in the *L. maculatus* genome, which were clustered into 42 sub-families (Figure S2B). Most *LmHsp40*s comprised 3 conserved motifs (motifs 1/2/3, Figure 2B) as the major DnaJ domain, while 10 *LmHsp40*s had extra motifs 4 or 5 in their sequences. Moreover, one *Hsp10* (*Hspe1*) and eight *Hsp60*s (*Hspd1* and 7 *Cct*s) were identified, with *Cct8* missing in the *L. maculatus* genome. The phylogeny supported the evolutionary relationship of *LmHspd1* and *LmHspe1*, respectively, clustered with orthologous genes of other teleost species (Figures 2C and S2C). There were three conserved motifs identified among *LmHsp60* members as the Cpn60_TCP1 domain, while none were found in *LmHsp10* (Figure 2C).

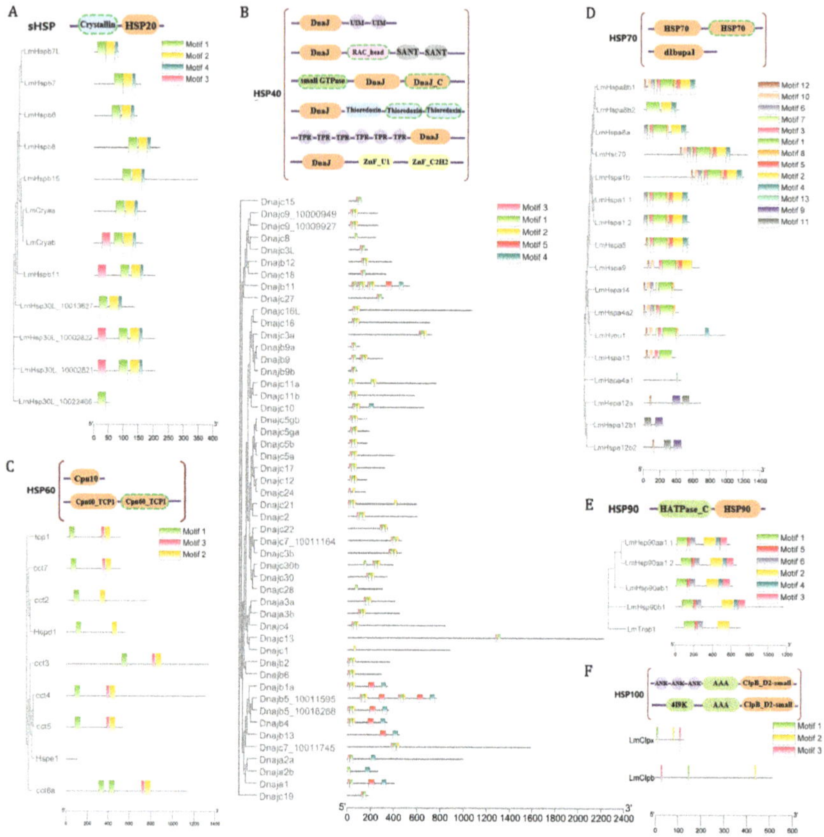

Figure 2. Conserved domain architecture of *LmHsp* families, as (**A**) *LmsHsps*, (**B**) *LmHsp40*s, (**C**) *LmHsp10/60*s, (**D**) *LmHsp70*s, (**E**) *LmHsp90*s, and (**F**) *LmHsp100*s, showing along their phylogenetic relationships. The horizontal grey bars represent amino acid sequences without predicted functional domains, whereas the colored boxes represent the regions with successfully predicted motifs. The orange boxes represent the corresponding HSP domain of each family, the green boxes represent the ATPase domains, and the other boxes represent specifically annotated domains. The domains that appear only in some family members are marked with dashed frames.

Seventeen *LmHsp70* genes were identified in the *L. maculatus* genome, which contained more duplicated copies, mainly from the *Hspa1*, *Hspa8*, and *Hspa12* sub-families (Table S1 and Figure 2D). Almost all the *Hsp70* sub-families were found between human and *L. maculatus* except for *Hspa2*, *Hspa6*, *Hspa7*, and *Hsph1*, which exist in human [45] but not in fish species apart from spotted gar, zebrafish, and stickleback (Figure S2D). Thirteen conserved motifs were identified in *LmHsp70*s, and genes with closer phylogenetic relationship had similar conserved motifs. For example, motifs 9 and 11 were only annotated in *Hspa12* sub-family (Figures 2D and S2D), which directly confirmed the fact that this sub-family was distantly related to other *Hsp70* genes in vertebrates [45]. Another noteworthy gene was *Hsp4a1*, which only contained one specific motif 13, and was significantly different from other *LmHsp70*s, such as *Hspa4a2*. The vertebrate *Hsp70*s were mainly scattered into nine clads, among which eight sub-families were easily distinguished, except for the sub-families *Hspa1-2-6-7-8* and *Hsc70* (Heat shock cognate 71 kDa), all clustering together (Figure S2D). This was in accordance with the fact that *Hspa1-2-6-7-8* and *Hsc70* shared a common domain called HSP1_2_6_8Nucleoid Binding Domain [46] with very similar conserved motifs (Figure S2D).

In addition, a total of five *Hsp90*s and two *Hsp100*s were collected from the *L. maculatus* genome, which was similar to that in other teleosts. The vertebrate *Hsp90* genes can be classified into four sub-families, supporting the four clads (*Hsp90aa1*, *Hsp90ab1*, *Hsp90b1* and *Trap1*) in their phylogeny (Figure S2E). Teleost *Hsp90aa1* genes were clustered into two sub-clads, *Hsp90aa1.1* and *Hsp90aa1.2*, further supporting that they were originated from the TSGD in the teleost lineage [42] (Figure S2E). Moreover, all the *LmHsp90* members contain multi-introns, and their detected six motifs are conserved with each other (Figure 2E). In total, all this information revealed the generally conserved sequence and function in teleost *Hsp* families.

3.2. Diverse Expression of LmHsp Genes among L. maculatus Normal Tissues

Gene expression profiling facilitates the understanding of the function and evolution of *Hsp* families. Here, we conducted transcriptome analysis with RNA-seq data from seven *L. maculatus* adult tissues (brain, stomach, liver, spleen, gill, ovary, and testis), which illustrated diverse expression profiles of the 95 *LmHsp* genes across tissues. In particular, *LmHsp10/60*s were mostly highly expressed among tissues, while *LmsHsp*s were lowly expressed (Figure 3). Interestingly, *Hspa8a* and *Hsp90ab1* were the most abundantly expressed across all tissues, suggesting their essential role in maintaining the normal physiological homeostasis in diverse tissues, whereas *Dnajb9*, *Dnajc5b*, and *Hspb7L* were not expressed in any tissues (Figure 3). Moreover, some *Hsp* genes showed biased expression in certain tissues. For example, *Cryab* was only expressed in brain, *Hspb11* was highly expressed in gill, and both spleen and liver had higher *Hspa1.1* and *Hspa1.2* expression than other tissues (Figure 3), which may indicate the tissue specific function of these *Hsp* members as chaperons. In addition, almost all *sHsp* genes (except *Hspb8*) were weakly expressed or even not expressed in ovary, while *Cryaa*, *Hspb6*, *Hspb7*, and *Hsp30L* were highly expressed in testis (Figure 3).

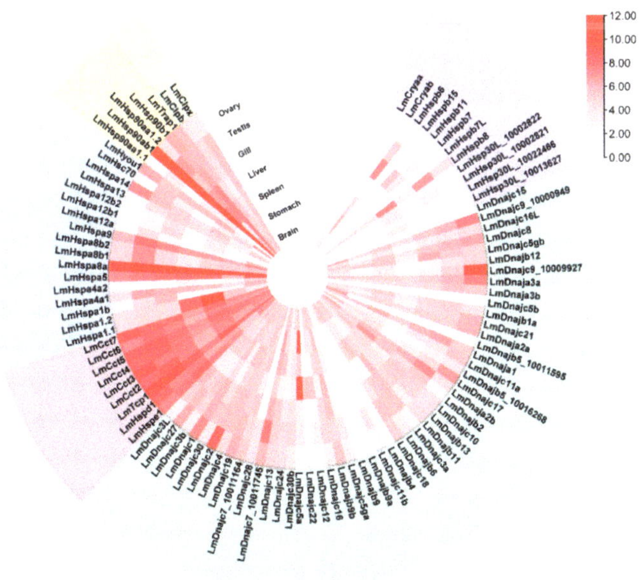

Figure 3. Expression profile of *LmHsp* genes among adult tissues of *L. maculatus*. The $\log_2$ (FPKM + 1) values are represented as 0–12 according to the colored scale bar, and the heatmap is ranked with each *LmHsp* families.

3.3. Regulated Expression of LmHsp Genes in Response to Salinity Change and Alkalinity Stress

To investigate the functional response of *LmHsps* to environmental changes, the regulated expression of 95 *LmHsp* genes was characterized under salinity or alkalinity stress in gills where the ion secretion and uptake happen. As chaperons, Hsps may help to maintain the homeostasis by interacting with the osmotic stress denatured proteins in gill cells. Firstly, *L. maculatus* could be well adapted to wide salinity ranges, therefore, with freshwater acclimation, the expression of *LmHsp* genes may be weakly regulated (Figure 4A and Table S3). For example, only *Hspa12b1* and three *Hsp40*s (*Dnajc3L, Dnajc3a,* and *Dnajc5b*) were significantly influenced in the brackish water group, and only *Dnajc5ga* was up-regulated in the sea water group (Figure 4A), which suggested that the gill tissues were already adapted to the stimulation caused by freshwater acclimation. Moreover, even though not significantly regulated, the expression of almost all *sHsp* members (except *Hspb7*) were reduced, and the expression of most *Hsp70*s were increased, whereas most *Hsp90* members (*Hsp90ab1, Hsp90b1,* and *Trap1*) did not respond to the stimulation (Figure 4A), suggesting possible functional diversification among the *LmHsp* families.

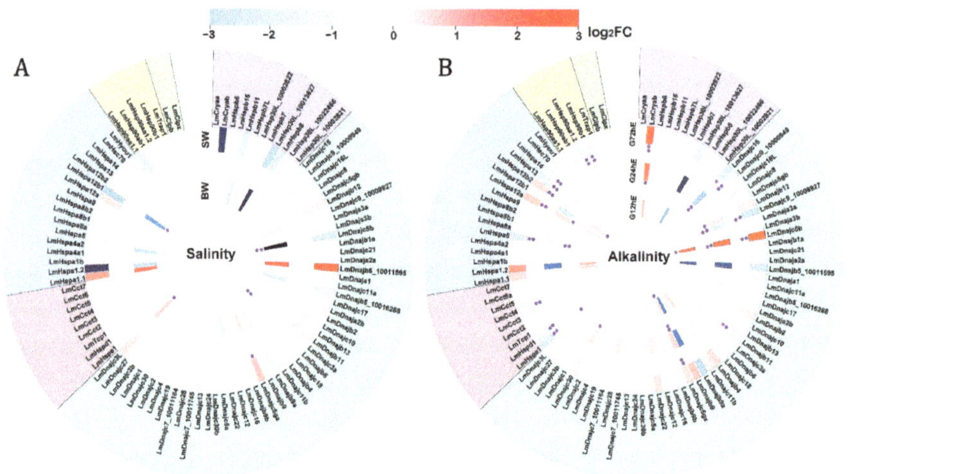

Figure 4. Regulated expression of *LmHsp* genes in gills of *L. maculatus* in response to (**A**) salinity change and (**B**) alkalinity stress. The heatmap was based on the $\log_2$FC values and ranked with each *LmHsp* family. BW and SW represent brackish and sew water groups compared to the freshwater group, while G12hE, G24hE, and G72hE indicate the alkalinity experiment duration for 12 h, 24 h, and 72 h, compared to the blank control 0 h, respectively. * indicates $|\log_2\text{FC}| > 0.7$ with $p < 0.05$, ** indicates $|\log_2\text{FC}| > 0.7$ with both p value and false discovery rate (FDR) < 0.05.

The alkalinity challenge revealed that *LmHsps* in gills were sensitive to alkalinity stress within a short period of time (Figure 4B and Table S3), which was most likely in accordance with the good adaptation of *L. maculatus* to diverse salinities rather than alkalinity. For example, under alkalinity stress, the number of significantly regulated *LmHsp* genes ($|\log_2\text{FC}| > 0.7$ and $p < 0.05$) steadily increased along the test time points (Figure 4B). In particular, only 3 *LmHsps* (*Hspa12b2, Dnajc5ga,* and *Dnajc5b*) were up-regulated at 12 h exposure, and 10 *LmHsps* (*Cryab, Hspe1, Dnajb12, Dnajc5b, Dnajc5ga, Dnajc9_10009927, Dnajc15, Dnajc30, Hspa5,* and *Hspa12a*) were significantly influenced at 24 h, while at 72 h, 17 *LmHsp* genes (*Cryab, Hspe1, Hspd1, Cct4, Dnajb11, Dnajc5b, Dnajc5ga, Dnajc9_10000949, Dnajc9_10009927, Dnajc15, Hspa5, Hspa12a, Hspa12b2, Hspa13, Hspa14, Hsp90aa1.2,* and *Hsp90b1*) were significantly regulated (Figure 4B). Of these regulated *LmHsps*, 12 were down-regulated (*Hspe1, Hspd1, Cct4, Dnajb11, Dnajc9_10000949, Dnajc9_10009927, Dnajc15, Hspa5, Hspa13, Hspa14, Hsp90aa1.2,* and *Hsp90b1*) with $\log_2$FC ranging from −0.72 to −2.59, whereas 7 were up-regulated (*Cryab, Dnajb12, Dnajc5b, Dnajc5ga, Dnajc30, Hspa12a,* and *Hspa12b2*) with $\log_2$FC ranging from 0.80 to 12.04 (Table S3), which was mostly in the *LmHsp40, LmHsp70,* and *LmHsp90* families but less in *LmsHsp* and *LmHsp10/60* families. *Hsp70* and *Hsp90* are the two highly conserved ATP-dependent molecular chaperones that fold and remodel proteins in almost every cellular process with co-chaperon *Dnaj*s [8,9]. The regulation of *LmHsp40*s (*Dnaj*s), *LmHsp70*s, and *LmHsp90*s indicated that they could play more essential roles in response to alkalinity stress than other *LmHsps*. Under alkalinity stress, the ionic balance disruption occurred due to the hyperosmotic environment with protein damage [47], which can increase the levels of some *LmHsp40*s and *LmHsp70*s (Figure 4B). However, there were more down-regulated *LmHsps* rather than up-regulated ones, possibly due to the certain degree of adaptation to alkalinity stress in *L. maculatus,* which could survive in alkaline water for more than two weeks to two months [27]. Several studies also reported down-regulation of *Hsp*s in marine organisms under diverse stimuli, such as scallops *Chlamys farreri* and *Patinopecten yessoensis Hsp70B2* genes in response to toxic dinoflagellates [16,17], ascidian *Ciona savignyi Hsp20/60/70/90* genes under both low and high salinity stresses [48], and European seabass *Dicentrarchus labrax Hsp70* in low salinity conditions [47]. Further investigation is warranted with a possible delayed response in *LmHsp* expression level as well as the regulatory heat shock factor (Hsf) of these *LmHsps*. In addition, the expression of *Hspa8a* and *Hsp90ab1* did not exhibit differential regulation (Figure 4B), although they showed high abundance in both normal and challenged gill tissues (Figure 3 and Table S3), indicating that these genes did not participate in the response to alkalinity stress but may be needed to maintain the physiological homeostasis.

3.4. Coordinated Regulation of LmHsp Co-Chaperones under Alkalinity Stress

When the *Hsps* repair misfolded proteins in a simplified model, *Hsp70* initially facilitates the re-folding of hydrophobic residue stretches into their native state together with *Hsp40*s (*Dnaj*s), whereas *Hsp90* catalyzes a step later in protein folding and activation [8,11]. With the possible co-chaperon among *Hsp40*s, *Hsp70*s, and *Hsp90*s, to further excavate their associated regulation, the expression trend and Pearson's correlation coefficient (PCC) of *LmHsp40-70-90* genes under alkalinity stress were further investigated. As a result, 24 *LmHsp40-70-90* genes were enriched into 4 expression profiles (Figure 5A and Table S4), among which 10 *LmHsp*s were enriched in profile 0 (declined expression) and 9 *LmHsp*s were enriched in profile 4 (raised expression) (Figure 5B). In profile 0, the expression of two *LmHsp70*s (*LmHspa5* and *LmHspa13*), two *LmHsp90*s (*LmHsp90aa1.2* and *LmHsp90b1*), and six *LmHsp40*s (*LmDnajc15, LmDnajc9_10000949, LmDnajc9_10009927, LmDnajb5_10016268, LmDnajb11,* and *LmDnajb9a*) were continuously down-regulated (Figure 5B), while in profile 4, the expression of three *LmHsp70*s (*LmHspa1.1, LmHspa12a,* and *LmHspa12b2*) and six *LmHsp40*s (*LmDnajb12, LmDnajc5b, LmDnajb4, LmDnajc5ga, LmDnajc22,* and *LmDnajc30*) were continuously up-regulated (Figure 5B). These transcriptional profiles may be suggestive of coordinated regulation of the *Hsp40-70-90* genes and could thus comprise the possible component of alkalinity adaptation in *L. maculatus*, which warrants further verification.

In addition, according to PCC > 0.7, 15 *LmHsp40-70-90* members from profile 0 (8 *Hsp*s) and profile 4 (7 *Hsp*s) may be coordinately regulated with each other (Figure 5C and Table S5). For example, the up-regulated *LmHspa12b2* was positively correlated with *LmDnajb12* and negatively correlated with *LmDnajc9,* while the down-regulated *LmHsp90b1* and *LmHsp90aa1.2* were positively correlated with *LmDnajc9* or *LmDnajb11* (Figure 5C and Table S5), which may indicate *LmDnajc9* as the key co-chaperon of *LmHspa12b2, LmHsp90b1,* and *LmHsp90aa1.2*. Moreover, the down-regulated *LmHspa5* also co-expressed with *LmDnajb11, LmDnajc9,* and *LmHsp90b1,* while negatively related to *LmHspa12b2* and *LmDnajc5ga.* These protein-protein interaction correlations among Hsp40-70-90 members could also be observed in model organisms such as zebrafish and human (Figure S3), indicating functional conservation of these Hsp co-chaperons. Moreover, the expression trends and PCC relationship also revealed that the down-regulated co-expression of *LmHsp40-70-90* members may reflect their corporation, whereas the up-regulation of *LmHsp40*s and atypical *LmHspa12*s may supplement the loss caused by down-regulation (Figure 5C).

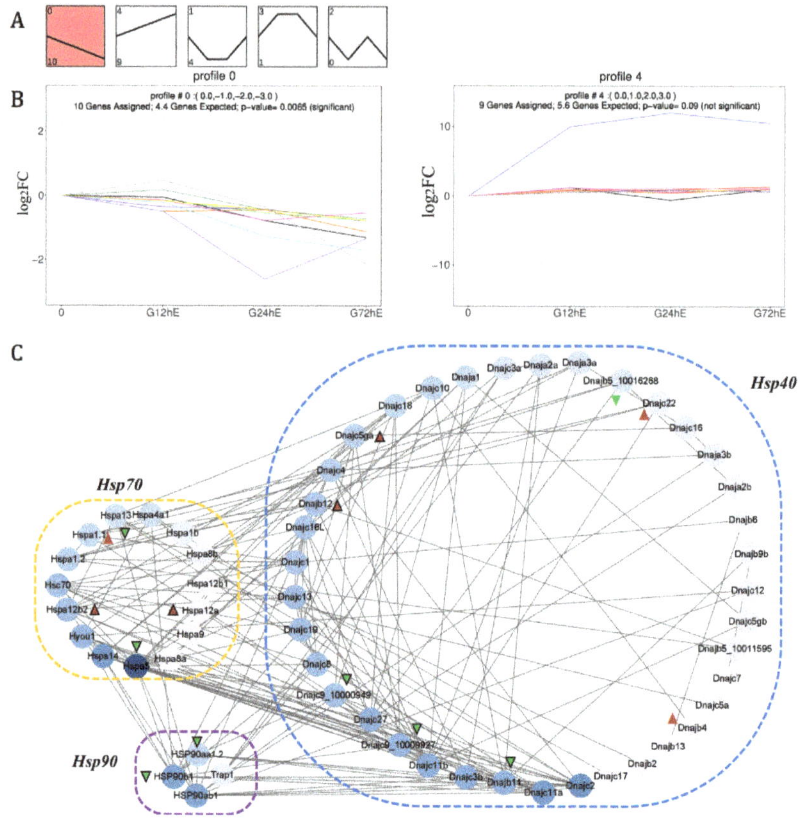

Figure 5. The expression trend and coordinated regulation of *LmHsp40-70-90* co-chaperons in response to alkalinity stress. (**A**) Expression patterns enriched in five profiles with the number of *LmHsps* indicated; (**B**) Expression trends of *LmHsp* genes enriched in the down-regulated profile 0 and up-regulated profile 4; (**C**) Protein-protein interaction (PPI) according to the PCC of *LmHsp* expression under alkalinity stress. Green and red triangles represent *Hsps* from profiles 0 and 4, respectively, and triangles with black frame are the significantly regulated *LmHsps* in response to alkalinity stress.

3.5. *Evolutionary Dynamics of LmHsp Families in L. maculatus*

To gain insights of whether the regulated *LmHsps* may function under adaptive evolution in dynamic aquatic environments, the molecular evolution of *LmHsp* families was evaluated through a series of model tests, including site model (SM), branch model (BM), and branch-site model (BSM) via PAML. Firstly, the ω values (dN/dS) of each *LmHsp* family were obtained via the SM tests, which revealed that the evolutionary rate (ω) of the *LmsHsp* family was generally higher than that of other *LmHsp* families (Figure 6A and Table S6). For example, the *LmsHsp* family exhibited rapid evolution with $\omega > 1$, while the *LmHsp70* and *LmHsp90* families were more conserved during evolution ($\omega < 0.1$) (Figure 6A). In addition, with BM tests for the 95 *LmHsp* genes labeling each family as the foreground branch (ω_1), respectively, it confirmed that in *L. maculatus*, *sHsps* represented faster molecular evolution rates than other *Hsp* families ($p < 0.01$), whereas *LmHsp70*s had the opposite evolutionary pattern with high sequence conservation (Figure 6B and Table S6). This was in line with the results obtained in the SM tests (Figure 6A). Moreover, *LmHsp90*s represented rapid evolution in the BM test but not in the SM test. These results suggested that although most *Hsp* genes were deemed to be highly conserved among teleost species, the molecular evolutionary rates of *sHsp*s and *Hsp90*s were generally faster in *L. maculatus*.

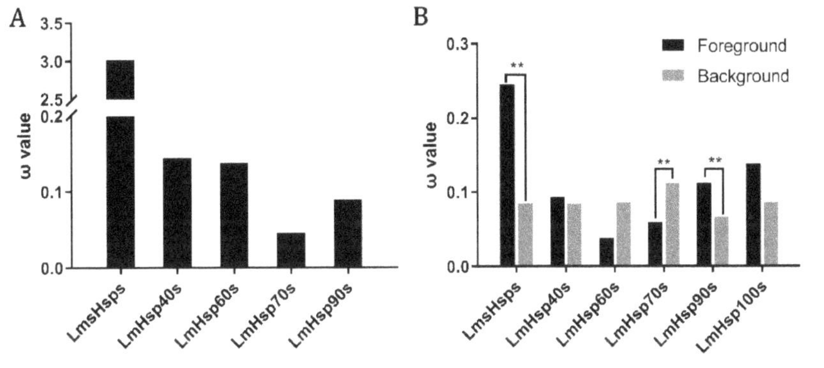

Figure 6. Molecular evolution of *LmHsp* genes in *L. maculatus*. (**A**) ω values of each *LmHsp* family from site model tests; (**B**) Branch-model tests of each *LmHsp* family from *L. maculatus*, black and gray columns indicate ω values of each *LmHsp* family (ω_1, foreground branch) and other *LmHsp* families (ω_0, background branch), respectively. ** indicates $p < 0.01$.

Accordingly, the BSM tests of *LmsHsps*, *LmHsp70*s and *LmHsp90*s, which represented significant altered evolutionary rates in *L. maculatus* (Figure 6B), were further conducted against *Hsp*s of other teleost lineages, and the putative positively selected sites (PSSs) were detected in 10 genes ($p < 0.01$, BEB probability > 0.95), including *Hspa4a1, Hspa4a2, Hspa8a, Hspa9, Hspa12b1, Hspa12b2, Hsc70, Hsp90aa1.2, Hsp90ab1,* and *Hsp90b1* (Table S7). The genes with PSS detected were mainly found in seven *Hsp70* and three *Hsp90* members but not in *sHsp*s (Table S7), among which *Hspa12b2, Hsp90aa1.2,* and *Hsp90b1* were also significantly regulated in response to alkalinity stress (Figure 5C). Therefore, in an adverse aquatic environment, the *sHsp* family may be under stronger evolutionary pressure in both *L. maculatus* and other teleosts (Figure 6), whereas in *L. maculatus*, some members of *Hsp70* and *Hsp90* families may evolve more rapidly than other teleosts (Table S7), possibly demonstrating the good adaptation of *L. maculatus* to the environmental changes, such as water temperature, salinity, and alkalinity.

In addition, with BSM tests of the 15 putative correlated *Hsp40-70-90* co-chaperons in response to alkalinity stress (Table 1 and Figure 5C), the PSSs in the four significantly regulated *Hsps*, *Hsp90aa1.2, Hsp90b1, Hspa12b2,* and *Dnajc9*, were detected in their conserved functional domains (Figure 7). Hsp90 chaperones normally exist as a dimer, and each part contains three domains: the N-terminal domain (NTD) with ATP hydrolytic activity, the middle domain (MD) to interact with Hsp70, and the C-terminal domain (CTD) to participate in identification and interaction of clients [49]. The eukaryotic Hsp90 proteins also contain a C-terminal extension of the MEEVD motif to help them bind to co-chaperones [8,49]. The fast-evolved PSSs of *Hsp90aa1.2* and *Hsp90b1* were mainly present in the MD and CTD (Figure 7A,B,E), which could most likely affect their interaction efficiency with Hsp70 proteins. Moreover, gene duplication is one of the most essential ways to develop functional divergence through environmental adaptation [42]. Both Hsp90 and Hsp70 function in an ATP-dependent scenario and require the involvement of a protein-named activator of Hsp90 ATPase homolog (Hsp90aa1), which interacts with Hsp90 and Hsp70 to stimulate ATPase activity [8,11]. Two putative *Hsp90aa1* genes were identified in *L. maculatus* due to TSGD, while only one (*LmHsp90aa1.2*) was significantly induced in response to alkalinity (Figure 4B). Therefore, there can be functional diversification between the duplicated *Hsp90aa1* genes in *L. maculatus*, with *Hsp90aa1.2* both evolving significantly fast and regulated under alkalinity stress, but not in *Hsp90aa1.1* (Figure 4 and Table 1).

Table 1. Branch-site model tests for the 15 coordinated regulated *LmHsp40-70-90* co-chaperons (Figure 5C) between *L. maculatus* and other teleost species.

Gene ID	*p* Value	Positive Selected Sites
Hspa1.1	1.0000	/
Hspa5	1.0000	/
Hspa12a	0.0534	8 S 0.919
Hspa12b2	**0.0000**	**32 P 0.958 ***, 36 T 0.918, 39 V 0.935, 41 L 0.907, 43 G 0.923, 46 P 0.936, 49 R 0.906, 112 C 0.903, **321 D 0.995 **, 323 T 1.000 ****, 327 I 0.907, **339 K 0.999 **, 340 A 0.998 ****, 341 S 0.929, **343 E 0.999 **, 344 L 0.963 *, 346 A 1.000 **, 347 K 1.000 **, 351 R 0.998 **, 353 V 0.999 **, 355 F 0.997 **, 366 P 0.999 **, 367 M 0.997 **, 368 L 0.998 **, 370 K 0.999 **, 371 A 0.998 **, 372 V 0.963 *, 374 K 1.000 **, 375 A 0.999 **, 377 G 0.999 **, 379 T 1.000 **, 384 I 0.963 *, 408 S 1.000 **, 409 Q 1.000 **, 411 H 0.997 **, 418 L 1.000 ****, 419 F 0.866, **420 D 0.997 ****
Hspa13	**0.0039**	/
Hsp90aa1.2	**0.0000**	16 G 0.908, 22 D 0.916, **431 M 0.978 *, 432 H 0.996 **, 488 D 0.974 *, 489 M 0.982 *, 491 F 1.000 **, 492 V 0.974 ***
Hsp90b1	**0.0000**	**723 S 0.983 *, 791 D 0.998 **, 965 H 0.981 ***, 1099 E 0.940
Dnajb4	0.6613	239 N 0.861
Dnajb5_10016268	1.0000	289 Y 0.507
Dnajb11	1.0000	/
Dnajb12	1.0000	/
Dnajc5ga	1.0000	/
Dnajc9_10000949	**0.0079**	77 R 0.825, **95 K 0.961 *, 96 E 0.990 ***, 97 A 0.522, **109 V 0.978 *, 130 V 0.980 ***, 141 K 0.930, 204 V 0.921, 205 Q 0.896, 206 H 0.944, 207 Q 0.791, 211 D 0.921, 216 S 0.596, 220 C 0.970*, 244 F 0.710, 269 A 0.614, 276 M 0.650, 283 D 0.911, **286 V 0.972 ***, 318 S 0.661, **320 D 0.976 *, 338 N 0.994 ****, 350 E 0.700
Dnajc9_10009927	1.0000	/
Dnajc22	1.0000	/

The ancestral branch leading to *L. maculatus* was set as the foreground branch (ω_1). Sites with the BEB posterior probabilities higher than 90% were presented, with those higher than 95% marked with * and higher than 99% marked with ** and in bold. *p* values < 0.05 were in bold.

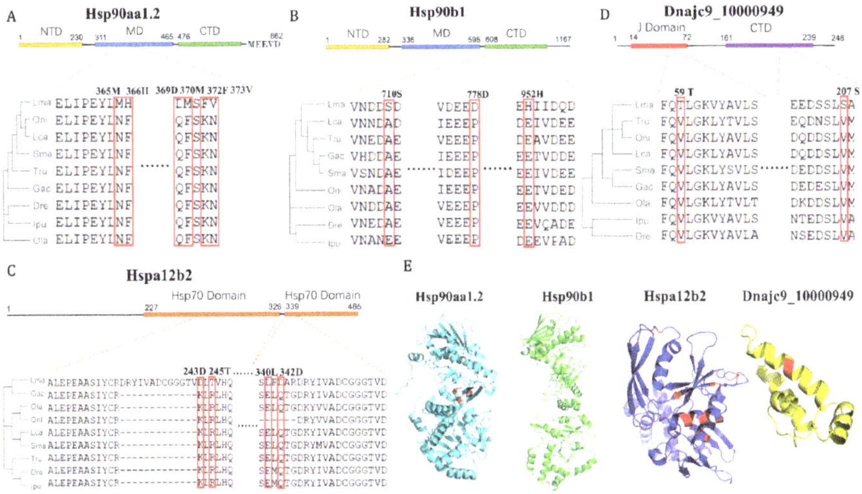

Figure 7. Examples of positively selected sites (PSSs) in (**A**) *Hsp90aa1.2*, (**B**) *Hsp90b1*, (**C**) *Hspa12b2*, and (**D**) *Dnajc9* functional domains from branch-site model tests. The partial aa sequence alignments of the selected vertebrates with their phylogeny are presented. Red frames indicate PSSs detected with their location at the bar above. Omitted multiple sequence alignment between the two segments are represented by dots. NTD, N-terminal domain; MD, middle domain; and CTD, C-terminal domain. (**E**) Schematic diagram of 3D structure of the above-mentioned LmHsp members with PSSs labeled in red. Only 710 S was predicted in the Hsp90b1 3D structure.

Moreover, Hsp70s are regarded as the most well-known member of the Hsp superfamily, which can act as co-chaperones of Hsp40s and Hsp90s involved in the folding and remodeling of various proteins [9]. Generally, the typical Hsp70 members consist of two regions, a conserved nucleotide-binding domain (NBD) with an ATP binding site, and a substrate-binding domain (SBD), which can mediate substrate or co-chaperone binding [50]. Those lacking NBD or SBD, such as Hspa4 and Hspa12 (Figure 2D), are considered atypical members of the Hsp70 family [45]. Here, only the atypical *Hspa12* genes in *L. maculatus* were up-regulated under alkalinity stress, while other typical *Hsp70*s (*LmHspa5*, *LmHspa13*, and *LmHspa14*) were mostly down-regulated together with their co-chaperon *Hsp90*s, indicating possible functional diversification. There are two Hsp70 domains detected in LmHspa12b2, and the PSSs mainly exist in its C-terminal domain (Figure 7C,E), indicating its possible binding specificity alternation. Moreover, PSSs were also detected in the J domain and CTD of LmDnajc9, which is essential in stimulating Hsp70 ATPase activity (Figure 7D,E) and also as the putative key co-chaperon with LmHspa12b2 and LmHsp90aa1.2 (Figure 5C). Therefore, these fast evolved loci in functional domains may contribute to the *Hsp*s functional co-evolution in *L. maculatus* (Figure 7E), and it may also suggest their vital correlated roles in response to diverse environmental stresses in *L. maculatus* through adaptive evolution. Further investigation is warranted among these Hsp40-70-90 co-chaperons to better understand their function and adaptive evolution in *L. maculatus*.

4. Conclusions

Hsps contribute to mediate stress responses by refolding damaged proteins to prevent their aggregation. This study provided a systematic genomic and transcriptomic survey of 95 *Hsp* genes in spotted sea bass (*L. maculatus*), which has good adaptability in saline-alkaline waters. Diverse expression regulation of *LmHsps* was observed in responses to alkalinity stress, which was majorly responded in coordinated *LmHsp40-70-90* families through the adaptive recruitment of these co-chaperons. These findings are useful for understanding the diverse functions of *Hsp* co-chaperon network and adaptive evolution in teleost species. Further investigation on the functions of teleost *Hsp* network will provide a more detailed explanation about their adaptation mechanisms against different environmental challenges and warrant a better feasibility regarding the cultivation of *L. maculatus* in alkaline water.

Supplementary Materials: The following supporting information can be downloaded at: https://www.mdpi.com/article/10.3390/biology11030353/s1, Table S1: Detailed copy number of *Hsp* family genes in selected vertebrate genomes; Table S2: Summarized information of *LmHsp* genes; Table S3: Differential expression of *LmHsps* under salinity change and alkalinity stress by edgeR analysis; Table S4: Trend analysis for the *LmHsp40-70-90* gene expression under alkalinity stress; Table S5: Protein-protein interaction (PPI) network for the *LmHsp40-70-90* gene expression under alkalinity stress; Table S6: Molecular evolution analysis of the *LmHsp* genes among teleost lineages via PAML; Table S7: Branch-site model tests of the *LmHsp* genes with other teleost species via PAML; Figure S1: Principal component analysis (PCA) with FPKM values of all transcriptome samples under salinity change and alkalinity stress; Figure S2: Structure, motif and phylogeny of *LmHsp* families, showing phylogenetic relationship along with their conserved motifs. (**A**) *LmsHsps*, (**B**) *LmHsp40s*, (**C**) *LmHsp10/60s*, (**D**) *LmHsp70s*, (**E**) *LmHsp90s*, and (**F**) *LmHsp100s*. Abbreviations: *Hsa: H. sapiens; Mmu: M. musculus; Gga: G. Gallus; Xtr: X. tropicalis; Loc: L. oculatus; Dre: D. rerio; Ipu: I. punctatus; Ola: O. latipes; Gac: G. aculeatus; Sma: S. maximus; Tru: T. rubripes; Oni: O. niloticus; Lca: L. calcarifer; Lma: L. maculatus;* Figure S3: Protein-protein interaction (PPI) of *Hsps* from zebrafish and human.

Author Contributions: J.C. and Q.Z. designed the study, X.L., S.L., Y.W. and W.L. analyzed the data. J.C. and X.L. conducted the manuscript writing. All authors have read and agreed to the published version of the manuscript.

Funding: This research was funded by the China Agriculture Research System (CARS-47-G06) and the National Marine Genetic Resource Center, China.

Institutional Review Board Statement: This study was conducted in accordance with the Institutional Animal Care and Use Committee of Ocean University of China (IACUC-OUC), and it does not contain any studies with human participants.

Informed Consent Statement: Not applicable.

Data Availability Statement: The transcriptome datasets used in this study can be found in the NCBI Sequence Read Archive (SRA) BioProject PRJNA611641 and PRJNA515986, as well as *L. maculatus* tissue transcriptomes, including brain-SRR7528887, stomach-SRR7528884, spleen-SRR7528888, liver-SRR7528886, gill-SRR7528883, testis-SRR7528885 and ovary-SRR2937376.

Conflicts of Interest: The authors declare that they have no conflict of interest.

References

1. Ritossa, F.M. A new puffing pattern induced by heat shock and DNP in *Drosophila*. *Experientia* **1962**, *18*, 515–523. [CrossRef]
2. Whitley, D.; Goldberg, S.P.; Jordan, W.D. Heat shock proteins: A review of the molecular chaperones. *J. Vasc. Surg.* **1999**, *29*, 748–751. [CrossRef]
3. Basu, N.; Todgham, A.E.; Ackerman, P.A.; Bibeau, M.R.; Nakano, K.; Schulte, P.M.; Iwama, G.K. Heat shock protein genes and their functional significance in fish. *Gene* **2002**, *295*, 173–183. [CrossRef]
4. Feder, M.E.; Hofmann, G.E. Heat-shock proteins, molecular chaperones, and the stress response: Evolutionary and ecological physiology. *Annu. Rev. Physiol.* **1999**, *61*, 243–282. [CrossRef] [PubMed]
5. Dabbaghizadeh, A.; Tanguay, R.M. Structural and functional properties of proteins interacting with small heat shock proteins. *Cell Stress Chaperones* **2020**, *25*, 629–637. [CrossRef]
6. Faust, O.; Abayev-Avraham, M.; Wentink, A.S.; Maurer, M.; Nillegoda, N.B.; London, N.; Bukau, B.; Rosenzweig, R. HSP40 proteins use class-specific regulation to drive HSP70 functional diversity. *Nature* **2020**, *587*, 489–494. [CrossRef] [PubMed]
7. Bie, A.S.; Cömert, C.; Körner, R.; Corydon, T.J.; Palmfeldt, J.; Hipp, M.S.; Hartl, F.U.; Bross, P. An inventory of interactors of the human HSP60/HSP10 chaperonin in the mitochondrial matrix space. *Cell Stress Chaperones* **2020**, *25*, 407–416. [CrossRef]
8. Genest, O.; Wickner, S.; Doyle, S.M. Hsp90 and Hsp70 chaperones: Collaborators in protein remodeling. *J. Biol. Chem.* **2019**, *294*, 2109–2120. [CrossRef]
9. Rosenzweig, R.; Nillegoda, N.B.; Mayer, M.P.; Bukau, B. The Hsp70 chaperone network. *Nat. Rev. Mol. Cell Biol.* **2019**, *20*, 665–680. [CrossRef]
10. Mayer, M.P.; Le Breton, L. Hsp90: Breaking the symmetry. *Mol. Cell* **2015**, *58*, 8–20. [CrossRef]

11. Jiang, F.; Chang, G.; Li, Z.; Abouzaid, M.; Du, X.; Hull, J.; Ma, W.; Lin, Y. The HSP/co-chaperone network in environmental cold adaptation of *Chilo suppressalis*. *Int. J. Biol. Macromol.* **2021**, *187*, 780–788. [CrossRef] [PubMed]
12. Chen, B.; Feder, M.E.; Kang, L. Evolution of heat-shock protein expression underlying adaptive responses to environmental stress. *Mol. Ecol.* **2018**, *27*, 3040–3054. [CrossRef] [PubMed]
13. Kim, B.M.; Rhee, J.S.; Jeong, C.B.; Seo, J.S.; Park, G.S.; Lee, Y.M.; Lee, J.S. Heavy metals induce oxidative stress and trigger oxidative stress-mediated heat shock protein (hsp) modulation in the intertidal copepod *Tigriopus japonicus*. *Comp. Biochem. Physiol. C Toxicol. Pharmacol.* **2014**, *166*, 65–74. [CrossRef] [PubMed]
14. Cantinha, R.D.S.; Borrely, S.I.; Oguiura, N.; Pereira, C.A.d.B.; Rigolon, M.M.; Nakano, E. HSP70 expression in *Biomphalaria glabrata* snails exposed to cadmium. *Ecotoxicol. Environ. Saf.* **2017**, *140*, 18–23. [CrossRef]
15. Ulaje, S.A.; Lluch-Cota, S.E.; Sicard, M.T.; Ascencio, F.; Cruz-Hernández, P.; Racotta, I.S.; Rojo-Arreola, L. *Litopenaeus vannamei* oxygen consumption and HSP gene expression at cyclic conditions of hyperthermia and hypoxia. *J. Therm. Biol.* **2020**, *92*, 102666. [CrossRef]
16. Cheng, J.; Xun, X.; Kong, Y.; Wang, S.; Yang, Z.; Li, Y.; Kong, D.; Wang, S.; Zhang, L.; Hu, X.; et al. *Hsp70* gene expansions in the scallop *Patinopecten yessoensis* and their expression regulation after exposure to the toxic dinoflagellate *Alexandrium catenella*. *Fish Shellfish Immunol.* **2016**, *58*, 266–273. [CrossRef]
17. Hu, B.; Li, M.; Yu, X.; Xun, X.; Lu, W.; Li, X.; Li, Y.; Lou, J.; Wang, S.; Zhang, L.; et al. Diverse expression regulation of *Hsp70* genes in scallops after exposure to toxic *Alexandrium* dinoflagellates. *Chemosphere* **2019**, *234*, 62–69. [CrossRef]
18. Lei, Q.N.; Wu, Y.Y.; Liang, H.Y.; Wang, Z.X.; Zheng, Z.; Deng, Y.W. Molecular cloning and expression analysis of *heat shock protein 20 (HSP20)* from the pearl oyster *Pinctada martensii*. *Genet. Mol. Res.* **2016**, *15*, 10. [CrossRef]
19. Xie, Y.; Song, L.; Weng, Z.; Liu, S.; Liu, Z. *Hsp90*, *Hsp60* and *sHsp* families of heat shock protein genes in channel catfish and their expression after bacterial infections. *Fish Shellfish Immunol.* **2015**, *44*, 642–651. [CrossRef]
20. Song, L.; Zhang, J.; Li, C.; Yao, J.; Jiang, C.; Li, Y.; Liu, S.; Liu, Z. Genome-wide identification of *hsp40* genes in channel catfish and their regulated expression after bacterial infection. *PLoS ONE* **2014**, *9*, e115752. [CrossRef]
21. Song, L.; Zhao, Y.; Song, Y.; Zhao, L.; Ma, C.; Zhao, J. Effects of saline-alkaline water on growth performance, nutritional processing, and immunity in Nile tilapia (*Oreochromis niloticus*). *Aquaculture* **2021**, *544*, 737036. [CrossRef]
22. Islam, M.J.; Kunzmann, A.; Slater, M.J. Extreme winter cold-induced osmoregulatory, metabolic, and physiological responses in European seabass (*Dicentrarchus labrax*) acclimatized at different salinities. *Sci. Total Environ.* **2021**, *771*, 145202. [CrossRef]
23. Islam, M.J.; Slater, M.J.; Kunzmann, A. What metabolic, osmotic and molecular stress responses tell us about extreme ambient heatwave impacts in fish at low salinities: The case of European seabass, *Dicentracus labrax*. *Sci. Total Environ.* **2020**, *749*, 141458. [CrossRef]
24. Zhang, M.; Li, L.; Liu, Y.; Gao, X. Effects of sudden drop in salinity on osmotic pressure regulation and antioxidant defense mechanism of *Scapharca subcrenata*. *Front. Physiol.* **2020**, *11*, 884.
25. Yokogawa, K.; Seki, S. Morphological and genetic differences between Japanese and Chinese sea bass of the genus *Lateolabrax*. *Jpn. J. Ichthyol.* **1995**, *41*, 437–445.
26. Tian, Y.; Wen, H.; Qi, X.; Zhang, X.; Li, Y. Identification of *mapk* gene family in *Lateolabrax maculatus* and their expression profiles in response to hypoxia and salinity challenges. *Gene* **2019**, *684*, 20–29. [CrossRef]
27. Wang, L.Y.; Tian, Y.; Wen, H.S.; Yu, P.; Liu, Y.; Qi, X.; Gao, Z.C.; Zhang, K.Q.; Li, Y. *Slc4* gene family in spotted sea bass. (*Lateolabrax maculatus*): Structure, evolution, and expression profiling in response to alkalinity stress and salinity changes. *Genes* **2020**, *11*, 1271. [CrossRef]
28. Schultz, J.; Copley, R.R.; Doerks, T.; Ponting, C.P.; Bork, P. SMART: A web-based tool for the study of genetically mobile domains. *Nucleic Acids Res.* **2000**, *28*, 231–234. [CrossRef]
29. El-Gebali, S.; Mistry, J.; Bateman, A.; Eddy, S.R.; Luciani, A.; Potter, S.C.; Qureshi, M.; Richardson, L.J.; Salazar, G.A.; Smart, A.; et al. The Pfam protein families database in 2019. *Nucleic Acids Res.* **2019**, *47*, D427–D432. [CrossRef]
30. Chen, C.; Chen, H.; Zhang, Y.; Thomas, H.R.; Frank, M.H.; He, Y.; Xia, R. TBtools: An integrative toolkit developed for interactive analyses of big biological data. *Mol. Plant* **2020**, *13*, 1194–1202. [CrossRef]
31. Edgar, R.C. MUSCLE: Multiple sequence alignment with high accuracy and high throughput. *Nucleic Acids Res.* **2004**, *32*, 1792–1797. [CrossRef]
32. Trifinopoulos, J.; Nguyen, L.T.; von Haeseler, A.; Minh, B.Q. W-IQ-TREE: A fast online phylogenetic tool for maximum likelihood analysis. *Nucleic Acids Res.* **2016**, *44*, W232–W235.
33. Letunic, I.; Bork, P. Interactive tree of life (iTOL) v3: An online tool for the display and annotation of phylogenetic and other trees. *Nucleic Acids Res.* **2016**, *44*, W242–W245. [CrossRef]
34. Pertea, M.; Kim, D.; Pertea, G.M.; Leek, J.T.; Salzberg, S.L. Transcript-level expression analysis of RNA-seq experiments with HISAT, StringTie and Ballgown. *Nat. Protoc.* **2016**, *11*, 1650–1667. [CrossRef]
35. Tian, Y.; Wen, H.; Qi, X.; Zhang, X.; Liu, S.; Li, B.; Sun, Y.; Li, J.; He, F.; Yang, W.; et al. Characterization of full-length transcriptome sequences and splice variants of *Lateolabrax maculatus* by single-molecule long-read sequencing and their involvement in salinity regulation. *Front. Genet.* **2019**, *10*, 1126. [CrossRef]
36. Robinson, M.D.; McCarthy, D.J.; Smyth, G.K. EdgeR: A Bioconductor package for differential expression analysis of digital gene expression data. *Bioinformatics* **2010**, *26*, 139–140. [CrossRef]

37. Shannon, P.; Markiel, A.; Ozier, O.; Baliga, N.S.; Wang, J.T.; Ramage, D.; Amin, N.; Schwikowski, B.; Ideker, T. Cytoscape: A software environment for integrated models of biomolecular interaction networks. *Genome Res.* **2003**, *13*, 2498–2504. [CrossRef]
38. Kumar, S.; Stecher, G.; Tamura, K. MEGA7: Molecular evolutionary genetics analysis version 7.0 for bigger datasets. *Mol. Biol. Evol.* **2016**, *33*, 1870–1874. [CrossRef]
39. Gao, F.; Chen, C.; Arab, D.A.; Du, Z.; He, Y.; Ho, S. EasyCodeML: A visual tool for analysis of selection using CodeML. *Ecol. Evol.* **2019**, *9*, 3891–3898. [CrossRef]
40. Yang, Z. PAML 4: Phylogenetic analysis by maximum likelihood. *Mol. Biol. Evol.* **2007**, *24*, 1586–1591. [CrossRef] [PubMed]
41. Kelley, L.A.; Mezulis, S.; Yates, C.M.; Wass, M.N.; Sternberg, M.J.E. The Phyre2 web portal for protein modeling, prediction and analysis. *Nat. Protoc.* **2015**, *10*, 845–858. [CrossRef] [PubMed]
42. Glasauer, S.M.; Neuhauss, S.C. Whole-genome duplication in teleost fishes and its evolutionary consequences. *Mol. Genet. Genom.* **2014**, *289*, 1045–1060. [CrossRef]
43. Kriehuber, T.; Rattei, T.; Weinmaier, T.; Bepperling, A.; Haslbeck, M.; Buchner, J. Independent evolution of the core domain and its flanking sequences in small heat shock proteins. *FASEB J.* **2010**, *24*, 3633–3642. [CrossRef]
44. Haslbeck, M.; Weinkauf, S.; Buchner, J. Small heat shock proteins: Simplicity meets complexity. *J. Biol. Chem.* **2019**, *294*, 2121–2132. [CrossRef] [PubMed]
45. Brocchieri, L.; Macario, E.C.d.; Macario, A.J. *Hsp70* genes in the human genome: Conservation and differentiation patterns predict a wide array of overlapping and specialized functions. *BMC Evol. Biol.* **2008**, *8*, 19. [CrossRef] [PubMed]
46. Song, L.; Li, C.; Xie, Y.; Liu, S.; Zhang, J.; Yao, J.; Jiang, C.; Li, Y.; Liu, Z. Genome-wide identification of *Hsp70* genes in channel catfish and their regulated expression after bacterial infection. *Fish Shellfish Immunol.* **2016**, *49*, 154–162. [CrossRef] [PubMed]
47. Kokou, F.; Con, P.; Barki, A.; Nitzan, T.; Slosman, T.; Mizrahi, I.; Cnaani, A. Short- and long-term low-salinity acclimation effects on the branchial and intestinal gene expression in the European seabass (*Dicentrarchus labrax*). *Comp. Biochem. Physiol. A* **2019**, *231*, 11–18. [CrossRef] [PubMed]
48. Huang, X.; Li, S.; Gao, Y.; Zhan, A. Genome-wide identification, characterization and expression analyses of heat shock protein-related genes in a highly invasive ascidian *Ciona savignyi*. *Front. Physiol.* **2018**, *9*, 1043. [CrossRef]
49. Hoter, A.; El-Sabban, M.E.; Naim, H.Y. The HSP90 Family: Structure, regulation, function, and implications in health and disease. *Int. J. Mol. Sci.* **2018**, *19*, 2560. [CrossRef]
50. Murphy, M.E. The HSP70 family and cancer. *Carcinogenesis* **2013**, *34*, 1181–1188. [CrossRef]

biology

MDPI

Article

Whole-Genome Transcript Expression Profiling Reveals Novel Insights into Transposon Genes and Non-Coding RNAs during Atlantic Salmon Seawater Adaptation

Valentina Valenzuela-Muñoz [1,2,3,*], Cristian Gallardo-Escárate [1,3], Bárbara P. Benavente [1,3], Diego Valenzuela-Miranda [1,3], Gustavo Núñez-Acuña [1,3], Hugo Escobar-Sepulveda [1,3] and Juan Antonio Váldes [1,2]

1 Interdisciplinary Center for Aquaculture Research (INCAR), University of Concepción, Concepcion 4030000, Chile; crisgallardo@udec.cl (C.G.-E.); Bbenavente@udec.cl (B.P.B.); divalenzuela@udec.cl (D.V.-M.); gustavonunez@udec.cl (G.N.-A.); hescobar.sepulveda@gmail.com (H.E.-S.); jvaldes@unab.cl (J.A.V.)
2 Laboratorio de Biotecnología Molecular, Facultad de Ciencias de la Vida, Universidad Andrés Bello, Santiago 8370035, Chile
3 Laboratory of Biotechnology and Aquatic Genomics, Department of Oceanography, University of Concepción, Concepcion 4030000, Chile
* Correspondence: valevalenzuela@udec.cl or valevalenzuela@gmail.com

Simple Summary: This study proposes a novel approach to analyze transcriptome data sets using the Atlantic salmon seawater adaptation process as a model. *Salmon salar* smolts were transferred to seawater under two strategies: (i) fish group exposed to gradual salinity changes (GSC) and (ii) fish group exposed to a salinity shock (SS). mRNA and miRNAs sequencing were performed for gills, intestine, and head kidney tissues. The whole-genome transcript expression profiling revealed specific gene expression patterns among the tissues and treatments. A great abundance of transposable elements was observed in chromosome regions differentially expressed under experimental conditions. Moreover, small RNA expression analysis suggested fewer of miRNAs associated with the smoltification process. However, target analysis of these miRNAs suggests a regulatory role of process such as growth, stress response, and immunity. The findings uncover whole-transcriptome modulation during seawater adaptation of Atlantic salmon, evidencing the interplaying among mRNAs and miRNAs.

Abstract: The growing amount of genome information and transcriptomes data available allows for a better understanding of biological processes. However, analysis of complex transcriptomic experimental designs involving different conditions, tissues, or times is relevant. This study proposes a novel approach to analyze complex data sets combining transcriptomes and miRNAs at the chromosome-level genome. Atlantic salmon smolts were transferred to seawater under two strategies: (i) fish group exposed to gradual salinity changes (GSC) and (ii) fish group exposed to a salinity shock (SS). Gills, intestine, and head kidney samples were used for total RNA extraction, followed by mRNA and small RNA illumina sequencing. Different expression patterns among the tissues and treatments were observed through a whole-genome transcriptomic approach. Chromosome regions highly expressed between experimental conditions included a great abundance of transposable elements. In addition, differential expression analysis showed a greater number of transcripts modulated in response to SS in gills and head kidney. miRNA expression analysis suggested a small number of miRNAs involved in the smoltification process. However, target analysis of these miRNAs showed a regulatory role in growth, stress response, and immunity. This study is the first to evidence the interplaying among mRNAs and miRNAs and the structural relationship at the genome level during Atlantic salmon smoltification.

Keywords: Atlantic salmon; smoltification; genome; mRNAs; miRNAs

Citation: Valenzuela-Muñoz, V.; Gallardo-Escárate, C.; Benavente, B.P.; Valenzuela-Miranda, D.; Núñez-Acuña, G.; Escobar-Sepulveda, H.; Váldes, J.A. Whole-Genome Transcript Expression Profiling Reveals Novel Insights into Transposon Genes and Non-Coding RNAs during Atlantic Salmon Seawater Adaptation. *Biology* **2022**, *11*, 1. https://doi.org/10.3390/biology11010001

Academic Editor: Roy Ambli Dalmo

Received: 28 September 2021
Accepted: 13 December 2021
Published: 21 December 2021

1. Introduction

Genomics tools have facilitated the elucidation of the global genomic changes under different conditions. Nowadays, the availability of the Atlantic salmon (*Salmo salar*) genome [1] allows for the identification of genome regions associated with pivotal biological processes and responses to the aquatic environmental. One of the most important biological processes during the salmon lifecycle is the parr-smolt transformation (smoltification), which is primarily influenced by water temperature and photoperiod [1]. This process involves physiological, morphological, endocrinal, and behavioral changes [2,3], which have been extensively studied, owing to their implications in salmon aquaculture [1,4–6].

Previous transcriptomic analyses have elucidated expression changes of genes associated with growth, metabolism, oxygen transport, osmoregulation, protein biosynthesis, and sensory reception during the Atlantic salmon smoltification process [7,8]. Recently, the role of non-coding RNAs (ncRNAs) during Atlantic salmon transition from freshwater (FW) to seawater (SW) has been proposed as a novel regulatory molecular mechanism involved in fish biology. For instance, previous studies have reported 2864 long non-coding RNAs (lncRNAs) differentially modulated in Atlantic salmon gills during the transition from FW to SW [9]. Among them, two putative lncRNAs with the potential to be used as smoltification-timing biomarkers were identified. One was highly regulated in FW, and the other was upregulated in SW. In addition, a putative regulation role of lncRNAs associated with Na^+/K^+-ATPase genes, hormone receptors, and thyroid hormone receptors was suggested [9].

Among ncRNAs, microRNAs (miRNAs) are crucial in post-transcriptional regulation, binding to target mRNAs in 3′UTR and repressing translation to proteins [10,11]. Different biological process, such as development, growth, cell division, metabolism, and apoptosis, are regulated by miRNAs [12–15]. For Atlantic salmon, 472 miRNAs have been reported in the miRBase [16]. In addition, a total of 521 miRNAs have been described for Atlantic salmon, with different expression patterns among kidney, head kidney, heart, brain, gills, white muscle, and intestine tissues [17]. Moreover, 71 miRNAs were reported to be differently modulated in head kidney of Atlantic salmon during the smoltification process [18]. Furthermore, the authors reported a negative-correlation expression pattern of miRNAs with genes associated with hormone biosynthesis, stress management, immune response, and ion transport. In addition, the study reported a cluster of 37 miRNAs highly regulated before the smoltification initiation and another cluster of 17 miRNAs that increased their expression until SW transfer [18]. Despite the evidence suggesting a role of ncRNAs in the regulation of the smoltification process, comprehensive transcriptome analyses linking physiological adaptation to seawater with the complexity of the salmon genome are unexplored. For instance, there is no evidence of how unduplicated and/or duplicated genes involved in the smoltification are modulated and which are the key molecular elements involved. Herein, the molecular interplaying among coding/non-coding genes expressed in different tissues during seawater adaptation is uncovered. The major obstacle to conducting extensive experimental trials is combining the transcriptome time-series with different sequencing approaches (e.g., mRNA vs. small RNA sequencing), and joining the analyzed target-tissues with the experimental conditions tested. The massive amount of transcriptome data is frequently completely unused or at least not straightforwardly used to identify the primary biological processes modulated and their molecular components. Furthermore, evidence that chromatin in the interphase nucleus has nonrandom localization supports the idea that the nuclear architecture is comprised of three-dimensional (3D) genome spaces. This spatial organization plays crucial roles in genome function and cellular processes, such as DNA replication [19,20], transcription [21], DNA-damage repair [22,23], development, and cell differentiation [24]. Chromosomes occupy distinct subnuclear territories, with transcriptionally active loci positioned at their surface [25–30]. Thus, the relationship between gene transcription, gene regulation, and spatial 3D genome structure for relevant biological processes requires further scientific investigation. This study aimed to explore global transcriptome modulation during Atlantic salmon seawater adaptation. Herein, a novel approach was developed to analyze time-series differential transcription data from head kidney, gills, and intestine tissues exposed to salinity changes. In parallel, transcriptional dynamics were associated with chromosome regions where differently expressed thresholds between salinity conditions were identified. Notably, transcriptome analyses revealed novel insights into transposon genes and non-coding RNAs involved in smoltification in Atlantic salmon. This study is the first to suggest putative chromosome regions transcriptionally activated in response to salinity stress in anadromous fish.

2. Materials and Methods

2.1. Smoltification and Seawater Transfer

Atlantic salmon smolts (60 ± 6.2 gr) were obtained from a commercial farm (Hendrix Genetics, Boxmeer, The Netherlands) and then transported to the Marine Biology Station, Universidad de Concepción, Dichato, Chile. Fish were maintained in ultraviolet-treated saltwater by single-pass flow-through tank systems on a 12:12 h light: dark cycle, fed daily with a commercial diet, dissolved oxygen level of 8.5 mg/L and pH = 8.0. After ten days of acclimation, a group of 30 smolts was exposed to a gradual salinity change (GSC) by increasing FW salinity to SW. The gradient was set at three salinity points, changing 10 PSU per week over a month. Meanwhile, another group of 30 smolts was exposed to a salinity shock (SS), directly from FW to 32 PSU. Gills, head kidney, and intestine samples were collected during the experiment trial. Samples were collected at FW, 10, 20, and 32 PSU for the GSC group and after a week of acclimation at 32 PSU for the SS group (Figure S1). Both processes were conducted in triplicate. The samples were fixed in RNAlater® (Thermofisher, Waltham, MA, USA) for subsequent RNA isolation. Furthermore, the salmon condition to SW transfer was evaluated by immune histochemistry analyses performed by the VEHICE company. In addition, RT-qPCR expression analysis of ATPase-α and ATPase-β was determined using comparative ΔΔCt relative expression analysis. Primers and qPCR conditions were similar, as described by Valenzuela-Muñoz, Váldes, and Gallardo-Escárate [9]. All animal procedures were carried out under the guidelines approved by the Ethics Committee of the University of Concepción. The experimental design for the current study considered the Three Rs (3Rs) guidelines for animal testing.

2.2. High-Throughput Transcriptome Sequencing

Total RNA was isolated from each experimental fish group using TRizol Reagent (Ambion®, Austin, TX, USA), following the manufacturer's instructions. The isolated RNA was evaluated by TapeStation 2200 (Agilent Technologies Inc., Santa Clara, CA, USA), using the R6K Reagent Kit. Three biological replicates were separately sequenced by tissue and sampling point from each experimental fish group. For each replicate, five individuals were used for the RNA extraction and then pooled for library preparation. RNAs with RIN > 8.0 were used for double-stranded cDNA library construction using the TruSeq RNA Sample Preparation Kit v2 (Illumina®, San Diego, CA, USA). The same RNA samples were used for small RNA library synthesis using the TruSeq Small RNA Library Prep Kit (Illumina®, San Diego, CA, USA). All libraries made for RNA and small RNAs were sequenced by the Hiseq (Illumina®, San Diego, CA, USA) platform in Macrogen Inc. Raw data used for the current study are available in SRA-NCBI (Bioproject # PRJNA761374).

2.3. Whole-Genome Transcript Expression Analysis

Raw data from each experimental group were separately trimmed and mapped to the Atlantic salmon genome (GCF_000233385.1) using CLC Genomics Workbench v21 software (Qiagen Bioinformatics, Hilden, Germany). Threshold values for mRNA and small RNA were calculated from the coverage analysis using the Graph Threshold Areas tool in CLC Genomics Workbench v21 software. Here, an index denoted as chromosome genome expression (CGE) was formulated to explore the whole-genome transcript expression profiling according to:

$$CGE = \frac{|X_1 - X_n| \times 100}{(X_1 - X_n)}$$

where X corresponds to mean coverage of transcripts mapped into a specific chromosome region and compared among experimental conditions (e.g., GSC cv. SS; tissues vs. experimental time-points, mRNAs vs. miRNAs). The transcript coverage values for each dataset were calculated using a threshold of 10,000 to 90,000 reads, where a window size of 10 positions was set to calculate and identify differentially transcribed chromosome regions. The CGE index represents the percentage of transcriptional variation between two or more groups for the same locus. This approach allows for visualization of actively transcribed chromosome regions, identification of differentially expressed loci, exploration of mRNA-miRNAs interactions in term of transcriptional activity, and observation of tissue-specific patterns in fish exposed to several experimental conditions. Finally, threshold values for each dataset and CGE index were visualized in Circos plots [31].

2.4. RNA-Seq Data Analysis

Raw sequencing reads were assembled de novo separately for each tissue using the CLC Genomics Workbench v21 software (CLC Bio, Aarhus, Denmark). Assembly was performed with overlap criteria of 70% and a similarity of 0.9 to exclude paralogous sequence variants (Renaut et al. 2010). The settings used were set as mismatch cost = 2, deletion cost = 3, insert cost = 3, minimum contig length = 200 base pairs, and trimming quality score = 0.05. After assembly, singletons were retained in the dataset as possible representatives of low-expression transcript fragments. Differential expression analysis was set with a minimum length fraction = 0.6 and a minimum similarity fraction (long reads) = 0.5.

Contig sequences obtained from each tissue were blasted to CGE regions to enrich the number of transcripts evaluated by RNA-Seq analysis (Figure 1). In addition, the sequences were extracted from the Atlantic salmon genome near to the threshold areas in a window of 10 kb for each transcriptome. The expression value was set as transcripts per million (TPM). The distance metric was calculated with the Manhattan method, with a mean expression level in 5–6 rounds of k-means, clustering subtracted. Finally, Generalized Linear Model (GLM) available in the CLC software was used for statistical analyses and to compare gene expression levels in terms of the $\log_2$ fold change ($p = 0.005$; FDR-corrected). In addition, k-means clustering was performed for transposable element (TE) expression values. The metric distance was calculated with the Manhattan method, where the mean expression level in 5–6 rounds of k-means clustering was subtracted.

2.5. Sequence Annotation and GO Enrichment Analysis

Differentially expressed contigs were annotated through BlastX analysis using a custom protein database constructed from GeneBank and UniProtKB/Swiss-Prot. The cutoff E-value was set at 1E-10. Transcripts were subjected to gene ontology (GO) analysis using the Blast2GO plugins included in the CLC Genomics Workbench v12 software (CLC Bio, Qiagen, Germantown, MA, USA). The results were plotted using the cluster Profiler R package [32]. GO enrichment analysis was conducted to identify the most represented biological processes among protein-coding genes located proximally to the identified lncRNAs. Enrichment of biological processes was identified using Fisher's exact test tool of Blast2GO among the different experimental groups against the control group (FW samples).

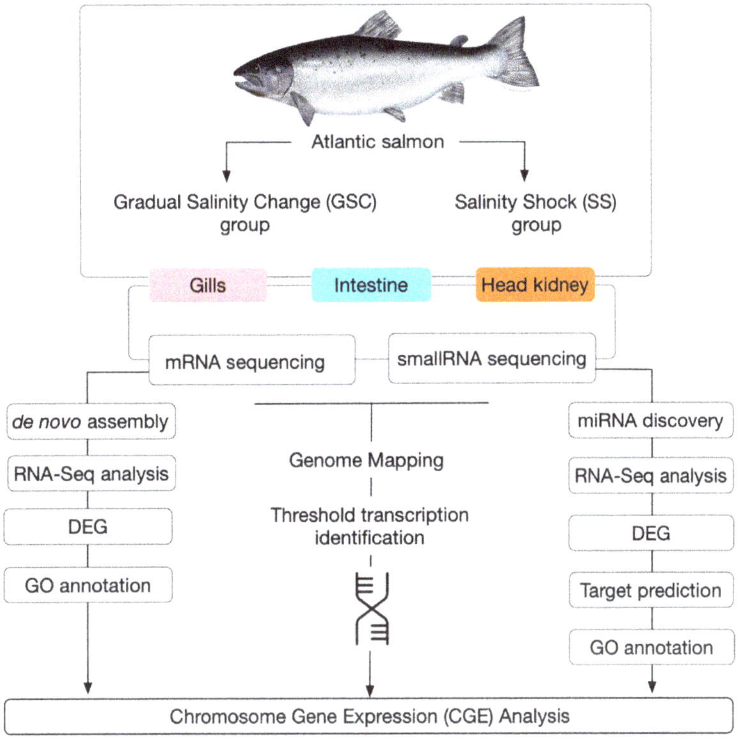

Figure 1. Whole-genome expression approach for transcriptome analysis during smoltification in Atlantic salmon.

2.6. miRNA Annotation and Expression Analysis in Response to Salinity Changes

Low-quality reads from illumina sequencing data, reads with a quality score of less than 0.05 on the Phred scale, with a short length, or with three or more ambiguous nucleotides were removed using CLC Genomics Workbench software (Version 21, CLC Bio, Aarhus, Denmark). Furthermore, any cleaned sequences matching metazoan mRNA, rRNA, tRNA, snRNA, snoRNA, repeat sequences, or other ncRNAs were deposited in the NCBI databases (http://www.ncbi.nlm.nih.gov/ (accessed on 27 July 2021)), RFam (http://rfam.janelia.org/ (accessed on 27 July 2021)), or Repbase (http://www.girinst.org/repbase/ (accessed on 27 July 2021)) were discarded. Then, the remaining transcripts were counted to generate a unique small RNA list. These sequences were annotated against pre-miRNA and mature miRNA (5′ and 3′) sequences listed for *Salmo salar* available in the miRbase (release 22) [16,33]. miRNA expression analysis followed a similar protocol, as previously described by our group [34].

2.7. miRNA Target Prediction and Expression Correlation

The computational target prediction algorithm used was RNA22 [35]. The datasets used were the differently modulated transcripts from both groups, GSC fish and SS fish. The RNA22 parameters were set at free energy < -15 kcal/mol and a score > 50. All target genes were annotated by GO analysis, following the protocol described above.

In addition, smoltification-related genes and differently expressed miRNAs were selected for expression-correlation analysis. Pearson's correlation among TPM values from both datasets, including all samples exposed and not exposed to each salinity condition, were calculated in R software [36]. Plots for correlation analyses were constructed using the Corrplot package [37], considering a correlations with p-value < 0.01 significant.

2.8. RT-qPCR Validation Analysis

Transcription expression profiles of genes associated with the smoltification process and ncRNAs were conducted by RT-qPCR. Briefly, 200 ng/μL of total RNA from three individuals per fish group obtained was used for cDNA synthesis using the RevertAid™ H Minus First Strand cDNA Synthesis Kit (Thermo Fisher Scientific™, USA), following the manufacturer's instructions. Four genes and their putative miRNAs were validated (Table S4). The comparative ΔΔCt relative expression analysis method was used. Selection of the housekeeping gene for the experiment was based on evaluation of the stability of elongation factor-α (EF-α), β-tubulin, and 18S genes by Normfinder. Here, EF-α was selected for gene normalization. Each RT-qPCR reaction was carried out in a final volume of 10 μL using the commercial PowerUp SYBR Green Master Mix kit (Applied Biosystems®, Waltham, MA, USA). RT-qPCR reactions were performed on the StepOnePlus™ (Applied Biosystems®, Life Technologies™, Carlsbad, CA, USA) using the following conditions: 95 °C for 10 min, 40 cycles at 95 °C for 15 s and 60 °C for 30 s, ending with 30 s at 72 °C. Statistical analyses were conducted through ANOVA-1 test and Student's *t*-test in the GraphPad Prims 8.4.7 package.

miRNA transcription-level validation was achieved by synthetizing cDNA from the same RNA samples using the miSCript II RT kit (Qiagen, Germany), with an incubation reaction at 37 °C for 60 min and 5 min at 95 °C. Specific primers were designed for bantam miRNA and were used for amplification by qPCR using the miScript SYBR Green PCR kit (Qiagen, Germany) in a QuanStudio 3 System (Life Technologies, USA). Thermal cycling conditions consisted of an initial denaturation and enzyme activation at 95 °C for 15 min, followed by 40 cycles of 94 °C for 15 s (denaturation), 55 °C for 30 s (annealing), and 70 °C for 45 s (extension). Ssa-mir-455e5p was used as an endogenous control for this reaction [38]. Gene and miRNA expression were quantified using the ΔΔCT comparative method, was previously described (Pfaffl, 2001).

3. Results

3.1. Atlantic Salmon Performance during the Experimental Trial

Immune histochemistry analysis performed in Atlantic salmon exposed to GSC and SS conditions showed chloride cell migration, indicating adequate salmon conditions for SW transfer (Figure S2). Furthermore, this condition was confirmed by RT-qPCR analysis of ATPase-α and ATPase-β subunits (Figure S3). No mortality was recorded in experimental groups.

3.2. Gill-Tissue-Transcription Modulation during Smoltification Process

Whole-genome modulation of Atlantic salmon tissues was evaluated in two groups: Atlantic salmon exposed to gradual salinity change (GSC) and salinity shock (SS). Atlantic salmon gill-tissue whole-genome expression showed low variation in MRNAs between the GSC and SS fish groups (Figure 2A). Moreover, miRNAs had an opposite expression pattern compared with mRNAs, with high threshold values in chromosome areas where mRNAs showed a downregulation. Chromosome expression variation between experimental groups was calculated by CGE index. Despite the low differences in mRNA threshold values between the GCS and SS groups, a group of chromosomes including chr1, chr2, chr4, chr9, chr12, chr13, chr14, chr17, chr18, chr22, and chr25 presented high CGE index values (Figure 2B). In addition, these chromosomes showed high threshold values for miRNAs, with a high CGE index. Furthermore, the synteny analysis of the selected chromosomes exhibits a section with high homology among the highlighting areas (Figure 2B). Notably, chr12 and chr22 showed a large synteny block and high mRNA modulation (red ribbons), and similar patterns were found in chr12 with chr2 in green ribbons. This also draws attention to chr4 and chr13, with high miRNA modulation linking a synteny block.

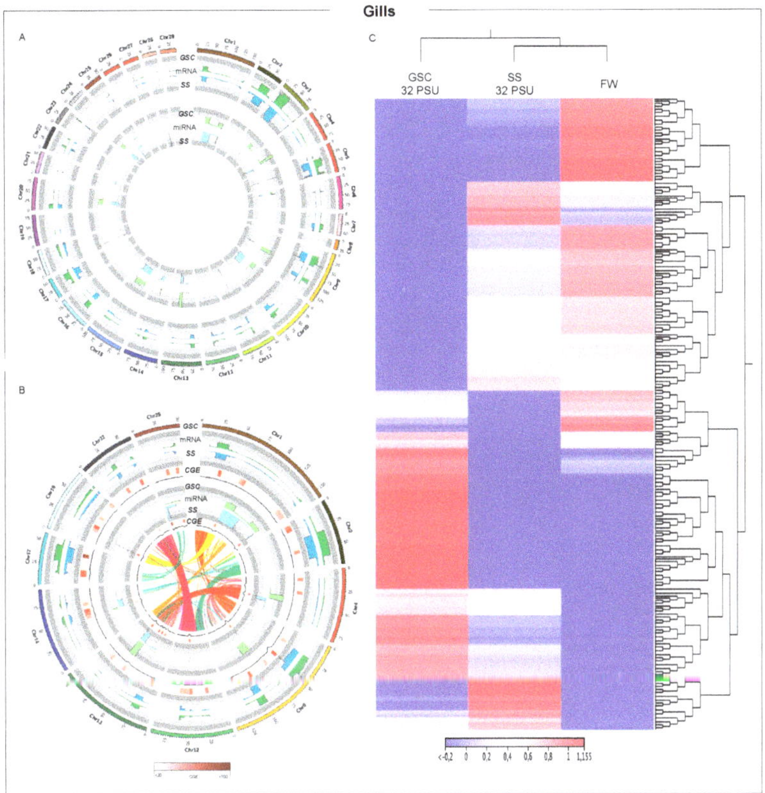

Figure 2. Whole-genome transcription of Atlantic salmon gills during smoltification process. (**A**) Threshold analysis of gills for GSC and SS fish groups. (**B**) Chromosome regions with high CGE index variation between GSC and SS fish groups. Heatmap in red shows the expression variation between both groups, CGE index. Black line graph indicates genome coverage of threshold areas. In the Circos plot, the ribbons represent the homoeologous regions in salmon genome. (**C**) RNA-Seq analysis of chromosome regions with high CGE index between experimental groups.

Putative differentially expressed chromosome regions were annotated by Blast analysis using a *Salmo salar* protein database. Notably, numerous transposase and transposable elements, *Tcb1*, *HSP70*, and *MCHII* genes were annotated in chromosomes showing high differential expression among experimental fish groups (Table S1). Moreover, RNA-Seq analyses of these selected chromosomes evidenced an interesting expression pattern, where gill samples of FW and fish from the SS group at 32 PSU were grouped in the same cluster, separated from the GSC 32PSU group (Figure 2C). Furthermore, differential expression analysis of transposable elements (TEs) showed that the median of the TEs expressed in fish exposed to GSC was downmodulated (Figure S5).

In addition, contigs annotated as TEs clustered in the Atlantic salmon gill transcriptome. Fish transfer from FW to SW gradually and by salinity shock exhibited different expression (Figure S5). Notably, clusters 1 and 4 were downmodulated after SW transfer at 32 PSU, GSC group. In contrast, cluster 2 showed contigs annotated as upmodulated TEs (Figure S5A). Furthermore, gill samples of Atlantic salmon from FW to 32 PSU by salinity sock showed four clusters, where cluster 1 and 3 showed different expression profiles, up- and down-modulated, respectively. Interestingly, TEs RT-qPCR validation showed a similar expression pattern to RNA-seq analysis in cluster 2 and cluster 3 for GSC and SS, respectively (Figure S5B).

3.3. Intestine Tissue-Transcription Modulation during Smoltification Process

A high regulation of the expression of chr2 to chr9 regions was observed from the whole-genome expression analysis. Moreover, miRNA transcriptional variation was observed in a reduced number of chromosomes, including chr4, chr13, chr15, and chr25 (Figure 3A), suggesting a low regulatory role of miRNAs in the intestine during the smoltification process. Notably, chromosomes with a high CGE index between mRNAs expressed in the GSC and SS fish group showed a high miRNA CGE index (Figure 3B), suggesting local miRNAs regulation, similar to the findings observed in gills. In addition, the synteny analysis among highlighting chromosomes showed an area with a high CGE index and mRNA homology between chr1-chr9 and chr12-chr22. Contrary to was observed in gills, heatmap representations of intestine transcripts group in the same cluster fish at 32 PSU (Figure 3C). In addition, intestine chromosomes with a high CGE index exhibited a great abundance of regulatory elements, such as transposase, transposable elements, and retro-transposable elements (Table S1).

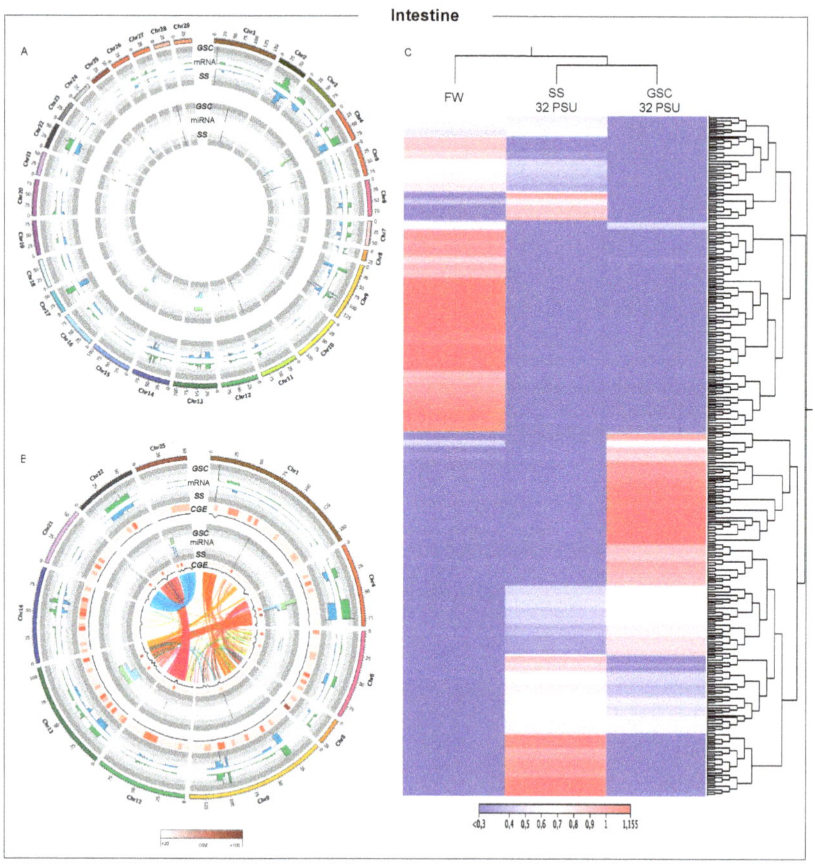

Figure 3. Whole-genome transcription in Atlantic salmon intestine during smoltification process. (**A**) Threshold analysis of gills for GSC and SS fish groups. (**B**) Chromosome regions with high CGE index variation between GSC and SS fish groups. Heatmap in red shows the expression variation between both groups, CGE index. Black line graph indicates genome coverage of threshold areas. In the Circos plot, the ribbons represent the homoeologous regions in salmon genome. (**C**) RNA-Seq analysis of chromosomes regions with high CGE index between experimental groups.

3.4. Head-Kidney Transcriptome Modulation during Smoltification Process

From chromosome analysis using read sequences obtained for head kidney tissue from Atlantic salmon exposed to GSC and SS, high modulated areas were observed in all Atlantic salmon genomes (Figure 4A). Additionally, head-kidney miRNA analysis of whole Atlantic salmon genome suggested an important regulatory function during the smoltification process in this tissue. Cromosomes chr2, chr3, chr4, chr9, chr13, chr15, chr17, chr20, chr24, and chr25 had high CGE indexes for mRNA and miRNA between head kidney tissue of fish exposed to GSC and SS (Figure 4B). Interestingly, while the synteny analysis of gill and intestine tissue showed homology among mRNAs areas, a synteny block in head kidney tissue was observed in regions with a high miRNA CGE index, such as chr13-chr15, chr20-chr24, chr3-chr6, and chr13-chr14 (Figure 4B). Interestingly, among the genes annotated in these chromosomes, *hemoglobin*, *HSP70*, *TLR8*, and a large number of transposase and transposable elements were found (Table S1). Notably, TEs expressed in head kidney were upregulated in both experimental conditions, GSC and SS (Figure S5).

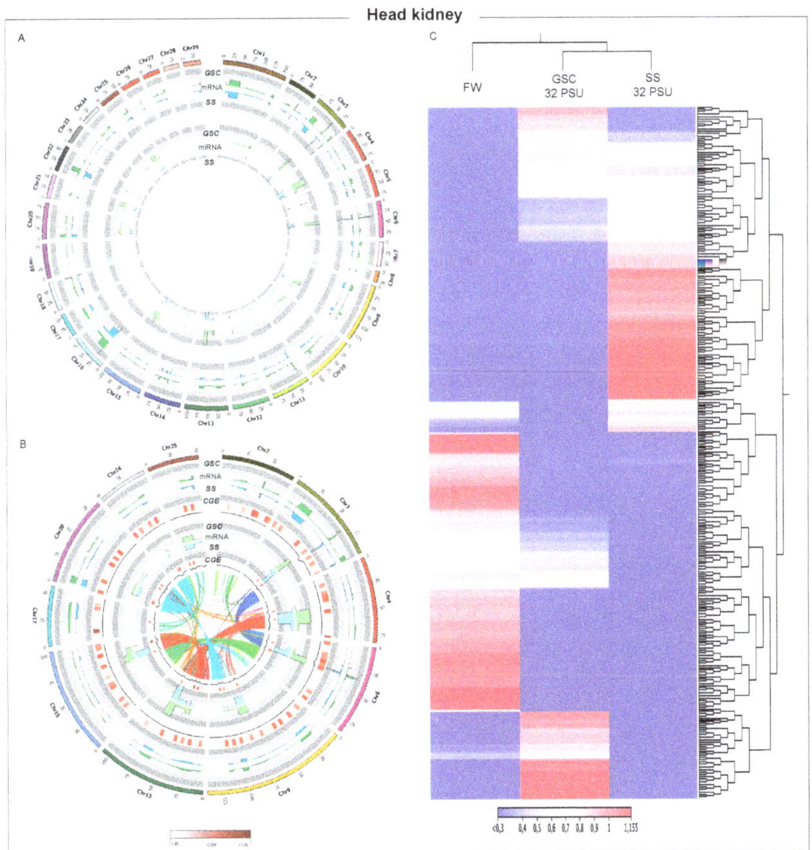

Figure 4. Whole-genome transcription in Atlantic salmon head kidney during smoltification process. (**A**) Threshold analysis of gills for GSC and SS fish groups. (**B**) Chromosome regions with high CGE index variation between GSC and SS fish groups. Heatmap in red shows the expression variation between both groups, CGE index. Black line graph indicates genome coverage of threshold areas. In the Circos plot, the ribbons represent the homoeologous regions in salmon genome. (**C**) RNA-Seq analysis of chromosome regions with high CGE index between experimental groups.

3.5. Differential Expression Analysis of Atlantic Salmon during Seawater Transfer

Nucleotide sequences near 10kb of each coverage threshold area were extracted from the Atlantic salmon genome. Later, sequences were blasted to the contigs obtained from the de novo assembly performed for each tissue. These sequences were used as a reference for RNA-Seq analysis using the filtered data of each tissue. Gill tissue of fish exposed to GSC and SS at 32 PSU showed similar expression patterns. On the other hand, samples obtained from the control group (FW) and fish exposed to 10 and 20 PSU were grouped in the same cluster (Figure S4A). From the differential expression analysis between gills of fish exposed to GSC and SS compared with the control group (FW), a larger number of transcripts differently modulated in response to the SS, with 2528 transcripts, compared to 646 transcripts differently expressed in the GSC group, suggesting a signification effect of SS in mRNA expression modulation (Figure 5A, Table S2). From GO analysis, the differentially expressed transcripts were annotated as biological processes, such as response to a steroid hormone, positive regulation of gene expression, muscle cell migration, localization of cell, insulin secretion, defense response, eye morphogenesis, and cell motility, with a large number of transcripts in the GSC group (Figure 5A).

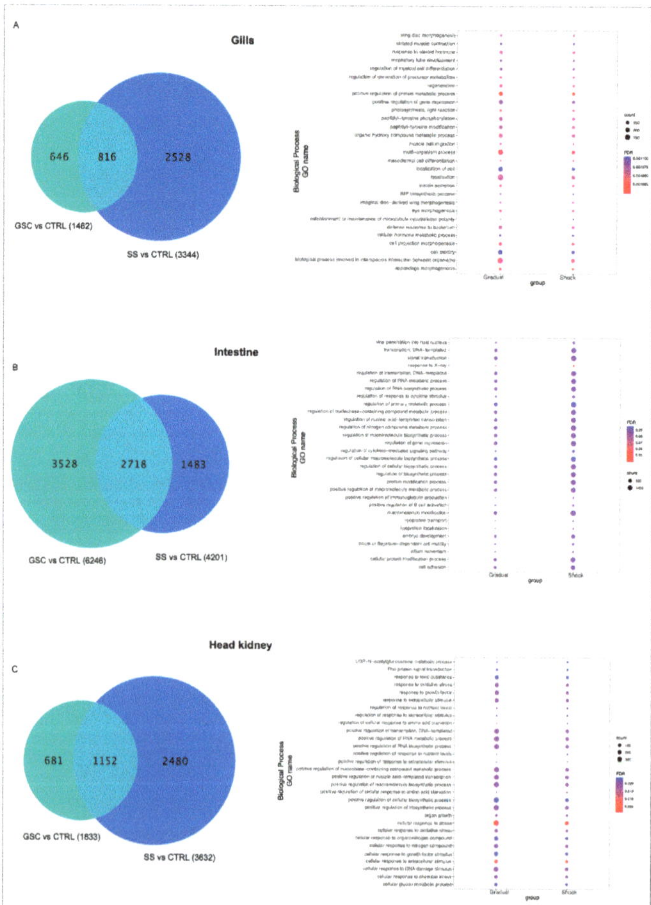

Figure 5. Differential expression analysis and GO enrichment of Atlantic salmon exposed to GSC and SS. (**A**) DEGs and GO enrichment of gills tissue. (**B**) DEGs and GO enrichment of intestine tissue. (**C**) DEGs and GO enrichment of head kidney tissue.

Similar expression patterns were found between intestine fish samples from GSC and SS under the 32 PSU condition. In the control group (FW), fish exposed to 10 and 20 PSU were clustered in the same group (Figure S4B). A total of 3528 transcripts were highly modulated in the intestine of the GSC group, while in SS fish group, 1483 DE transcripts in intestine tissue were obtained (Figure 5B, Table S2). GO enrichment analysis of differently modulated transcripts in intestine tissue showed BP processes associated with response to hormones, organic substance metabolic processes, metabolic processes, ion transmembrane transport, cellular processes, and some processes associate with immune response, with a greater number of transcripts in the GSC group (Figure 5B).

Transcriptome expression analyses showed a similar expression profile between head-kidney samples of fish at 32 PSU from GSC and SS groups. Notably, the head-kidney samples obtained from the GSC fish group at 10 PSU showed high expression levels of transcripts downregulated under the other evaluated conditions (Figure S4C). Interestingly, differential expression analysis showed a large number of transcripts modulated in the SS group, with 2480, compared with to 681 transcripts modulated in response to the GSS condition (Figure 5C, Table S2). GO analysis showed that the most relevant modulated BP were in response to oxidative stress, response to growth factor, positive regulation of transcription, and positive regulation of response to extracellular stimulus, among others (Figure 5C).

3.6. *miRNA Regulation in Atlantic Salmon during Smoltification*

The small RNAs obtained from Illumina sequencing for each tissue were annotated using the *Salmo salar* miRNA database published in miRBase release 22.1 [16,33]. A total of 478 miRNAs were annotated, like those reported in the miRBase for Atlantic salmon. In gills, two clusters of transcriptional profiles were identified, the first one grouping the control group (FW) with fish exposed to 10 PSU, and a second cluster grouping samples obtained from fish exposed to GSC at 20, 32 PSU and the SS group at 32 PSU (Figure 6A). Unlike gills, the transcriptional patterns of miRNAs of SS and GSC were grouped by salinity (32 PSU) in the same cluster. Furthermore, in intestine and head kidney tissue, fish exposed to GSC (32 PSU) evidenced an upregulation compared to fish from the SS group. In contrast, GSC fish at 20 and 10 PSU were fish sampled in freshwater (Figure 6B,C).

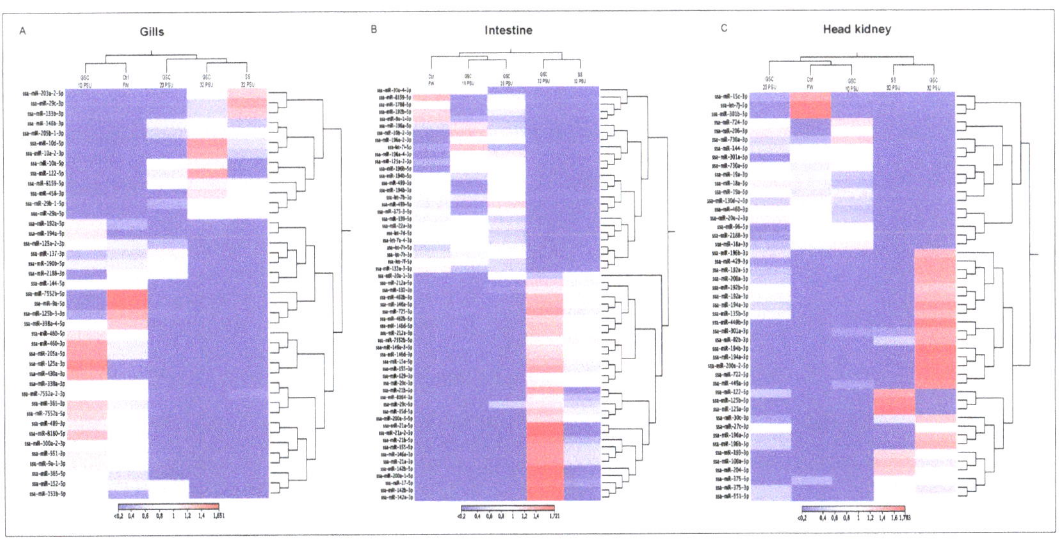

Figure 6. miRNA expression profile during gradual salinity changes and salinity shock in gills, intestine, and head kidney of Atlantic salmon.

Notably, clustering analysis showed different miRNAs with expression profile changes related to the FW condition between GSC and SS groups. Interestingly, miRNAs with expression changes between FW and GSC showed an upregulation in gill samples after transferring to 32 PSU (Figure S6A). Furthermore, these miRNAs expressed in response to GSC, ssa-miR-143, ssa-miR-21b, and ssa-miR-10d, were validated by RT-qPCR. Expression evaluation showed an up-modulation during salinity changes from FW to 32 PSU, similar to the in silico analysis (Figure S6B). Furthermore, four cluster exhibited significant expression changes in fish groups exposed to SS (Figure S6A). In addition, RT-qPCR analysis in gill samples obtained at each sample point (Figure S1) showed similar expression patterns of ssa-miR-181 and ssa-miR-30d from cluster 1, and ssa-miR-10b from cluster 2 (Figure S6B).

Differential expression analysis was performed between GSC and SS salmon groups using sampled tissues in FW as a control. Differences in the number and class of miRNAs among tissues were found. For instance, gill samples showed a large number of miRNAs differently expressed in response to SS, including SSA-miR-122-5p, ssa-miR-122-3p, and ssa-miR-122-2-3p upregulated in GSC group. Moreover, according to prediction target analysis, SSA-miR-122-5p has as target gene the *sodium/potassium-transporting ATPase subunit beta-3-like* and *thyroid hormone receptor interactor 11*, with delta G values of −17.3 and −16.2, respectively, While SSA-miR-365-5p is the most downregulated miRNA in this tissue (Figure 7).

A larger number of miRNAs in the intestine were differently expressed in samples obtained from fish exposed to GSC. However, no significant differences in the expression change values were observed (Figure 7). However, upregulated SSA-miR-155-3p miRNA showed a putative binding site to *myosin-11-like*, *probable cation-transporting ATPase*, and *collagen Type XI Alpha2* in the GSC group, with delta G values of −14.5, −13.5, and −14.7, respectively (Table S3). Moreover, ssa-miR-20a-1-3p was upregulated, and among its target genes, the *membrane heat shock 70 kDa protein* and *toll-like receptor 13* were identified, with delta G values of −12.54 and −18.2, respectively (Table S3).

The head kidney showed a similar number of differently expressed miRNAs between CGS and SS fish groups. Nevertheless, the miRNAs differently modulated between groups were different. For instance, SSA-miR-499a-5p, ssa-miR-192a-3p, ssa-miR-192b-3p, and ssa-miR-200b-5p were overexpressed in the GSC group. Target gene analysis determined that cathelicidin antimicrobial peptide has a putative binding site to ssa-miR-192a-3p, with a delta G value of −15.2, while ssa-miR-122-5p and ssa-miR-122-2-3p were upregulated in response to SS, evidencing a putative binding site to *immunoglobulin tau heavy chain* (delta G −15.2) (Table S3).

Interestingly, some miRNAs are expressed in two tissues, suggesting a role in salinity changes. For instance, ssa-miR-499a-5p miRNAs was overexpressed in gills and head kidney tissues in response to GSC. Another example is ssa-miR-196b-3p, which is overexpressed in the intestine and head kidney. Regarding the response to SS, the miRNAs ssa-miR-27d-5p is upmodulate in the gills and intestine. Others interesting miRNAs are ssa-miR-144-3p and ssa-miR-301a-5p, downregulated in gills and head kidney from fish exposed to SS (Figure 7). Furthermore, target analysis of commonly expressed miRNAs evidenced a putative role in the modulation of collagen genes in the case of ssa-miR-499a-5p and ssa-miR-27d-5p. Additionally, it was a gene associated with cell differentiation was observed as *G-protein-signaling modulator 2-like*, with a target site to ssa-miR-144-3b. Finally, among the SSA-miR-301a-5p target genes was identified *clathrin heavy chain 1-like*. (Table S3).

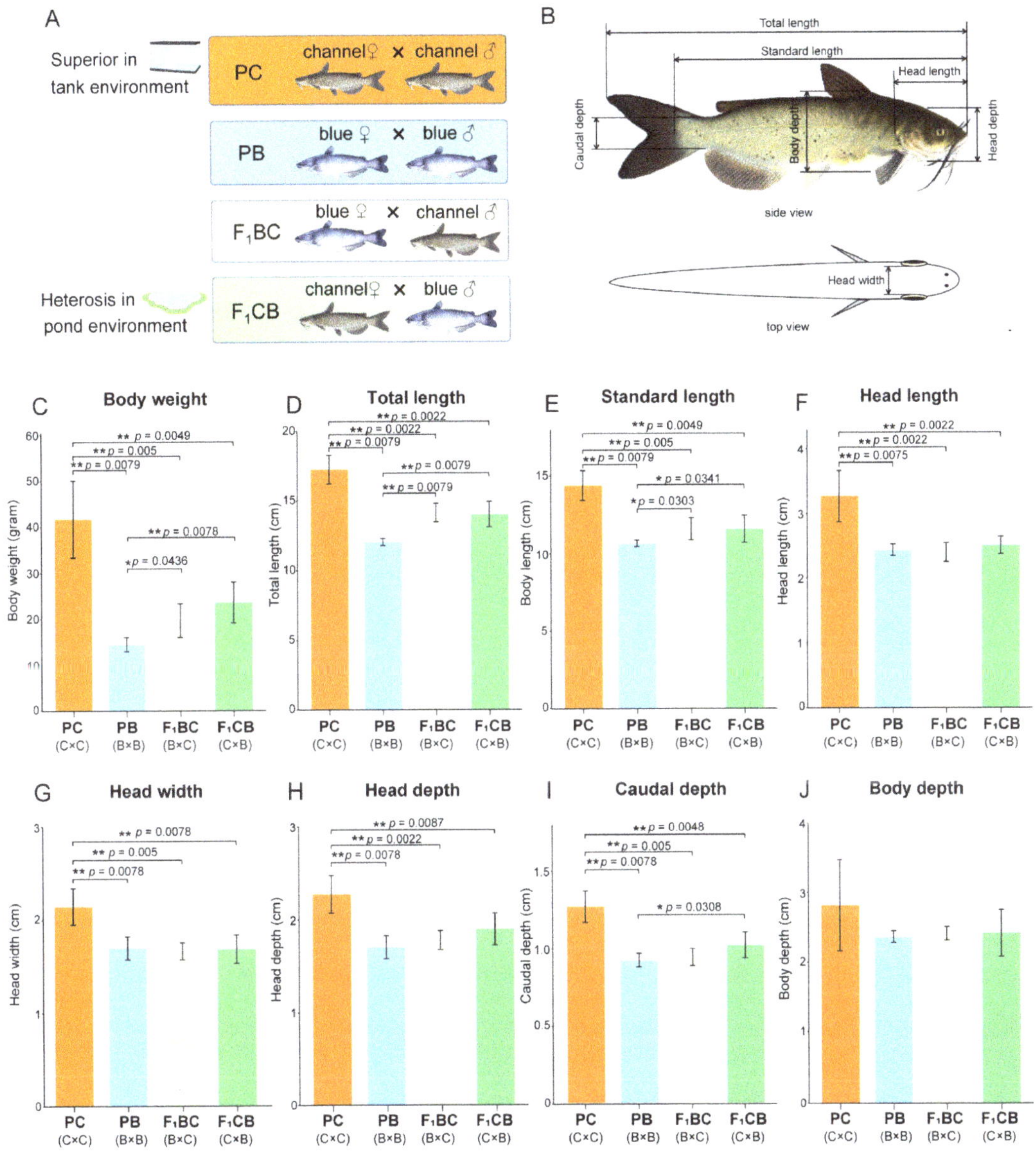

Figure 1. Morphometric measurements of channel catfish (C), *Ictalurus punctatus*, blue catfish (B), *I. furcatus*, and their reciprocal F1 hybrids raised in the tank environment. (**A**) Schematic illustration of four genetic cross types: channel catfish (C) parental cross (PC), blue catfish (B) parental cross (PB), blue catfish female × channel catfish male hybrids (F_1BC), and channel catfish female × blue catfish male hybrids (F_1CB) (**B**); Morphometric traits measured in this study: body weight (**C**); total length (**D**); body length (**E**); head length (**F**); head width (**G**); head depth (**H**) caudal depth (**I**); and body depth (**J**). Statistical significance was assessed by nonparametric Mann-Whitney U test (*, $p < 0.05$; **, $p < 0.01$;). The different colors representing the four genetic types were used consistently in this and subsequent figures.

2.4. Total RNA Extraction, RNA-Seq Library Preparation, and Sequencing

Two biological replicates were included for each of the four genetic types at 12 months of age (PB, PC, F_1BC, and F_1CB). RNA extraction was conducted using AllPrep DNA/RNA Mini Kit (Qiagen, Redwood City, CA, USA) following the manufacturer's protocol. RNA concentrations were quantified using a NanoDrop OneC Microvolume Spectrophotometer (Thermo Scientific, Waltham, MA, USA). The RNA integrity was evaluated with the LabChip GX Touch HT (PerkinElmer, Hopkinton, MA, USA). The RNA library preparation was performed using NEBNext Poly(A) mRNA Magnetic Isolation Module and NEBNext Ultra II RNA Library Prep Kit for Illumina (New England BioLabs, Ipswich, MA, USA) with 1 µg of total RNA input. Purified mRNA samples were fragmented for 10 min at 94 °C. The first-strand cDNA synthesis, second-strand cDNA synthesis, end repair, and adaptor ligation were performed according to the manufacturer's protocol. The library PCR amplification was performed with 16 cycles. The average size of the RNA libraries was approximately 350 bp (including the sequencing adapters). The RNA sequencing libraries were checked using the LabChip GX Touch HT (PerkinElmer, Hopkinton, MA, USA) and quantified using a Qubit 3.0 Fluorometer (Thermo Fisher Scientific, Waltham, MA, USA). The libraries were sequenced using a 2×150 Paired-End configuration in an Illumina NovoSeq 6000 lane at Novogene (Novogene Corporation Inc., Sacramento, CA, USA).

2.5. RNA-Seq Data Analysis and Identification of Differentially Expressed Genes among Channel Catfish, Blue Catfish, and Their Reciprocal F1 Hybrids

The quality of raw sequence data was checked by FastQC (version 0.11.5) [53]. Low-quality base and adapter sequences were trimmed by Trimmomatic (version 0.36) with default parameters [54]. Trimmed reads of 30 bp or longer were retained and mapped to the channel catfish reference genome [55] using TopHat (version 2.1.1) [56]. BedTools (version 2.29.0) [57] was used to quantify read counts that mapped to gene models. Differentially-expressed genes (DEGs) among four genetic types were identified using the edgeR package in R (version 3.6.3) [58]. The expression values of each gene were calculated as Reads Per Kilobase of transcript, per Million mapped reads (RPKM). Adjusted p-values were computed using the Benjamini and Hochberg method [59] with a threshold of 0.05. The thresholds for detecting significant DEGs were $|$log2FC (fold change)$|$ > 1.5 and an adjusted p-value < 0.05.

2.6. Gene Ontology and Functional Enrichment Analysis for DEGs among PC, PB, and Reciprocal Hybrids F_1BC and F_1CB

For the DEGs from each pairwise comparison, Gene Ontology (GO) terms and KEGG (Kyoto Encyclopedia of Genes and Genomes) pathways enrichment analysis were performed by Metascape [60] with default parameters. The gene IDs were determined according to the homology to zebrafish annotations. GO analyses on biological processes, cellular components, and molecular functions were performed at an adjusted p-value cutoff of 0.01.

2.7. Identification and Functional Pathway Analysis of Transgressive Genes in Reciprocal F_1 Hybrids

Transgressive genes were identified as genes with F_1 hybrid expression levels at least 20% higher (or lower) than that in both parents (blue catfish and channel catfish). These transgressive genes were further classified into concordant and discordant transgressive genes based on the expression pattern in the reciprocal hybrids. Genes with higher (or lower) expression levels than the channel (PC) and blue catfish parents (PB) in both reciprocal hybrids (F_1BC and F_1CB) are defined as concordant transgressive genes, and the remaining genes are discordant. Among the discordant genes, the Discordant I category includes genes that are only transgressive in one hybrid cross but not the other. Discordant II genes show the opposite directions in the expression level in reciprocal F_1 hybrids. These three subtypes of transgressive genes and transgressive genes in F_1BC or F_1CB hybrid catfish were subject to GO term and KEGG pathway enrichment analyses using Metascape.

All analyses are carried out based on the knowledge base associated with the zebrafish at an adjusted *p*-value cutoff of 0.01.

2.8. Quantitative Reverse Transcription PCR Validation of DEGs and Transgressive Genes

A 400 ng aliquot of total RNA from each liver sample was reverse-transcribed using the LunaScript® RT SuperMix Kit (New England BioLabs, Ipswich, MA, USA) with Oligo dT Primer in a 20 μL reaction, according to the manufacturer's instructions. Six candidate genes were selected to be verified from DEGs and transgressive genes. Primer sequences were designed using the Oligo 7.0 software (Molecular Biology Insights Inc., Cascade, CO, USA). The primers were synthesized by Eurofins (Eurofins Genomics LLC., Louisville, KY, USA), and the amplification performance was checked by agarose gel electrophoresis. The qRT-PCR was performed in 96-well plates on a Bio-Rad C1000 Touch Thermal Cycler with CFX96 Real-Time PCR Detection Systems (Bio-Rad Laboratories, Hercules, CA, USA). The PCR reaction was performed in 20 μL systems using Luna® Universal qPCR Master Mix (New England BioLabs, Ipswich, MA, USA). Each well contained 10 μL of Luna Universal qPCR Master Mix, 8 μL of nuclease-free water, 0.5 μL of each primer (10 μmol/L), and 1 μL of cDNA template. The reaction conditions were 95 °C for 60 s, followed by 40 cycles at 95 °C for 15 s and 60 °C for 30 s. After PCR amplification, a melting curve was generated by heating from 65 to 95 °C with 0.5 °C increments, 3 s dwell time, and a plate read at each temperature. All qRT-PCR assays were carried out with two technical replicates.

2.9. Statistical Analysis

For morphometric measurements, the one-way analysis of variance (ANOVA) was conducted to test for differences among four types of catfish. A Tukey post hoc test was carried out to compare morphometric measurements and hematological parameters (glucose, lysozyme, lactate, and ACH50) between genetic types. Mann–Whitney U test was used to compare morphometric data, hematological parameters level, and gene expression level measured by qRT-PCR for pairwise comparisons of two genetic types in the main figures. Statistical significances were determined at the $p < 0.05$ level.

3. Results

3.1. Environment-Dependent Heterobeltiosis—Channel Catfish Is Superior in Aquarium Culture

Channel catfish female × blue catfish male cross displayed a series of heterobeltiosis characteristics in pond culture [42]. To investigate whether the heterosis pattern holds in aquarium culture, we measured eight morphometric traits (Table S1 and Data S1) of 10-month old fish for each of the four genetic types (PB, PC, F_1BC, and F_1CB). The body weight of the channel catfish (41.3 g) was 2~3 fold higher than the other three genotypes (14.4 g, 19.5 g, and 23.3 g in PC, F_1BC, and F_1CB), indicating the channel catfish was superior in growth ($p < 0.01$; Figure 1C). Total length and standard length were also measured (Figure 1D,E), and channel catfish grew significantly faster in length than blue catfish and the reciprocal hybrids within 10 months of age in tank culture ($p < 0.01$). The head shape metrics (head length, head width, and head depth) and caudal depth showed the same pattern ($p < 0.01$; Figure 1F–I), which were known to be highly correlated. There was also a trend of wider body depth in channel catfish (2.83 ± 0.65 cm) than PC, F_1BC, and F_1CB, but the results did not achieve statistical significance ($p > 0.05$; Figure 1J).

When the body metrics were standardized by total length (TL), PB was significantly higher than F_1BC (adjusted *p*-value < 0.05, Mann–Whitney U test; Figure S1) in all six body metrics (standard length, body depth, head length, head width, head depth, and body weight). PB also had a slightly elevated standard length/TL than all three other genetic types, and an increased head length/TL than F_1CB (Figure S1). All other pairwise comparisons of normalized body shape parameters were not significant. As shown in the radar chart, the values of seven body metrics in channel catfish were higher than the three other genetic types (Figure S2). The results clearly demonstrated that heterosis in growth was not observed in C×B hybrid as expected in the pond environment, and channel catfish was the superior genomic configuration in tank culture.

3.2. Low Level of Innate Immunity and Complement Activities in Channel Catfish Raised in the Aquarium Environment

Lysozyme activity is a key index used to evaluate fish innate immune system activity [61]. Channel catfish plasma lysozyme activity was ~24-fold lower than blue catfish (p = 0.004), ~6-fold lower than the C×B hybrid (p = 0.008), and 2.4-fold lower than the B×C hybrid (p = 0.012; Figure 2A, Table S2, and Data S2). The complement system is an important component of the innate immune system, enhancing the ability to clear microbes and foreign cells [62], and the alternative pathway of this system is commonly measured. Blue catfish had the highest ACH50, 11-fold higher than channel catfish (p = 0.016; Figure 2B, Table S2, and Data S2). The reciprocal F_1 hybrids were near the mid-parent value (Figure 2B, Table S2, and Data S2).

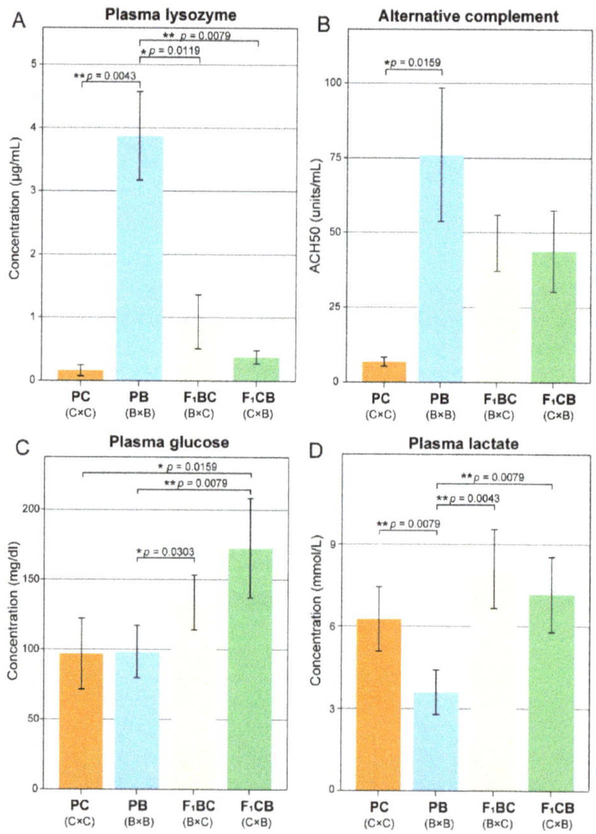

Figure 2. Plasma biochemical and immunological measurements in channel catfish (C), *Ictalurus punctatus*, blue catfish (B), *I. furcatus*, and their reciprocal F_1 hybrids raised in the tank environment. Plasma lysozyme activity (**A**); alternative complement pathway hemolytic activity (**B**); plasma glucose level (**C**); and plasma lactate level (**D**) were measured in channel catfish parental cross (PC), blue catfish parental cross (PB), blue catfish female × channel catfish male hybrids (F_1BC), and channel catfish female × blue catfish male hybrids (F_1CB). Statistical significance was assessed by non-parametric Mann–Whitney U test (*, $p < 0.05$; **, $p < 0.01$).The different colors representing the four genetic types were used consistently in this and subsequent figures.

3.3. Transgressive Effects in Metabolism—F_1 Hybrids Have Significantly Higher Blood Glucose Lactate Levels Than both Channel and Blue Catfish Parents

The blood glucose levels in the channel catfish and blue catfish were around 100 mg/dl (Figure 2C). The reciprocal F_1 hybrids had significantly higher blood glucose levels than both parents: F_1BC (133.6 mg/dl) was 38% higher than the parental species, and F_1CB (172.5 mg/dl) was 78% higher ($p < 0.05$; Figure 2C and Table S2), indicating a transgressive effect in which the F_1s had significant upregulation in blood glucose. Blood lactate was known to correlate with hyperactivity, stress level, and mortality in fish species. A similar transgressive effect was observed for the blood lactate level: both F_1BC and F_1CB were significantly higher than blue catfish ($p < 0.01$; Figure 2D and Table S2), and they were also higher than channel catfish, but it did not achieve statistical significance (Figure 2D).

3.4. Transcriptome Analysis in the Channel and Blue Catfish Parents and the Reciprocal F_1 Hybrids Revealed >2000 Differentially Expressed Genes

Liver RNA-seq analyses were performed on channel catfish (PC), blue catfish (PB), C×B hybrid catfish (F_1CB), and B×C hybrid catfish (F_1BC) (Table S3). A total of 13,420 expressed genes were identified with Reads Per Kilobase of transcript, per Million mapped reads (RPKM) value greater than 1.0 in at least one genetic type. Pairwise differential gene expression analysis was conducted to identify the differentially expressed genes (DEGs) between two genetic types (Figure 3 and Table S4). There were more DEGs between the two parental species ($n = 2308$; FDR < 0.05 and $|\log2FoldChange| > 1.5$) than the parent-hybrid comparisons (458~810 DEGs), which was consistent with the fact that PC and PB had the lowest transcriptome-wide gene expression correlation (spearman correlation coefficient $\rho = 0.76$; Figure 3). The reciprocal hybrids F_1CB and F_1BC only had 98 DEGs with the highest expression correlation ($\rho = 0.93$; Figure 3). Since the F_1 hybrids had identical nuclear genome configurations, and they were expected to display similar gene expression profiles. The results confirmed that this was the case. The evolutionary divergence and genetic distance between the channel and blue catfish resulted in gene expression changes of >2000 genes (Data S3), which account for 17.2% of expressed genes in the liver transcriptome.

3.5. Fatty Acid Metabolism and Transport Genes Were Significantly Upregulated in Channel Catfish Compared to Blue Catfish and Hybrids

Gene ontology (GO) and Kyoto Encyclopedia of Genes and Genomes (KEGG) pathways enrichment analyses were performed to identify the enriched functional pathways among DEGs. Among the identified terms, a third of the PC-PB significant GO categories were also significant in PC-F_1 comparisons (Figure 4A,B), including cellular lipid metabolic processes (GO:0044255), cellular lipid catabolic processes (GO:0044242), secondary alcohol metabolic process (GO:1902652), and oxoacid metabolic process (GO:0043436; Figure 4A). Specific non-overlapping terms enriched only in PC-F_1 comparisons included phospholipid efflux (GO: 0033700) in the PC-F_1CB comparison, as well as phosphatidate phosphatase activity (GO: 0008195) in the PC-F_1BC comparison (Figure 4A). These findings suggested that the lipid metabolism pathway activities were altered in PC.

Since channel catfish is superior in aquarium growth than PB and hybrids, we focused our analysis on the significantly upregulated and downregulated genes in PC compared to PB ($p < 0.001$; Figure 4C,D). Interestingly, the PC upregulated genes were enriched for four carboxylic acid pathways (metabolic, catabolic, transport, and binding; Figure 4C) and three lipid metabolism pathways (cellular lipid metabolic process, glycerophospholipid metabolism, and glycerolipid catabolic process; Figure 4C). The network analyses revealed that the fatty acids and lipid metabolism genes were interconnected (Figure 4E), which was separated from the transporter-related terms (Figure 4E), suggesting these two broader functional categories were significantly overrepresented in PC upregulated genes. With regard to the genes that were significantly highly expressed in PB, the top three enriched functional terms are small-molecule biosynthesis, organic acid metabolic process, and carbon metabolism (Figure 4D), which are in sharp contrast to PC upregulated genes.

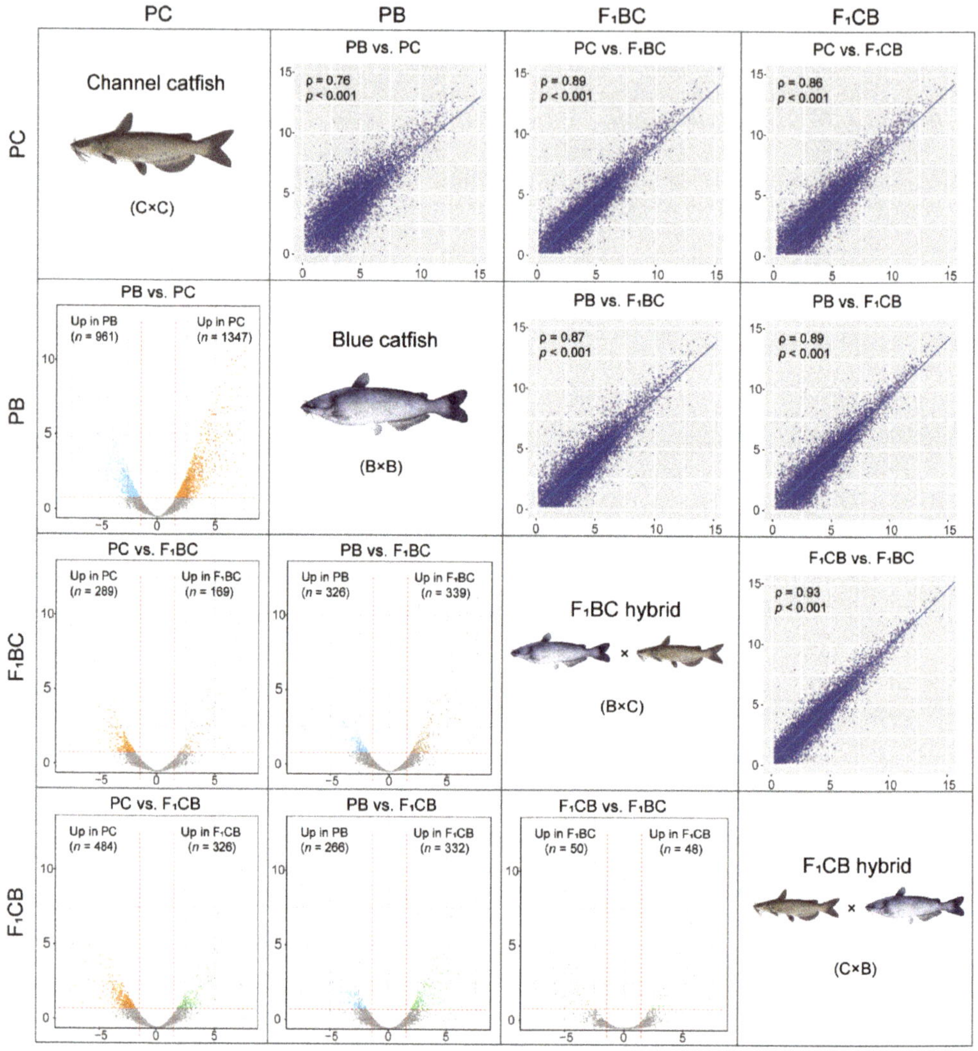

Figure 3. Transcriptome-wide gene expression correlation and differentially expressed genes in the liver among channel catfish (C), *Ictalurus punctatus*, blue catfish (B), *I. furcatus*, and their reciprocal F_1 hybrids raised in the tank environment (***diagonal panels***). ***Bottom-left*** panels: volcano plots of six pairwise comparisons among the four genetic types from channel catfish parental cross (PC), blue catfish parental cross (PB), blue catfish female × channel catfish male hybrids (F_1BC), and channel catfish female × blue catfish male hybrids (F_1CB). Differentially expressed genes (DEGs) are highlighted (FDR < 0.05). The x-axis stands for $\log_2$ fold changes, and the y-axis represents $-\log_{10}(p\text{-value})$. The vertical lines indicate $|\log_2\text{FoldChange}| = 1.5$. ***Upper-right*** panels: scatterplots of the $\log_2$ (RKPM) values for six pairwise comparisons among the four genetic types. Spearman's rank correlation coefficient ρ and the corresponding p-values are labeled.

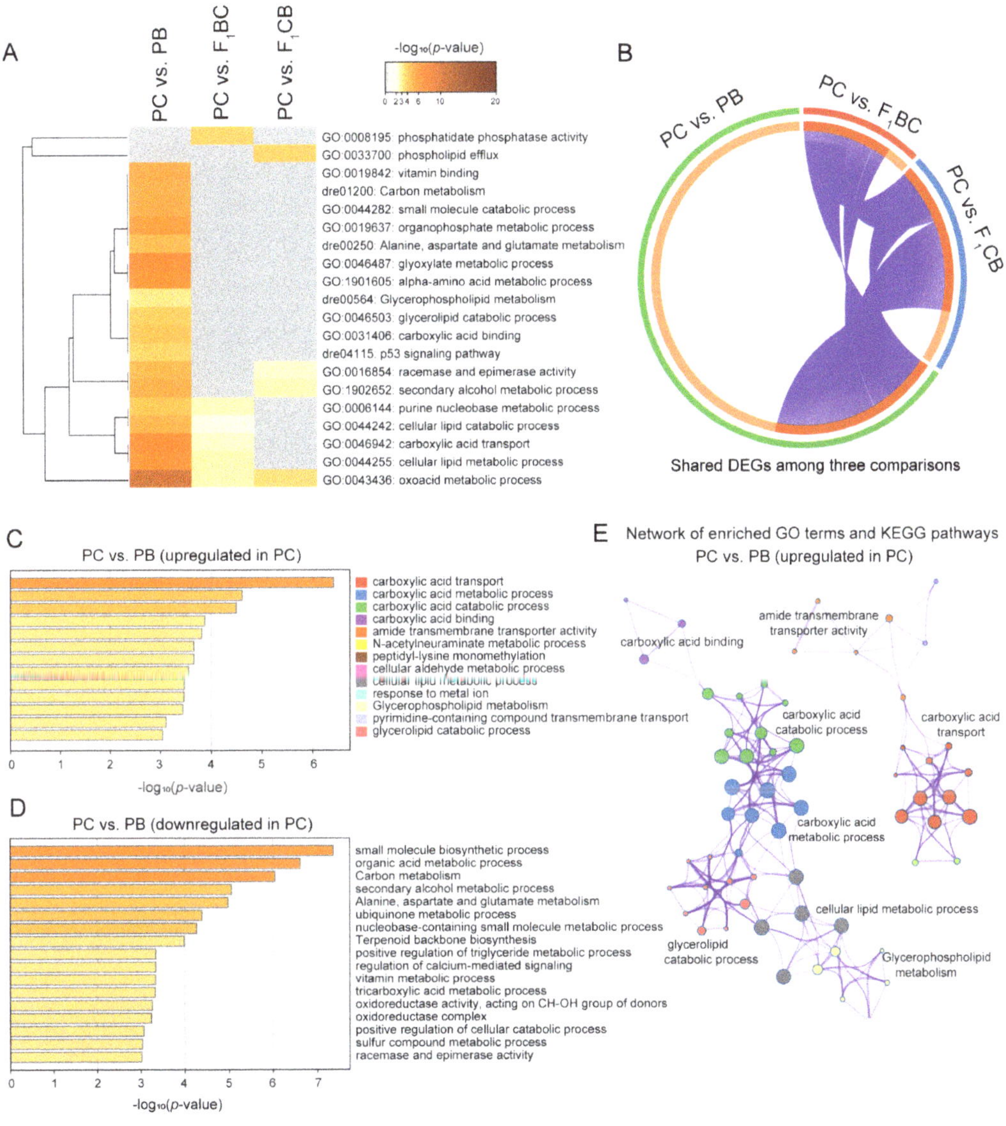

Figure 4. Pathway enrichment analysis of liver Differentially Expressed Genes (DEGs) between channel catfish *Ictalurus punctatus* parental cross (PC) and three other genetic types (PB: blue catfish *I. furcatus* parental cross, blue catfish female × channel catfish male hybrids (F_1BC), and channel catfish female × blue catfish male hybrids (F_1CB). (**A**) Hierarchical clustering of significant gene ontology terms shared in at least two of the three comparisons (PC vs. PB, PC vs. F_1BC, and PC vs. F_1CB); (**B**) A circular plot of shared DEGs in the three comparisons; (**C**,**D**) Enriched functional categories for upregulated genes in PC compared to PB (**C**) and downregulated genes in PC compared to PB (**D**); Enrichment scores measured by $-\log_{10}(p\text{-value})$ were shown on the *x*-axis; (**E**) A plot of enriched term network for upregulated genes in PC. GO terms were represented by the same color dots as in (**C**), and the interconnectivity was represented by the edges.

3.6. One Thousand Genes Displayed Transgressive Expression Patterns in the Liver of F_1 Hybrid Catfish

To elucidate the molecular basis of heterobeltiosis vs. hybrid incompatibility, a group of genes called transgressive genes were investigated, which were defined as genes that show higher or lower expression levels in both parents (see Materials and Methods). According to the transgressive pattern in reciprocal hybrids, the transgressive genes were further classified as (1) concordant: transgressive genes in both F_1 hybrids with the same direction, which are higher than both parents (upregulated concordant genes) or lower than both parents (downregulated concordant genes); (2) discordant I: genes that are transgressive in only one reciprocal F_1, but not the other (Figure 5A); (3) discordant II: genes that are transgressive in both F_1s, but the expression directions are opposite (Figure 5A). A total of 1140 transgressive genes were identified in F_1 hybrids, which count for 8.5% of all expressed genes in the liver transcriptome (Figure 5B and Data S4). Over 90% of the transgressive genes were shared in F_1CB and F_1BC (Figure 5C), suggesting that the cross direction-dependent transgressive effect (discordant I genes) only occurred in less than 10% of transgressive genes (Figure 5A,B). Even fewer genes displayed the opposite pattern in gene expression changes compared to the parental species, and these discordant II genes (n = 49) accounted for 4.3% of all transgressive genes (Figure 5B). To validate the transgressive genes detected in RNA-seq experiments, qRT-PCR experiments for gene expression quantifications for six genes were performed (Table S5), and all of them were confirmed, including two non-transgressive genes *cyp2k19* and *fgf1b* (Figure 6A,B), two concordant genes *hmox* and *irf7* (Figure 6C,D), as well as two discordance genes *tm4sf4* and *creg1* (Figure 6E,F).

3.7. Concordant Transgressive Genes in Hybrid Catfish Were Enriched for Cytoskeleton Functions, Stress, and Immune-Related Pathways

Six cytoskeleton and extracellular matrix-related terms were enriched in concordant transgressive genes, including microtubule cytoskeleton (GO: 0015630), cytoskeletal protein binding (GO: 0008092), polymeric cytoskeletal fiber (GO: 0099513), focal adhesion (dre04510), gamma–tubulin complex (GO: 0000930), and lamellipodium (GO: 0030027). Oxidative stress and immune functions were also significantly overrepresented in concordant genes. The most significantly enriched term was AGE-RAGE signaling (dre04933; $p < 0.0001$; Figure 5D), which plays an important role in inflammation in diabetes. The Nod-like receptors (dre04521; $p < 0.001$; Figure 5D) are master regulators of inflammation and defense, which activate innate and adaptive immunity by recognizing pathogen patterns. CARD domain binding genes (GO: 0050700; $p < 0.001$; Figure 5D) are often associated with inflammation and apoptosis. NAD+ binding (GO: 0070403; $p < 0.001$; Figure 5D) function is involved in cellular energy metabolism. Discordant I transgressive genes were enriched for ribosome function ($p < 0.001$; Figure 5D), the term regulation of cell cycle process (GO: 0010564) was overrepresented in discordance II genes ($p < 0.01$; Figure 5D).

3.8. Overrepresentation of Glycan Degradation Function among Upregulated Transgressive Genes Provided a Potential Mechanism for the Blood Glucose Elevation in F_1 Hybrids

For the concordant transgressive genes (n = 985), the majority of them (94%) were downregulated in both F_1 hybrids, whereas only 60 concordant genes were upregulated compared to the channel catfish and blue catfish parents (Figure 5A). These overexpressed genes may explain the transgressive phenotypes observed in the F_1 hybrids (Figure 2). GO and KEGG analysis of upregulated transgressive genes in F_1BC and F_1CB were performed. Interestingly, the top enriched term was the same, which was glycan degradation (dre00511; $p < 0.0001$; Figure 5E). Glycan breakdown in the liver will result in an elevation in blood glucose level in both hybrids F_1BC and F_1CB compared to the channel catfish and blue catfish parents, which was what we observed (Figure 2C).

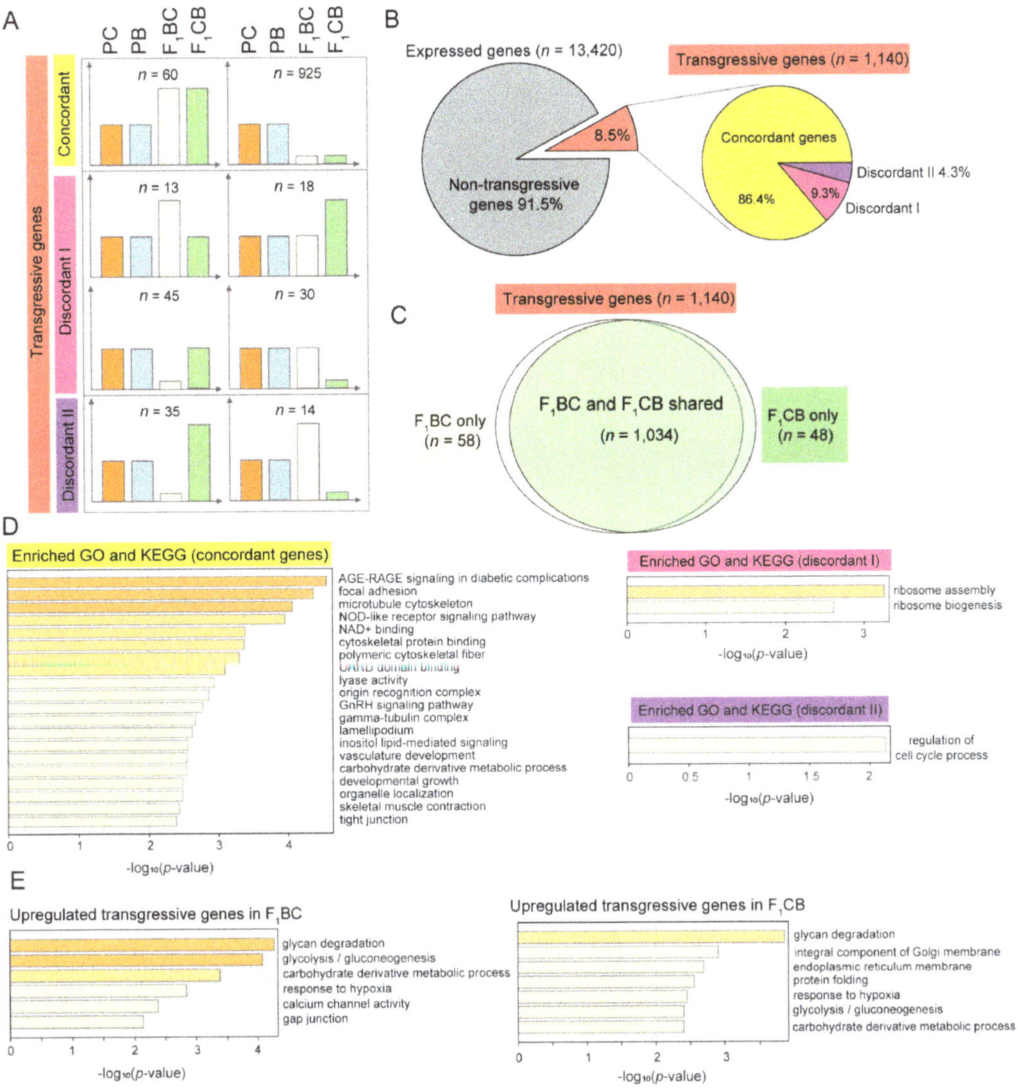

Figure 5. Identification and functional enrichment analysis of liver transgressive genes in the reciprocal hybrids of channel catfish, *Ictalurus punctatus*, and blue catfish, *I. furcatus*. (**A**) Definition of different classes of transgressive genes based on the gene expression levels in channel catfish parental cross (PC), blue catfish parental cross (PB), blue catfish female × channel catfish male hybrids (F_1BC), and channel catfish female × blue catfish male hybrids (F_1CB). The *y*-axis represents relative gene expression levels. The gene counts were labeled for each class; (**B**) Piechart of transgressive gene distributions in hybrid catfish; (**C**) Venn diagram of transgressive genes in F_1BC and F_1CB hybrids; (**D**) Enriched functional categories for concordant and discordant transgressive genes; (**E**) Enriched functional categories for upregulated transgressive genes in F_1BC and F_1CB hybrids. Enrichment scores measured by $-\log_{10}(p\text{-value})$ were shown on the *x*-axis.

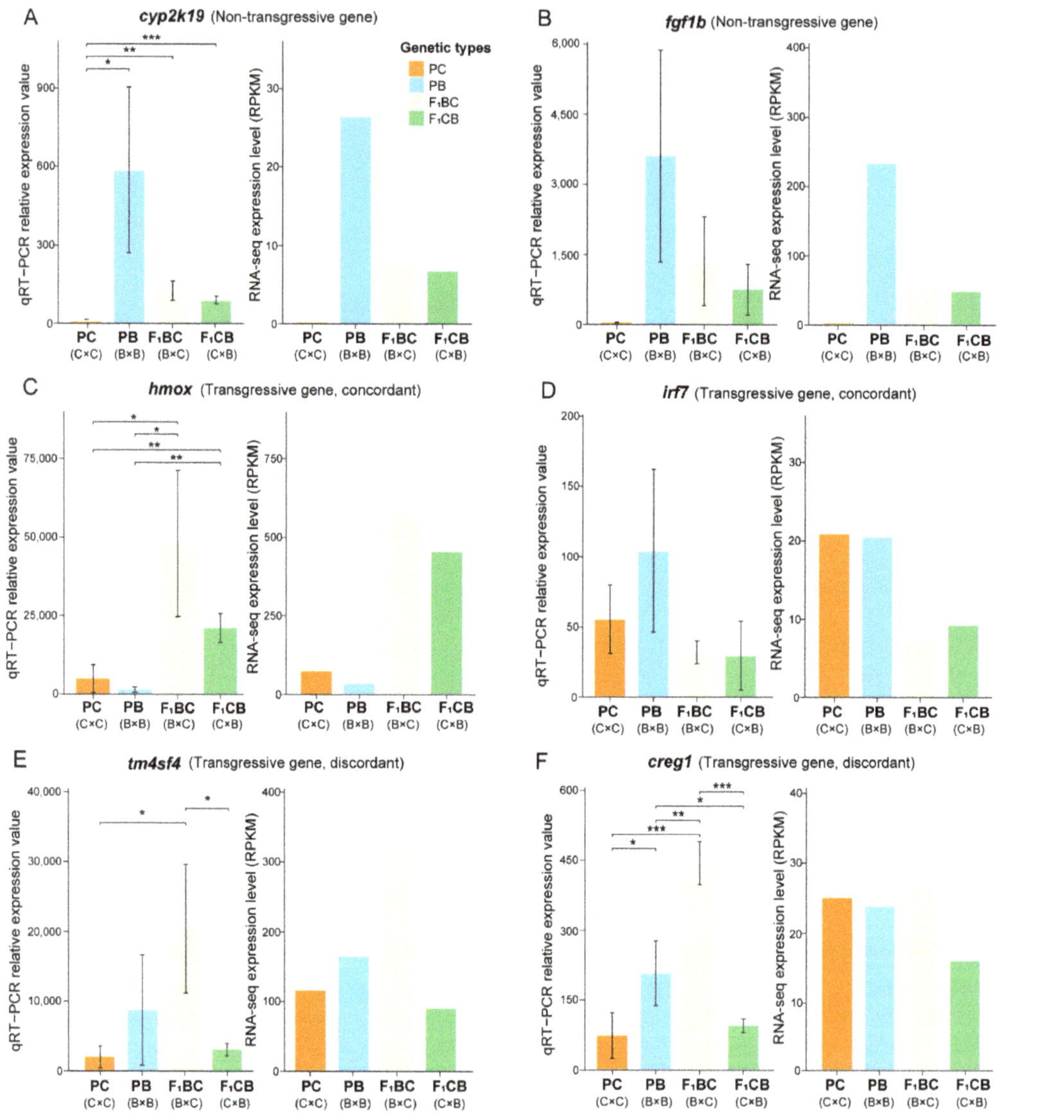

Figure 6. Quantitative reverse transcription PCR validation of differentially expressed genes and transgressive genes. PC: channel catfish *Ictalurus punctatus* parental cross; PB: blue catfish I. furcatus parental cross; F_1BC: blue catfish female × channel catfish male hybrids; F_1CB: channel catfish female × blue catfish male hybrids. Barplots of qRT-PCR relative quantification and RNA-seq RPKM values (Reads Per Kilobase of transcript per Million mapped reads) for non-transgressive genes *cyp2k1* (**A**) and *fgf1* (**B**), concordant transgressive genes *hmox* (**C**) and *irf7* (**D**), and discordant transgressive genes *tm4sf4* (**E**) and *creg1* (**F**). Mann–Whitney U test was used to assess the statistical significance (*, $p < 0.05$; **, $p < 0.01$; ***, $p < 0.001$).

4. Discussion

4.1. The Phenomenon of Environment-Dependent Heterosis in Hybrid Catfish

The degree of heterosis was known to be affected by the environment and genotype-environment interactions [63]. Environment-influenced heterosis has been reported in cattle, in which the advantage of milk and protein yields was suppressed in a high HLI (summer heat load index) environment [64]. Examples of environment-influenced heterosis also can be found in aquatic species. Crossbred offspring from two silver perch (*Bidyanus bidyanus*) strains from Murray River and Cataract Dam had the best performance in growth under pond environment, whereas heterosis decreased when reared in cages and tanks [65]. In these situations, the degree of heterosis was affected by the rearing environment, but heterosis did not disappear entirely. We define this phenomenon as environment-influenced heterosis.

Environment-dependent heterosis was first discovered in *Drosophila* in 1987. Crosses among five geographically diverse *D. melanogaster* inbred lines identified significant heterosis in fecundity [50]. When the flies were maintained at a higher than optimal density, heterosis was still observed for all hybrid crosses with a slightly lesser degree, indicating that crowdedness can affect hybrid vigor, but it does not abolish heterosis [50]. However, when the flies were reared in the lower temperature (17 °C rather than 24 °C), heterosis was only present in two hybrid crosses, and all other hybrid line pairs had lower fecundity than the inbred crosses [50], suggesting heterosis disappeared for most hybrid crosses under low-temperature environment. The hybridization of the channel catfish and blue catfish is a vertebrate example of environment-dependent heterosis. The hybrid cross grew 30–121% faster than the channel catfish in pond environments with different densities, whereas the channel catfish's mean body weight was 49.8% higher than hybrid fish when they were grown in cages [66]. Both environment-influenced and environment-dependent heterosis had been observed in hybrid catfish [66]. In this study, we discovered that the channel catfish was superior in body weight (Figure 1C) and all other morphometric traits (Figure S2) than blue catfish and the reciprocal hybrids, at 10 months of age in tank culture, although channel catfish did not differ from the reciprocal hybrids in standardized shape parameters (Figure S1). The reciprocal hybrids between the channel and blue catfish may serve as an excellent system to investigate environment-dependent heterosis and their molecular mechanisms.

4.2. The Biological Robustness and Prevalence of Transgressive Genes in Channel-Blue Catfish Hybrid System

Interspecific hybrids can display heterosis, but they could also suffer from hybrid necrosis [67] or hybrid incompatibility [68]. If the two parental genomes were too distantly related, DNA sequence and gene expression divergence would result in misregulation of protein-coding gene expression in F_1s. Uneven chromosome numbers can also cause the hybrid breakdown. As a classic example of heterosis in mammals, a mule is an offspring of a female horse ($2n = 64$) and a male donkey ($2n = 62$). Although mules outperform both parents in many aspects, including strength, stamina, temper, and longevity, mules are infertile because the odd number of chromosomes will affect proper segregation during meiosis. Hybridization can also lead to the aberrant activation of transposable elements (TEs), which was known as "genome shock" discovered in maize by Barbara McClintock [69]. Genes with transgressive expression patterns may explain the heterosis vs. hybrid incompatibility. After merging the two parental genomes, the F_1 gene expression level is expected to be near the mid-parental value (additive effect), or close to either parent (dominant or incomplete dominant effects), or within the parental range (both effects). In contrast, transgressive genes have higher (or lower) expression levels compared to both parents, which can explain the superior phenotypes or hybrid misregulation in F_1s. In our study, over 90% of expressed genes were non-transgressive genes, and transgressive genes only accounted for 8%, which is consistent with the level of divergence between the channel catfish and blue catfish (13–15 SNPs per Kb) estimated from EST data [70]. The channel-blue hybrids are fully viable with strain-dependent variable fertility [71], suggesting major biological functions and metabolic/developmental pathways can tolerate transgressive genes in some crosses, exhibiting robustness at the organism level.

4.3. The Superiority in Tank Growth May Be Associated with Low Immune Activity and Stress Levels in Channel Catfish

In the pond culture, heterobeltiosis in growth, disease resistance, and harvestability was only observed in C×B hybrids [42]. In the tank environment, channel catfish is the fast-growing genetic type compared to all other three genomic configurations. Based on previous literature on the density-influenced heterosis in pond vs. cage culture in other fish species, the degree of crowding and/or accompanying stress levels was believed to play important roles in the variation in heterosis. However, density-dependent stress alone cannot explain the fact that hybrid catfish did show heterobeltiosis in tanks of larger sizes (>1 cubic meter), even if the density was extremely high. Therefore, our current results only apply to the smaller aquarium environment and cannot be generalized to larger tanks.

As important defense mechanisms in fish's innate immune system [72], lysozyme and complement activity (including the alternative pathway) are widely used to evaluate the immunity and ability of the fish to clear pathogens [61,73]. In channel catfish, lysozyme activity was reported to correlate with blood bacteria concentrations after exposure to pathogenic *Edwardsiella ictaluri* [74]. Additionally, it has been previously reported that alternative complement activity can vary across fish species, and as a whole, catfish were found to have relatively low ACH50 values in comparison to barramundi (*Lates calcarifer*) and rabbitfish (*Siganus rivulatus*) [75]. Thus, innate immune parameters need to be compared within closely related fish species or strains.

In this research, lysozyme and alternative complement levels in channel catfish were more than 10-fold lower than blue catfish and significantly lower than the hybrids (Figure 2A,B), suggesting a dramatic decrease in innate immune activity in channel catfish in the absence of pathogenic infections under aquarium environment. This was consistent with previous findings using in vivo pathogen challenge experiments, in which channel catfish were found to be least resistant to bacterial pathogens overall compared to the blue catfish and F_1CB (F_1BC not tested). For the three major infectious diseases in catfish production, blue catfish was almost completely resistant to *Edwardsiella ictaluri*, the pathogen for Enteric Septicemia of Catfish (ESC), with 0.7–10.5% mortality [76,77]. F_1CB had a 26% mortality under ESC [77], whereas the mortality for channel catfish was up to 72.3% [76] (resistance to ESC: PB >> F_1CB > PC). For *Aeromonas* spp. infections, blue catfish (32% mortality) [78] were reported to be more resistant than F_1CB [79], with channel catfish (90% mortality in [80] and 78% mortality in [81]) as the least resistant genetic type (resistance to Aeromonas disease: PB > F_1CB >> PC). For columnaris disease, the hybrid F_1CB was observed to be much more resistant (32% mortality) to *Flavobacterium columnare* than channel catfish (74% mortality) and blue catfish (87% mortality) [45] (resistance to columnaris disease: F_1CB >> PC > PB). Similarly, Zhang et al. (2020) also demonstrated increased lysozyme levels in hybrid yellow catfish (female yellow catfish *Pelteobagrus fulvidraco* × male darkbarbel catfish *P. vachelli*) compared to all-male yellow catfish [82]. Since maintaining a highly-activated immune system is both energetically and nutritionally expensive, which will inhibit growth and development, the lowered innate immunity activity discovered in the channel catfish may explain the superior phenotype in weight gain under aquarium culture in small tanks.

Elevated plasma glucose and lactate levels were observed in hybrid catfish in this study. Increased glucose levels in rainbow trout were reported to be attributed to environmental changes, such as season [83], density stocking [84], as well as nutritional changes or stressors [85]. Interestingly, a recent study in channel catfish evaluated the role of gastric peptides on glucose levels and discovered complex regulation of glucose levels [86]. Plasma lactate is also an important stress-related parameter and has been shown to increase in channel catfish when confined or deprived of oxygen [87]. With respect to stress management, stressed fish typically have higher plasma lactate levels, which could be the result of hyperactivity [64,88]. Metabolic activities, such as lipolysis and glycolysis, provide the energy required to meet the demands of the stress response, potentially resulting in a negative impact on growth. In summary, the lower plasma glucose and lactate levels may also contribute to the fast-growing phenotype of channel catfish in tank culture.

4.4. Fatty Acid and Lipid Metabolism Are Enriched in Channel Catfish

Among the upregulated genes in the channel catfish compared to blue catfish, the top four significant gene ontology terms were all related to carboxylic acid ($p < 0.001$; Figure 4C). Three lipid metabolism terms were also significant ($p < 0.001$; Figure 4C). Hydroxy-carboxylic acids (HCAs) are intermediates in animal energy metabolism, and HCA receptors play an important role in homeostasis by regulating energy metabolism, lipolysis, inflammation, and immunity [89]. HCAs were understudied in fish species, but HCA metabolic process was reported to be the most significant GO term between non-alcoholic fatty liver patients and controls [90], and research has shown that activated hydroxy-carboxylic acids (HCAs) inhibit adipocyte lipolysis [91,92], suggesting its relevance to lipid metabolism and storage. For many fish species, fatty acid oxidation is the primary source of energy. The enriched HCA and lipid metabolism-related terms in channel catfish upregulated genes may mediate the difference in immune function and growth phenotypes.

4.5. High Blood Glucose in Hybrids Is Likely to Be due to Glycan Degradation in the Liver

The liver is a crucial organ for maintaining glucose homeostasis. The blood glucose levels were found to be significantly higher in both F_1s than the channel catfish and blue catfish (Figure 2C), and they also exceeded the normal range. Hyperglycemia is common in teleost species [93], but this transgressive effect may result in increased glucose utilization and reduced weight gain in F_1s. To explore the molecular basis, transgressive genes that were upregulated in F_1BC and F_1CB were investigated. Interestingly, glycan degradation was identified as the top enriched pathway for both hybrids (Figure 5E). The glycolysis pathway was also overrepresented in the upregulated transgressive genes. These findings suggested that rapid glycan degradation and glucose utilization may explain the higher blood glucose level and reduced weight gain in the F_1 hybrids in the small tank/aquarium environment.

4.6. Toward a Better Understanding of the Environment-Dependent Heterobeltiosis in Hybrid Catfish

The environment-dependent heterobeltiosis is an intriguing phenomenon. It is the first reported case in any vertebrate species, and it is of great interest in both evolutionary biology of heterosis and agriculture practice to enhance the catfish genetic stock. However, this is a complicated problem, and many aspects of it warrant further research. First of all, environmental variations can affect the status of heterosis. We have shown that in small tanks, channel catfish was the superior genetic type for growth and development. However, from the previous literature discussed in 4.3, this may not be the case under pathogenic infections, under which blue catfish and the C$\times$B hybrids have better survival rates overall. Second, a "tank size" effect exists. In tanks larger than one cubic meter, heterobeltiosis was observed just as in the pond culture, independent of the fish density. Third, there might be an age, size, and density-dependent effect as well, since channel catfish grow the fastest during year 1 in low-density ponds, but hybrids grow faster than channel catfish in year 1 in high-density ponds. In contrast, this density-related relationship does not occur during year 1 growth in aquaria. To further complicate the situation, these factors interact with the genetic and epigenetic backgrounds of the four types, as well as neurological differences in complex social behavior, level of stress, physical activities, hormonal changes, innate immune attributes, feed intake and frequency, and flighting or tank hierarchy among individuals. Further studies are needed to disentangle these complex factors to elucidate the underlying mechanisms of this fascinating phenomenon.

5. Conclusions

Heterosis and heterobeltiosis are the genetic basis for production enhancement using interspecific hybrid breeding techniques. Hybrid catfish are superior in a number of production and disease-resistant traits, and they grow much faster than both channel catfish and blue catfish parents. Interestingly, this heterobeltiosis only occurs in pond culture, and channel catfish are superior in growth in smaller culturing units, such as tanks and aquaria. This research investigated this intriguing environment-dependent heterosis, and identified three potential mechanisms of this phenomenon: (1) significantly lower lysozyme activity and alternative complement activity discovered in channel catfish may reduce the energy cost of immune function to promote growth; (2) fatty acid metabolism/transport pathways were enriched in channel catfish upregulated genes, which may explain the faster growth of channel catfish than the other three genetic types; (3) F_1 hybrids had elevated blood glucose levels compared to channel catfish, which may result from liver glycan degradation. Collectively, these gene expression and physiological differences contributed to the lack of heterosis in the tank culture environment. For the first time, this study provided insights into the regulation of environment-dependent heterosis in a vertebrate model. Further studies are needed to determine when and how the heterobeltiosis happened after the tank-pond environment transition.

Supplementary Materials: The following supporting information can be downloaded at: https://www.mdpi.com/article/10.3390/biology11010117/s1. Figure S1: Morphometric measurements of channel catfish (C), *Ictalurus punctatus*, blue catfish (B), *I. furcatus*, and their reciprocal F1 hybrids raised in the tank environment, standardized by total length (TL); Figure S2: Radar plot of seven body metrics in channel catfish (C), *Ictalurus punctatus*, blue catfish (B), *I. furcatus*, and their reciprocal F1 hybrids raise in the tank environment; Table S1: Comparison of morphometric measurements for blue catfish, channel catfish, and their reciprocal hybrids in the tank environment; Table S2: Biochemical and immunological assays in blue catfish, channel catfish, and their reciprocal hybrids cultured in the tank environment; Table S3: Liver sample RNA sequencing yield, quality control, and alignment statistics; Table S4: Numbers of significantly differentially expressed genes (DEGs) in pairwise comparisons among four genetic types (PB, PC, F_1BC, and F_1CB); Table S5: List of primer sequences for quantitative reverse transcription PCR validation; Data S1: Raw measurements of morphometric characteristics for 10-month blue catfish, channel catfish, and their reciprocal hybrids cultured in the tank environment; Data S2: Raw biochemical and immunological measurements in blue catfish, channel catfish, and their reciprocal hybrids cultured in the tank environment; Data S3: List of differentially expressed genes from pairwise comparisons among PB, PC, F_1BC, and F_1CB; Data S4: List of transgressive genes in blue-channel catfish hybrids. Supplemental Data files are available at github.com/XuWangLab/2021_catfish_hybrid_heterosis_sppData (last accessed on 5 January 2022).

Author Contributions: X.W., R.A.D., T.J.B. and B.S. contributed to the conception and design of the study. H.W., B.S. and S.L. performed the animal work, sample collection, and measurements. H.W. performed the RNA experiments and qPCR validation. H.W. and X.W. performed the bioinformatic data analysis. X.W., R.A.D. and T.J.B. provided samples, resources, and analysis tools. H.W. and X.W. wrote the first draft of the manuscript. T.J.B., B.S. and R.A.D. wrote sections of the manuscript. All authors contributed to manuscript revision, read, and approved the submitted version. All authors have read and agreed to the published version of the manuscript.

Funding: This research was funded by USDA National Institute of Food and Agriculture Hatch project 1018100, National Science Foundation EPSCoR RII Track-4 Research Fellowship (1928770). X.W. is supported by an Alabama Agriculture Experiment Station Agriculture Research Enhancement, Exploration, and Development (AAES-AgR-SEED) award, and a laboratory start-up fund from Auburn University College Veterinary Medicine. H.W. is supported by the Auburn University Presidential Graduate Research Fellowship, College of Veterinary Medicine Dean's Fellowship, and the China Scholarship Council.

Institutional Review Board Statement: All experimental animal protocols, including animal care and tissue sample collections, were approved by the Auburn University Institutional Animal Care and Use Committee (IACUC) under PRN# 2019-3520.

Informed Consent Statement: Not applicable.

Data Availability Statement: The raw RNA-seq data is available at NCBI GEO (Gene Expression Omnibus) databases under the accession number GSE186603.

Acknowledgments: The authors would like to thank Allison Wise for help with determining lysozyme activity and ACH50. We acknowledge the Auburn University Easley Cluster for support of this work.

Conflicts of Interest: The authors declare no conflict of interest.

References

1. Darwin, C. *The Works of Charles Darwin, Volume 25: The Effects of Cross and Self Fertilization in the Vegetable Kingdom*; NYU Press: New York, NY, USA, 2010.
2. Shull, G.H. What is " heterosis"? *Genetics* **1948**, *33*, 439. [CrossRef]
3. Batte, M.; Nyine, M.; Uwimana, B.; Swennen, R.; Akech, V.; Brown, A.; Hovmalm, H.P.; Geleta, M.; Ortiz, R. Significant progressive heterobeltiosis in banana crossbreeding. *BMC Plant Biol.* **2020**, *20*, 489. [CrossRef] [PubMed]
4. Dunham, R.A.; Smitherman, R.O. Crossbreeding channel catfish for improvement of body weight in earthen ponds. *Growth* **1983**, *47*, 97–103. [PubMed]
5. Dunham, R.A. *Aquaculture and Fisheries Biotechnology: Genetic Approaches*, 2nd ed.; CABI: Wallingford, UK, 2011; pp. 1–495.
6. Chen, Z.J. Molecular mechanisms of polyploidy and hybrid vigor. *Trends Plant Sci.* **2010**, *15*, 57–71. [CrossRef] [PubMed]
7. Troyer, A.F. Development of hybrid corn and the seed corn industry. In *Handbook of Maize*; Springer: Berlin/Heidelberg, Germany, 2009; pp. 87–114.
8. Williams, J.; Aguilar, I.; Rekaya, R.; Bertrand, J. Estimation of breed and heterosis effects for growth and carcass traits in cattle using published crossbreeding studies. *J. Anim. Sci.* **2010**, *88*, 460–466. [CrossRef] [PubMed]
9. Buckley, F.; Lopez-Villalobos, N.; Heins, B. Crossbreeding: Implications for dairy cow fertility and survival. *Animal* **2014**, *8*, 122–133. [CrossRef] [PubMed]
10. Ferreira, V.; Rosa, G.; Berger, Y.; Thomas, D. Survival in crossbred lambs: Breed and heterosis effects. *J. Anim. Sci.* **2015**, *93*, 912–919. [CrossRef] [PubMed]
11. Lalev, M.; Mincheva, N.; Oblakova, M.; Hristakieva, P.; Ivanova, I. Estimation of heterosis, direct and maternal additive effects from crossbreeding experiment involving two White Plymouth Rock lines of chickens. *Biotechnol. Anim. Husb.* **2014**, *30*, 103–114. [CrossRef]
12. Dagilis, A.J.; Kirkpatrick, M.; Bolnick, D.I. The evolution of hybrid fitness during speciation. *PLoS Genet.* **2019**, *15*, e1008125. [CrossRef]
13. Carroll, S.B. Evo-devo and an expanding evolutionary synthesis: A genetic theory of morphological evolution. *Cell* **2008**, *134*, 25–36. [CrossRef]
14. Hodgins-Davis, A.; Rice, D.P.; Townsend, J.P. Gene expression evolves under a house-of-cards model of stabilizing selection. *Mol. Biol. Evol.* **2015**, *32*, 2130–2140. [CrossRef]
15. Mack, K.L.; Nachman, M.W. Gene regulation and speciation. *Trends Genet.* **2017**, *33*, 68–80. [CrossRef] [PubMed]
16. Haerty, W.; Singh, R.S. Gene regulation divergence is a major contributor to the evolution of Dobzhansky–Muller incompatibilities between species of Drosophila. *Mol. Biol. Evol.* **2006**, *23*, 1707–1714. [CrossRef] [PubMed]
17. Banho, C.A.; Merel, V.; Oliveira, T.Y.K.; Carareto, C.M.A.; Vieira, C. Comparative transcriptomics between Drosophila mojavensis and D. arizonae reveals transgressive gene expression and underexpression of spermatogenesis-related genes in hybrid testes. *Sci. Rep.* **2021**, *11*, 9844. [CrossRef]
18. Crow, J.F. Dominance and overdominance. *Genet. Exploit. Heterosis Crops* **1999**, 49–58. [CrossRef]
19. Malone, J.H.; Michalak, P. Gene expression analysis of the ovary of hybrid females of Xenopus laevis and X. muelleri. *BMC Evol. Biol.* **2008**, *8*, 82. [CrossRef]
20. Bolnick, D.I.; Turelli, M.; Lopez-Fernández, H.; Wainwright, P.C.; Near, T.J. Accelerated mitochondrial evolution and "Darwin's corollary": Asymmetric viability of reciprocal F_1 hybrids in Centrarchid fishes. *Genetics* **2008**, *178*, 1037–1048. [CrossRef]
21. Davidson, J.H.; Balakrishnan, C.N. Gene regulatory evolution during speciation in a songbird. *G3 Genes Genomes Genet.* **2016**, *6*, 1357–1364. [CrossRef]
22. Turelli, M.; Moyle, L.C. Asymmetric postmating isolation: Darwin's corollary to Haldane's rule. *Genetics* **2007**, *176*, 1059–1088. [CrossRef]
23. Brideau, N.J.; Flores, H.A.; Wang, J.; Maheshwari, S.; Wang, X.; Barbash, D.A. Two Dobzhansky-Muller genes interact to cause hybrid lethality in Drosophila. *Science* **2006**, *314*, 1292–1295. [CrossRef]
24. Lippman, Z.B.; Zamir, D. Heterosis: Revisiting the magic. *Trends Genet.* **2007**, *23*, 60–66. [CrossRef]

25. Dobzhansky, T. Studies on hybrid sterility. II. Localization of sterility factors in Drosophila pseudoobscura hybrids. *Genetics* **1936**, *21*, 113. [CrossRef]
26. Wei, X.; Zhang, J. The optimal mating distance resulting from heterosis and genetic incompatibility. *Sci. Adv.* **2018**, *4*, eaau5518. [CrossRef] [PubMed]
27. Mendel, G. *Experiments in Plant Hybridisation*; Harvard University Press: Cambridge, MA, USA, 1965.
28. Larièpe, A.; Mangin, B.; Jasson, S.; Combes, V.; Dumas, F.; Jamin, P.; Lariagon, C.; Jolivot, D.; Madur, D.; Fievet, J. The genetic basis of heterosis: Multiparental quantitative trait loci mapping reveals contrasted levels of apparent overdominance among traits of agronomical interest in maize (Zea mays L.). *Genetics* **2012**, *190*, 795–811. [CrossRef] [PubMed]
29. Li, L.; Lu, K.; Chen, Z.; Mu, T.; Hu, Z.; Li, X. Dominance, overdominance and epistasis condition the heterosis in two heterotic rice hybrids. *Genetics* **2008**, *180*, 1725–1742. [CrossRef] [PubMed]
30. Shen, G.; Zhan, W.; Chen, H.; Xing, Y. Dominance and epistasis are the main contributors to heterosis for plant height in rice. *Plant Sci.* **2014**, *215*, 11–18. [CrossRef] [PubMed]
31. Swanson-Wagner, R.A.; Jia, Y.; DeCook, R.; Borsuk, L.A.; Nettleton, D.; Schnable, P.S. All possible modes of gene action are observed in a global comparison of gene expression in a maize F1 hybrid and its inbred parents. *Proc. Natl. Acad. Sci. USA* **2006**, *103*, 6805–6810. [CrossRef]
32. Crow, J.F. Mid-century controversies in population genetics. *Annu. Rev. Genet.* **2008**, *42*, 1–16. [CrossRef]
33. Bartley, D.M.; Rana, K.; Immink, A.J. The use of inter-specific hybrids in aquaculture and fisheries. *Rev. Fish Biol. Fish.* **2000**, *10*, 325–337. [CrossRef]
34. Hulata, G. A Review of Genetic-Improvement of the Common Carp (Cyprinus carpio L.) and Other Cyprinids by Crossbreeding, Hybridization and Selection. *Aquaculture* **1995**, *129*, 143–155. [CrossRef]
35. Steffens, W.; Jahnichen, H.; Fredrich, F. Possibilities of Sturgeon Culture in Central-Europe. *Aquaculture* **1990**, *89*, 101–122. [CrossRef]
36. Amiri, B.M.; Maebayashi, M.; Hara, A.; Adachi, S.; Yamauchi, K. Ovarian development and serum sex steroid and vitellogenin profiles in the female cultured sturgeon hybrid, the bester. *J. Fish Biol.* **1996**, *48*, 1164–1178. [CrossRef]
37. Henderson-Arzapalo, A.; Colura, R.L.; Maciorowski, A.F. A Comparison of Black Drum, Red Drum, and their Hybrid in Saltwater Pond Culture. *J. World Aquac. Soc.* **1994**, *25*, 289–296. [CrossRef]
38. Tripp-Valdez, M.A.; Cicala, F.; Galindo-Sanchez, C.E.; Chacon-Ponce, K.D.; Lopez-Landavery, E.; Diaz, F.; Re-Araujo, D.; Lafarga-De la Cruz, F. Growth Performance and Transcriptomic Response of Warm-Acclimated Hybrid Abalone Haliotis rufescens (female symbol) × H. corrugata (male symbol). *Mar. Biotechnol.* **2021**, *23*, 62–76. [CrossRef] [PubMed]
39. Harrell, R.M.; Webster, D.W. An overview of Morone culture. *Dev. Aquac. Fish. Sci.* **1997**, *30*, 1–10.
40. Dunham, R.A.; Elaswad, A. Catfish biology and farming. *Annu. Rev. Anim. Biosci.* **2018**, *6*, 305–325. [CrossRef]
41. Dunham, R.A.; Umali, G.M.; Beam, R.; Kristanto, A.H.; Trask, M. Comparison of production traits of NWAC103 channel catfish, NWAC103 channel catfish× blue catfish hybrids, Kansas Select 21 channel catfish, and blue catfish grown at commercial densities and exposed to natural bacterial epizootics. *N. Am. J. Aquac.* **2008**, *70*, 98–106. [CrossRef]
42. Brown, T.W.; Chappell, J.A.; Boyd, C.E. A commercial-scale, in-pond raceway system for Ictalurid catfish production. *Aquac. Eng.* **2011**, *44*, 72–79. [CrossRef]
43. Bosworth, B.G. Effects of winter feeding on growth, body composition, and processing traits of co-cultured Blue Catfish, Channel Catfish, and Channel Catfish × Blue Catfish hybrids. *N. Am. J. Aquac.* **2012**, *74*, 553–559. [CrossRef]
44. Dunham, R.A.; Smitherman, R.O.; Webber, C. Relative tolerance of channel x blue hybrid and channel catfish to low oxygen concentrations. *Progress. Fish-Cult.* **1983**, *45*, 55–57. [CrossRef]
45. Arias, C.R.; Cai, W.; Peatman, E.; Bullard, S.A. Catfish hybrid Ictalurus punctatus× I. furcatus exhibits higher resistance to columnaris disease than the parental species. *Dis. Aquat. Org.* **2012**, *100*, 77–81.
46. Dunham, R.; Masser, M. *Production of Hybrid Catfish*; Southern Regional Aquaculture Center: Stoneville, MI, USA, 2012; Volume 436.
47. Bott, L.B.; Roy, L.A.; Hanson, T.R.; Chappell, J.; Whitis, G.N. Research verification of production practices using intensive aeration at a hybrid catfish operation. *N. Am. J. Aquac.* **2015**, *77*, 460–470. [CrossRef]
48. Kumar, G.; Engle, C.R.; Hanson, T.R.; Tucker, C.S.; Brown, T.W.; Bott, L.B.; Roy, L.A.; Boyd, C.E.; Recsetar, M.S.; Park, J.; et al. Economics of alternative catfish production technologies. *J. World Aquac. Soc.* **2018**, *49*, 1039–1057. [CrossRef]
49. Torrans, L.; Ott, B. Effect of grading fingerling hybrid catfish (♀channel catfish × ♂blue catfish) on growth, production, feed conversion, and food fish size distribution. *N. Am. J. Aquac.* **2018**, *80*, 187–192. [CrossRef]
50. Domínguez, A.; Albornoz, J. Environment-dependent heterosis in Drosophila melanogaster. *Génétique Sélection Évolution* **1987**, *19*, 37–48. [CrossRef] [PubMed]
51. Sankaran, K.; Gurnani, S. On the variation in the catalytic activity of lysozyme in fishes. *Indian J. Biochem. Biophys.* **1972**, *9*, 162–165. [PubMed]
52. Welker, T.L.; Lim, C.; Klesius, P.; Liu, K. Evaluation of Distiller's Dried Grains with Solubles from Different Grain Sources as Dietary Protein for Hybrid Tilapia, Oreochromis niloticus (♀) × Oreochromis aureus (♂). *J. World Aquac. Soc.* **2014**, *45*, 625–637. [CrossRef]
53. Andrews, S. FastQC: A Quality Control Tool for High Throughput Sequence Data. 2010. Available online: http://www.bioinformatics.babraham.ac.uk/projects/fastqc/ (accessed on 5 January 2022).

54. Bolger, A.M.; Lohse, M.; Usadel, B. Trimmomatic: A flexible trimmer for Illumina sequence data. *Bioinformatics* **2014**, *30*, 2114–2120. [CrossRef]
55. Liu, Z.; Liu, S.; Yao, J.; Bao, L.; Zhang, J.; Li, Y.; Jiang, C.; Sun, L.; Wang, R.; Zhang, Y. The channel catfish genome sequence provides insights into the evolution of scale formation in teleosts. *Nat. Commun.* **2016**, *7*, 11757. [CrossRef]
56. Kim, D.; Pertea, G.; Trapnell, C.; Pimentel, H.; Kelley, R.; Salzberg, S.L. TopHat2: Accurate alignment of transcriptomes in the presence of insertions, deletions and gene fusions. *Genome Biol.* **2013**, *14*, R36. [CrossRef]
57. Quinlan, A.R.; Hall, I.M. BEDTools: A flexible suite of utilities for comparing genomic features. *Bioinformatics* **2010**, *26*, 841–842. [CrossRef]
58. Robinson, M.D.; McCarthy, D.J.; Smyth, G.K. edgeR: A Bioconductor package for differential expression analysis of digital gene expression data. *Bioinformatics* **2010**, *26*, 139–140. [CrossRef]
59. Benjamini, Y.; Hochberg, Y. Controlling the False Discovery Rate—A Practical and Powerful Approach to Multiple Testing. *J. R. Stat. Soc. Ser. B-Stat. Methodol.* **1995**, *57*, 289–300. [CrossRef]
60. Zhou, Y.; Zhou, B.; Pache, L.; Chang, M.; Khodabakhshi, A.H.; Tanaseichuk, O.; Benner, C.; Chanda, S.K. Metascape provides a biologist-oriented resource for the analysis of systems-level datasets. *Nat. Commun.* **2019**, *10*, 1523. [CrossRef]
61. Saurabh, S.; Sahoo, P.K. Lysozyme: An important defence molecule of fish innate immune system. *Aquac. Res.* **2008**, *39*, 223–239. [CrossRef]
62. Holland, M.C.; Lambris, J.D. The complement system in teleosts. *Fish Shellfish Immunol.* **2002**, *12*, 399–420. [CrossRef] [PubMed]
63. Falconer, D. Selection in different environments: Effects on environmental sensitivity (reaction norm) and on mean performance. *Genet. Res.* **1990**, *56*, 57–70. [CrossRef]
64. Bryant, J.; López-Villalobos, N.; Pryce, J.; Holmes, C.; Johnson, D.; Garrick, D. Effect of environment on the expression of breed and heterosis effects for production traits. *J. Dairy Sci.* **2007**, *90*, 1548–1553. [CrossRef]
65. Guy, J.A.; Jerry, D.R.; Rowland, S.J. Heterosis in fingerlings from a diallel cross between two wild strains of silver perch (Bidyanus bidyanus). *Aquac. Res.* **2009**, *40*, 1291–1300. [CrossRef]
66. Dunham, R.A.; Brummett, R.E.; Ella, M.O.; Smitherman, R.O. Genotype-environment interactions for growth of blue, channel and hybrid catfish in ponds and cages at varying densities. *Aquaculture* **1990**, *85*, 143–151. [CrossRef]
67. Bomblies, K.; Weigel, D. Hybrid necrosis: Autoimmunity as a potential gene-flow barrier in plant species. *Nat. Rev. Genet.* **2007**, *8*, 382–393. [CrossRef] [PubMed]
68. Vaid, N.; Laitinen, R.A.E. Diverse paths to hybrid incompatibility in Arabidopsis. *Plant J.* **2019**, *97*, 199–213. [CrossRef] [PubMed]
69. McClintock, B. The significance of responses of the genome to challenge. *Science* **1984**, *226*, 792–801. [CrossRef] [PubMed]
70. Wang, S.; Peatman, E.; Abernathy, J.; Waldbieser, G.; Lindquist, E.; Richardson, P.; Lucas, S.; Wang, M.; Li, P.; Thimmapuram, J. Assembly of 500,000 inter-specific catfish expressed sequence tags and large scale gene-associated marker development for whole genome association studies. *Genome Biol.* **2010**, *11*, R8. [CrossRef] [PubMed]
71. Dunham, R.A.; Argue, B.J. Reproduction among Channel Catfish, Blue Catfish, and Their F_1 and F_2 Hybrids. *Trans. Am. Fish. Soc.* **2000**, *129*, 222–231. [CrossRef]
72. Sahoo, P. Innate/natural resistance and non-lymphoid defence mechanisms in fish. In *Fish and Shellfish Immunology, An Introduction*; Swain, P., Sahoo, P.K., Ayyappan, S., Eds.; Narendra Publishing House: Dehli, India, 2006; pp. 37–46.
73. Sunyer, J.O.; Tort, L. Natural hemolytic and bactericidal activities of sea bream Sparus aurata serum are effected by the alternative complement pathway. *Vet. Immunol. Immunopathol.* **1995**, *45*, 333–345. [CrossRef]
74. Small, B.C.; Bilodeau, A.L. Effects of cortisol and stress on channel catfish (Ictalurus punctatus) pathogen susceptibility and lysozyme activity following exposure to Edwardsiella ictaluri. *Gen. Comp. Endocrinol.* **2005**, *142*, 256–262. [CrossRef]
75. Nayak, S.; Portugal, I.; Zilberg, D. Analyzing complement activity in the serum and body homogenates of different fish species, using rabbit and sheep red blood cells. *Vet. Immunol. Immunopathol.* **2018**, *199*, 39–42. [CrossRef]
76. Wolters, W.R.; Johnson, M.R. Enteric Septicemia Resistance in Blue Catfish and Three Channel Catfish Strains. *J. Aquat. Anim. Health* **1994**, *6*, 329–334. [CrossRef]
77. Wolters, W.R.; Wise, D.J.; Klesius, P.H. Survival and Antibody Response of Channel Catfish, Blue Catfish, and Channel Catfish Female × Blue Catfish Male Hybrids after Exposure to Edwardsiella ictaluri. *J. Aquat. Anim. Health* **1996**, *8*, 249–254. [CrossRef]
78. Li, C.; Beck, B.; Su, B.; Terhune, J.; Peatman, E. Early mucosal responses in blue catfish (Ictalurus furcatus) skin to Aeromonas hydrophila infection. *Fish Shellfish Immunol.* **2013**, *34*, 920–928. [CrossRef] [PubMed]
79. Zhou, T.; Yuan, Z.; Tan, S.; Jin, Y.; Yang, Y.; Shi, H.; Wang, W.; Niu, D.; Gao, L.; Jiang, W.; et al. A Review of Molecular Responses of Catfish to Bacterial Diseases and Abiotic Stresses. *Front. Physiol.* **2018**, *9*, 1113. [CrossRef]
80. Zhang, D.; Xu, D.-H.; Shoemaker, C. Experimental induction of motile Aeromonas septicemia in channel catfish (Ictalurus punctatus) by waterborne challenge with virulent Aeromonas hydrophila. *Aquac. Rep.* **2016**, *3*, 18–23. [CrossRef]
81. Zhang, D.; Moreira, G.S.; Shoemaker, C.; Newton, J.C.; Xu, D.H. Detection and quantification of virulent Aeromonas hydrophila in channel catfish tissues following waterborne challenge. *FEMS Microbiol. Lett.* **2016**, *363*, fnw080. [CrossRef] [PubMed]
82. Zhang, Y.-L.; Zhang, C.-M.; Fan, Q.-X.; Shen, Z.-G. Comparative Study of Growth, Feed Utilization, and Immunity of Hybrid Catfish and All-Male Yellow Catfish. *N. Am. J. Aquac.* **2020**, *82*, 71–74. [CrossRef]
83. Soengas, J.; Fuentes, J.; Otero, J.; Andrés, M.; Aldegunde, M. Seasonal changes in carbohydrate metabolism in the rainbow trout (Oncorhynchus mykiss) and their relationship to changes in gill (Na+-K+)-ATPase activity. *Aquaculture* **1992**, *108*, 369–380. [CrossRef]

84. Conde-Sieira, M.; Aguilar, A.J.; López-Patiño, M.A.; Míguez, J.M.; Soengas, J.L. Stress alters food intake and glucosensing response in hypothalamus, hindbrain, liver, and Brockmann bodies of rainbow trout. *Physiol. Behav.* **2010**, *101*, 483–493. [CrossRef]
85. Polakof, S.; Panserat, S.; Soengas, J.L.; Moon, T.W. Glucose metabolism in fish: A review. *J. Comp. Physiol. B* **2012**, *182*, 1015–1045. [CrossRef]
86. Schroeter, J.C.; Fenn, C.M.; Small, B.C. Elucidating the roles of gut neuropeptides on channel catfish feed intake, glycemia, and hypothalamic NPY and POMC expression. *Comp. Biochem. Physiol. A Mol. Integr. Physiol.* **2015**, *188*, 168–174. [CrossRef]
87. Small, B.C. Effect of isoeugenol sedation on plasma cortisol, glucose, and lactate dynamics in channel catfish Ictalurus punctatus exposed to three stressors. *Aquaculture* **2004**, *238*, 469–481. [CrossRef]
88. Caillouet, C.W., Jr. Hyperactivity, blood lactic acid and mortality in channel catfish. *Iowa Agric. Home Econ. Exp. Stn. Res. Bull.* **1967**, *35*, 1.
89. Offermanns, S. Hydroxy-Carboxylic Acid Receptor Actions in Metabolism. *Trends Endocrinol. Metab.* **2017**, *28*, 227–236. [CrossRef] [PubMed]
90. Jia, X.; Zhai, T. Integrated analysis of multiple microarray studies to identify novel gene signatures in non-alcoholic fatty liver disease. *Front. Endocrinol.* **2019**, *10*, 599. [CrossRef] [PubMed]
91. Cai, T.-Q.; Ren, N.; Jin, L.; Cheng, K.; Kash, S.; Chen, R.; Wright, S.D.; Taggart, A.K.; Waters, M.G. Role of GPR81 in lactate-mediated reduction of adipose lipolysis. *Biochem. Biophys. Res. Commun.* **2008**, *377*, 987–991. [CrossRef] [PubMed]
92. Liu, C.; Wu, J.; Zhu, J.; Kuei, C.; Yu, J.; Shelton, J.; Sutton, S.W.; Li, X.; Yun, S.J.; Mirzadegan, T. Lactate inhibits lipolysis in fat cells through activation of an orphan G-protein-coupled receptor, GPR81. *J. Biol. Chem.* **2009**, *284*, 2811–2822. [CrossRef] [PubMed]
93. Moon, T.W. Glucose intolerance in teleost fish: Fact or fiction? *Comp. Biochem. Physiol. B Biochem. Mol. Biol.* **2001**, *129*, 243–249. [CrossRef]

22/11/466
biology

MDPI

Article

Association Analysis between Genetic Variants of *elovl5a* and *elovl5b* and Poly-Unsaturated Fatty Acids in Common Carp (*Cyprinus carpio*)

Yan Zhang [1,2,3,†], Qing-Song Li [1,†], Yu-Qing Ye [1], Qi Wang [2], Xiao-Qing Sun [2], Ran Zhao [2] and Jiong-Tang Li [2,3,*]

[1] National Demonstration Center for Experimental Fisheries Science Education, Shanghai Ocean University, Shanghai 201306, China; zhangy@cafs.ac.cn (Y.Z.); m190110454@st.shou.edu.cn (Q.-S.L.); yeyuqing95@163.com (Y.-Q.Y.)
[2] Key Laboratory of Aquatic Genomics, Ministry of Agriculture and Rural Affairs, Chinese Academy of Fishery Sciences, Beijing 100141, China; wangqi@cafs.ac.cn (Q.W.); sunxiaoqing@cafs.ac.cn (X.-Q.S.); zhaoran@cafs.ac.cn (R.Z.)
[3] Beijing Key Laboratory of Fishery Biotechnology, Beijing 100141, China
* Correspondence: lijt@cafs.ac.cn; Tel.: +86-010-68691136
† These authors contributed equally to this work.

Simple Summary: PUFAs have an essential impact on human health, but their availability constitutes a critical bottleneck in food production. Although fish is the traditional source of PUFAs, it is limited by the stagnation of fisheries. Many studies aim to increase the PUFA products of fish. Genetic markers are efficient in aquaculture breeding. Fatty acid desaturase 2 (*fads2*) and elongase 5 (*elovl5*) are the rate-limiting enzymes in the synthesis of PUFAs. The allo-tetraploid common carp is able to biosynthesize endogenous PUFAs. However, selective breeding common carp with high PUFA contents was hindered due to a lack of effective molecular markers. For future breeding common carp capable of producing endogenous PUFAs more effectively, we previously identified the polymorphisms in the coding regions of two duplicated *fads2*, *fads2a* and *fads2b*. However, the polymorphisms in the duplicated *elovl5*, *elovl5a* and *elovl5b*, were not detected. This study screened the genetic variants in the coding regions of *elovl5a* and *elovl5b*. Moreover, the joint effects of multiple coding SNPs in *fads2b* and *elovl5b*, two major genes regulating the PUFA biosynthesis, were evidenced with the increased explained percentages of the PUFA contents. These polymorphisms in these two genes were used to evaluate the breeding values of PUFAs. These SNPs would be potential markers for future selection to improve the PUFA contents in common carp.

Citation: Zhang, Y.; Li, Q.-S.; Ye, Y.-Q.; Wang, Q.; Sun, X.-Q.; Zhao, R.; Li, J.-T. Association Analysis between Genetic Variants of *elovl5a* and *elovl5b* and Poly-Unsaturated Fatty Acids in Common Carp (*Cyprinus carpio*). *Biology* **2022**, *11*, 466. https://doi.org/10.3390/biology11030466

Academic Editor: Patricia Pereiro

Received: 26 January 2022
Accepted: 14 March 2022
Published: 18 March 2022

Abstract: The allo-tetraploid common carp, one widely cultured food fish, is able to produce poly-unsaturated fatty acids (PUFAs). The genetic markers on the PUFA contents for breeding was limited. The polymorphisms in *elovl5a* and *elovl5b*, the rate-limiting enzymes in the PUFA biosynthesis, have not been investigated yet. Herein, we identified one coding SNP (cSNP) in *elovl5a* associated with the content of one PUFA and two cSNPs in *elovl5b* with the contents of eight PUFAs. The heterozygous genotypes in these three loci were associated with higher contents than the homozygotes. Together with previously identified two associated cSNPs in *fads2b*, we found the joint effect of these four cSNPs in *fads2b* and *elovl5b* on the PUFA contents with the increased explained percentages of PUFA contents. The genotype combinations of more heterozygotes were associated with higher PUFA contents than the other combinations. Using ten genomic selection programs with all cSNPs in *fads2b* and *elovl5b*, we obtained the high and positive correlations between the phenotypes and the estimated breeding values of eight PUFAs. These results suggested that *elovl5b* might be the major gene corresponding to common carp PUFA contents compared with *elovl5a*. The cSNP combinations in *fads2b* and *elovl5b* and the optimal genomic selection program will be used in the future selection breeding to improve the PUFA contents of common carp.

Keywords: common carp; poly-unsaturated fatty acid; fatty acid elongase; association study; genomic selection

1. Introduction

Poly-unsaturated fatty acids (PUFAs) with at least 18 carbons [1] play physiologically important roles and are essential for human health because they are the major components of complex lipid molecules involved in numerous critical biological processes [2–4]. In general, fish are the main available source of long-chain PUFAs (LC-PUFAs, fatty acids with at least 20 carbons [5]) for human dietary [6]. In diploid fish, fatty acid desaturase 2 (*fads2*) and elongase 5 (*elovl5*) are two rate-limiting enzymes in the PUFA biosynthesis pathway [7,8]. Therefore, the polymorphisms in these two genes are hypothesized to be associated with the PUFA contents. Indeed, the polymorphisms of *fads2* in bovine and fish were reported to be significantly associated with the PUFA content [9–11]. Genome-wide association studies revealed that single nucleotide polymorphisms (SNPs) of *elovl5* were associated with omega-6 (n-6) and omega-3 (n-3) fatty acid (FA) levels in human, sheep, and bovine [12–14]. Identifying the mutants in *fads2* and *elovl5* in fish of economic value would benefit for the future breeding fish having high contents of PUFAs.

The allo-tetraploid common carp is widely cultured in the world and is able to convert dietary 18-Carbon PUFAs to LC-PUFAs including arachidonic acid, eicosapentaenoic acid, and docosahexaenoic acid [7,15]. As an allo-tetraploid fish, it maintains the tetraploidization status and encodes almost twice that the diploid Cyprininae fish. Previously we cloned two *fads2* genes (*fads2a* and *fads2b*) and identified the polymorphisms in these two genes associated with the PUFA contents in common carp [9]. These findings not only supported that there existed two homoeologues of *fads2* but also suggested that the PUFA biosynthesis pathway in common carp was more complex than diploid Cyprininae fish. Although two *elovl5* genes (*elovl5a* and *elovl5b*) in common carp were sequenced [16], the questions of whether there exist the polymorphisms in these two genes and whether these polymorphisms are associated with the PUFA contents have not been studied.

Recent study revealed that two subgenomes in common carp performed balancing of differential expression in response to different conditions, dampening the stimulus impact to the expression of the duplicated genes [17]. As for the PUFA biosynthesis, which of two duplicated *elovl5* genes is the major effect gene is still unknown. Furthermore, how duplicated *fads2* genes and duplicated *elovl5* genes coordinate to regulate the PUFA biosynthesis is less studied.

To answer these questions, in this work, we sequenced the coding regions of common carp *elovl5a* and *elovl5b* and detected the polymorphisms in these two genes. With the association study, we identified the SNPs significantly associated with the PUFA contents and found more associated SNPs in *elovl5b* than *elovl5a*. We further examined the joint effects of the associated SNPs in *fads2b* and *elovl5b* on the PUFA contents. Finally, we obtained the high and positive correlations between the contents and the predicted breeding values of eight PUFAs using the cSNPs in these two genes. These cSNPs would be used as the biomarkers to facilitate the selective breeding of common carp with high PUFA contents.

2. Materials and Methods

2.1. Sampling and Measuring PUFA Contents

We collected the juveniles of three bred strains of common carp, including 'HuangHe' (HHC) strain, 'FuRui' (FRC) strain and 'Jian' (JC) strain in May 2018, described in our previous study [9]. These strains were sampled from different provinces of China and had different morphology traits, which were shown in detail before. We had cultivated these juveniles for one year in one pond at the Chinese Academy of Fishery Sciences (Fangshan, Beijing, China) with the same commercial diet. In May 2019, we randomly selected 124 individuals of FRC, 98 JC fish, and 47 HHC fish. The tissue collection, liver RNA extraction, reverse-transcription of RNA to cDNA, and the muscles PUFA content calculation were described before.

Briefly, FAs were converted into the fatty acid methyl esters (FAMEs), which were further extracted and purified by thin-layer chromatography following the strategy of Li et al. [18]. A total of 25 types of FAs were identified using the 7890A GC System (Agilent Technologies, Wilmington, Delaware, USA) by comparing their GC retention time with the time of the peaks of a Supelco 37 Component FAMEs standard mix (Nu-chek Prep Inc., Elysian, MN, USA). The relative proportion of each among 4 types of 18-Carbon PUFAs and 8 types of LC-PUFAs was calculated as (area of one PUFA/total area of 25 types of FAs) $\times$ 100.

2.2. Sequencing and Genotyping

Based on the reference full-length sequence of *elovl5a* and *elovl5b* in common carp (MK893918.1 and MK893919.2), the gene-specific primers were designed to amplify the complete coding sequence (CDS) regions (Supplementary Table S1). PCR amplification was carried out according to the protocol as described previously [9]. The annealing temperatures were set as 56 °C and 60 °C, respectively. After sequencing the PCR products with the Sanger method, we aligned them to the reference sequences of *elovl5a* and *elovl5b* using Blastn, respectively [9]. Theoretically, one sequence expected to be from *elovl5a* should have a higher identity value to *elovl5a* than *elovl5b* and vice versa. To call cSNPs, the confirmed sequences were aligned to the corresponding reference sequences using the novoSNP software [19]. The SNPs were identified with F-scores $\geq$ 30, and the homozygotes and heterozygotes were auto-detected with this software. If a site in one individual had one sequencing peak, then this site was homozygous. If this site had two peaks, it was heterozygous. If this site in one sample had three or more peaks, it was discarded.

2.3. Genetic Diversity of Common Carp elovl5a and elovl5b

The genetic distances and population structures among three strains were calculated with all retained genotypes in *elovl5a* and *elovl5b* together. The genetic distances and the population structures of all samples were calculated with Tassel 5 [20] and the admixture function [21] of the package LEA [22] in R 4.1.0, respectively. The first two eigenvectors of the PCA result were plotted. The population structures (K value ranging from 2 to 6) were displayed with pophelper v2.3.1 [23].

We grouped three strains into one population and calculated the genetic diversities of the cSNPs, which had a frequency over 0.04. The diversity indicators included the observed heterozygosity (Ho), the expected heterozygosity (He), the minor allele frequency (MAF), and the polymorphism information content (PIC). The former three indicators were measured with the Genepop software 4.7 [24]. The PIC was estimated using PICcalc 0.6 [25]. Using TBtools [26], we classified the effects of cSNPs on the coding sequences into the stop loss, stop gain, non-synonymous substitution, and synonymous substitution. We also calculated the linkage disequilibrium (LD) between any two SNPs in each gene. The LD was measured using the LDheatmap function in R [27] and represented with D′ value. We clustered cSNPs into one haplotype block if the D′ of any two compared cSNPs in this gene was over 0.8.

2.4. Associations of cSNPs in elovl5a and elovl5b with the Contents of 12 PUFAs

To identify the cSNPs in these two genes associated with the PUFA contents, the general linear model (GLM) and the analysis of variance (ANOVA) were used to study the association between each PUFA content and the genotypes, respectively. We ran the GLM model with the parameters of the genetic distance matrix and 100,000 permutations using Tassel 5 [20]. This method was widely applied in the association study between the polymorphisms in candidate genes and the phenotypes [28,29]. To perform ANOVA, we classified all individuals into different groups based on their genotypes in one SNP. The pairwise comparison between any two groups on each PUFA content with ANOVA.

We corrected the ANOVA p values using the false discovery rate (FDR) method for multiple hypothesis testing. One cSNP was deemed to be significantly associated with one PUFA content when it had a p value < 0.05 in the GLM method and an FDR-corrected p value < 0.05 in the ANOVA. The explained percentage of phenotypic variation (PV) of each cSNP was measured using Tassel 5.

2.5. Joint Effects of Significant SNPs in elovl5b and fads2b on the PUFA Contents

Previously, we identified one cSNP in *fads2a* associated with the content of C20:3n-6 PUFA. Another two cSNPs in *fads2b* were significantly associated with the contents of seven PUFAs and six PUFAs, respectively. Herein, we identified three cSNPs associated with the contents of multiple PUFAs. Since *fads2b* and *elovl5b* were two major effect genes on the contents of multiple PUFAs, we hypothesized that the joint analysis of multiple associated SNPs could detect a larger effect than single SNP and identify the optimal genotype combinations associated with higher PUFA contents. Hence, we estimated the joint effects of four significantly associated SNPs (two in *fads2b* and two in *elovl5b*) on the contents of multiple PUFAs. We generated different genotype combinations from all these SNPs. If one genotype combination was observed in at least three individuals, this combination was used in the comparison. For each PUFA, we performed the pairwise comparisons of the PUFA contents among different retained combinations using ANOVA in R software. The explained percentages of PV of the genotype combination to the content of each PUFA was estimated with the function of 'lm' [30] in R.

2.6. Estimating the Breeding Values with the cSNPs in elovl5b and fads2b on the PUFA Contents

Further, we were interested in whether all identified cSNPs in *fads2b* and *elovl5b* would be applied into estimating the breeding values (BVs) of the PUFA contents. The cSNPs with MAFs over 0.03 were used to estimate the BVs with BWGS [31]. This package integrates multiple programs available for the genomic BV prediction, including GBLUP [32], EG-BLUP [33], Ridge regression (RR) [34], LASSO [35], Elastic Net (EN) [36], Bayesian ridge regression (BRR) [37], Bayesian LASSO (BL) [38], Bayes A (BA) [39], Bayes B (BB) [40], and Bayes C (BC) [41]. We used each program to estimate the BV of each PUFA in each validated individual. The maximum proportion of missing value for filtering marker column was set as 0.2 with the minimum allele frequency for filtering markers as 0.03. We performed 20 independent replicates in the cross validation for each program on each PUFA. First, in each replicate of one cross validation, these individuals were separated into the reference group and the validation group, respectively. The randomly sampled 90% of all individuals having the amplified sequences from both *fads2b* and *elovl5b* were treated as the reference group to train the breeding models and the remaining 10% of all individuals as the validation group. For each PUFA, the BV of each individual in the validation group by each program was represented as the predicted content. Further, in each replicate of one cross validation by each program, we calculated the Pearson correlation coefficient value (CV) between the actual contents and the predicted contents of each PUFA across individuals in the validation group. Second, we calculated the mean BV of each PUFA and its standard deviation (SD) of all individuals predicted by each program during 20 replicates. Third, to estimate the BV accuracy of each program, we calculated the mean CV and its SD of each program for each PUFA during 20 replicates. The mean squared error of prediction (MSEP) and corresponding standard deviation (SD-MSEP) were also computed.

3. Results

3.1. Genetic Diversities of Common Carp elovl5a and elovl5b

We confirmed that the entire CDS regions of *elovl5a* and *elovl5b* were successfully sequenced in 204 and 269 individuals, respectively. Both amplified lengths of *elovl5a* and *elovl5b* cDNA sequences were 876 bp, corresponding to seven exons. The base contents of A, G, C, and T in the *elovl5a* CDS were 28.9%, 24.5%, 21.4%, and 25.2%, respectively. For *elovl5b*, the base contents of A, G, C, and T in the CDS were 25.9%, 25.4%, 23.0%, and 25.4%, respectively. These two reference mRNAs were highly identical with a similarity of 95%, higher than that of common carp *fads2a* and *fads2b* (89.86%) [9].

Ten cSNPs including one non-synonymous cSNP (ns-cSNP) and nine synonymous cSNPs (s-cSNP) were identified in six exons of *elovl5a* (Table 1 and Supplementary Table S2). These cSNPs had 25 genotypes where five SNPs had three genotypes per locus. The MAFs of the cSNPs ranged from 0.0147 to 0.3769 with seven SNPs having MAF less than 0.1. The Ho values of ten cSNPs were from 0.0294 to 0.4724 where the Ho values of five cSNPs were smaller than 0.1. The He values of ten cSNPs were from 0.029 to 0.4697 and six cSNPs had He values smaller than 0.1. The PIC values of these SNPs had a range from 0.0286 to 0.3594. All the MAF, Ho and He values of five cSNPs, half of SNPs in *elovl5a*, were lower than 0.1. These data suggested their low polymorphic levels.

Table 1. Genetic diversities of SNPs in the coding sequences of *elovl5a* and *elovl5b*.

Locus	Gene	Position */Exon	Ho	He	PIC	MAF	Genotype	Amino Acid Change
E5a.29	*elovl5a*	29/1	0.1225	0.1649	0.1513	0.0907	CC CT TT	T-I
E5a.87	*elovl5a*	87/2	0.0493	0.0480	0.0469	0.0246	AG GG	L-L
E5a.180	*elovl5a*	180/2	0.1029	0.0976	0.0929	0.0515	AC CC	S-S
E5a.351	*elovl5a*	351/4	0.0735	0.0708	0.0683	0.0368	CC CT	Y-Y
E5a.429	*elovl5a*	429/4	0.0588	0.0843	0.0808	0.0441	CC CT TT	H-H
E5a.552	*elovl5a*	552/5	0.4608	0.4538	0.3508	0.348	CC CT TT	Y-Y
E5a.582	*elovl5a*	582/5	0.1569	0.2001	0.1801	0.1127	AA AG GG	P-P
E5a.651	*elovl5a*	651/6	0.0294	0.029	0.0286	0.0147	AA GA	T-T
E5a.798	*elovl5a*	798/7	0.0837	0.0802	0.077	0.0419	AG GG	S-S
E5a.810	*elovl5a*	810/7	0.4724	0.4697	0.3594	0.3769	AA AT TT	I-I
E5b.172	*elovl5b*	172/2	0.1822	0.1656	0.1519	0.0911	CC CT	P-S
E5b.174	*elovl5b*	174/2	0.1413	0.1313	0.1227	0.0706	CA CC	P-S
E5b.195	*elovl5b*	195/2	0.2156	0.1924	0.1739	0.1078	AA AC	L-L
E5b.333	*elovl5b*	333/4	0.9071	0.4988	0.3744	0.4758	CC CT TT	N-N
E5b.424	*elovl5b*	424/4	0.4387	0.3587	0.2944	0.2342	CC CT TT	L-L

Table 1. *Cont.*

Locus	Gene	Position */Exon	Ho	He	PIC	MAF	Genotype	Amino Acid Change
E5b.711	*elovl5b*	711/6	0.5576	0.435	0.3404	0.3197	CC TC TT	T-T
E5b.782	*elovl5b*	782/7	0.145	0.1345	0.1254	0.0725	GA GG	R-Q
E5b.813	*elovl5b*	813/7	0.2193	0.2181	0.1943	0.1245	CC CT TT	N-N

* The base position of SNP related to the start codon in the reference mRNA.

Eight cSNPs including three ns-cSNPs and five synonymous ones were identified in five exons of *elovl5b* (Table 1 and Supplementary Table S3). We found 20 genotypes in these eight cSNPs, four of which had three genotypes per locus. The Ho, He, PIC, and MAF values of these SNPs were in the ranges of 0.1413~0.9071, 0.1313~0.4988, 0.1227~0.3744, and 0.0706~0.4758, respectively. Comparing the diversities of these two genes revealed that *elovl5b* had higher polymorphic levels than *elovl5a*. Although three cSNPs in *elovl5b* were observed with MAF smaller than 0.1, all SNPs had PIC, Ho, and He higher than 0.1, supporting the higher polymorphisms in *elovl5b*.

These three common carp strains were grouped together based on all genotypes in *elovl5a* and *elovl5b* in the PCA analysis (Supplementary Figure S1A). The first two eigenvectors accounted for 41.61% and 10.76% of the total genetic variances. The population structures, plotted with different K values using all genotypes in *elovl5a* and *elovl5b*, also suggested that three strains had similar genetic components (Supplementary Figure S1B). These two data revealed that there were no strain-specific SNPs in *elovl5a*/*elovl5b*. Hence, these three strains were grouped into one population in the association study.

3.2. cSNPs in elovl5a and elovl5b Associated with the PUFA Contents

Three cSNPs, one in *elovl5a* (E5a.87) and two in *elovl5b* (E5b.172 and E5b.782), were identified to be associated with the contents of multiple PUFAs. In these three loci, we only observed the homozygous wild-types and the heterozygous genotypes while the homozygous mutations were not found. The synonymous SNP in *elovl5a* was significantly associated with the C20:5n-3 PUFA content (Table 2 and Figure 1A). The heterozygote of this SNP corresponded to higher content of C20:5n-3 than the homozygote with a fold change of 2.19.

Two ns-cSNPs in *elovl5b*, E5b.172 and E5b.782, were significantly associated with several PUFA contents by using both GLM and ANOVA (Table 2). The ns-cSNP of E5b.172 led to the mutation from proline to serine. This SNP was significantly associated with nine PUFAs contents, including four n-3 PUFAs (C18:3n-3, C20:3n-3, C20:4n-3, and C22:5n-3) and five n-6 PUFAs (C18:2n-6, C20:3n-6, C20:4n-6, C22:4n-6, and C22:5n-6) (Figure 1B–J). The heterozygote had higher contents of the above nine PUFAs than the homozygote. The fold changes of the mean contents of the associated PUFAs were from 1.43 to 2.13 (Table 2). The ns-cSNP of E5b.782 led to the mutation from arginine to glutamine. It was associated with only four PUFAs contents, including C20:3n-3, C20:3n-6, C22:4n-6, and C22:5n-6 (Figure 1K–N). Likewise, the heterozygote of this SNP had higher contents of the above four PUFAs than the homozygote. The fold changes of the mean associated PUFA contents were from 1.43 to 1.66 (Table 2). The contents of four PUFAs (C20:3n-3, C20:3n-6, C22:4n-6, and C22:5n-6) were associated with both cSNPs.

Two haplotype blocks were detected in each of the CDS regions of *elovl5a* and *elovl5b*, respectively (Supplementary Figure S2). E5b.172 and E5b.782 were distributed in two different blocks, respectively. The contents of four PUFAs (C20:3n-3, C20:3n-6, C22:4n-6, and C22:5n-6) were associated with both cSNPs in *elovl5b*. These data suggested these two cSNPs might coordinate to regulate the biosynthesis of these four PUFAs. However, the SNP of E5b.172 was associated with the contents of the other five PUFAs, suggesting that this cSNP solely affected these PUFAs contents. Since the cSNP of E5b.172 had more associated PUFAs and explained more percentages of PV than E5b.782 (except C20:3n-6), suggesting that the former might be the major SNP.

Table 2. Association analysis of the PUFA contents in common carp.

Trait	Marker	Perm_*p* Value of GLM	MarkerR^2	FDR of ANOVA	MM	Mm
C18:3n-3	E5b.172	4.44×10^{-3}	6.90	1.29×10^{-9}	0.84 ± 0.59 *	1.79 ± 1.46
C20:3n-3	E5b.172	2.01×10^{-3}	3.96	5.61×10^{-11}	0.24 ± 0.11	0.41 ± 0.2
	E5b.782	3.54×10^{-2}	2.15	2.69×10^{-4}	0.26 ± 0.13	0.38 ± 0.19
C20:4n-3	E5b.172	1.54×10^{-3}	7.12	1.07×10^{-11}	0.28 ± 0.11	0.46 ± 0.22
C20:5n-3	E5a.87	3.12×10^{-3}	4.13	1.89×10^{-7}	3.45 ± 1.46	7.57 ± 6.88
C22:5n-3	E5b.172	7.71×10^{-3}	3.70	1.58×10^{-8}	1.58 ± 0.82	2.61 ± 1.36
C18:2n-6	E5b.172	3.60×10^{-4}	2.86	1.73×10^{-15}	23.21 ± 5.75	37.02 ± 17
C20:3n-6	E5b.172	6.70×10^{-4}	3.37	1.71×10^{-13}	2.63 ± 1.12	4.82 ± 2.71
	E5b.782	3.51×10^{-2}	4.48	2.50×10^{-4}	2.83 ± 1.51	4.25 ± 2.53
C20:4n-6	E5b.172	3.67×10^{-2}	2.10	4.06×10^{-4}	8.11 ± 3.79	11.61 ± 8.04
C22:4n-6	E5b.172	7.23×10^{-3}	3.44	1.21×10^{-8}	0.65 ± 0.28	1.05 ± 0.62
	E5b.782	4.51×10^{-2}	1.83	1.57×10^{-3}	0.68 ± 0.33	0.97 ± 0.62
C22:5n-6	E5b.172	2.30×10^{-3}	4.07	1.22×10^{-10}	2.15 ± 1.2	4.23 ± 3.02
	E5b.782	3.46×10^{-2}	1.83	2.26×10^{-4}	2.32 ± 1.51	3.84 ± 3.06

Perm_*p* value of GLM: *p* value corrected with 100,000 permutations test. FDR of ANNOVA: the corrected *p* value of ANOVA using the FDR method. Marker R^2: the explained percentage of PVs by markers. M: major allele; m: minor allele. * The content of each PUFA is displayed as the mean ± SD value.

3.3. Joint Effect of *elovl5b* and *fads2b* on the PUFA Contents

The cSNPs in *elovl5b* were associated with more PUFA contents than that in *elovl5a*, suggesting that *elovl5b* might be the major effect gene regulating the PUFA biosynthesis in common carp. Previously, we found that *fads2b* might be another major gene responding to common carp PUFA contents and that two cSNPs in *fads2b* were associated with the contents of multiple PUFAs [9].

Interestingly, the cSNPs in *fads2b* were associated with four n-3 PUFAs (C18:3n-3, C20:3n-3, C20:4n-3, and C22:5n-3) and four n-6 PUFAs (C18:2n-6, C20:3n-6, C22:4n-6, and C22:5n-6). The cSNPs in *elovl5b* had significant association with four n-3 PUFAs (C18:3n-3, C20:3n-3, C20:4n-3, and C22:5n-3) and five n-6 PUFAs (C18:2n-6, C20:3n-6, C20:4n-6, C22:4n-6, and C22:5n-6). The cSNPs in these two genes were involved in the contents of eight common PUFAs, suggesting the coordinative regulation of the PUFA biosynthesis through the mutations in *fads2b* and *elovl5b*. Hence, we estimated the joint effects of four significantly associated cSNPs in two major effect genes (two in *fads2b*, fads2b.751 and fads2b.1197; and two in *elovl5b*, E5b.172 and E5b.782) on the contents of 12 PUFAs. Theoretically, we should observe 16 genotype combinations with four cSNPs. In practice, among 223 individuals having the genotypes of both *fad2b* and *elovl5b*, ten types were detected in at least three individuals (Figure 2 and Supplementary Table S4). The most frequent combination (H1), where all four loci were the homozygotes of the reference bases, was observed in 162 individuals. The other combinations were observed in much fewer individuals. We did not detect the combination including four homozygotes of the mutation bases.

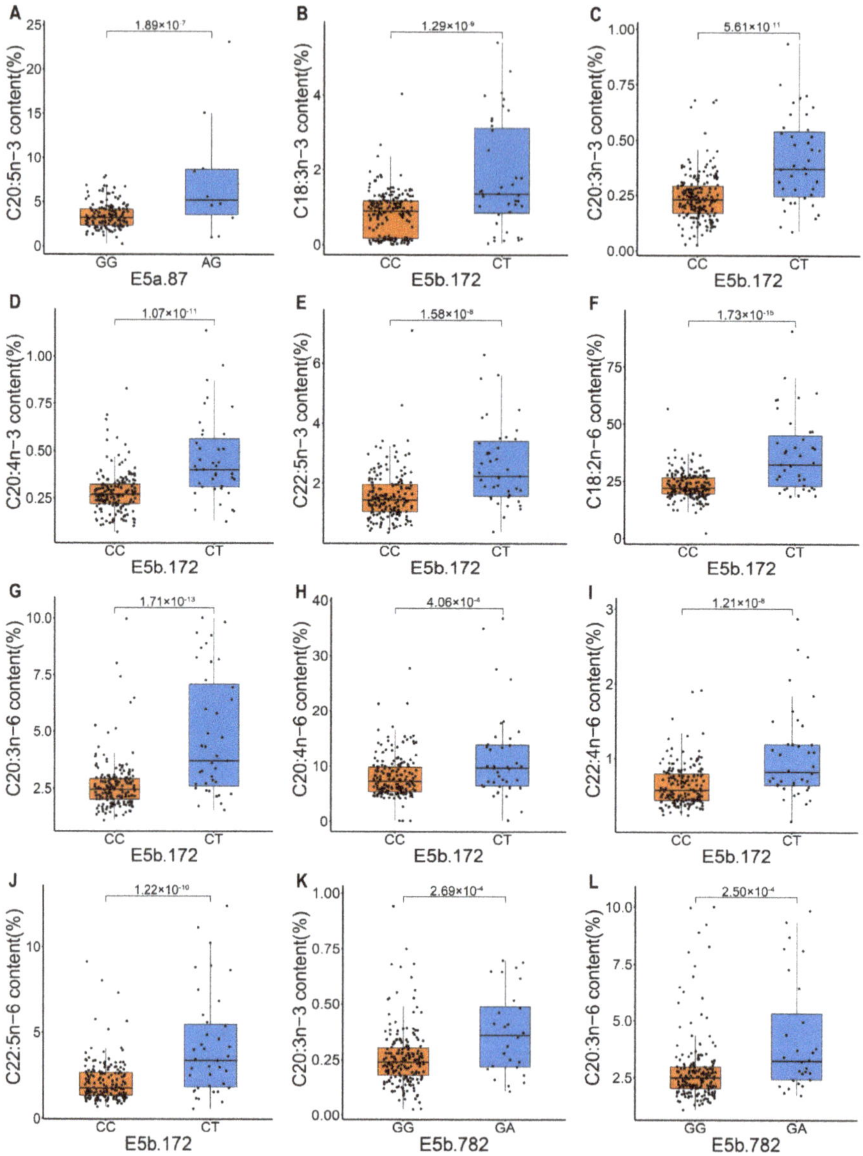

Figure 1. *Cont.*

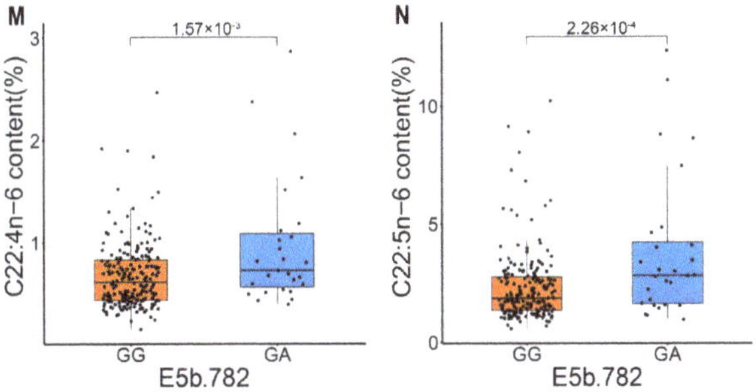

Figure 1. Box plots showing significant PUFA content differences between two genotypes at three loci, E5a.87, E5b.172, and E5b.782. The corrected *p* values were calculated with ANOVA. (**A**): PUFA content values of genotype GG and AG atE5a.87 locus; (**B–J**): PUFA content values of genotype CC and CT at E5b.172 locus; (**K–N**): PUFA content values of genotype GG and GA at E5b.782 locus.

The multi-cSNPs combinations greatly improved the explained percentages of PV compared with the single SNP (Table 3). For three PUFAs (C20:3n-3, C20:3n-6, and C22:5n-6), the contents had four associated SNPs in both two genes. For C20:3n-3, the explained percentages of PV by each SNP ranged between 2.15~13% while the percentage increased to 32.73% with the combination of four cSNPs. For C20:3n-6 and C22:5n-6, the genotype combination improved the explained percentages to 37.59% and 28.19%, respectively. For three PUFAs (C20:4n-3, C22:5n-3, and C22:4n-6), the contents had three associated SNPs in both these two genes. The genotype combinations also increased the explained percentages to 33.22%, 22.63%, and 21.26%, respectively. Although there were only two associated cSNPs in *fads2b* and *elovl5b* in C18:3n-3 and C18:2n-6, the SNP combination explained 33.6% and 54.95% of PV, respectively, much higher than single SNP. For the C20:4n-6 PUFA content, only E5b.172 was associated with 2.1% of PV whereas the combination of four cSNPs had 13.06% of PV. Intriguingly, for the remaining three PUFAs (C18:4n-3, C18:3n-6, and C20:5n-3) having no associated SNPs in these two genes, the genotype combination also contributed to the percentages of PV with a range between 4.24%~13.06%. The increased explained percentages of PV indicated the joint effects of these four cSNPs on the contents of 12 PUFAs.

Table 3. The explained percentage of PV for each PUFA by each genotype combination from four SNP loci in *fads2b* and *elovl5b* across 223 individuals.

PUFA	fads2b.751	fads2b.1197	E5b.172	E5b.782	Genotype Combination
C18:3n-3	5	NA	6.9	NA	33.6
C20:3n-3	13	4.3	3.96	2.15	32.73
C20:4n-3	11	5	7.12	NA	33.22
C22:5n-3	11	5	3.7	NA	22.63
C18:2n-6	NA	4	2.86	NA	54.95
C20:3n-6	11	4	3.37	4.48	37.59
C22:4n-6	7	NA	3.44	1.83	21.26
C22:5n-6	9	3	4.07	1.83	28.19
C18:4n-3	NA	NA	NA	NA	4.24
C18:3n-6	NA	NA	NA	NA	5.02
C20:5n-3	NA	NA	NA	NA	8.02
C20:4n-6	NA	NA	2.1	NA	13.06

The explained percentage of PV is represented as the Marker R^2. NA means that this SNP is not associated with the PUFA content.

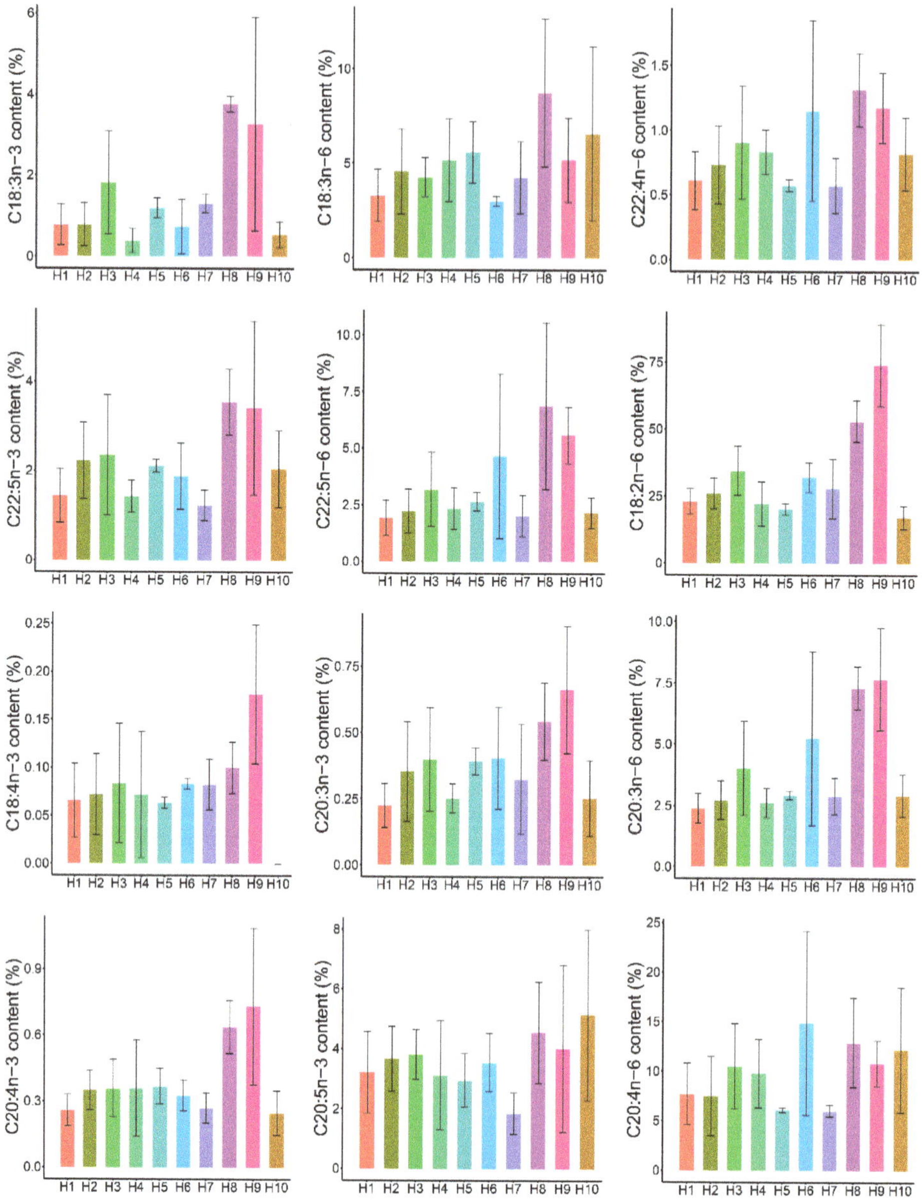

Figure 2. Bar plots showing the mean contents of PUFAs among ten genotype combinations. The genotype combination information and corresponding PUFA contents were shown in Supplementary Table S4. The bar heights represent the mean contents of PUFAs except the outliers. The whiskers are the standard deviations.

The joint effect analysis also clearly showed which genotype combination had higher contents of PUFAs than the other combinations. For eight PUFAs having the associated cSNPs in both *fads2b* and *elovl5b*, the genotype combination with higher heterozygous levels corresponded to higher contents of PUFAs. The combination (H8) including four heterozygotes corresponded to the highest contents of five PUFAs including C18:3n-3, C18:3n-6, C22:4n-6, C22:5n-3, and C22:5n-6 (Figure 2 and Supplementary Table S4). The genotype combination corresponded to higher contents of five PUFAs than the homozygote combination (H1) with fold changes ranging between 2.16 and 4.83. Another combination (H9) including three heterozygotes and one homozygote corresponded to the highest contents of five PUFAs including C18:2n-6, C18:4n-3, C20:3n-3, C20:3n-6, and C20:4n-3 (Figure 2 and Supplementary Table S4). This combination corresponded to higher contents of these five PUFAs than the homozygote combination (H1) with fold changes between 2.57 and 3.22. Taken together, these two genotype combinations could be effective markers to select common carp of high PUFAs contents.

3.4. The cSNPs in fads2b and elovl5b to Predict the Breeding Values

Because the cSNPs in *fads2b* and *elovl5b* had joint effects on the PUFA contents, we tried to answer whether these cSNPs had the potentials to be used for selection breeding. A total of 35 cSNPs were detected in *fads2b* and *elovl5b*. The MAFs of 25 cSNPs were over 0.03 and hence were used to estimate the BVs. These 25 cSNPs existed in 223 individuals. Ten programs in BWGS tool provided the prediction results (Supplementary Table S5).

Since there was only one cSNP or no cSNP in these two genes associated with the contents of C18:3n-6, C18:4n-3, C20:5n-3, and C20:4n-6, the highest mean CV between the actual contents and the BVs were only 0.26. Even in C18:4n-3, the mean CVs of nine programs except BRR were negative (Figure 3). However, for the remaining eight PUFAs having at least two cSNPs in these two genes, the BB program had the best mean CVs in C18:2n-6 (0.5824), C20:3n-3 (0.52605), C20:3n-6 (0.53285), and C22:5n-6 (0.4558). For C18:3n-3 and C 22:4n-6, the BB program was the second-best tool with slight lower CVs (0.4245 and 0.33805) than the top best tool, BC (0.43135 and 0.3413). Likewise, for 22:5n-3, the performance of the BB program also ranked the second with a slight lower CV (0.38595) than the best tool, LASSO (0.39055). For C20:4n-3, the CV of BB program (0.50045) was much close to that of the top best (LASSO, 0.50395) and the second-best tool (BC, 0.50605). These results indicated that the cSNPs in *fads2b* and *elovl5b* could be used to predict the selection breeding potential of high contents of eight PUFAs and that the BB program would be the optimal tool for breeding if taken all eight PUFAs into consideration.

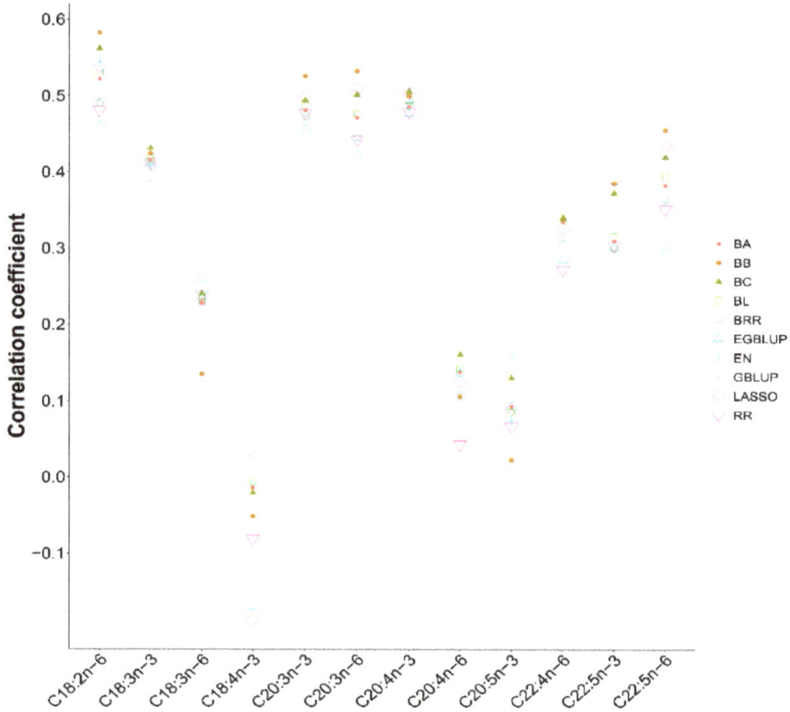

Figure 3. Distribution of the mean Pearson correlation CVs for each PUFA using ten methods with 25 cSNPs in *fads2b* and *elovl5b*. The CV was calculated between the actual contents and the predicted contents of each PUFA across individuals in the validation group.

4. Discussion

The availability of PUFAs has become important in food production. Although fish is the traditional source of PUFAs, fish PUFA product is limited. Many studies aim to increase fish PUFA contents. Although feeding different ingredients could improve the fish PUFA content [42], this strategy is feed-consuming. This goal could be alternatively achieved by breeding fish capable of biosynthesizing endogenous PUFAs more effectively. Genetic markers were evidenced to be useful in selective breeding of aquaculture species. To date, few effective molecular markers associated with high PUFA contents were identified, hindering the selective breeding of common carp with high PUFA contents. Therefore, it would be necessary to develop the genetic markers applicable to breeding.

The species-specific whole genome duplication and parallel subgenome evolution in common carp generated two homoeologues of *fads2* (*fads2a* and *fads2b*) and two homoeologues of *elovl5* (*elovl5a* and *elovl5b*). We explored that *fads2b* was the major effect gene associated with the common carp PUFA contents compared with *fads2a* [9]. However, the question which gene in two duplicated *elovl5* genes makes more contributions to the PUFA biosynthesis has not been answered yet. *Elovl5b* was associated with more PUFAs and had higher explained percentage of PVs than *elovl5a*, suggesting that the former might be the major effect gene to improve the PUFA contents. More functional studies including yeast heterologous expression system and expression patterns of these two genes are required for further validation of their effects.

In our current study and previous study [9], three strains of common carp showed no significant inter-strain genetic differences in *fads2a*/*fads2b*/*elovl5a*/*elovl5b*, demonstrated by both PCA and population structure analysis. The individuals from one strain were grouped together with the samples from the other strains, suggesting that they might be sampled from multiple families. The grouping also increased the genetic diversities of the studied population and improved the statistical power of the association analysis

We observed one synonymous cSNP in *elovl5a* associated with one PUFA content. One possible reason is that this synonymous cSNP might lead to the changes in DNA methylation and gene expression of *elovl5a*. In humans, the polymorphisms in the FADS region modify the epigenetic methylation, which further regulate the fatty acid metabolism [43] and contribute to the non-alcoholic fatty liver disease [44]. Whether this cSNP possibly modifies the *elovl5a* DNA methylation needs further research. Two ns-cSNPs in *elovl5b* were associated with several PUFA contents. The enzymes activity would be affected by either amino acid mutations in the active sites [45] or distal mutations site away from the active site [46]. Thus, we speculate that E5b.172 and E5b.782 lead to amino acid change and further possibly changed the elongase activities.

In our previous study, the heterozygote advantage on the PUFA contents was observed in the polymorphisms of *fads2b* [9]. Herein, we confirmed the heterozygote advantages in the polymorphisms in *elovl5a* and *elovl5b*. We did not find the homozygotes of the minor alleles in both *elovl5a* and *elovl5b*. The heterozygotes of three identified cSNPs corresponded to higher PUFA contents than the homozygotes of the reference bases, suggesting that the heterozygosity of *elovl5a* and *elovl5b* might increase their elongase activities. We also revealed the heterozygote advantage in the genotype combination from *fads2b* and *elovl5b*. The combinations of H8 and H9 having four or three heterozygotes had higher contents of the PUFAs than the homozygote combination.

How duplicated *fads2* and *elovl5* genes coordinate to regulate the PUFA biosynthesis is less studied. Our studies showed that *fads2b* and *elovl5b* were two major effect genes on the PUFA contents compared with *fads2a* and *elovl5a*, respectively. The cSNPs in *fads2b* and *elovl5b* were associated with the contents of eight common PUFAs. However, in six PUFAs, fads2b.751 made higher explained percentages of PVs than the cSNPs in *elovl5b* (Table 3), suggesting that improving the fatty acid desaturase activity would be more efficient to increase the PUFA contents than the elongase activity.

None has studied the joint effects of multiple SNPs on the PUFA contents. The explained percentages of PVs by four cSNPs were higher than the sum percentage of each SNP, indicating the additive effects by these cSNPs on the PUFA contents. The coordination to improve the PUFA contents required the simultaneous mutations in these two genes. We also used the cSNPs in these two genes to predict the BVs of the PUFA contents. The best prediction accuracies of the contents of eight PUFAs ranged between 0.3413 and 0.5824. These values were in the BV reliability range from 0.04 to 0.72 for economic traits in sheep, pig, and cattle [47–49].

We comprehensively investigated the polymorphisms in *fads2a*, *fads2b*, *elovl5a*, and *elovl5b*. Although *fads2* and *elovl5* were two rate-limiting enzymes in the PUFA biosynthesis [18], few cSNPs in *fads2a* and *elovl5a* were identified and four cSNPs in *fads2b* and *elovl5b* explained 21.26~54.95% of PVs for the contents of eight PUFAs, suggesting that there might exist other regulatory elements or genes involved in the common carp PUFA biosynthesis. The SNPs in the promoter region of *elovl5* significantly related to fatty acid content in many animals [13,50–54]. Moreover, it was reported that *elovl4*, *elovl6*, and other genes also participated in the PUFA biosynthesis [51,54,55]. Thus, it is necessary to scan the polymorphisms in the promoter of *elovl5a* and *elovl5b* and the genomic regions of the related genes and then identify their associations with the PUFA contents in the future.

5. Conclusions

We identified the polymorphisms in the CDSs of two duplicated common carp *elovl5*, *elovl5a* and *elovl5b*. The association study identified three cSNPs in these two genes significantly related to the PUFA contents in common carp. *Elovl5b* might be the major gene regulating common carp n-3 and n-6 PUFA biosynthesis, and two ns-cSNPs in this gene might be the main effect SNPs. The joint effects of four cSNPs in *fads2b* and *elovl5b* improved the explained percentages of PVs of the PUFA contents. The individuals having more heterozygotes of four cSNPs had higher PUFA contents than the ones having the homozygotes, suggesting that the former would be used as the parents for selective breeding of offspring having higher PUFA contents. The cSNPs in these two genes could be applied to estimating the breeding values of the PUFA contents with the optimal tool of the BB program. In sum, our results highlight the importance of the polymorphisms in *elovl5a* and *elovl5b*, other critical factors in the PUFA biosynthesis in common carp. These cSNPs would be useful markers for selection to improve the PUFA contents in common carp.

Supplementary Materials: The following supporting information can be downloaded at: https://www.mdpi.com/article/10.3390/biology11030466/s1. Figure S1: Genetic relationship among three strains; Figure S2: Haplotype blocks in the coding regions of *elovl5a* and *elovl5b*; Table S1: The primers used for amplification of the coding sequence of *elovl5a* and *elovl5b*; Table S2: *Elovl5a* genotypes; Table S3: *Elovl5b* genotypes; Table S4: Genotype combination of two SNPs in *elovl5b* and two SNPs in *fads2b* and the corresponding PUFA contents for joint analysis; Table S5: Pearson correlation between the breeding values and the PUFAs contents based on all cSNPs in *fads2b* and *elovl5b*.

Author Contributions: Conceptualization, J.-T.L.; methodology, Y.Z., J.-T.L. and Y.-Q.Y.; software, Q.-S.L.; validation, Y.Z.; formal analysis, Q.-S.L. and Q.W.; investigation, Y.Z. and J.-T.L.; resources, J.-T.L., Y.-Q.Y. and X.-Q.S.; data curation, R.Z.; original draft preparation, Y.Z.; review and editing, J.-T.L.; visualization, Q.-S.L.; supervision, J.-T.L.; project administration, Y.Z. and J.-T.L.; funding acquisition, Y.Z. and J.-T.L. All authors have read and agreed to the published version of the manuscript.

Funding: This research was funded by the Beijing Municipal Natural Science Foundation (grant number 6212033), the National Key Research and Development Program (grant numbers 2018YFD0900102), the National Natural Science Foundation of China (grant number 31872557), and the Special Scientific Research Funds for Central Non-profit Institutes, Chinese Academy of Fishery Sciences (grant numbers 2020XT0103, 2021A005, and 2020TD24).

Institutional Review Board Statement: The study was conducted according to the guidelines of the Declaration of China and approved by the Animal Care and Use Committee of the Chinese Academy of Fishery Sciences (protocol code ACUC-CAFS-20191202 and date of approval is 27 December 2018).

Informed Consent Statement: Not applicable.

Data Availability Statement: The data presented in this study are available in supplementary material.

Conflicts of Interest: The authors declare no conflict of interest.

References

1. Bentsen, H. Dietary polyunsaturated fatty acids, brain function and mental health. *Microb. Ecol. Health Dis.* **2017**, *28*, 1281916. [CrossRef]
2. Hoppenbrouwers, T.; Cvejić Hogervorst, J.H.; Garssen, J.; Wichers, H.J.; Willemsen, L.E.M. Long Chain Polyunsaturated Fatty Acids (LCPUFAs) in the Prevention of Food Allergy. *Front. Immunol.* **2019**, *10*, 1118. [CrossRef] [PubMed]
3. Soo, H.J.; Sam, K.K.; Chong, J.; Lau, N.S.; Ting, S.Y.; Kuah, M.K.; Kwang, S.Y.; Ranjani, M.; Shu-Chien, A.C. Functional characterisation of fatty acyl desaturase, Fads2, and elongase, Elovl5, in the Boddart's goggle-eyed goby *Boleophthalmus boddarti* (Gobiidae) suggests an incapacity for long-chain polyunsaturated fatty acid biosynthesis. *J. Fish Biol.* **2020**, *97*, 83–99. [CrossRef] [PubMed]
4. Xie, D.; Chen, C.; Dong, Y.; You, C.; Wang, S.; Monroig, O.; Tocher, D.R.; Li, Y. Regulation of long-chain polyunsaturated fatty acid biosynthesis in teleost fish. *Prog. Lipid Res.* **2021**, *82*, 101095. [CrossRef] [PubMed]
5. Wijendran, V.; Brenna, J.T.; Wang, D.H.; Zhu, W.; Meng, D.; Ganguli, K.; Kothapalli, K.S.D.; Requena, P.; Innis, S.; Walker, W.A. Long-chain polyunsaturated fatty acids attenuate the IL-1β-induced proinflammatory response in human fetal intestinal epithelial cells. *Pediatr. Res.* **2015**, *78*, 626–633. [CrossRef] [PubMed]
6. Guo, F.; Bunn, S.E.; Brett, M.T.; Kainz, M.J. Polyunsaturated fatty acids in stream food webs—High dissimilarity among producers and consumers. *Freshw. Biol.* **2017**, *62*, 1325–1334. [CrossRef]
7. Castro, L.F.C.; Tocher, D.R.; Monroig, O. Long-chain polyunsaturated fatty acid biosynthesis in chordates: Insights into the evolution of Fads and Elovl gene repertoire. *Prog. Lipid Res.* **2016**, *62*, 25–40. [CrossRef] [PubMed]
8. Monroig, O.; Tocher, D.R.; Navarro, J.C. Biosynthesis of polyunsaturated fatty acids in marine invertebrates: Recent advances in molecular mechanisms. *Mar. Drugs* **2013**, *11*, 3998–4018. [CrossRef] [PubMed]

9. Zhang, Y.; Sun, X.Q.; Ye, Y.Q.; Wang, Q.; Li, Q.S.; Zhao, R.; Wang, H.W.; Li, J.T. Association between the Polymorphisms of fads2a and fads2b and Poly-Unsaturated Fatty Acids in Common Carp (*Cyprinus carpio*). *Animals* **2021**, *11*, 1780. [CrossRef] [PubMed]
10. Renaville, B.; Tulli, F.; Bruno, M.; Tibaldi, E.; Messina, M. Fatty acid desaturase 2 (FADS2) insertion/deletion polymorphism impact on muscle fatty acid profile in European grayling (*Thymallus thymallus*). *Br. J. Nutr.* **2013**, *110*, 1559–1564. [CrossRef]
11. Proskura, W.S.; Liput, M.; Zaborski, D.; Sobek, Z.; Yu, Y.H.; Cheng, Y.H.; Dybus, A. The effect of polymorphism in the FADS2 gene on the fatty acid composition of bovine milk. *Arch. Anim. Breed.* **2019**, *62*, 547–555. [CrossRef] [PubMed]
12. Li, X.; Gan, Z.W.; Ding, Z.; Wu, Y.X.; Chen, X.Y.; Tian, H.M.; Liu, G.L.; Yang, Y.T.; Xie, L. Genetic Variants in the ELOVL5 but not ELOVL2 Gene Associated with Polyunsaturated Fatty Acids in Han Chinese Breast Milk. *Biomed. Environ. Sci.* **2017**, *30*, 64–67. [CrossRef] [PubMed]
13. Matsumoto, H.; Shimizu, Y.; Tanaka, A.; Nogi, T.; Tabuchi, I.; Oyama, K.; Taniguchi, M.; Mannen, H.; Sasazaki, S. The SNP in the promoter region of the bovine ELOVL5 gene influences economic traits including subcutaneous fat thickness. *Mol. Biol. Rep.* **2013**, *40*, 3231–3237. [CrossRef] [PubMed]
14. Zhao, L.; Li, F.; Liu, T.; Yuan, L.; Zhang, X.; Zhang, D.; Li, X.; Zhang, Y.; Zhao, Y.; Song, Q.; et al. Ovine ELOVL5 and FASN genes polymorphisms and their correlations with sheep tail fat deposition. *Gene* **2022**, *807*, 145954. [CrossRef]
15. Monroig, Ó.; Tocher, D.R.; Hontoria, F.; Navarro, J.C. Functional characterisation of a Fads2 fatty acyl desaturase with Δ6/Δ8 activity and an Elovl5 with C16, C18 and C20 elongase activity in the anadromous teleost meagre (*Argyrosomus regius*). *Aquaculture* **2013**, *412–413*, 14–22. [CrossRef]
16. Ren, H.T.; Huang, Y.; Tang, Y.K.; Yu, J.H.; Xu, P. Two Elovl5-like elongase genes in Cyprinus carpio var. Jian: Gene characterization, mRNA expression, and nutritional regulation. *Mol. Biol.* **2015**, *49*, 592–600. [CrossRef]
17. Li, J.T.; Wang, Q.; Huang Yang, M.D.; Li, Q.S.; Cui, M.S.; Dong, Z.J.; Wang, H.W.; Yu, J.H.; Zhao, Y.J.; Yang, C.R.; et al. Parallel subgenome structure and divergent expression evolution of allo-tetraploid common carp and goldfish. *Nat. Genet.* **2021**, *53*, 1493–1503. [CrossRef]
18. Li, Y.; Monroig, O.; Zhang, L.; Wang, S.; Zheng, X.; Dick, J.R.; You, C.; Tocher, D.R. Vertebrate fatty acyl desaturase with Delta4 activity. *Proc. Natl. Acad. Sci. USA* **2010**, *107*, 16840–16845. [CrossRef]
19. Weckx, S.; Del-Favero, J.; Rademakers, R.; Claes, L.; Cruts, M.; De Jonghe, P.; Van Broeckhoven, C.; De Rijk, P. novoSNP, a novel computational tool for sequence variation discovery. *Genome Res.* **2005**, *15*, 436–442. [CrossRef] [PubMed]
20. Bradbury, P.; Zhang, Z.; Kroon, D.; Casstevens, T.; Ramdoss, Y.; Buckler, E. TASSEL: Software for Association Mapping of Complex Traits in Diverse Samples. *Bioinformatics* **2007**, *23*, 2633–2635. [CrossRef] [PubMed]
21. Alexander, D.H.; Novembre, J.; Lange, K. Fast model-based estimation of ancestry in unrelated individuals. *Genome Res.* **2009**, *19*, 1655–1664. [CrossRef] [PubMed]
22. Frichot, E.; François, O.; O'Meara, B. LEA: An R package for landscape and ecological association studies. *Methods Ecol. Evol.* **2015**, *6*, 925–929. [CrossRef]
23. Francis, R.M. pophelper: An R package and web app to analyse and visualize population structure. *Mol. Ecol. Resour.* **2017**, *17*, 27–32. [CrossRef]
24. Dereeper, A.; Nicolas, S.; Le Cunff, L.; Bacilieri, R.; Doligez, A.; Peros, J.P.; Ruiz, M.; This, P. SNiPlay: A web-based tool for detection, management and analysis of SNPs. Application to grapevine diversity projects. *BMC Bioinform.* **2011**, *12*, 134. [CrossRef] [PubMed]
25. Nagy, S.; Poczai, P.; Cernák, I.; Gorji, A.M.; Hegedűs, G.; Taller, J. PICcalc: An online program to calculate polymorphic information content for molecular genetic studies. *Biochem. Genet.* **2012**, *50*, 670–672. [CrossRef] [PubMed]
26. Chen, C.; Chen, H.; Zhang, Y.; Thomas, H.R.; Frank, M.H.; He, Y.; Xia, R. TBtools: An Integrative Toolkit Developed for Interactive Analyses of Big Biological Data. *Mol. Plant* **2020**, *13*, 1194–1202. [CrossRef] [PubMed]
27. Shin, J.-H.; Blay, S.; Graham, J.; McNeney, B. LDheatmap: AnRFunction for Graphical Display of Pairwise Linkage Disequilibria between Single Nucleotide Polymorphisms. *J. Stat. Softw.* **2006**, *16*, 1–9. [CrossRef]
28. Porth, I.; Klapšte, J.; Skyba, O.; Hannemann, J.; McKown, A.D.; Guy, R.D.; DiFazio, S.P.; Muchero, W.; Ranjan, P.; Tuskan, G.A.; et al. Genome-wide association mapping for wood characteristics in Populus identifies an array of candidate single nucleotide polymorphisms. *New Phytol.* **2013**, *200*, 710–726. [CrossRef] [PubMed]
29. Shao, Y.; Jin, L.; Zhang, G.; Lu, Y.; Shen, Y.; Bao, J. Association mapping of grain color, phenolic content, flavonoid content and antioxidant capacity in dehulled rice. *Theor. Appl. Genet.* **2011**, *122*, 1005–1016. [CrossRef] [PubMed]
30. Wilkinson, G.N.; Rogers, C.E. Symbolic Description of Factorial Models for Analysis of Variance. *Appl. Stat.* **1973**, *22*, 392–399. [CrossRef]
31. Charmet, G.; Tran, L.G.; Auzanneau, J.; Rincent, R.; Bouchet, S. BWGS: A R package for genomic selection and its application to a wheat breeding programme. *PLoS ONE* **2020**, *15*, e0222733. [CrossRef] [PubMed]
32. Clark, S.A.; van der Werf, J. Genomic best linear unbiased prediction (gBLUP) for the estimation of genomic breeding values. *Methods Mol. Biol.* **2013**, *1019*, 321–330. [CrossRef] [PubMed]
33. Jiang, Y.; Reif, J.C. Modeling Epistasis in Genomic Selection. *Genetics* **2015**, *201*, 759–768. [CrossRef] [PubMed]
34. Endelman, J.B. Ridge Regression and Other Kernels for Genomic Selection with R Package rrBLUP. *Plant Genome* **2011**, *4*, 250–255. [CrossRef]
35. Li, Z.; Sillanpää, M.J. Overview of LASSO-related penalized regression methods for quantitative trait mapping and genomic selection. *Theor. Appl. Genet.* **2012**, *125*, 419–435. [CrossRef] [PubMed]

36. Zou, H.; Hastie, T. Addendum: Regularization and variable selection via the elastic net. *J. R. Stat. Soc. Ser. B* **2005**, *67*, 768. [CrossRef]
37. De Los Campos, G.; Hickey, J.M.; Pong-Wong, R.; Daetwyler, H.D.; Calus, M.P. Whole-genome regression and prediction methods applied to plant and animal breeding. *Genetics* **2013**, *193*, 327–345. [CrossRef]
38. Park, T.; Casella, G. The Bayesian Lasso. *J. Am. Stat. Assoc.* **2008**, *103*, 681–686. [CrossRef]
39. Meuwissen, T.H.; Hayes, B.J.; Goddard, M.E. Prediction of total genetic value using genome-wide dense marker maps. *Genetics* **2001**, *157*, 1819–1829. [CrossRef] [PubMed]
40. Habier, D.; Fernando, R.L.; Kizilkaya, K.; Garrick, D.J. Extension of the bayesian alphabet for genomic selection. *BMC Bioinform.* **2011**, *12*, 186. [CrossRef]
41. Iheshiulor, O.O.M.; Woolliams, J.A.; Svendsen, M.; Solberg, T.; Meuwissen, T.H.E. Simultaneous fitting of genomic-BLUP and Bayes-C components in a genomic prediction model. *Genet. Sel. Evol.* **2017**, *49*, 63. [CrossRef] [PubMed]
42. Psofakis, P.; Karapanagiotidis, I.T.; Malandrakis, E.E.; Golomazou, E.; Exadactylos, A.; Mente, E. Effect of fishmeal replacement by hydrolyzed feather meal on growth performance, proximate composition, digestive enzyme activity, haematological parameters and growth-related gene expression of gilthead seabream (*Sparus aurata*). *Aquaculture* **2020**, *521*, 735006. [CrossRef]
43. He, Z.; Zhang, R.; Jiang, F.; Zhang, H.; Zhao, A.; Xu, B.; Jin, L.; Wang, T.; Jia, W.; Jia, W.; et al. FADS1-FADS2 genetic polymorphisms are associated with fatty acid metabolism through changes in DNA methylation and gene expression. *Clin. Epigenet.* **2018**, *10*, 113. [CrossRef] [PubMed]
44. Walle, P.; Männistö, V.; de Mello, V.D.; Vaittinen, M.; Perfilyev, A.; Hanhineva, K.; Ling, C.; Pihlajamäki, J. Liver DNA methylation of FADS2 associates with FADS2 genotype. *Clin. Epigenet.* **2019**, *11*, 10. [CrossRef]
45. Walkiewicz, K.; Benitez Cardenas, A.S.; Sun, C.; Bacorn, C.; Saxer, G.; Shamoo, Y. Small changes in enzyme function can lead to surprisingly large fitness effects during adaptive evolution of antibiotic resistance. *Proc. Natl. Acad. Sci. USA* **2012**, *109*, 21408–21413. [CrossRef] [PubMed]
46. Wang, X.; Zhang, X.; Peng, C.; Shi, Y.; Li, H.; Xu, Z.; Zhu, W. D3DistalMutation: A Database to Explore the Effect of Distal Mutations on Enzyme Activity. *J. Chem. Inf. Model.* **2021**, *61*, 2499–2508. [CrossRef]
47. Baby, S.; Hyeong, K.E.; Lee, Y.M.; Jung, J.H.; Oh, D.Y.; Nam, K.C.; Kim, T.H.; Lee, H.K.; Kim, J.J. Evaluation of genome based estimated breeding values for meat quality in a berkshire population using high density single nucleotide polymorphism chips. *Asian-Australas. J. Anim. Sci.* **2014**, *27*, 1540–1547. [CrossRef]
48. Brito, L.F.; Clarke, S.M.; McEwan, J.C.; Miller, S.P.; Pickering, N.K.; Bain, W.E.; Dodds, K.G.; Sargolzaei, M.; Schenkel, F.S. Prediction of genomic breeding values for growth, carcass and meat quality traits in a multi-breed sheep population using a HD SNP chip. *BMC Genet.* **2017**, *18*, 7. [CrossRef]
49. Liu, J.J.; Liang, A.X.; Campanile, G.; Plastow, G.; Zhang, C.; Wang, Z.; Salzano, A.; Gasparrini, B.; Cassandro, M.; Yang, L.G. Genome-wide association studies to identify quantitative trait loci affecting milk production traits in water buffalo. *J. Dairy Sci.* **2018**, *101*, 433–444. [CrossRef]
50. Hwang, J.Y.; Lee, H.J.; Go, M.J.; Jang, H.B.; Choi, N.H.; Bae, J.B.; Castillo-Fernandez, J.E.; Bell, J.T.; Spector, T.D.; Lee, H.J.; et al. Genome-wide methylation analysis identifies ELOVL5 as an epigenetic biomarker for the risk of type 2 diabetes mellitus. *Sci. Rep.* **2018**, *8*, 14862. [CrossRef]
51. Jin, M.; Monroig, Ó.; Navarro, J.C.; Tocher, D.R.; Zhou, Q.C. Molecular and functional characterisation of two *elovl4* elongases involved in the biosynthesis of very long-chain (>C_{24}) polyunsaturated fatty acids in black seabream *Acanthopagrus schlegelii*. *Comp. Biochem. Physiol. B Biochem. Mol. Biol.* **2017**, *212*, 41–50. [CrossRef] [PubMed]
52. Ryan, M.T.; Hamill, R.M.; O'Halloran, A.M.; Davey, G.C.; McBryan, J.; Mullen, A.M.; McGee, C.; Gispert, M.; Southwood, O.I.; Sweeney, T. SNP variation in the promoter of the PRKAG3 gene and association with meat quality traits in pig. *BMC Genet.* **2012**, *13*, 66. [CrossRef] [PubMed]
53. Wu, Y.; Wang, Y.; Tian, H.; Lu, T.; Yu, M.; Xu, W.; Liu, G.; Xie, L. DHA intake interacts with ELOVL2 and ELOVL5 genetic variants to influence polyunsaturated fatty acids in human milk. *J. Lipid Res.* **2019**, *60*, 1043–1049. [CrossRef] [PubMed]
54. Zhu, K.C.; Song, L.; Guo, H.Y.; Guo, L.; Zhang, N.; Liu, B.S.; Jiang, S.G.; Zhang, D.C. Elovl4a participates in LC-PUFA biosynthesis and is regulated by PPARαβ in golden pompano *Trachinotus ovatus* (Linnaeus 1758). *Sci. Rep.* **2019**, *9*, 4684. [CrossRef] [PubMed]
55. Corominas, J.; Marchesi, J.A.; Puig-Oliveras, A.; Revilla, M.; Estelle, J.; Alves, E.; Folch, J.M.; Ballester, M. Epigenetic regulation of the ELOVL6 gene is associated with a major QTL effect on fatty acid composition in pigs. *Genet. Sel. Evol.* **2015**, *47*, 20. [CrossRef] [PubMed]

biology

MDPI

Article

Influence of Varying Dietary ω6 to ω3 Fatty Acid Ratios on the Hepatic Transcriptome, and Association with Phenotypic Traits (Growth, Somatic Indices, and Tissue Lipid Composition), in Atlantic Salmon (*Salmo salar*)

Tomer Katan [1,*], Xi Xue [1], Albert Caballero-Solares [1,*], Richard G. Taylor [2], Christopher C. Parrish [1] and Matthew L. Rise [1]

[1] Department of Ocean Sciences, Memorial University of Newfoundland, St. John's, NL A1C 5S7, Canada; xi.xue@mun.ca (X.X.); cparrish@mun.ca (C.C.P.); mrise@mun.ca (M.L.R.)
[2] Cargill Animal Nutrition, 10383 165th Avenue NW, Elk River, MN 55330, USA; richard_taylor@cargill.com
* Correspondence: tkatan@mun.ca (T.K.); acaballeroso@mun.ca (A.C.-S.); Tel.: +1-709-7703846 (T.K.); Tel.: +1-709-3251598 (A.C.-S.)

Citation: Katan, T.; Xue, X.; Caballero-Solares, A.; Taylor, R.G.; Parrish, C.C.; Rise, M.L. Influence of Varying Dietary ω6 to ω3 Fatty Acid Ratios on the Hepatic Transcriptome, and Association with Phenotypic Traits (Growth, Somatic Indices, and Tissue Lipid Composition), in Atlantic Salmon (*Salmo salar*). *Biology* **2021**, *10*, 578. https://doi.org/10.3390/biology10070578

Academic Editor: Patricia Pereiro

Received: 20 April 2021
Accepted: 21 June 2021
Published: 24 June 2021

Simple Summary: Plant oils are routinely used in fish feeds as a fish oil replacement. However, these terrestrial alternatives typically contain high levels of ω6 fatty acids (FA) and, thus, high ω6 to ω3 (ω6:ω3) FA ratios, which influence farmed fish and their consumers. The ω6:ω3 ratio is known to affect many biological processes (e.g., inflammation, FA metabolism) and human diseases; however, its impacts on fish physiology and the underlying molecular mechanisms are less well understood. In this study, we used 44 K microarrays to examine which genes and molecular pathways are altered by variation in dietary ω6:ω3 in Atlantic salmon. Our microarray study showed that several genes related to immune response, lipid metabolism, cell proliferation, and translation were differentially expressed between the two extreme ω6:ω3 dietary treatments. We also revealed that the PPARα activation-related transcript *helz2* is a potential novel molecular biomarker of tissue variation in ω6:ω3. Further, correlation analyses illustrated the relationships between liver transcript expression and tissue (liver, muscle) lipid composition, and other phenotypic traits in salmon fed low levels of fish oil. This nutrigenomic study enhanced the current understanding of Atlantic salmon gene expression response to varying dietary ω6:ω3.

Abstract: The importance of dietary omega-6 to omega-3 (ω6:ω3) fatty acid (FA) ratios for human health has been extensively examined. However, its impact on fish physiology, and the underlying molecular mechanisms, are less well understood. This study investigated the influence of plant-based diets (12-week exposure) with varying ω6:ω3 (0.4–2.7) on the hepatic transcriptome of Atlantic salmon. Using 44 K microarray analysis, genes involved in immune and inflammatory response (*lect2a*, *itgb5*, *helz2a*, *p43*), lipid metabolism (*helz2a*), cell proliferation (*htra1b*), control of muscle and neuronal development (*mef2d*) and translation (*eif2a*, *eif4b1*, *p43*) were identified; these were differentially expressed between the two extreme ω6:ω3 dietary treatments (high ω6 vs. high ω3) at week 12. Eight out of 10 microarray-identified transcripts showed an agreement in the direction of expression fold-change between the microarray and qPCR studies. The PPARα activation-related transcript *helz2a* was confirmed by qPCR to be down-regulated by high ω6 diet compared with high ω3 diet. The transcript expression of two *helz2* paralogues was positively correlated with ω3, and negatively with ω6 FA in both liver and muscle, thus indicating their potential as biomarkers of tissue ω6:ω3 variation. *Mef2d* expression in liver was suppressed in the high ω6 compared to the balanced diet (ω6:ω3 of 2.7 and 0.9, respectively) fed fish, and showed negative correlations with ω6:ω3 in both tissues. The hepatic expression of two *lect2* paralogues was negatively correlated with viscerosomatic index, while *htra1b* correlated negatively with salmon weight gain and condition factor. Finally, *p43* and *eif2a* were positively correlated with liver Σω3, while these transcripts and *eif4b2* showed negative correlations with 18:2ω6 in the liver. This suggested that some aspects of protein synthesis were influenced by dietary ω6:ω3. In summary, this nutrigenomic study identified hepatic transcripts responsive to dietary variation in ω6:ω3, and relationships of transcript expression with tissue (liver, muscle) lipid composition and other phenotypic traits.

Keywords: hepatic transcript expression; lipid metabolism; salmon; microarray; omega-6/omega-3 ratio; nutrigenomics; fatty acids; liver; muscle

1. Introduction

Plant-based oils are commonly used in aquafeeds to replace fish oil (FO), due to decreasing global availability, rising market price, and concerns regarding the ecological sustainability of the finite fishery resources upon which FO production depends [1,2]. Indeed, plant oils (PO) were shown to be more economical and environmentally sustainable [3], and their inclusion as alternatives to FO in many experimental diets did not affect the growth and survival of farmed Atlantic salmon (*Salmo salar*) [4–6]. However, terrestrial oils are devoid of long-chain polyunsaturated fatty acids (LC-PUFA), such as eicosapentaenoic acid (EPA, 20:5ω3), docosahexaenoic acid (DHA, 22:6ω3), and arachidonic acid (ARA, 20:4ω6), which are abundant in FO. These LC-PUFA have important functions in vertebrate health, reproduction, neural development, and growth, among other biological processes [5,7]. This has resulted in decreased fillet EPA and DHA levels in farmed fish that were fed with PO as a partial or full replacement for FO, compromising their nutritional quality for human consumers [8–10]. Further, previous studies reported impacts on fish health and physiology with the dietary replacement of FO by PO (e.g., liver steatosis, altered complement pathway and phagocytic activity, and modulated expression of genes involved in immune response) [11–15]. Another concern is that most terrestrial oils used in aquafeeds, and the farmed seafood consuming them, may not provide adequate ratios of ω6 to ω3 (ω6:ω3) fatty acids (FA) due to high ω6 FA content [10,16–18]. Previous human nutrition studies reported that high dietary ω6:ω3 promotes the pathogenesis of many diseases, including cardiovascular, inflammatory, autoimmune, and cognitive diseases, as well as obesity and cancer [19–22]. An optimal ratio of ω6:ω3 is important for maintaining the homeostasis of many biological processes such as cell apoptosis, inflammation, fatty acid and cholesterol metabolism, and others [23–25]. However, the underlying molecular mechanisms are still poorly understood in fish, and it is not known which genes are involved in variation in dietary and tissue ω6:ω3 in salmon fed high levels of terrestrial-based oils.

A feeding trial was performed to examine the impact of five plant-based diets with varying ω6:ω3 on salmon growth, tissue (i.e., muscle, liver) lipid composition, liver LC-PUFA synthesis, and transcript expression (targeted qPCR) of lipid metabolism and eicosanoid synthesis-related genes [26]. The objective of our current study was to utilize a 44 K salmonid oligonucleotide microarray [27–29] for the examination of the impact of the two extreme ω6:ω3 diets (i.e., high ω6 and high ω3) on the hepatic transcriptome at week 12. We hypothesized that salmon fed the two diets with the most extreme lipid compositions (i.e., High ω3 and High ω6) would show the most extensive transcriptomic differences. The current study used the same fish as in Katan et al. [26]. The aim was to identify novel biomarker genes and molecular pathways that are altered by variation in ω6:ω3. To aid in the elucidation of the relationships between liver transcripts and phenotypic traits (i.e., growth parameters, somatic indices, tissue FA and lipid class composition), correlation analyses were also performed.

2. Materials and Methods

2.1. Fish and Experimental Diets

Five experimental diets with varying ratios of ω6:ω3 were formulated and manufactured by Cargill Innovation Center (Dirdal, Norway). The diets had ω6:ω3 of 1:3 (high ω3), 1:2 (medium ω3), 1:1 (balanced), 2:1 (medium ω6) and 3:1 (high ω6). Dietary formulations and their lipid composition were published previously [26]. However, as they pertain to the current study, the formulation and lipid profiles of the relevant diets (i.e., high ω3, balanced, and high ω6) are also included as supplementary material herein (Tables S1 and S2). All diets contained the same sources and equal levels of marine and plant proteins, but had different mixes of plant-based oils (i.e., linseed (flax), soy, and palm). All diets were formulated to be isonitrogenous and isoenergetic (Table S1), and to meet the nutritional requirements of salmonids [30].

Atlantic salmon pre-smolts were transported from Northern Harvest Sea Farms (Stephenville, NL, Canada) in October 2015, and held in the Dr. Joe Brown Aquatic Research Building (Ocean Sciences Centre, Memorial University of Newfoundland, St. John's, NL, Canada) in 3800-L tanks. After their arrival, fish were graded in order to select the most uniform population, and this was followed by PIT (Passive Integrated Transponder; Easy AV, Avid Identification Systems, Norco, CA, USA)-tagging for individual identification. Then, post-smolts (203 ± 24 g mean initial weight ± SE) were randomly distributed into twenty 620-L tanks (40 fish $tank^{-1}$), and subjected to a 2.5-week acclimation period. After the completion of the acclimation period, fish were switched from the commercial diet (Nutra Transfer NP, 3 mm, Skretting Canada, St. Andrews, NB, Canada), and fed with the experimental diets (4 tanks $diet^{-1}$) for 12 weeks. The photoperiod was maintained at 24 h light. Fish were fed overnight using automatic feeders, and apparent feed intake was recorded throughout the trial. Mortalities were also recorded during the trial. For additional details regarding the rearing conditions and recordings, refer to Katan et al. [26].

2.2. Sample Collection

Growth performance parameters (e.g., fork-length, weight, organ indices) were measured at the beginning and the end of the 12-week feeding trial [26]. At the end of the trial, salmon were starved for 24 h, and then 5 fish per tank were euthanized with an overdose of MS-222 (400 mg L^{-1}; Syndel Laboratories, Vancouver, BC, Canada) and dissected for tissue collection. For gene expression analyses, liver samples (50–100 mg) were collected in 1.5 mL nuclease-free tubes, flash-frozen in liquid nitrogen, and stored at −80 °C until RNA extractions were performed. Liver and muscle samples, for lipid analyses, were collected, processed, and stored as described in Katan et al. [26]. Only liver samples from fish that showed weight gains within one standard deviation below and above the mean value of each tank were utilized for this study, in order to reduce biological variability in the gene expression data among fish. Tank means rather than dietary treatment means, were chosen for sample selection, so that variability between tanks could be included in the statistical analysis.

2.3. RNA Extraction, DNase Treatment, Column Purification and cDNA Synthesis

The TissueLyser II system (at 25 Hz for 2.5 min) with 5 mm stainless steel beads, QIAGEN, Mississauga, ON, Canada) was used to homogenize liver samples in TRIzol reagent (Invitrogen, Carlsbad, CA, USA). Samples were subjected to RNA extraction according to manufacturer's instructions. Due to low 260/230 ratios (i.e., 1.0–1.6) following TRIzol extraction, all RNA samples were then re-extracted (phenol-chloroform) and precipitated following standard methods [31]. This was followed by DNaseI treatment and column purification using RNase-free DNase Set and RNeasy Mini Kit (QIAGEN). All procedures were conducted according to manufacturer instructions, and as described in Xue et al. [29]. RNA integrity was verified by 1% agarose gel electrophoresis, and RNA purity and quantity were assessed by NanoDrop UV spectrophotometry (NanoDrop, Thermo Scientific, Mississauga, ON, Canada). DNased and column-purified RNA samples had A260/280 and A260/230 ratios of 1.8–2.2. All cDNAs were synthesized by reverse transcription of 1 μg of DNaseI-treated, column-purified total RNA from each sample, with 1 μL of random primers (250 ng; Invitrogen), 1 μL of dNTPs (0.5 mM final concentration; Invitrogen), 4 μL of 5× first-strand buffer (1× final concentration; Invitrogen), 2 μL of DTT (10 mM final concentration; Invitrogen) and 1 μL of Moloney murine leukemia virus (M-MLV) reverse transcriptase (RT) (200 U; Invitrogen) at 37 °C for 50 min, following the manufacturer's instructions, and as described in Xue et al. [29]. The total reaction volume was 20 μL. Finally, all cDNAs were diluted 40× with nuclease-free water (Invitrogen) prior to the qPCR.

2.4. Microarray Hybridization and Data Acquisition

Eight fish (2 from each of the 4 dietary tanks) from each of the 2 extreme ω6:ω3 treatments (high ω6 or high ω3) were used in the microarray analysis (i.e., 16 fish total), using a common reference design. Four array slides were used in the current study, and each array contained 2 fish per treatment, which were randomly selected. The common reference was made by an equal quantity of each DNase I-treated, column-purified total RNA liver sample. The microarray experiment was performed as described in Xue et al. [29]. Briefly, anti-sense amplified RNA (aRNA) was in vitro transcribed from 1 μg of each column-purified RNA or reference pooled RNA using Ambion's Amino Allyl MessageAmp II aRNA Amplification kit (Life Technologies, Burlington, ON, Canada), following the manufacturer's instructions. The quantity and quality of aRNA were assessed using NanoDrop spectrophotometry and 1% agarose gel electrophoresis, respectively. Then, 20 μg of each aRNA were precipitated overnight, following standard molecular biology procedures, and re-suspended in coupling buffer. Each individual aRNA sample was labeled with Cy5 (i.e., experimental samples), whereas the reference pool was labeled with Cy3 (i.e., common reference) fluor (GE HealthCare, Mississauga, ON, Canada), following the manufacturer's instructions. The "microarray" function of the NanoDrop spectrophotometer was used to measure the labeling efficiency of the aRNA. The labeled aRNA (825 ng) from each experimental sample (i.e., Cy5) was mixed with an equal quantity of labeled aRNA from the common reference (i.e., Cy3), for each array, and the resulting pool was fragmented, following the manufacturer's instructions (Agilent, Mississauga, ON, Canada). Each pool was co-hybridized to a consortium for Genomic Research on All Salmonids Project (cGRASP)-designed 4 × 44 K salmonid oligonucleotide microarray (GEO accession # GPL11299) [27] (Agilent). Finally, the arrays were hybridized at 65 °C for 17 h with rotation (10 rpm), using an Agilent hybridization oven. The microarray slides were washed immediately after hybridization as per the manufacturer's instructions.

Each microarray slide was scanned at 5 μm resolution with 90% laser power using a ScanArray Gx Plus scanner and ScanExpress v4.0 software (Perkin Elmer, Waltham, MA, USA), and the Cy3 and Cy5 channel photomultiplier tube (PMT) settings were adjusted to balance the fluorescence signal between channels. The resulting raw data were saved as TIFF images, and the signal intensity data were extracted using Imagene 9.0 (BioDiscovery, El Segundo, CA, USA). Removal of low-quality or flagged spots on the microarray, as well as log_2-transformation and Loess-normalization of the data, were performed using R and the Bioconductor package mArray [32]. Features absent in more than 25% (i.e., 4 out of 16 arrays) of the arrays were omitted, and the missing values were imputed using the EM_array method and the LSimpute package [33,34]. The final dataset used for statistical analyses consisted of 10,264 probes for all arrays (GEO accession number: GSE139418; https://www.ncbi.nlm.nih.gov/geo/query/acc.cgi?acc=GSE139418.

2.5. Microarray Data Analysis

The Significance Analysis of Microarrays (SAM) algorithm [35] was performed to identify genes that were significantly differentially expressed between the two extreme ω6:ω3 treatments. A false discovery rate (FDR) threshold of 10% was used with the Bioconductor package siggenes [36] in R. For the identification of additional transcripts that were differentially expressed between the two dietary treatments, the Rank Products (RP) method was also used, as this method is less sensitive to high biological variability [37,38]. The latter analysis was performed at a percentage of false-positives (PFP) threshold of 10%, using the Bioconductor package RankProd [39]. Due to high background signal in the first slide (i.e., slide # 11,502), no genes were initially identified as significantly differentially expressed; therefore, this slide was removed from the analyses. In order to maximize our capacity to identify differentially expressed genes, gene lists were obtained with 2 and 3 of the remaining slides (consisting of 4 and 6 fish per treatment, respectively). Each slide is composed of 4 arrays, i.e., 4 biological replicates analyzed per slide.

The resulting gene lists were annotated using the contiguous sequences (contigs) that were used for the design of the 60 mer oligonucleotide probes of the array [27]. Annotations were performed manually with a BLASTx alignment against the NCBI non-redundant (nr) amino acid database using an E-value threshold of 10^{-5}. The best BLASTx hits corresponding to putative *Homo sapiens* orthologues were used to obtain gene ontology (GO) terms manually from the UniProt Knowledgebase (http://www.uniprot.org/, accessed on 4 November 2020).

2.6. qPCR Study and Data Analysis

Transcript expression levels of 10 genes of interest (GOI) (Table 1), identified as differentially expressed in the microarray analyses, were assayed by qPCR. In addition to the high ω6 and high ω3 treatments, the qPCR analysis also included liver samples from fish fed the balanced diet. In addition to the microarray-identified GOI, BLASTn searches using publicly available Atlantic salmon cDNA sequences (i.e., in NCBI's non-redundant nucleotide (nt) and expressed sequence tags (EST) databases) were used to identify paralogues for each GOI, as described in Caballero-Solares et al. [40].

Table 1. qPCR primers.

Gene Name (Symbol) [a]	Nucleotide Sequence (5′-3′) [b]	Amplification Efficiency (%)	Amplicon Size (bp)	GenBank Accession Number
Serine protease HTRA1 a (*htra1a*) [c]	F:GCTGATGTGGTGGAGGAGAT	113.3	127	NM001141717
	R:TCAAGCCGTCCTCTGACAC	-	-	-
Serine protease HTRA1 b (*htra1b*) [c]	F:ATGATGACTCTCACACCAATGC	95.4	104	EG831192
	R:GTTTTTGGGATGACCTCGATT	-	-	-
Aminoacyl tRNA synthase complex-interacting multifunctional protein 1 (*p43*)	F:GGAAGACGAATGCAGAGGAC	97.2	82.0	[illegible]
	R:GGAGCGGTCATTCACACTTT	-	-	-
Eukaryotic translation initiation factor 2A (*eif2a*)	F:TAAACCCAGATGCCCTTGAG	94.9	143	NM001140088
	R:GGCTTTCAGCTCGTCGATAG	-	-	-
Eukaryotic translation initiation factor 4B 1 (*eif4b1*)	F:CGCAGGGACCGGGATGAT	85.3	123	BT072661
	R:TCGGTCCTC5CTGTCCGC	-	-	-
Eukaryotic translation initiation factor 4B 2 (*eif4b2*)	F:CACATCCAGGAAGTACCTCT	87.4	94.0	DY739566
	R:TCGTCCTCCTTACCGCTGA	-	-	-
Cytochrome c oxidase subunit 2 (*mtco2*) [d]	F:CACCGATTACGAAGACTTAGGC	107.9	136	DW554935
	R:TGAAACTAGGACCCGGATTG	-	-	-
Leukocyte cell-derived chemotaxin 2 precursor a (*lect2a*) [c]	F:CAGATGGGGACAAGGACACT	94.6	150	BT059281
	R:GCCTTCTTCGGGTCTGTGTA	-	-	-
Leukocyte cell-derived chemotaxin 2 precursor b (*lect2b*) [c]	F:ACAACTGGGGACAAGGACAG	84.8	125	DV106130
	R:CACTTTGCCGTTGAGTTTCA	-	-	-
60S ribosomal protein L18 (*rpl18*)	F:AGTTCCACGACTCGAAGATC	93.8	143	DW535031
	R:TTTTATTGTGCCGCACAAGGT	-	-	-
Myocyte-specific enhancer factor 2D (*mef2d*)	F:GCAGCAACATCAACAACAGC	89.5	160	XM014177143
	R:CTCATCTCTACCCAAGAGGA	-	-	-
Helicase with zinc finger domain 2 a (*helz2a*, alias *pric285a*) [e]	F:GCAAGGTTGGGTATGAGGAA	91.3	149	BT072427
	R:TTCGGAGTTGCTCCAGTCTT	-	-	-
Helicase with zinc finger domain 2 b (*helz2b*, alias *pric285b*) [e]	F:AGACGTAGTGGTTCGGATCG	82.0	145	EG928625
	R:GACCGTGATTTCGTCCAGTT	-	-	-
Integrin beta-5-like (*itgb5*) [f]	F:CCTGCCAGCGGCTATGCAA	94.1	147	DW540995/
	R:AGGACTGACATGCCGTTGG	-	-	XM014165323
Elongation factor 1 alpha-2 (*eef1α-2*) [g]	F:GCACAGTAACACCGAAACGA	86.4	132	BG933853
	R:ATGCCTCCGCACTTGTAGAT	-	-	-
60S ribosomal protein 32 (*rpl32*) [g]	F:AGGCGGTTTAAGGGTCAGAT	96.1	119	BT043656
	R:TCGAGCTCCTTGATGTTGTG	-	-	-

[a] Bolded gene symbols refer to microarray-identified transcripts. [b] F is forward and R is reverse primer. [c] Primers that were previously published in Caballero-Solares et al. [40]. [d] The *Salmo salar* sequence of *mtco2* used in the qPCR assay showed 87% identity with the 60 mer microarray probe (C060R108) affiliated with rainbow trout (*Oncorhynchus mykiss*). [e] Primers that were previously published in Caballero-Solares et al. [15] (annotated as VHSV-induced protein in that study). Alias *pric285* stands for peroxisomal proliferator-activated receptor A interacting complex 285. [f] The Atlantic salmon sequences of *itgb5* used in the qPCR assay showed 86% identity with the 60mer microarray probe (C002R106) affiliated with rainbow trout. Primers were designed based on common regions between DW540995 and XM014165323. [g] Primers that were previously published in Katan et al. [26].

Paralogue-specific primers were used for *eif4b*, *htra1*, *lect2* and *helz2* (Table S3 and Figures S1–S4). The sequences of the primer pairs used in qPCR, GenBank accession number of sequences used for primer design, and other details are presented in Table 1. Notably, primers for the transcript *lhpl4* (GenBank accession number NM_001146670) failed quality testing, and thus, this transcript was not included in the qPCR study. In addition, the 60 mer microarray probe for *mtco1* (C188R069) is affiliated with a rainbow trout sequence, and had relatively low identity (i.e., <85%) with available *Salmo salar* sequences (using NCBI's EST and nt databases) and, therefore, was excluded from the qPCR study. The program Primer 3 (http://frodo.wi.mit.edu, accessed on 19 October 2019) was used for primer design. Each primer pair was quality-tested, including standard curve and dissociation curve to ensure that a single product was amplified with no primer dimers [32,41]. Primer pairs were quality-tested using the 7500 Fast Real Time PCR system (Applied Biosystems/Life Technologies, Foster City, CA, USA). The amplification efficiency [42] of each primer pair was determined using a 5-point 1:3 dilution series starting with cDNA representing 10 ng of input total RNA. Two pools were generated (i.e., high ω3 pool and high ω6 pool), with each pool consisting of 8 fish (and each fish contributing an equal quantity to the pool). The reported primer pair amplification efficiencies are an average of the two pools, except if one pool showed poor efficiency or spacing (i.e., *p43*, *eif2a*, *htra1a*, *helz2a* and *helz2b*, where one pool was used due to low expression levels). A 5-point 1:2 dilution series was used for the primers *mtco2* and *helz2b* as these transcripts had lower expression levels (fluorescence threshold cycle (C_T) values of ~30 and 31, respectively). Furthermore, amplicons were checked by 1.5% agarose gel electrophoresis and compared with the 1 kb plus DNA Ladder (Invitrogen) to ensure that the correct size fragment was amplified.

To select the most suitable normalizer genes, six candidate normalizers were tested based on our previous qPCR studies (*rpl32*, *actb*, *eef1α-1*, *eef1α-2*, *abcf2*, *pabpc1*) [15,29], and salmon literature on reference genes (*actb*, *eef1α-1*, *eef1α-2*) [43]. Their qPCR primers were quality-tested as described above. Then, their C_T values were measured using cDNA (corresponding to 5 ng of input total RNA) of 6 randomly selected fish per treatment (18 total). The geNorm algorithm [44] was used to analyze their expression stability. *Rpl32* and *eef1α-2* were shown to be the most stable (i.e., geNorm M-values of 0.30 and 0.25, respectively) among the 6 candidate reference genes and, therefore, were selected as normalizers.

All PCR amplifications were performed in a total reaction volume of 13 μL and consisted of 4 μL of cDNA (5 ng input total RNA), 50 nM each of forward and reverse primer and 1× Power SYBR Green PCR Master Mix (Applied Biosystems), and nuclease-free water (Invitrogen). The qPCR reactions, including no-template controls, were performed in technical triplicates using the ViiA 7 Real-Time PCR System (384-well format) (Applied Biosystems, Foster City, CA, USA) and the Power SYBR Green I dye chemistry. The Real-Time analysis program consisted of 1 cycle of 50 °C for 2 min, 1 cycle of 95 °C for 10 min, followed by 40 cycles (of 95 °C for 15 s and 60 °C for 1 min), with the fluorescence signal data collection after each 60 °C step. When a C_T value within a triplicate was greater than 0.5 cycles from the other two values, it was considered to be an outlier, discarded and the average C_T of the remaining two values was calculated. The relative quantity (RQ) of each transcript was calculated using ViiA 7 Software v1.2 (Applied Biosystems) for Comparative CT ($\Delta\Delta C_T$) analysis [45], with primer amplification efficiencies incorporated (Table 1). The expression levels of each GOI were normalized to both normalizer genes, and the sample with the lowest normalized expression was used as the calibrator sample (i.e., RQ = 1.0) for each GOI, as in [46]. Transcript expression data are presented as RQ values (mean ± SE) relative to the calibrator.

2.7. Statistical Analyses

2.7.1. qPCR Data

A general linear model with tank nested in diet, followed by a Tukey pairwise comparison ($p < 0.05$), was used to identify significant differences among dietary treatments at week 12. In cases where significant tank effect was observed ($p < 0.05$), a one-way ANOVA followed by a Tukey pairwise comparison post-hoc test was performed (Minitab 17 Statistical Software, State College, PA, USA). The RQ data are presented as mean ± SE. Each dietary treatment group was tested for outliers using Grubb's test ($p < 0.05$). In total, 9 RQ values were identified as statistical outliers in the entire dataset (i.e., out of 322 values), and excluded from the study. Each GOI had a minimum of 6 samples per dietary treatment, while most GOI had a sample size of 7–8 per dietary treatment. The qPCR fold-changes were calculated by dividing the mean RQ value of the high ω6 fish by that of the high ω3 fish. Finally, residuals were tested to verify normality, independence, and homogeneity of variance. Normality was examined using the Anderson–Darling test. If the test failed ($p < 0.05$), a one-way ANOVA on ranks was performed, which was followed by the Kruskal–Wallis test (SigmaPlot, Systat Software, Inc., Version 13, San Jose, CA, USA). In all cases, differences were considered statistically significant when $p < 0.05$.

2.7.2. Correlation Analyses of qPCR and Lipid Composition Data

Tissue lipid composition (muscle and liver) and growth performance of salmon fed varying ω6:ω3 diets were previously published by Katan et al. [26]. Pearson correlation analyses were performed in the current study to identify the relationships between hepatic transcript expression (i.e., qPCR data), tissue composition (i.e., % FA and lipid classes), and growth parameters (i.e., weight gain (WG), condition factor (CF)), using individual fish. All GOI in the qPCR study were used in the correlation analysis in order to identify differences between the liver and muscle tissue. Only ω3 and ω6 FA that accounted for >0.5% of the total FA in the tissue (average of each treatment) were included in the analyses. Furthermore, hierarchical clustering was used to group transcripts and lipid composition (using group average in PRIMER (Version 6.1.15, Ivybridge, UK)). IBM SPSS Statistics was used for the correlation analyses.

3. Results

3.1. Liver Microarray Analysis

RP analysis detected nine differentially expressed features (PFP < 10%; Table 2). Eight of these features (i.e., *lhpl4*, *htra1b*, *mtco2*, *lect2a*, *rpl18*, *helz2a*, *itgb5*, and *mtco1*) were identified analyzing data from two slides (slides # 11,504–11,505; comprising four fish per treatment), and one (i.e., *mef2d*) was identified analyzing data from three slides (slides # 11,503–11,505; comprising six fish per treatment). Two features (i.e., *lhpl4* and *htra1b*) showed higher expression in the high ω6 fish (4.78- and 3.57-fold change, respectively), while the other seven RP-identified features (i.e., *mtco2*, *lect2a*, *rpl18*, *mef2d*, *helz2a*, *itgb5*, and *mtco1*) showed down-regulation in the high ω6 fish (fold-change ranged from −3.27 to −7.11).

SAM analysis identified *p43*, *eif2a*, *eif4b1*, and *itgb5* as differentially expressed genes (FDR < 10%) between the high ω6 and high ω3 fed salmon, using three slides (slides # 11,503–11,505) (Table 2). These genes were down-regulated in the high ω6 compared with the high ω3 fed fish (fold-change values ranged from −2.79 to −5.12). One feature (*itgb5*) was represented in both SAM and RP analysis, and was down-regulated in the high ω6 compared to the high ω3 fed fish, in both analyses (−5.12 and −5.25- fold-change, respectively).

Table 2. Microarray-identified transcripts that were significantly differentially expressed in the liver of salmon fed high ω6 compared to high ω3 diet.

Probe ID [a]	BLASTx Identification [b]				Gene Ontology (GO) of Putative Human Orthologues [d]	Fold-Change [e]
	Best Named BLASTx Hit (Species) [c]	Accession No.	E-Value	% ID (AA)		
C187R103	Lipoma HMGIC fusion partner-like 4 protein (*lhpl4*) (*Salmo salar*)	NP_001140142	0	272/272 (100%)	BP: regulation of inhibitory synapse assembly, gamma-aminobutyric acid receptor clustering. MF: protein binding, GABA receptor binding. CC: inhibitory synapse, postsynaptic membrane, cell projection, plasma membrane, cell junction.	4.78
C231R170	Serine protease HTRA1 (*htra1b*) (*Salvelinus alpinus*)	XP_023864611	4e−171	248/256 (97%)	BP: proteolysis, extracellular matrix disassembly, negative regulation of transforming growth factor beta receptor signaling pathway, negative regulation of defense response to virus, positive regulation of epithelial cell proliferation. MF: serine-type endopeptidase and peptidase activity, insulin-like growth factor binding, hydrolase activity. CC: collagen-containing extracellular matrix, extracellular space, plasma membrane, cytoplasm.	3.57
C103R052	Aminoacyl tRNA synthase complex-interacting multifunctional protein 1 (*p43*) (*Salmo trutta*)	XP_029622221	0	321/326 (98%)	BP: inflammatory response, apoptotic process, response to wounding, tRNA aminoacylation for protein translation, defense response to virus, leukocyte migration, angiogenesis, chemotaxis, positive regulation of glucagon secretion. MF: RNA binding, tRNA binding, protein binding, cytokine activity, protein homodimerization activity. CC: aminoacyl-tRNA synthetase multienzyme complex, nucleus, cytosol, endoplasmic reticulum, extracellular region.	−2.79
C067R040	Eukaryotic translation initiation factor 2A (*eif2a*) (*Salmo salar*)	NP_001133560	0	576/576 (100%)	BP: translational initiation, ribosome assembly, protein phosphorylation, SREBP signaling pathway, response to amino acid starvation. MF: translation initiation factor activity, cadherin binding, ribosome binding, tRNA binding, protein binding. CC: blood microparticle, extracellular space, cytosolic small ribosomal subunit.	−3.13
C253R093	Eukaryotic translation initiation factor 4B (*eif4b1*) (*Salvelinus alpinus*)	XP_023852969	6e−11	37/40 (93%)	BP: translational initiation, eukaryotic translation initiation factor 4F complex assembly. MF: RNA binding, protein binding, translation initiation factor activity, RNA strand annealing activity. CC: polysome, cytosol, eukaryotic translation initiation factor 4F complex.	−3.23
C060R108	Cytochrome c oxidase subunit 2 (*mtco2*) (*Oncorhynchus masou masou*)	ASB29545	7e−74	115/182 (63%)	BP: electron transport chain, oxidation-reduction process. MF: cytochrome-c oxidase activity, copper ion binding, metal ion binding, oxidoreductase activity. CC: membrane, respirasome, mitochondrion.	−3.27
C159R112	Leukocyte cell-derived chemotaxin 2 precursor (*lect2a*) (*Salmo salar*)	ACI67916	6e−102	155/156 (99%)	BP: chemotaxis, skeletal system development. MF: protein binding, metal ion binding. CC: cytoplasm, extracellular space.	−3.48
C152R057	60S ribosomal protein L18 (*rpl18*) (*Salmo trutta*)	XP_029599741	3e−122	172/173 (99%)	BP: translation, viral transcription, SRP-dependent cotranslational protein targeting to membrane. MF: structural constituent of ribosome, protein binding, RNA binding. CC: ribosome, cytosolic large ribosomal subunit, cytosol.	−4.37
C133R018	Myocyte-specific enhancer factor 2D (*mef2d*) (*Oncorhynchus mykiss*)	XP_021427816	3e−70	193/193 (100%)	BP: positive regulation of vascular smooth muscle cell proliferation, muscle organ development, skeletal muscle and neuronal cell differentiation, apoptotic process, positive regulation of transcription by RNA polymerase II, adult heart development, nervous system development. MF: DNA-binding transcription factor activity, RNA polymerase II-specific, protein binding, histone deacetylase binding, protein heterodimerization activity. CC: nucleus, nuclear chromatin, intracellular membrane-bounded organelle, nucleoplasm.	−4.54
C065R088	Helicase with zinc finger domain 2 (*helz2a* alias, *pric285a*) (*Salmo trutta*)	XP_029548942	0	694/714 (97%)	BP: regulation of lipid metabolic process, positive regulation of transcription by RNA polymerase II, nuclear-transcribed mRNA catabolic process, nonsense-mediated decay. MF: nuclear receptor transcription activity, helicase activity, ribonuclease activity, hydrolase activity, RNA binding, ATP binding, protein binding, metal ion binding. CC: nucleus, membrane, nucleoplasm.	−4.71

Table 2. *Cont.*

Probe ID [a]	BLASTx Identification [b]				Gene Ontology (GO) of Putative Human Orthologues [d]	Fold-Change [e]
	Best Named BLASTx Hit (Species) [c]	Accession No.	E-Value	% ID (AA)		
C002R106 *	Integrin beta-5-like (*itgb5*) (*Oncorhynchus mykiss*)	XP_021453113	0	283/315 (90%)	BP: cell adhesion mediated by integrin, integrin-mediated signaling pathway, muscle contraction, antigen processing and presentation of exogenous peptide antigen via MHC class I, TAP-dependent, viral process, transforming growth factor beta receptor signaling pathway. MF: protein binding, signaling receptor activity, virus receptor activity. CC: cell surface, extracellular exosome, phagocytic vesicle, plasma membrane, integrin complex.	−5.12
C188R069	Cytochrome c oxidase subunit 1 (*mtco1*) (*Oncorhynchus tshawytscha*) *	NP_148940	0	410/437 (94%)	BP: oxidation-reduction process, oxidative phosphorylation, electron transport chain, aerobic respiration. MF: oxidoreductase activity, cytochrome-c oxidase activity, heme binding, metal ion binding. CC: mitochondrial inner membrane, respiratory chain complex IV, respirasome.	−7.11

[a] Refers to the identity of the probe on the 44 K array. Probes that are shown in bold font are features that were identified by SAM (FDR < 10%), and the remaining features were identified by RP analysis (PFP < 10%). The probe with an asterisk represents a feature that was identified in both SAM and RP analysis. Two 4 × 44 K array slides (slides # 11,504–11,505; representing 4 fish per treatment) were used in the RP analysis. However, the RP-identified *mef2d* was obtained using 3 slides (slides # 11,503–11,505; representing 6 fish per treatment). SAM-identified features were obtained using three slides (slides # 11,503–11,505). [b] Genes were identified by BLASTx, using the contig from which the microarray probe was designed against the NCBI non-redundant database. The best BLASTx hit with E-value < 10^{-5} and an informative protein name was used, and presented with species name, GenBank accession number, E-value and % amino acid (AA) identity. [c] All microarray-identified genes, with the exception of *lhpl4* and *mtco1*, were quantified by qPCR (see Materials and Methods). [d] Gene Ontology (GO) terms were selected from putative *Homo sapiens* orthologues (i.e., best BLASTx hit). Representative GO terms were identified (i.e., redundancies were not included), and divided into the categories: biological process (BP), molecular function (MF) and cellular component (CC). [e] Fold-change values between the 2 dietary treatments (high ω6/high ω3) for each of the significant microarray features. Down-regulated transcripts are shown with negative values (−(1/fold-change)) The SAM- and RP-identified *itgb5* showed fold-changes of −5.12 and −5.25, respectively.

Putative identities were determined for the 12 microarray-identified features, and functional annotations (i.e., GO terms) were collected for them (Table 2). The microarray-identified gene *lhpl4* (4.78-fold up-regulated) is involved in the nervous system, with GO annotations "regulation of inhibitory synapse assembly" and "GABA receptor binding". The feature *htra1b* (3.57-fold up-regulated) was classified as a gene involved in cell proliferation and showed the functional annotations "positive regulation of epithelial cell proliferation", "proteolysis" and "extracellular space" (Table 2). Several informative microarray features represented genes involved in translation, such as *p43*, *eif2a*, *eif4b1*, and *rpl18* (−2.79 to −4.37-fold down-regulated), with the associated GO terms "tRNA binding", "aminoacyl-tRNA synthetase multienzyme complex", "translational initiation", "ribosome assembly", and "structural constituent of ribosome". Furthermore, the GO terms "defense response to virus", "inflammatory response" and "response to wounding" were also identified with *p43*. The features *mtco2* and *mtco1* (−3.27- and −7.11-fold down-regulated, respectively) were classified as mitochondrion respiratory chain components, and showed the GO terms "electron transport chain", "oxidation-reduction process" and "cytochrome-c oxidase activity". Other microarray-identified features corresponded to immune- and inflammation-related genes such as *lect2a* (with the GO terms "chemotaxis" and "metal ion binding") and *itgb5* ("antigen processing and presentation" and "phagocytic vesicle"), and showed down-regulation in the high ω6 fed fish (−3.48 to −5.25-fold-change, respectively). The gene *mef2d* (−4.54-fold down-regulated) is involved in muscle cell proliferation, and in neuronal cell differentiation and survival, with the associated GO terms "muscle organ development", "skeletal muscle cell differentiation" "nervous system development", "apoptotic process" and "DNA-binding transcription factor activity" (Table 2). Further, the microarray-identified feature *itgb5* was associated with the GO term "muscle contraction". Finally, the gene *helz2a* (−4.71-fold down-regulated) was classified as a gene involved in lipid metabolism regulation by peroxisome proliferator-activated receptor alpha (PPARα), and showed the functional annotations "regulation of lipid metabolic process", "nuclear receptor transcription activity", "ATP binding", "metal ion binding", "hydrolase activity" and "ribonuclease activity".

3.2. qPCR Study

Ten microarray-identified genes were used in the qPCR study. All genes, with the exception of *mtco2* and *rpl18*, showed an agreement in the direction of expression fold-change (i.e., up- or down-regulation) between the microarray and qPCR studies (Table 3). The microarray-identified *helz2a* showed significantly lower transcript expression in the high ω6 compared to the high ω3 fed fish (−1.49-fold; $p = 0.04$). The paralogue *helz2b* showed significantly lower expression in both the high ω6 and balanced groups compared to the high ω3 fed fish (−1.61-fold; $p = 0.002$). The transcript *mef2d* showed significantly lower expression in the high ω6 compared to the balanced fed fish (−1.27-fold; $p = 0.03$), and a lower expression trend in the high ω6 compared to the high ω3 fish (−1.22-fold; $p = 0.06$). Both paralogues of *htra1* were numerically higher (although not statistically significant) in the high ω6 compared to the balanced and high ω3 fish (1.34–1.57-fold and 3.75–2.09-fold; $p = 0.25$ and 0.07, respectively) (Table 3).

3.3. Correlations between Hepatic qPCR Transcript Expression and Liver Lipid Composition

Hierarchical clustering of the qPCR transcripts showed four separate clusters (Figures 1 and 2). The first cluster consisted of both paralogues of *htra1*. The second cluster comprised *rpl18* only. The third cluster included some of the immune- and inflammation-related transcripts such as *lect2*, *p43*, and *helz2a*, as well as *mef2d*, *eif2a*, *eif4b*, and *mtco2*. The transcripts *helz2b* and *itgb5* composed the fourth cluster.

Table 3. Hepatic qPCR analysis of microarray-identified transcripts, and comparison between the microarray and qPCR results.

Microarray Probe [a]	Transcript Name	qPCR RQ Values [b]			*p*-Value (qPCR) [c]	Fold-Change [d]	
		High ω3	Balanced	High ω6		Microarray	qPCR
N/A	*htra1a*	2.2 ± 0.41	1.9 ± 0.29	3.0 ± 0.65	0.25	N/A	1.34
C231R170	*htra1b*	6.0 ± 2.14	10.7 ± 3.56	22.4 ± 7.43	0.07	3.57	3.75
C103R052	*p43*	3.4 ± 0.66	2.9 ± 0.41	2.2 ± 0.47	0.24	−2.79	−1.59
C067R040	*eif2a*	5.2 ± 0.40	5.3 ± 0.74	3.5 ± 0.92	0.19	−3.13	−1.47
C253R093	*eif4b1*	8.8 ± 1.78	6.7 ± 1.43	5.5 ± 1.53	0.29	−3.23	−1.59
N/A	*eif4b2*	2.7 ± 0.30	2.7 ± 0.40	2.2 ± 0.30	0.55	N/A	−1.22
C060R108	*mtco2*	1.4 ± 0.10	1.2 ± 0.06	1.4 ± 0.14	0.43	−3.27	1.07
C159R112	*lect2a*	7.6 ± 2.58	4.0 ± 1.20	4.2 ± 0.96	0.38	−3.48	−1.79
N/A	*lect2b*	3.4 ± 0.74	3.7 ± 0.89	3.8 ± 0.61	0.96	N/A	1.12
C152R057	*rpl18*	2.0 ± 0.17	2.1 ± 0.25	2.2 ± 0.16	0.88	−4.37	1.08
C133R018	*mef2d*	1.9 ± 0.10 [a,b]	2.0 ± 0.11 [a]	1.5 ± 0.13 [b]	0.03	−4.54	−1.22
C065R088	*helz2a*	2.3 ± 0.32 [a]	1.6 ± 0.15 [a,b]	1.5 ± 0.14 [b]	0.04	−4.71	**−1.49**
N/A	*helz2b*	2.3 ± 0.21 [a]	1.4 ± 0.09 [b]	1.4 ± 0.11 [b]	0.002	N/A	**−1.61**
C002R106	*itgb5*	2.1 ± 0.23	1.9 ± 0.08	1.6 ± 0.08	0.13	−5.25	−1.34

[a] Refers to the identity of the probe on the 44 K array. Transcripts with no probe ID are paralogues of microarray-identified transcripts. [b] Mean relative quantity (RQ) ± standard error (n = 6–8). RQ values were normalized to *elongation factor 1 alpha-2* (*eef1α-2*) and *60S ribosomal protein 32* (*rpl32*), and calibrated to the lowest expressing individual for each gene of interest. Different letters indicate significant differences among treatments (General linear model followed by Tukey pairwise comparison). [c] *p*-values obtained in the qPCR study. Differences were considered statistically significant when $p < 0.05$. [d] Microarray and qPCR comparison of fold-changes (i.e., high ω6/high ω3). Down-regulated transcripts are negative values (−(1/fold-change)). qPCR fold-changes corresponding to GOI with significant differences between the high ω6 and high ω3 treatments are bolded.

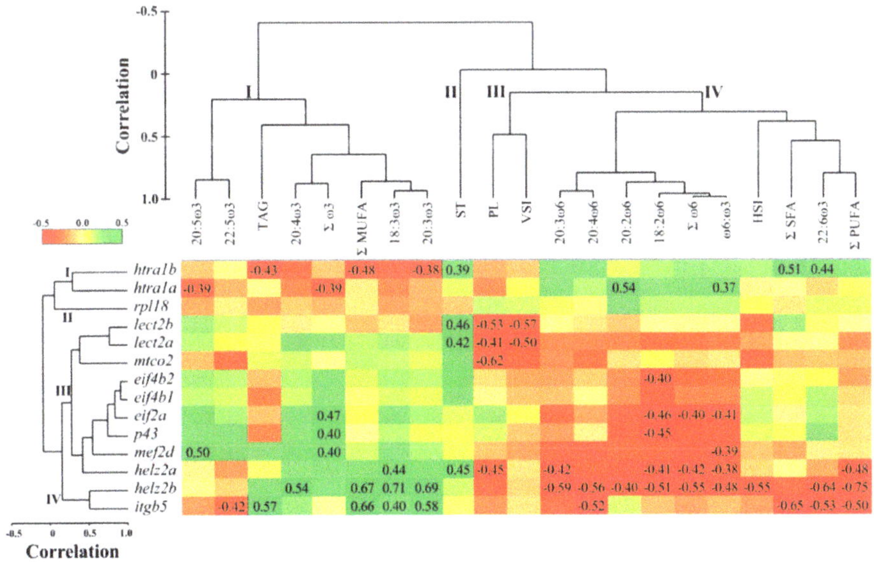

Figure 1. Pearson correlation matrix and hierarchical clustering of liver transcript expression (qPCR relative quantity values (RQ)), liver lipid composition, and somatic indices in Atlantic salmon fed diets with varying ω6 to ω3 fatty acid ratios. Correlation coefficients were described when correlations were statistically significant ($p < 0.05$). Red signifies negative and green signifies positive relationships. ΣSFA, ΣMUFA, and ΣPUFA represents total saturated, monounsaturated and polyunsaturated fatty acids, respectively. 20:5ω3, 22:6ω3, and 20:4ω6 represent EPA, DHA, and ARA, respectively. TAG, ST and PL represent triacylglycerols, sterols, and phospholipids, respectively. HSI and VSI represent hepatosomatic and viscerosomatic indices, respectively.

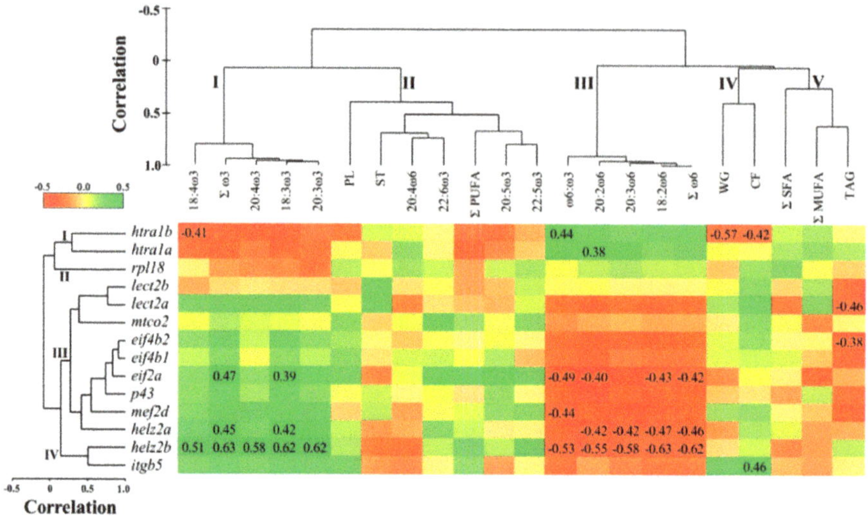

Figure 2. Pearson correlation matrix and hierarchical clustering of liver transcript expression (qPCR relative quantity values (RQ)), muscle lipid composition, and growth in Atlantic salmon fed diets with varying ω6 to ω3 fatty acid ratios. Correlation coefficients were described when correlations were statistically significant ($p < 0.05$). Red signifies negative and green signifies positive relationships. ΣSFA, ΣMUFA, and ΣPUFA represent total saturated, monounsaturated and polyunsaturated fatty acids, respectively. 20:5ω3, 22:6ω3, and 20:4ω6 represent EPA, DHA, and ARA, respectively. TAG, ST and PL represent triacylglycerols, sterols, and phospholipids, respectively. WG and CF represent weight gain and condition factor, respectively.

Cluster analysis of liver lipid composition and somatic indices showed four clusters (Figure 1). Cluster one consisted of the ω3 FA: 18:3ω3, 20:3ω3, 20:4ω3, 20:5ω3, 22:5ω3, as well as the sums of ω3 (Σω3) and monounsaturated FA (ΣMUFA), and the lipid class triacylglycerols (TAG). The lipid class sterols (ST) represented cluster two, while total phospholipids (PL) segregated with viscerosomatic index (VSI) in cluster three. Cluster four consisted of the ω6 FA: 18:2ω6, 20:2ω6, 20:3ω6, 20:4ω6, Σω6, and the ratio ω6:ω3; in addition, 22:6ω3, the sums of PUFA (ΣPUFA) and saturated fatty acids (ΣSFA), and the hepatosomatic index (HSI) were associated with this cluster.

The hepatic transcript expression of *htra1b* was negatively correlated with TAG, ΣMUFA, and 20:3ω3, and positively with ST, ΣSFA, and 22:6ω3 (p = 0.009–0.043; Figure 1). *Htra1a* showed negative correlations with 20:5ω3 and Σω3, and positive with 20:2ω6, and ω6:ω3 (p = 0.006–0.047). Both paralogues of *lect2* were correlated negatively with PL and VSI, and positively with ST (p = 0.004–0.032). Transcript expression of *mtco2* was negatively correlated with PL (p = 0.001), while that of *eif4b2* was negatively correlated with 18:2ω6 (p = 0.036; Figure 1). Both *eif2a* and *p43* transcript expression correlated negatively with 18:2ω6 (p = 0.019 and 0.021, respectively) and positively with Σω3, whereas *eif2a* alone correlated negatively with Σω6 and ω6:ω3 (p = 0.037 and 0.033, respectively). *Mef2d* was correlated negatively with ω6:ω3 and positively with 20:5ω3 (p = 0.042, and 0.011, respectively), while the three transcripts *eif2a*, *p43*, and *mef2d* showed positive correlations with Σω3 (p = 0.016–0.037). Furthermore, both paralogues of *helz2* correlated negatively with ω6 PUFA (i.e., 18:2ω6, 20:3ω6, Σω6), ω6:ω3, and ΣPUFA, and positively with 18:3ω3 (p = 0.0001–0.047). However, *helz2b* had negative correlations with additional ω6 (i.e., 20:2ω6, 20:4ω6; p = 0.038 and 0.004, respectively), and positive correlations with ω3 PUFA (i.e., 20:3ω3, 20:4ω3; p = 0.0001 and 0.005, respectively). In addition, *helz2b* correlated negatively with 22:6ω3 and HSI, and positively with ΣMUFA (p = 0.0001–0.005). In contrast, *helz2a* correlated negatively with PL, and positively with ST (p = 0.021). Finally, *itgb5* had negative correlations with 22:5ω3, 20:4ω6, ΣSFA, 22:6ω3, and ΣPUFA, and positive correlations with TAG, ΣMUFA, 18:3ω3, and 20:3ω3 (p =0.001–0.034; Figure 1).

3.4. Correlations between Hepatic qPCR Transcript Expression and Muscle Lipid Composition

Muscle tissue lipid composition and growth showed five separate clusters (Figure 2). Cluster one consisted of the ω3 FA: 18:3ω3, 18:4ω3, 20:3ω3, 20:4ω3, and Σω3. Cluster two included ΣPUFA, and the LC-PUFA: 20:5ω3, 22:5ω3. 22:6ω3, and 20:4ω6. Furthermore, the lipid classes ST and PL were grouped in cluster two. Cluster three grouped the ω6 FA: 18:2ω6, 20:2ω6, 20:3ω6, as well as Σω6 and the ratio ω6:ω3. Cluster four showed the growth parameters (i.e., WT and CF), while cluster five grouped TAG, ΣSFA and ΣMUFA.

The hepatic transcript expression of *htra1b* was correlated negatively with muscle 18:4ω3 and growth parameters WG and CF, and positively with ω6:ω3 (p = 0.004–0.037), while that of *htra1a* showed positive correlation with 20:2ω6 (p = 0.049; Figure 2). The transcript expression of *lect2a* and *eif4b2* was negatively correlated with TAG (p = 0.021 and 0.049, respectively). *Eif2a* was correlated negatively with muscle ω6 FA (i.e., 18:2ω6, 20:2ω6, Σω6) and ω6:ω3, and positively with ω3 FA (18:3ω3 and Σω3) (p = 0.014–0.045; Figure 2). *Mef2d* was negatively correlated with ω6:ω3 (p = 0.026). Further, both paralogues of *helz2* were negatively correlated with ω6 FA (18:2ω6, 20:2ω6, 20:3ω6, Σω6), and positively correlated with ω3 (i.e., 18:3ω3 and Σω3) (p = 0.001–0.032). *Helz2b* alone correlated negatively with ω6:ω3, and positively with 18:4ω3, 20:3ω3, and 20:4ω3 (p = 0.002–0.01). Finally, *itgb5* showed a positive correlation with CF (p = 0.024).

3.5. Overlapping Lipid–Gene Correlations between the Liver and Muscle Analyses

Some significant correlations showed an overlap between the liver and muscle analyses (Figure 3). The hepatic transcript expression of *htra1a* was correlated positively with 20:2ω6 in both tissues. The expression of *eif2a* was correlated positively with Σω3, and negatively with 18:2ω6, Σω6 and ω6:ω3, while that of *mef2d* showed negative correlations with ω6:ω3 in both liver and muscle. The transcript expression of *helz2a* was correlated positively with 18:3ω3, and negatively with 18:2ω6, 20:3ω6 and Σω6, while that of *helz2b* correlated positively with 18:3ω3, 20:3ω3 and 20:4ω3, and negatively with 18:2ω6, 20:2ω6, 20:3ω6, Σω6 and ω6:ω3 (Figure 3). The hepatic transcript expression of most genes (all except *helz2b* with 20:4ω3) showed stronger positive correlations with liver compared to muscle FA. However, negative correlations were mostly (all except *eif2a* with 18:2ω6, and both *helz2* paralogues with 20:3ω6) more significant with muscle compared with the liver FA (Figure 3).

A	18:3ω3	20:3ω3	20:4ω3	Σ ω3	20:2ω6	18:2ω6	20:3ω6	ω6	ω6:ω3
htra1a					0.54, 0.38				
eif2a				0.47, 0.47		-0.46, -0.43		-0.40, -0.42	-0.41, -0.49
mef2d									-0.39, -0.44
helz2a	0.44, 0.42					-0.41, -0.47	-0.42, -0.42	-0.42, -0.46	
helz2b	0.71, 0.62	0.69, 0.62	0.54, 0.58		-0.40, -0.55	-0.51, -0.63	-0.59, -0.58	-0.55, -0.62	-0.48, -0.53
B *htra1a*					0.006, 0.049				
eif2a				0.016, 0.019		0.019, 0.031		0.037, 0.033	0.033, 0.014
mef2d									0.042, 0.026
helz2a	0.023, 0.032					0.033, 0.019	0.031, 0.032	0.029, 0.021	
helz2b	0.0001, 0.002	0.0001, 0.002	0.005, 0.004		0.038, 0.006	0.010, 0.002	0.003, 0.004	0.005, 0.002	0.015, 0.008

Figure 3. Overlapping Pearson correlations between the liver and muscle analyses. Liver transcript expression (qPCR relative quantity values (RQ)) of GOIs was correlated with liver and muscle lipid composition in Atlantic salmon fed diets with varying ω6 to ω3 fatty acid ratios. Statistically significant ($p < 0.05$) correlations are shown. Green and red cells signify positive and negative relationships, respectively. Upper panel shows correlation coefficients (**A**), and lower panel depicts p-values (**B**). Commas separated values from the liver and muscle analyses, respectively (**A**,**B**).

4. Discussion

The microarray study indicated that dietary variation in ω6:ω3 resulted only in small changes in the liver transcriptome of salmon fed plant-based diets. This can partly be explained by the fact that growth performance and somatic indices were not significantly affected by diet [26]. It was previously shown that different replacements of FO with camelina oil had no impact on Atlantic cod (*Gadus morhua*) growth, and resulted in only one microarray-identified gene that showed a significant difference in spleen basal expression between treatments [47]. Furthermore, our results are in line with previous microarray studies, which demonstrated that dietary replacement of fish meal (FM) and FO with terrestrial ingredients resulted in subtle gene expression changes in Atlantic salmon distal intestine [48], head kidney [49], and liver [50]. However, Atlantic salmon fed soy and linseed oils showed large alterations in hepatic gene expression compared to those fed FO [51]. Differences in the numbers of responsive transcripts between Leaver et al. [51] and the current study could be related to dietary lipid sources and studied time points.

Several transcripts that play important roles in immune and inflammatory response (*lect2a, itgb5, helz2a, p43*), lipid metabolism (*helz2a*), cell proliferation (*htra1b*), muscle and neuronal cell development (*mef2d*), and translation (*eif2a, eif4b1, p43*) were identified by our microarray study as diet-responsive. All transcripts, with the exception of *mtco2* and *rpl18*, showed an agreement in the direction of expression fold-change between the microarray and the qPCR analyses (Table 3). The 60mer microarray probe representing the transcript *mtco2*, which was designed using a rainbow trout cDNA sequence, showed only 87% similarity (see Materials and Methods) with the *Salmo salar* cDNA sequence that was used in the qPCR study (and other *S. salar* sequences in NCBI databases). This fact, as well as other limitations (e.g., mRNA regions targeted by the qPCR primers and microarray probe may not be the same; possibility of contig misassembly) could have contributed to the disagreement between microarray and qPCR results [32].

Hepatic *helz2a* showed a significant differential expression between the high ω6 and high ω3 fed fish in the microarray experiment, and both paralogues of this transcript (i.e., *helz2a* and *helz2b*) were significantly down-regulated in the high ω6 compared to the high ω3 fed fish in the qPCR analysis. Interestingly, the transcript expression of *helz2b* was positively correlated with ω3 (i.e., 18:3ω3, 20:3ω3, 20:4ω3), and negatively with ω6 PUFA (i.e., 18:2ω6, 20:2ω6, 20:3ω6, 20:4ω6), Σω6 and ω6:ω3, in the liver tissue (Figure 1). In the muscle tissue, these PUFA (with the exception of 20:4ω6) were also correlated with hepatic *helz2b* expression (Figure 2). These data suggest that *helz2* is a potential novel molecular biomarker of tissue variation in ω6:ω3. The protein encoded by this gene is a nuclear transcriptional co-activator for PPARα [52–54], which is a master regulator of numerous genes involved in lipid metabolism processes (e.g., FA oxidation, and metabolism of bile acids, triacylglycerols, and retinoids) [55]. Additionally, HELZ2 was shown to have an antiviral function in mammals [56,57], and its gene was identified as an ancestral (between mammals and fish) interferon stimulated gene (ISG) with conserved components of antiviral immunity [58]. *Helz2* transcripts (referred to as VHSV-induced protein (*vig1*)) showed up-regulation in the head kidney of Atlantic salmon exposed to the viral mimic polyriboinosinic polyribocytidylic acid (pIC) [15]. The negative correlation between liver *helz2b* and HSI is not surprising, given the involvement of PPARα in hepatic FA β-oxidation, and in liver steatosis [59,60]. In addition, the observed positive correlations with ω3 PUFA are in line with the anti-inflammatory properties of PPARα [60]. In our previous study [26], it was observed that the fatty acid binding protein-encoding transcript *fabp10* showed an upregulation trend ($p = 0.06$) in the high ω3 compared to the balanced and high ω6 fed fish. Thus, this suggests that the high ω3 diet may have influenced the transport of ω3 FA in liver cells, and played a role in the activation of PPARα. The interaction between liver fatty acid transport and PPARα activation has been shown in previous mammalian studies [61,62]. Another potential mechanism that could explain the positive correlation between *helz2b* expression with ω3 PUFA, is that ω3 PUFA bind to PPARα with higher affinity than ω6 PUFA [63]. However, there is still a lack of knowledge about the interaction between dietary ω3 and ω6 PUFA, and the mechanisms by which they regulate PPARα activators [64].

Mef2d was identified in the microarray as down-regulated by the high ω6 diet, with qPCR showing significantly lower expression in the high ω6 compared to the balanced diet fed fish. Furthermore, hepatic *mef2d* expression was correlated positively with liver 20:5ω3 and Σω3, and negatively with the ratio of ω6:ω3 in both liver and muscle tissues. The gene *mef2*, characterized in zebrafish (*Danio rerio*) [65] and common carp (*Cyprinus carpio*) [66], is involved in skeletal and cardiac muscle development and differentiation, as well as in neuronal cell development [67–70]. Wei et al. [71] reported a significant increase in skeletal muscle *mef2c* transcript expression in pigs fed with linseed-enriched (10%) as compared with a control diet (0%). Additionally, a study with Atlantic salmon revealed that feeding with a synthetic FA (i.e., tetradecylthioacetic acid (0.25%) compared to a control diet (0%) increased the cardiosomatic index, and the cardiac expression of *mef2c* [72]. In relation to the liver, previous studies showed that members of the MEF2 family regulate the activation of hepatic stellate cells –a type of cell involved in liver fibrosis– in mice [73] and rats [74], and ω3 PUFA inhibited the proliferation and activation of these cells in mouse liver [75]. Further, Wang et al. [74] reported that *mef2d* was induced during hepatic stellate cells activation.These data may support the idea that the transcript *mef2* is responding to dietary FA (particularly ω3 PUFA) in vertebrates. However, as most studies examined the role of *mef2* expression in skeletal [71,76] and cardiac [72,77] muscle development, the interactions between *mef2d* and liver physiology are less understood in fish. Future studies should investigate the influence of dietary ω6:ω3 on liver *mef2d* expression, and their interaction with hepatic stellate cells in fish.

Serine protease HTRA1-encoding transcript (*htra1b*) was up-regulated in the high ω6 compared to the high ω3 fed fish, in the microarray study. *Htra1b* showed a similar trend of higher expression in the high ω6 fed fish (p = 0.07) in the qPCR analysis, and a similar fold-change (i.e., high ω6/high ω3) in the microarray and qPCR studies (Table 3). Further, hepatic *htra1b* was positively correlated with ΣSFA, 22:6ω3 and ST, and negatively with ΣMUFA, 20:3ω3 and TAG in the liver, while, in the muscle, it showed positive and negative correlations with ω6:ω3 and 18:4ω3, respectively. The transcript *htra1a* was positively correlated with 20:2ω6 and ω6:ω3, and negatively with ω3 PUFA (i.e., 20:5ω3 and Σω3) in the liver, and showed positive correlation with 20:2ω6 in the muscle. Serine protease HTRA1 function was linked to cell growth and apoptosis, as well as immune and inflammatory responses (by inhibiting the TGF-beta pathway) in mammalian tissues (e.g., eye, bone and liver) [78–81]. It was previously shown that dietary FM replacement with terrestrial plant alternatives induced higher hepatic *htra1* transcript expression in Atlantic salmon [82]. Conversely, replacing both dietary FM and FO by terrestrial plant alternatives down-regulated the transcription of *htra1a* and *htra1b* in Atlantic salmon liver [40]. Discrepancies between Caballero-Solares et al. [40] and the present study extend to the FA–transcript correlation analyses; while, in the present study, the transcript expression of *htra1* paralogues correlated positively and negatively with tissue ω6 and ω3 FA levels, respectively, the opposite tendency was observed in Caballero-Solares et al. [40]. However, unlike the previous studies [40,82], our study tested different mixes of vegetable oils while keeping FM/FO inclusion levels equal across diets. Therefore, although the studies cannot be directly compared, such discrepancies suggest that the regulation of HTRA1-mediated processes in the liver of Atlantic salmon depends on the combination of protein and lipid sources included in the diet. Finally the negative correlation observed between *htra1b* and growth parameters (i.e., WG and CF) in the current study (Figure 2) is interesting, as mammalian HTRA1 was negatively linked to skeletal muscle development and bone formation [80,83,84].

The immune related microarray features *lect2a* and *itgb5* were down-regulated in the high ω6 compared to the high ω3 fed fish, and an agreement was observed in the direction of expression fold-change between the microarray and qPCR studies. LECT2 is a multifunctional protein that plays a role in cell growth, neutrophil chemotactic activity, and innate immune response against pathogens in fish [85–88]. LECT2 also functions as a hepatokine that modulates the inflammatory response in mammals [89,90]. Earlier microarray studies reported down-regulation of hepatic *lect2* in Atlantic salmon fed terrestrial as compared with marine diets [29,40]. This may indicate that pro-inflammatory plant-based diets suppress the constitutive transcript expression of hepatic *lect2*. The observed negative correlation between both paralogues of *lect2* and VSI suggests that *lect2* suppression is related to higher lipid deposition. However, the interaction between fat deposition and immune response is very complex, and requires further investigation in fish [90,91]. Additionally, we showed a positive correlation between the transcript expression of both paralogues of *lect2* and liver sterol content. Interestingly, a previous study reported down-regulation of hepatic *lect2* in Atlantic salmon fed a cholesterol-supplemented diet as compared with a non-supplemented plant-based diet, and this coincided with reduced plasma phytosterols (i.e., sitosterol and campesterol) [92]. Although dietary sterol levels were not significantly different in our study, liver sterol concentration did vary among treatments [26]. Clearly, more studies are required in order to elucidate the impact of dietary and tissue cholesterol and phytosterols on the constitutive transcript expression of *lect2* in fish. Finally, the correlations observed in our study between *itgb5* and liver FA are in line with the notion that FA can regulate the mRNA and protein levels of integrins and other adhesion proteins in leukocytes and endothelial cells [93,94]. Interestingly, European seabass (*Dicentrarchus labrax*) fed a plant-based diet showed a down-regulation in hepatic *Integrin beta-2* compared to those fed a marine diet [95], while a reduction in the ω6:ω3 ratio of human lung cancer cells resulted in a delayed adhesion, and down-regulation of *integrin-α2* [96]. Taken together, these data suggest that the transcript expression of integrins may be impacted by dietary or tissue ω6:ω3.

Similar to *lect2a* and *itgb5*, the transcripts *p43*, *eif2a*, and *eif4b1* were down-regulated in the high ω6 fed fish in the microarray experiment, and they showed an agreement in the direction of dietary modulation with the qPCR study. The transcript expression of *p43*, *eif2a*, and *eif4b2* was negatively correlated with 18:2ω6, and both *p43* and *eif2a* were positively correlated with liver Σω3. Further, *eif2a* expression was positively correlated with 18:3ω3 and Σω3, and negatively correlated with 18:2ω6, 20:2ω6, Σω6 and ω6:ω3 in the muscle. The protein p43 is associated with a multi-tRNA synthetase complex, and regulates tRNA channeling in mammals [97]. In addition, p43 also encodes an apoptosis-induced cytokine, which regulates inflammation, wound healing, and angiogenesis [98–100]. Phosphorylation of the protein eIF4B was shown to stimulate translation in zebrafish [101,102] and yeast [103]. However, phosphorylation of eIF2A repressed translation in response to accumulation of misfolded proteins in the ER of several fish species [104–106]. Thus, changes in the expression patterns of *p43*, *eif2a* and *eif4b*, and the correlations observed with tissue lipid composition, suggest that some aspects of protein synthesis were influenced by dietary and tissue ω6:ω3. Previous mammalian studies demonstrated that translation is inhibited in apoptotic cells, and this was correlated with enhanced cleavage of the eukaryotic translation initiation factors eIF4B, eIF2, and others [107,108]. Thus, the fact that the high ω6 fed fish showed up-regulation of *htra1b* (Table 2) may suggest that apoptosis was associated with the observed modulation of translation-related transcripts (Table 2), and their correlations with tissue FA (Figures 1 and 2). However, as our microarray study did not identify other well-known apoptosis biomarkers (e.g., genes encoding caspases and Bcl-2 family members), this can only be postulated. Further, the stimulatory effects of ω3 PUFA on protein synthesis [109] could be another potential mechanism explaining the positive correlations observed between *p43*, *eif2a*, and liver ω3 FA. Research examining the impact of replacing FO/FM with plant-based diets on protein synthesis in salmonids has been contradictory. Some authors showed an induction [82], while others showed a suppression [110] of these and other translation-related transcripts. Indeed, protein synthesis regulation in fish is a dynamic process, and is influenced by dietary formulations, genetic [50] and abiotic factors, protein requirement, growth, and the tissues examined [111,112].

5. Conclusions

Our 44 K microarray study demonstrated that high ω6 and high ω3 plant-based diets with varying ratios of ω6:ω3 (i.e., 2.7 and 0.4, respectively) resulted in a relatively low number of differentially expressed transcripts in salmon liver. However, the identified transcripts and/or their functional annotations suggested important roles in lipid metabolism (*helz2a*), cell proliferation (*htra1b*), immune and inflammatory response (*lect2a, itgb5, helz2a, p43*), control of muscle and neuronal cell development (*mef2d*), and translation (*eif2a, eif4b1, p43*). Two paralogues of *helz2* were down-regulated in the high ω6 compared to the high ω3 fed fish in the qPCR study. Significant positive correlations were observed between the hepatic transcript expression of *helz2b* and ω3 PUFA, while negative correlations were identified with ω6 PUFA and ω6:ω3, in both the liver and muscle tissues. This indicated that the PPARα activation-related transcript *helz2* is a potential novel molecular biomarker of tissue variation in ω6:ω3. Given these data and the importance of *helz2* as an ancestral vertebrate interferon stimulated gene, future studies should investigate the dietary ω6:ω3 impact on Atlantic salmon anti-viral response. The transcript *mef2d* was suppressed in the high ω6 compared to the balanced fed fish, and was negatively correlated with ω6:ω3 in both tissues. Our microarray study revealed that the upregulation of hepatic *htra1b* concurred with the suppression of immune- and inflammatory-related transcripts (i.e., *lect2a, p43, helz2a, helz2b*, and *itgb5*). This supported the idea proposed by other researchers [40,82] of a link between the dietary modulation of *htra1* and that of immune-related transcripts. Finally, the transcripts *p43, eif2a*, and *eif4b1* were significantly down-regulated in the high ω6 compared to the high ω3 fed fish in the microarray, and showed an agreement in the direction of expression fold-change between the microarray and qPCR studies. These data, along with the significant correlations observed between *p43, eif2a* and *eif4b2* expression and tissue PUFA, suggested that the molecular regulation of protein synthesis in the liver may have been impacted by dietary ω6:ω3.

Supplementary Materials: The following are available online at https://www.mdpi.com/article/10.3390/biology10070578/s1, Table S1: Formulation and nutrient composition (%) of experimental diets fed to Atlantic salmon. This diet information was published in Katan et al. [26]. However, it is included here as this information is pertinent to the current study as well, Table S2: Lipid and FA composition (%) of experimental diets fed to Atlantic salmon. This diet information was published in Katan et al. [26]. However, it is included here as this information is pertinent to the current study as well, Table S3: Paralogues of genes involved in the qPCR, and their identity (%) over the aligned nucleotide regions, Figure S1: Alignment of the nucleotide sequences of *htra1a* (GenBank accession number NM_001141717) and *htra1b* (GenBank accession number EG831192). Conserved regions are highlighted in yellow. Alignments were performed using AlignX (Vector NTI Advance 11). Forward primers are bolded and underlined, whereas reverse primers are bolded without an underline, Figure S2: Alignment of the nucleotide sequences of *eif4b1* (GenBank accession number BT072661) and *eif4b2* (accession number DY739566). Conserved regions are highlighted in yellow. Alignments were performed using AlignX (Vector NTI Advance 11). Forward primers are bolded and underlined, whereas reverse primers are bolded without an underline, Figure S3: Alignment of the nucleotide sequences of *lect2a* (GenBank accession number BT059281) and *lect2b* (GenBank accession number DV106130). Conserved regions are highlighted in yellow. Alignments were performed using AlignX (Vector NTI Advance 11). Forward primers are bolded and underlined, whereas reverse primers are bolded without an underline, Figure S4: Alignment of the nucleotide sequences of *helz2a* (GenBank accession number BT072427) and *helz2b* (GenBank accession number EG928625). Conserved regions are highlighted in yellow. Alignments were performed using AlignX (Vector NTI Advance 11). Forward primers are bolded and underlined, whereas reverse primers are bolded without an underline.

Author Contributions: Conceptualization, M.L.R., C.C.P. and R.G.T.; Methodology, T.K., X.X. and A.C.-S.; Software, T.K., X.X. and A.C.-S.; Validation, T.K., X.X., A.C.-S., R.G.T., C.C.P. and M.L.R.; Formal Analysis, T.K., X.X. and A.C.-S.; Investigation, T.K. and X.X.; Resources, R.G.T., C.C.P. and M.L.R.; Data Curation, T.K. and X.X.; Writing—Original Draft Preparation, T.K.; Writing—Review and Editing, T.K., X.X., A.C.-S., R.G.T., C.C.P. and M.L.R.; Visualization, T.K., X.X. and A.C.-S.;

Supervision, M.L.R. and C.C.P.; Project Administration, T.K., X.X., C.C.P. and M.L.R.; Funding Acquisition, M.L.R., C.C.P. and R.G.T. All authors have read and agreed to the published version of the manuscript.

Funding: This study was conducted within the Biomarker Platform for Commercial Aquaculture Feed Development project. This research was funded by the Government of Canada through Genome Canada and Genome Atlantic (Genomic Applications Partnership Program, GAPP), grant number: 6604. This project was also funded by Innovate NL (Government of Newfoundland and Labrador Department of Tourism, Culture, Industry and Innovation), award number: 211219, as well as Atlantic Canada Opportunities Agency (ACOA), grant number: 206200, to M.L.R. and C.C.P. M.L.R. was also supported by Natural Sciences and Engineering Research Council of Canada (NSERC) Discovery Grants, grant numbers: 341304-2012 and 2020-04519. T.K. was supported by a Postgraduate Scholarship-Doctoral (PGS D) from NSERC, Ocean Industries Student Research Award (OISRA) from Innovate NL, and a SGS fellowship from Memorial University Newfoundland.

Institutional Review Board Statement: Ethical treatment of fish in this experiment was carried out in accordance with the guidelines of the Canadian Council on Animal Care, and approved by the Institutional Animal Care Committee of Memorial University of Newfoundland (protocol # 16-74-MR; date of approval: 5 April 2016).

Informed Consent Statement: Not applicable.

Data Availability Statement: The data presented in this study are available within the article. The microarray data were submitted to NCBI's Gene Expression Omnibus (GEO) repository (GSE139418, https://www.ncbi.nlm.nih.gov/geo/query/acc.cgi?acc=GSE139418. If required, any additional data are available on request from the authors.

Acknowledgments: The authors would like to acknowledge Cara Kirkpatrick for managing the project, Dominic Nanton for feed formulations, Umasuthan Navaneethaiyer for assistance with microarray analysis, and Jeanette Wells for her help with the analytical work. We would also like to thank the sampling teams in the Matthew L. Rise and Christopher C. Parrish laboratories. Finally, we would like to thank Danny Boyce, and the Dr. Joe Brown Aquatic Research Building (JBARB) staff, for their assistance with fish husbandry and sampling.

Conflicts of Interest: The authors declare no conflict of interest. R.G.T. was employed by Cargill Innovation at the time this research was conducted. R.G.T., in the representation of Cargill Innovation, participated in the formulation of the experimental diets, the design of the trial, and reviewed the manuscript. However, he had no role in the design of the gene expression experiments and correlation analyses, in the collection, analyses, or interpretation of data; in the writing of the manuscript, or in the decision to publish the results.

References

1. Wilberg, M.J.; Miller, T.J. Comment on "Impacts of biodiversity loss on ocean ecosystem services". *Science* **2007**, *316*, 787–790. [CrossRef] [PubMed]
2. Naylor, R.L.; Hardy, R.W.; Bureau, D.P.; Chiu, A.; Elliott, M.; Farrell, A.P.; Forster, I.; Gatlin, D.M.; Goldburg, R.J.; Hua, K.; et al. Feeding aquaculture in an era of finite resources. *Proc. Natl. Acad. Sci. USA* **2009**, *106*, 15103–15110. [CrossRef] [PubMed]
3. Turchini, G.M.; Francis, D.S. Fatty acid metabolism (desaturation, elongation and β-oxidation) in rainbow trout fed fish oil- or linseed oil-based diets. *Br. J. Nutr.* **2009**, *102*, 69–81. [CrossRef] [PubMed]
4. Bransden, M.P.; Carter, C.G.; Nichols, P.D. Replacement of fish oil with sunflower oil in feeds for Atlantic salmon (*Salmo salar* L.): Effect on growth performance, tissue fatty acid composition and disease resistance. *Comp. Biochem. Physiol. B Biochem. Mol. Biol.* **2003**, *135*, 611–625. [CrossRef]
5. Tocher, D.R. Fatty acid requirements in ontogeny of marine and freshwater fish. *Aquac. Res.* **2010**, *41*, 717–732. [CrossRef]
6. Liland, N.S.; Rosenlund, G.; Berntssen, M.H.G.; Brattelid, T.; Madsen, L.; Torstensen, B.E. Net production of Atlantic salmon (FIFO, Fish in Fish out <1) with dietary plant proteins and vegetable oils. *Aquac. Nutr.* **2013**, *19*, 289–300. [CrossRef]
7. Tocher, D.R. Omega-3 long-chain polyunsaturated fatty acids and aquaculture in perspective. *Aquaculture* **2015**, *449*, 94–107. [CrossRef]
8. Alhazzaa, R.; Bridle, A.R.; Nichols, P.D.; Carter, C.G. Replacing dietary fish oil with Echium oil enriched barramundi with C18 PUFA rather than long-chain PUFA. *Aquaculture* **2011**, *312*, 162–171. [CrossRef]
9. Calder, P.C. Functional roles of fatty acids and their effects on human health. *J. Parenter. Enter. Nutr.* **2015**, *39*, 18S–32S. [CrossRef]
10. Sprague, M.; Dick, J.R.; Tocher, D.R. Impact of sustainable feeds on omega-3 long-chain fatty acid levels in farmed Atlantic salmon, 2006–2015. *Sci. Rep.* **2016**, *6*, 21892. [CrossRef]

11. Montero, D.; Kalinowski, T.; Obach, A.; Robaina, L.; Tort, L.; Caballero, M.J.; Izquierdo, M.S. Vegetable lipid sources for gilthead seabream (*Sparus aurata*): Effects on fish health. *Aquaculture* **2003**, *225*, 353–370. [CrossRef]
12. Ruyter, B.; Moya-Falcón, C.; Rosenlund, G.; Vegusdal, A. Fat content and morphology of liver and intestine of Atlantic salmon (*Salmo salar*): Effects of temperature and dietary soybean oil. *Aquaculture* **2006**, *252*, 441–452. [CrossRef]
13. Jordal, A.E.O.; Lie, Ø.; Torstensen, B.E. Complete replacement of dietary fish oil with a vegetable oil blend affect liver lipid and plasma lipoprotein levels in Atlantic salmon (*Salmo salar* L.). *Aquac. Nutr.* **2007**, *13*, 114–130. [CrossRef]
14. Liland, N.S. Atlantic Salmon (*Salmo salar* L.) Sterol Metabolism and Metabolic Health Impact of Dietary Lipids. Ph.D. Thesis, Department of Biology, University of Bergen, National Institute of Nutrition and Seafood Research, Bergen, Norway, December 2014.
15. Caballero-Solares, A.; Hall, J.R.; Xue, X.; Eslamloo, K.; Taylor, R.G.; Parrish, C.C.; Rise, M.L. The dietary replacement of marine ingredients by terrestrial animal and plant alternatives modulates the antiviral immune response of Atlantic salmon (*Salmo salar*). *Fish Shellfish Immunol.* **2017**, *64*, 24–38. [CrossRef]
16. Pickova, J.; Mørkøre, T. Alternate oils in fish feeds. *Eur. J. Lipid Sci. Technol.* **2007**, *109*, 256–263. [CrossRef]
17. Weaver, K.L.; Ivester, P.; Chilton, J.A.; Wilson, M.D.; Pandey, P.; Chilton, F.H. The content of favorable and unfavorable polyunsaturated fatty acids found in commonly eaten fish. *J. Am. Diet. Assoc.* **2008**, *108*, 1178–1185. [CrossRef]
18. Young, K. Omega-6 (n-6) and omega-3 (n-3) fatty acids in tilapia and human health: A review. *Int. J. Food Sci. Nutr.* **2009**, *60* (Suppl. 5), 203–211. [CrossRef]
19. Simopoulos, A.P. Importance of the ratio of omega-6/omega-3 essential fatty acids: Evolutionary aspects. *World Rev. Nutr. Diet.* **2003**, *92*, 1–22. [CrossRef]
20. Wijendran, V.; Hayes, K.C. Dietary n-6 and n-3 fatty acid balance and cardiovascular health. *Annu. Rev. Nutr.* **2004**, *24*, 597–615. [CrossRef]
21. Simopoulos, A.P. The importance of the omega-6/omega-3 fatty acid ratio in cardiovascular disease and other chronic diseases. *Exp. Biol. Med.* **2008**, *233*, 674–688. [CrossRef]
22. Gómez Candela, C.; Bermejo López, L.M.; Loria Kohen, V. Importance of a balanced omega 6/omega 3 ratio for the maintenance of health: Nutritional recommendations. *Nutr. Hosp.* **2011**, *26*, 323–329. [CrossRef]
23. De Pablo Martinez, M.A.; Álvarez De Cienfuegos, G. Modulatory effects of dietary lipids on immune system functions. *Immunol. Cell Biol.* **2000**, *78*, 31–39. [CrossRef]
24. Wymann, M.P.; Schneiter, R. Lipid signalling in disease. *Nat. Rev. Mol. Cell Biol.* **2008**, *9*, 162–176. [CrossRef]
25. Duan, Y.; Li, F.; Li, L.; Fan, J.; Sun, X.; Yin, Y. n-6:n-3 PUFA ratio is involved in regulating lipid metabolism and inflammation in pigs. *Br. J. Nutr.* **2014**, *111*, 445–451. [CrossRef]
26. Katan, T.; Caballero-Solares, A.; Taylor, R.G.; Rise, M.L.; Parrish, C.C. Effect of plant-based diets with varying ratios of ω6 to ω3 fatty acids on growth performance, tissue composition, fatty acid biosynthesis and lipid-related gene expression in Atlantic salmon (*Salmo salar*). *Comp. Biochem. Physiol. Part D Genom. Proteom.* **2019**, *30*, 290–304. [CrossRef]
27. Jantzen, S.G.; Sanderson, D.S.; Von Schalburg, K.R.; Yasuike, M.; Marass, F.; Koop, B.F. A 44K microarray dataset of the changing transcriptome in developing Atlantic salmon (*Salmo salar* L.). *BMC Res. Notes* **2011**, *4*, 88. [CrossRef]
28. Sahlmann, C.; Sutherland, B.J.G.; Kortner, T.M.; Koop, B.F.; Krogdahl, Å.; Bakke, A.M. Early response of gene expression in the distal intestine of Atlantic salmon (*Salmo salar* L.) during the development of soybean meal induced enteritis. *Fish Shellfish Immunol.* **2013**, *34*, 599–609. [CrossRef]
29. Xue, X.; Hixson, S.M.; Hori, T.S.; Booman, M.; Parrish, C.C.; Anderson, D.M.; Rise, M.L. Atlantic salmon (*Salmo salar*) liver transcriptome response to diets containing *Camelina sativa* products. *Comp. Biochem. Physiol. Part D Genom. Proteom.* **2015**, *14*, 1–15. [CrossRef]
30. National Research Council. *Nutrient Requirements of Fish and Shrimp*; National Academies Press: Washington, DC, USA, 2011. [CrossRef]
31. Xu, Q.; Feng, C.Y.; Hori, T.S.; Plouffe, D.A.; Buchanan, J.T.; Rise, M.L. Family-specific differences in growth rate and hepatic gene expression in juvenile triploid growth hormone (GH) transgenic Atlantic salmon (*Salmo salar*). *Comp. Biochem. Physiol. Part D Genom. Proteom.* **2013**, *8*, 317–333. [CrossRef]
32. Booman, M.; Borza, T.; Feng, C.Y.; Hori, T.S.; Higgins, B.; Culf, A.; Léger, D.; Chute, I.C.; Belkaid, A.; Rise, M.L.; et al. Development and experimental validation of a 20K Atlantic cod (*Gadus morhua*) oligonucleotide microarray based on a collection of over 150,000 ESTs. *Mar. Biotechnol.* **2011**, *13*, 733–750. [CrossRef]
33. Celton, M.; Malpertuy, A.; Lelandais, G.; de Brevern, A.G. Comparative analysis of missing value imputation methods to improve clustering and interpretation of microarray experiments. *BMC Genom.* **2010**, *11*, 15. [CrossRef] [PubMed]
34. Bø, T.H.; Dysvik, B.; Jonassen, I. LSimpute: Accurate estimation of missing values in microarray data with least squares methods. *Nucleic Acids Res.* **2004**, *32*, e34. [CrossRef] [PubMed]
35. Tusher, V.G.; Tibshirani, R.; Chu, G. Significance analysis of microarrays applied to the ionizing radiation response. *Proc. Natl. Acad. Sci. USA* **2001**, *98*, 5116–5121. [CrossRef] [PubMed]
36. Schwender, H.; Krause, A.; Ickstadt, K. Identifying interesting genes with siggenes. *RNews* **2006**, *6*, 45–50.
37. Breitling, R.; Armengaud, P.; Amtmann, A.; Herzyk, P. Rank products: A simple, yet powerful, new method to detect differentially regulated genes in replicated microarray experiments. *FEBS Lett.* **2004**, *573*, 83–92. [CrossRef]
38. Jeffery, I.B.; Higgins, D.G.; Culhane, A.C. Comparison and evaluation of methods for generating differentially expressed gene lists from microarray data. *BMC Bioinform.* **2006**, *7*, 359. [CrossRef]

39. Hong, F.; Breitling, R.; McEntee, C.W.; Wittner, B.S.; Nemhauser, J.L.; Chory, J. RankProd: A bioconductor package for detecting differentially expressed genes in meta-analysis. *Bioinformatics* **2006**, *22*, 2825–2827. [CrossRef]
40. Caballero-Solares, A.; Xue, X.; Parrish, C.C.; Foroutani, M.B.; Taylor, R.G.; Rise, M.L. Changes in the liver transcriptome of farmed Atlantic salmon (*Salmo salar*) fed experimental diets based on terrestrial alternatives to fish meal and fish oil. *BMC Genom.* **2018**, *19*, 796. [CrossRef]
41. Rise, M.L.; Hall, J.R.; Rise, M.; Hori, T.S.; Browne, M.J.; Gamperl, A.K.; Hubert, S.; Kimball, J.; Bowman, S.; Johnson, S.C. Impact of asymptomatic nodavirus carrier state and intraperitoneal viral mimic injection on brain transcript expression in Atlantic cod (*Gadus morhua*). *Physiol. Genom.* **2010**, *42*, 266–280. [CrossRef]
42. Pfaffl, M.W. A new mathematical model for relative quantification in real-time RT-PCR. *Nucleic Acids Res.* **2001**, *29*, e45. [CrossRef]
43. Olsvik, P.A.; Lie, K.K.; Jordal, A.E.O.; Nilsen, T.O.; Hordvik, I. Evaluation of potential reference genes in real-time RT-PCR studies of Atlantic salmon. *BMC Mol. Biol.* **2005**, *6*, 21. [CrossRef]
44. Vandesompele, J.; De Preter, K.; Pattyn, F.; Poppe, B.; Van Roy, N.; De Paepe, A.; Speleman, F. Accurate normalization of real-time quantitative RT-PCR data by geometric averaging of multiple internal control genes. *Genome Biol.* **2002**, *3*, 34. [CrossRef]
45. Livak, K.J.; Schmittgen, T.D. Analysis of relative gene expression data using real-time quantitative PCR and the 2-ΔΔCT method. *Methods* **2001**, *25*, 402–408. [CrossRef]
46. Rise, M.L.; Hall, J.R.; Nash, G.W.; Xue, X.; Booman, M.; Katan, T.; Gamperl, A.K. Transcriptome profiling reveals that feeding wild zooplankton to larval Atlantic cod (*Gadus morhua*) influences suites of genes involved in oxidation-reduction, mitosis, and selenium homeostasis. *BMC Genom.* **2015**, *16*, 1016. [CrossRef]
47. Booman, M.; Xu, Q.; Rise, M.L. Evaluation of the impact of camelina oil-containing diets on the expression of genes involved in the innate anti-viral immune response in Atlantic cod (*Gadus morhua*). *Fish Shellfish Immunol.* **2014**, *41*, 52–63. [CrossRef]
48. Brown, T.D.; Hori, T.S.; Xue, X.; Ye, C.L.; Anderson, D.M.; Rise, M.L. Functional genomic analysis of the impact of Camelina (*Camelina sativa*) meal on Atlantic salmon (*Salmo salar*) distal intestine gene expression and physiology. *Mar. Biotechnol.* **2016**, *18*, 418–435. [CrossRef]
49. Eslamloo, K.; Xue, X.; Hall, J.R.; Smith, N.C.; Caballero-Solares, A.; Parrish, C.C.; Taylor, R.G.; Rise, M.L. Transcriptome profiling of antiviral immune and dietary fatty acid dependent responses of Atlantic salmon macrophage-like cells. *BMC Genom.* **2017**, *18*, 706. [CrossRef]
50. Morais, S.; Pratoomyot, J.; Taggart, J.B.; Bron, J.E.; Guy, D.R.; Bell, J.G.; Tocher, D.R. Genotype-specific responses in Atlantic salmon (*Salmo salar*) subject to dietary fish oil replacement by vegetable oil: A liver transcriptomic analysis. *BMC Genom.* **2011**, *12*, 255. [CrossRef]
51. Leaver, M.J.; Villeneuve, L.A.N.; Obach, A.; Jensen, L.; Bron, J.E.; Tocher, D.R.; Taggart, J.B. Functional genomics reveals increases in cholesterol biosynthetic genes and highly unsaturated fatty acid biosynthesis after dietary substitution of fish oil with vegetable oils in Atlantic salmon (*Salmo salar*). *BMC Genom.* **2008**, *9*, 1–15. [CrossRef]
52. Surapureddi, S.; Yu, S.; Bu, H.; Hashimoto, T.; Yeldandi, A.V.; Kashireddy, P.; Cherkaoui-Malki, M.; Qi, C.; Zhu, Y.J.; Rao, M.S.; et al. Identification of a transcriptionally active peroxisome proliferator-activated receptor α-interacting cofactor complex in rat liver and characterization of PRIC285 as a coactivator. *Proc. Natl. Acad. Sci. USA* **2002**, *99*, 11836–11841. [CrossRef]
53. Katano-Toki, A.; Satoh, T.; Tomaru, T.; Yoshino, S.; Ishizuka, T.; Ishii, S.; Ozawa, A.; Shibusawa, N.; Tsuchiya, T.; Saito, T.; et al. THRAP3 interacts with HELZ2 and plays a novel role in adipocyte differentiation. *Mol. Endocrinol.* **2013**, *27*, 769–780. [CrossRef]
54. Yoshino, S.; Satoh, T.; Yamada, M.; Hashimoto, K.; Tomaru, T.; Katano-Toki, A.; Kakizaki, S.; Okada, S.; Shimizu, H.; Ozawa, A.; et al. Protection against high-fat diet-induced obesity in Helz2-deficient male mice due to enhanced expression of hepatic leptin receptor. *Endocrinology* **2014**, *155*, 3459–3472. [CrossRef]
55. Kersten, S. Integrated physiology and systems biology of PPARα. *Mol. Metab.* **2014**, *3*, 354–371. [CrossRef]
56. Fusco, D.N.; Pratt, H.; Kandilas, S.; Cheon, S.S.Y.; Lin, W.; Cronkite, D.A.; Basavappa, M.; Jeffrey, K.L.; Anselmo, A.; Sadreyev, R.; et al. HELZ2 is an IFN effector mediating suppression of dengue virus. *Front. Microbiol.* **2017**, *8*, 240. [CrossRef]
57. Fu, M.; Blackshear, P.J. RNA-binding proteins in immune regulation: A focus on CCCH zinc finger proteins. *Nat. Rev. Immunol.* **2017**, *17*, 130–143. [CrossRef]
58. Levraud, J.; Jouneau, L.; Briolat, V.; Laghi, V.; Boudinot, P. IFN-stimulated genes in zebrafish and humans define an ancient arsenal of antiviral immunity. *J. Immunol.* **2019**, *203*, 3361–3373. [CrossRef]
59. Kersten, S.; Desvergne, B.; Wahli, W. Roles of PPARS in health and disease. *Nature* **2000**, *405*, 421–424. [CrossRef]
60. Van Raalte, D.H.; Li, M.; Pritchard, P.H.; Wasan, K.M. Peroxisome proliferator-activated receptor (PPAR)-α: A pharmacological target with a promising future. *Pharm. Res.* **2004**, *21*, 1531–1538. [CrossRef]
61. Huang, H.; Starodub, O.; McIntosh, A.; Kier, A.B.; Schroeder, F. Liver fatty acid-binding protein targets fatty acids to the nucleus. Real time confocal and multiphoton fluorescence imaging in living cells. *J. Biol. Chem.* **2002**, *277*, 29139–29151. [CrossRef]
62. Hostetler, H.A.; McIntosh, A.L.; Atshaves, B.P.; Storey, S.M.; Payne, H.R.; Kier, A.B.; Schroeder, F. L-FABP directly interacts with PPARα in cultured primary hepatocytes. *J. Lipid Res.* **2009**, *50*, 1663–1675. [CrossRef]
63. Desvergne, B.; Wahli, W. Peroxisome proliferator-activated receptors: Nuclear control of metabolism. *Endocr. Rev.* **1999**, *20*, 649–688. [CrossRef] [PubMed]
64. Schroeder, F.; Petrescu, A.D.; Huang, H.; Atshaves, B.P.; McIntosh, A.L.; Martin, G.G.; Hostetler, H.A.; Vespa, A.; Landrock, D.; Landrock, K.K.; et al. Role of fatty acid binding proteins and long chain fatty acids in modulating nuclear receptors and gene transcription. *Lipids* **2008**, *43*, 1–17. [CrossRef] [PubMed]

65. Ticho, B.S.; Stainier, D.Y.R.; Fishman, M.C.; Breitbart, R.E. Three zebrafish MEF2 genes delineate somitic and cardiac muscle development in wild-type and mutant embryos. *Mech. Dev.* **1996**, *59*, 205–218. [CrossRef]
66. He, M.; Zhou, D.; Ding, N.Z.; Teng, C.B.; Yan, X.C.; Liang, Y. Common carp MEF2 genes: Evolution and expression. *Genes* **2019**, *10*, 588. [CrossRef]
67. Black, B.L.; Olson, E.N. Transcriptional control of muscle development by myocyte enhancer factor-2 (MEF2) proteins. *Annu. Rev. Cell Dev. Biol.* **1998**, *14*, 167–196. [CrossRef]
68. Flavell, S.W.; Cowan, C.W.; Kim, T.K.; Greer, P.L.; Lin, Y.; Paradis, S.; Griffith, E.C.; Hu, L.S.; Chen, C.; Greenberg, M.E. Activity-dependent regulation of MEF2 transcription factors suppresses excitatory synapse number. *Science* **2006**, *311*, 1008–1012. [CrossRef]
69. Haberland, M.; Arnold, M.A.; McAnally, J.; Phan, D.; Kim, Y.; Olson, E.N. Regulation of HDAC9 gene expression by MEF2 establishes a negative-feedback loop in the transcriptional circuitry of muscle differentiation. *Mol. Cell. Biol.* **2007**, *27*, 518–525. [CrossRef]
70. Flavell, S.W.; Kim, T.K.; Gray, J.M.; Harmin, D.A.; Hemberg, M.; Hong, E.J.; Markenscoff-Papadimitriou, E.; Bear, D.M.; Greenberg, M.E. Genome-wide analysis of MEF2 transcriptional program reveals synaptic target genes and neuronal activity-dependent polyadenylation site selection. *Neuron* **2008**, *60*, 1022–1038. [CrossRef]
71. Wei, H.; Zhou, Y.; Jiang, S.; Huang, F.; Peng, J.; Jiang, S. Transcriptional response of porcine skeletal muscle to feeding a linseed-enriched diet to growing pigs. *J. Anim. Sci. Biotechnol.* **2016**, *7*, 6. [CrossRef]
72. Grammes, F.; Rørvik, K.A.; Takle, H. Tetradecylthioacetic acid modulates cardiac transcription in Atlantic salmon, *Salmo salar* L., suffering heart and skeletal muscle inflammation. *J. Fish Dis.* **2012**, *35*, 109–117. [CrossRef]
73. Xiaomeng, M.; Zhang, L.; Lixin, D. Research progress on MEF2B gene in human and animals. *Agric. Sci. Technol.* **2016**, *17*, 2477–2482.
74. Wang, X.; Tang, X.; Gong, X.; Albanis, E.; Friedman, S.L.; Mao, Z. Regulation of hepatic stellate cell activation and growth by transcription factor myocyte enhancer factor 2. *Gastroenterology* **2004**, *127*, 1174–1188. [CrossRef]
75. Zhang, K.; Chang, Y.; Shi, Z.; Han, X.; Han, Y.; Yao, Q. ω-3 PUFAs ameliorate liver fibrosis and inhibit hepatic stellate cells proliferation and activation by promoting YAP/TAZ degradation. *Sci. Rep.* **2016**, *6*, 1–14. [CrossRef]
76. Fuentes, E.N.; Valdés, J.A.; Molina, A.; Björnsson, B.T. Regulation of skeletal muscle growth in fish by the growth hormone—Insulin-like growth factor system. *Gen. Comp. Endocrinol.* **2013**, *192*, 136–148. [CrossRef]
77. Lien, C.L.; Schebesta, M.; Makino, S.; Weber, G.J.; Keating, M.T. Gene expression analysis of zebrafish heart regeneration. *PLoS Biol.* **2006**, *4*, 1386–1396. [CrossRef]
78. Clausen, T.; Southan, C.; Ehrmann, M. The HtrA family of proteases: Implications for protein composition and cell fate. *Mol. Cell* **2002**, *10*, 443–455. [CrossRef]
79. Oka, C.; Tsujimoto, R.; Kajikawa, M.; Koshiba-Takeuchi, K.; Ina, J.; Yano, M.; Tsuchiya, A.; Ueta, Y.; Soma, A.; Kanda, H.; et al. HtrA1 serine protease inhibits signaling mediated by Tgfβ family proteins. *Development* **2004**, *131*, 1041–1053. [CrossRef]
80. Graham, J.R.; Chamberland, A.; Lin, Q.; Li, X.J.; Dai, D.; Zeng, W.; Ryan, M.S.; Rivera-Bermúdez, M.A.; Flannery, C.R.; Yang, Z. Serine protease HTRA1 antagonizes transforming growth factor-β signaling by cleaving its receptors and loss of HTRA1 in vivo enhances bone formation. *PLoS ONE* **2013**, *8*, e74094. [CrossRef]
81. Zhu, F.; Jin, L.; Luo, T.P.; Luo, G.H.; Tan, Y.; Qin, X.H. Serine protease HtrA1 expression in human hepatocellular carcinoma. *Hepatobiliary Pancreat. Dis. Int.* **2010**, *9*, 508–512.
82. Tacchi, L.; Secombes, C.J.; Bickerdike, R.; Adler, M.A.; Venegas, C.; Takle, H.; Martin, S.A.M. Transcriptomic and physiological responses to fishmeal substitution with plant proteins in formulated feed in farmed Atlantic salmon (*Salmo salar*). *BMC Genom.* **2012**, *13*, 363. [CrossRef]
83. Bakay, M.; Zhao, P.; Chen, J.; Hoffman, E.P. A web-accessible complete transcriptome of normal human and DMD muscle. *Neuromuscul. Disord.* **2002**, *12*, S125–S141. [CrossRef]
84. Tiaden, A.N.; Richards, P.J. The emerging roles of HTRA1 in musculoskeletal disease. *Am. J. Pathol.* **2013**, *182*, 1482–1488. [CrossRef] [PubMed]
85. Kokkinos, P.A.; Kazantzi, A.; Sfyroera, G.; Zarkadis, I.K. Molecular cloning of leukocyte cell-derived chemotaxin 2 in rainbow trout. *Fish Shellfish Immunol.* **2005**, *18*, 371–380. [CrossRef] [PubMed]
86. Wei, J.; Guo, M.; Cui, H.; Yan, Y.; Ouyang, Z.; Qin, Q. A new leukocyte cell-derived chemotaxin-2 from marine fish grouper, *Epinephelus coioides*: Molecular cloning and expression analysis. *Fish Shellfish Immunol.* **2011**, *31*, 600–605. [CrossRef]
87. Chen, J.; Chen, Q.; Lu, X.J.; Li, C.H. LECT2 improves the outcomes in ayu with *Vibrio anguillarum* infection via monocytes/macrophages. *Fish Shellfish Immunol.* **2014**, *41*, 586–592. [CrossRef]
88. Fu, G.H.; Bai, Z.Y.; Xia, J.H.; Liu, X.J.; Liu, F.; Wan, Z.Y.; Yue, G.H. Characterization of the LECT2 gene and its associations with resistance to the big belly disease in Asian seabass. *Fish Shellfish Immunol.* **2014**, *37*, 131–138. [CrossRef]
89. Lan, F.; Misu, H.; Chikamoto, K.; Takayama, H.; Kikuchi, A.; Mohri, K.; Takata, N.; Hayashi, H.; Matsuzawa-Nagata, N.; Takeshita, Y.; et al. LECT2 functions as a hepatokine that links obesity to skeletal muscle insulin resistance. *Diabetes* **2014**, *63*, 1649–1664. [CrossRef]
90. Jung, T.W.; Chung, Y.H.; Kim, H.; Elaty, A.M.A.; Jeong, J.H. LECT2 promotes inflammation and insulin resistance in adipocytes via P38 pathways. *J. Mol. Endocrinol.* **2018**, *61*, 37–45. [CrossRef]

91. Todorčević, M.; Škugor, S.; Krasnov, A.; Ruyter, B. Gene expression profiles in Atlantic salmon adipose-derived stromo-vascular fraction during differentiation into adipocytes. *BMC Genom.* **2010**, *11*, 39. [CrossRef]
92. Kortner, T.M.; Björkhem, I.; Krasnov, A.; Timmerhaus, G.; Krogdahl, Å. Dietary cholesterol supplementation to a plant-based diet suppresses the complete pathway of cholesterol synthesis and induces bile acid production in Atlantic salmon (*Salmo salar* L.). *Br. J. Nutr.* **2014**, *111*, 2089–2103. [CrossRef]
93. Calder, P.C. Immunoregulatory and anti-inflammatory effects of n-3 polyunsaturated fatty acids. *Braz. J. Med. Biol. Res.* **1998**, *31*, 467–490. [CrossRef]
94. Pompéia, C.; Lopes, L.R.; Miyasaka, C.K.; Procópio, J.; Sannomiya, P.; Curi, R. Effect of fatty acids on leukocyte function. *Braz. J. Med. Biol. Res.* **2000**, *33*, 1255–1268. [CrossRef]
95. Geay, F.; Ferraresso, S.; Zambonino-Infante, J.L.; Bargelloni, L.; Quentel, C.; Vandeputte, M.; Kaushik, S.; Cahu, C.L.; Mazurais, D. Effects of the total replacement of fish-based diet with plant-based diet on the hepatic transcriptome of two European sea bass (*Dicentrarchus labrax*) half-sibfamilies showing different growth rates with the plant-based diet. *BMC Genom.* **2011**, *12*, 522. [CrossRef]
96. Xia, S.H.; Wang, J.; Kang, J.X. Decreased n-6/n-3 fatty acid ratio reduces the invasive potential of human lung cancer cells by downregulation of cell adhesion/invasion-related genes. *Carcinogenesis* **2005**, *26*, 779–784. [CrossRef]
97. Ivakhno, S.S.; Kornelyuk, A.I. Cytokine-like activities of some aminoacyl-tRna synthetases and auxiliary p43 cofactor of aminoacylation reaction and their role in oncogenesis. *Exp. Oncol.* **2004**, *26*, 250–255.
98. Ko, Y.G.; Park, H.; Kim, T.; Lee, J.W.; Park, S.G.; Seol, W.; Kim, J.E.; Lee, W.H.; Kim, S.H.; Park, J.E.; et al. A cofactor of mRNA synthetase, p43, is secreted to up-regulate proinflammatory genes. *J. Biol. Chem.* **2001**, *276*, 23028–23033. [CrossRef]
99. Park, H.; Park, S.G.; Kim, J.; Ko, Y.G.; Kim, S. Signaling pathways for TNF production induced by human aminoacyl-tRNA synthetase-associating factor, p43. *Cytokine* **2002**, *20*, 148–153. [CrossRef]
100. Sang, G.P.; Shin, H.; Young, K.S.; Lee, Y.; Choi, E.C.; Park, B.J.; Kim, S. The novel cytokine p43 stimulates dermal fibroblast proliferation and wound repair. *Am. J. Pathol.* **2005**, *166*, 387–398. [CrossRef]
101. Le, X.; Pugach, E.K.; Hettmer, S.; Storer, N.Y.; Liu, J.; Wills, A.A.; DiBiase, A.; Chen, E.Y.; Ignatius, M.S.; Poss, K.D.; et al. A novel chemical screening strategy in zebrafish identifies common pathways in embryogenesis and rhabdomyosarcoma development. *Development* **2012**, *140*, 2354–2364. [CrossRef]
102. Faught, E.; Vijayan, M.M. Loss of the glucocorticoid receptor in zebrafish improves muscle glucose availability and increases growth. *Am. J. Physiol. Endocrinol. Metab.* **2019**, *316*, E1093–E1104. [CrossRef]
103. Sen, N.D.; Zhou, F.; Harris, M.S.; Ingolia, N.T.; Hinnebusch, A.G. EIF4B stimulates translation of long mRNAs with structured 5′ UTRs and low closed-loop potential but weak dependence on eIF4G. *Proc. Natl. Acad. Sci. USA* **2016**, *113*, 10464–10472. [CrossRef]
104. Harding, H.P.; Zhang, Y.; Bertolotti, A.; Zeng, H.; Ron, D. Perk is essential for translational regulation and cell survival during the unfolded protein response. *Mol. Cell* **2000**, *5*, 897–904. [CrossRef]
105. Howarth, D.L.; Lindtner, C.; Vacaru, A.M.; Sachidanandam, R.; Tsedensodnom, O.; Vasilkova, T.; Buettner, C.; Sadler, K.C. Activating transcription factor 6 is necessary and sufficient for alcoholic fatty liver disease in zebrafish. *PLoS Genet.* **2014**, *10*, e1004335. [CrossRef]
106. Kavaliauskis, A.; Arnemo, M.; Rishovd, A.L.; Gjøen, T. Activation of unfolded protein response pathway during infectious salmon anemia virus (ISAV) infection in vitro and in vivo. *Dev. Comp. Immunol.* **2016**, *54*, 46–54. [CrossRef]
107. Saelens, X.; Kalai, M.; Vandenabeele, P. Translation inhibition in apoptosis: Caspase-dependent PKR activation and eIF2-α phosphorylation. *J. Biol. Chem.* **2001**, *276*, 41620–41628. [CrossRef]
108. Morley, S.J.; Coldwell, M.J.; Clemens, M.J. Initiation factor modifications in the preapoptotic phase. *Cell Death Differ.* **2005**, *12*, 571–584. [CrossRef]
109. Smith, G.I.; Atherton, P.; Reeds, D.N.; Mohammed, B.S.; Rankin, D.; Rennie, M.J.; Mittendorfer, B. Dietary omega-3 fatty acid supplementation increases the rate of muscle protein synthesis in older adults: A randomized controlled trial. *Am. J. Clin. Nutr.* **2011**, *93*, 402–412. [CrossRef] [PubMed]
110. Panserat, S.; Hortopan, G.A.; Plagnes-Juan, E.; Kolditz, C.; Lansard, M.; Skiba-Cassy, S.; Esquerré, D.; Geurden, I.; Médale, F.; Kaushik, S.; et al. Differential gene expression after total replacement of dietary fish meal and fish oil by plant products in rainbow trout (*Oncorhynchus mykiss*) liver. *Aquaculture* **2009**, *294*, 123–131. [CrossRef]
111. Houlihan, D.F.; Carter, C.G. Protein synthesis. *Fish Physiol.* **2001**, *20*, 31–75. [CrossRef]
112. Kaushik, S.J.; Seiliez, I. Protein and amino acid nutrition and metabolism in fish: Current knowledge and future needs. *Aquac. Res.* **2010**, *41*, 322–332. [CrossRef]

biology

MDPI

Article

Brains in Metamorphosis: Temporal Transcriptome Dynamics in Hatchery-Reared Flatfishes

Laura Guerrero-Peña [1], Paula Suarez-Bregua [1,*], Luis Méndez-Martínez [1], Pablo García-Fernández [2], Ricardo Tur [2], Juan A. Rubiolo [3,4], Juan J. Tena [5] and Josep Rotllant [1,*]

[1] Aquatic Biotechnology Lab., Institute of Marine Research, Spanish National Research Council (IIM-CSIC), 36208 Vigo, Spain; lguerrero@iim.csic.es (L.G.-P.); lmendez@iim.csic.es (L.M.-M.)

[2] Nueva Pescanova Biomarine Center, S.L., 36980 O Grove, Spain; pgarciaf@nuevapescanova.com (P.G.-F.); rtur@nuevapescanova.com (R.T.)

[3] Facultad de Ciencias Bioquímicas y Farmacéuticas-Centro Científico y Tecnológico Acuario del Río Paraná, Universidad Nacional de Rosario, Rosario S2002LRK, Argentina; ja.rubiolo@usc.es

[4] Departamento de Genética, Facultad de Veterinaria, Universidad de Santiago de Compostela, 27002 Lugo, Spain

[5] Centro Andaluz de Biología del Desarrollo (CABD), CSIC-Universidad Pablo de Olavide, 41013 Sevilla, Spain; jjtenagu@upo.es

* Correspondence: paulasuarez@iim.csic.es (P.S.-B.); rotllant@iim.csic.es (J.R.)

Simple Summary: Metamorphosis, the process by which the young forms of some animals develop into adult forms, has long kept scientists on tenterhooks. Flatfish undergo one of the most dramatic metamorphoses described in the entire animal kingdom, in which a symmetrical larva that swims upright and looks just like a typical baby fish becomes a completely flat, asymmetrical juvenile that will live its entire adult life attached to the bottom. To answer the question of how the same organism can generate two completely different body plans associated with different phases of the life cycle, we investigated the dynamics of the brain transcriptome, which is the regulatory center of specific endocrine-activated developmental processes during metamorphosis. Our results show, for the first time, a temporary immune system reorganization during flatfish metamorphic remodelling process. Therefore, characterizing and understanding all the developmental changes that take place during metamorphosis will assist in the understanding the importance of each of these processes in the normal development of an individual and therefore, facilitate the transfer of knowledge to prevent abnormal development or developmental pathologies.

Abstract: Metamorphosis is a captivating process of change during which the morphology of the larva is completely reshaped to face the new challenges of adult life. In the case of fish, this process initiated in the brain has traditionally been considered to be a critical rearing point and despite the pioneering molecular work carried out in other flatfishes, the underlying molecular basis is still relatively poorly characterized. Turbot brain transcriptome of three developmental stages (pre-metamorphic, climax of metamorphosis and post-metamorphic) were analyzed to study the gene expression dynamics throughout the metamorphic process. A total of 1570 genes were differentially expressed in the three developmental stages and we found a specific pattern of gene expression at each stage. Unexpectedly, at the climax stage of metamorphosis, we found highly expressed genes related to the immune response, while the biological pathway enrichment analysis in pre-metamorphic and post-metamorphic were related to cell differentiation and oxygen carrier activity, respectively. In addition, our results confirm the importance of thyroid stimulating hormone, increasing its expression during metamorphosis. Based on our findings, we assume that immune system activation during the climax of metamorphosis stage could be related to processes of larval tissue inflammation, resorption and replacement, as occurs in other vertebrates.

Keywords: turbot; metamorphosis; brain; RNA; sequencing; transcriptome

Citation: Guerrero-Peña, L.; Suarez-Bregua, P.; Méndez-Martínez, L.; García-Fernández, P.; Tur, R.; Rubiolo, J.A.; Tena, J.J.; Rotllant, J. Brains in Metamorphosis: Temporal Transcriptome Dynamics in Hatchery-Reared Flatfishes. *Biology* **2021**, *10*, 1256. https://doi.org/10.3390/biology10121256

Academic Editor: Arne Ludwig

Received: 28 October 2021
Accepted: 29 November 2021
Published: 2 December 2021

1. Introduction

Metamorphosis is a post-embryonic process that involves radical changes in morphology, physiology and habitat, leading to stage-specific organs and structures [1]. Metamorphosis is widely distributed throughout the animal kingdom [2], which results in varying types of metamorphosis that probably do not share the same evolutionary mechanisms [3]. The well-known metamorphoses are those carried out by insects [4,5] and amphibians [6], but many other invertebrates and vertebrates have to face this process [7].

Most teleost fish undergo metamorphosis in the transition from larva to juvenile stage [8,9]; however, the most dramatic metamorphosis known in fish is that of flatfish (order Pleuronectiformes) [10], which includes species of high economic value such as turbot (*Scophthalmus maximus*), Senegalese sole (*Solea senegalensis*) or Atlantic halibut (*Hippoglossus hippoglossus*). During flatfish metamorphosis, a pelagic and bilateral symmetric larva becomes a benthonic asymmetric juvenile. Profound internal and external changes, such as eye migration to the opposite side of the body, remodelling of the craniofacial complex [11–13] and redistribution of the skin pigmentation [14], lead to an asymmetric juvenile fish adapted to benthic life.

Metamorphosis is an energy demanding process that has an important impact on the feeding, growth and, in some cases, a higher mortality rate in flatfishes [15]. The unsuccessful larva-to-juvenile transition can lead to malformed tissues due to incomplete migration of the eye, bone deformity and/or abnormal pigmentation deposition, ultimately affecting the commercial value of flatfishes [16].

As occurs in amphibians [17], thyroid hormones (THs) have an important role as endocrine regulators of flatfish metamorphosis. In fact, TH action has been shown as mandatory to successfully carry out this process [18]. Treatments with inhibitors of THs arrest metamorphosis, interrupting eye migration alongside development and growth rate [19]. TH production is determined by the hypothalamus-pituitary-thyroid (HPT) axis. The pituitary, located in the brain and ventral to the hypothalamus [20], prompts the thyroid gland by releasing thyroid stimulating hormone (TSH). Thyroid glands are the only component of the HPT axis that is located outside the brain and secretes THs (triiodothyronine [T3] and tetraiodothyronine [T4]) to act on target tissues [21,22] via thyroid hormone receptors (i.e., TH receptor alpha [TRα] and TH receptor beta [TRβ]) [9]. Then, the functional adaptive changes associated with post-embryonic development are supposed to be tightly regulated by the expression of specific genes.

Although one of the two major components of the HPT axis is found in the brain, this organ does not seem to suffer an obvious morphological remodelling during the flatfish metamorphosis. Despite the fact that the horizontal semicircular canals, with respect to the eyes, change from parallel to perpendicular in the transition from larva to juvenile, the symmetry of the brain is not affected [23]. The only regions of the brain known to become asymmetric after metamorphosis are the olfactory lobes [24] and telencephalic hemispheres [25]. Neither has evidence of asymmetric remodelling been found in the neuronal nerve pathways [10].

The present study aims to characterize the gene expression profile and molecular mechanisms involved in the flatfish brain during metamorphosis. Turbot brain transcriptome of three developmental stages (pre-metamorphic, climax of metamorphosis and post-metamorphic) were analyzed to study the gene expression dynamics throughout the metamorphic process. The brain was selected as the target tissue since it is the regulatory center of specific endocrine-activated developmental processes during metamorphosis. We focused on the differential gene expression profile throughout metamorphosis and the enriched pathways in each developmental stage.

2. Materials and Methods

2.1. Fish Collection and Sampling

Newborn turbots (*Scophthalmus maximus*) reared under a standard commercial production cycle were supplied by the company Insuiña SL, Grupo Nueva Pescanova (Pontevedra,

Spain). Fish from a single-pair mating were collected at different stages by experienced company staff. The number of fish sampled for all experimental procedures was estimated according to the minimum number of animals necessary to provide reliable and robust statistical results. The metamorphic stages were defined following the morphological criteria described by Al-Maghazachi and Gibson [26] and Suarez-Bregua [27], based on eye migration, whole body symmetry and rearing temperature (18 °C): pre-metamorphic stage (stage 3b: 15 days post fertilization [dpf]), before eye migration on a symmetrical larva; metamorphic climax (stage 4d: 30 dpf), larva exhibiting asymmetrical features with upper edge of right eye visible from left side; and post-metamorphic stage (stage 5c: 57 dpf), asymmetric juveniles that achieved complete eye migration. Individual fish samples at each metamorphic stage ($N = 3$ independent biological replicates per stage) were euthanized using a lethal dose of MS-222 (250 mg/L for 30–40 min) [28] (Sigma-Aldrich, Saint Louis, MO, USA), photographed and dissected with a Leica M165FC stereomicroscope equipped with a DFC310FX camera (Leica, Wetzlar, Germany).

Ethical approval (AGL2017-89648P) for all studies was obtained from the Institutional Animal Care and Use Committee of the IIM-CSIC Institute in accordance with the National Advisory Committee for Laboratory Animal Research Guidelines licensed by the Spanish Authority (RD53/2013). This work was in conformance with the European animal directive (2010/63/UE) for the protection of experimental animals.

2.2. RNA Isolation and Sequencing

Fish brains ($N = 9$) were dissected. Briefly, the turbot head was cut off and the top of skull was opened to remove all brain tissue, including the pituitary gland. Samples were then fixed in RNAlater (Thermo Fisher Scientific, Waltham, MA, USA) for 24 h at 4 °C and stored at −80 °C until use. Brain samples were removed from RNAlater solution (Invitrogen, Waltham, MA, USA) and homogenized in RLT buffer (RNeasy Mini Kit, Qiagen, Venlo, Germany). Total RNA was extracted and purified using the RNeasy Mini Kit (Qiagen) with on-column DNase digestion (Qiagen) according to the manufacturer's instructions. RNA concentration was quantified on a Qubit 4 fluorometer (Thermo Fisher Scientific) and RNA integrity (RIN (RNA integrity number) > 8) was verified on an Agilent 2100 bioanalyzer (Agilent Technologies, Santa Clara, CA, USA).

Approximately 1 μg of total RNA was initially used for BGISEQ-500 standard library construction at BGI (Beijing Genomics Institute, Shenzhen, China). Prepared cDNA libraries were sequenced on a BGISEQ-500 platform and single-end reads of 50 base pairs (bp) length were generated per sample.

2.3. Transcriptome Analysis and Annotation

Reads were quality checked (phred score > 30) using FastQC v0.11.8 (http://www.bioinformatics.babraham.ac.uk/projects/fastqc/; accesed 18 November 2019) and mapped to the turbot genome assembly (ASM318616v1) [29] using STAR v2.7.0e alignment software [30]. The turbot reference genome and respective annotation file were downloaded from Ensembl Genome Browser (ftp://ftp.ensembl.org/pub/release-101/fasta/scophthalmus_maximus/; accesed 25 November 2019). HTseq v0.10.0 [31] was used to transform uniquely mapped reads into counts and assign them to genes.

We performed a functional annotation of the whole turbot genome using Sma3s v2 software [32]. We first obtained the predicted amino acid sequences from Ensembl REST server through the API with a custom script, and then, query sequences were compared with non-redundant protein sequences of the Swissprot and TrEMBL vertebrate databases using an E-value threshold of 1×10^{-6}. Subsequently, the annotated genes were assigned to the three main categories of Gene Ontology (GO): biological process (BP), molecular function (MF), and cell component (CC). The large computational operations were carried out using the resources of the Supercomputing Center of Galicia (CESGA).

2.4. Differential Expression Analysis, Clustering and GO Enrichment

Gene count data were normalized and pairwise comparisons were performed to identify differentially expressed genes (DEGs) with the DESeq2 R package v1.26.0 [33]. *p*-values were adjusted (padj) by false discovery rate (FDR) [34]. Only genes with a padj < 0.01 and Log2 fold change (Log2FC) ≤ -2 or ≥ 2 were considered as DEGs. DEGs were analyzed according to their expression pattern throughout three different approaches: (1) Hierarchical clustering via heatmap; (2) Soft clustering using Mfuzz software [35]; (3) overlapping clustered up and downregulated genes using custom Venn diagram.

GO enrichment analysis was carried out using clusterProfiler R package v3.14.3 [36], based on hypergeometric distribution and FDR control [34]. We used the previously functionally annotated turbot genome as a background.

2.5. Quantitative Real-Time PCR (qRT-PCR)

To validate the RNA sequencing and transcriptome analysis, a set of five selected genes were evaluated by qRT-PCR. Genes related to immune system response (chitinase 3, *chit3;* and Interferon-induced helicase C domain-containing protein, *ifih1*) and brain-driven metamorphic remodeling (thyroid hormone receptor alpha, *thra;* thyroid-stimulating hormone subunit beta a, *tshba;* and ependymin, *epd*) were selected.

For cDNA synthesis, 200 ng of the total RNA isolated from each sample was reverse-transcribed according to the Maxima First Strand cDNA Synthesis Kit (Thermo Fisher Scientific) protocol. Samples were amplified in duplicate containing 10 µL of PowerUp SYBR Green Master Mix (2×) (Thermo Fisher Scientific), 1 µL of 0.5 µM of each primer, 7 µL nuclease-free water, and 1 µL of cDNA template. qRT-PCR reactions were analyzed with a QuantStudio3 Real-Time PCR System (Thermo Fisher Scientific) under the following cycling conditions: initial uracil-DNA-glycosylase step at 50 °C for 2 min, Dual-Lock™ DNA polymerase activation at 95 °C for 2 min, followed by 40 cycles of denaturation at 95 °C for 15 s and annealing/extension at 60 °C for 1 min. Gene expression of *chit3, epd, thra, tshba* and *ifih1* genes was assessed in two independent experiments by using the efficiency-calibrated method, as previously described [37]. Relative mRNA expression levels were normalized to the housekeeping 18S ribosomal gene. Primer sets used for each gene are listed in Supplementary Table S1.

3. Results

3.1. Transcriptome Assembly and Annotation

We sequenced the brain transcriptomes in three key developmental stages across turbot metamorphosis (pre-metamorphic, climax and post-metamorphic stages) (Figure 1). Nine cDNA libraries (three replicates per each metamorphic stage) were sequenced and more than 208 million 50 bp single-end reads were generated. Reads were processed for subsequent transcriptome analysis (Table 1). The reads were mapped to the turbot reference genome, obtaining average mapping rates of 88.98%, 88.78% and 89.23% for pre-metamorphic, climax and post-metamorphic stages, respectively. The average number of genes assigned was 19,119, 19,045 and 18,996 for pre-metamorphic, climax and post-metamorphic stages, respectively (Table 1, Figure 2). Reads assigned to each gene of turbot genome are listed in Supplementary Table S2.

Table 1. Summary of RNA sequencing, assembly and annotation data of the turbot brain samples at three key metamorphic stages.

	Pre-Metamorphic Stage			Climax Stage			Post-Metamorphic Stage			Total
	R1	R2	R3	R1	R2	R3	R1	R2	R3	
Reads	24,000,160	22,065,907	23,172,275	23,261,839	23,167,451	23,191,562	23,275,068	23,173,993	23,358,252	208,666,507
Mapped reads	21,366,876	19,668,853	20,569,271	20,699,424	20,540,205	20,569,248	20,731,453	20,676,181	20,882,737	185,704,248
Mapping rate (%)	89.03	89.14	88.77	88.98	88.66	88.69	89.07	89.22	89.40	-
Number of genes	19,137	19,058	19,162	19,042	19,045	19,048	18,977	18,993	19,017	-

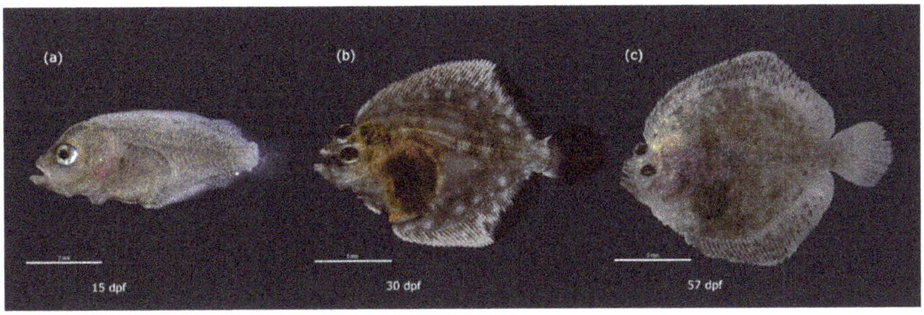

Figure 1. Turbot developmental stages selected during the metamorphosis process for transcriptome analysis. From left to right: (**a**) pre-metamorphic symmetrical larva prior to eye migration (stage 3b: 15dpf) (**b**) asymmetrical larva at the metamorphic climax with upper edge of migrating right eye visible from left side (stage 4d: 30dpf) (**c**) post-metamorphic asymmetrical juvenile with upper eye entirely placed on the left side (stage 5c: 57dpf). Fish were reared at 18 °C. Scale bars (**a**) 2mm, (**b**,**c**) 5 mm.

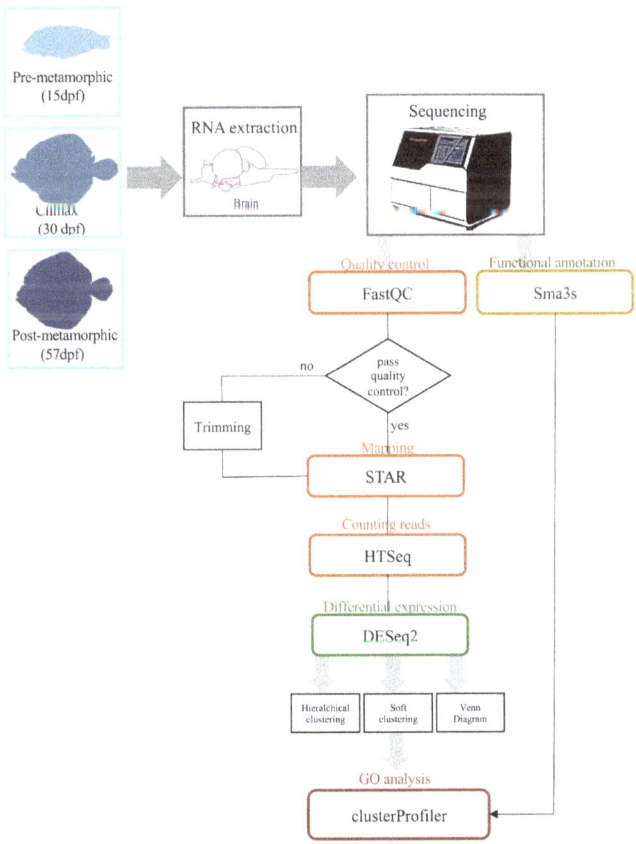

Figure 2. Overview of pipeline strategy followed in this work, from brain dissection to GO analysis of DEGs.

On the lookout for a subsequent successful GO enrichment analysis, a functional annotation of the turbot genome was carried out. 95.30% of the genes in the turbot genome were functionally annotated, summarized in three categories: BP, MF and CC. A total of

286,299 different GO terms were assigned, with 53.15%, 22.46% and 24.38%, corresponding to the BP, MF and CC categories, respectively. This functionally annotated turbot genome was used as a background to perform GO enrichment for selected genes of interest.

3.2. Gene Expression Dynamics in Metamorphosing Turbot Brains

To identify DEGs in the turbot brain during metamorphosis, we performed pairwise comparisons among three postembryonic developmental stages (pre-metamorphic vs. climax, climax vs. post-metamorphic and pre-metamorphic vs. post-metamorphic). A total of 1570 genes were differentially expressed in the three metamorphic developmental stages. A high proportion of DEGs were found when pre-metamorphic vs. climax stages (338 up- and 221 down-regulated) and pre-metamorphic vs. post-metamorphic stages (415 up- and 487 down-regulated) were compared (Figure 3a,c, respectively; Supplementary Tables S3 and S4). However, a significantly lower number of DEGs (33 up- and 76 down-regulated) was found after comparison between the climax and post-metamorphic stages (Figure 3b; Supplementary Table S5). All DEGs from the three developmental stages were combined into a single set and hierarchically clustered within a heatmap in order to produce an overview of the gene expression profiles across metamorphosis (Figure 3d). In the heatmap, DEGs were clustered according to gene expression level. Overall, two major clusters could be observed. Most genes clustered on the top half of the heatmap displayed decreased expression throughout the metamorphosis process (Figure 3d), while groups of genes on the bottom half showed increased expression over time. Specifically, a set of tightly clustered DEGs showed a marked expression peak at the metamorphic climax, while exhibiting down-regulated gene expression at the pre- and post-metamorphic stages (Figure 3d).

We next focused on the selected gene clusters that exhibited specific expression at pre-metamorphic stage (Supplementary Table S6), climax stage (Supplementary Table S7) and post-metamorphic stage (Supplementary Table S8) and a GO term enrichment analysis was performed to investigate the biological functions associated with each gene dataset (Figure 4), using the previous functional annotation of the turbot genome as a background. In this analysis, the most significant GO terms (padj < 0.05) were identified. Genes from the cluster showing overexpression at the pre-metamorphic stage revealed a significant enrichment of BP GO terms related to the developmental process and the development of anatomical structures, such as anterior/posterior pattern specification (GO:0009952), embryonic skeletal system (GO:0048704) and multicellular organism development (GO:0007275). Regarding the MF ontology, the most significant GO terms included heme binding (GO:0020037) and monooxygenase activity (GO:0004497) and in the CC category, we found ontologies such as RNA polymerase II transcription regulator complex (GO:009075) and apical plasma membrane (GO:0016324) (Figure 4a; Supplementary Table S9). Genes that were predominantly expressed during the climax of metamorphosis enrich BP ontologies related to immune system processes, such as antigen processing and presentation of endogenous peptide antigen via MHC class I (GO:00019885), innate immune response (GO:0045087) and immune response (GO:0006958), in MF category the most significant GO terms were NAD+ ADP-ribosyltransferase activity (GO:0003950), extracellular matrix structural constituent (GO:0005201) and protein transmembrane transporter activity (GO:0008320) and in the CC category we found ontologies such as collagen trimer (GO:0005581), collagen type I primer (GO:0005584) and extracellular matrix (GO:0031012) (Figure 4b; Supplementary Table S10). We observed that clustered genes with a higher expression in the post-metamorphosis stage also enrich the BP ontologies related to immune system process, such as immune response (GO:0006955), although we can also highlight ontologies related to cellular components organization, such as extracellular matrix organization (GO:0030198) or muscle system processes, like the cardiac muscle contraction (GO:0060048) and skeletal muscle contraction (GO:0003009) (Figure 4c; Supplementary Table S11).

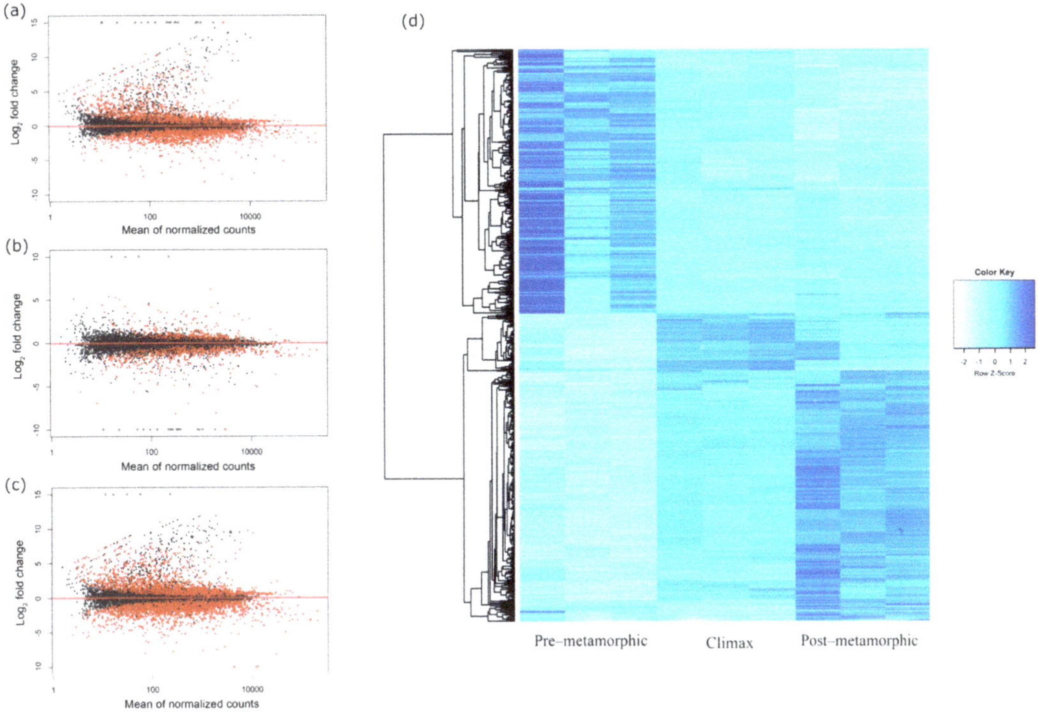

Figure 3. Visualization of gene expression datasets in turbot brains throughout the developmental metamorphosis process. MA plot of all transcriptome genes from pairwise comparisons: (**a**) pre-metamorphic vs. climax, (**b**) climax vs. post-metamorphic and (**c**) pre-metamorphic vs. post-metamorphic. The red dots plotted represent genes with an adjusted p-value < 0.1, while gray dots are those genes that do not show the established significance between the different stages. (**d**) Heatmap displaying the DEGs hierarchically clustered according to the expression profiles throughout metamorphosis. Each column represents an individual triplicate from each metamorphic stage (pre-metamorphic, climax and post-metamorphic) and each row represents different DEGs. The colors from light blue to dark blue indicate gene expression from low to high, respectively.

3.3. qPCR Validation of Gene Expression Patterns

Real-time qPCR was used to validate the expression levels of *chit3*, *epd*, *thra*, *tshba* and *ifih1* genes across turbot metamorphosis (Figure 5d). The qPCR results of the analyzed genes were consistent with the RNA sequencing data. We found that *chit3*, *ifih1* and *thra* expression levels peaked at the metamorphic climax and decreased after overcoming metamorphosis (Figure 5d). In addition, the expression of *tshba* increased at the climax stage, but remained high in the post-metamorphic stage. The only gene that exhibited a gradual increase of expression from the pre-metamorphic to post-metamorphic stage was *epd*. (Figure 5d).

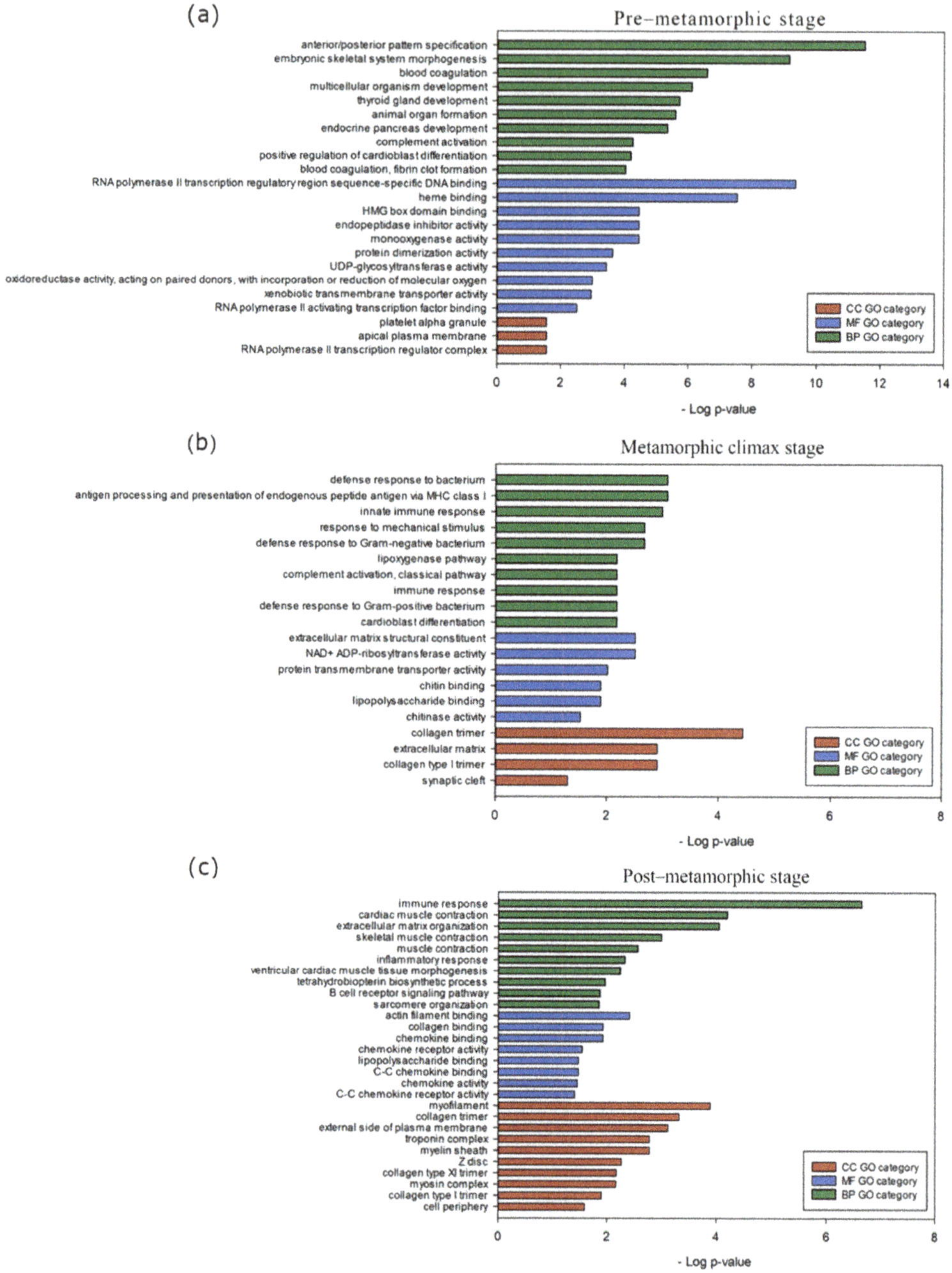

Figure 4. Main biological processes, molecular functions and cellular components boosted during metamorphosis in turbot brain. Gene Ontology (GO) enrichment of genes from hierarchical clustering showing high expression at (**a**) pre-metamorphic, (**b**) climax or (**c**) post-metamorphic stages. The most enriched GO terms (top 10, adjusted p-value < 0.05) belonging to biological process (BP), molecular function (MF) and cellular component (CC) categories are represented.

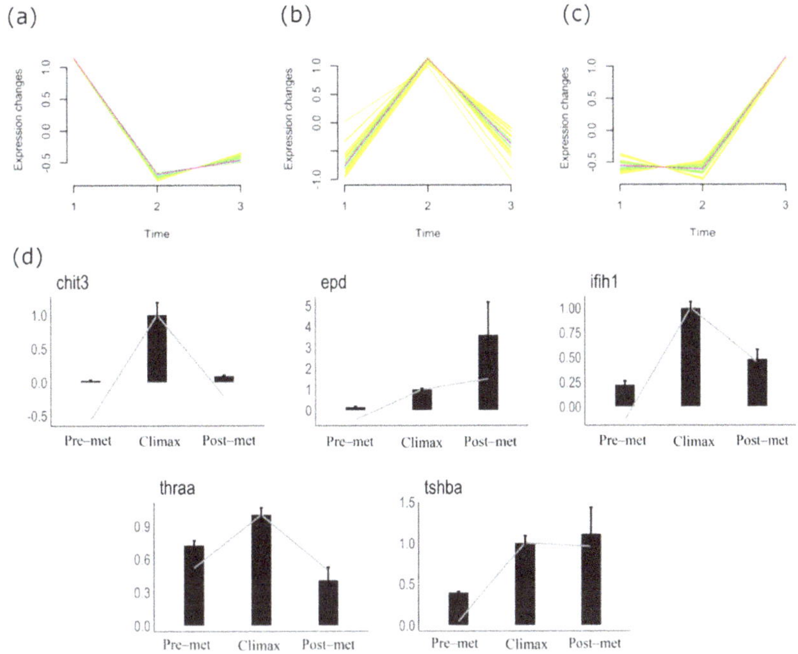

Figure 5. Selected clusters from Mfuzz soft clustering analysis of DEGs in turbot brains throughout the developmental metamorphosis process. DEGs show up-regulated genes corresponding to the: (**a**) pre-metamorphic stage, (**b**) climax stage, (**c**) post-metamorphic stage. Color code, from magenta to yellow, denote high or low Mfuzz membership values, respectively. Time 1, 2 and 3 in *X*-axis corresponds to pre-metamorphic, climax and post-metamorphic stage, respectively. (**d**) Validation of expression patterns of *chit3, epd, thra, tshba,* and *ifih1* genes in the turbot brain using qPCR. In the bar plots, the trend lines represent the fold change obtained by analyzing the RNAseq values during the metamorphic stages. The bars represent the fold change values obtained by qPCR. Results were normalized to *18S* gene and expressed as the mean ± SEM of two independent experiments. Data from climax stage was set at 1.

3.4. Stage-Specific Gene Expression and GO Enrichment Analysis

To further identify genes in the turbot brain with potentially important roles throughout the developmental metamorphosis process, we first explored the temporal expression patterns of DEGs by soft clustering analysis. The 1570 DEGs were divided into 28 clusters according to the similarity of their expression patterns throughout the three selected key stages. We then selected clusters containing genes that showed consistent and specific up-regulated expression at each metamorphic stage (Figure 5a–c). Clusters with genes highly expressed in the pre-metamorphic stage and then down-regulated displayed a total of 236 genes (Figure 5a; Supplementary Table S12), while 23 genes from a single cluster were specifically expressed in the post-metamorphic stage (Figure 5c; Supplementary Table S13). Interestingly, we found two clusters contained a total of 63 genes whose expression peaks at the metamorphic climax (Figure 5b; Supplementary Table S14), which supports the data represented on the heatmap (Figure 3d).

Additionally, we performed a GO enrichment analyses of the selected clusters (Figure 6). In the pre-metamorphic stage, the gene set from clusters selected showed significant enrichment of early development-related BP GO terms, such as embryonic skeletal system morphogenesis (GO:0048704), cholesterol homeostasis (GO:0042632) and anterior/posterior pattern specification (GO:0009952). For MF ontology, the most significant GO terms were UDP-glycosyltransferase activity (GO:0008194), hormone activity (GO:0005179) and heme binding (GO:0020037) (Figure 6a; Supplementary Table S15). Clusters with genes with an overexpressed expression pattern at climax were associated with immune response-related BP GO terms, such as chitin catabolic process (GO:0006032), immune system process (GO:0002376) and positive regulation of T cell migration (GO:2000406). NAD+ ADP-ribosyltransferase activity (GO:0003950), chitinase activity (GO:0004568) and chitin binding (GO:0008061) were the most highly represented MF GO terms (Figure 6b; Supplementary Table S16). Finally, the post-metamorphic clustered genes exhibited a significant enrichment in GO terms, including dopamine biosynthetic process (GO:0042416), regulation of removal of superoxide radicals (GO:2000121) and tetrahydrofolate biosynthetic process (GO:0046654) in the BP category; oxygen carrier activity (GO:0005344), oxygen binding (GO:0019825) and mitogen-activated protein kinase binding (GO:0051019) in the MF category; and nuclear membrane (GO:0031965) in the CC category (Figure 6c; Supplementary Table S17).

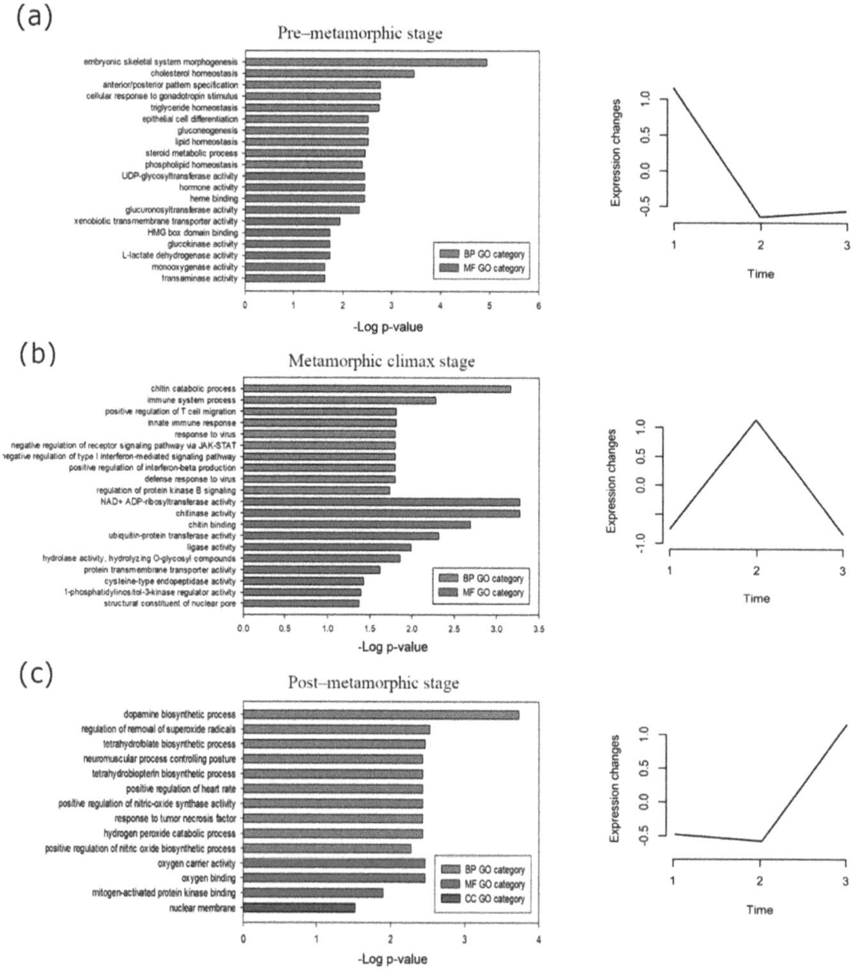

Figure 6. Main biological processes, molecular functions and cellular components boosted during metamorphosis in turbot brains. Gene Ontology (GO) enrichment of genes from Mfuzz clusters showing high expression at (**a**) pre-metamorphic, (**b**) climax or (**c**) post-metamorphic stages. The most enriched GO terms (top 10, adjusted p-value < 0.05) belonging to biological process (BP), molecular function (MF) and cellular component (CC) categories are represented.

3.5. Snapshot of Up- and Down-Regulated Gene Expression at the Metamorphic Climax

A tight transcriptional control of crucial genes is supposed to occur during the turbot metamorphic climax, which is a critical metamorphosis stage. To gain information on the differentially expressed genes in the turbot brain at the climax of metamorphosis, we identified the most significantly up-regulated and down-regulated genes (Figure 7). The Venn diagram in Figure 7a shows 14 overlapping genes that increased in expression to a peak at the metamorphic climax (i.e., DEGs down-regulated by pre-metamorphic vs. climax comparison), and subsequently, gene expression decreased in the post-metamorphic stage to the same level as the pre-metamorphic stage (i.e., DEGs up-regulated by climax vs. post-metamorphic comparison). In contrast, only four overlapping genes were significantly down-regulated at the climax of metamorphosis when compared to the pre-metamorphic stage (i.e., DEGs up-regulated by pre-metamorphic vs. climax comparison) and the post-metamorphic stage (i.e., DEGs down-regulated by climax vs. post-metamorphic comparison), where the expression level was gradually recovered prior to metamorphosis (Figure 7b). Based on the annotation results, the set of up-regulated genes included neurotrophic factors (e.g., meteorin and MYCBP-associated protein), cellular signaling regulators (e.g., tetraspanin, interferon regulatory factor 3 and interferon-gamma 1) and immune-response activators (e.g., chitinase and TFN 2 domain-containing protein), among others (Figure 7c). Down-regulated genes were associated with biological processes, such as osmoregulation (e.g., aquaporin 8a) and iron transport (e.g., hemopexin, Figure 7c).

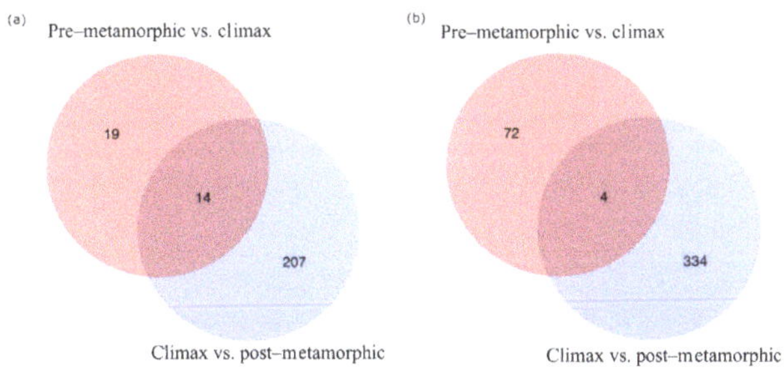

(c)

Ensembl ID	Gene Symbol	Gene Name	log2FC		padj		Read Count [1]		
Up-regulated			Pre vs Cli.	Cli vs Post	Pre vs Cli.	Cli vs Post	Pre	Cli	Post
ENSSMAG00000015091	Metrnl (*)/zgc:92275 (**)	Meteorin like, glial cell differentiation regulator (*)	−2.46	2.29	2.09×10^{-5}	0.00030	84	443.3	80
ENSSMAG00000004351	cd9	Tetraspanin	−5.71	3.51	3.29×10^{-38}	8.22×10^{-18}	199.3	6702	508
ENSSMAG00000009806	-	-	−4.65	2.65	2.61×10^{-43}	1.93×10^{-3}	693.3	17530	2418
ENSSMAG00000007192	loxe3	Epidermis-type lipoxygenase 3	−4.45	3.60	3.58×10^{-10}	1.89×10^{-6}	20.6	429.3	31
ENSSMAG00000011856	chi	Chitinase	−4.7	3.63	1.59×10^{-24}	3.04×10^{-14}	79.3	2049	141.3
ENSSMAG00000011681	-	-	−2.6	2.23	0.00016	0.0043	100.3	623	115
ENSSMAG00000000373	FASLG (#)	TNF_2 domain-containing protein (#)	−2.35	3.13	7.18×10^{-6}	5.81×10^{-8}	20.3	101.3	10
ENSSMAG00000015932	irf3	Interferon regulatory factor 3	−2.34	2.43	3.92×10^{-6}	6.83×10^{-6}	62.3	314	50.6
ENSSMAG00000007460	nt5c1ab	5′-nucleotidase, cytosolic IAb	−2.46	2.00	1.07×10^{-49}	2.66×10^{-32}	525	2837.6	616.3
ENSSMAG00000003312	mycbpap	MYCBP-associated protein	−2.56	2.03	6.11×10^{-15}	2.87×10^{-9}	32	185.3	39.3
ENSSMAG00000016022	-	-	−3.92	3.12	1.02×10^{-5}	0.002	220.3	3407	339.3
ENSSMAG00000010402	mx1	MX dynamin like GTPase 1	−4.8	4.15	1.20×10^{-9}	9.76×10^{-7}	164	4662	228.3
ENSSMAG00000009138	ifng1 (#)	Interferon-gamma 1(#)	−7.08	2.46	3.28×10^{-8}	0.0014	0.6	89.6	14
ENSSMAG00000014583	CCL19L1 (#)	SCY domain-containing protein (#)	−8.15	3.07	3.73×10^{-7}	5.21×10^{-21}	12.3	3433.3	352.3
Down-regulated									
ENSSMAG00000008469	aqp8.2	aquaporin 8a, tandem duplicate 2	7.08	−6.92	0.00030	0.0013	854	6.66	704.6
ENSSMAG00000003294	znf395a	Zinc finger protein 395a	2.35	−2.26	0.0011	0.005	110.3	22	91.3
ENSSMAG00000005388	hpxa	Hemopexin a	15.59	−10.34	2.50×10^{-6}	0.0075	8704	0	204
ENSSMAG00000016108	Mpx (*)/Epx (**)	C1q domain-containing protein (*)/Eosinophil peroxidase (**)	2.57	−4.26	2.63×10^{-5}	5.48×10^{-13}	65	10.6	177.6

(*) (**) Controversial annotation: different gene annotation proposed by Sma3s v2 software (*) and Ensembl (**).
(#) Gene annotation solely proposed by Sma3s v2 software, no matches were found by Ensembl.
[1] Mean of three replicates.

Figure 7. The most significantly up-regulated and down-regulated DEGs in the turbot brain at the climax stage of metamorphosis. Venn diagrams show (**a**) up-regulated and (**b**) down-regulated DEGs after pairwise comparisons between metamorphic stages (pre-metamorphic vs. climax and climax vs. post-metamorphic). (**c**) Gene annotation proposed by Sma3s v2 software3.5, fold change values (Log2FC), *p*-values adjusted (padj) and read count of each sample.

4. Discussion

Some marine fish species have complex life cycles, in which one or more free-living developmental stages eventually transform into morphologically, ecologically and physiologically distinct juvenile stages. However, the molecular and cellular processes underlying the regulation of this dramatic transformation process remains a mystery.

The purpose of this study was to investigate the dynamics of the brain transcriptome and determine the potential signaling pathways involved in the flatfish developmental metamorphosis process. To the best of our knowledge, this paper is the first to study the whole transcriptomic profile of the brain during metamorphosis in a flatfish species, identifying differential expression in 1570 of the analyzed 21,000 different protein coding genes.

Our results show a specific gene expression profile for each characterized developmental stage. In addition, we observed larger differences in the quantity of genes expressed in the pre-metamorphic stage with respect to the other stages, while the climax and post-metamorphosis stages presented a lower number of genes differentially expressed between them. This is supported because the pre-metamorphic stage is a larval development period that requires a tight regulation of gene expression for specific ontogenesis [38].

Comparative analysis of the different approaches to cluster and analyze the DEGs obtained (i.e., hard clustering, soft clustering and overlapping clustered DEGs by custom Venn diagram), revealed that both hierarchical and soft clustering highlight genes mostly involved in developmental processes during the pre-metamorphosis stage. At the climax stage, the three approaches revealed up-regulated genes associated to immune system functions. However, comparisons between hierarchical and soft clustering at the post-metamorphic stage showed different enriched ontologies. From hierarchical clustering, significant GO terms at post-metamorphic stage were related to immune system processes as found in the metamorphic climax. Soft clustering exhibited stage-specific GO terms and, thus, this approach led to an increased resolution to identify stage-specific gene expression and enriched ontologies across turbot metamorphosis. For this reason, we focused on the data analyzed by soft clustering approach.

At the pre-metamorphic stage (15 dpf), which corresponds to a larva with symmetrical morphology [26], the main biological processes affected were, as expected, those related to embryonic skeletal system morphogenesis, epithelial cell differentiation, gluconeogenesis, steroid metabolism and cholesterol, triglyceride and lipid homeostasis, among others. Thus, the up-regulation of several Hox genes at this stage suggests that it is still an active stage of morphogenesis [39,40]. As previously stated, our results show an enrichment of ontologies related to lipid homeostasis and steroid metabolism. Therefore, during early larva development and the time of mouth opening, lipids located in the yolk or oil drop play an essential structural and energetic role in turbot development. As a result, high activation of lipid metabolism is observed, but after this stage, lipid levels greatly decrease and regulation of lipid homeostasis occurs. Sterols are the only lipids that remain stable even after this event [41,42].

At the onset of the metamorphic climax (30 dpf), a strong morphological remodelling occurs, including the beginning of the migration of the right eye. As expected, our results showed significant transcriptional activation of thyroid hormone receptor alpha-A *(thraa)* and thyroid-stimulating hormone beta subunit *(tshba)* genes at this specific stage. All studies to date suggest that THs play a key role in the induction of the teleost metamorphosis process. Thus, metamorphosis in teleosts is triggered by the hypothalamic-pituitary-thyroid (HPT) axis, which is constituted of brain neuropeptide thyrotropin-releasing hormone (TRH), brain neuropeptide corticotropin-releasing hormone (CRH), pituitary glycoprotein hormone (thyrotropin, TSH) and THs (thyroxime [T4] and triiodothyronine [T3]). In some teleosts, such as coho salmon, it has been suggested that CRH, rather than TRH, plays a key role as a stimulator of TSH secretion by the pituitary gland [21,43]. Our results corroborate the critical role of THs in the regulation of the flatfish metamorphosis process. Interestingly, during this stage, one of the most enriched GO categories was innate immune system.

A significant up-regulation was detected for the chitinase family genes [44,45]. In addition, several genes of innate immune response were also up-regulated, including the *dhx58*, *ifih1*, *irf3* and *irf7*genes. *ifih1* gene encodes the melanoma differentiation-associated protein 5 (Mda5). This protein increases the phosphorylation levels of the transcription interferon regulatory factors 3 and 7 (*irf3* and *irf7*), activating the expression of the type 1 interferon genes (*ifnα* and *ifnβ*) and initiating the processes of inflammation and cell death [46]. Mda5 helicase is activated by both endogenous and exogenous double-stranded RNA and several studies demonstrate that the regulation of Mda5 expression is linked to the induction of autoimmunity [47]. This suggests that Mda5 activation is not only due to the antiviral response but may also be stimulated by endogenous factors. It is well known that diverse innate immunity-related molecules are also expressed in the brain and play important roles in brain development [48]. We also observed an overexpression of *casp10* and *ripk3*, which stimulate cell apoptosis and inflammation [49]. The results obtained by applying more restrictive statistical conditions also show a significant up-regulation of genes that enrich the immune response ontologies, such as *mx*, which promotes cell apoptosis [50], or *faslg*, which induces apoptosis in T cells [51]. Another enriched GO term at the onset of the metamorphic climax stage was regulation of T-cell migration. It is generally believed that the development of an immune response involves T-cell activation in lymphoid organs and subsequent migration to peripheral tissues to mediate tissue damage inflammation [52]. However, it has recently been shown that, in addition to the defense function of cells and immune molecules, they also play a key role in neurodevelopmental processes [48].

Our data show that during the climax stage of the developmental metamorphosis process in turbot brains, components of the immune system and THs could play an important role, as well as the processes of inflammation and cell death in the metamorphosis process at the brain level. Other studies in *Xenopus* also highlight the reorganization of the immune system during metamorphosis due to the need to replace larval tissue in adults and to support all the new chemical and biological products generated during this transformation [53,54].

At the post-metamorphic stage (57dpf), and after successfully completing the metamorphosis process, both morphological and behavioral changes were observed in the juvenile turbot. At this stage, one of the most enriched GO terms was dopamine biosynthetic process. A significant up-regulation was detected for nuclear receptor subfamily 4 (*nr4a1*) and GTP cyclohydrolase 1 (*gch1*) family genes. Dopamine is a neurotransmitter involved in the inhibitory control of TSH secretion [55]. The timeline profile of *tshb* expression during the metamorphosis process shows a significant increase during climax stage; however, significant high levels of *tshb* remain after the end of this stage. This may be because sufficiently high levels of dopamine have not yet been reached to inhibit *tshb* expression. Another enriched GO term induced at the post-metamorphic stage was regulation of removal of superoxide radicals. It is well known that the active processes of inflammation and apoptosis found at the climax stage generate free radicals that must be removed to avoid further damage in the organism at later stages [56].

In conclusion, the generated transcripts expression patterns provide a framework of novel developmental process-responsive genes in the brain during turbot metamorphosis. This molecular response entails the initial activation of signaling networks, mainly related to morphogenesis and cell differentiation at the pre-metamorphic stage, thus suggesting an active stage of embryonic development. Subsequent activation of the signaling network was mainly related to immune response, inflammation and cell apoptosis at the climax stage and finally, a signaling network related to a different mechanism of biosynthesis and homeostasis at the post-metamorphic stage. Caution should be taken, however, in the interpretation of these results, because some studies have shown that the different protocols used in the different rearing procedures of the animals in intensive aquaculture systems could significantly stimulate different signaling network systems [57]. Likewise, environmental factors, such as temperature, can also be determining factors at the level of transcriptomic characterization, since, in poikilothermic animals, temperature variations are directly related to metabolism and growth [58]. Nevertheless, this study is the first genome-wide transcriptome analysis of flatfish brain tissue during the developmental metamorphosis process and it is an important resource for future research on the molecular characterization of vertebrate metamorphosis.

5. Conclusions

Our results show a clear and evident reorganization of the immune system during the metamorphosis process of flatfish. By studying the gene expression profile through transcriptomic analysis and throughout turbot development (pre-metamorphic, climax of metamorphosis and post-metamorphic stages), we observed that during the metamorphic climax there is an overexpression of genes related to immune system processes, such as inflammatory processes or cell apoptosis, among others. This suggest that this overexpression could be related to processes linked to the generation of new juvenile tissues and reorganization and/or destruction of larval tissues and to support all the new chemical and biological products generated during this transformation process [53]. Thus, the need for further studies related to the immune system during metamorphosis in flatfish to determine and describe its specific function is emphasized.

Supplementary Materials: The following are available online at https://www.mdpi.com/article/10.3390/biology10121256/s1: Table S1: Primer sequences used for quantitative real-time PCR (qRT-PCR) gene expression analyses, Table S2: Read counts of genes in three key developmental stages. Table S3: List of DEGs in turbot brain between pre-metamorphic and climax stages (DEseq2 results, padj < 0.01 and Log2FC ≤ -2 or ≥ 2), Table S4: List of DEGs in turbot brain between pre-metamorphic and post-metamorphic stages (DEseq2 results, padj < 0.01 and Log2FC ≤ -2 or ≥ 2), Table S5: List of DEGs in turbot brain between climax and post-metamorphic stages (DEseq2 results, padj < 0.01 and Log2FC ≤ -2 or ≥ 2), Table S6: Norm counts of genes highly expressed in turbot brain at pre-metamorphic stage. Gene set from hierarchical clustering, Table S7: Norm counts of genes highly expressed in turbot brain at climax stage. Gene set from hierarchical clustering, Table S8: Norm counts of genes highly expressed in turbot brain at post-metamorphic stage. Gene set from hierarchical clustering. Table S9: GO enrichment of genes highly expressed in pre-metamorphic stage. Gene set from Hierarchical clustering, Table S10: GO enrichment of genes highly expressed in climax stage. Gene set from Hierarchical clustering, Table S11: GO enrichment of genes highly expressed in post-metamorphic stage. Gene set from Hierarchical clustering, Table S12: Genes highly expressed in turbot brain at the pre-metamorphic stage. Gene set from Mfuzz clusters 6, 20, 22 and 25, Table S13: Genes highly expressed in turbot brain at the post-metamorphic stage. Gene set from Mfuzz cluster 14, Table S14: Genes highly expressed in turbot brain at the metamorphic climax. Gene set from Mfuzz clusters 1 and 8, Table S15: GO enrichment of genes highly expressed in pre-metamorphic stage. Gene set from Mfuzz clusters, Table S16: GO enrichment of genes highly expressed in metamorphic stage. Gene set from Mfuzz clusters, Table S17: GO enrichment of genes highly in post-metamorphic stage. Gene set from Mfuzz clusters.

Author Contributions: Conceptualization, P.S.-B., J.J.T. and J.R.; methodology, L.G.-P., P.S.-B., L.M.-M., J.A.R., J.J.T.; validation, L.G.-P., P.S.-B.; formal analysis, L.G.-P., P.S.-B., J.A.R., J.J.T.; resources, P.G.-F., R.T. and J.R.; data curation, L.G.-P., P.S.-B., J.A.R., J.J.T.; writing—original draft preparation, L.G.-P., P.S.-B. and J.R.; writing—review and editing, J.J.T., J.R.; supervision, P.S.-B. and J.R.; project administration, J.R.; and funding acquisition, J.R. All authors have read and agreed to the published version of the manuscript.

Funding: This research was funded by the MCIN/AEI/10.13039/501100011033 grant number AGL2017-89648P to JR and by "ERDF A way of making Europe". L. Guerrero-Peña was supported by a pre-doctoral fellowship of the Spanish FPI Research Training Program funded by Spanish Economy and Competitiveness Ministry, grant number (PRE2018-085475). L. Méndez-Martínez was supported by a pre-doctoral fellowship of the Xunta de Galicia, grant number (IN606A-2020/006).

Institutional Review Board Statement: The study was conducted according to the guidelines of the Declaration of Helsinki, and approved by the Ethics Committee of IIM-CSIC Institute and Xunta de Galicia (protocol code ES360570202001/18/FUN. 01/BIOL AN. 08/JRM).

Data Availability Statement: Not applicable.

Acknowledgments: We would like to thank Tirso Martínez Fondo for developing custom scripts and providing IT support.

Conflicts of Interest: The authors declare no conflict of interest.

References

1. Bishop, C.D.; Erezyilmaz, D.F.; Flatt, T.; Georgiou, C.D.; Hadfield, M.G.; Heyland, A.; Hodin, J.; Jacobs, M.W.; Maslakova, S.A.; Pires, A.; et al. What is metamorphosis? *Integr. Comp. Biol.* **2006**, *46*, 655–661. [CrossRef]
2. Ten Brink, H.; de Roos, A.M.; Dieckmann, U. The evolutionary ecology of metamorphosis. *Am. Nat.* **2019**, *193*, E116–E131. [CrossRef]
3. Nielsen, C. The origin of metamorphosis. *Evol. Dev.* **2000**, *2*, 127–129. [CrossRef] [PubMed]
4. Rolff, J.; Johnston, P.R.; Reynolds, S. Complete metamorphosis of insects. *Philos. Trans. R. Soc. B Biol. Sci.* **2019**, *374*. [CrossRef] [PubMed]
5. Truman, J.W. The Evolution of Insect Metamorphosis. *Curr. Biol.* **2019**, *29*, R1252–R1268. [CrossRef]
6. Fritzsch, B. The evolution of metamorphosis in amphibians. *J. Neurobiol.* **1990**, *21*, 1011–1021. [CrossRef]
7. Laudet, V. The Origins and Evolution of Vertebrate Metamorphosis. *Curr. Biol.* **2011**, *21*, R726–R737. [CrossRef]
8. Campinho, M.A. Teleost metamorphosis: The role of thyroid hormone. *Front. Endocrinol.* **2019**, *10*, 1–12. [CrossRef]
9. McMenamin, S.K.; Parichy, D.M. Chapter Five–Metamorphosis in Teleosts. *Curr. Top. Dev. Biol.* **2013**, *103*, 127–165. [CrossRef]
10. Schreiber, A.M. Chapter Six—Flatfish: An Asymmetric Perspective on Metamorphosis. *Curr. Top. Dev. Biol.* **2013**, *103*, 167–194. [CrossRef] [PubMed]
11. Sun, M.; Wei, F.; Liu, H.; Hu, J.; Chen, X.; Gong, X.; Tian, Y.; Chen, S.; Bao, B. Distortion of frontal bones results from cell apoptosis by the mechanical force from the up-migrating eye during metamorphosis in Paralichthys olivaceus. *Mech. Dev.* **2015**, *136*, 87–98. [CrossRef]
12. Brewster, B. Eye migration and cranial development during flatfish metamorphosis: A reappraisal (Teleostei: Pleuronectiformes). *J. Fish Biol.* **1987**, *31*, 805–833. [CrossRef]
13. Sæle, Ø.; Smáradóttir, H.; Pittman, K. Twisted story of eye migration in flatfish. *J. Morphol.* **2006**, *267*, 730–738. [CrossRef] [PubMed]
14. Chen, Q.; Sato, K.; Yokoi, H.; Suzuki, T. Developmental regulatory system of ocular-side-specific asymmetric pigmentation in flounder: Critical role of retinoic acid signaling. *J. Exp. Zool. Part B Mol. Dev. Evol.* **2020**, *334*, 156–167. [CrossRef]
15. Geffen, A.J.; van der Veer, H.W.; Nash, R.D.M. The cost of metamorphosis in flatfishes. *J. Sea Res.* **2007**, *58*, 35–45. [CrossRef]
16. Power, D.M.; Einarsdóttir, I.E.; Pittman, K.; Sweeney, G.E.; Hildahl, J.; Campinho, M.A.; Silva, N.; Sæle, Ø.; Galay-Burgos, M.; Smáradóttir, F.; et al. The molecular and endocrine basis of flatfish metamorphosis. *Rev. Fish. Sci.* **2008**, *16*, 93–109. [CrossRef]
17. Tata, J.R. Amphibian metamorphosis as a model for studying the developmental actions of thyroid hormone. *Cell Res.* **1998**, *8*, 259–272. [CrossRef] [PubMed]
18. De Jesus, E.G.; Hirano, T.; Inui, Y. Flounder metamorphosis: Its regulation by various hormones. *Fish Physiol. Biochem.* **1993**, *11*, 323–328. [CrossRef] [PubMed]
19. Gomes, A.S.; Alves, R.N.; Rønnestad, I.; Power, D.M. Orchestrating change: The thyroid hormones and GI-tract development in flatfish metamorphosis. *Gen. Comp. Endocrinol.* **2015**, *220*, 2–12. [CrossRef] [PubMed]
20. Einarsdóttir, I.E.; Silva, N.; Power, D.M.; Smáradóttir, H.; Björnsson, B.T. Thyroid and pituitary gland development from hatching through metamorphosis of a teleost flatfish, the Atlantic halibut. *Anat. Embryol.* **2006**, *211*, 47–60. [CrossRef]
21. Deal, C.K.; Volkoff, H. The Role of the Thyroid Axis in Fish. *Front. Endocrinol.* **2020**, *11*, 1–25. [CrossRef] [PubMed]
22. Ortiga-Carvalho, T.M.; Chiamolera, M.I.; Pazos-Moura, C.C.; Wondisford, F.E. Hypothalamus-pituitary-thyroid axis. *Compr. Physiol.* **2016**, *6*, 1387–1428. [CrossRef]
23. Graf, W.; Baker, R. Neuronal adaptation accompanying metamorphosis in the flatfish. *J. Neurobiol.* **1990**, *21*, 1136–1152. [CrossRef]
24. Campinho, M.A.; Silva, N.; Martins, G.G.; Anjos, L.; Florindo, C.; Roman-Padilla, J.; Garcia-Cegarra, A.; Louro, B.; Mnachado, M.; Power, D.M. A thyroid hormone regulated asymmetric responsive centre is correlated with eye migration during flatfish metamorphosis. *Sci. Rep.* **2018**, *8*, 12267. [CrossRef]
25. Briñón, J.G.; Médina, M.; Arévalo, R.; Alonso, J.R.; Lara, J.M.; Aijón, J. Volumetric Analysis of the Telencephalon and Tectum During Metamorphosis in a Flatfish, the Turbot Scophthalmus maximus. *Brain. Behav. Evol.* **1993**, *41*, 1–5. [CrossRef] [PubMed]
26. Al-Maghazachi, S.J.; Gibson, R. The developmental stages of larval turbot, *Scophthalmus maximus* (L.). *J. Exp. Mar. Bio. Ecol.* **1984**, *82*, 35–51. [CrossRef]
27. Suarez-Bregua, P.; Pérez-Figueroa, A.; Hernández-Urcera, J.; Morán, P.; Rotllant, J. Temperature-independent genome-wide DNA methylation profile in turbot post-embryonic development. *J. Therm. Biol.* **2020**, *88*, 1–7. [CrossRef] [PubMed]
28. Part II—Methods of Euthanasia. Available online: https://www.avma.org/sites/default/files/2020-01/2020_Euthanasia_Final_1-15-20.pdf (accessed on 27 October 2021).
29. Figueras, A.; Robledo, D.; Corvelo, A.; Hermida, M.; Pereiro, P.; Rubiolo, J.A.; Gomez-Garrido, J.; Carrete, L.; Bello, X.; Gut, M.; et al. Whole genome sequencing of turbot (Scophthalmus maximus; Pleuronectiformes): A fish adapted to demersal life. *DNA Res.* **2016**, *23*, 181–192. [CrossRef]

30. Dobin, A.; Davis, C.A.; Schlesinger, F.; Drenkow, J.; Zaleski, C.; Jha, S.; Batut, P.; Chaisson, M.; Gingeras, T.R. STAR: Ultrafast universal RNA-seq aligner. *Bioinformatics* **2013**, *29*, 15–21. [CrossRef] [PubMed]
31. Anders, S.; Pyl, P.T.; Huber, W. HTSeq-A Python framework to work with high-throughput sequencing data. *Bioinformatics* **2015**, *31*, 166–169. [CrossRef]
32. Muñoz-Mérida, A.; Viguera, E.; Claros, M.G.; Trelles, O.; Pérez-Pulido, A.J. Sma3s: A three-step modular annotator for large sequence datasets. *DNA Res.* **2014**, *21*, 341–353. [CrossRef]
33. Love, M.I.; Huber, W.; Anders, S. Moderated estimation of fold change and dispersion for RNA-seq data with DESeq2. *Genome Biol.* **2014**, *15*, 550. [CrossRef] [PubMed]
34. Benjamini, Y.; Hochberg, Y. Controlling the False Discovery Rate: A Practical and Powerful Approach to Multiple Testing. *J. R. Stat. Soc. Ser. B* **1995**, *57*, 289–300. [CrossRef]
35. Kumar, L.; Futschik, M.E. Mfuzz: A software package for soft clustering of microarray data. *Bioinformation* **2007**, *2*, 5–7. [CrossRef]
36. Yu, G.; Wang, L.G.; Han, Y.; He, Q.Y. ClusterProfiler: An R package for comparing biological themes among gene clusters. *Omi. A J. Integr. Biol.* **2012**, *16*, 284–287. [CrossRef]
37. Pfaffl, M.W. A new mathematical model for relative quantification in real-time RT-PCR. *Nucleic Acids Res.* **2001**, *29*, 2002–2007. [CrossRef]
38. Zoologie, R.C.; De Mazurais, D.; Darias, M.; Cahu, C.L. Transcriptomics for understanding marine fish larval development. *Can. J. Zool.* **2011**, *89*, 599–611.
39. Kavouras, M.; Malandrakis, E.E.; Golomazou, E.; Konstantinidis, I.; Blom, E.; Palstra, A.P.; Anastassiadis, K.; Panagiotaki, P.; Exadactylos, A. Hox gene expression profiles during embryonic development of common sole. *Anim. Biol.* **2019**, *69*, 183–198. [CrossRef]
40. Mallo, M.; Wellik, D.M.; Deschamps, J. Hox genes and regional patterning of the vertebrate body plan. *Dev. Biol.* **2010**, *344*, 7–15. [CrossRef] [PubMed]
41. Cunha, I.; Galante-Oliveira, S.; Rocha, E.; Urbatzka, R.; Castro, L.F.C. Expression of intercellular lipid transport and cholesterol metabolism genes in eggs and early larvae stages of turbot, Scophthalmus maximus, a marine aquaculture species. *Mar. Biol.* **2015**, *162*, 1673–1683. [CrossRef]
42. Cunha, I.; Galante-Oliveira, S.; Rocha, E.; Planas, M.; Urbatzka, R.; Castro, L.F.C. Dynamics of PPARs, fatty acid metabolism genes and lipid classes in eggs and early larvae of a teleost. *Comp. Biochem. Physiol. B Biochem. Mol. Biol.* **2013**, *164*, 247–258. [CrossRef]
43. Larsen, D.A.; Swanson, P.; Dickey, J.T.; Rivier, J.; Dickhoff, W.W. In vitro thyrotropin-releasing activity of corticotropin-releasing hormone-family peptides in coho salmon, Oncorhynchus kisutch. *Gen. Comp. Endocrinol.* **1998**, *109*, 276–285. [CrossRef]
44. Gao, C.; Cai, X.; Zhang, Y.; Su, B.; Song, H.; Wenqi, W.; Li, C. Characterization and expression analysis of chitinase genes (CHIT1, CHIT2 and CHIT3) in turbot (*Scophthalmus maximus* L.) following bacterial challenge. *Fish Shellfish Immunol.* **2017**, *64*, 357–366. [CrossRef]
45. Teng, Z.; Sun, C.; Liu, S.; Wang, H.; Zhang, S. Functional characterization of chitinase-3 reveals involvement of chitinases in early embryo immunity in zebrafish. *Dev. Comp. Immunol.* **2014**, *46*, 489–498. [CrossRef]
46. Brisse, M.; Ly, H. Comparative structure and function analysis of the RIG-I-like receptors: RIG-I and MDA5. *Front. Immunol.* **2019**, *10*, 1–27. [CrossRef]
47. Sadler, A.J. The role of MDA5 in the development of autoimmune disease. *J. Leukoc. Biol.* **2017**, *103*, 185–192. [CrossRef]
48. Morimoto, K.; Nakajima, K. Role of the Immune System in the Development of the Central Nervous System. *Front. Neurosci.* **2019**, *13*. [CrossRef] [PubMed]
49. Orozco, S.; Oberst, A. RIPK3 in cell death and inflammation: The Good, the Bad, and the Ugly. *Immunol. Rev.* **2017**, *277*, 102–112. [CrossRef] [PubMed]
50. Numajiri Haruki, A.; Naito, T.; Nishie, T.; Saito, S.; Nagata, K. Interferon-inducible antiviral protein MxA enhances cell death triggered by endoplasmic reticulum stress. *J. Interf. Cytokine Res.* **2011**, *31*, 847–856. [CrossRef]
51. Andersen, M.H.; Schrama, D.; Thor Straten, P.; Becker, J.C. Cytotoxic T cells. *J. Invest. Dermatol.* **2006**, *126*, 32–41. [CrossRef] [PubMed]
52. Yu, J.; Zhou, X.; Chang, M.; Nakaya, M.; Chang, J.-H.; Xiao, Y.J.; Lindsey, W.; Dorta-Estremera, S.; Cao, W.; Zal, N.; et al. Regulation of T-cell activation and migration by the kinase TBK1 during neuroinflammation. *Nat. Commun.* **2015**, *6*, 6074. [CrossRef]
53. Rollins-Smith, L.A. Metamorphosis and the amphibian immune system. *Immunol. Rev.* **1998**, *166*, 221–230. [CrossRef] [PubMed]
54. Pollet, N. Expression of immune genes during metamorphosis of Xenopus: A survey. *Front. Biosci.* **2010**, *15*, 348–358. [CrossRef]
55. Roelfsema, F.; Boelen, A.; Kalsbeek, A.; Fliers, E. Regulatory aspects of the human hypothalamus-pituitary-thyroid axis. *Best Pract. Res. Clin. Endocrinol. Metab.* **2017**, *31*, 487–503. [CrossRef] [PubMed]
56. Khansari, N.; Shakiba, Y.; Mahmoudi, M. Chronic Inflammation and Oxidative Stress as a Major Cause of Age- Related Diseases and Cancer. *Recent Pat. Inflamm. Allergy Drug Discov.* **2009**, *3*, 73–80. [CrossRef]
57. Carbone, D.; Faggio, C. Importance of prebiotics in aquaculture as immunostimulants. Effects on immune system of Sparus aurata and Dicentrarchus labrax. *Fish Shellfish Immunol.* **2016**, *54*, 172–178. [CrossRef] [PubMed]
58. Remen, M.; Nederlof, M.A.J.; Folkedal, O.; Thorsheim, G.; Sitjà-Bobadilla, A.; Pérez-Sánchez, J.; Oppedal, F.; Olsen, R.E. Effect of temperature on the metabolism, behaviour and oxygen requirements of Sparus aurata. *Aquac. Environ. Interact.* **2015**, *7*, 115–123. [CrossRef]

biology

MDPI

Article

Characterization of miRNAs in Embryonic, Larval, and Adult Lumpfish Provides a Reference miRNAome for *Cyclopterus lumpus*

Setu Chakraborty [1], Nardos T. Woldemariam [2], Tina Visnovska [3], Matthew L. Rise [4], Danny Boyce [5], Javier Santander [1,*] and Rune Andreassen [2,*]

[1] Marine Microbial Pathogenesis and Vaccinology Laboratory, Department of Ocean Sciences, Memorial University of Newfoundland, 0 Marine Lab Rd, St. John's, NL A1C 5S7, Canada; schakraborty@mun.ca

[2] Department of Life Sciences and Health, Faculty of Health Sciences, OsloMet–Oslo Metropolitan University, Pilestredet 50, N-0130 Oslo, Norway; nardostwoldemariam@gmail.com

[3] Bioinformatics Core Facility, Oslo University Hospital, 0372 Oslo, Norway; tina.visnovska@rr-research.no

[4] Department of Ocean Sciences, Faculty of Sciences, Memorial University of Newfoundland, 0 Marine Lab Rd, St. John's, NL A1C 5S7, Canada; mrise@mun.ca

[5] Dr. Joe Brown Aquatic Research Building (JBARB), Department of Ocean Sciences, Memorial University of Newfoundland, 0 Marine Lab Rd, St. John's, NL A1C 5S7, Canada; dboyce@mun.ca

* Correspondence: jsantander@mun.ca (J.S.); rune.andreassen@oslomet.no (R.A.)

Simple Summary: Lumpfish (*Cyclopterus lumpus*) is an emergent aquaculture species, and its miRNA repertoire is still unknown. miRNAs are critical post-transcriptional modulators of teleost gene expression. Therefore, a lumpfish reference miRNAome was characterized by small RNA sequencing and miRDeep analysis of samples from different organs and developmental stages. The resulting miRNAome, an essential reference for future expression analyses, consists of 443 unique mature miRNAs from 391 conserved and eight novel miRNA genes. Enrichment of specific miRNAs in particular organs and developmental stages indicates that some conserved lumpfish miRNAs regulate organ and developmental stage-specific functions reported in other teleosts.

Abstract: MicroRNAs (miRNAs) are endogenous small RNA molecules involved in the post-transcriptional regulation of protein expression by binding to the mRNA of target genes. They are key regulators in teleost development, maintenance of tissue-specific functions, and immune responses. Lumpfish (*Cyclopterus lumpus*) is becoming an emergent aquaculture species as it has been utilized as a cleaner fish to biocontrol sea lice (e.g., *Lepeophtheirus salmonis*) infestation in the Atlantic Salmon (*Salmo salar*) aquaculture. The lumpfish miRNAs repertoire is unknown. This study identified and characterized miRNA encoding genes in lumpfish from three developmental stages (adult, embryos, and larvae). A total of 16 samples from six different adult lumpfish organs (spleen, liver, head kidney, brain, muscle, and gill), embryos, and larvae were individually small RNA sequenced. Altogether, 391 conserved miRNA precursor sequences (discovered in the majority of teleost fish species reported in miRbase), eight novel miRNA precursor sequences (so far only discovered in lumpfish), and 443 unique mature miRNAs were identified. Transcriptomics analysis suggested organ-specific and age-specific expression of miRNAs (e.g., miR-122-1-5p specific of the liver). Most of the miRNAs found in lumpfish are conserved in teleost and higher vertebrates, suggesting an essential and common role across teleost and higher vertebrates. This study is the first miRNA characterization of lumpfish that provides the reference miRNAome for future functional studies.

Keywords: conserved miRNA; high-throughput sequencing; lumpfish; novel miRNA; RT-qPCR

Citation: Chakraborty, S.; Woldemariam, N.T.; Visnovska, T.; Rise, M.L.; Boyce, D.; Santander, J.; Andreassen, R. Characterization of miRNAs in Embryonic, Larval, and Adult Lumpfish Provides a Reference miRNAome for *Cyclopterus lumpus*. *Biology* **2022**, *11*, 130. https://doi.org/10.3390/biology11010130

Academic Editor: Patricia Pereiro

Received: 23 November 2021
Accepted: 7 January 2022
Published: 13 January 2022

1. Introduction

The discovery of microRNAs (miRNAs) in 1993 in *Caenorhabditis elegans* and further identification in humans and many other animals significantly alters the longstanding dogmas that defined gene regulation [1]. These studies revealed that miRNAs were a class of small noncoding RNAs that function as guide molecules in RNA silencing machinery, often termed the RNA-induced silencing complex (RISC). RISC regulates gene expression at the messenger RNA level either by degrading mRNAs targeted by the miRNAs or preventing their translation [1–3]. miRNA constitute a large family of post-transcriptional regulators with ~22 nucleotides in length and present in animals, plants, and some viruses [3,4]. Functional studies indicate that miRNAs have diverse expression patterns and regulate almost every cellular process, including developmental, physiological, and pathophysiological processes [3,5,6].

miRNA biogenesis involves multiple steps; first, miRNAs are processed from primary molecules (pri-miRNAs), which are transcribed by RNA-specific endoribonuclease (Drosha) and processed into an ~70-nucleotide pre-miRNA in the nucleus [2,3,6–9]. Pre-miRNAs are then transported to the cytoplasm for further processing by the enzyme Dicer to an ~22-bp miRNA/miRNA duplex [2,3,6–9]. The miRNA duplex is loaded into the RISC. Only one of the mature miRNAs (guide miRNA) is incorporated in RISC, and the other is degraded (passenger miRNA). The guide miRNA directs the RISC to target mRNAs, where the mature miRNA usually binds in the 3′ untranslated region (UTR) of the target mRNAs [2,3,6–9].

Teleosts are an essential component of aquatic ecosystems and a primary source of proteins for human and animal consumption worldwide. Teleosts are one of the most diverse vertebrates on the earth [10]. The exploration of the role of miRNA in teleost development, organogenesis, tissue differentiation, growth, regeneration, reproduction, endocrine system, and responses to environmental stimuli, as well as their role in the maturation of the immune system and response to infectious diseases, is still under investigation [11–17]. miRNA characterization is the first step in any investigation of their regulatory roles. Such characterizations have been carried out in some economically important fish species such as Atlantic salmon (*Salmo salar*), Atlantic cod (*Gadus morhua*), rainbow trout (*Oncorhynchus mykiss*), Atlantic halibut (*Hippoglossus hippoglossus*), channel catfish (*Ictalurus punctatus*), turbot (*Scophthalmus maximus*), Asian seabass (*Lates calcarifer*), olive flounder (*Paralichthys olivaceus*) [18–26], and fish models like zebrafish (*Danio rerio*) and three-spined stickleback (*Gasterosteus aculeatus*) [27,28].

Global production of farmed Atlantic salmon is estimated at just over 2.6 million tonnes in 2019. This growth was mainly driven by Norway and Chile, the two leading producing countries. The Norwegian salmon industry alone earned some NOK 19–20 billion (USD 2.1–2.2 billion) in profits before tax in 2019 (The Food and Agriculture Organization of United Nations (FAO), 2021) (https://www.fao.org/in-action/globefish/market-reports/resource-detail/en/c/1268636/) (Last accesed: 21 January 2021). Farmed salmon is the main species of the Atlantic Canadian aquaculture industry, which represents approximately 80% of Atlantic Canada's total aquaculture value [29].

Infectious diseases are a challenge for the aquaculture industry. Globally losses due to diseases in the aquaculture industry exceed US$6 billion annually [30,31]. One of the most prominent disease challenges currently restraining Atlantic salmon aquaculture is the infestation by the parasite sea lice, specifically *Lepeophtheirus salmonis* and *Caligus* spp. [32–36]. Sea lice are a group of visible host-dependent ectoparasite copepods with vast reproductive potential [32–36]. The attached sea lice feed on salmon mucus, blood, and skin, which leads to significant physical damage and immunosuppression [32,36–39]. In addition, these effects on fish health lead to substantial economic impacts due to production losses and treatment costs [32,37,40]. The salmon industry in the North Atlantic region has adopted cleaner fish, e.g., Lumpfish (*Cyclopterus lumpus*), for biological control of sea-lice infestations [32,41–44]. However, several aspects of lumpfish biology remain unknown, including their miRNA repertoire.

This study aimed to identify and characterize miRNA encoding genes in lumpfish by small RNAs high-throughput sequencing (HTS) followed by miRDeep2 analysis. The identification was carried out in two early developmental stages, embryos and larvae, and six organs of adult lumpfish—brain, muscle, gill, liver, spleen, and head kidney. A combination of HTS and computational analytical approaches (e.g., miRNA precursor prediction) has been successfully used for miRNA characterization in particular two early developmental stages and adult lumpfish organs. Therefore, here we provide the first reference miRNAome for lumpfish.

2. Materials and Methods

2.1. Fish Holding

Five adult lumpfish (1100 g ± 99.5, male = 4, female = 1) were obtained from the Joe Brown Aquatic Research Building (JBARB) at the Department of Ocean Sciences (DOS), Memorial University of Newfoundland (MUN), Canada. The animals were kept in a 23,000 L tank, with flow-through (140 L min^{-1}) of UV-treated seawater (6 °C), ambient photoperiod (winter–spring), and 95–110% air saturation. Biomass density was maintained at 20 kg m^{-3}. The fish were fed daily using a commercial diet (Skretting—Europa 18 (6.0–9.0 mm pellet), Vancouver, BC, Canada) with a ratio of 0.25% of their body weight per day. Additionally, lumpfish embryos (300 degree days) and lumpfish larvae (one week posthatch) were obtained from the JBARB. Lumpfish egg masses were fertilized and maintained with flow-through in 5 L upwelling black nontranslucent incubators at 8–10 °C supplied with 95–110% air saturated and 5 μm UV-treated filtered flow-through seawater (spring–summer) [45]. After completing the development of the embryo, the larvae hatch are maintained at 10 °C [46].

2.2. Ethics Statement

The fish dissection and tissue sample collection were performed following the Canadian Council on Animal Care guidelines (https://ccac.ca/en/standards/guidelines/) (Last accessed: 10 January 2022) and approved by Memorial University of Newfoundland's Institutional Animal Care Committee (https://www.mun.ca/research/about/acs/) (Last accessed: 10 January 2022) under the protocols #18-1-JS and #18-03-JS.

2.3. Sample Collection

Fish were euthanized with 400 mg of MS222 (Syndel Laboratories, Vancouver, BC, Canada) per litre of seawater and dissected immediately after death confirmation. Tissue samples were collected from adult lumpfish brain, gill, skeletal muscle, liver, spleen, and head kidney. Tissue samples from five adult lumpfish, two pools of lumpfish embryos, and two pools of lumpfish larvae were immediately flash-frozen in liquid nitrogen (Air Liquide Canada Atlantic, St. John's, NL, Canada) and stored at −80 °C until further processing.

2.4. RNA Extraction

Total RNA was extracted using the mirVana RNA isolation kit (Invitrogen, Carlsbad, CA, USA) following the manufacturer's protocol. RNA concentration and integrity were quantified using spectrophotometry (Genova-nano, Jenway, Stone, Staffordshire, England) and 1% agarose gel electrophoresis. The total RNA concentrations of 32 samples (four adult individuals with brain and muscle samples, five adult individuals with gill, liver, spleen, and head kidney samples, and four samples from two early life stages) ranged from 100–3250 ng μL^{-1} (total volume 100 μL) (Table S1). Sixteen samples from two adult lumpfish, two embryos, and two larvae were used for high-throughput sequencing (HTS) by independent library preparation and sequencing of each sample (Table S2), whereas all the 32 samples were used for qPCR analysis.

2.5. High-Throughput Sequencing (HTS)

The library construction and small RNA sequencing were performed by the Genomics Core Facility Oslo (Oslo University Hospital, Oslo, Norway). The NEBNext Small RNA Library Prep Set for Illumina (New England Biolabs, Inc., Ipswich, MA, USA) was used to prepare the libraries according to the manufacturer's protocol. One µg of total RNA from each sample was used as input to prepare the libraries, and a final size selection of 140–150 bp fragments using 6% polyacrylamide gel was used to enrich small RNAs. The adapter sequences (5′ AGATCGGAAGAGCACACGTCTGAACTCCAGTCAC 3′) were used in the library preparation process. Following library preparation, next-generation RNA sequencing was carried out using the Illumina Genome Analyzer IIx sequencing platform described in Woldemariam et al. [19], generating single-end reads of length 75 bp.

2.6. Analysis of Sequencing Data

The sequencing of raw and processed data quality was checked before miRNA discovery analysis. FastQC v0.11.9 (http://www.bioinformatics.babraham.ac.uk/projects/fastqc/) (Last accessed: 20 November 2021) was used to check the quality of both the raw sequencing data and the data obtained after adapter removal and size filtering. The adapter removal and size filtering were carried out using cutadapt v1.8.3 [47]. Reads shorter than 18 bp and longer than 25 bp after the adapter removal were discarded. The reads passing the filtering step were converted to fasta format with fastq_to_fasta from FASTX toolkit v0.0.14 (http://hannonlab.cshl.edu/fastx_toolkit/) (Last accessed: 20 November 2021).

Each of the 16 samples were analyzed independently to detect miRNAs highly expressed in particular adult organs/tissues and early developmental stages. The lumpfish reference genome [48] and bowtie v1.0.0 [49] were used for mapping the reads to the reference genome. The workflow applied to identify novel lumpfish miRNA sequences is illustrated in Figure 1.

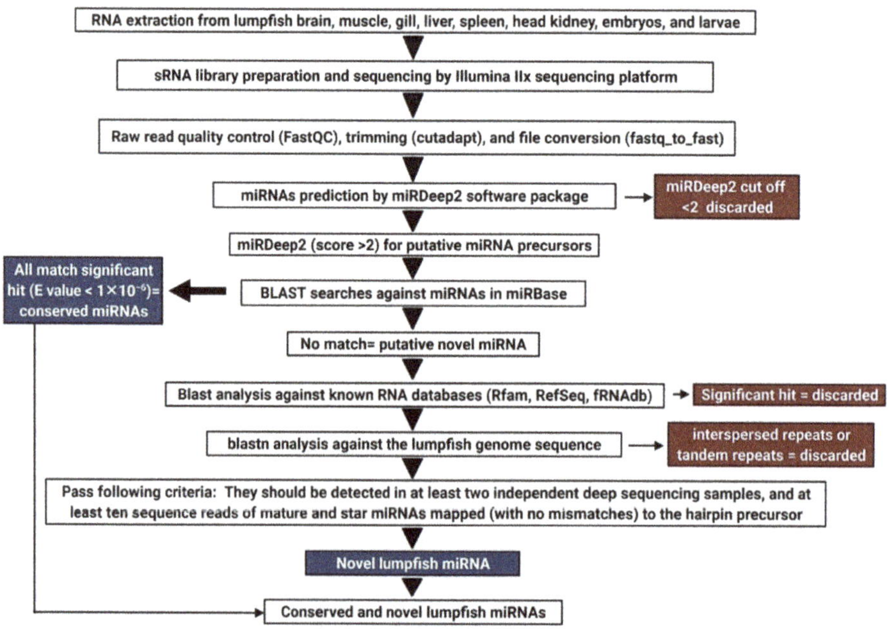

Figure 1. Experimental workflow used for characterization of lumpfish miRNAome.

High-quality trimmed reads were used to discover lumpfish miRNA using the miRDeep2 software package v0.0.7 (mapper and miRDeep2 analysis modules) applying default commands [50,51]. The miRDeep2 tools assign a log-odds score (the miRDeep2 score) based on an algorithm that integrates the statistics of the read positions, the frequencies of reads within hairpins, and the posterior probability that the hairpin was derived from a true miRNA gene [50]. A miRDeep2 score of ≥ 2 was used as a cutoff to prevent false positive detection of miRNA precursors. In addition, they were inspected regarding the following criteria: (i) reads between $5'$ and $3'$ end of a precursor should be aligned perfectly in a discrete manner; (ii) miRNA precursors should be detected in at least two independent deep sequencing samples, and (iii) at least ten sequence reads of mature and miRNAs mapped to the hairpin precursor [52]. We further analyzed these putative precursor sequences by BLAST searches against known precursor sequences deposited in miRBase, (http://www.mirbase.org/index.shtml) (last accessed on 22 November 2021). Any putative miRNA precursor sequence having a significant hit (E-value $< 1 \times 10^{-6}$) in the BLAST analyses was regarded as a true evolutionarily conserved lumpfish ortholog of the miRNA gene in miRBase that retrieved the best hit and annotated as the evolutionarily conserved lumpfish ortholog of the miRNA gene according to the miRbase nomenclature guidelines (clu-prefix and same number as in other teleosts) [53,54]. The putative miRNA precursor sequences that were identified by miRDeep2 and passed the additional criteria but did not show any significant match to the existing precursors in miRBase were considered putative novel miRNAs. All those sequences were further analyzed by blastn searches against RNA databases in GenBank (http://blast.ncbi.nlm.nih.gov/Blast) (Last accessed: 20 November 2021), the small RNA databases Rfam (https://rfam.xfam.org/search) (Last accessed: 20 November 2021), and the functional RNA database fRNAdb (https://dbarchive.biosciencedbc.jp/en/frnadb/desc.html) (Last accessed: 20 November 2021). Sequences that had a significant hit against these databases were considered other kinds of small RNA and discarded from the analysis. The remaining precursors were used as queries in blastn analysis against the lumpfish genome sequence. Sequences with a significant BLAST hit (E value $< 1 \times 10^{-6}$) against multiple loci (>10) in the lumpfish genome reference sequence were considered to be interspersed repeats or tandem repeats and discarded from the analysis. Sequences that passed all these filtering steps were regarded as true novel lumpfish miRNAs. A reference miRNAome of unique mature miRNA sequences (5p or 3p) for expression analysis of HTS data were generated by aligning all mature miRNAs using Sequencher software 5.3 (Gene Codes Corporation, Ann Arbor, MI, USA). The identical mature miRNAs from the same families were aligned applying strict settings, and the final reference, thus, consisted of only the unique, different mature miRNAs.

2.7. Disclosing Putative Differentially Expressed and Organ and Developmental Stage Enriched miRNAs

The HTS data from 16 tissue samples were used to estimate the expression of individual miRNAs across the different organs and developmental stages. The adapters were trimmed from the raw reads, and the resulting reads were filtered based on size. The filtered reads from all 16 samples were mapped to the reference applying STAR aligner software (v.2.5.2b) [47]. The index for mapping was generated from the unique mature lumpfish miRNAs (see 2.6) with parameters genomeSAindexNbases 6. STAR aligner software (v.2.5.2b) with alignIntronMax1 and default parameters was then used for the mapping. Next, the output files of STAR mapping (BAM format) were processed further in R-Studio by using the feature Counts function from the Rsubread package (v.1.34.2) to produce count matrices [47]. The count tables were used as input in the DESeq2 R package (v.1.24.0) for differential expression analysis. Samples from an organ or developmental stage ($n = 2$) were compared to all other tissues sampled ($n = 14$). Putative differentially expressed miRNAs were defined as those with Benjamini-Hochberg adjusted $p \leq 0.05$, $\log_2$ fold change threshold value of at least ≤ -3.0 or ≥ 3.0. The miRNA abundance of the different miRNAs within a particular organ or developmental stage was estimated as the percentage of a specific miRNA out of the total based on the average of normalized read counts from duplicated samples (reads less than 20 were filtered out). Enriched miRNAs were analyzed for each organ and developmental stage.

2.8. RT-qPCR

We selected eight different miRNAs that were suggested as differentially expressed in literature and enriched in one of the organs by the DESeq2 analysis for further expression analysis with qPCR. These miRNAs had previously shown similar organ-specific enrichment in other teleosts [19,20]. The RNA-seq read numbers of these eight miRNAs are provided in Table S8. Those eight miRNAs (clu-miR-135c-5p, clu-miR-9b-3p, clu-miR-133ab-3p, clu-miR-205-1-5b, clu-miR-203-3p, clu-miR-203a-5p, clu-miR-192a-5p, clu-miR-122-1-5p) were analysed by RT-qPCR to verify the DESeq2 results. All forward primer sequences used for qPCR were retrieved from the mature sequences of these miRNAs in the characterization step (methods 2.6). The primer sequences are listed in Table 1. The cDNA synthesis and qPCR were carried out applying the miScript (miScript II RT Kit and miScript SYBR Green PCR Kit) assays following the manufacturer's instructions (Qiagen, Hilden, Germany). The qPCR reaction mixture contained 12.5 µL 2 × QuantiTect SYBR Green Master Mix, 2.5 µL 10× miScript Universal Primer, 2.5 µL of 10 µM forward miRNA-specific primer, 5 µL RNase free water, and 2.5 µL cDNA. The qPCR analysis was carried out by Mx3000p (Stratagene, Agilent Technologies, LA Jolla, CA, USA) using the following cycle, 95 °C for 15 min followed by 40 cycles of 94 °C for 15 s, 55 °C for 30 s and 70 °C for 30 s as described in Andreassen et al., 2016 [20]. The mature sequences of clu-mir-25-3p and clu-mir-17-5p were used as reference genes [20,55]. The instrument-provided ct values were applied to the LinRegPCR (v2021.1) software to calculate efficiency in all assays, and then the efficiency-adjusted Ct-values were provided [56]. The efficiency adjusted values were also used in the normalization (geomean from the two reference genes) to provide the dCt-values. The relative change in expression in each miRNA's target organ was calculated using the comparative Ct method (ΔΔCt-method) [57]. All the comparisons were relative to the lowest expressed organ/tissue for the particular miRNA. All relative quantity (RQ) data are presented as mean ± standard error (SE). The RQ values for each target gene were subjected to a one-way ANOVA with Tukey post-tests to compare gene expression across tissues. All statistical tests were performed using GraphPad Prism 7.04 (San Diego, CA, USA) with the p-value threshold set at $\leq$0.05. The number of organ samples was four or five for each group, while the early developmental stages had two biological replicates in each group (Table S1).

Table 1. Primers used in qPCR analysis of mature miRNAs.

miRNAs	Primer Sequences (5′ to 3′)
clu-miR-25-3p	CATTGCACTTGTCTCGGTCTGA
clu-miR-17-1-5p	CAAAGTGCTTACAGTGCAGGTA
clu-miR-122-1-5p	TGGAGTGTGACAATGGTGTTTG
clu-miR-133ab-3p	TTTGGTCCCCTTCAACCAGCTGT
clu-miR-205-1-5p	TCCTTCATTCCACCGGAGTCTG
clu-miR-135c-5p	TATGGCTTTTTATTCCTATGTG
clu-miR-203-3p	GTGAAATGTTTAGGACCACTTG
clu-miR-203a-5p	AGTGGTTCTCAACAGTTCAACA
clu-miR-192a-5p	ATGACCTATGAATTGACAGCCA
clu-miR-9b-3p	TAAAGCTAGAGAACCGAATGTA

3. Results

3.1. Total RNA Extraction, Library Preparation, and Small RNA Sequencing

Total RNA extracted from 32 samples (brain, muscle, gill, liver, spleen, and head kidney from five adult fish and two samples each from larvae and embryos) showed concentrations ranging from 100 to 3250 ng/µL (Table S1) and intact 28S and 18S bands in 1% agarose gel indicated that they were of high quality. All these samples qualified for further analysis by HTS and qPCR. Small RNA libraries were successfully generated for

16 samples (twelve tissue samples from two adult fish and two samples from each early developmental stage). The HTS resulted in a total of 147,972,041 raw reads, ranging from 6.6 to 13.3 million reads per sample. After adapter trimming, there were a total of 86,054,423 reads ranging from 4.5 to 6.9 million reads per sample (Table S2). All raw HTS results were submitted to NCBI with BioProject accession number PRJNA679415. The individual SRA accession numbers are given in Table S2.

3.2. Characterization of Lumpfish miRNA

The processed reads from each sample were analyzed with miRDeep2 software for miRNA gene discovery (Figure 1). Subsequent BLAST homology searches of all putative miRNA precursor sequences against miRbase revealed a total of 391 miRNA genes from 104 different families that were lumpfish orthologs to evolutionarily conserved miRNAs. They were subsequently annotated as the lumpfish orthologs of these miRNAs. The miRDeep2 analysis also revealed 5p or 3p arm domination (most abundant mature miRNA from a given precursor) and the genome location of each miRNA gene. An overview of all precursor sequences along with their corresponding 5p and 3p mature sequences is given for all evolutionarily conserved miRNA genes in Table S3.

A total of 98 precursors identified by miRDeep2 did not show significant matches in the homology analyses against miRBase. These were considered as putative novel miRNA precursor sequences. They were further analyzed by blastn searches against RNA databases in GenBank, small RNA databases Rfam, functional RNA database fRNAdb, and lumpfish genome sequence (GenBank Accession: PRJNA625538). Sequences that had a significant hit against these databases were discarded from the analysis, as described in the methods section. Following this filtering process, eight precursor and corresponding mature sequences showed characteristics expected from true miRNAs. These eight miRNA precursor sequences are likely to represent novel lumpfish miRNAs, and all these novel miRNA genes along with their corresponding 5p and 3p mature sequences, the observed arm dominance of mature sequences, and the genome location of each miRNA gene are given in the last part of Table S3. Finally, the mature miRNAs were aligned using Sequencher software to identify all unique mature miRNAs (many mature miRNAs from the same families were identical). There were 443 unique mature miRNAs. These unique miRNAs representing the lumpfish miRNAome are given in Supplementary File S4.

3.3. Abundance of miRNAs within Organs and Developmental Stages

We determined the diversity of miRNAs within the lumpfish tissues/organs and developmental stages based on the normalized read counts. The normalized read counts for all samples are shown in Table S5, while the average normalized read counts for each tissue/organ or developmental stage are shown in Table S6.

Our results show the presence of 340 unique mature miRNAs in the lumpfish brain, 328 in muscle, 289 in gill, 288 in the liver, 268 in the spleen, 289 in the head kidney, 328 in embryos, and 327 in larvae (Figure 2). Two hundred forty-one mature miRNAs were expressed in all six organs of adult lumpfish, 324 mature miRNAs were expressed commonly in embryos and larvae, and 223 mature miRNAs were expressed across all three developmental stages. All the miRNAs expressed in the early life stages, such as embryos and larvae, were also expressed in at least one organ of adult fish. The exceptions were clu-miR-19a-2-5p, which was only expressed in embryos, and clu-miR-137-1-5p, only in larvae.

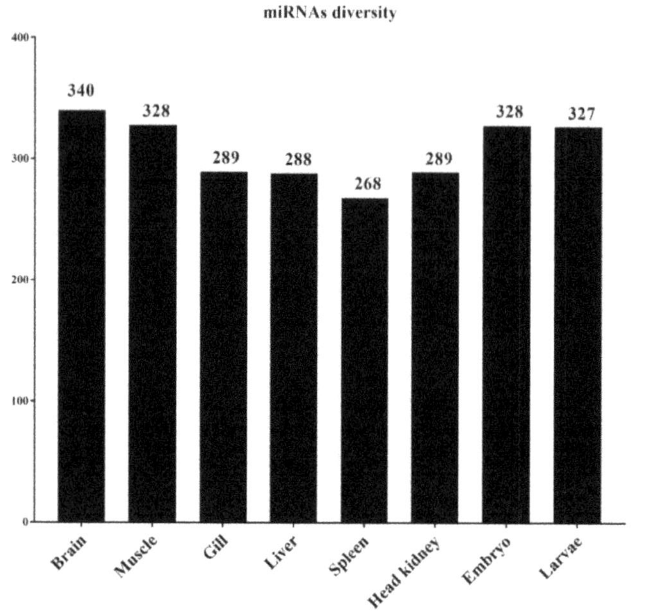

Figure 2. miRNA diversity in lumpfish tissue/organs and early developmental stages.

The abundance of most common mature miRNAs within each organ and developmental stage is shown in Figures 3 and 4, respectively. These figures show the distribution of the top 20 enriched mature miRNAs within each of the six organs and in the two early developmental stages. The abundances for all miRNAs within each of the organs and early developmental stages are shown in Table S6. Five of the top 20 enriched mature miRNAs, clu-miR-21a-5p, clu-miR-22ab-3p, clu-miR-26-1-5p, clu-miR-100-2-5p, and clu-let-7g-5p were highly abundant within all organ and early developmental stages. While the five mature miRNAs clu-miR-146a-5p, clu-let-7a-3-5p, clu-miR-126-3p, clu-let-7e-5p, and clu-miR-143-3p were highly abundant miRNAs within all six organs of adult lumpfish, but not among the highly expressed miRNAs within lumpfish embryos and larvae (Figures 3 and 4). Additionally, several miRNAs were highly abundant within one of the tissue/organs from adult fish compared to others. For example, Clu-miR-122-1-5p, clu-miR-192a-5p, clu-miR-152ab-3p, and one novel miRNA (clu-miR-nov-5-5p) were also among the top 20 most abundant miRNAs in the liver, but with much lower abundance when comparing expression of miRNAs within other organs. Likewise, clu-miR-1-1-3p, clu-miR-206-3p, and clu-miR-133ab-3p were abundant only in muscle, clu-miR-451a-5p only in spleen, clu-miR-142-2-3p only in head kidney, and clu-miR-9-2-5p and clu-miR-7-3-5p only in brain (Figure 3, Table S6). Two miRNAs, clu-miR-217b-5p and clu-miR-181b-3-5p, were common in the two early developmental stages while having relatively low expression within adult organs. In addition, there were some miRNAs common in one organ and early developmental stage. These were clu-miR-9-2-5p and clu-miR-7-3-5p (brain and early developmental stages), clu-miR-1-1-3p and clu-miR-206-3p (brain and early developmental stages) and clu-miR-192a-5p (liver and early developmental stages).

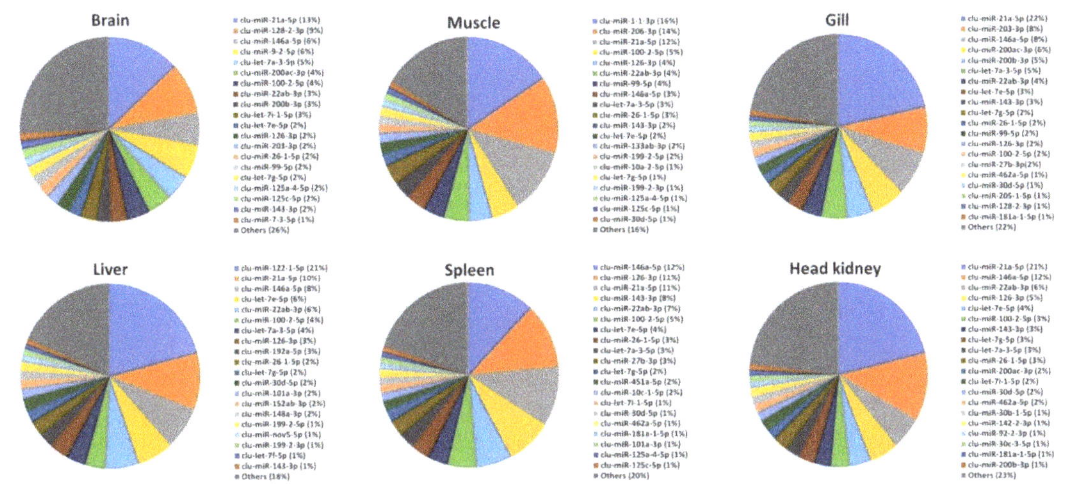

Figure 3. Twenty most abundant miRNAs in lumpfish brain, muscle, gill, liver, spleen, and head kidney.

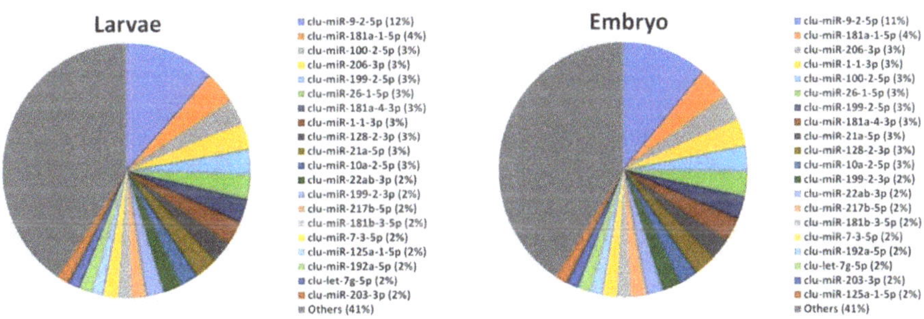

Figure 4. Twenty most abundant miRNAs in lumpfish embryos and larvae.

3.4. Comparison of Mature miRNA Expression between Organs and Early Developmental Stages

To further explore whether some miRNAs (any of the miRNAs, not only top common ones) were differentially expressed between adult organs or early developmental stages, we carried out expression analysis of the HTS data and additional RT-qPCR of selected miRNAs. DESeq2 analysis of the HTS data was conducted by comparing one organ or early developmental stage ($n = 2$) to all other samples ($n = 14$). The results (Table S7) suggested that several miRNAs have higher or lower expression in one organ or early developmental stages compared to all other samples. The suggested miRNAs with an increased expression ($\log_2$ fold change > 3.0) in a particular organ or early developmental stage compared to expression in all others are given in Tables 2 and 3, respectively. The numbers of such miRNAs were 9 in the brain, 5 in muscle, 8 in gill, 15 in the liver, 3 in the spleen, 13 in embryos, and 22 in larvae. However, our DESeq2 analysis did not suggest any enrichment of miRNAs in the lumpfish head kidney.

Table 2. Mature miRNAs suggested as highly expressed in one organ compared to others.

Organ [1]	miRNAs [2]	Log_2FC [3]
Brain	clu-miR-31-3p	6.14
Brain	clu-miR-153c-3p	5.96
Brain	clu-miR-153a-3p	5.33
Brain	clu-miR-1788-5p	4.91
Brain	clu-miR-212b-1-5p	4.10
Brain	clu-miR-212b-1-3p	3.14
Brain	clu-miR-128-2-3p	3.49
Brain	clu-miR-338-1-3p	3.40
Brain	clu-miR-132-1-5p	3.08
Muscle	clu-miR-133b-3p	6.08
Muscle	clu-miR-133ab-3p	5.45
Muscle	clu-miR-1-1-3p	5.23
Muscle	clu-miR-1-3-5p	3.56
Gill	clu-miR-31-5p	6.91
Gill	clu-miR-1788-3p	6.21
Gill	clu-miR-203-3p	5.13
Gill	clu-miR-203a-5p	4.61
Gill	clu-miR-375-1-3p	4.82
Gill	clu-miR-205-1-3p	4.16
Gill	clu-miR-200b-3p	3.8
Gill	clu-miR-200b-5p	3.36
Liver	clu-miR-122-1-5p	8.23
Liver	clu-miR-122-1-3p	7.65
Liver	clu-miR-nov3-3p	6.78
Liver	clu-miR-nov3-5p	4.68
Liver	clu-miR-nov1-5p	5.58
Liver	clu-miR-101b-3p	4.83
Liver	clu-miR-101b-5p	4.41
Liver	clu-miR-722-3p	4.71
Liver	clu-miR-722-5p	4.38
Liver	clu-miR-92b-3p	4.04
Liver	clu-miR-92b-5p	3.91
Liver	clu-miR-192a-5p	3.75
Liver	clu-miR-94a-5p	3.43
Liver	clu-miR-152ab-3p	3.37
Liver	clu-miR-nov5-5p	3.36
Spleen	clu-miR-2187b-5p	5.10
Spleen	clu-miR-2187b-3p	3.47
Spleen	clu-miR-460-5p	3.27

[1] Organ samples were obtained from adult lumpfish. [2] The names are in a few cases with different lettered/numbered suffixes than in miRBase as several mature family members are identical. The miRNAs in the table are grouped in families, and the family member with the highest FC is used to list families in descending order. [3] Log_2-transformed fold-change (FC) as determined by DESeq2 analysis.

RT-qPCR was applied to verify the findings from the DESeq2 analysis in a few selected miRNAs. The two conserved mature miRNAs clu-mir-25-3p and clu-mir-17-5p, shown as suitable reference genes in other teleosts [20,55], revealed stable expression across all samples in this study (mean Ct values were 22.8 ± 0.9 (SD), 22.4 ± 1.1 (SD), respectively) and were, consequently, used as reference genes in the RT-qPCR analysis. Eight miRNAs known to be highly expressed in certain organs [19,20] were selected for RT-qPCR (Table 1). These selected miRNAs showed significantly increased expression levels in the expected tissue/organs (Figure 5) that align with the literature [19,20]. For instance, clu-miR-135c-5p and clu-miR-9b-3p expression was significantly higher in brain, clu-miR-205b-5b, clu-miR-203a-3p, and clu-miR-203b expression was significantly higher in gill, clu-miR-133-3p expression was significantly higher in muscle, and clu-miR-122-5p and clu-miR-192a-5p expression was significantly higher in liver compared with other tissue/organs. These qPCR results agreed with the DESeq2 results for six organ samples, while the increases observed for clu-miR-135c-5p and clu-miR-9b-3p in the brain were similar in the DESeq2 analysis but not significant. However, we utilized four or five biological replicates for the RT-qPCR analysis, whereas two were used in the DESeq2 analysis. Additionally, the significance levels were adjusted according to a large number of tests in the DESeq2 analysis. This could explain why the increases did not reach the significant thresholds in the DESeq2 analysis for these two miRNAs.

Table 3. Mature miRNAs suggested as highly expressed in embryos or larvae.

Embryos/Larvae [1]	miRNAs [2]	Log_2FC [3]
Embryos	clu-miR-430b-5-5p	5.61
Embryos	clu-miR-430b-4-3p	4.51
Embryos	clu-miR-430b-1-3p	4.37
Embryos	clu-miR-190b-5p	5.45
Embryos	clu-miR-726-5p	4.91
Embryos	clu-miR-184ab-2-3p	4.77
Embryos	clu-miR-184ab-3p	4.77
Embryos	clu-miR-301b-5p	4.73
Embryos	clu-miR-301b-1-5p	4.40
Embryos	clu-miR-124-1-5p	4.40
Embryos	clu-miR-217b-5p	4.23
Embryos	clu-miR-217a-5p	4.13
Embryos	clu-miR-216a-1-5p	4.20
Larvae	clu-miR-124-1-5p	4.41
Larvae	clu-miR-130-1-5p	3.15
Larvae	clu-miR-130-6-5p	3.62
Larvae	clu-miR-183-5p	4.09
Larvae	clu-miR-184ab-2-3p	4.71
Larvae	clu-miR-184ab-3p	4.71
Larvae	clu-miR-190b-5p	5.44
Larvae	clu-miR-194b-3p	3.46
Larvae	clu-miR-196a-1-5p	3.95
Larvae	clu-miR-216a-1-5p	4.00
Larvae	clu-miR-217a-5p	4.11
Larvae	clu-miR-217b-5p	4.11
Larvae	clu-miR-301b-1-5p	4.38
Larvae	clu-miR-301b-3p	3.44
Larvae	clu-miR-301b-5p	4.70
Larvae	clu-miR-430a-12-3p	3.97
Larvae	clu-miR-430a-3-3p	3.97
Larvae	clu-miR-430b-1-3p	4.28
Larvae	clu-miR-430b-4-3p	4.40
Larvae	clu-miR-430b-5-5p	5.41
Larvae	clu-miR-459-3p	4.01
Larvae	clu-miR-726-5p	4.64

[1] Lumpfish embryos were obtained at 300 degree days, and lumpfish larvae were obtained after one week post-hatch. [2] The names are in a few cases with different lettered/numbered suffixes than in miRBase as several mature family members are identical. The miRNAs in the table are grouped in families, and the family member with the highest FC is used to list families in descending order. [3] Log_2-transformed fold-change (FC) as determined by DESeq2 analysis.

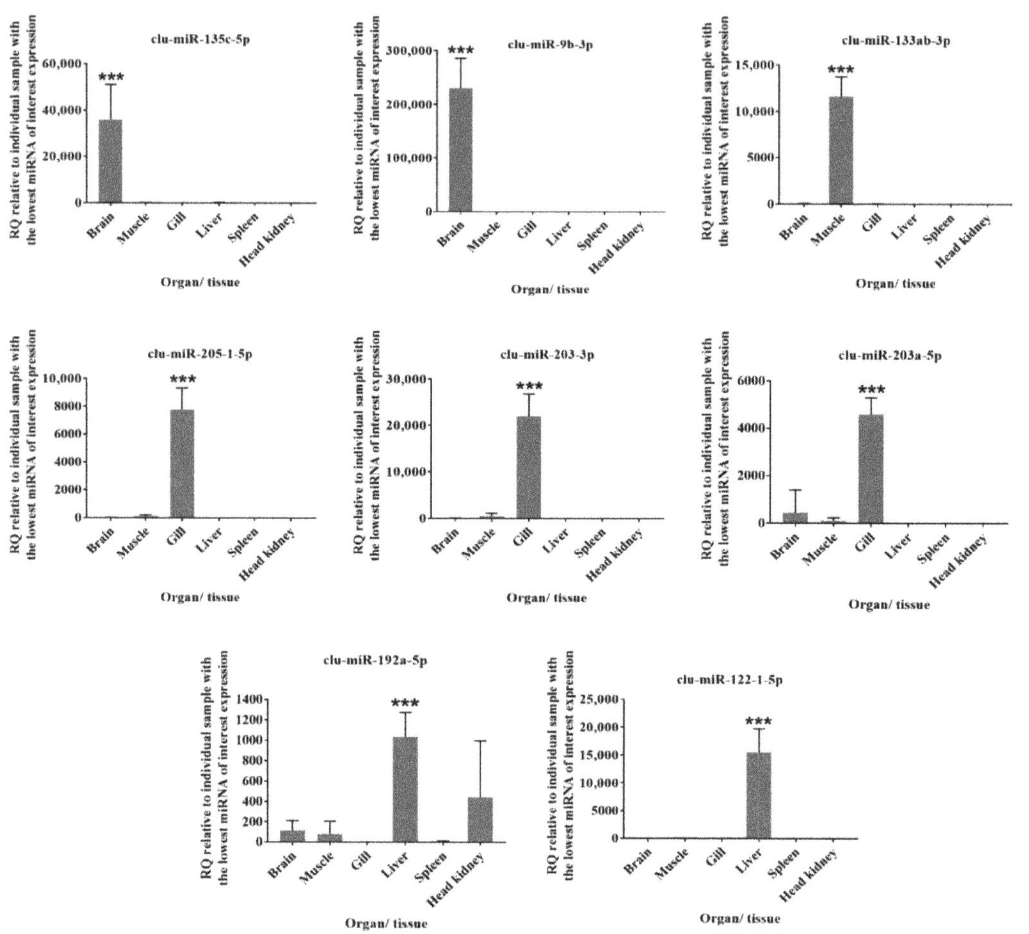

Figure 5. Verification of tissue-specific expression of conserved miRNAs. RT-qPCR results show the relative expression of eight miRNAs (clu-miR-135c-5p, clu-miR-9b-3p, clu-miR-133ab-3p, clu-miR-205-1-5b, clu-miR-203-3p, clu-miR-203a-5p, clu-miR-192a-5p, clu-miR-122-1-5p) across lumpfish organs (brain, muscle, gill, liver, spleen, and head kidney). Number of replicates for tissue samples were five (n = 5) except brain (n = 4) and muscle (n = 4). RQ: relative quantity normalized to clu-miR-25-3p and clu-miR-17-1-5p and calibrated to the individual sample with the lowest miRNA of interest expression. *** on the top of a particular sample indicates that the expression of the particular miRNA is significantly higher when compared to others by one-way ANOVA (p < 0.001).

4. Discussion

miRNAs play a significant role in embryonic development, determination of cell fate, and control of cell proliferation, differentiation, and death. Their dysregulation has a significant impact on critical cellular pathways and is linked to a variety of diseases [11,13,16,48]. A species-specific and well-characterized miRNAome generated from small RNA sequencing of different developmental stages is required to study miRNA expression by analysis of HTS data. Characterization of miRNAs in multiple organs and developmental stages in a new aquaculture species like lumpfish will also facilitate further studies to determine their role in development, whether they regulate organ developmental stage-specific functions, immune responses to infectious diseases, and disease progression. Therefore, this study was undertaken to define and characterize miRNAs expressed in the brain, muscle, gill, liver, spleen, and head kidney of adult lumpfish, as well as the two developmental stages, embryos and larvae. Together this resulted in a miRNAome consisting of 443 unique mature miRNAs that were used as lumpfish miRNA reference for analysis of HTS data and primer design (RT-qPCR analysis) of single miRNAs.

The expression of different miRNAs within an organ or developmental stage would reveal which ones were highly abundant and likely to have essential regulatory functions. Comparisons between adult organs and early developmental stages could further reveal the highly expressed ones in a few or single organs. We applied DESeq2 analysis to demonstrate that the miRNAome worked well as a reference in such HTS analysis. However, as there were two biological replicates of each adult organ (or early developmental stages) compared to all other HTS samples ($n = 14$) in these analyses, we report them as suggestive expression differences. Ideally, there should be three or more biological replicates in each group compared in such analysis. However, we did choose a rather conservative $\log_2$FC (3 or more) to suggest them as differently expressed between organs (Table S7), and some of the miRNAs increased in particular organs were also supported in the additional RT-qPCR analysis (Figure 5).

Our analysis identified that 10 mature miRNAs were highly abundant and among the top 20 enriched miRNAs within all six organs (five were also among the top 20 enriched in the early developmental stages). These 10 mature miRNAs (clu-let-7a-3-5p, clu-let-7e-5p, clu-let-7g-5p, clu-miR-21a-5p, clu-miR-22ab-3p, clu-miR-26-1-5p, clu-miR-100-2-5p, clu-miR-126-3p, clu-miR-143-3p, and clu-miR-146a-5p) are conserved miRNA families discovered in the majority of vertebrates in miRBase [19,20,58]. Their high expression within all adult organs could suggest that these miRNAs play a critical role in lumpfish cellular homeostasis. Still, as they are highly abundant in all adult organs, they are not likely to regulate organ-specific functions.

The brain receives information from sense organs that monitor conditions both within and around the fish. In the brain, the immune cells and the central nervous system interactions allow the immune system to fight against infection and enable the nervous system to regulate immune functioning [59,60]. Any change in these interaction pathways can cause many pathological conditions attributed to organ dysfunction [59,60]. However, miRNAs are critical brain development and function regulators, such as neuronal activity [11,61]. Our miRDeep2 analysis identified 340 conserved mature miRNAs in the lumpfish brain. Among highly enriched in the brain are clu-miR-9-2-5p and clu-miR-7-3-5p. These two miRNAs do not have similar high relative expression levels within any other adult organs but are similarly enriched in the two early developmental stages, indicating they could be important in developing neural tissue in lumpfish (Table S6). A similar enrichment pattern of miR-9-5p is seen in Atlantic salmon, cod, halibut, three-spined stickleback, and zebrafish brain [19,20,22,28]. Enrichment of miR-7 in the brain is also observed across vertebrates [62]. Several studies have shown that these two miRNAs are crucial for brain development in zebrafish and other vertebrates [63–65], and it is likely that clu-miR-9-2-5p and clu-miR-7-3-5p may have similar functions in lumpfish. The DESeq2 analysis also suggested that clu-miR-128-2-3p, clu-miR-153c-3p, clu-miR-212b-1-5p, and clu-miR-338-1-3p were more than 10-times higher expressed in the brain than other organs (Table S7a). Similar findings were observed in Atlantic cod, three-spined stickleback, and zebrafish [19,66]. In higher vertebrates, miR-128 controls neural motor behaviours by regulating the expression of various ion channels [67]. The three other miRNAs have also all been reported as having important brain functions in higher vertebrates [68–70].

Fish muscles are the major edible parts worldwide, determining the nutritional and the market value. The teleost muscle is also an immunologically active organ, playing an important role against pathogens [71]. MicroRNAs are established modulators of muscle cell proliferation, differentiation, regeneration, and diseases [72]. Our miRDeep2 analysis identified 328 conserved mature miRNAs in lumpfish muscle. The muscle-specific top enriched miRNAs were clu-miR-1-1-3p and clu-miR-133ab-3p (present in other organs but much less abundant). Similar to our study, miR-133 and miR-1 were enriched in zebrafish, Atlantic salmon, and cod, suggesting the maintenance of muscle-specific miRNAs expression and function [18–20,73]. For example, miR-133 is one of the foremost studied and best-characterized miRNAs in vertebrates. It is required for proper skeletal and cardiac muscle development and function in mammals and fish [74,75]. On the other hand, miR-1 is a conserved miRNA in the muscle tissue that plays a crucial role in maintaining muscle integrity [76].

Because of direct exposure to the water, teleost gills are the main mucosal surfaces for the entrance of pathogens, which trigger an immune response [77]. miRNAs are important regulators of immune response to those infections in the gills of fish [20,78,79]. However, our DESEq2 analysis was on apparently healthy organs and suggested the enrichment of clu-miR-200 and clu-miR-203 family members and clu-miR-205-1-3p, clu-miR-375-1-3p, clu-miR-31-5p, and clu-miR-1788-3p in lumpfish gill. RT-qPCR results confirmed the enrichment of clu-miR-203-3p, clu-miR-203a-5p, and clu-miR-205-1-5p. Some of these miRNAs, such as miR-200, miR-205, and miR-375, were enriched in cod gill as well [20], while miR-200, miR-203, and miR-205 were enriched in gills of tilapia [80]. One of these miRNAs, miR-200, has been shown as important to gill function in cell studies of fish [81]. However, no study has been conducted to decipher the gill-associated role of the remaining five lumpfish miRNAs suggested as differentially expressed in teleost gill.

The liver is involved in various vital functions in controlling biochemical processes, including detoxification and metabolism [82]. miRNAs are essential for regulating liver development and functions, and alterations in intrahepatic miRNA networks have been associated with liver disease in humans [83]. They are also associated with hepatic lipid metabolism in Atlantic salmon [84]. Our miRDeep analysis identified 288 conserved mature miRNAs in the lumpfish liver (Table S6).

Four of the top 20 enriched miRNAs in liver—clu-miR-122-1-5p, clu-miR-152ab-3p, clu-miR-192a-5p and clu-miR-nov5-5p—did not show similar enrichment in any other adult organs (Figure 3). DESEq2 analysis also suggested these as having significantly increased expression in the liver, and this was confirmed by RT-qPCR for clu-miR-122-1-5p and clu-miR-192a-5p (Figure 5). This finding is similar to other teleosts and mammals [20,85,86]. miR-122 is the most abundant miRNA in the liver of many species. In mammals, miR-122 is studied extensively and is known to be involved in lipid metabolism [85]. Furthermore, miR-192 is involved in cell growth, lipid synthesis, and apoptosis [87] and having such roles also aligns with this miRNA being among the top 20 miRNAs expressed in the early developmental stage samples (Figure 4). Dysregulation of miR-152 is associated with liver disease in higher vertebrates indicating they are important hepatic miRNAs [84,88]. Based on the high conservation of these miRNAs among vertebrates (miRBase 22.1) [58] and with a similar enrichment pattern observed in lumpfish, we could assume they also play a similar liver-specific role in lumpfish.

As the body's primary blood filter, the spleen plays a major role in detecting cell damage during infection [89]. The spleen is the home of different types of immune cells that trigger different immune responses [89–91]. Splenic miRNAs have been identified to modulate immune responses during diseases in humans, mice, chickens, dogs, and fishes [92–99]. Our miRDeep analysis identified 268 conserved mature miRNAs in the lumpfish spleen. One mature miRNA, clu-miR-451a-5p, was only among the top 20 enriched miRNAs in the spleen, and this particular miRNA has been shown to regulate erythroid maturation in zebrafish [100]. Furthermore, our DESeq2 analysis suggested the enrichment of clu-miR-2187b-5p and clu-miR-460-5p in the lumpfish spleen. These two miRNAs are also enriched in Atlantic salmon and cod spleen [19,20], but their particular function in the spleen has not yet been investigated.

The anteriormost part of the kidney in the teleost is referred to as the head kidney. It is predominantly a lymphoid compartment. The head kidney is an essential hematopoietic organ and serves as a secondary lymphoid organ, a lymph node analog, vital in inducing and elaborating immune responses [90,91]. Assessing changes in the expression of miRNAs in the head kidney could provide more comprehensive insight into the immune response to infection. Our miRDeep2 analysis identified 289 conserved mature miRNAs in the lumpfish head kidney. Our DESeq2 analysis did not suggest any enrichment of miRNA in the head kidney.

The embryos and larvae samples did reveal several miRNAs suggested as early developmental stage enriched. Notably, the miR-430 family was suggested as enriched by the DESeq2 analysis. These are known as highly expressed in early development, and among suggested functions is maternal RNA clearance during early embryogenesis in zebrafish [101,102]. Another miRNA that was highly enriched and expressed only in the early developmental stages was clu-miR-217b-5p. This miRNA, as well as mature miRNAs from miR-124, miR-184, and miR-216 families that were also enriched in the lumpfish early developmental stages, have all been shown as important in zebrafish development [73].

5. Conclusions

In conclusion, this study represents the first characterization of a lumpfish miRNA transcriptome produced by independent analysis of small RNA sequences from several adult organs and early developmental stages. We identified 391 conserved and eight novel miRNA precursor sequences, which account for 443 unique mature miRNAs. Our results demonstrate that most of the lumpfish miRNAs are highly conserved with highly similar precursor sequences to those observed in other teleosts. Many miRNAs also appear to have similar tissue-specific expression patterns as in other vertebrates. Thus the miRNAs profile of lumpfish suggested a similar organ-specific expression pattern as other vertebrates. It is possible that these conserved miRNAs are regulating essential and conserved genes in vertebrates. Furthermore, the identification and characterization of lumpfish-specific novel miRNAs repertoire in this study will be crucial for further functional studies of the novel miRNAs in this species.

Supplementary Materials: The following are available online at https://www.mdpi.com/article/10.3390/biology11010130/s1, Table S1: Concentration and quality of RNA samples, Table S2: Descriptive data from small RNA sequenced samples, Table S3: All evolutionarily conserved and novel lumpfish miRNA genes discovered in different organs and developmental stages, File S4: Unique mature miRNAs, Table S5, Normalized read counts of individual mature miRNAs from sequencing data in each sample sequenced, Table S6: Relative distribution of miRNAs within organs and within developmental stages, Table S7: Mature miRNAs suggested as differentially expressed by DESeq2 analysis, Table S8: miRNAs analyzed by qPCR.

Author Contributions: Conceptualization, M.L.R., J.S., R.A.; formal analysis, S.C., N.T.W. and T.V.; data curation, N.T.W. and S.C.; writing—original draft, S.C., R.A., J.S.; writing—review & editing S.C., R.A., M.L.R., D.B., J.S.; investigation, S.C.; project administration, J.S., R.A. and D.B. supervision, J.S. and R.A. All authors have read and agreed to the published version of the manuscript.

Funding: This research was funded by Norwegian Research Council grant 280839/E40 to R.A, Canada First—Ocean Frontier Institute (Module J), Atlantic Fisheries Funds, Genome Atlantic RP3 fund, Cooke Aquaculture division Cold Ocean Salmon-Newfoundland, Atlantic Canada Oportunities Agency (ACOA) and NSERC-Discovery grant (RGPIN-2018-05942) to J.S.

Institutional Review Board Statement: The fish dissection and tissue sample collection were performed following the Canadian Council on Animal Care guidelines and approved by Memorial University of Newfoundland's Institutional Animal Care Committee under protocols #18-1-JS and #18-03-JS.

Informed Consent Statement: Not applicable.

Data Availability Statement: All raw HTS results were submitted to NCBI with bio project accession number PRJNA679415. The individual SRA accession numbers are given in Table S2.

Acknowledgments: We thank Sanatander lab members (Memorial University of Newfoundland), JBARB personnel (Memorial University of Newfoundland), and Andreassen lab (Oslo Metropolitan University) for their logistical support. We are thankful for the sequencing services performed by the Genomics Core Facility Oslo (Oslo University Hospital, Norway).

Conflicts of Interest: The authors declare no conflict of interest.

References

1. Bhaskaran, M.; Mohan, M. MicroRNAs: History, biogenesis, and their evolving role in animal development and disease. *Vet. Pathol.* **2014**, *51*, 759–774. [CrossRef] [PubMed]
2. Michlewski, G.; Caceres, J.F. Post-transcriptional control of miRNA biogenesis. *RNA* **2019**, *25*, 1–16. [CrossRef] [PubMed]
3. Krol, J.; Loedige, I.; Filipowicz, W. The widespread regulation of microRNA biogenesis, function and decay. *Nat. Rev. Genet.* **2010**, *11*, 597–610. [CrossRef]
4. Saliminejad, K.; Khorram Khorshid, H.R.; Soleymani Fard, S.; Ghaffari, S.H. An overview of microRNAs: Biology, functions, therapeutics, and analysis methods. *J. Cell. Physiol.* **2019**, *234*, 5451–5465. [CrossRef]
5. Bronevetsky, Y.; Ansel, K.M. Regulation of miRNA biogenesis and turnover in the immune system. *Immunol. Rev.* **2013**, *253*, 304–316. [CrossRef]
6. Matsuyama, H.; Suzuki, H.I. Systems and Synthetic microRNA Biology: From Biogenesis to Disease Pathogenesis. *Int. J. Mol. Sci.* **2019**, *21*, 132. [CrossRef]
7. De Sousa, M.C.; Gjorgjieva, M.; Dolicka, D.; Sobolewski, C.; Foti, M. Deciphering miRNAs' Action through miRNA Editing. *Int. J. Mol. Sci.* **2019**, *20*, 6249. [CrossRef] [PubMed]
8. Ha, M.; Kim, V.N. Regulation of microRNA biogenesis. *Nat. Rev. Mol. Cell Biol.* **2014**, *15*, 509–524. [CrossRef]
9. Hammond, S.M. An overview of microRNAs. *Adv. Drug Deliv. Rev.* **2015**, *87*, 3–14. [CrossRef] [PubMed]
10. Ravi, V.; Venkatesh, B. Rapidly evolving fish genomes and teleost diversity. *Curr. Opin. Genet. Dev.* **2008**, *18*, 544–550. [CrossRef]
11. Bizuayehu, T.T.; Babiak, I. MicroRNA in Teleost Fish. *Genome Biol. Evol.* **2014**, *6*, 1911–1937. [CrossRef]
12. Xue, X.; Woldemariam, N.T.; Caballero-Solares, A.; Umasuthan, N.; Fast, M.D.; Taylor, R.G.; Rise, M.L.; Andreassen, R. Dietary Immunostimulant CpG Modulates MicroRNA Biomarkers Associated with Immune Responses in Atlantic Salmon (Salmo salar). *Cells* **2019**, *8*, 1592. [CrossRef] [PubMed]
13. Smith, N.C.; Christian, S.L.; Woldemariam, N.T.; Clow, K.A.; Rise, M.L.; Andreassen, R. Characterization of miRNAs in Cultured Atlantic Salmon Head Kidney Monocyte-Like and Macrophage-Like Cells. *Int. J. Mol. Sci.* **2020**, *21*, 3989. [CrossRef] [PubMed]
14. Woldemariam, N.T.; Agafonov, O.; Sindre, H.; Høyheim, B.; Houston, R.D.; Robledo, D.; Bron, J.E.; Andreassen, R. miRNAs Predicted to Regulate Host Anti-viral Gene Pathways in IPNV-Challenged Atlantic Salmon Fry Are Affected by Viral Load, and Associated with the Major IPN Resistance QTL Genotypes in Late Infection. *Front. Immunol.* **2020**, *11*, 2113. [CrossRef] [PubMed]
15. Andreassen, R.; Woldemariam, N.T.; Egeland, I.O.; Agafonov, O.; Sindre, H.; Hoyheim, B. Identification of differentially expressed Atlantic salmon miRNAs responding to salmonid alphavirus (SAV) infection. *BMC Genom.* **2017**, *18*, 349. [CrossRef]
16. Andreassen, R.; Høyheim, B. miRNAs associated with immune response in teleost fish. *Dev. Comp. Immunol.* **2017**, *75*, 77–85. [CrossRef] [PubMed]
17. Eslamloo, K.; Inkpen, S.M.; Rise, M.L.; Andreassen, R. Discovery of microRNAs associated with the antiviral immune response of Atlantic cod macrophages. *Mol. Immunol.* **2018**, *93*, 152–161. [CrossRef]
18. Andreassen, R.; Worren, M.M.; Høyheim, B. Discovery and characterization of miRNA genes in atlantic salmon (Salmo salar) by use of a deep sequencing approach. *BMC Genom.* **2013**, *14*, 482. [CrossRef] [PubMed]
19. Woldemariam, N.T.; Agafonov, O.; Høyheim, B.; Houston, R.D.; Taggart, J.B.; Andreassen, R. Expanding the miRNA Repertoire in Atlantic Salmon; Discovery of IsomiRs and miRNAs Highly Expressed in Different Tissues and Developmental Stages. *Cells* **2019**, *8*, 42. [CrossRef]
20. Andreassen, R.; Rangnes, F.; Sivertsen, M.; Chiang, M.; Tran, M.; Worren, M.M. Discovery of miRNAs and Their Corresponding miRNA Genes in Atlantic Cod (Gadus morhua): Use of Stable miRNAs as Reference Genes Reveals Subgroups of miRNAs That Are Highly Expressed in Particular Organs. *PLoS ONE* **2016**, *11*, e0153324. [CrossRef]
21. Ma, H.; Hostuttler, M.; Wei, H.; Rexroad, C.E., 3rd; Yao, J. Characterization of the rainbow trout egg microRNA transcriptome. *PLoS ONE* **2012**, *7*, e39649.
22. Bizuayehu, T.T.; Fernandes, J.M.O.; Johansen, S.D.; Babiak, I. Characterization of Novel Precursor miRNAs Using Next Generation Sequencing and Prediction of miRNA Targets in Atlantic Halibut. *PLoS ONE* **2013**, *8*, e61378. [CrossRef]
23. Xu, Z.; Chen, J.; Li, X.; Ge, J.; Pan, J.; Xu, X. Identification and Characterization of MicroRNAs in Channel Catfish (Ictalurus punctatus) by Using Solexa Sequencing Technology. *PLoS ONE* **2013**, *8*, e54174. [CrossRef] [PubMed]
24. Fu, Y.; Shi, Z.; Wu, M.; Zhang, J.; Jia, L.; Chen, X. Identification and Differential Expression of MicroRNAs during Metamorphosis of the Japanese Flounder (Paralichthys olivaceus). *PLoS ONE* **2011**, *6*, e22957. [CrossRef]
25. Xia, J.H.; He, X.P.; Bai, Z.Y.; Yue, G.H. Identification and Characterization of 63 MicroRNAs in the Asian Seabass Lates calcarifer. *PLoS ONE* **2011**, *6*, e17537. [CrossRef] [PubMed]
26. Robledo, D.; Martin, A.P.; Álvarez-Dios, J.A.; Bouza, C.; Pardo, B.G.; Martínez, P. First characterization and validation of turbot microRNAs. *Aquaculture* **2017**, *472*, 76–83. [CrossRef]

27. Chen, P.Y.; Manninga, H.; Slanchev, K.; Chien, M.; Russo, J.J.; Ju, J.; Sheridan, R.; John, B.; Marks, D.S.; Gaidatzis, D.; et al. The developmental miRNA profiles of zebrafish as determined by small RNA cloning. *Genes Dev.* **2005**, *19*, 1288–1293. [CrossRef]
28. Desvignes, T.; Batzel, P.; Sydes, J.; Eames, B.F.; Postlethwait, J.H. miRNA analysis with Prost! reveals evolutionary conservation of organ-enriched expression and post-transcriptional modifications in three-spined stickleback and zebrafish. *Sci. Rep.* **2019**, *9*, 3913. [CrossRef]
29. DFO. Aquaculture in Atlantic Canada—Atlantic Salmon. 2021. Available online: https://www.canada.ca/en/atlantic-canada-opportunities/services/factsheetsandbrochures35.html (accessed on 21 January 2021).
30. DFO. *Evaluation of Bacterial Kidney Disease (BKD) Impacts on the Canadian Salmon Aquaculture Industry Final Report*; Fisheries and Oceans Canada (DFO): Ottawa, ON, Canada, 2010; Available online: https://www.cahs-bc.ca/wp-content/uploads/2019/03/BKDWhitePaper_FinalApr2010_WithAppendices.pdf (accessed on 21 January 2021).
31. Akazawa, N.; Alvial, A.; Blanc, P.P.; Burgos, J.; Chamberlain, G.; Forster, J.; Hoang, T.; Ibarra, R.; Khoa, L.; Kibenge, F.; et al. *Reducing Disease Risk in Aquaculture—World Bank Report Number 88257-GLB*; World Bank: Washington, DC, USA, 2014.
32. Brooker, A.J.; Papadopoulou, A.; Gutierrez, C.; Rey, S.; Davie, A.; Migaud, H. Sustainable production and use of cleaner fish for the biological control of sea lice: Recent advances and current challenges. *Vet. Rec.* **2018**, *183*, 383. [CrossRef]
33. Lam, C.T.; Rosanowski, S.M.; Walker, M.; St-Hilaire, S. Sea lice exposure to non-lethal levels of emamectin benzoate after treatments: A potential risk factor for drug resistance. *Sci. Rep.* **2020**, *10*, 932–938. [CrossRef]
34. Karbowski, C.; Finstad, B.; Karbowski, N.; Hedger, R. Sea lice in Iceland: Assessing the status and current implications for aquaculture and wild salmonids. *Aquac. Environ. Interact.* **2019**, *11*, 149–160. [CrossRef]
35. Aaen, S.M.; Helgesen, K.O.; Bakke, M.J.; Kaur, K.; Horsberg, T.E. Drug resistance in sea lice: A threat to salmonid aquaculture. *Trends Parasitol.* **2015**, *31*, 72–81. [CrossRef] [PubMed]
36. Abolofia, J.; Asche, F.; Wilen, J.E. The Cost of Lice: Quantifying the Impacts of Parasitic Sea Lice on Farmed Salmon. *Mar. Resour. Econ.* **2017**, *32*, 329–349. [CrossRef]
37. Overton, K.; Dempster, T.; Oppedal, F.; Kristiansen, T.S.; Gismervik, K.; Stien, L.H. Salmon lice treatments and salmon mortality in Norwegian aquaculture: A review. *Rev. Aquac.* **2019**, *11*, 1398–1417. [CrossRef]
38. Barker, S.E.; Bricknell, I.R.; Covello, J.; Purcell, S.; Fast, M.D.; Wolters, W.; Bouchard, D.A. Sea lice, Lepeophtheirus salmonis (Krøyer 1837), infected Atlantic salmon (Salmo salar L.) are more susceptible to infectious salmon anemia virus. *PLoS ONE* **2019**, *14*, e0209178. [CrossRef]
39. Torrissen, O.; Jones, S.; Asche, F.; Guttormsen, A.; Skilbrei, O.T.; Nilsen, F.; Horsberg, T.E.; Jackson, D. Salmon lice—Impact on wild salmonids and salmon aquaculture. *J. Fish Dis.* **2013**, *36*, 171–194. [CrossRef]
40. Costello, M.J. The global economic cost of sea lice to the salmonid farming industry. *J. Fish Dis.* **2009**, *32*, 115–118. [CrossRef]
41. Barrett, L.T.; Overton, K.; Stien, L.H.; Oppedal, F.; Dempster, T. Effect of cleaner fish on sea lice in Norwegian salmon aquaculture: A national scale data analysis. *Int. J. Parasitol.* **2020**, *50*, 787–796. [CrossRef]
42. Imsland, A.K.D.; Hanssen, A.; Nytrø, A.V.; Reynolds, P.; Jonassen, T.M.; Hangstad, T.A.; Elvegård, T.A.; Urskog, T.C.; Mikalsen, B. It works! Lumpfish can significantly lower sea lice infestation in large-scale salmon farming. *Biol. Open* **2018**, *7*, bio036301. [CrossRef]
43. Vasquez, I.; Cao, T.; Chakraborty, S.; Gnanagobal, H.; O'Brien, N.; Monk, J.; Boyce, D.; Westcott, J.D.; Santander, J. Comparative Genomics Analysis of *Vibrio anguillarum* Isolated from Lumpfish (*Cyclopterus lumpus*) in Newfoundland Reveal Novel Chromosomal Organizations. *Microorganisms* **2020**, *8*, 1666. [CrossRef]
44. Chakraborty, S.; Cao, T.; Hossain, A.; Gnanagobal, H.; Vasquez, I.; Boyce, D.; Santander, J. Vibrogen-2 vaccine trial in lumpfish (Cyclopterus lumpus) against Vibrio anguillarum. *J. Fish Dis.* **2019**, *42*, 1057–1064. [CrossRef]
45. Dang, M.; Cao, T.; Vasquez, I.; Hossain, A.; Gnanagobal, H.; Kumar, S.; Hall, J.; Monk, J.; Boyce, D.; Westcott, J.; et al. Oral Immunization of Larvae and Juvenile of Lumpfish (*Cyclopterus lumpus*) against *Vibrio anguillarum* Does Not Influence Systemic Immunity. *Vaccines* **2021**, *9*, 819. [CrossRef]
46. Boyce, D.; Ang, K.P.; Prickett, R. Cunner and lumpfish as cleaner fish species in Canada. In *Cleaner Fish Biology and Aquaculture Applications*; Treasurer, J.W., Ed.; 5M Publishing: Sheffield, UK, 2018.
47. Shwe, A.; Østbye, T.-K.K.; Krasnov, A.; Ramberg, S.; Andreassen, R. Characterization of Differentially Expressed miRNAs and Their Predicted Target Transcripts during Smoltification and Adaptation to Seawater in Head Kidney of Atlantic Salmon. *Genes* **2020**, *11*, 1059. [CrossRef]
48. Knutsen, T.M.; Kirubakaran, G.T.; Mommens, M.; Moen, T. *Lumpfish (Cyclopterus lumpus) Draft Genome Assembly*; Figshare: Iasi, Romania, 2018.
49. Langmead, B.; Trapnell, C.; Pop, M.; Salzberg, S.L. Ultrafast and memory-efficient alignment of short DNA sequences to the human genome. *Genome Biol.* **2009**, *10*, R25. [CrossRef] [PubMed]
50. Friedlander, M.R.; Chen, W.; Adamidi, C.; Maaskola, J.; Einspanier, R.; Knespel, S.; Rajewsky, N. Discovering microRNAs from deep sequencing data using miRDeep. *Nat. Biotechnol.* **2008**, *26*, 407–415. [CrossRef]
51. Friedländer, M.R.; Mackowiak, S.D.; Li, N.; Chen, W.; Rajewsky, N. miRDeep2 accurately identifies known and hundreds of novel microRNA genes in seven animal clades. *Nucleic Acids Res.* **2012**, *40*, 37–52. [CrossRef]
52. Kozomara, A.; Griffiths-Jones, S. miRBase: Annotating high confidence microRNAs using deep sequencing data. *Nucleic Acids Res.* **2014**, *42*, D68–D73. [CrossRef]

53. Griffiths-Jones, S.; Grocock, R.J.; Van Dongen, S.; Bateman, A.; Enright, A.J. miRBase: MicroRNA sequences, targets and gene nomenclature. *Nucleic Acids Res.* **2006**, *34* (Supplement S1), D140–D144. [CrossRef] [PubMed]
54. Ambros, V.; Bartel, B.; Bartel, D.P.; Burge, C.B.; Carrington, J.C.; Chen, X.; Dreyfuss, G.; Eddy, S.R.; Griffiths-Jones, S.; Marshall, M.; et al. A uniform system for microRNA annotation. *RNA* **2003**, *9*, 277–279. [CrossRef] [PubMed]
55. Johansen, I.; Andreassen, R. Validation of miRNA genes suitable as reference genes in qPCR analyses of miRNA gene expression in Atlantic salmon (Salmo salar). *BMC Res. Notes* **2014**, *7*, 945. [CrossRef] [PubMed]
56. Ruijter, J.M.; Ramakers, C.; Hoogaars, W.M.H.; Karlen, Y.; Bakker, O.; van den Hoff, M.J.B.; Moorman, A.F.M. Amplification efficiency: Linking baseline and bias in the analysis of quantitative PCR data. *Nucleic Acids Res.* **2009**, *37*, e45. [CrossRef] [PubMed]
57. Schmittgen, T.D.; Livak, K.J. Analyzing real-time PCR data by the comparative C(T) method. *Nat. Protoc.* **2008**, *3*, 1101–1108. [CrossRef] [PubMed]
58. miRBase: The MicroRNA Database. Available online: http://www.mirbase.org (accessed on 15 November 2012).
59. Dantzer, R.; Wollman, E.E. Relationships between the brain and the immune system. *J. Société Biol.* **2003**, *197*, 81–88. [CrossRef]
60. Dantzer, R. Neuroimmune Interactions: From the Brain to the Immune System and Vice Versa. *Physiol. Rev.* **2018**, *98*, 477–504. [CrossRef]
61. Brennan, G.; Henshall, D.C. MicroRNAs as regulators of brain function and targets for treatment of epilepsy. *Nat. Rev. Neurol.* **2020**, *16*, 506–519. [CrossRef]
62. Tessmar-Raible, K.; Raible, F.; Christodoulou, F.; Guy, K.; Rembold, M.; Hausen, H.; Arendt, D. Conserved Sensory-Neurosecretory Cell Types in Annelid and Fish Forebrain: Insights into Hypothalamus Evolution. *Cell* **2007**, *129*, 1389–1400. [CrossRef]
63. Giraldez, A.J.; Cinalli, R.M.; Glasner, M.E.; Enright, A.; Thomson, J.M.; Baskerville, S.; Hammond, S.M.; Bartel, D.P.; Schier, A.F. MicroRNAs Regulate Brain Morphogenesis in Zebrafish. *Science* **2005**, *308*, 833–838. [CrossRef]
64. Radhakrishnan, B.; Anand, A.A.P. Role of miRNA-9 in Brain Development. *J. Exp. Neurosci.* **2016**, *10*, JEN.S32843. [CrossRef]
65. Zhao, J.; Zhou, Y.; Guo, M.; Yue, D.; Chen, C.; Liang, G.; Xu, L. MicroRNA-7: Expression and function in brain physiological and pathological processes. *Cell Biosci.* **2020**, *10*, 1–12. [CrossRef]
66. Vaz, C.; Wee, C.W.; Lee, G.P.S.; Ingham, P.W.; Tanavde, V.; Mathavan, S. Deep sequencing of small RNA facilitates tissue and sex associated microRNA discovery in zebrafish. *BMC Genom.* **2015**, *16*, 950. [CrossRef]
67. Tan, C.L.; Plotkin, J.L.; Venø, M.T.; von Schimmelmann, M.; Feinberg, P.; Mann, S.; Handler, A.; Kjems, J.; Surmeier, D.J.; O'Carroll, D.; et al. MicroRNA-128 Governs Neuronal Excitability and Motor Behavior in Mice. *Science* **2013**, *342*, 1254–1258. [CrossRef] [PubMed]
68. Tsuyama, J.; Bunt, J.; Richards, L.J.; Iwanari, H.; Mochizuki, Y.; Hamakubo, T.; Shimazaki, T.; Okano, H. MicroRNA-153 Regulates the Acquisition of Gliogenic Competence by Neural Stem Cells. *Stem Cell Rep.* **2015**, *5*, 365–377. [CrossRef]
69. Wanet, A.; Tacheny, A.; Arnould, T.; Renard, P. miR-212/132 expression and functions: Within and beyond the neuronal compartment. *Nucleic Acids Res* **2012**, *40*, 4742–4753. [CrossRef] [PubMed]
70. Howe, J.; Li, E.S.; Streeter, S.E.; Rahme, G.; Chipumuro, E.; Russo, G.B.; Litzky, J.F.; Hills, L.B.; Rodgers, K.; Skelton, P.; et al. MiR-338-3p regulates neuronal maturation and suppresses glioblastoma proliferation. *PLoS ONE* **2017**, *12*, e0177661. [CrossRef] [PubMed]
71. Valenzuela, C.A.; Zuloaga, R.; Poblete-Morales, M.; Vera-Tobar, T.; Mercado, L.; Avendaño-Herrera, R.; Valdés, J.A.; Molina, A. Fish skeletal muscle tissue is an important focus of immune reactions during pathogen infection. *Dev. Comp. Immunol.* **2017**, *73*, 1–9. [CrossRef] [PubMed]
72. Brzeszczyńska, J.; Brzeszczyński, F.; Hamilton, D.F.; McGregor, R.; Simpson, A.H.R.W. Role of microRNA in muscle regeneration and diseases related to muscle dysfunction in atrophy, cachexia, osteoporosis, and osteoarthritis. *Bone Jt. Res.* **2020**, *9*, 798–807. [CrossRef]
73. Bhattacharya, M.; Sharma, A.R.; Sharma, G.; Patra, B.C.; Nam, J.-S.; Chakraborty, C.; Lee, S.-S. The crucial role and regulations of miRNAs in zebrafish development. *Protoplasma* **2017**, *254*, 17–31. [CrossRef]
74. Yu, H.; Lu, Y.; Li, Z.; Wang, Q. microRNA-133: Expression, Function and Therapeutic Potential in Muscle Diseases and Cancer. *Curr. Drug Targets* **2014**, *15*, 817–828. [CrossRef]
75. Zhang, R.; Li, R.; Lin, Y. Identification and characterization of microRNAs in the muscle of Schizothorax prenanti. *Fish Physiol. Biochem.* **2017**, *121*, 207–1064. [CrossRef]
76. Horak, M.; Novak, J.; Bienertova-Vasku, J. Muscle-specific microRNAs in skeletal muscle development. *Dev. Biol.* **2016**, *410*, 1–13. [CrossRef]
77. Yu, Y.; Wang, Q.; Huang, Z.; Ding, L.; Xu, Z. Immunoglobulins, Mucosal Immunity and Vaccination in Teleost Fish. *Front. Immunol.* **2020**, *11*, 567941. [CrossRef]
78. Zhang, Q.-L.; Dong, Z.-X.; Luo, Z.-W.; Jiao, Y.-J.; Guo, J.; Deng, X.-Y.; Wang, F.; Chen, J.-Y.; Lin, L.-B. MicroRNA profile of immune response in gills of zebrafish (Danio rerio) upon Staphylococcus aureus infection. *Fish Shellfish. Immunol.* **2019**, *87*, 307–314. [CrossRef] [PubMed]
79. Juanchich, A.; Bardou, P.; Rué, O.; Gabillard, J.-C.; Gaspin, C.; Bobe, J.; Guiguen, Y. Characterization of an extensive rainbow trout miRNA transcriptome by next generation sequencing. *BMC Genom.* **2016**, *17*, 164. [CrossRef]
80. Su, H.; Fan, J.; Ma, D.; Zhu, H. Identification and Characterization of Osmoregulation Related MicroRNAs in Gills of Hybrid Tilapia Under Three Types of Osmotic Stress. *Front. Genet.* **2021**, *12*, 361. [CrossRef]

81. Ng, H.M.; Ho, J.C.H.; Nong, W.; Hui, J.H.L.; Lai, K.P.; Wong, C.K.C. Genome-wide analysis of MicroRNA-messenger RNA interactome in ex-vivo gill filaments, Anguilla japonica. *BMC Genom.* **2020**, *21*, 208. [CrossRef]
82. Sales, C.F.; Silva, R.F.; Amaral, M.G.C.; Domingos, F.F.T.; Ribeiro, R.I.M.A.; Thomé, R.G.; Santos, H.B. Comparative histology in the liver and spleen of three species of freshwater teleost. *Neotrop. Ichthyol.* **2017**, *15*. [CrossRef]
83. Dai, Z.; Yang, T.; Song, G. The roles of miRNAs in liver diseases. *Non-Coding RNA Investig.* **2019**, *3*, 25. [CrossRef]
84. Østbye, T.K.; Woldemariam, N.T.; Lundberg, C.E.; Berge, G.M.; Ruyter, B.; Andreassen, R. Modulation of hepatic miRNA expression in Atlantic salmon (Salmo salar) by family background and dietary fatty acid composition. *J. Fish Biol.* **2021**, *98*, 1172–1185. [CrossRef] [PubMed]
85. Lagos-Quintana, M.; Rauhut, R.; Yalcin, A.; Meyer, J.; Lendeckel, W.; Tuschl, T. Identification of Tissue-Specific MicroRNAs from Mouse. *Curr. Biol.* **2002**, *12*, 735–739. [CrossRef]
86. Trattner, S.; Vestergren, A.S. Tissue distribution of selected microRNA in Atlantic salmon. *Eur. J. Lipid Sci. Technol.* **2013**, *115*, 1348–1356. [CrossRef]
87. Liu, X.-L.; Cao, H.-X.; Wang, B.-C.; Xin, F.-Z.; Zhang, R.-N.; Zhou, D.; Yang, R.-X.; Zhao, Z.-H.; Pan, Q.; Fan, J.-G. miR-192-5p regulates lipid synthesis in non-alcoholic fatty liver disease through SCD-1. *World J. Gastroenterol.* **2017**, *23*, 8140–8151. [CrossRef]
88. Su, H.; Yang, J.-R.; Xu, T.; Huang, J.; Xu, L.; Yuan, Y.; Zhuang, S.-M. MicroRNA-101, Down-regulated in Hepatocellular Carcinoma, Promotes Apoptosis and Suppresses Tumorigenicity. *Cancer Res.* **2009**, *69*, 1135–1142. [CrossRef]
89. Bronte, V.; Pittet, M.J. The Spleen in Local and Systemic Regulation of Immunity. *Immunity* **2013**, *39*, 806–818. [CrossRef] [PubMed]
90. Kaattari, S.L.; Irwin, M.J. Salmonid spleen and anterior kidney harbor populations of lymphocytes with different B cell repertoires. *Dev. Comp. Immunol.* **1985**, *9*, 433–444. [CrossRef]
91. Press, C.; Evensen, Ø. The morphology of the immune system in teleost fishes. *Fish Shellfish Immunol.* **1999**, *9*, 309–318. [CrossRef]
92. Li, G.; Zhao, Y.; Wen, L.; Liu, Z.; Yan, F.; Gao, C. Identification and Characterization of MicroRNAs in the Spleen of Common Carp Immune Organ. *J. Cell. Biochem.* **2014**, *115*, 1768–1778. [CrossRef]
93. Cao, Y.; Wang, D.; Li, S.; Zhao, J.; Xu, L.; Liu, H.; Lu, T.; Mou, Z. A transcriptome analysis focusing on splenic immune-related mciroRNAs of rainbow trout upon Aeromonas salmonicida subsp. salmonicida infection. *Fish Shellfish. Immunol.* **2019**, *91*, 350–357. [CrossRef] [PubMed]
94. Zhou, Y.; Wang, Y.-Y.; Fu, H.-C.; Huang, H.-Z. MicroRNA expression and analysis of immune-related putative target genes in ISKNV-infected spleen of mandarin fish (Siniperca chuatsi). *Aquaculture* **2021**, *547*, 737450. [CrossRef]
95. Grimes, J.A.; Prasad, N.; Levy, S.; Cattley, R.; Lindley, S.; Boothe, H.W.; Henderson, R.A.; Smith, B.F. A comparison of microRNA expression profiles from splenic hemangiosarcoma, splenic nodular hyperplasia, and normal spleens of dogs. *BMC Vet. Res.* **2016**, *12*, 272. [CrossRef] [PubMed]
96. Li, Z.J.; Zhang, Y.P.; Li, Y.; Zheng, H.W.; Zheng, Y.S.; Liu, C.J. Distinct expression pattern of miRNAs in Marek's disease virus infected-chicken splenic tumors and non-tumorous spleen tissues. *Res. Vet. Sci.* **2014**, *97*, 156–161. [CrossRef] [PubMed]
97. He, J.-J.; Ma, J.; Wang, J.-L.; Xu, M.-J.; Zhu, X.-Q. Analysis of miRNA expression profiling in mouse spleen affected by acute Toxoplasma gondii infection. *Infect. Genet. Evol.* **2016**, *37*, 137–142. [CrossRef] [PubMed]
98. Arribas, A.J.; Gomez-Abad, C.; Sánchez-Beato, M.; Martinez, N.; DiLisio, L.; Casado, F.; Cruz, M.A.; Algara, P.; Piris, M.A.; Mollejo, M. Splenic marginal zone lymphoma: Comprehensive analysis of gene expression and miRNA profiling. *Mod. Pathol.* **2013**, *26*, 889–901. [CrossRef]
99. Huang, L.; Ma, J.; Sun, Y.; Lv, Y.; Lin, W.; Liu, M.; Tu, C.; Zhou, P.; Gu, W.; Su, S.; et al. Altered splenic miRNA expression profile in H1N1 swine influenza. *Arch. Virol.* **2015**, *160*, 979–985. [CrossRef] [PubMed]
100. Pase, L.; Layton, J.E.; Kloosterman, W.P.; Carradice, D.; Waterhouse, P.M.; Lieschke, G.J. miR-451 regulates zebrafish erythroid maturation in vivo via its target gata2. *Blood* **2009**, *113*, 1794–1804. [CrossRef]
101. Giraldez, A.J.; Mishima, Y.; Rihel, J.; Grocock, R.J.; Van Dongen, S.; Inoue, K.; Enright, A.J.; Schier, A.F. Zebrafish MiR-430 Promotes Deadenylation and Clearance of Maternal mRNAs. *Science* **2006**, *312*, 75–79. [CrossRef] [PubMed]
102. Takacs, C.M.; Giraldez, A.J. miR-430 regulates oriented cell division during neural tube development in zebrafish. *Dev. Biol.* **2016**, *409*, 442–450. [CrossRef]

biology

MDPI

Article

De novo Assembly and Analysis of Tissue-Specific Transcriptomes of the Edible Red Sea Urchin *Loxechinus albus* Using RNA-Seq

Paulette Antiqueo [1,2], Rodrigo Zuloaga [1,2], Macarena Bastias-Molina [1,3], Claudio Meneses [1,3], Juan Manuel Estrada [4], Alfredo Molina [1,2,4] and Juan Antonio Valdés [1,2,4,*]

1 Departamento Ciencias Biológicas, Facultad de Ciencias de la Vida, Universidad Andres Bello, Santiago 8370186, Chile; p.antiqueo.a@gmail.com (P.A.); r.zuloaga@uandresbello.edu (R.Z.); macarena.bastias@unab.cl (M.B.-M.); claudio.meneses@unab.cl (C.M.); amolina@unab.cl (A.M.)
2 Interdisciplinary Center for Aquaculture Research (INCAR), Concepción 4030000, Chile
3 Centro de Biotecnología Vegetal, FONDAP Center for Genome Regulation, Facultad de Ciencias de la Vida, Universidad Andres Bello, Santiago 8370146, Chile
4 Centro de Investigación Marina Quintay (CIMARQ), Facultad de Ciencias de la Vida, Universidad Andres Bello, Valparaíso 2340000, Chile; mestrada@unab.cl
* Correspondence: jvaldes@unab.cl; Tel.: +56-2661-8363; Fax: +56-2661-8415

Simple Summary: Edible red sea urchin (*Loxechinus albus*) is an endemic species of echinoderm distributed along the Chilean coasts. This resource has been overexploited in recent years, depleting their natural populations. At present, there are few reported gene sequences available in public databases, restricting the molecular studies associated with aquaculture for this species. The aim of this study was to present the first annotated reference transcriptome of *L. albus* using NGS technologies and the differential expression transcripts analysis of the evaluated tissues. The transcriptome data obtained in this study will serve as a reference for future molecular research in the edible red sea urchin and other sea urchin species.

Abstract: Edible red sea urchin (*Loxechinus albus*) is an endemic echinoderm species of the Chilean coasts. The worldwide demand for high-quality gonads of this species has addressed the depletion of its natural populations. Studies on this sea urchin are limited, and genomic information is almost nonexistent. Hence, generate a transcriptome is crucial information that will considerably enrich molecular data and promote future findings for the *L. albus* aquaculture. Here, we obtained transcriptomic data of the edible red sea urchin by Illumina platform. Total RNA was extracted from gonads, intestines, and coelomocytes of juvenile urchins, and samples were sequenced using MiSeq Illumina technology. A total of 91,119,300 paired-end reads were *de novo* assembled, 185,239 transcripts produced, and a reference transcriptome created with 38.8% GC content and an N50 of 1769 bp. Gene ontology analysis revealed notable differences in the expression profiles between gonads, intestines, and coelomocytes, allowing the detection of transcripts associated with specific biological processes and KEGG pathways. These data were validated using 12 candidate transcripts by real-time qPCR. This dataset will provide a valuable molecular resource for *L. albus* and other species of sea urchins.

Keywords: edible red sea urchin; *Loxechinus albus*; RNA-seq; reference transcriptome

Citation: Antiqueo, P.; Zuloaga, R.; Bastias-Molina, M.; Meneses, C.; Estrada, J.M.; Molina, A.; Valdés, J.A. *De novo* Assembly and Analysis of Tissue-Specific Transcriptomes of the Edible Red Sea Urchin *Loxechinus albus* Using RNA-Seq. *Biology* **2021**, *10*, 995. https://doi.org/10.3390/biology10100995

Academic Editor: Patricia Pereiro

Received: 25 August 2021
Accepted: 29 September 2021
Published: 2 October 2021

1. Introduction

The *Loxechinus albus* (Molina, 1782), or edible red sea urchin, is an echinoderm species of the Chilean and Peruvian coasts, distributed along ca. Cape Horn, Chile (56°70′ S) to the Isla Lobos de Afuera, Peru (6°53′ S) [1]. The worldwide demand for high-quality gonads of this sea urchin has addressed a vast overexploitation of its natural populations [2]. Harvesting of *L. albus* represents the major sea urchin fishery among world urchin fisheries [3].

The aquaculture of this species, involving the rearing tank production of larvae, juvenile, and later fattening in natural environments, are important approaches to aquaculture diversification in Chile and to restore the overexploited coastal areas [4].

One of the main difficulties in the study of biological and molecular mechanisms associated with the farming of this species is the limited genomic information available [5,6]. In this context, transcriptome sequencing is useful to identify genes participating certain biological processes when genomic data are not available [7]. This analysis allows a broad comprehension of molecular mechanisms involved in biological processes from data on predicted function of genes [8]. Progress in the characterization of the transcriptome in commercial sea urchins is achievable due to advances in next-generation sequencing (NGS) technologies. NGS has allowed the research of sea urchin transcriptomes and other non-model species in brief periods of time at a low cost [9–11]. The molecular information achieved has provided significant value regarding the physiological responses to adaptation in a variety of commercial sea urchins under fluctuating environmental conditions [12,13].

At this time, the existing information on *L. albus* biology is limited and is related to with oxidative metabolism [14], growth patterns [15], the performance of early juveniles under food type and feeding frequency [16], and cryopreservation of embryos and larvae [17]. However, biological studies with molecular bases carried out in this species are scarce, mainly due to the low amount of genomic information available [11,18]. Although some advances have been made in the transcriptome characterization and mitogenome of this species in recent years, the low coverage of the technology used, as well as the use of gonads as the only target tissue, has limited the obtainment of a high-quality reference transcriptome [5,6,9,19]. Therefore, we present here the first annotated transcriptome of juvenile edible red sea urchin using NGS technologies based on three critical tissues for physiological homeostasis of echinoderms and the expression analysis of the transcripts present in each tissue: (i) gonads, involved in reproduction and exportation product for aquaculture, (ii) intestine, involved in food digestion and nutrient uptake, and (iii) coelomocytes, involved mainly in immune surveillance and inflammatory process. The transcriptome data obtained here will provide a reference for molecular studies in the farming of *L. albus* and other sea urchin species.

2. Materials and Methods

2.1. Experimental Design and Sampling

Loxechinus albus specimens were obtained from the Centro de Investigación Marina de Quintay (CIMARQ; 33°13′ S, 71°38′ O, Valparaiso, Chile). Briefly, fertilization was performed using a pool of gametes from four females and four males stimulated to spawn by injection of 3 mL of 0.5 M KCl. The embryos generated were cultured in 200 L larval rearing containers and larvae developed were fed with *Chaetoceros gracilis* microalgae. The larvae were grown in 50 L tanks in filtrated and aerated seawater and then preconditioned to settle in post-larval condition. Juvenile sexually immature sea urchins were maintained under natural conditions (13 ± 1 °C) in the spring season. The sea urchins were three years old and weighed 30 ± 5 g. The animals were fed macroalgae ad libitum (*Lessonia sp., Macrocystis sp., Durvillea sp.*). A total of 10 sea urchins were selected, dissected, and three different tissues were collected: intestines, gonads, and coelomocytes. Intestines were cleaned with phosphate buffer solution (PBS 1×) before storage. In immature gonads, germ cells were undifferentiated, revealing no sex differentiation. The coelomic fluid was collected by cutting the peristomal membrane, mixed with anticoagulant (20 mM Tris–HCl, 0.5 M NaCl, and 30 mM EDTA; pH 7.4), centrifuged for 5 min at 5000× *g*, and then coelomocyte pellet was collected. Samples were rapidly frozen in liquid nitrogen and deposited at −80 °C until use.

2.2. Isolation of RNA and Sequencing

Total RNA was obtained using columns of the RNeasy Mini Kit (Qiagen, Austin, TX, USA). The genomic DNA from RNA samples with removed by DNase I treatment. RNA was quantified by fluorometry using a Qubit 2.0 Fluorometer (Life Technology, Carlsbad, CA, USA), and the integrity of RNA was measured using the Fragment Analyzer (Analytical Advanced Technologies, Ames, IA, USA). Total RNA from five sea urchins were pooled in equal quantities by tissue, in duplicate, and then used to mRNA libraries construction. These libraries were generated by the TruSeq RNA Sample Preparation Kit v2 (Illumina, San Diego, CA, USA). Finally, libraries were sequenced (2 × 250 bp) utilizing the MiSeq technology (Illumina) at the Center for Plant Biotechnology (Universidad Andrés Bello, Santiago, Chile). The raw reads of the present study were uploaded to the NCBI SRA database under BioProject PRJNA475570, with accession number SRP150640.

2.3. Processing of Raw Data, De novo Assembly, and Validation of Assembly

First, the raw sequence reads were quality checked using FASTQC software. Adapters were removed, and raw data were trimmed using FlexBar [20] with Phred scores below 38 and 250 bp reads. The *de novo* transcriptome was assembled using all libraries (two libraries per tissue) with the Trinity program using default parameters [21]. Transcripts were filtered based on the minimal number of mapped reads with the Corset program using default parameters [22]. To evaluate *de novo* assembly integrity, the assembled transcriptome by Benchmarking Universal Single-Copy Orthologs (BUSCO) was compared against the OrthoDBv9 database (Vertebrata and Eukaryota) to identify orthologous genes that were highly conserved [23].

2.4. Functional Annotation and Analysis of Differentially Expressed Transcripts

For transcriptome annotation, a search in BlastX against the UniProt (https://www.uniprot.org/blast/; accessed 18 March 2021), Nonredundant (NR; https://blast.ncbi.nlm.nih.gov/Blast.cgi; accessed 18 March 2021), and Clusters of orthologous groups for eukaryotic complete genomes (COG; https://www.ncbi.nlm.nih.gov/research/cog; accessed 18 March 2021) databases was performed. The Blast2GO program was employed to acquire gene ontology (GO) annotation [24], and the WEGO application [25] was used to carry out GO functional classification for all transcripts. Recognition of differentially expressed transcripts (DETs) from the gonads, intestines, and coelomocytes were realized using Bowtie by mapping against the assembled *L. albus* transcriptome [26]. The RSEM software was used to assess expression values of fragments per kilobase million (FPKM) [27]. EdgeR was employed to determine differential expression between intestine vs. gonad, coelomocytes vs. intestine, and coelomocytes vs. gonads [28]. Transcripts detected with false discovery rate (FDR)-corrected p values < 0.001 and absolute values of fold-change > 4.0 were incorporated in the GO and Kyoto encyclopedia of genes and genomes (KEGG) enrichment analyses.

2.5. Gene Ontology and KEGG Enrichment Analysis

The DETs were examined against the DAVID resource [29] and then categorized based on GO terms for molecular functions, biological processes, cellular components, and KEGG pathways. To determine a relationship between the DAVID background and *L. albus* DETs, a search in BLASTx was performed against *Strongylocentrotus purpuratus* Ensembl proteins for major matches with the *L. albus* transcriptome. Ensembl Gene IDs of *S. purpuratus* were acquired from the resultant Ensembl protein entries. Custom IDs set were selected for DAVID analysis as the "Background" Standard settings for ease (0.1) and gene count (2). The cut off p value used for molecular functions and cellular components was 1×10^{-3}, and for biological processes was 1×10^{-6}.

2.6. *Validation of RNA-Seq by Real-Time qPCR*

All quantitative real-time polymerase chain reaction (qPCR) assays were performed according to MIQE recommendations [30]. Total RNA isolation from gonads, intestines, and coelomocytes was realized using columns of RNeasy Mini Kit (Qiagen). RNA quantification was measured by NanoDrop technology with an Epoch Multivolume Spectrophotometer System (BioTek, Winooski, VT, USA). For complementary DNA (cDNA) synthesis, only RNA with an A260/280 ratio between 1.9 and 2.1 was selected. This procedure was performed using 1 μg of RNA by QuantiTect Reverse Transcription Kit (Qiagen), eliminating first genomic DNA with the wipeout buffer included and then reverse transcribed into cDNA at 42 °C for 30 min. qPCR reaction was made using Brilliant II SYBR Green QPCR Master Mix (Agilent Technologies, Santa Clara, CA, USA), a 1/dilution of cDNA, and 5 μM primers (Supplementary Table S1). Real-time PCRs were run on an Mx3000P qPCR System (Agilent Technologies) in triplicate. The PCR amplification program included 95 °C for 10 min; 40 cycles of 95 °C for 30 s, Tm for 30 s, 95 °C for 32 s, and 72 °C for 30 s. The QGene application was utilized to analyze gene expression [31] and data were normalized with 18S ribosomal subunit RNA as housekeeping gene.

2.7. *Statistical Analysis*

All statistical analyses were performed using GraphPad Prism v8.0 (GraphPad Software, La Jolla, CA, USA). The generated data were presented as the mean ± standard error of the mean (SEM). A one-way analysis of variance (ANOVA) followed by the Bonferroni post hoc test was used to differentiate the means between groups. A value of $p < 0.001$ was accepted as significative data. Finally, correlation analysis between real-time qPCR and RNA-seq were evaluated by multiple linear regression through p value and coefficients of determination (R^2).

3. Results

3.1. *Raw Data Sequencing, De novo Assembly Transcriptome and Functional Annotation*

The cDNA libraries were generated from intestines, gonads, and coelomocytes of pooled edible red sea urchins (Figure 1).

Figure 1. Juvenile edible red sea urchins (*Loxechinus albus*).

The Illumina MiSeq sequencing produced 95,745,640 paired end reads of cDNA library replicates for each tissue (Table 1). The obtained raw data were trimmed by eliminating adapters, contaminant sequences and filtering base pairs with low-quality, the high-quality reads were reduced to 91,119,300 base pairs (Table 1). The Trinity software was used to *de novo* assembly using all libraries, resulting in 278,803 transcripts. The high-quality reads were mapped against the transcriptome generated and reduced by Corset software. The newly reduced transcriptome had 185,239 transcripts, with N50 = 1769 bp and GC% = 38.81 (Table 1). The length distribution of the obtained transcripts was detailed in Supplementary Figure S1. The BUSCO database was selected to compare the assembled transcriptome, which includes information of orthologous genes highly conserved. The assembly found 248 BUSCO genes of the eukaryotic core gene, 222 complete (89.5%), 8 fragmented (3.2%), and 18 missing genes (7.3%). The annotation of *L. albus* transcripts was carried out by BlastX searches against the non-redundant (NR), UniProt, and cluster of orthologous groups (COG) databases. The statistics of similarity and target species of all transcripts are shown in Figure 2, revealing that 91.2% of sequences matched *S. purpuratus*.

Table 1. Summary of transcriptome sequencing for the edible red sea urchin (*L. albus*) tissue and assembly statistics. bp: base pair.

Name	Number of Reads	Number of Reads after Trim
Coelomocyte	20,682,190	19,948,624
Coelomocyte (replicate)	16,865,448	15,699,186
Intestine	12,145,212	11,748,696
Intestine (replicate)	11,348,164	10,620,928
Gonad	18,495,858	17,901,954
Gonad (replicate)	16,208,768	15,199,912
Total	95,745,640	91,119,300
Transcriptome	*De novo* Assembly (Trinity)	After Filter (Corset)
Total contigs	278,803	185,239
Average large contig (bp)	326	929
Coverage contig	708.32	-
%GC	38.2	38.81
N10 (bp)	5015	5328
N30 (bp)	2645	2945
N50 (bp)	1418	1769
Total bases	197,480,887	172,122,576

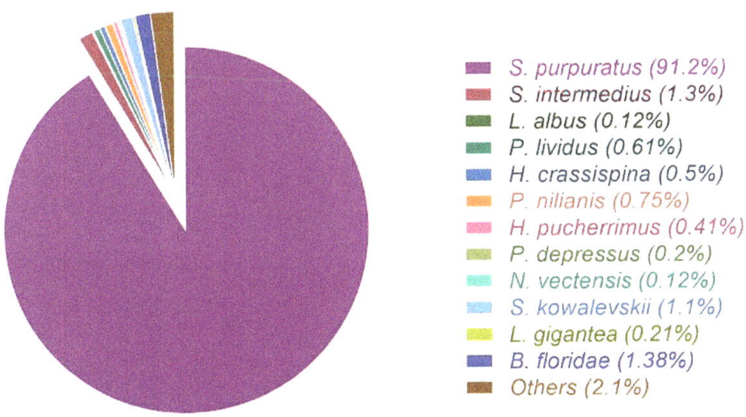

Figure 2. A species-based BlastX comparative analysis revealed the major match with *Strongylocentrotus purpuratus*.

The functional annotation of transcripts was realized using the Blast2GO platform, applying a GO term search through transcripts with BLAST hits matched against NR database. We detected significant similarity with a total of 57,106 (31%) transcripts. The GO analysis revealed 38,265 GO outcomes for biological processes (20.2%), 36,046 for molecular functions (19.1%) and 35,909 GO for cellular components (19.0%). A significant proportion of the annotated transcripts in biological process were assigned to cellular component biogenesis (GO:0044085) and cellular localization (GO:0051641) (Figure 3).

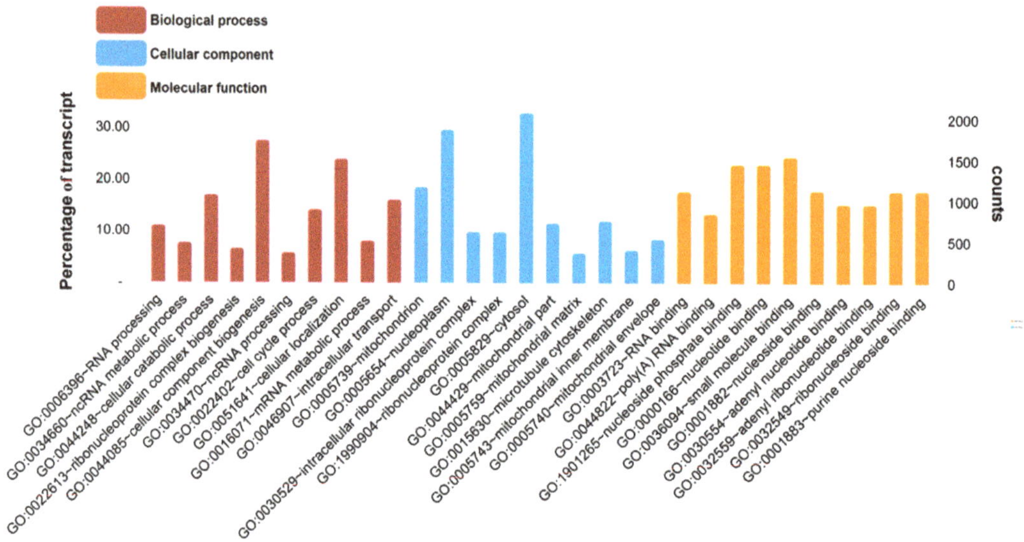

Figure 3. GO functional classification assigned the most percentage of the annotated transcripts to cellular component biogenesis term for biological process; cytosol term for cellular component; and small molecule binding term for molecular function, respectively. Analysis was carried out with the WEGO program for the edible sea urchin (*L. albus*) reference transcriptome.

For cellular components and molecular functions, several annotated transcripts were allocated to cytosol (GO:0005829) and nucleoplasm (GO:0005654); and nucleotide binding (GO:0000166), nucleoside phosphate binding (GO:1901265), and small molecule binding (GO:0036094) terms, respectively (Figure 3). Based on sequence homology, 32,231 sequences were classified into 25 functional categories (Figure 4). The most represented categories were General Functional Prediction only, followed by Signal Transduction. These results indicate that we generated a reference transcriptome for the edible sea urchin based on 91,119,300 high-quality reads that assembled *de novo* into 185,239 transcripts with an N50 of 1769 bp and 38.8% GC content.

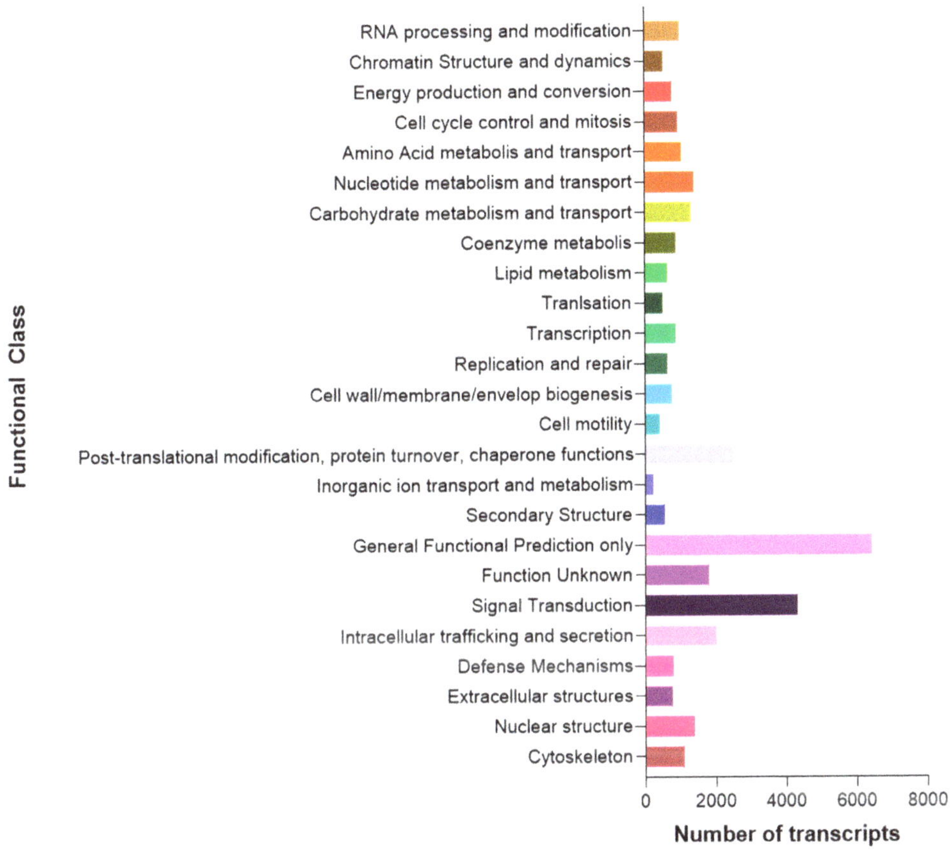

Figure 4. The General Functional Prediction only was the most represented category for the 25 cluster of orthologous groups (COG) functional classification in the *L. albus* reference transcriptome.

3.2. Assessment of Differentially Expressed Transcripts

About ~95.0% of the reads were mapped to the assembled transcriptome by Bowtie, and then RSEM and EdgeR were used to estimate the expression values and to calculate differential expression transcripts (DETs) between tissues, respectively. Transcripts with false discovery rate (FDR)-corrected p values < 0.001 and absolute fold-change values > 4.0 were defined as DETs. The mapping of *L. albus* transcriptome indicated that 26,864 transcripts were differentially expressed among coelomocytes, intestine, and gonad. These DETs were clustered using hierarchy by comparisons between patterns of gene expression (Figure 5).

The correlation analysis of heat map clustered samples together in the same group, representing a high reproducibility of RNA-seq data among replicates (Supplementary Figure S2). To identified DETs, tissues were exposed to a succession of paired comparations (Figure 6): In coelomocytes vs. gonad, 16,406 DETs were identified (8394 upregulated and 8012 downregulated); in coelomocytes vs. intestine, 14,025 DETs were identified (5558 upregulated and 8467 downregulated); intestine vs. gonad, 10,015 DETs were identified (6179 upregulated and 3836 downregulated).

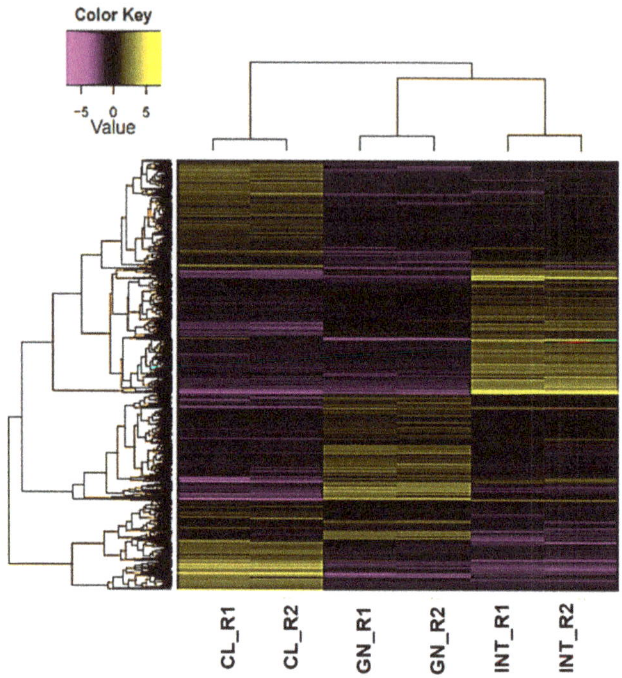

Figure 5. The heat map showed differentiated clustering of expressed transcripts according to respective expression values across tissues. Parameters: fold change (absolute values > 4.0) and FDR corrected p value ($p < 0.001$). Abbreviations: CL: coelomocyte, INT: intestine, GN: gonad. R1: Replicate 1, R2: Replicate 2.

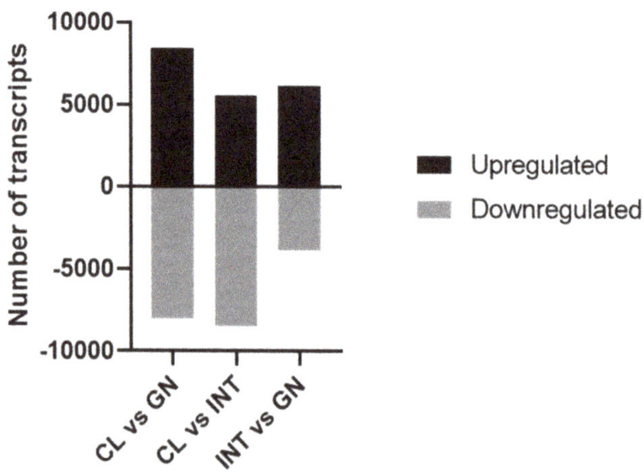

Figure 6. Paired comparisons of differentially expressed transcripts between tissues of *L. albus*. Upregulated and downregulated transcripts measured are indicated by black and gray columns, respectively. Parameters: fold change (absolute values > 4.0) and FDR corrected p value ($p < 0.001$). Abbreviations: CL: coelomocyte, INT: intestine, GN: gonad.

The analysis of Venn diagram revealed that 737 DETs (2.7%) were commonly expressed among the three tissues (Figure 7). Also, 7061 DETs (26.3%) were specifically expressed between coelomocytes and gonad, specific 4280 DETs (15.9%) among coelomocytes and intestine, and 2678 DETs (10%) between intestine and gonad. A complete list of the DETs for each tissue was included in Supplementary Table S2. The tissue-specific DETs were examined in DAVID resource and categorized as biological processes, molecular functions, and cellular components. The coelomocyte DETs were considerably enriched in positive regulation of apoptotic process (GO:0043065) and intracellular signal transduction (GO:0035556) for biological processes (Figure 8a). The most enriched GO terms for molecular functions and cellular components were assigned to poly(A) RNA binding (GO:0044822) and ATP binding (GO:0005524), and cytosol (GO:0005829), respectively (Supplementary Figure S3). In contrast, DNA repair (GO:0006281) and microtubule-based process (GO:0007017) were the major enriched biological processes in gonads (Figure 8b). The main transcripts assigned to molecular functions and cellular components were microtubule motor activity (GO:0003777) and ATP binding (GO:0005524) GO terms, and dynein complex and cilium (GO:0005929) GO terms, respectively (Supplementary Figure S4). For biological processes, transmembrane transport (GO:0055085) and microtubule-based process (GO:0007017) were the most enriched in intestinal DETs (Figure 8c). The majority assigned transcripts for molecular functions and cellular components were GTP binding (GO:0005525) and motor activity (GO:0003774), and extracellular exosome (GO:0070062) and membrane (GO:0016020), respectively (Supplementary Figure S5).

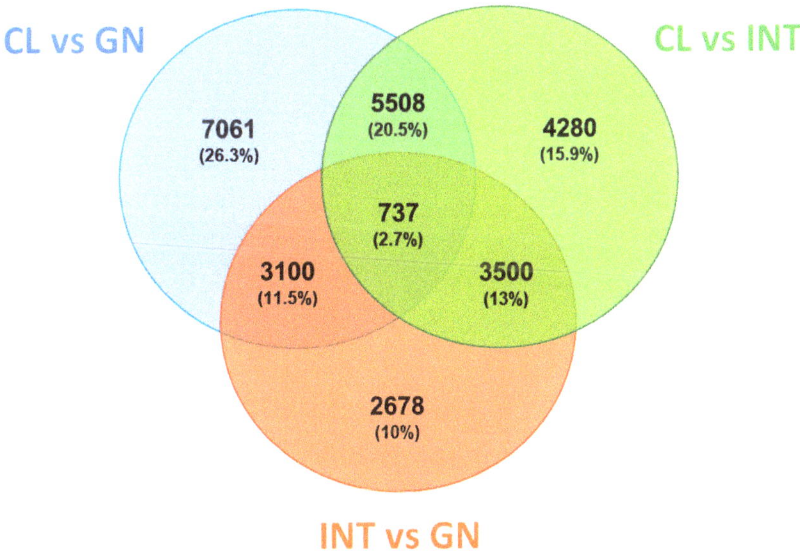

Figure 7. Differentially expressed transcripts between gonads, intestines, and coelomocytes of the edible sea urchin. Venn diagram of the total differentially expressed transcripts in *L. albus*. Each color indicates the comparison between tissue and the numbers of genes that were differentially expressed. Parameters: fold change (absolute values > 4.0) and FDR corrected p value ($p < 0.001$). Abbreviations: CL: coelomocyte, INT: intestine, GN: gonad.

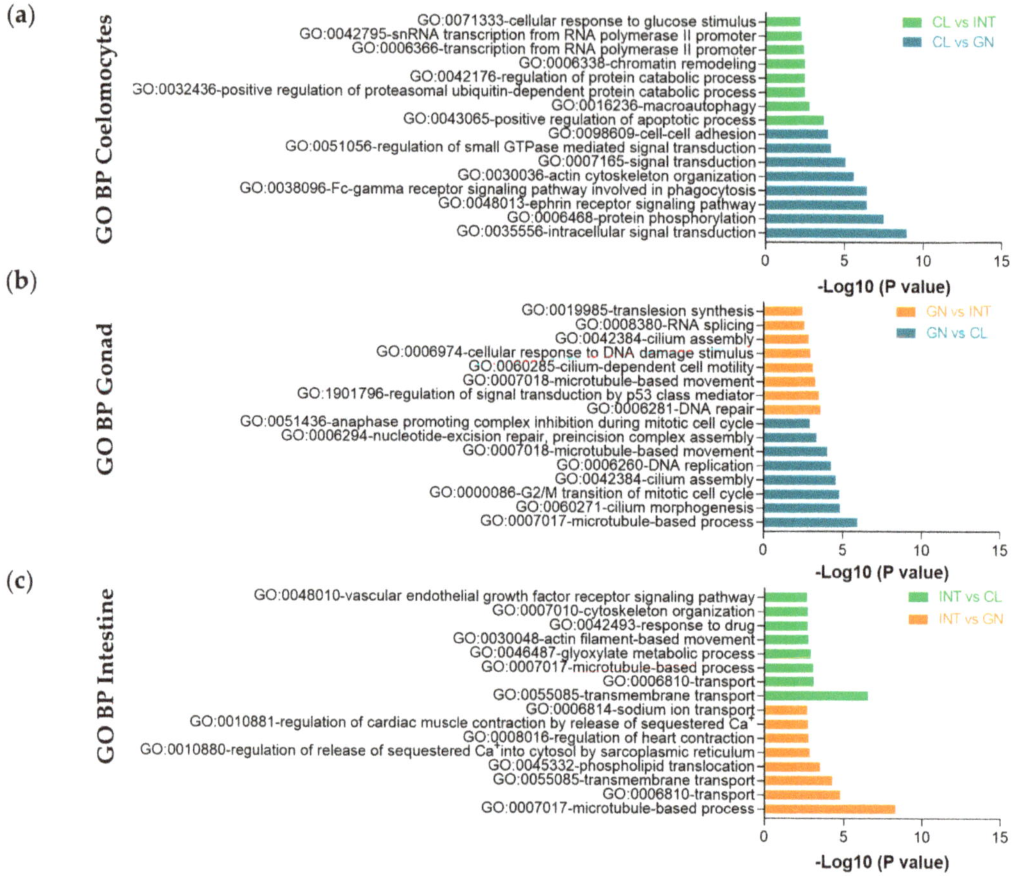

Figure 8. The Top-16 Gene Ontology biological process (BP) enrichment of up-regulated transcripts from *L. albus* tissues. (**a**) In coelomocytes compared to intestine (CL vs. INT), the most enriched term was positive regulation of apoptotic process, and compared to the gonad (CL vs. GN) was intracellular signal transduction; (**b**) in the gonad compared to intestine (GN vs. INT), the most enriched term was DNA repair, and compared to coelomocytes (GN vs. CL) was microtubule-based process; (**c**) in intestine compared to coelomocytes (INT vs. CL) the most enriched term was transmembrane transport, and compared to the gonad (INT vs. GN) was microtubule-based process.

The analysis of KEGG pathway indicated that various transcripts related to cAMP signaling pathway, Platelet activation, and Neurotrophin signaling pathway were over-represented in coelomocytes vs. intestine analysis (Table 2). Other relevant pathways represented in coelomocytes vs. gonads correspond to Fc gamma R-mediated phagocytosis, Platelet activation, and Pathogenic Escherichia coli infection (Table 2). In gonad vs. intestine, the Purine metabolism was the main pathway overrepresented, followed by the Spliceosome and Huntington's disease (Table 3). We also found that Biosynthesis of antibiotics, Gap junction, and Fatty acid degradation were highly represented in gonads vs. coelomocytes (Table 3). Regarding the main signaling pathways overrepresented in intestine (intestine vs. coelomocytes), we highlight adherens junction, Pathogenic Escherichia coli infection, and ABC transporters (Table 4). Finally, the comparison of intestine vs. gonad confirmed that pathways such ABC transporters and Pathogenic Escherichia coli infection were relevant overrepresented pathways (Table 4). In summary, we detected 26,864 transcripts were differentially expressed among the three tissues, which GO analysis revealed several processes and pathways that were expressed in common and, most importantly, tissue specific.

Table 2. The Top-16 KEGG pathways enrichment of up-regulated transcripts of coelomocytes in the edible red sea urchin (*L. albus*). In comparison to intestine, cAMP signaling pathway was the most overrepresented and Fc gamma R-mediated phagocytosis compared to gonad.

KEGG Term (ID)	Number of Genes	Percentage of Genes	Fold Enrichment	*p* Value
Coelomocytes vs. Intestine				
cAMP signaling pathway (4024)	13	2.9%	2.4	8.1×10^{-3}
Platelet activation (4611)	10	2.2%	2.8	9.3×10^{-3}
Neurotrophin signaling pathway (4722)	9	2.0%	2.7	1.7×10^{-2}
Regulation of lipolysis in adipocytes (4923)	6	1.4%	3.8	1.8×10^{-2}
Proteoglycans in cancer (5205)	12	2.7%	2.2	2.1×10^{-2}
Focal adhesion (4510)	12	2.7%	2.1	2.6×10^{-2}
Insulin secretion (4911)	7	1.6%	3.0	2.9×10^{-2}
cGMP-PKG signaling pathway (4022)	10	2.2%	2.3	3.0×10^{-2}
Coelomocytes vs. Gonad				
KEGG Term (ID)	**Number of Genes**	**Percentage of Genes**	**Fold Enrichment**	***p* Value**
Fc gamma R-mediated phagocytosis (4666)	15	2.5%	4.9	1.7×10^{-6}
Platelet activation (4611)	16	2.6%	3.6	7.0×10^{-5}
Pathogenic *Escherichia coli* infection (5130)	10	1.6%	5.4	7.9×10^{-5}
cGMP-PKG signaling pathway (4022)	17	2.8%	2.9	1.9×10^{-4}
Gap junction (4540)	12	1.9%	3.7	3.3×10^{-4}
Proteoglycans in cancer (5205)	19	2.0%	2.6	3.5×10^{-4}
Focal adhesion (4510)	19	3.1%	2.5	4.9×10^{-4}
Bacterial invasion of epithelial cells (5100)	11	1.8%	3.8	4.9×10^{-4}

Table 3. The Top-16 KEGG pathways enrichment of up-regulated transcripts of gonad in the edible red sea urchin (*L. albus*). In comparison to intestine, purine metabolism was the most overrepresented and Biosynthesis of antibiotics compared to coelomocytes.

KEGG Term (ID)	Number of Genes	Percentage of Genes	Fold Enrichment	*p* Value
Gonad vs. Intestine				
Purine metabolism (230)	10	2.9%	2.4	6.9×10^{-3}
Spliceosome (3040)	8	2.3%	2.8	1.4×10^{-2}
Huntington's disease (5016)	9	2.6%	2.7	3.2×10^{-2}
p53 signaling pathway (4115)	5	1.5%	3.8	4.0×10^{-2}
Focal adhesion (4510)	9	2.6%	2.1	4.6×10^{-2}
Metabolic pathways (1100)	32	9.2%	3.0	5.9×10^{-2}
Pathogenic *Escherichia coli* infection (5130)	4	1.5%	2.3	7.5×10^{-2}
Gap junction (4540)	5	1.7%	2.4	9.0×10^{-2}
Gonad vs. Coelomocytes				
KEGG Term (ID)	**Number of Genes**	**Percentage of Genes**	**Fold Enrichment**	***p* Value**
Biosynthesis of antibiotics (1130)	20	4.5%	3.4	4.5×10^{-6}
Gap junction (4540)	10	2.3%	4.1	6.2×10^{-4}
Fatty acid degradation (71)	7	1.6%	6.1	8.9×10^{-4}
Valine, leucine, and isoleucine degradation (280)	7	1.6%	5.4	1.6×10^{-3}
Fatty acid metabolism (1212)	7	1.6%	5.3	1.8×10^{-3}
Phagosome (4145)	12	2.7%	2.9	2.6×10^{-3}
Metabolic pathways (1100)	47	10.6%	1.4	1.1×10^{-2}
DNA replication (3030)	5	1.1%	5.1	1.6×10^{-2}

Table 4. The Top-16 KEGG pathways enrichment of up-regulated transcripts of intestine in the edible red sea urchin (*L. albus*). In comparison to coelomocytes, adherens junction was the most overrepresented and ABC transporters compared to gonad.

	Intestine vs. Coelomocytes			
KEGG Term (ID)	**Number of Genes**	**Percentage of Genes**	**Fold Enrichment**	***p* Value**
Adherens junction (4520)	13	2.1%	5.0	7.7×10^{-6}
Pathogenic *Escherichia coli* infection (5130)	11	1.7%	5.9	1.2×10^{-5}
ABC transporters (2010)	10	1.6%	6.2	2.2×10^{-5}
Lysosome (4142)	16	2.5%	3.6	2.9×10^{-5}
Metabolic pathways (1100)	68	10.7%	1.5	1.8×10^{-4}
Gap junction (4540)	11	1.7%	3.4	1.2×10^{-3}
Chemical carcinogenesis (5204)	9	1.4%	3.1	8.2×10^{-3}
cAMP signaling pathway (4024)	15	2.3%	2.1	1.3×10^{-2}
	Intestine vs. Gonad			
KEGG Term (ID)	**Number of Genes**	**Percentage of Genes**	**Fold Enrichment**	***p* Value**
ABC transporters (2010)	11	1.9%	7.9	7.1×10^{-7}
Chemical carcinogenesis (5204)	12	2.2%	4.8	3.4×10^{-5}
Pathogenic *Escherichia coli* infection (5130)	9	1.6%	5.6	1.6×10^{-4}
Lysosome (4142)	12	2.2%	3.2	1.3×10^{-3}
Gap junction (4540)	10	1.8%	3.6	1.6×10^{-3}
Galactose metabolism (52)	6	1.1%	6.4	2.1×10^{-3}
Starch and sucrose metabolism (500)	6	1.1%	5.8	3.3×10^{-3}
Amino sugar and nucleotide sugar metabolism (520)	7	1.3%	4.6	3.6×10^{-3}

3.3. *Transcriptomic Data Validation*

We used real-time qPCR to validate RNA-seq and assay upregulated transcripts in each tissue compared to the others. According to *p* values and fold enrichments indicated in GO analysis, we selected four transcripts from coelomocytes: heat shock protein 70 kDa 1 A (HSP70), lysosomal trafficking regulator (LYST), B-cell lymphoma 2 (BCL2), and ubiquitin A-52 residue ribosomal protein fusion product 1 (UBA52) (Figure 9a). Four transcripts from gonad were identified: testis-specific serine/threonine-protein kinase 3 (TSSK3), centrin 2 (CETN2), cation channel sperm associated 3 (CATSPER3), and sperm surface protein 17 (SPA17) (Figure 9b). Four transcripts from intestine were identified: notch homolog 1 (NOTCH1), toll-like receptor 3 (TLR3), glutathione s-transferase theta 1 (GSTT1), and caspase 3 (CASP3) (Figure 9c). The transcript expression revealed a high statistical correlation between RNA-seq and qPCR measuring (R^2 = 0.7259, $p < 0.001$) (Figure 9d). The data indicate that high correlation between RNA-seq and qPCR was obtained, validating the selected transcripts.

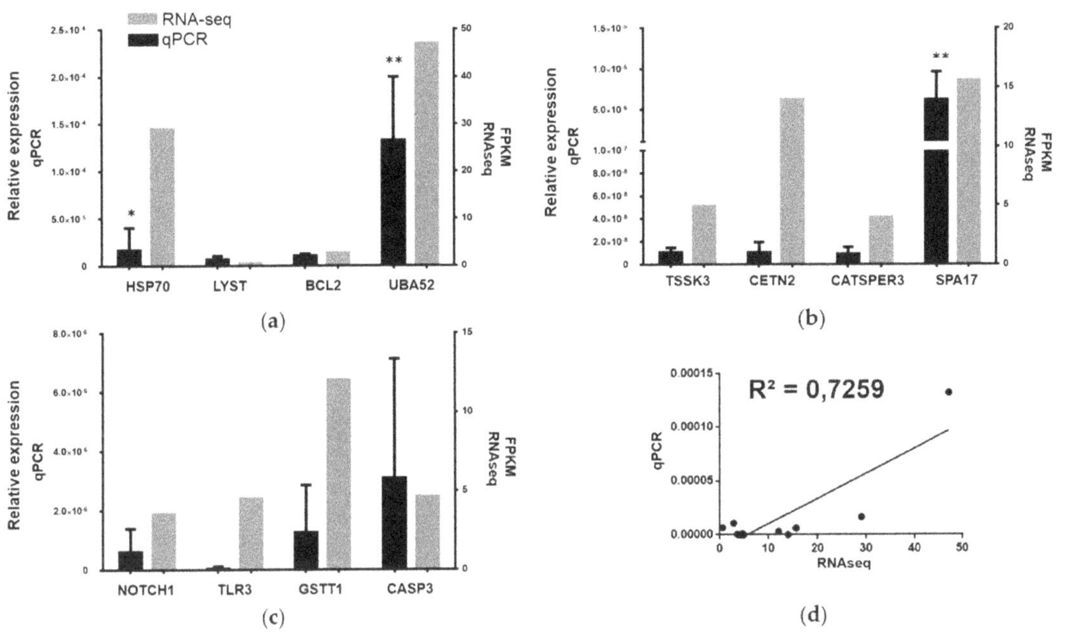

Figure 9. The quantitative real-time PCR validation of differentially expressed transcripts was highly correlated to RNA-seq. (**a**) coelomocyte; (**b**) gonad; (**c**) intestine; and (**d**) statistical correlation analysis. Expression fold changes measured by RT-qPCR and RNA-seq are indicated by black and gray columns, respectively. Abbreviations: heat shock protein 70 kDa 1 A (HSP70), lysosomal trafficking regulator (LYST), B-cell lymphoma 2 (BCL2), ubiquitin A-52 residue ribosomal protein fusion product 1 (UBA52), testis-specific serine/threonine-protein kinase 3 (TSSK3), centrin 2 (CETN2), cation channel sperm associated 3 (CATSPER3), sperm surface protein 17 (SPA17), notch homolog 1 (NOTCH1), toll-like receptor 3 (TLR3), glutathione s-transferase theta 1 (GSTT1), and caspase 3 (CASP3). * $p < 0.05$, ** $p < 0.01$.

4. Discussion

In this study, the transcriptome of the edible red sea urchin (*L. albus*) was sequenced and annotated by using NGS technology. This is the first report on the RNA sequencing, transcriptome assembly, and functional annotation from juvenile *L. albus* based on three different tissues: intestines, gonads, and coelomocytes. This transcriptome contains 185,239 transcripts with similar features to existing transcriptomes of other sea urchins. For instance, the GC content of *L. albus* (38.8%) has a similar value in relation to the sea urchin transcriptomes of *Evechinus chloroticus* (39.0%) [32], *Sterechinus neumayeri* (38.6%) [33], *Strongylocentrotus intermedius* (39.6%) [34], and *Mesocentrotus nudus* (39.9%) [35]. However, our results exhibit a GC content slightly lower than values reported for the testis transcriptome of *L. albus* (40.4%) [6]. This small difference may be attributed to the tissue used in the study (mature gonad), as well as the sequencing technology (Roche 454 GS-FLX Titanium). This group obtained 1062,716 raw reads with a mean length of 309.8 bp, generating a reference transcriptome of 42,530 transcripts with an N50 of 645 bp [6], in contrast to our study, which presents 91,119,300 paired end reads with a mean length of 250 bp, generating a reference transcriptome of 185,239 transcripts with an N50 of 1769 bp.

A detailed analysis of tissue expression reveal that 26,864 transcripts are differentially expressed among the three tissues. Coelomocytes are the cells responsible for immunity in sea urchins, of which 50% to 70% are motile cells with a high energy requirement and are considered equivalent to human macrophages, which are cellular immune system components with high catabolic activity and are part of the innate immune response involved in pathogen digestion and autophagy [36,37]. Besides coelomocytes are predominantly in the coelomic fluid, they also function as wandering cells and infiltrate all tissues [36]. Consequently, the differentially expressed transcripts in the coelomocytes are mainly associated with biological processes, such as positive regulation of apoptotic process and intracellular signal transduction, and KEGG pathways associated with Platelet activation, Fc gamma R-mediated phagocytosis, and Pathogenic *Escherichia coli* infection. Studies in sea urchin coelomocytes transcriptomes describe similar observations. In *S. intermedius*, the expression of 546 unique transcripts in coelomocytes is associated with lysozyme, lectin, pattern recognition receptors (PRRs), and the complement system [38]. In a related study carried out in coelomocytes of *Arbacia lixula*, the expression of transcripts is associated to lipid metabolism and the immune response [39]. In addition, an RNA-seq analysis in coelomocytes reveal key functions of NOD-like receptor pathway and phagosomes in spotting diseased *S. intermedius* [40]. Recently, the immune response of *L. albus* coelomocytes by poly I:C, bacterial lipopolysaccharides (LPS), and temperature reveal a dynamic expression of TLR genes (*tlr3* including), as well as *strongylocin-1* and *strongylocin-2* [18].

Among the transcripts identified with a high expression in coelomocytes and validated by RT-qPCR stand out the heat shock protein 70 kDa 1 A (HSP70), the lysosomal trafficking regulator (LYST), the B-cell lymphoma 2 (BCL2) and the ubiquitin A-52 residue ribosomal protein fusion product 1 (UBA52). HSP70 is a chaperone protein responsible for protein folding to protect cells against stressors or presenting antigens for immune response [41]. Interestingly, a recent study has shown that LPS can induce a stress response by increasing the protein levels of HSP70 in *Paracentrotus lividus* coelomocytes, suggesting a relevant role in the sea urchin immune response [42]. LYST plays a role in the transport of materials into structures called lysosomes, acting as recycling centers within cells [43]. Although there are no reports of the importance of this gene in sea urchin coelomocytes, in mammalian macrophages has been linked as a key regulator of membrane trafficking to inflammatory responses mediated by TLRs [44]. BCL2 is a member of protein regulators for cell death, through inhibition of apoptosis [45]. The participation of BCL2 as an important mediator of the immune response in marine organisms has recently been described in *Apostichopus japonicus* challenged with *Vibrio splendidus* [46]. The UBA52 gene encodes to 60S ribosomal protein L40 (RPL40) and, together with ubiquitin, has a main function of targeting proteins for degradation by the 26S proteosome. Additionally, UBA52 can regulate gene expression, chromatin structure, and the stress response [47]. Although there are no reports of the relevance of UBA52 in the immune response of sea urchin coelomocytes, its participation in the immune response of higher vertebrates has been described [48]. These observations suggest a permanent activity of protein catabolism in sea urchin coelomocytes, as sentinel organisms of the immune response.

In sea urchins, the gonads are considered a dual organ since they regulate reproduction and nutrient storage [49]. Both functions are carried out by two specific cell populations, somatic and germ cells, the latter also called nutritive phagocytes [50]. Several transcriptomic analyses have been described in gonads of edible sea urchin species, obtaining results in agreement with ours. Particularly, a *de novo* assembly of *M. nudus* gonad transcriptome, key genes associated with biological processes such lipid metabolism, and the biosynthesis of polyunsaturated fatty acids have been identified [35]. Furthermore, a comparative analysis between reproductive tissues of this specie, reveal upregulated GO categories related to energy generation in the testis and negative regulation of nucleotide metabolism in the ovary [39]. Additionally, a transcriptomic analysis indicates important differences in the expression of genes of *M. nudus* related to the biosynthesis of polyunsaturated fatty acids and metabolism in relation to high-quality gonads, similar to our observations [34,51]. In addition, a gonadal transcriptomic of *M. nudus* present candidate sex-related genes that could be involved in significant roles in spermatogenesis, oogenesis, and germ cell development [52]. More recently, integrated analyses of metabolomic and transcriptomic reveal key genes for metabolism and eicosapentaenoic acid biosynthesis in the sea urchin *S. intermedius*, identifying six accumulated metabolites and several differentially expressed genes associated with polyunsaturated fatty acids in the testis compared with the ovary [53].

The testis-specific serine/threonine-protein kinase 3 (TSSK3), centrin 2 (CETN2), cation channel sperm associated 3 (CATSPER3) and sperm surface protein 17 (SPA17) were upregulated specifically in gonad. TSSK3 is a protein involved in the development and maturation of male germ cell [54]. TSSK3 role as a key regulator in spermiogenesis of the Scallop *Argopecten irradians* has recently been described [55]. CETN2 is a member of calcium-binding proteins and is also a component of the centrosome [56]. However, there are no previous reports of its importance in the gonadal function of marine echinoderms. CATSPER3 is a sperm-specific ion channel that plays a central role for the successful fertilization, which involves sperm hyperactivation, acrosome reaction, and chemotaxis towards the egg [57]. In mammals, it has been determined that CATSPER3 is exclusively expressed in the testis, which should be validated in future analyzes in adult sea urchins [58]. SPA17 encodes a sperm surface zona pellucida binding protein. In mammals, SPA17 may be implicated in fertilization through zona pellucida attaching of the oocyte [59]. Members of this group of genes have shown an important participation in the maturation and fertilization of echinoderm eggs [60]. However, it is difficult to speculate if differentially expressed transcripts of gonads are relevant in the function of the testis or ovary, because the present study was performed with sexually immature juveniles *L. albus*.

Finally, the intestine is the organ responsible for the digestion of food [61]. In echinoderms, it also has a key role in the immune response, where it has been described that it presents a high number of sequences related to this process [62,63]. Consistently, we observe an overrepresentation of GO and KEGG categories similar to that observed in coelomocytes, showing high enrichment in the sequences related to SSCR and Cargo Receiver. This is consistent with the importance of the intestine in the immune response; therefore, it has been postulated that the control of normal flora or pathogen-associated gut response is responsible for the increase in the expression of the pattern recognition receptors NLR and TLR [64]. Similarly, we also detect an overrepresentation of lysosomal pathways, which was also has been found in the transcriptomic analysis in *S. intermedius* [38].

The notch homolog 1 (NOTCH1), toll-like receptor 3 (TLR3), glutathione s-transferase theta 1 (GSTT1) and caspase 3 (CASP3) were upregulated specifically in the intestine. NOTCH1 is a Type 1 transmembrane protein involves in numerous processes of development by regulating cellular fate. The Notch signaling pathway is highly conserved among species and regulates interactions between adjacent cells [65]. Interestingly, the importance of intestinal epithelial NOTCH1 as a protector for the development of colorectal adenocarcinoma in a murine model has been described [66]. TLR3 is a member of pattern recognition receptors that recognizes specific molecules of pathogens during innate immune response [67]. Recently our group described the upregulation of TLR3 expression in the coelomocyte response to bacterial LPS, poly I:C and temperature in *L. albus* [18]. GSTT1 is an enzyme that catalyzes the coupling of reduced glutathione to diverse hydrophobic and electrophilic compounds [68]. Remarkably, the glutathione S-transferase activity in the anterior portion of *P. lividus* intestine has been employed as biomarker of environmental contamination, revealing an important role in the homeostasis of this tissue [69]. Finally, CASP3 gene encodes for a cysteine-aspartic acid protease (caspase) that has a crucial role in the execution-phase of apoptosis [70]. This protein has been described as an essential element if the innate immunity of sea urchins [64]. These observations suggest the presence of coelomocytes in the sea urchin intestinal tract, as a fundamental part of the crosstalk between intestinal microbiota and the inflammatory response.

5. Conclusions

This is the first evidence on the RNA sequencing, *de novo* assembly, and functional annotation of the edible red sea urchin (*L. albus*) transcriptome, as well as the differential expression in the gonads, intestines, and coelomocytes. The *de novo* assembly produced 185,239 transcripts, creating a reference transcriptome with an N50 of 1769 bp and 38.8% GC content. Gene ontology analysis of transcripts revealed notable differences in the expression profiles between gonads, intestines, and coelomocytes, allowing the detection of transcripts associated with specific biological processes. In coelomocytes, DETs were mostly associated to positive regulation of apoptotic process and intracellular signal transduction. In the gonad, DETs were associated to DNA repair and microtubule-based process. In Intestine, DETs were associated to transmembrane transport and microtubule-based process. The dataset generated in this work contribute to enrich the molecular resources of *L. albus*, improvement futures biological studies of this species. The acquired information is also relevant to discovering novel candidate genes that could be employed to estimate the physiological condition of edible red sea urchins under aquaculture rearing.

Supplementary Materials: The following are available online at https://www.mdpi.com/article/10.3390/biology10100995/s1, Table S1: Primer sequences for qPCR assays, including amplicon size and PCR efficiencies for the assessed genes, Table S2: The complete list of differentially expressed transcripts in the edible sea urchin (*L. albus*), Figure S1: Length distribution of assembled transcripts, Figure S2: The heat map analysis of transcript expression across tissues in *L. albus* was highly correlated among replicates, Figure S3: The Top-16 Gene Ontology molecular function (MF) and cellular component (CC) enrichment of up-regulated transcripts from *L. albus* coelomocytes, Figure S4: The Top-16 Gene Ontology molecular function (MF) and cellular component (CC) enrichment of up-regulated transcripts from *L. albus* gonad, Figure S5: The Top-16 Gene Ontology molecular function (MF) and cellular component (CC) enrichment of up-regulated transcripts from *L. albus* intestine.

Author Contributions: Conceptualization, J.M.E. and J.A.V.; methodology, P.A., J.M.E. and J.A.V.; software, P.A.; validation, J.M.E., C.M., A.M. and J.A.V.; formal analysis, P.A. and M.B.-M.; investigation, P.A. and J.A.V.; resources, J.M.E., A.M. and J.A.V.; data curation, P.A.; writing—original draft preparation, R.Z. and J.A.V.; writing—review and editing, R.Z. and J.A.V.; visualization, J.M.E., C.M., A.M. and J.A.V.; supervision, J.M.E., C.M., A.M. and J.A.V.; project administration, A.M. and J.A.V.; funding acquisition, J.A.V. All authors have read and agreed to the published version of the manuscript.

Funding: This research was funded by Fondo Nacional de Desarrollo Científico y Tecnológico (FONDECYT), grant number 1201498 (to Juan Antonio Valdés), Fondo de Financiamiento de Centros de Investigación en Áreas Prioritarias (FONDAP) grant INCAR 15110027 and Fondo de Innovación para la Competitividad Regional (FIC), grant BIP 30480912-0.

Institutional Review Board Statement: The study was conducted according to the guidelines of the Declaration of Helsinki and approved by the Ethics Committee of Universidad Andres Bello and the National Commission for Scientific and Technological Research (CONICYT) of the Chilean government (N° 010P/2021).

Informed Consent Statement: Not applicable.

Data Availability Statement: The raw read sequences obtained from sequencing were deposited in the Sequence Read Archive (SRA) under BioProject PRJNA475570, with accession number SRP150640. The datasets generated and/or analyzed during the current study are not publicly available due to privacy or ethical restrictions but are available from the corresponding author upon reasonable request.

Conflicts of Interest: The authors declare no conflict of interest. The funders had no role in the design of the study, in the collection, analyses, or interpretation of data, in the writing of the manuscript, or in the decision to publish the results.

References

1. Oyarzún, S.T.; Marín, S.L.; Valladares, C.; Iriarte, J.L. Reproductive Cycle of *Loxechinus albus* (Echinodermata: Echinoidea) in Two Areas of the Magellan Region (53oS, 70-72oW), Chile. *Sci. Mar.* **1999**, *63*, 439–449. [CrossRef]
2. Vásquez, J.A.; Donoso, G.A. Loxechinus albus. In *Developments in Aquaculture and Fisheries Science*; Elsevier: Amsterdam, The Netherlands, 2013; Volume 38, pp. 285–296, ISBN 978-0-12-396491-5.
3. Vásquez, J.A. Loxechinus albus. In *Developments in Aquaculture and Fisheries Science*; Elsevier: Amsterdam, The Netherlands, 2020; Volume 43, pp. 431–445, ISBN 978-0-12-819570-3.
4. Olave, S.; Bustos, E.; Lawrence, J.M.; Carcamo, P. The Effect of Size and Diet on Gonad Production by the Chilean Sea Urchin *Loxechinus albus*. *JWAS* **2001**, *32*, 210–214. [CrossRef]

5. Cea, G.; Gaitán-Espitia, J.D.; Cárdenas, L. Complete Mitogenome of the Edible Sea Urchin *Loxechinus albus*: Genetic Structure and Comparative Genomics within Echinozoa. *Mol. Biol. Rep.* **2015**, *42*, 1081–1089. [CrossRef]
6. Gaitán-Espitia, J.D.; Sánchez, R.; Bruning, P.; Cárdenas, L. Functional Insights into the Testis Transcriptome of the Edible Sea Urchin *Loxechinus albus*. *Sci. Rep.* **2016**, *6*, 36516. [CrossRef]
7. Chandhini, S.; Rejish Kumar, V.J. Transcriptomics in Aquaculture: Current Status and Applications. *Rev. Aquac.* **2018**, *11*, 1379–1397. [CrossRef]
8. Inbakandan, D. Transcriptomics in Aquaculture. In *Encyclopedia of Marine Biotechnology*; Kim, S., Ed.; Wiley: Hoboken, NJ, USA, 2020; pp. 1919–1936, ISBN 978-1-119-14377-2.
9. Jung, G.; Lee, Y.-H. Complete Mitochondrial Genome of Chilean Sea Urchin: *Loxechinus albus* (Camarodonta, Parechinidae). *Mitochondrial DNA* **2015**, *26*, 883–884. [CrossRef]
10. Oulas, A.; Pavloudi, C.; Polymenakou, P.; Pavlopoulos, G.A.; Papanikolaou, N.; Kotoulas, G.; Arvanitidis, C.; Iliopoulos, L. Metagenomics: Tools and Insights for Analyzing Next-Generation Sequencing Data Derived from Biodiversity Studies. *Bioinform. Biol. Insights* **2015**, *9*, 75–88. [CrossRef] [PubMed]
11. Vergara-Amado, J.; Silva, A.X.; Manzi, C.; Nespolo, R.F.; Cárdenas, L. Differential Expression of Stress Candidate Genes for Thermal Tolerance in the Sea Urchin *Loxechinus albus*. *J. Therm. Biol.* **2017**, *68*, 104–109. [CrossRef] [PubMed]
12. Wong, J.M.; Johnson, K.M.; Kelly, M.W.; Hofmann, G.E. Transcriptomics Reveal Transgenerational Effects in Purple Sea Urchin Embryos: Adult Acclimation to Upwelling Conditions Alters the Response of Their Progeny to Differential pCO_2 Levels. *Mol. Ecol.* **2018**, *27*, 1120–1137. [CrossRef]
13. Zhan, Y.; Li, J.; Sun, J.; Zhang, W.; Li, Y.; Cui, D.; Hu, W.; Chang, Y. The Impact of Chronic Heat Stress on the Growth, Survival, Feeding, and Differential Gene Expression in the Sea Urchin Strongylocentrotus Intermedius. *Front. Genet.* **2019**, *10*, 301. [CrossRef]
14. Malanga, G.; Perez, A.; Calvo, J.; Puntarulo, S. The Effect of Seasonality on Oxidative Metabolism in the Sea Urchin *Loxechinus albus*. *Mar. Biol.* **2009**, *156*, 763–770. [CrossRef]
15. Flores, L.; Ernst, B.; Parma, A.M. Growth Pattern of the Sea Urchin, *Loxechinus albus* (Molina, 1782) in Southern Chile: Evaluation of Growth Models. *Mar. Biol.* **2010**, *157*, 967–977. [CrossRef]
16. Cárcamo, P.F. Effects of Food Type and Feeding Frequency on the Performance of Early Juveniles of the Sea Urchin *Loxechinus albus* (Echinodermata: Echinoidea): Implications for Aquaculture and Restocking. *Aquaculture* **2015**, *436*, 172–178. [CrossRef]
17. Dupré, E.; Carvajal, J. Cryopreservation of Embryos and Larvae of the Edible Sea Urchin *Loxechinus albus* (Molina, 1782). *Cryobiology.* **2019**, *86*, 84–88. [CrossRef]
18. Dettleff, P.; Villagra, M.; González, J.; Fuentes, M.; Estrada, J.M.; Valenzuela, C.; Molina, A.; Valdés, J.A. Effect of Bacterial LPS, Poly I:C and Temperature on the Immune Response of Coelomocytes in Short Term Cultures of Red Sea Urchin (*Loxechinus albus*). *Fish Shellfish Immunol.* **2020**, S1050464820306513. [CrossRef]
19. Arévalo, A.; Bruning, P.; Sanchez, R.; Cárdenas, L. Development and Characterization of EST-Microsatellites for the Edible Sea Urchin *Loxechinus albus* Using next Generation Sequencing. *Conserv. Genet. Resour.* **2014**, *6*, 433–435. [CrossRef]
20. Dodt, M.; Roehr, J.; Ahmed, R.; Dieterich, C. FLEXBAR—Flexible Barcode and Adapter Processing for Next-Generation Sequencing Platforms. *Biology.* **2012**, *1*, 895–905. [CrossRef] [PubMed]
21. Grabherr, M.G.; Haas, B.J.; Yassour, M.; Levin, J.Z.; Thompson, D.A.; Amit, I.; Adiconis, X.; Fan, L.; Raychowdhury, R.; Zeng, Q.; et al. Full-Length Transcriptome Assembly from RNA-Seq Data without a Reference Genome. *Nat. Biotechnol.* **2011**, *29*, 644–652. [CrossRef] [PubMed]
22. Davidson, N.M.; Oshlack, A. Corset: Enabling Differential Gene Expression Analysis for *de Novoassembled* Transcriptomes. *Genome Biol.* **2014**, *15*, 410. [CrossRef] [PubMed]
23. Simão, F.A.; Waterhouse, R.M.; Ioannidis, P.; Kriventseva, E.V.; Zdobnov, E.M. BUSCO: Assessing Genome Assembly and Annotation Completeness with Single-Copy Orthologs. *Bioinformatics.* **2015**, *31*, 3210–3212. [CrossRef]
24. Conesa, A.; Gotz, S.; Garcia-Gomez, J.M.; Terol, J.; Talon, M.; Robles, M. Blast2GO: A Universal Tool for Annotation, Visualization and Analysis in Functional Genomics Research. *Bioinformatics.* **2005**, *21*, 3674–3676. [CrossRef]
25. Ye, J.; Zhang, Y.; Cui, H.; Liu, J.; Wu, Y.; Cheng, Y.; Xu, H.; Huang, X.; Li, S.; Zhou, A.; et al. WEGO 2.0: A Web Tool for Analyzing and Plotting GO Annotations, 2018 Update. *Nucleic Acids Res.* **2018**, *46*, W71–W75. [CrossRef]
26. Langmead, B.; Trapnell, C.; Pop, M.; Salzberg, S.L. Ultrafast and Memory-Efficient Alignment of Short DNA Sequences to the Human Genome. *Genome Biol.* **2009**, *10*, R25. [CrossRef] [PubMed]
27. Li, B.; Dewey, C.N. RSEM: Accurate Transcript Quantification from RNA-Seq Data with or without a Reference Genome. *BMC Bioinform.* **2011**, *12*, 323. [CrossRef]
28. Robinson, M.D.; McCarthy, D.J.; Smyth, G.K. EdgeR: A Bioconductor Package for Differential Expression Analysis of Digital Gene Expression Data. *Bioinformatics* **2010**, *26*, 139–140. [CrossRef] [PubMed]
29. Huang, D.W.; Sherman, B.T.; Lempicki, R.A. Systematic and Integrative Analysis of Large Gene Lists Using DAVID Bioinformatics Resources. *Nat. Protoc.* **2009**, *4*, 44–57. [CrossRef] [PubMed]
30. Bustin, S.A.; Benes, V.; Garson, J.A.; Hellemans, J.; Huggett, J.; Kubista, M.; Mueller, R.; Nolan, T.; Pfaffl, M.W.; Shipley, G.L.; et al. The MIQE Guidelines: Minimum Information for Publication of Quantitative Real-Time PCR Experiments. *Clin. Chem.* **2009**, *55*, 611–622. [CrossRef] [PubMed]
31. Simon, P. Q-Gene: Processing Quantitative Real-Time RT-PCR Data. *Bioinformatics* **2003**, *19*, 1439–1440. [CrossRef] [PubMed]

32. Gillard, G.B.; Garama, D.J.; Brown, C.M. The Transcriptome of the NZ Endemic Sea Urchin Kina (*Evechinus chloroticus*). *BMC Genom.* **2014**, *15*, 45. [CrossRef]
33. Dilly, G.F.; Gaitán-Espitia, J.D.; Hofmann, G.E. Characterization of the Antarctic Sea Urchin (*Sterechinus neumayeri*) Transcriptome and Mitogenome: A Molecular Resource for Phylogenetics, Ecophysiology and Global Change Biology. *Mol. Ecol. Resour.* **2015**, *15*, 425–436. [CrossRef]
34. Wang, H.; Ding, J.; Ding, S.; Chang, Y. Transcriptome Analysis to Characterize the Genes Related to Gonad Growth and Fatty Acid Metabolism in the Sea Urchin *Strongylocentrotus intermedius*. *Genes Genom.* **2019**, *41*, 1397–1415. [CrossRef] [PubMed]
35. Wei, Z.; Liu, X.; Zhou, Z.; Xu, J. De Novo Transcriptomic Analysis of Gonad of *Strongylocentrotus nudus* and Gene Discovery for Biosynthesis of Polyunsaturated Fatty Acids. *Genes Genom.* **2019**, *41*, 583–597. [CrossRef] [PubMed]
36. Da Silva, J.R.M.C. Immunology in sea urchins. In *Developments in Aquaculture and Fisheries Science*; Elsevier: Amsterdam, The Netherlands, 2020; Volume 43, pp. 227–236. ISBN 978-0-12-819570-3.
37. Smith, L.C.; Hawley, T.S.; Henson, J.H.; Majeske, A.J.; Oren, M.; Rosental, B. Methods for Collection, Handling, and Analysis of Sea Urchin Coelomocytes. *Methods Cell. Biol.* **2019**, *150*, 357–389. [CrossRef] [PubMed]
38. Chen, Y.; Chang, Y.; Wang, X.; Qiu, X.; Liu, Y. De Novo Assembly and Analysis of Tissue-Specific Transcriptomes Revealed the Tissue-Specific Genes and Profile of Immunity from *Strongylocentrotus intermedius*. *Fish Shellfish Immunol.* **2015**, *46*, 723–736. [CrossRef]
39. Pérez-Portela, R.; Turon, X.; Riesgo, A. Characterization of the Transcriptome and Gene Expression of Four Different Tissues in the Ecologically Relevant Sea Urchin *Arbacia lixula* Using RNA-Seq. *Mol. Ecol. Resour.* **2016**, *16*, 794–808. [CrossRef] [PubMed]
40. Zhang, W.; Lv, Z.; Li, C.; Sun, Y.; Jiang, H.; Zhao, M.; Zhao, X.; Shao, Y.; Chang, Y. Transcriptome Profiling Reveals Key Roles of Phagosome and NOD-like Receptor Pathway in Spotting Diseased *Strongylocentrotus intermedius*. *Fish Shellfish Immunol.* **2019**, *84*, 521–531. [CrossRef]
41. Lubkowska, A.; Pluta, W.; Strońska, A.; Lalko, A. Role of Heat Shock Proteins (HSP70 and HSP90) in Viral Infection. *IJMS* **2021**, *22*, 9366. [CrossRef]
42. Chiaramonte, M.; Inguglia, L.; Vazzana, M.; Deidun, A.; Arizza, V. Stress and Immune Response to Bacterial LPS in the Sea Urchin *Paracentrotus lividus* (Lamarck, 1816). *Fish Shellfish Immunol.* **2019**, *92*, 384–394. [CrossRef]
43. Sepulveda, F.E.; Burgess, A.; Heiligenstein, X.; Goudin, N.; Ménager, M.M.; Romao, M.; Côte, M.; Mahlaoui, N.; Fischer, A.; Raposo, G.; et al. LYST Controls the Biogenesis of the Endosomal Compartment Required for Secretory Lysosome Function: LYST Controls the Biogenesis of Secretory Lysosome. *Traffic* **2015**, *16*, 191–203. [CrossRef]
44. Westphal, A.; Cheng, W.; Yu, J.; Grassl, G.; Krautkrämer, M.; Holst, O.; Föger, N.; Lee, K.-H. Lysosomal Trafficking Regulator Lyst Links Membrane Trafficking to Toll-like Receptor–Mediated Inflammatory Responses. *J. Exp. Med.* **2017**, *214*, 227–244. [CrossRef]
45. Banjara, S.; Suraweera, C.D.; Hinds, M.G.; Kvansakul, M. The Bcl-2 Family: Ancient Origins, Conserved Structures, and Divergent Mechanisms. *Biomolecules* **2020**, *10*, 128. [CrossRef]
46. Guo, M.; Chen, K.; Lv, Z.; Shao, Y.; Zhang, W.; Zhao, X.; Li, C. Bcl-2 Mediates Coelomocytes Apoptosis by Suppressing Cytochrome c Release in *Vibrio splendidus* Challenged Apostichopus Japonicus. *Dev. Comp. Immunol.* **2020**, *103*, 103533. [CrossRef] [PubMed]
47. Kobayashi, M.; Oshima, S.; Maeyashiki, C.; Nibe, Y.; Otsubo, K.; Matsuzawa, Y.; Nemoto, Y.; Nagaishi, T.; Okamoto, R.; Tsuchiya, K.; et al. The Ubiquitin Hybrid Gene *UBA52* Regulates Ubiquitination of Ribosome and Sustains Embryonic Development. *Sci. Rep.* **2016**, *6*, 36780. [CrossRef] [PubMed]
48. Ebner, P.; Versteeg, G.A.; Ikeda, F. Ubiquitin Enzymes in the Regulation of Immune Responses. *Crit. Rev. Biochem. Mol. Biol.* **2017**, *52*, 425–460. [CrossRef] [PubMed]
49. Walker, C.W.; Unuma, T.; Lesser, M.P. Chapter 2 Gametogenesis and reproduction of sea urchins. In *Edible Sea Urchins: Biology and Ecology*; Lawrence, J.M., Ed.; *Developments in Aquaculture and Fisheries Science*; Elsevier: Amsterdam, The Netherlands, 2007; Volume 37, pp. 11–33.
50. Walker, C.W.; Harrington, L.M.; Lesser, M.P.; Fagerberg, W.R. Nutritive Phagocyte Incubation Chambers Provide a Structural and Nutritive Microenvironment for Germ Cells of *Strongylocentrotus droebachiensis*, the Green Sea Urchin. *Biol. Bull.* **2005**, *209*, 31–48. [CrossRef]
51. Jia, Z.; Wang, Q.; Wu, K.; Wei, Z.; Zhou, Z.; Liu, X. De Novo Transcriptome Sequencing and Comparative Analysis to Discover Genes Involved in Ovarian Maturity in *Strongylocentrotus nudus*. *Comp. Biochem. Physiol. D Genom. Proteom.* **2017**, *23*, 27–38. [CrossRef] [PubMed]
52. Sun, Z.-H.; Zhang, J.; Zhang, W.-J.; Chang, Y.-Q. Gonadal Transcriptomic Analysis and Identification of Candidate Sex-Related Genes in *Mesocentrotus nudus*. *Gene* **2019**, *698*, 72–81. [CrossRef]
53. Wang, H.; Ding, J.; Ding, S.; Chang, Y. Integrated Metabolomic and Transcriptomic Analyses Identify Critical Genes in Eicosapentaenoic Acid Biosynthesis and Metabolism in the Sea Urchin *Strongylocentrotus intermedius*. *Sci. Rep.* **2020**, *10*, 1697. [CrossRef]
54. Salicioni, A.M.; Gervasi, M.G.; Sosnik, J.; Tourzani, D.A.; Nayyab, S.; Caraballo, D.A.; Visconti, P.E. Testis-Specific Serine Kinase Protein Family in Male Fertility and as Targets for Non-Hormonal Male Contraception. *Biol. Reprod.* **2020**, *103*, 264–274. [CrossRef]
55. Xue, X.; Zhang, L.; Li, Y.; Wei, H.; Wu, S.; Liu, T.; Liu, L.; Xing, Q.; Wang, S.; Bao, Z. Expression of the Testis-Specific Serine/Threonine Kinases Suggests Their Role in Spermiogenesis of Bay Scallop *Argopecten irradians*. *Front. Physiol.* **2021**, *12*, 657559. [CrossRef]

56. Mullee, L.I.; Morrison, C.G. Centrosomes in the DNA Damage Response—The Hub Outside the Centre. *Chromosome Res.* **2016**, *24*, 35–51. [CrossRef]
57. Singh, A.P.; Rajender, S. CatSper Channel, Sperm Function and Male Fertility. *Reprod. Biomed. Online* **2015**, *30*, 28–38. [CrossRef]
58. Jin, J.-L.; O'Doherty, A.M.; Wang, S.; Zheng, H.; Sanders, K.M.; Yan, W. Catsper3 and Catsper4 Encode Two Cation Channel-Like Proteins Exclusively Expressed in the Testis. *Biol. Reprod.* **2005**, *73*, 1235–1242. [CrossRef]
59. Chiriva-Internati, M. Sperm Protein 17: Clinical Relevance of a Cancer/Testis Antigen, from Contraception to Cancer Immunotherapy, and Beyond. *Int. Rev. Immunol.* **2011**, *30*, 138–149. [CrossRef]
60. Santella, L.; Limatola, N.; Vasilev, F.; Chun, J.T. Maturation and Fertilization of Echinoderm Eggs: Role of Actin Cytoskeleton Dynamics. *Biochem. Biophys. Res. Commun.* **2018**, *506*, 361–371. [CrossRef] [PubMed]
61. Holland, N.D. Digestive system in regular sea urchins. In *Developments in Aquaculture and Fisheries Science*; Elsevier: Amsterdam, The Netherlands, 2020; Volume 43, pp. 147–163. ISBN 978-0-12-819570-3.
62. Ramírez-Gómez, F.; Ortíz-Pineda, P.A.; Rojas-Cartagena, C.; Suárez-Castillo, E.C.; García-Arrarás, J.E.; García-Ararrás, J.E. Immune-Related Genes Associated with Intestinal Tissue in the Sea Cucumber *Holothuria glaberrima*. *Immunogenetics* **2008**, *60*, 57–71. [CrossRef]
63. Yang, A.-F.; Zhou, Z.-C.; He, C.-B.; Hu, J.-J.; Chen, Z.; Gao, X.-G.; Dong, Y.; Jiang, B.; Liu, W.-D.; Guan, X.-Y.; et al. Analysis of Expressed Sequence Tags from Body Wall, Intestine and Respiratory Tree of Sea Cucumber (*Apostichopus japonicus*). *Aquaculture* **2009**, *296*, 193–199. [CrossRef]
64. Hibino, T.; Loza-Coll, M.; Messier, C.; Majeske, A.J.; Cohen, A.H.; Terwilliger, D.P.; Buckley, K.M.; Brockton, V.; Nair, S.V.; Berney, K.; et al. The Immune Gene Repertoire Encoded in the Purple Sea Urchin Genome. *Dev. Biol.* **2006**, *300*, 349–365. [CrossRef] [PubMed]
65. Garis, M.; Garrett-Sinha, L.A. Notch Signaling in B Cell Immune Responses. *Front. Immunol.* **2021**, *11*, 609324. [CrossRef] [PubMed]
66. Dunkin, D.; Iuga, A.C.; Mimouna, S.; Harris, C.L.; Haure-Mirande, J.-V.; Bozec, D.; Yeretssian, G.; Dahan, S. Intestinal Epithelial Notch-1 Protects from Colorectal Mucinous Adenocarcinoma. *Oncotarget* **2018**, *9*, 33536–33548. [CrossRef] [PubMed]
67. Habib, Y.J.; Zhang, Z. The Involvement of Crustaceans Toll-like Receptors in Pathogen Recognition. *Fish Shellfish Immunol.* **2020**, *102*, 169–176. [CrossRef] [PubMed]
68. Vaish, S.; Gupta, D.; Mehrotra, R.; Mehrotra, S.; Basantani, M.K. Glutathione S-Transferase: A Versatile Protein Family. *3 Biotech.* **2020**, *10*, 321. [CrossRef] [PubMed]
69. Cunha, I.; García, L.M.; Guilhermino, L. Sea-Urchin (*Paracentrotus lividus*) Glutathione S-Transferases and Cholinesterase Activities as Biomarkers of Environmental Contamination. *J. Environ. Monit.* **2005**, *7*, 288–294. [CrossRef] [PubMed]
70. Asadi, M.; Taghizadeh, S.; Kaviani, E.; Vakili, O.; Taheri-Anganeh, M.; Tahamtan, M.; Savardashtaki, A. Caspase-3: Structure, Function, and Biotechnological Aspects. *Biotechnol. Appl. Biochem.* **2021**, 1–13. [CrossRef]

biology

MDPI

Article

Comparison of Myosepta Development and Transcriptome Profiling between Blunt Snout Bream with and Tilapia without Intermuscular Bones

Jia-Jia Zhou [1,2], Yong-Jie Chang [1,2], Yu-Long Chen [1,2], Xu-Dong Wang [1,2], Qing Liao [1,2], Rui-Hui Shi [1,2,*] and Ze-Xia Gao [1,2,3,*]

1 College of Fisheries, Key Lab of Freshwater Animal Breeding, Ministry of Agriculture, Key Lab of Agricultural Animal Genetics, Breeding and Reproduction of Ministry of Education, Huazhong Agricultural University, No. 1 Shizishan Street, Hongshan District, Wuhan 430070, China; J-2020@webmail.hzau.edu.cn (J.-J.Z.); changyj@webmail.hzau.edu.cn (Y.-J.C.); cyl76@webmail.hzau.edu.cn (Y.-L.C.); wangxudong@webmail.hzau.edu.cn (X.-D.W.); liangqwang@webmail.hzau.edu.cn (Q.L.)

2 Guangdong Laboratory for Lingnan Modern Agriculture, Guangzhou 510000, China

3 Engineering Research Center of Green Development for Conventional Aquatic Biological Industry in the Yangtze River Economic Belt, Ministry of Education, Wuhan 430070, China

* Correspondence: Shirh@mail.hzau.edu.cn (R.-H.S.); gaozx@mail.hzau.edu.cn (Z.-X.G.); Tel.: +86-027-8728-2113 (Z.-X.G.); Fax: +86-027-8728-2114 (Z.-X.G.)

Simple Summary: The presence or absence of intermuscular bones (IBs) is directly related to the economic and edible value of fish. The specific regulatory mechanism of IB formation is not completely known yet. Here, we explored the molecular mechanisms that regulate the formation of IBs based on histological analysis, transcriptome profiling, and gene expression quantification using *M. amblycephala* (with IBs) and *O. niloticus* (without IBs) as models. As a result, we identified several bone-related genes and elucidated their regulatory roles in the development of IBs.

Citation: Zhou, J.-J.; Chang, Y.-J.; Chen, Y.-L.; Wang, X.-D.; Liao, Q.; Shi, R.-H.; Gao, Z.-X. Comparison of Myosepta Development and Transcriptome Profiling between Blunt Snout Bream with and Tilapia without Intermuscular Bones. *Biology* **2021**, *10*, 1311. https://doi.org/10.3390/biology10121311

Academic Editor: Patricia Pereiro

Received: 26 October 2021
Accepted: 6 December 2021
Published: 10 December 2021

Abstract: Intermuscular bones (IBs) are small spicule-like bones located in the myosepta of basal teleosts, which negatively affect the edibleness and economic value of fish. Blunt snout bream (*Megalobrama amblycephala*, with epineural and epipleural IBs) and tilapia (*Oreochromis niloticus*, without epineural and epipleural IBs) are two major aquaculture species and ideal models for studying the formation mechanisms of fish IBs. Here, we compared myosepta development between *M. amblycephala* and *O. niloticus*, based on histological analysis, transcriptome profiling, and expression analysis of bone-related genes. The histological results showed that dye condensation began to appear in the myosepta 20 days post hatching (dph) in *M. amblycephala*, and IBs could be clearly observed 50 dph in the myosepta, based on different staining methods. However, in *O. niloticus*, dye condensation was not observed in the myosepta from 10 to 60 dph. Differentially expressed genes (DEGs) at different developmental stages were screened by comparing the transcriptomes of *M. amblycephala* and *O. niloticus*, and KEGG analysis demonstrated that these DEGs were enriched in many bone-related pathways, such as focal adhesion, calcium, and Wnt signaling pathways. Quantitative PCR was performed to further compare the expression levels of some bone-related genes (*scxa*, *scxb*, *runx2a*, *runx2b*, *bgp*, *sp7*, *col1a2*, *entpd5a*, *entpd5b*, *phex*, *alpl*, and *fgf23*). All the tested genes (except for *alpl*) exhibited higher expression levels in *M. amblycephala* than in *O. niloticus*. A comparison of the dorsal and abdominal muscle tissues between the two species also revealed significant expression differences for most of the tested genes. The results suggest that *scxa*, *scxb*, *runx2a*, *runx2b*, *entpd5a*, *col1a2*, and *bgp* may play important roles in IB development. Our findings provide some insights into the molecular mechanisms of IB formation, as well as clues for further functional analysis of the identified genes to better understand the development of IBs.

Keywords: intermuscular bone; development; *Megalobrama amblycephala*; *Oreochromis niloticus*; histological structure; transcriptome; gene expression

1. Introduction

Intermuscular bones (IBs), which only occur in the myosepta of basal teleosts, are small spicule-like bones generated from tendon differentiation [1]. According to the statistics of FAO in 2018, nearly half of the top farmed finfish species worldwide have IBs, such as cyprinids and salmonids [2]. The potential risk of harm to the throat or digestive organs upon consumption greatly reduces the attractiveness of fish with IBs to producers and consumers, and causes obstacles for deep processing, value enhancement, and consumption of such fish [3]. Since the 1960s, IBs have received increasing attention, and extensive research has been carried out on the morphology, ossification process, and number of IBs in different species [3]. Previous studies have also revealed the possibility of reducing the IB number based on ploidy change [4], distant hybridization [5,6], and selective breeding [7]. Notably, an IB-deficient grass carp (*Ctenopharyngodon idella*) was found in an artificial gynogenetic group, and some specimens of tambaqui (*Colossoma macropomum*) without IBs were found in a culture population (normal individuals possess a significant number of IBs) [5,8], indicating the feasibility of genetic improvement of the IB trait. Therefore, the clarification of the molecular mechanisms of IB formation would contribute to the genetic improvement of fish in aquaculture.

Recent advance in high-throughput sequencing technology has greatly facilitated the research on the expression of RNA transcripts in specific tissues or cells, which has significantly improved the understanding of the molecular regulatory mechanisms of the formation of IBs. Recently, comparative transcriptome studies have revealed the molecular characteristics of IB formation and the distinction between IBs and other tissues [9,10]. Additionally, a previous study with histological–transcriptomic–proteomic data suggested that IBs are gradually formed through intramembranous ossification without a cartilaginous phase [11]. Another study compared the orthologous gene families of fish with and without IBs and demonstrated the functional importance of key signaling pathways associated with IB formation [12]. Some studies have characterized the expression of some genes related to IB development, such as *sclerostin* (*sost*) in *Carassius auratus* [13], *tenomodulin* (*tnmd*) and *bone morphogenetic proteins* (*bmps*) in *M. amblycephala* [14], and *muscle segment homeobox 3* (*msx3*) in *Hemibarbus labeo* [15], and the influence of some genes on IB formation has also been verified. For example, scleraxis bHLH transcription factor a (*scxa*) mutation obtained by on the CRISPR/Cas9 system resulted in a clear reduction of mineralized IBs in zebrafish [16], and sp7 transcription factor (*sp7*) mutation in common carp led to shorter IBs [17]. However, many other genes related to bone formation remain uncharacterized during IB development. The formation of IBs may undergo several processes including differentiation of tendon stem/progenitor cells (TSPCs) into osteoblasts, ossification, and mineralization [3,18,19]. Therefore, the genes involved in these processes are worth of exploration to better understand the formation of IBs. It has also been reported that the differentiation of the osteoblast lineage is coordinately regulated by various signaling pathways such as Hedgehog, Notch, Wnt, BMP, and FGF signaling pathways [20,21].

Currently, most studies of IB development, morphology, and transcriptome have been focused on specific fish species with IBs. There is still a lack of comparative studies of myosepta development and transcriptome profiling in fish species with and without IBs. In this study, a comparison analysis was conducted to better understand IB development by using two typical fish species: Blunt snout bream (*Megalobrama amblycephala*), a typical aquaculture species in China belonging to the Cyprinidae family, with a certain number of IBs (epineurals and epipleurals) [11], and Nile tilapia (*Oreochromis niloticus*), an important economic fish belonging to the Perciformes order, without epineurals and epipleurals [22]. We compared the histological structure of the myosepta of these two species from 5 to 60 dph with different staining methods (alcian blue–nuclear red, alizarin red, hematoxylin–eosin (HE), and toluidine blue). Transcriptome sequencing was performed at four key developmental stages of IBs to compare the gene expression between *M. amblycephala* and *O. niloticus*. Then, the expression levels of possible regulatory genes (*scxa, scxb,* RUNX family transcription factor 2 (*runx2a, runx2b),* bone gamma-carboxyglutamate protein (*bgp*), *sp7*, collagen type I alpha 2 (*col1a2*), ectonucleoside triphosphate diphosphohydrolase 5 (*entpd5a*, *entpd5b*), phosphate regulating endopeptidase homolog, X-linked (*phex*), alkaline phosphatase (*alpl*), and fibroblast growth factor 23 (*fgf23*)) during IB development were analyzed. Our findings may contribute to a better understanding of IB development in teleosts.

2. Materials and Methods

2.1. Ethics Statement

All experiments were conducted in accordance with the guidelines of the National Institute of Health Guide for the Care and Use of Laboratory Animals and approved by the Research Ethics Committee, Huazhong Agricultural University, Wuhan, China (HZAUDO-2016-005, 2016-10-26). All efforts were made to minimize fish suffering.

2.2. Sample Collection

M. amblycephala and *O. niloticus* were obtained from the Fish Breeding Base of the College of Fisheries, Huazhong Agricultural University. The specimens were anesthetized by MS-222 (Sigma, Saint Louis, MO, USA; 100 mg/L) and sterilized with 75% alcohol before tissue collection. After manual removal of the skins, fins, and vertebra, the tail muscles containing IBs were immediately collected under an anatomical lens at eight stages (10, 15, 20, 25, 30, 40, 50, and 60 dph) as described in our previous study [23]. The whole fish larvae at 5 dph were collected as they were too small for the collection of tail muscle tissue, and the dorsal (with epineurals) and abdominal (with epipleurals) muscle tissues were collected from one-year-old *M. amblycephala* and *O. niloticus*. Each group included three samples, and each sample contained muscle tissues from three *M. amblycephala* or *O. niloticus* specimens. All samples were rapidly placed in RNAiso Plus (TaKaRa, Dalian China), refrigerated at 4 °C for 24 h, and stored at –80 °C before total RNA extraction.

2.3. Histological Analysis

The caudal peduncle at different developmental stages (5, 10, 15, 20, 25, 30, 40, 50, 60 dph) of *M. amblycephala* and *O. niloticus* was collected and fixed in 4% paraformaldehyde. Then, the post-fixed tissues were decalcified in 0.5 M EDTA until full decalcification (3–7 d) for histological processing. The decalcified tissues were processed by dehydration through a graded series of ethanol (70–100%), cleared in xylene, and then embedded in paraffin blocks. The 5 μm slices were sectioned using a rotary microtome and then stained with HE, alcian blue–nuclear red, alizarin red, and toluidine blue, among which alizarin red was used to detect bone mineralization, alcian blue–nuclear red to visualize the cartilage, and HE and toluidine blue to observe the overall structure of the tissue [24–26].

2.4. RNA Library Construction and Sequencing

Total RNA was isolated from samples of *M. amblycephala* and *O. niloticus* at 11 stages using RNAiso Plus Reagent (TaKaRa, Dalian, China) according to the manufacturer's instructions. RNA quality and quantity were measured using 1% agarose gels and NanoDrop 2000 (Thermo Scientific, Waltham, MA, USA), respectively. The RNA samples from four developmental stages of *M. amblycephala* as indicated in our previous published study [23] and the corresponding stages of *O. niloticus* from 15, 25, 30, and 40 dph were used for sequencing analysis. A total of 2 μg of RNA per sample with RIN controlled at 9.6–10 was used as input material for RNA sample preparation. Sequencing libraries were generated using the NEBNext® UltraTM RNA Library Prep Kit for Illumina® (NEB, Ipswich, MA, USA) following the manufacturer's recommendations, and index codes were added to attribute sequences to each sample. Subsequently, clustering of the index-coded samples was performed on a cBot cluster generation system using the HiSeq PE Cluster Kit v4-cBot-HS (Illumina, San Diego, CA, USA) according to the manufacturer's instructions. After cluster generation, the libraries were sequenced on an Illumina platform, and 150 bp paired-end reads were generated. Finally, clean data were obtained by filtering the sequencing adapter sequence and low-quality sequencing data.

2.5. Differential Expression and Functional Enrichment Analysis

In order to compare the gene expression levels at different developmental stages in *M. amblycephala* and *O. niloticus*, the raw data of four developmental stages of *M. amblycephala* from our previous study [23] were filtered and aligned to the *M. amblycephala* reference genome [9], while the data of *O. niloticus* were aligned to the *O. niloticus* reference genome (https://asia.ensembl.org/Oreochromis_niloticus/Info/Index, accessed on 15 September 2019). The DESeq2 in R packages were used to identify differentially expressed mRNAs with q-value < 0.05 and Log2FoldChange > 1. Besides, Gene Ontology (GO) and Kyoto Encyclopedia of Genes and Genomes (KEGG) enrichment analysis of differentially expressed genes (DEGs) were implemented by KOBAS (http://kobas.cbi.pku.edu.cn/kobas3/genelist/, accessed on 24 February 2021).

2.6. Gene Expression Analysis

Reverse-transcription PCR was conducted to synthesize cDNA from 1 μg of total RNA by using the HiScript®IIQ RT SuperMix for qPCR (+gDNA wiper) (Vazyme, Nanjing, China) following the manufacturer's protocol. Quantitative real-time PCR (qRT-PCR) was carried out on a QuantStudio™ 6 Flex real-time PCR System (Applied Biosystems, Carlsbad, CA, USA) according to the manufacturer's instructions. Primer pairs are shown in Table S2. qRT-PCR was performed using SYBR® Premix DimerEraser™ (TaKaRa, Shiga, Japan). The qRT-PCR program consisted of pre-denaturation at 95 °C for 5 min and 40 cycles of amplification at 95 °C for 15 s, 60 °C for 30 s, and 72 °C for 30 s. Each experiment was conducted with three replicates. A housekeeping gene (*β-actin*) was used as a reference gene in quantification analysis. A melting curve analysis was performed at the end of the reaction to demonstrate the specificity of the reaction. For each data analysis, the expression was quantified based on the comparative C_T method ($2^{-\triangle\triangle C_T}$ formula), and the developmental stage or gene corresponding to the largest C_T value in the same species was chosen as the reference sample. All values are presented as means ± standard error (SE). One-way ANOVA was conducted using GraphPad Prism8 software to examine the differential expression of genes at different stages in *M. amblycephala* and *O. niloticus*. Statistical significance was accepted at the level of $p < 0.05$. Trend analysis was performed on the hiplot (https://hiplot.com.cn//, accessed on 2 September 2021) using the average value of relative expression level. The entire research scheme diagram is shown in Figure 1.

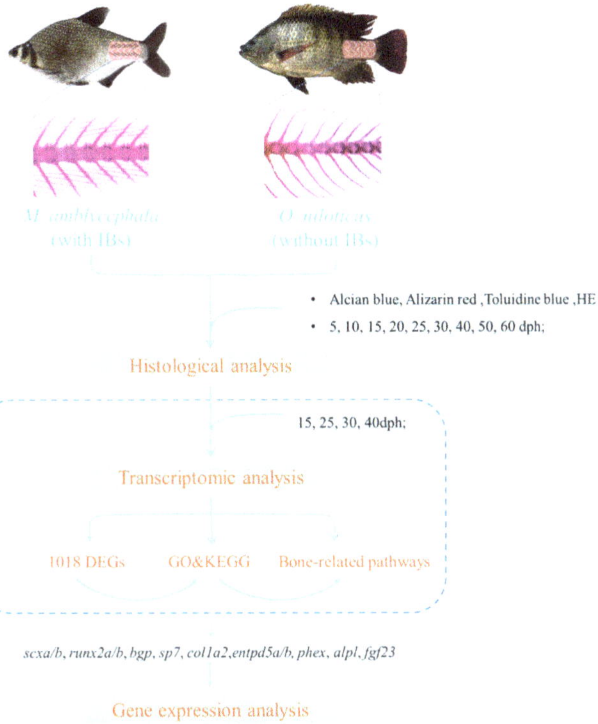

Figure 1. Research scheme diagram.

3. Results

3.1. Histological Structure

To identify histological characteristics during IB development, four different staining methods were used to examine changes in tissue structure during the ossification process of IBs. Samples of nine developmental stages were collected from *M. amblycephala* and *O. niloticus*. In *M. amblycephala,* dye condensation was observed in the myosepta 20 dph using toluidine blue staining (Figure 2A). Then, bone-like staining nodules became clearer and larger from 20 to 50 dph with the growth of IBs (Figure 2B–D). At 50 dph, the staining nodules of IBs were clearly visible in the myosepta when performing alizarin red, HE, and toluidine blue staining; however, alcian blue–nuclear red failed to stain the IBs (Figure 3A–D). As for *O. niloticus,* that does not possess IBs, dye condensation and bone-like nodules were not observed in the myosepta at any stage (Figure 2E–H).

3.2. Comparative Transcriptome Analysis

To investigate the genetic regulation of IB development, we compared the transcriptome profiles of *M. amblycephala* and *O. niloticus* at different developmental stages. Figure 4A presents the results from the preliminary analysis of DEGs at different developmental stages of *M. amblycephala*, while Figure 4B presents the analysis results for *O. niloticus*.

In the S2-vs-S1 comparison, 53 DEGs were uniquely found in *M. amblycephala* and not in *O. niloticus* at any stage; these genes may play an important role in the early development of IBs. The Venn diagram revealed that 1074 DEGs were uniquely present in S4-vs-S1 and S3-vs-S1 in *M. amblycephala* (Figure 4C), while 1884 DEGs were uniquely found in S4-vs-S1 and S3-vs-S1 in *O. niloticus* (Figure 4E). Finally, 1018 DEGs present in *M. amblycephala* but not in *O. niloticus* were selected for further analysis (Figure 4D); these genes are very likely associated with the development of IBs. Then, GO enrichment analysis revealed that many of these genes are involved in nucleus (GO:0005634), cytoplasm (GO:0005737), and ATP binding (GO:0005730) (Table S5). The KEGG analysis results demonstrated that these genes were enriched in 133 pathways, including many bone-related pathways, such as focal adhesion, calcium signaling, tight junction, and Wnt signaling pathways (Table S6) (Figure 4F). There were 15, 11, 11, and 10 DEGs in calcium signaling, mTOR signaling, Wnt signaling, MAPK signaling pathways, respectively, which were all related to osteoclast and osteoblast differentiation.

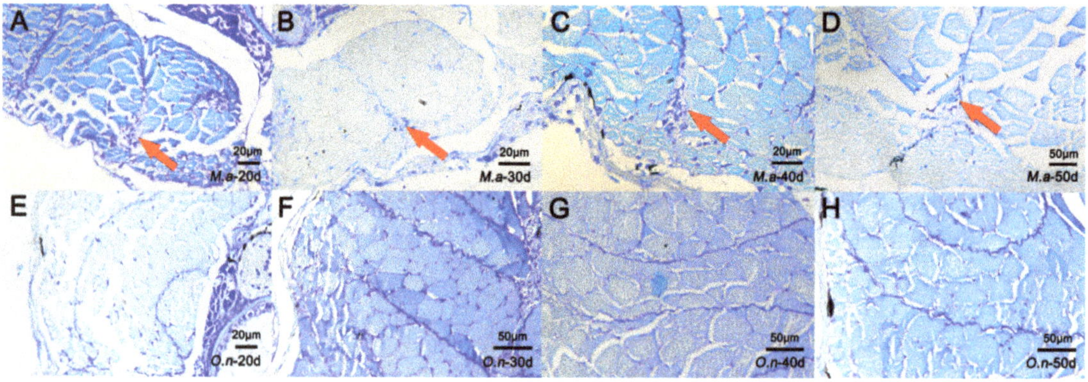

Figure 2. Histological characteristics during IB development based on toluidine blue staining in *M. amblycephala* and *O. niloticus*. (**A**–**D**) Staining at 20, 30, 40, and 50 dph of *M. amblycephala*; (**E**–**H**) staining at 20, 30, 40, and 50 dph of *O. niloticus*. The IBs in *M. amblycephala* are marked by red arrows. No IBs were identified in *O. niloticus*.

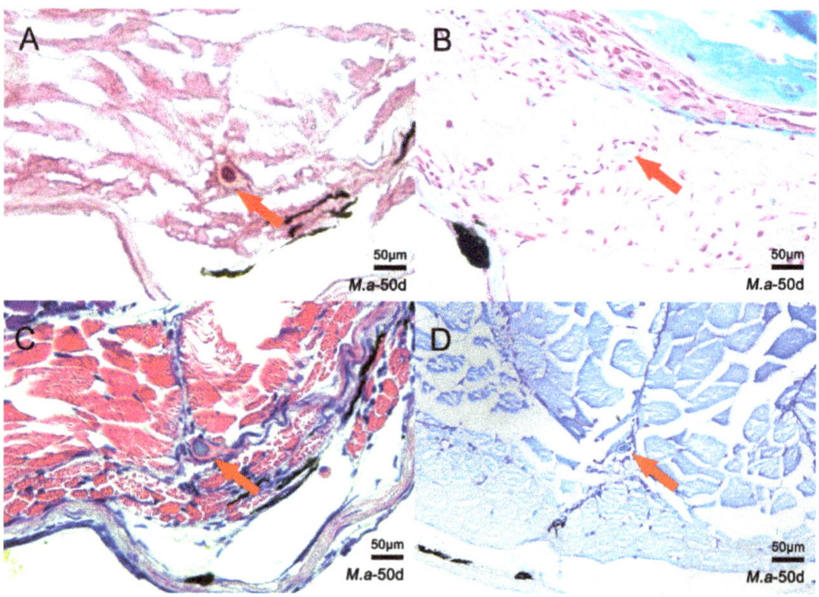

Figure 3. Histological characteristics of *M. amblycephala* IBs based on four staining methods at 50 dph. (**A**) Alizarin red; (**B**) alcian blue–nuclear red; (**C**) HE; (**D**) toluidine blue. The IBs are marked by red arrows.

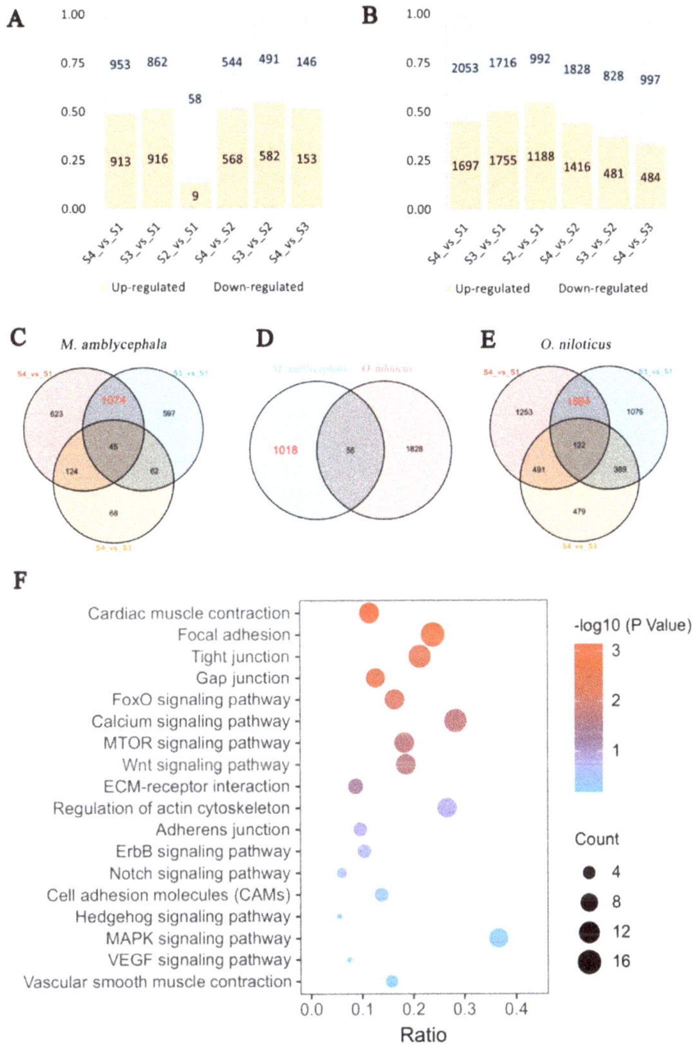

Figure 4. Comparative transcriptome analysis. (**A**) Number of DEGs in *M. amblycephala*; (**B**) number of DEGs in *O. niloticus*; (**C**) Venn diagram for DEGs in three comparison groups in *M. amblycephala*; (**D**) Venn diagram of shared and unique genes between *M. amblycephala* and *O. niloticus*; (**E**) Venn diagram for DEGs in three comparison groups in *O. niloticus*; (**F**) signaling pathways related to bone development enriched in 1018 DEGs identified by comparative transcriptome analysis.

3.3. Gene Expression Analysis

The gene expression profiles during the development of *M. amblycephala* and *O. niloticus* were analyzed by qRT-PCR (Figure 5A–K). The relative expression levels of all the tested genes in *M. amblycephala* showed an increasing tendency from 10 dph to 20 dph, particularly those of *scxa*, *scxb*, *bgp*, *col1a2*, and *alpl*. Moreover, from 25 dph to 40 dph, the expression levels of all tested genes still showed an increasing tendency in *M. amblycephala*, except for *alpl*, while they remained unchanged or even declined in *O. niloticus*, suggesting that most of these genes are involved in the formation of IBs. At 40 dph, the relative expression of *runx2a*, *runx2b*, *entpd5a*, *sp7*, and *fgf23* in *M. amblycephala* reached a peak, and the expression of *col1a2* was 103 times that at 10 dph. From 40 dph to 60 dph, the expression of *runx2a*, *runx2b*, *scxa*, *scxb*, *entpd5b*, *bgp*, and *col1a2* was maintained at relatively high levels in *M. amblycephala*, which was not observed in *O. niloticus*. It was thus confirmed that most of the tested genes had higher expression levels in *M. amblycephala* than in *O. niloticus* at the same stage, particularly *scx*, *runx2*, *entpd5*, *col1a2*, and *bgp*. However, *O. niloticus* showed higher expression levels of *alpl*, *sp7*, and *phex* than *M. amblycephala*. There was no significant difference in the expression level of *fgf23* between *M. amblycephala* and *O. niloticus* for most of the tested stages. Taken together, the qRT-PCR analysis revealed that *M. amblycephala* had higher relative expression levels of *scx*, *runx2*, *entpd5*, *col1a2*, and *bgp* at most stages, especially at the late stage of IB development. Interestingly, these genes showed generally the same changing pattern of expression in *M. amblycephala* (Figure 5J), which was completely different from that in *O. niloticus* (Figure 5K).

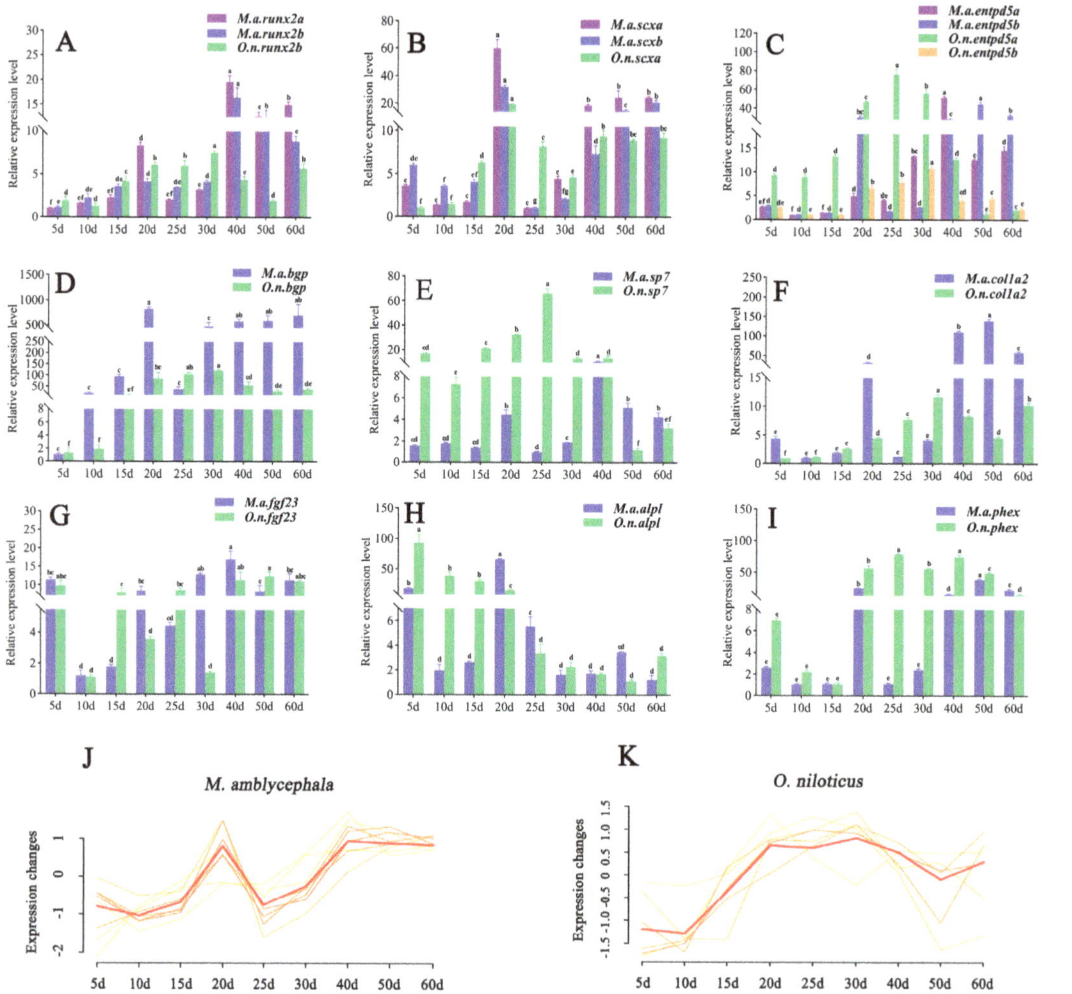

Figure 5. Expression analysis of the tested genes during IB development. (**A–I**) Relative expression levels of the tested genes (*runx2, scx, entpd5, bgp, sp7, col1a2, fgf23, alpl, phex*) at nine developmental stages in *M. amblycephala* and *O. niloticus*. The same letters above the columns for each gene mean no significant difference ($p > 0.05$); (**J,K**) expression trends of *scx*, *runx2*, *entpd5*, *bgp*, and *col1a2* in *M. amblycephala* and *O. niloticus* at nine developmental stages.

We further characterized the expression of the tested genes in the dorsal and abdominal muscle tissues of *M. amblycephala* and *O. niloticus*. Figure 6A–D shows that there was no significant difference in the expression of most genes between the dorsal and the abdominal part in one-year-old *M. amblycephala* ($p < 0.05$), except for the *bgp* gene, suggesting that *bgp* has a greater impact on epipleurals than on epineurals. In both the dorsal and the abdominal parts, the *scxa, scxb, runx2a, col1a2, bgp, entpd5b,* and *phex* genes had relatively higher expression levels than *runx2b, entpd5a, fgf23, alpl,* and *sp7* in *M. amblycephala*. Surprisingly, both *scxa* and *col1a2* showed high expression levels in the dorsal and abdominal parts, and their expression in the dorsal part was significantly higher than that in the abdominal part in one-year old *O. niloticus*. The expression trend of these genes in the dorsal and abdominal parts was basically the same in *M. amblycephala* and *O. niloticus*.

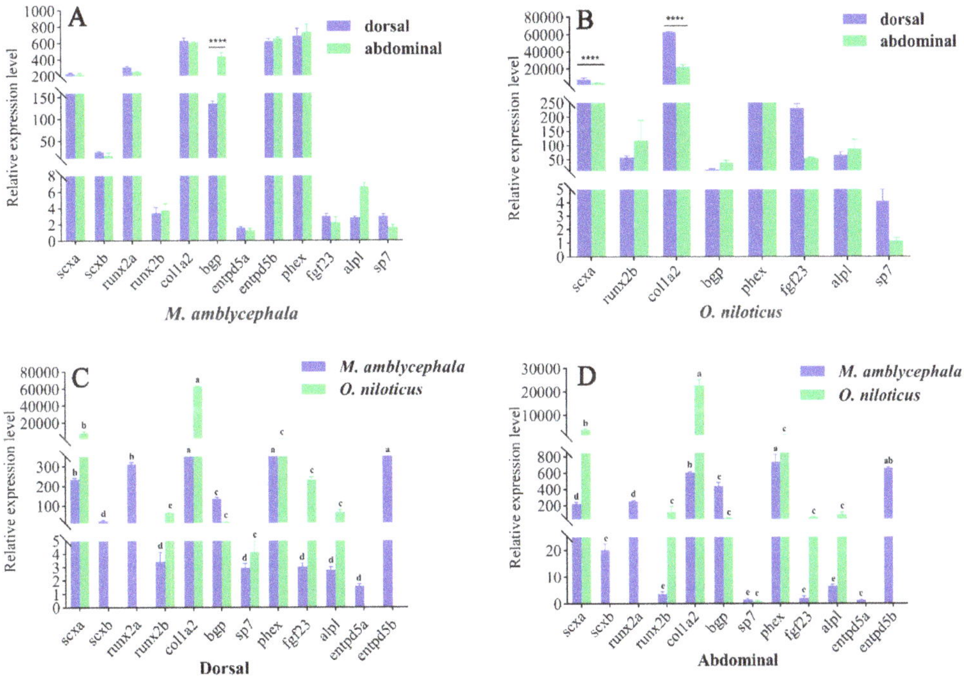

Figure 6. Relative expression levels of the tested genes (*runx2, scx, entpd5, bgp, sp7, col1a2, fgf23, alpl, phex*) in the dorsal and abdominal parts in one-year-old *M. amblycephala* and *O. niloticus*. **** indicates an extremely significant difference ($p < 0.0001$); the same letters above the columns for each gene mean no significant difference ($p > 0.05$). (**A**) *M. amblycephala*; (**B**) *O. niloticus*; (**C**) comparison between *M. amblycephala* and *O. niloticus* in the dorsal part; (**D**), comparison between *M. amblycephala* and *O. niloticus* in the abdominal part.

4. Discussion

It has been reported that IBs are ossified from myosepta [1]. To better understand the difference in myosepta development between fish with and without IBs, a histological analysis was carried out to compare myosepta development in *M. amblycephala* (with IBs) and *O. niloticus* (without IBs). Toluidine blue staining showed significant differences in myosepta development between *M. amblycephala* and *O. niloticus*, and dye condensation and bone-like nodules were found during myosepta development in *M. amblycephala* but not in *O. niloticus*. The results of alizarin red and alcian blue–nuclear red staining at different developmental stages of *M. amblycephala* further support the conclusion that IBs are membrane bones formed from intramembranous ossification without a cartilaginous phase [11].

In this study, a temporal differential expression analysis was performed to compare gene expression at different developmental stages between *M. amblycephala* and *O. niloticus*, and a total of 53 DEGs were screened. These genes may play important roles in the initial formation of IBs. A preliminary analysis of the transcriptome data in the late period of IB development revealed that 1018 DEGs may be related to IB development. GO and KEGG analyses identified some pathways involved in osteoblast and/or osteoclast development, including calcium, mTOR, MAPK, VEGF, FoxO, and ErbB signaling pathways [27–30]. A previous review has particularly explored the functions of several molecular signaling pathways during bone formation and development, including Hedgehog signaling, Notch signaling, and Wnt signaling pathways [20]. Three pathways, respectively, for cardiac muscle contraction, regulation of actin cytoskeleton, and vascular smooth muscle contraction have certain correlations with muscle contraction. Previous research has demonstrated that IBs may be associated with the regulation of muscle contraction [12,31,32]. Bone is isolated by a layer of osteoblasts connected by tight and gap junctions with unique cellular functions and complex molecular composition [33]. During the formation of the skeleton structure, junction-associated proteins, which only allow the regulated transport and limit the free diffusion of molecules, are expressed in osteoblasts, generating significant resistance across osteoblast monolayers, while gap junctions may play important roles in the communication between cells through connexins as well as in growth, development, and tissue homeostasis [34–36]. Tight junction proteins can transmit signals to the cell interior either directly or by recruiting other signaling molecules to regulate cell proliferation, migration, survival, and differentiation [37]. Cellular interaction requires the support of signaling pathways such as those associated with focal adhesions, ECM–receptor interactions, adherens junctions, tight junctions, Notch, and cell adhesion molecules [38–41]. Hence, it can be speculated that some of the genes screened in our research may participate in bone formation and development. Although numerous studies have shown that many pathways are involved in skeletal development, no pathway or gene has been found to be specifically essential for IB development, and our study may provide a reference for future research.

As a type of pluripotent stem cells, TSPCs can form tendon-like, bone-like, and tendon-bone junction-like tissues in rat models [42,43]. The development of IBs is a complex biological process, which may involve condensation of pluripotent stem cells, differentiation of osteoblasts, and mineralization [3]. This study analyzed the expression of relevant genes (*scxa, scxb, runx2a, runx2b, bgp, sp7, col1a2, entpd5a, entpd5b, phex, alpl,* and *fgf23*) during the development of IBs, with the aim to improve our understanding of the molecular mechanisms of IB development.

Scx, a basic helix–loop–helix (bHLH) transcription factor, is relevant to tendon differentiation and maturation as a well-established marker of tendon cells [44,45]. As reported in mammals, the *bgp* gene could promote the differentiation of preosteoblasts into mature osteoblasts [46]. As a Type I collagen, COL1A2 is particularly abundant in tendon and bone tissues [47]. *Alpl*, that encodes alkaline phosphatase (ALP), is highly expressed in the cells of mineralized tissues and plays a critical role in the formation of hard tissues [48]. The expression of these genes showed a dramatic increase at 20 dph in *M. amblycephala*, which is consistent with the histological results showing that dye condensation occurred in the myosepta at 20 dph. These results suggest that these genes may be involved in the early formation of IBs. This finding is consistent with that of a previous study, which reported that the knockout of *scxa* would lead to an obvious reduction of IBs in zebrafish [16].

IBs are mostly composed of osteocytes, bone collagen, and calcium hydroxyapatite [49]. Osteocytes are derived from osteoblasts and then trapped and surrounded by bone matrix. The bone matrix is composed of bone collagen protein synthesized by the osteoblasts and requiring the expression of *col1a2*, non-collagen proteins (such as osteocalcin encoded by *bgp*), which may help to regulate matrix mineralization and affect bone cell activity, and minerals (mainly hydroxyapatite) [50]. The *runx2* gene can promote the differentiation of multipotent mesenchymal cells into osteoblasts [21,51]. Besides, *runx2* induces the expression of *bgp*, encoding osteocalin, to induce the differentiation of preosteoblasts into mature osteoblasts [52,53]. The *entpd5* gene plays an important role in phosphate homeostasis and has been demonstrated to be essential for skeletal mineralization in zebrafish [54,55]. In this study, these genes showed high expression levels from 40 dph to 60 dph (the period of rapid growth and perfection of IBs) in *M. amblycephala*, which may be of great significance to the development of IBs. The *osterix* (*Sp7*) gene can activate the differentiation of preosteoblast cells into mature osteoblast cells and osteocytes [56,57]. However, it is surprising that the expression of *sp7* did not show much change during the formation and development of IBs and, in *M. amblycephala*, was even much lower than in *O. niloticus*.

The PPi/Pi ratio is important for normal bone mineralization. *Alpl* encodes the tissue-non-specific isoenzyme of alkaline phosphatase (TNSALP), which dephosphorylates PPi into inorganic Pi [58]. The *phex* gene can increase the concentration of Pi in the ECM by promoting renal Pi reabsorption, while *fgf23* has an opposite effect [59,60]. There was no significant change in the expression levels of these genes during 40–60 dph in *M. amblycephala* and *O. niloticus*, but the expression of these genes in the dorsal and abdominal muscles of one-year-old *O. niloticus* was higher than in the corresponding parts of one-year-old *M. amblycephala*. However, in the dorsal and abdominal parts of one-year-old *M. amblycephala*, the *phex*/*fgf23* expression ratios were 229 and 324, respectively, much higher than in *O. niloticus* (6 and 22, respectively). The higher relative expression of *phex* with respect to *fgf23* is conducive to Pi deposition and may promote the production of hydroxyapatite crystals during IB development. These genes may have important effects on the formation of IBs, which needs to be further validated considering their higher expression in *O. niloticus*.

In general, our quantitative results showed that *scx*, *runx2*, *entpd5*, *col1a2*, and *bgp* have higher relative expression levels at most developmental stages of *M. amblycephala*, particularly at the late stage of IB development.

5. Conclusions

IBs are characterized as spicule-like bones existing in basal teleosts, with uncertain functions. In this study, histological staining was combined with RNA-Seq to reveal differences in myosepta development between *M. amblycephala* and *O. niloticus*. First, histological changes in IBs during the development of *M. amblycephala* were revealed, and it was found that IB development does not involve the cartilage phase. Then, the dynamics of molecular features in developing IBs were analyzed, and a number of candidate genes were screened according to the literature. Finally, the functions of these candidate genes in IB development were characterized. Our results suggest that the *scxa*, *scxb*, *runx2a*, *runx2b*, *entpd5a*, *col1a2*, and *bgp* genes may play important roles in the formation, ossification, and mineralization of IBs. Overall, our findings contribute to a better understanding of IB formation and development and provide an important reference for further understanding the molecular mechanisms of IB formation. In future research, gene knockout or RNA interference may be employed to elucidate gene functions during IB development in fish.

Supplementary Materials: The following are available online at https://www.mdpi.com/article/10.3390/biology10121311/s1, Table S1: Amino acid sequence GeneBank accession number/Ensembl ID of bone-related genes in different species, Table S2: Primer sequences for genes qRT-PCR of *M. amblycephala* and *O. niloticus*, Table S3: List of DEGs in different comparison groups of *M. amblycephala*, Table S4: List of DEGs in different comparison groups of *O. niloticus*, Table S5: GO terms of 1018 DEGs uniquely shared in S4_vs_S1 and S3_vs_S1 in *M. amblycephala*, Table S6: KEGG pathway information of 1018 DEGs uniquely shared in S4_vs_S1 and S3_vs_S1 in *M. amblycephala*.

Author Contributions: J.-J.Z., Y.-J.C. and R.-H.S., methodology, investigation, formal analysis, writing—original draft preparation. Q.L. and X.-D.W., methodology, investigation, formal analysis. Y.-L.C., formal analysis, Writing. Z.-X.G., conceptualization, project administration, funding acquisition, writing—review and editing. All authors have read and agreed to the published version of the manuscript.

Funding: This work was financially supported by the Laboratory for Lingnan Modern Agriculture (NZ2021011), National Key Research and Development Program (Grant No. 2018YFD0900102), the China Agriculture Research System of MOF and MARA (Grant No. CARS-46-08) and State Key Laboratory of Developmental Biology of Freshwater Fish (2018KF005), and Wuhan Applied Foundational Frontier Project (2020020601012253).

Institutional Review Board Statement: All experiments were conducted in accordance with the guidelines of the National Institute of Health Guide for the Care and Use of Laboratory Animals and approved by the Research Ethics Committee, Huazhong Agricultural University, Wuhan, China (HZAUDO-2016-005, 26 October 2016).

Informed Consent Statement: Not applicable.

Data Availability Statement: All transcriptome data of tilapia are available in the NCBI database under Accession number PRJNA729911. All transcriptome data of blunt snout bream are available in the NCBI database under Accession number PRJNA263980.

Acknowledgments: We thank all the researchers at our laboratories for their help.

Conflicts of Interest: The authors declare no conflict of interest.

References

1. Gemballa, S.; Britz, R. Homology of intermuscular bones in acanthomorph fishes. *Am. Mus. Novit.* **1988**, *3241*, 1–25. [CrossRef]
2. FAO. The State of World Fisheries and Aquaculture 2020. In *Sustainability in Action*; FAO: Rome, Italy, 2020; p. 206. [CrossRef]
3. Nie, C.H.; Hilsdorf, A.W.S.; Wan, S.M.; Gao, Z.X. Understanding the development of intermuscular bones in teleost: Status and future directions for aquaculture. *Rev. Aquac.* **2020**, *12*, 759–772. [CrossRef]
4. Li, L.; Zhong, Z.Z.; Zeng, M.; Liu, S.J.; Zhou, Y.; Xiao, J.; Wang, J.; Liu, Y. Comparative analysis of intermuscular bones in fish of different ploidies. *Sci. China Life Sci.* **2013**, *56*, 341–350. [CrossRef]
5. Qian, Y.; Zheng, J.; Xu, X.; Luo, C. Normally grown and developed intermuscular bone-deficient mutant in grass carp, *Ctenopharyngodon idellus*. *Chin. Sci. Bull.* **2015**, *60*, 52–57. [CrossRef]
6. Jiang, W.P.; Jia, Y.Y.; Liu, S.L.; Li, Q.; Li, T.; Gu, Z.M. Comparative analysis of intermuscular bone in hybrid F1, F2 of (*C. alburnus*) (♀) × (*M. amblycephala*) (♂) and its parents. *Acta Hydrobiol. Sin.* **2016**, *40*, 227–286.
7. Xiong, X.-M.; Robinson, N.A.; Zhou, J.-J.; Chen, Y.-L.; Wang, W.M.; Wang, X.-B.; Gao, Z.-X. Genetic parameter estimates for intermuscular bone in blunt snout bream (*Megalobrama amblycephala*) based on a microsatellite-based pedigree. *Aquaculture* **2019**, *502*, 371–377. [CrossRef]
8. Perazza, C.A.; de Menezes, J.T.B.; Ferraz, J.B.S.; Pinaffi, F.L.V.; Silva, L.A.; Hilsdorf, A. Lack of intermuscular bones in specimens of *Colossoma macropomum*: An unusual phenotype to be incorporated into genetic improvement programs. *Aquaculture* **2017**, *472*, 57–60. [CrossRef]
9. Liu, H.; Chen, C.H.; Gao, Z.X.; Min, J.M.; Gu, Y.M.; Jian, J.B.; Jiang, X.W.; Cai, H.M.; Ebersberger, I.; Xu, M.; et al. The draft genome of blunt snout bream (*Megalobrama amblycephala*) reveals the development of intermuscular bone and adaptation to herbivorous diet. *GigaScience* **2017**, *6*, gix039. [CrossRef]
10. Nie, C.-H.; Wan, S.-M.; Tomljanovic, T.; Treer, T.; Hsiao, C.-D.; Wang, W.-M.; Gao, Z.-X. Comparative proteomics analysis of teleost intermuscular bones and ribs provides insight into their development. *BMC Genom.* **2017**, *18*, 147. [CrossRef] [PubMed]
11. Nie, C.-H.; Wan, S.-M.; Liu, Y.; Liu, H.; Wang, W.-M.; Gao, Z.-X. Development of teleost intermuscular bones undergoing intramembranous ossification based on histological-transcriptomic-proteomic data. *Int. J. Mol. Sci.* **2019**, *20*, 4698. [CrossRef] [PubMed]
12. Chen, J.; Chen, X.L.; Huang, X.; Huang, G.H.; Gao, Z.X.; Wang, W.M.; Liu, H. Genome-wide analysis of intermuscular bone development reveals changes of key genes expression and signaling pathways in blunt snout bream (*Megalobrama amblycephala*). *Genomics* **2021**, *113*, 654–663. [CrossRef] [PubMed]
13. Tian, X.; Wang, L.Y.; Chen, L.; Wang, L.; Ma, X.; Hu, C.C.; Kong, X.H.; Nie, G.X.; Li, X.J. The mRNA and protein expression of gene SOST during ossification process of intermuscular bone in crucian carp (*Carassius auratus*) in Qihe River. *J. Fish. China* **2016**, *40*, 673–680.
14. Wang, X.D.; Nie, C.H.; Gao, Z.X. Research progress on molecular regulation mechanism and genetic selection of intermuscular bones in teleosts. *Acta Hydrobiol. Sin.* **2021**, *45*, 680–691.
15. Lv, Y.P.; Yao, W.J.; Chen, J.; Bao, B.L. Newly identified gene muscle segment homeobox C may play a role in intermuscular bone development of *Hemibarbus labeo*. *Genet. Mol. Res.* **2015**, *14*, 11324–11334. [CrossRef]
16. Nie, C.H.; Wan, S.M.; Chen, Y.L.; Zhu, D.J.; Wang, X.D.; Dong, X.R.; Gao, Z.-X. Loss of scleraxis leads to distinct reduction of mineralized intermuscular bone in zebrafish. *Aquac. Fish.* **2021**, *6*, 169–177. [CrossRef]
17. Zhong, Z.; Niu, P.; Wang, M.; Huang, G.; Xu, S.; Sun, Y.; Xu, X.; Hou, Y.; Sun, X.; Yan, Y.; et al. Targeted disruption of *sp7* and myostatin with CRISPR-Cas9 results in severe bone defects and more muscular cells in common carp. *Sci. Rep.* **2016**, *6*, 22953. [CrossRef]

18. Walia, B.; Huang, A.H. Tendon stem progenitor cells: Understanding the biology to inform therapeutic strategies for tendon repair. *J. Orthop. Res.* **2018**, *37*, 1270–1280. [CrossRef]
19. Huang, Z.; Yin, Z.; Xu, J.; Fei, Y.; Heng, B.C.; Jiang, X.; Chen, W.; Shen, W. Tendon stem/progenitor cell subpopulations and their implications in tendon biology. *Front. Cell Dev. Biol.* **2021**, *9*. [CrossRef]
20. Salhotra, A.; Shah, H.N.; Levi, B.; Longaker, M.T. Mechanisms of bone development and repair. *Nat. Rev. Mol. Cell Biol.* **2020**, *21*, 696–711. [CrossRef]
21. Komori, T. Runx2, A multifunctional transcription factor in skeletal development. *J. Cell. Biochem.* **2002**, *87*, 1–8. [CrossRef]
22. Su, S.; Dong, Z. Comparative expression analyses of bone morphogenetic protein 4 (BMP4) expressions in muscles of tilapia and common carp indicate that BMP4 plays a role in the intermuscular bone distribution in a dose-dependent manner. *Gene Expr. Patterns* **2018**, *27*, 106–113. [CrossRef] [PubMed]
23. Wan, S.M.; Yi, S.K.; Zhong, J.; Nie, C.H.; Guan, N.N.; Zhang, W.Z.; Gao, Z.X. Dynamic mRNA and miRNA expression analysis in response to intermuscular bone development of blunt snout bream (*Megalobrama amblycephala*). *Sci Rep.* **2016**, *6*, 31050. [CrossRef] [PubMed]
24. Walker, M.B.; Kimmel, C.B. A two-color acid-free cartilage and bone stain for zebrafish larvae. *Biotech. Histochem.* **2007**, *82*, 23–28. [CrossRef]
25. Witten, P.E.; Huysseune, A. A comparative view on mechanisms and functions of skeletal remodelling in teleost fish, with special emphasis on osteoclasts and their function. *Biol. Rev.* **2009**, *84*, 315–346. [CrossRef] [PubMed]
26. Eames, B.F.; Yan, Y.-L.; Swartz, M.E.; Levic, D.S.; Knapik, E.W.; Postlethwait, J.H.; Kimmel, C.B. Mutations in *fam20b* and *xylt1* reveal that cartilage matrix controls timing of endochondral ossification by inhibiting chondrocyte maturation. *PLoS Genet.* **2011**, *7*, e1002246. [CrossRef]
27. Fisher, M.C.; Clinton, G.M.; Maihle, N.J.; Dealy, C.N. Requirement for ErbB2/ErbB signaling in developing cartilage and bone. *Dev. Growth Differ.* **2007**, *49*, 503–513. [CrossRef]
28. Liu, Y.; Olsen, B.R. Distinct VEGF functions during bone development and homeostasis. *Arch. Immunol. Ther. Exp.* **2014**, *62*, 363–368. [CrossRef] [PubMed]
29. Majidinia, M.; Sadeghpour, A.; Yousefi, B. The roles of signaling pathways in bone repair and regeneration. *J. Cell. Physiol.* **2018**, *233*, 2937–2948. [CrossRef]
30. Ma, X.; Su, P.; Yin, C.; Lin, X.; Wang, X.; Gao, Y.; Patil, S.; War, A.R.; Qadir, A.; Tian, Y.; et al. The roles of Foxo transcription factors in regulation of bone cells function. *Int. J. Mol. Sci.* **2020**, *21*, 692. [CrossRef] [PubMed]
31. Charvet, B.; Malbouyres, M.; Pagnon-Minot, A.; Ruggiero, F.; Le Guellec, D. Development of the zebrafish myoseptum with emphasis on the myotendinous junction. *Cell Tissue Res.* **2011**, *346*, 439–449. [CrossRef]
32. Fiedler, I.A.K.; Zeveleva, S.; Duarte, A.; Zhao, X.; Depalle, B.; Cardoso, L.; Jin, S.; Berteau, J.P. Microstructure, mineral and mechanical properties of teleost intermuscular bones. *J. Biomech.* **2019**, *94*, 59–66. [CrossRef]
33. Schlesinger, P.H.; Blair, H.C.; Stolz, D.B.; Riazanski, V.; Ray, E.C.; Tourkova, I.L.; Nelson, D.J. Cellular and extracellular matrix of bone, with principles of synthesis and dependency of mineral deposition on cell membrane transport. *Am. J. Physiol. Cell Physiol.* **2020**, *318*, C111–C124. [CrossRef]
34. Sharrow, A.C.; Li, Y.; Micsenyi, A.; Griswold, R.D.; Wells, A.; Monga, S.S.P.; Blair, H.C. Modulation of osteoblast gap junction connectivity by serum, TNFα, and TRAIL. *Exp. Cell Res.* **2008**, *314*, 297–308. [CrossRef]
35. Wongdee, K.; Pandaranandaka, J.; Teerapornpuntakit, J.; Tudpor, K.; Thongbunchoo, J.; Thongon, N.; Jantarajit, W.; Krishnamra, N.; Charoenphandhu, N. Osteoblasts express claudins and tight junction-associated proteins. *Histochem. Cell Biol.* **2008**, *130*, 79–90. [CrossRef]
36. Dbouk, H.A.; Mroue, R.M.; El-Sabban, M.E.; Talhouk, R.S. Connexins: A myriad of functions extending beyond assembly of gap junction channels. *Cell Commun. Signal.* **2009**, *7*, 4. [CrossRef]
37. Bhat, A.A.; Uppada, S.; Achkar, I.W.; Hashem, S.; Yadav, S.K.; Shanmugakonar, M.; Al-Naemi, H.A.; Haris, M.; Uddin, S. Tight junction proteins and signaling pathways in cancer and inflammation: A functional crosstalk. *Front. Physiol.* **2019**, *9*, 1942. [CrossRef] [PubMed]
38. Bauer, M.S.; Baumann, F.; Daday, C.; Redondo, P.; Durner, E.; Jobst, M.A.; Milles, L.F.; Mercadante, D.; Pippig, D.A.; Gaub, H.E.; et al. Structural and mechanistic insights into mechanoactivation of focal adhesion kinase. *Proc. Natl. Acad. Sci. USA* **2019**, *116*, 6766–6774. [CrossRef] [PubMed]
39. Tonnesen, M.G.; Feng, X.; Clark, R.A. Angiogenesis in wound healing. *J. Investig. Dermatol. Symp. Proc.* **2000**, *5*, 40–46. [CrossRef]
40. Campbell, H.K.; Maiers, J.L.; DeMali, K.A. Interplay between tight junctions & adherens junctions. *Exp. Cell Res.* **2017**, *358*, 39–44. [CrossRef] [PubMed]
41. Cohen, D.J.; Nelson, W.J. Secret handshakes: Cell–cell interactions and cellular mimics. *Curr. Opin. Cell Biol.* **2018**, *50*, 14–19. [CrossRef]
42. Lui, P.P.Y.; Chan, K.M. Tendon-Derived Stem Cells (TDSCs): From basic science to potential roles in tendon pathology and tissue engineering applications. *Stem Cell Rev. Rep.* **2011**, *7*, 883–897. [CrossRef] [PubMed]
43. Zhang, Q.; Cheng, B. Tendon-derived stem cells as a new cell source for tendon tissue engineering. *Front. Biosci.* **2013**, *18*, 756–764.
44. Cserjesi, P.; Brown, D.; Ligon, K.L.; Lyons, G.E.; Copeland, N.G.; Gilbert, D.J.; Jenkins, N.A.; Olson, E.N. Scleraxis: A basic helix-loop-helix protein that prefigures skeletal formation during mouse embryogenesis. *Development* **1995**, *121*, 1099–1110. [CrossRef]

45. Schweitzer, R.; Chyung, J.H.; Murtaugh, L.C.; Brent, A.E.; Rosen, V.; Olson, E.N.; Lassar, A.; Tabin, C.J. Analysis of the tendon cell fate using Scleraxis, a specific marker for tendons and ligaments. *Development* **2001**, *128*, 3855–3866. [CrossRef]
46. Maruyama, Z.; Yoshida, C.A.; Furuichi, T.; Amizuka, N.; Ito, M.; Fukuyama, R.; Miyazaki, T.; Kitaura, H.; Nakamura, K.; Fujita, T.; et al. *Runx2* determines bone maturity and turnover rate in postnatal bone development and is involved in bone loss in estrogen deficiency. *Dev. Dyn.* **2007**, *236*, 1876–1890. [CrossRef] [PubMed]
47. Fisher, S.; Jagadeeswaran, P.; Halpern, M.E. Radiographic analysis of zebrafish skeletal defects. *Dev. Biol.* **2003**, *264*, 64–76. [CrossRef]
48. Vimalraj, S. Alkaline phosphatase: Structure, expression and its function in bone mineralization. *Gene* **2020**, *754*, 144855. [CrossRef]
49. Xu, B.; Ju, Y.; Cui, Y.; Song, G. Carbon nanotube array inducing osteogenic differentiation of human mesenchymal stem cells. *Mater. Sci. Eng. C* **2015**, *51*, 182–188. [CrossRef] [PubMed]
50. Clarke, B. Normal Bone Anatomy and Physiology. *Clin. J. Am. Soc. Nephrol.* **2008**, *3*, S131–S139. [CrossRef]
51. Liang, W.; Li, X.; Gao, B.; Gan, H.; Lin, X.; Liao, L.; Li, C. Observing the development of the temporomandibular joint in embryonic and post-natal mice using various staining methods. *Exp. Ther. Med.* **2015**, *11*, 481–489. [CrossRef]
52. Pico, M.J.; Hashemi, S.; Xu, F.; Nguyen, K.H.; Donnelly, R.; Moran, E.; Flowers, S. Glucocorticoid receptor-mediated cis-repression of osteogenic genes requires BRM-SWI/SNF. *Bone Rep.* **2016**, *5*, 222–227. [CrossRef]
53. Komori, T. Runx2, an inducer of osteoblast and chondrocyte differentiation. *Histochem. Cell Biol.* **2018**, *149*, 313–323. [CrossRef] [PubMed]
54. Huitema, L.F.A.; Apschner, A.; Logister, I.; Spoorendonk, K.M.; Bussmann, J.; Hammond, C.; Schulte-Merker, S. *Entpd5* is essential for skeletal mineralization and regulates phosphate homeostasis in zebrafish. *Proc. Natl. Acad. Sci. USA* **2012**, *109*, 21372–21377. [CrossRef] [PubMed]
55. Apschner, A.; Huitema, L.F.; Ponsioen, B.; Peterson-Maduro, J.; Schulte-Merker, S. Zebrafish enpp1 mutants exhibit pathological mineralization, mimicking features of generalized arterial calcification of infancy (GACI) and pseudoxanthoma elasticum (PXE). *Dis. Model. Mech.* **2014**, *7*, 811–822. [CrossRef] [PubMed]
56. Chen, Z.J.; Song, Z.Y.; Yang, J.J.; Huang, J.; Jiang, H.B. *Sp7/osterix* positively regulates *dlx2b* and *bglap* to affect tooth development and bone mineralization in zebrafish larvae. *J. Biosci.* **2019**, *44*, 127. [CrossRef]
57. Zhang, C.; Tang, W.; Li, Y.; Yang, F.; Dowd, D.R.; Macdonald, P.N. Osteoblast-specific transcription factor osterix increases vitamin d receptor gene expression in osteoblasts. *PLoS ONE* **2011**, *6*, e26504. [CrossRef]
58. Simon, S.; Resch, H.; Klaushofer, K.; Roschger, P.; Zwerina, J.; Kocijan, R. Hypophosphatasia: From diagnosis to treatment. *Curr. Rheumatol. Rep.* **2018**, *20*, 69. [CrossRef]
59. Marsell, R.; Jonsson, K.B. The phosphate regulating hormone fibroblast growth factor-23. *Acta Physiol.* **2010**, *200*, 97–106. [CrossRef]
60. Takeda, E.; Taketani, Y.; Sawada, N.; Sato, T.; Yamamoto, H. The regulation and function of phosphate in the human body. *BioFactors* **2004**, *21*, 345–355. [CrossRef] [PubMed]

biology

MDPI

Article

Bulked Segregant Analysis and Association Analysis Identified the Polymorphisms Related to the Intermuscular Bones in Common Carp (*Cyprinus carpio*)

Ming-Shu Cui [1,2], Ran Zhao [1,2], Qi Wang [1,2], Yan Zhang [1,2], Qing-Song Li [1,3], Mei-Di Huang Yang [1,3], Xiao-Qing Sun [1,2] and Jiong-Tang Li [1,2,*]

1 Key Laboratory of Aquatic Genomics, Ministry of Agriculture and Rural Affairs, Chinese Academy of Fishery Sciences, Chinese Academy of Agricultural Sciences, Beijing 100141, China; cms851860@163.com (M.-S.C.); zhaoran@cafs.ac.cn (R.Z.); wangqi@cafs.ac.cn (Q.W.); zhangy@cafs.ac.cn (Y.Z.); liqingsong686@163.com (Q.-S.L.); huangymd@163.com (M.-D.H.Y.); sunxiaoqing@cafs.ac.cn (X.-Q.S.)
2 Beijing Key Laboratory of Fishery Biotechnology, Beijing 100141, China
3 National Demonstration Center for Experimental Fisheries Science Education, Shanghai Ocean University, Shanghai 201306, China
* Correspondence: lijt@cafs.ac.cn; Tel.: +86-010-6869-1136

Simple Summary: Many widely cultured freshwater fish species, such as common carp, belong to the Cyprinidae family. However, most cyprinids have numerous and complex intermuscular bones (IBs), resulting in an adverse effect on cyprinid fish meat processing and consumption. Numerous studies have been trying to understand the development mechanism of IBs and to identify the SNPs associated with the total IB number. However, the SNPs associated with different forms of IBs have been studied less thoroughly. The joint effects of the SNPs on IB development also remain poorly understood. The common carp has numerous geographical populations and domesticated strains, diversifying its phenotypes. The question of whether consensus IB-related SNPs or genes exist among multiple strains of common carp has also not yet been answered. Selective breeding of IB-reduced common carp has been hindered due to a lack of effective molecular markers. To answer these questions, we performed bulked segregant analysis (BSA) to detect the consensus SNPs in three strains. The consensus BSA-SNPs and the other SNPs in their flanking regions were validated in additional individuals. The SNPs associated with the frequency of different IB types were identified. We examined the joint effects of significant SNPs on the numbers of different types of IBs. The identified genetic markers may benefit future selective breeding and reduce the IB number in common carp.

Abstract: The allotetraploid common carp is one of the most important freshwater food fish. However, the IBs found in allotetraploid common carp increase the difficulty in fish meat processing and consumption. Although candidate genes associated with the total IB number have been identified, the SNPs associated with the numbers of the total IBs and different forms of IBs have not yet been identified, hindering the breeding of IB-reduced common carp. Herein, the numbers of different types of IBs in three common carp strains were measured. Using whole-genome resequencing and bulked segregant analysis in three pairs of IB-more and IB-less groups, we identified the consensus nonsynonymous SNPs in three strains of common carp. Screening the flanking regions of these SNPs led to the detection of other SNPs. Association study detected 21 SNPs significantly associated with the number of total IBs, epineural-IBs, and ten detailed types of IBs. We observed the joint effects of multiple SNPs on each associated IB number with an improved explained percentage of phenotypic variation. The resulting dataset provides a resource to understand the molecular mechanisms of IB development in different common carp strains. These SNPs are potential markers for future selection to generate IB-reduced common carp.

Citation: Cui, M.-S.; Zhao, R.; Wang, Q.; Zhang, Y.; Li, Q.-S.; Huang Yang, M.-D.; Sun, X.-Q.; Li, J.-T. Bulked Segregant Analysis and Association Analysis Identified the Polymorphisms Related to the Intermuscular Bones in Common Carp (*Cyprinus carpio*). *Biology* **2022**, *11*, 477. https://doi.org/10.3390/biology11030477

Academic Editor: Patricia Pereiro

Received: 1 February 2022
Accepted: 14 March 2022
Published: 21 March 2022

Keywords: intermuscular bone; bulked segregant analysis; SNP; association analysis; joint effect; common carp

1. Introduction

Most cyprinids have numerous and complex intermuscular bones (IBs) [1], which are embedded into the myosepta [2]. The IBs were classified based on their locations into the epineural bones (en-IBs), epipleural IBs (ep-IBs), and epicentral IBs (ec-IBs) [3]. The IBs were also sorted based on their morphology into seven forms, including I type, Y type, one-end-unequal-bi-fork type (OEUBF), one-end-multi-fork type (OEMF), two-end-bi-fork type (TEBF), two-end-multi-fork type (TEMF), and tree-branch type (TB) [4]. The IBs increase the labor required to process cyprinids meat, hindering the meat consumption. Therefore, to reduce the IB count in cyprinids and understand the mechanisms of IB development, comparative genomic, transcriptomic, and proteomic analyses were performed to identify the genes and pathways associated with IB development [5–7]. Numerous protein-coding genes [8–10] and noncoding RNAs [6] have been found to be involved in IB development in various fishes, suggesting the complex regulation of IB development. Some genes were knocked out by using the clustered regularly interspaced short palindromic repeats (CRISPR)-Cas9 technology, decreasing the IB count in fish [11]. Although genome editing provides promise of generating the IB-free or IB-reduced fish, the edited fish exhibited developmental defects [12,13]. Generating IB-deficient fish without any defects by genetic breeding or ploidy breeding may be more effective [4,14]. However, most current research focused on the identification of SNPs associated with the number of total IBs. The SNPs associated with different forms of IBs have been studied less thoroughly. Analyzing multiple SNPs is a promising approach to finding genetic effects beyond single-locus associations. Nevertheless, the joint effects of the SNPs on IB development remain poorly understood.

Common carp, an allotetraploid fish [15], is cultured worldwide. In China, common carp accounts for about 11.2% of annual freshwater fish production. Over 20 strains of common carp have been bred and cultivated in China. In addition to serving as a food fish, it is also a common ornamental fish species [16]. Research on common carp IB development is ongoing. Quantitative trait locus (QTL) analysis and comparative transcriptomic study have been used to analyze candidate genes associated with the total IB count in common carp [17,18]. However, the SNPs associated with the numbers of the total IBs and different forms of IBs have not been identified yet. Common carp has numerous geographical populations and domesticated strains, diversifying the phenotypes [19]. Whether consensus IB-associated SNPs or genes exist among multiple strains of common carp also requires further study. Finally, our recent study revealed that two subgenomes in the allotetraploid common carp underwent the differential expression balance in response to different conditions, dampening the stimulus' impact on the homoeologous expression [15]. Whether both homoeologous subgenomes participate in IB development is still unknown.

Herein, we performed bulked segregant analysis (BSA) to detect SNPs in each of three strain pairs consisting of IB-less and IB-more bulks. The consensus BSA-SNPs and their neighboring SNPs among three strain pairs were identified in additional individuals. Then the SNPs associated with the numbers of different types of IBs were identified. We examined the joint effects of significant SNPs on IB numbers. Our study advanced knowledge in this field, allowing a better understanding of the genetic diversity of IB numbers in common carp, and provided genetic markers for future selection to reduce the IB number in common carp.

2. Materials and Methods

2.1. Ethics Statement

This study was performed under the recommendations on animal care and use for scientific purposes established by the Animal Care and Use Committee of the Chinese

Academy of Fishery Sciences (ACUC-CAFS). All the fish in this study were euthanized with MS222 solution (a concentration of 40 mg/L).

2.2. Counting and Classifying the IBs

In November 2020, we collected the six-month-old juvenile common carp including the variety Hebao (HB, Wuyuan, China), the variety Wenqing (WQ, Zhaoqing, China), and the variety Hehua (HH, Quanzhou, China) (Supplementary Figure S1). The samples had been cultivated in one pond together at the experimental fish station of the Chinese Academy of Fishery Sciences (Fangshan, Beijing, China) for half a year. The fish were fed with the same commercial diet (Tongwei, China) in the pond. In total, 122 common carp individuals were randomly selected, including 38 HH individuals, 50 HB individuals, and 34 WQ individuals.

Before counting the IB number, the back muscles were dissected from each individual. We followed the method of Lv et al. [20] to separate the IBs from the muscle. The IBs were photographed in order with a digital camera. The IBs were first classified into two major types:epineural bones (en-IBs) and epipleural bones (ep-IBs). The en-IBs and ep-IBs were further sorted into seven types: I type, Y type, OEUBF, OEMF, TEBF, TEMF, and TB (Supplementary Figure S2). The numbers of total IBs, en-IBs, ep-IBs, and 14 types of IBs were applied to the association study.

We investigated the phenotypic distances among three strains using principal component analysis (PCA) with the 'pcrcomp' function in R package with the above 14 types of IBs. The first three eigenvectors were plotted. The Spearman correlation between any two traits across the 122 individuals was calculated with the 'correl' function in R. The 14 types of IBs were clustered on the basis of the correlations using the 'single method' and visualized using the R function 'heatmap'.

2.3. Genome Resequencing and Bulk Segregant Analysis

Genomic DNA from the back muscle of each individual was extracted using a commercial DNA extraction kit (TIANGEN Biotech, Beijing, China), which was based on the traditional phenol-chloroform extraction method. The DNA was treated with RNase and then stored at −20 °C until further analysis. The DNA quality was checked using gel electrophoresis and the quantity was measured with a UV spectrophotometer, NanoVue Plus (GE healthcare, Boston, MA, USA). In each strain, seven samples with the fewest total IBs were clustered together as an IB-fewer group (F-pool) and another seven samples with the most total IBs as an IB-more group (M-pool). DNA from each group with a concentration of at least 100 ng/ul was used to construct a paired-end genome library, which was sequenced on an Illumina Novaseq 6000 (Illumina, San Diego, CA, USA) platform with the 150-bp PE mode and coverage of at least 37-fold.

The raw reads of each group were filtered using Trimmomatic v0.35 [21] and mapped to the latest common carp reference genome (GenBank no: GCA_018340385.1 [15]) using BWA v0.7.17 [22]. Picard in the Genome Analysis Tool Kit (GATK, v3.8) [23] was used to filter out the low-quality alignments and PCR-duplicated fragments. Variants were called using HaplotypeCaller with the parameters of '-ERC GVCF -stand-call-conf 20-mbq 20′ and then GenotypeGVCFs with default parameters. The SNPs were identified using SelectVariants with the parameters of '-select-type SNP' and filtered using VariantFiltration with the parameters of '-cluster 4—window 10—mask-extension 3—filter-name LowQual—filter QUAL <20′ in GATK. In each strain, to identify the SNP with different frequencies between two groups, the VCF files from the IB-fewer (F-pool) and IB-more (M-pool) groups were inputted to MutMap [24] with the minimum base quality of 20 and 5000 simulation replicates. An SNP-index of one SNP in each pool was represented as the proportion of reads harboring it. The Δ(SNP-index) of this SNP was the difference between the SNP-index in the M-pool and that in the F-pool. If the Δ(SNP-index) was over the 95% confidence interval of all Δ(SNP-index), this SNP was considered as M-pool-enriched. Likewise, we identified the F-pool-enriched SNPs. The SNPs identified by the BSA analysis were designated as the BSA-SNPs.

In each strain, we examined the genomic locations and putative protein-coding changes by these BSA-SNPs using the package ANNOVAR version 2020-06-07 [25]. Based on the common carp gene annotation, the genomic locations of these BSA-SNPs were categorized into the exonic regions, splicing sites, 5′ untranslated regions (UTRs), 3′ UTRs, intronic regions, upstream and downstream regions, and intergenic regions. The protein-coding changes affected by the exonic SNPs were further classified into synonymous mutation, nonsynonymous mutation, stop-gain, and stop-loss. The latter three types of SNPs were attributed to deleterious mutations. For each strain, we detected the Gene Ontology (GO) terms overrepresented in the genes with the deleterious exonic BSA-SNPs using TBtools v1.075 [26] with p values < 5%. This software performs the GO term enrichment analysis based on hypergeometric distribution.

We investigated whether consensus SNPs would be observed in all three strains. The consensus SNPs were also categorized using ANNOVAR [25]. We only found two types of exonic SNPs, including synonymous SNPs (sSNPs) and nonsynonymous SNPs (nsSNPs). The nsSNPs were retained in the association study of IB traits.

2.4. Amplification, SNP Calling, and Genotyping

The specific primers for four BSA-nsSNPs were designed to amplify the loci and their flanking regions of 500 bp in 122 individuals (Supplementary Table S1). The PCR amplifications were performed in a volume of 50 μL containing 1 μL of DNA template, 2 μL of each primer (10 umol), 25 μL of easyTaq PCR SuperMix (TransGen Biotech, Beijing, China), and 20 μL of nuclease-free water. The amplification profile included 3 min at 94 °C, 40 cycles of 30 s at 94 °C, 30 s at the annealing temperature of 55 °C, 40 s at 72 °C, and a final extension of 10 min at 72 °C. The amplification products were purified with a commercial DNA fragment purification kit (Taihegene, Beiing, China) and then sequenced using the the Sanger method. Since the amplified region of each SNP was, at most, 1 kb, two ends of one round of Sanger sequencing covered the complete region.

Since the protein-coding genes in the common carp A and B subgenomes have high nucleotide similarities [15], to avoid misalignments and putative artificial SNPs, all amplified sequences were mapped to the genome assembly using blastn analysis with an e-value of 10^{-5}. If the best aligned region of each sequence was not the expected region, this sequence was discarded.

Before genotyping, we identified other neighboring non-BSA-SNPs around the BSA-SNPs. The retained amplified sequences were aligned to their corresponding reference genomic regions with novoSNP [27], with the reliable region ranging from 50 bp to 650 bp. In each individual, if one locus had up to two sequencing peaks and an F-score > 30, it was reserved. Those sites in one sample having at least three peaks, possibly resulting from the sequencing errors, were not included in the analysis. We identified the high-quality non-BSA-SNPs and heterozygote/homozygote following Zhang et al.'s protocol [28]. In brief, a homozygote had only one sequencing peak and a heterozygote had two sequencing peaks. The effects of non-BSA-SNPs on the coding sequences were also classified into synonymous substitution, nonsynonymous substitution, stop-gain, and stop-loss.

2.5. Examining the Genetic Diversities of the Examined SNPs

If one genotype of this locus was observed in fewer than four samples, this genotype was not included in the genetic diversity analysis. We calculated the genetic distances and population structures among three strains with all retained genotypes of the BSA-SNP and non-BSA-SNP loci. The genetic distances were estimated using principal component analysis (PCA) in Tassel 5.0 [29]. The first two eigenvectors were plotted. With the genotypes of each locus, the population structure among three strains was analyzed using Admixture 1.3.0 [30]. K was set from 2 to 6. The population structures and the genetic compositions of each individual were displayed with pophelper v2.3.1 [31]. As three strains were grouped together in the PCA analysis and had similar structures, we combined these strains into one population. If one SNP was successfully genotyped in at least 70 samples, we measured its genetic diversity with four indicators, including the observed heterozygosity (Ho), expected heterozygosity (He), polymorphism information content (PIC), and minor allele frequencies (MAF), using Genepop [32]. Mann-Whitney U tests were used to examine the genetic diversity distributions among three gene loci.

2.6. Identifying the SNPs Associated with the Numbers of IBs

Associations between the numbers of each type of IBs and the genotypes among all individuals were analyzed using Tassel 5 [29], with the general linear model (GLM) considering the PCA-matrix, which had the top five principal components. A test of 100,000 permutations was performed. The GLM model was widely used in the association study between the polymorphisms and the growth traits of common carp, including body weight and body length [33–36]. The percentage of phenotypic variation (PV) explained by each SNP was represented as Marker R^2, which was calculated with Tassel 5. We also performed the association study for each IB trait using analysis of variance (ANOVA). For the count of each type of IB, we grouped the individuals on the basis of their genotypes of one SNP. A pairwise comparison of the counts among the genotypes of this SNP with ANOVA was performed. If a significant difference in the counts between two compared groups was found with a p value < 0.05, this SNP was significantly related to the number of this IB type with ANOVA. For each type of IB, one SNP was considered significantly associated with the count if it had a p value < 0.05 in the GLM model and a p value < 0.05 in the ANOVA.

2.7. Joint Effects of Significant SNPs on the IB Numbers

Analyzing the joint effects of multiple SNPs on the phenotype is a promising approach to finding genetic effects beyond single-locus associations [37]. Hence, we performed the joint analysis of multiple SNPs to detect a larger effect on the numbers of IBs than that of the individual SNP. The analysis also helped to detect the genotype combination associated with fewer IBs. To estimate the joint effects of the significantly associated SNPs on the number of each type of IB, all SNP loci associated with this type of IB were selected, and we generated different combinations of genotypes from all of these SNPs. Only the genotype combinations observed in at least three individuals were retained. We compared the IB numbers among all genotype combinations with ANOVA. For each type of IB, the percentages of PV explained by the SNP combination were estimated with the function 'lm' [38] in the R package.

3. Results

3.1. Comparison of the IB Numbers in Different Strains

The numbers of total IBs in the HB, HH, and WQ strains fell within the range of 80–108 (median = 99), 79–101 (median = 95), and 76–104 (median = 95), respectively (Supplementary Table S2). PCA analysis based on the numbers of 14 types of IBs grouped these three strains together (Figure 1). The first two principal components explained 39.21% and 27.18% of the total genetic variance, respectively. The third component accounted for 11.94% of the variances.

In all strains, we did not observe the epicentral IBs. The en-IBs (median = 66) were the dominant type of IBs compared with the ep-IBs (median = 30.5). Among the 14 different types of IBs, the en-I and ep-I were the two main types of IBs (both median = 20), followed by the en-Y IBs (median = 16). We did not observe the ep-TB IBs in all samples and found only one sample with the en-TB IBs.

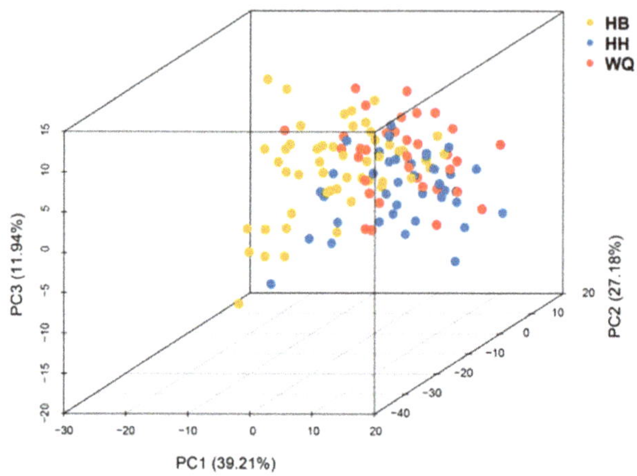

Figure 1. The PCA plot clustering three common carp strains with the numbers of 14 types of IBs.

Excluding the ep-TB and en-TB IBs, we clustered the other 12 types of IBs into two groups based on a matrix of pairwise number comparisons between each of the two forms of IBs (Figure 2). The first group consisted of the en-Y, en-I, and ep-I IBs, which were the dominant types of IBs. The en-I and ep-I IBs were significantly positively correlated, while they were significantly negatively correlated with the en-Y IBs. The second group included the other nine types of IBs, the numbers of which were lower than the first group. Most members in this group were significantly positively correlated with one another, except for the en-OEUB and ep-TEMF IBs. We also observed that the number of total IBs had a significantly positive correlation with the numbers of en-IBs and ep-IBs (Pearson's correlations = 0.822 and 0.852, $p < 0.05$, respectively, Supplementary Figure S3), suggesting that the number of total IBs was affected by both numbers of en-IBs and ep-IBs.

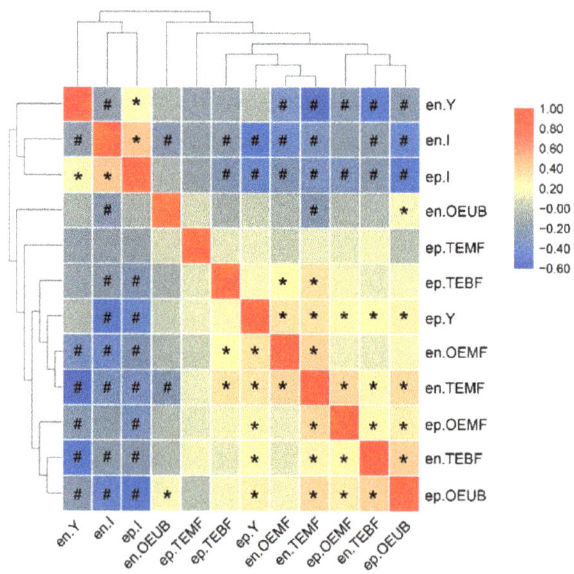

Figure 2. Heat map showing count correlations among 14 types of IBs. * two compared traits are significantly positively correlated. # two compared traits are significantly negatively correlated.

3.2. Identification of Segregant SNPs by BSA

In each strain, we sorted the samples with the fewest IBs and the most IBs into two extreme phenotypic groups, respectively. In the HB strain, the IB-fewer and IB-more groups had total IB number means of 87.3 ± 4.29 and 104.5 ± 1.99, respectively, with a ratio of 1.197 (Supplementary Figure S4 and Table S2). The IB mean ratios of two extreme groups in the other two strains were 1.16 (HH strain) and 1.17 (WQ strain), respectively. We sequenced six pools of the IB-fewer and IB-more fish in three strains and obtained about 60 Gb of sequence data with approximately 37-fold depth (ranging from 35.9–41.8×) per pool. Over 98.8% of these reads were aligned to the common carp genome (Table 1).

Table 1. Summary of genome sequencing and alignment.

Sample	Q20 (%)	Clean Read Pairs	Depth	Mapping Ratio (%)
HBF	96.83	205,678,671	37.24	99.20
HBM	96.77	211,458,150	38.28	98.88
WQF	96.90	206,924,782	37.46	99.29
WQM	96.70	198,379,639	35.92	99.20
HHF	97.02	231,095,221	41.84	99.31
HHM	97.01	228,265,661	41.33	99.27

HBF: the individuals with the fewest IBs in the 'HB' strain; HBM: the individuals with the most IBs in the 'HB' strain; HHF: the individuals with the fewest IBs in the 'HH' strain; HHM: the individuals with the most IBs in the 'HH' strain; WQF: the individuals with the fewest IBs in the 'WQ' strain; WQM: the individuals with the most IBs in the 'WQ' strain.

We obtained a basic set of 20,718,699 SNPs (1.23% of genome size, one SNP per 81.1 bp), including 12,531,928 SNPs in the HH strain, 10,835,374 SNPs in the HB strain, and 11,477,507 SNPs in the WQ strain (Supplementary Figure S5). In the basic set, the ratio of transition to transversion (Ts/Tv) was 2.2. The C/T transition accounted for 34.4% of all SNPs. A total of 87.5% of SNPs were located in the intergenic regions or intronic regions, and 3.7% were exonic (Supplementary Figure S6a). The percentage of sSNPs (61.4%) was higher than that of nsSNPs (38.1%) (Supplementary Figure S6b). A core set of 4,812,113 SNPs was shared among the three strains. In this core SNP set, the Ts/Tv ratio (2.25), C/T transition proportion (34.6%), genomic distribution of SNPs, and percentage of sSNPs were almost equivalent to those in the basic set (Supplementary Figure S7). The genomic distribution of SNPs and the percentages of different types of exonic SNPs in each strain also had similar distributions to those in the basic and core SNP sets (Supplementary Figures S8 and S9).

The BSA analysis identified 114,909 significant segregant SNPs in the HB strain, 118,240 SNPs in the HH strain, and 230,667 sites in the WQ strain, respectively (Supplementary Table S3). In each strain, the proportion of segregant SNPs accounted for 0.9–2.0% of all SNPs. The genomic and exonic distributions of the BSA-SNPs were subject to those of all SNPs in each strain (Supplementary Figures S10 and S11). The Ts/Tv ratio and C/T transition proportion of the BSA-SNPs were slightly higher than those of all SNPs (Supplementary Table S4). In each strain, 4.2–4.57% of the BSA-SNPs were exonic and the majority (59.66–59.79%) of these exonic BSA-SNPs were synonymous. The exonic BSA-SNPs were located in 6008 genes in the HB strain, 5761 genes in the HH strain, and 7073 genes in the WQ strain (Supplementary Figure S12). Most genes with the exonic BSA-SNPs were strain-specific. The Venn diagram showed that only 596 genes with the exonic BSA-SNPs occurred in three strains.

In the HB strain, 2307 genes with the deleterious exonic BSA-SNPs were enriched in 400 GO terms (Supplementary Figure S13 and Table S5). A total of 2192 genes in the HH strain were enriched in 330 GO terms, and in the WQ strain, 2676 genes with the satisfied exonic BSA-SNPs were enriched in 393 GO terms. In total, 22 GO terms were covered in three strains.

A deeper investigation found 207 BSA-SNPs shared by all three strains (Supplementary Figure S14), suggesting that most BSA-SNPs were strain-specific. They were distributed close to or in 318 genes (Supplementary Table S6). Most of them (193) were located in the intergenic or intronic regions (Table 2). Only eight SNPs, including four sSNPs and four nsSNPs, were located in the coding exons of eight genes. Four nsSNPs were encoded in four genes, which were in the chromosomes of A1, A16, A18, and B25, respectively. The first nsSNP in the A1 chromosome was found in the exon of *Mtap*, an enzyme converting the phosphorylation of S-methyl-5′-thioadenosine to adenine and 5-methylthioribose-1-phosphate [39]. The second nsSNP in the A16 chromosome belonged to the *Samd3* (sterile alpha motif domain-containing protein 3) gene [40]. The third nsSNP in the A18 chromosome was located in the *Ripor1* gene (also named FAM65A), Rho family-interacting cell polarization regulator 1. The final nsSNP identified in the B25 chromosome was detected in the *Neogenin* gene, a receptor for bone morphogenetic proteins [41].

Table 2. The genomic distributions of the BSA-SNPs shared by three strains.

Genomic Location		Number
downstream		3
exonic	nonsynonymous	4
	synonymous	4
intergenic		108
intronic		85
upstream		4

3.3. *Genetic Diversity of Three Sequenced BSA-SNP Regions*

Four exonic ns-BSA-SNPs in four genes observed in all three strains were genotyped for the association studies. Due to the assay capacity and the limitations of primer design for the genotyping analysis, the ns-BSA-SNP in *Ripor1* was not amplified. The other three ns-BSA-SNPs were successfully genotyped. Among the 122 samples, the A1:3023753 locus (*Mtap*) was successfully sequenced in 120 samples with the A16:2710463 site (*Samd3*) in 93 samples and the B25:3800828 locus (*Neogenin*) in 101 samples. Except for the ns-BSA-SNP in *Samd3*, the polymorphisms of two ns-BSA-SNP loci in *Mtap* and *Neogenin* were successfully validated. Besides the ns-BSA-SNPs, we also called another 47 non-BSA SNPs neighboring these loci (Supplementary Table S7). In the reliable amplified region of *Mtap* with a length of 516 bp, we detected 17 SNPs with a polymorphism rate of 3.29%. In the A16 region with a size of 564 bp, we called only five SNPs with a polymorphism rate of 0.89%. Another 27 SNPs were identified in *Neogenin* with a higher polymorphism rate of 3.95%.

A total of 25 SNP loci had only two genotypes with 23 loci and 1 locus having three and four genotypes, respectively (Supplementary Table S8). Among the 49 SNPs, 25 were located in the introns and 24 were in the exons. We identified five ns-SNPs in the A1 (three SNPs) and B25 loci (two SNPs), including two BSA-SNPs and three non-BSA-SNPs. The Ho distributions in these three gene loci were not significantly different (p value = 0.809, 0.109, and 0.120, respectively). Likewise, the He distributions in these three gene loci were not significantly different (p value = 0.181, 0.283, and 0.614, respectively). PIC analysis showed that 11 SNP loci (6 and 5 in the A1 and B25 locus, respectively) displayed low polymorphism levels (PIC < 0.1) and that the remaining loci had a polymorphism rate over 0.1. The PIC values in these three gene loci were not significantly distributed (Mann–Whitney U test, p value = 0.197, 0.614, and 0.283, respectively). No significant MAF distribution differences among these three regions were observed (p value = 0.101, 0.189, and 0.579, respectively).

PCA based on all A1 genotypes clustered three common carp strains together (Supplementary Figure S15a). The first two principal components explained 25.39% and 17.82% of the total genetic variances, respectively. Similar phenomena of clustering together were observed with the genotypes of the A16 and B25 loci (Supplementary Figure S15b,c). These data proved that these three strains had a similar genetic background. In the admixture plots with different K values (K from two to six), three strains had similar genetic components of these three loci (Supplementary Figure S16). There were no strain-specific SNPs in these three loci, as evidenced by both PCA and population structure analyses. Taken together with the similar IB number distribution among the three strains, we combined these three strains into one population in the association analyses.

3.4. SNPs Associated with the IB Numbers

Among the 49 SNPs, 21 were identified to be significantly associated with the numbers of the total IBs, en-IBs, and ten different types of IBs using GLM and ANOVA (Table 3 and Supplementary Figure S17). The remaining types of IBs, including ep-IBs, en-TB, ep-OEUB, ep-TEMF, and ep-TB, were not found to have associated SNPs. The mean values of the latter four types of IBs were 0.01 ± 0.09, 5.60 ± 2.47, 0.25 ± 0.58, and 0 ± 0, respectively. Few IBs of these types were observed among individuals, possibly hindering the identification of the associated SNPs. Two SNPs in *Samd3* were associated with the numbers of three types of IBs (Supplementary Figure S18a–c). However, 8 SNPs in *Mtap* and 11 in *Neogenin* were the major SNPs related to IB numbers. Eight SNPs were identified to be associated with the numbers of at least two types of IBs. Except for the en-IBs, ep-I, and ep-OEMF with only one SNP, the total IBs and other eight types of IBs had at least two associated SNPs. Intriguingly, seven types of IBs had SNPs from at least two genes.

Among 17 SNPs in *Mtap*, 8 were significantly associated with the numbers of seven types of IBs, including total IBs, en-I, en-OEUB, en-TEBF, en-TEMF, en-Y, and ep-I (Supplementary Figure S18d–n). Three SNPs were identified to be associated with the numbers of two types of IBs. For the total IBs, en-OEUB, en-TEBF, and en-TEMF, the heterozygotes of the associated SNP loci were significantly associated with fewer numbers than the homozygotes. However, the trends were reversed for the en-I, en-Y, and ep-I. Most associated SNPs in this region were synonymous and only one nsSNP of *Mtap* was associated with the number of en-TEBF IBs.

Among 27 SNPs in *Neogenin*, 12 were significantly associated with the numbers of 10 types of IBs, excluding en-OEUB and ep-I (Supplementary Figure S18o–af). Two SNPs were identified to be associated with the numbers of two types of IBs and two SNPs with three types of IBs. For the total IBs, en-IBs, en-TEMF, and en-Y, the heterozygotes of the associated SNP loci were significantly associated with fewer numbers than the homozygotes. However, the trend was reversed for the en-OEMF, en-Y, ep-OEMF, and ep-TEMF. Interestingly, we observed the coexistence of dominant heterozygotes and homozygotes of *Neogenin* associated with the numbers of en-I and en-TEBF. Most associated SNPs in *Neogenin* were synonymous and only one ns-SNP of B25.3800710 was significantly related to the numbers of en-I, en-TEBF, and en-TEMF IBs.

Two SNPs in *Mtap* and *Neogenin* were found to be significantly associated with the total IB number using both GLM and ANOVA (Table 3). The homozygotes of these two loci were associated with more IBs than the heterozygotes. These two loci could explain 4.72% and 5.9% of the total IB number variation, respectively. The SNP of B25.3800532 was also significantly associated with the number of en-IBs. Likewise, the homozygote of this locus corresponded to more IBs than the heterozygote.

Table 3. Association analysis of IB numbers using GLM and ANOVA.

Trait	SNP	GLM		ANOVA		
		p Value	Marker R^2 (%) #	MM	Mm	p Value
IB	A1.3023796	0.0487	4.72	96.0 ± 5.35	90.2 ± 8.79	0.0236
	B25.3800532	0.0224	5.90	96.1 ± 5.41	90.1 ± 5.87	0.0059
en-IB	B25.3800532	0.0228	5.57	66.1 ± 3.05	62.6 ± 2.64	0.0039
en-I	A1.3023490	0.0167	6.76	20.5 ± 5.45	27.2 ± 12.80	0.0022
	B25.3800904	0.0369	5.49	21.4 ± 6.22	15.9 ± 4.26	0.0213
	A1.3023694	0.0382	5.47	20.6 ± 5.90	22.2 ± 6.95	0.0208
	B25.3800710 *	0.0498	5.04	20.6 ± 5.46	27.3 ± 12.58	0.0055

Table 3. *Cont.*

Trait	SNP	GLM		ANOVA		
		p Value	Marker R^2 (%) #	MM	Mm	*p* Value
en-OEMF	B25.3800526	0.0019	7.51	3.1 ± 2.94	6.00 ± 3.44	0.0014
	B25.3800904	0.00001	16.39	3.1 ± 2.80	8.6 ± 2.07	0.00002
	B25.3800999	0.0059	8.02	2.9 ± 2.40	4.5 ± 3.72	0.0342
	A16.2710728	0.0297	5.65	4.6 ± 3.50	3.0 ± 2.85	0.0418
en-OEUB	A1.3023466	0.0221	6.26	11.4 ± 4.26	8.3 ± 2.80	0.0126
	A16.2710488	0.0450	5.13	9.9 ± 3.15	11.9 ± 4.57	0.0301
en-TEBF	A1.3023753 *	0.0333	4.80	7.9 ± 3.89	7.7 ± 3.87	0.0005
	A1.3023814	0.0414	2.96	9.7 ± 4.57	6.9 ± 3.78	0.0014
	B25.3800710 *	0.0425	4.47	8.7 ± 4.39	5.1 ± 5.79	0.0433
	B25.3800929	0.0206	3.79	8.7 ± 4.50	13.0 ± 4.12	0.0401
	A1.3023730	0.0358	4.66	9.0 ± 4.35	5.3 ± 3.13	0.0469
	B25.3801033	0.0268	5.06	8.5 ± 4.49	8.7 ± 3.95	0.0148
en-TEMF	A1.3023814	0.0138	4.64	6.9 ± 6.57	3.2 ± 3.68	0.0013
	A16.2710488	0.0183	6.09	4.6 ± 5.79	7.7 ± 6.37	0.0198
	B25.3800685	0.0276	5.48	5.8 ± 5.99	3.0 ± 4.07	0.0488
	B25.3800710 *	0.0349	5.14	5.6 ± 5.85	1.00 ± 1.91	0.0414
en-Y	A1.3023447	0.00001	15.84	14.8 ± 6.67	19.2 ± 7.38	0.0013
	A1.3023694	0.0155	6.02	14.7 ± 6.92	16.4 ± 7.49	0.0395
	B25.3801016	0.0330	6.32	15.5 ± 6.63	19.8 ± 7.42	0.0137
ep-I	A1.3023730	0.0302	5.22	19.2 ± 3.81	22.0 ± 2.14	0.0177
ep-OEMF	B25.3801033	0.0165	4.65	0.4 ± 0.72	0.9 ± 1.19	0.0357
ep-TEMF	B25.3800904	0.0149	7.02	0.2 ± 0.53	0.9 ± 1.07	0.0047
	B25.3800911	0.0380	3.64	0.2 ± 0.40	0.5 ± 0.76	0.0118
ep-Y	B25.3801016	0.0399	4.87	3.6 ± 1.62	2.1 ± 1.37	0.0073
	B25.3801028	0.0306	3.55	3.7 ± 1.47	1.9 ± 1.37	0.0019

* This SNP was nonsynonymous. # percentage of phenotypic variation explained by markers in GLM method was represented as Marker R^2; M: major allele; m: minor allele. The data are presented as the mean ± SE representing the IB numbers in one genotype group. The SNPs marked with the asterisks are nonsynonymous.

3.5. Joint Effects of Significant SNPs on the IB Numbers

The numbers of nine types of IBs had at least two associated SNPs. The SNPs in *Mtap* and *Neogenin* were significantly associated with the numbers of four types of IBs (total IBs, en-I, en-TEBF, and en-Y). The en-OEMF number had associated SNPs in *Samd3* and *Neogenin* and the en-OEUB number was associated with the SNPs in *Mtap* and *Samd3*. We also found that the SNPs in these three genes were associated with the number of en-TEMF. These data suggest that these seven types of IBs might be regulated by these genes. For nine types of IBs with at least two associated SNPs, we were interested in testing two simple joint effects that failed to be identified by single-marker analysis: (a) whether a genotype combination corresponding to the lowest IB number existed, and (b) whether the explained percentage of PV could be improved by multiple SNPs compared with a single SNP.

For the number of total IBs, we observed only three combinations in two SNP loci in the population (Supplementary Table S9 and Figure S19a), including GG/TT (both homozygotes in two SNP loci, named as hap1), GT/TT (one homozygote in A1.3023796 and one heterozygote in B25.3800532, hap2), and GG/TG (one heterozygote in A1.3023796 and one homozygote in B25.3800532, hap3). We did not find the combination of both heterozygotes in two SNP loci. The first combination was observed in 101 individuals, and was dominant in the population. The explained percentage of PV was increased from 4.72% with one SNP to 10% with two loci (Table 4).

In the other eight types of IBs, we found higher number deviations among different genotype combinations than those in the single-marker analysis (Supplementary Figure S19b–i). For the en-I number, we observed seven genotype combinations from four SNP loci. In the single-SNP analysis, the mean IB number ratio between the homozygote and the heterozygote of each single SNP ranged from 0.742 to 0.927 and the lowest mean IB number was 15.9 ± 4.26, observed in the homozygotes of B25.3800904. As shown in Table S9, these combinations had different numbers of samples where most individuals had a mean en-I number of 21.61 ± 4 (observed in the hap3 combinations). The lowest mean en-I IB number (14 ± 5.2) and the highest number (25 ± 2) corresponded to two extreme genotype combinations (hap7 and hap1, respectively), with a ratio of 0.56. The explained percentage of PV was improved from 5.04% with one SNP to 15.93% with multiple loci. For the remaining seven types of IBs, we also observed the joint effects of all associated SNPs, demonstrated by both the decreased mean IB number and the improved explained percentage of PV.

Table 4. The explained percentage of PV of each type of IBs by genotype combination.

Trait	SNP	Each SNP Marker R^2 (%)	Marker R^2 of Genotype Combination (%)
IB	A1.3023796	4.72	10
	B25.3800532	5.90	
en-I	A1.3023490	6.76	15.9
	B25.3800904	5.49	
	A1.3023694	5.47	
	B25.3800710 *	5.04	
en-OEMF	B25.3800526	7.51	34.3
	B25.3800904	16.39	
	B25.3800999	8.02	
	A16.2710728	5.65	
en-OEUB	A1.3023466	6.26	7.44
	A16.2710488	5.13	
	A1.3023753 *	4.80	
en-TEBF	A1.3023814	2.96	42.07
	B25.3800710 *	4.47	
	B25.3800929	3.79	
	A1.3023730	4.66	
	B25.3801033	5.06	
en-TEMF	A1.3023814	4.64	17.57
	A16.2710488	6.09	
	B25.3800685	5.48	
	B25.3800710 *	5.14	
en-Y	A1.3023447	15.84	9.99
	A1.3023694	6.02	
	B25.3801016	6.32	
ep-TEMF	B25.3800904	7.02	7.95
	B25.3800911	3.64	
ep-Y	B25.3801016	4.87	18.21
	B25.3801028	3.55	

* This SNP was nonsynonymous.

4. Discussions

IBs widely exist in cyprinids and have become a critical factor affecting cyprinid consumption and processing. In total, 12 different types of IBs were observed in common carp. Few en-TB and ep-TB IBs were also observed by Li et al. [4]. We clustered these 12 types of IBs into two groups. In general, the numbers of the first group were negatively correlated with those of the second group, suggesting that the former group possibly had an antagonistic effect on the second group. On the contrary, most intra-group members were significantly positively correlated with one another. These data all suggested the complexity of IB morphology in common carp.

One notable avenue to decrease the IB number is to knock out IB-related genes by using Cripsr/Cas9 technology, generating IB-reduced fish. However, gene-edited fish are still not allowed to be cultivated or marketed. Another possible strategy to generate the IB-few fish is genome selection breeding. Current studies on common carp IB development have narrowed the IB-related genomic regions down to the gene level or the QTL level [17,18].

Nevertheless, SNPs associated with the numbers of IBs have not been identified yet. Previous studies mainly focused on comparing the number of total IBs or limited types of IBs. Herein, we executed a comprehensive association with the numbers of total IBs, en-IBs, ep-IBs, and 14 other types of IBs. We observed many IB-related SNPs in all three strains. These SNPs were associated with the numbers of different types of IBs, suggesting that they might be involved in the development of different types of IBs separately. The SNP combinations had joint effects on the numbers of multiple types of IBs, supported by both the decreased mean IB number and the improved explained percentage of PV. For instance, the smallest mean numbers of en-OEMF and en-TEMF were decreased to 0.833 ± 1.33 and 0 ± 0, respectively, if considering multiple associated SNPs together, suggesting that it is possible to select the individuals having the fewest en-OEMF and en-TEMF with the SNP sets.

Many reported IB-development-related genes, including SAM and SH3 domain containing protein 1 (*Sash1*) [5], fibroblast growth factor receptors *Fgfr2* and *Fgfr3* [42], SRY-box transcription factor 9 (*Sox9*) [43], and Osteopontin (*Opn*) [44], were identified in our study (Supplementary Table S6). We identified the BSA-SNPs in the genes of the bone morphogenetic proteins (BMP) signaling pathway [45] in different strains (Supplementary Table S10). A total of 15, 10, and 13 genes involved in the BMP signaling pathway had deleterious exonic BSA-SNPs in the HB, HH, and WQ strains, respectively (Supplementary Figure S20). We performed consensus comparisons in three different levels, including SNPs, genes, and GO processes. In three strains, only eight exonic BSA-SNPs located in eight genes were covered, whereas 596 genes with the deleterious exonic SNPs were shared. These data suggested that each strain had different exonic polymorphisms in the same gene related to regulating IB development. These 596 genes might be common regulatory genes of IB generation in these three strains. We also observed that 22 GO biological processes were found in all three strains, although not all genes in these processes had deleterious mutations. One enriched category that was common in the three strains included the assembly, movement, and organization of cilium. These processes have been reported to regulate osteoblast differentiation, polarity, and alignment to reduce bone formation [46]. The cilia prompt chondrocyte differentiation and production of a cartilage matrix. Cilia are required for increased bone formation in response to mechanical forces. The primary cilium is also associated with many signaling pathways, including primary cilium-related Ca^{2+} signaling, BMP/Smad1/5/8 signaling, and Wnt signaling [47].

Our analysis provided hints that three genes had consensus nsSNPs in three strains to regulate IB development. The homologs of these three genes participate in bone development. The dysfunction of human *Mtap* homolog causes diaphyseal medullary stenosis with malignant fibrous histiocytoma [48]. Another nsSNP was located in the *Samd3* gene. Although there was no direct evidence to prove that *Samd3* was related to bone development, a deleterious mutation in *Samd9*, a homolog of *Samd3*, causes familial tumoral calcinosis [49], representing painful ulcerative lesions and severe skin and bone infections. The protein of *Neogenin* was found to bind directly with BMP2, BMP4, BMP6, and BMP7, and to negatively regulate the functions of BMPs [41]. This gene had more associated SNPs and IB types than *Mtap* and *Samd3*, suggesting that this gene might make a greater contribution to IB development. Studying the functions of these three genes would contribute to further elucidation of IB development mechanisms in common carp and other cyprininae fish.

Different from diploid fish, the allotetraploid common carp underwent a species-specific whole-genome duplication [16]. Whether both homeologues generated by whole-genome duplication participate in the IB development requires further study. In polyploid plant, Cheng et al. detected strong subgenome parallel selection linked to the domestication of the tuberous morphotypes [50]. However, our previous expression analysis across conditions in common carp revealed differential expression balance in two subgenomes, i.e., the majority of homoeologous pairs had only one differentially expressed homoeologue [50]. Therefore, it is hypothesized that two common carp subgenomes might be under unparalleled selection on the economic traits. In this study, the homologues of these gene regions were not found to have SNPs associated with IB number. This result provided evidence that two subgenomes performed flexible and unparalleled processes related to IB development and morphology.

The association analysis of the common carp IB development could be improved in the future. Firstly, two SNPs associated with the number of total IBs explained at most 10% of PV, suggesting that other SNPs that have not yet been identified could possibly exist. Secondly, the BSA analysis detected SNPs significantly associated with the numbers of nine types of IBs. The other five IB traits, including the numbers of ep-IBs, en-TB, ep-OEUB, ep-TEMF, and ep-TB, had no associated SNPs. Thirdly, whether polymorphisms exist in the promoters, miRNA binding sites, and other functional elements of these four genes will be studied in the future.

5. Conclusions

We classified and measured the numbers of total IBs, ep-IBs, en-IBs, and 14 other types of IBs in three common carp strains. Using BSA analysis, we detected polymorphisms of different frequencies between two pools in each strain and identified the consensus SNPs in three strains. Fine genotyping of three polymorphic regions in *Mtap*, *Samd3*, and *Neogenin* showed similar genetic backgrounds in these three regions across the three strains. Association studies on the polymorphisms of these three genes and the numbers of different IB types identified three SNPs significantly related to IB number. The joint effect analysis indicated that these SNP combinations improved the explained percentage of PV and suggested the optimal combinations related to the reduced IB numbers. These SNPs and their optimal genotype combinations will be potential markers applied in future marker-assisted selective breeding of common carp to decrease IB numbers.

Supplementary Materials: The following supporting information can be downloaded at: https://www.mdpi.com/article/10.3390/biology11030477/s1, Figure S1: the photos of the studied common carp strains; Figure S2: illustration of the seven types of IBs observed in common carp; Figure S3: the correlation matrix among the numbers of IB, en-IB, and ep-IB; Figure S4: illustration of the IBs in the six sequenced groups of common carp types; Figure S5: the Venn diagram of identified SNPs among three strains; Figure S6: the genomic and exonic distribution of the basic SNP set; Figure S7: the genomic and exonic distribution of the core SNP set; Figure S8: the function effects of the exonic SNPs in each strain; Figure S9: the function effects of the exonic SNPs in each strain; Figure S10: the genomic distributions of the identified BSA-SNPs in three strains; Figure S11: the function effects of the exonic BSA-SNPs in three strains; Figure S12: the Venn diagram of the shared BSA-genes in three strains; Figure S13: the Venn diagram of the shared GO terms enriched by the genes with the deleterious exonic BSA-SNPs in three strains; Figure S14: the Venn diagram of the shared BSA-SNPs in three strains; Figure S15: PCA plots detected with the SNPs in three regions, respectively; Figure S16: population structures detected with the SNPs in three regions, respectively; Figure S17: the Manhattan plots showing the SNPs and their associated IB numbers; Figure S18: the homozygote and heterozygote of each SNP and corresponding IB numbers; Figure S19: the genotype combinations and corresponding IB numbers; Figure S20: the BMP-related genes identified with BSA in each strain; Table S1: the primers used for amplification of the BSA-SNP regions; Table S2: the IB numbers in 122 common carp individuals from three strains; Table S3: the identified BSA-SNPs in each strain; Table S4: slight higher Ts/Tv ratio and C/T transition in the BSA-SNPs than all SNPs in each strain; Table S5: the enriched GO terms by the genes with the deleterious exonic BSA-SNPs in three strains; Table S6: the functional annotations of 318 genes neighboring the shared 207 BSA-SNPs in three strains; Table S7: the genotypes of three loci in 122 common carp individuals from three strains; Table S8: genetic diversities of three sequenced regions; Table S9: the joint effects of the associated SNPs; Table S10: the BMP-related genes having the deleterious exonic BSA-SNPs in each strain.

Author Contributions: Conceptualization, J.-T.L.; methodology, M.-S.C., R.Z., Q.W., Y.Z. and X.-Q.S.; software, M.-S.C., Q.W., Y.Z., Q.-S.L. and M.-D.H.Y.; validation, J.-T.L. and Y.Z.; resources, M.-S.C., X.-Q.S. and J.-T.L.; writing, M.-S.C. and J.-T.L.; funding acquisition, J.-T.L. All authors have read and agreed to the published version of the manuscript.

Funding: We acknowledge the grant supports from the National Key Research and Development Program of China (2021YFD1200804 to Y.Z.), Special Scientific Research Funds for Central Non-profit Institutes, Chinese Academy of Fishery Sciences (2020XT0103 and 2020TD24 to J.T.L.), and National Freshwater Genetic Resource Centre (FGRC:18537 to Y.Z.).

Institutional Review Board Statement: The study was conducted according to the guidelines of the Declaration of China, and approved by the Animal Care and Use Committee of the Chinese Academy of Fishery Sciences (protocol code ACUC-CAFS-20200101 and date of approval is January 2020).

Data Availability Statement: The genome resequencing data of six common carp stocks were deposited in the SRA database (project number PRJNA751998), respectively. The other analysis data presented in this study are available in Supplementary Materials.

Conflicts of Interest: The authors declare no conflict of interest.

References

1. Yang, K.; Jiang, W.; Wang, X.; Zhang, Y.; Pan, X.; Yang, J. Evolution of the intermuscular bones in the Cyprinidae (Pisces) from a phylogenetic perspective. *Ecol. Evol.* **2019**, *9*, 8555–8566. [CrossRef] [PubMed]
2. Li, B.; Zhang, Y.W.; Liu, X.; Ma, L.; Yang, J.X. Molecular mechanisms of intermuscular bone development in fish: A review. *Zool. Res.* **2021**, *42*, 362–376. [CrossRef] [PubMed]
3. Nie, C.-H.; Hilsdorf, A.W.S.; Wan, S.-M.; Gao, Z.-X. Understanding the development of intermuscular bones in teleost: Status and future directions for aquaculture. *Rev. Aquac.* **2020**, *12*, 759–772. [CrossRef]
4. Li, L.; Zhong, Z.; Zeng, M.; Liu, S.; Zhou, Y.; Xiao, J.; Wang, J.; Liu, Y. Comparative analysis of intermuscular bones in fish of different ploidies. *Sci. China Life Sci.* **2013**, *56*, 341–350. [CrossRef]
5. Wan, S.-M.; Xiong, X.-M.; Tomljanović, T.; Chen, Y.-L.; Liu, H.; Treer, T.; Gao, Z.-X. Identification and mapping of SNPs associated with number of intermuscular bone in blunt snout bream. *Aquaculture* **2019**, *507*, 75–82. [CrossRef]
6. Wan, S.M.; Yi, S.K.; Zhong, J.; Nie, C.H.; Guan, N.N.; Chen, B.X.; Gao, Z.X. Identification of MicroRNA for Intermuscular Bone Development in Blunt Snout Bream (*Megalobrama amblycephala*). *Int. J. Mol. Sci.* **2015**, *16*, 10686–10703. [CrossRef]
7. Wan, S.M.; Yi, S.K.; Zhong, J.; Nie, C.H.; Guan, N.N.; Zhang, W.Z.; Gao, Z.X. Dynamic mRNA and miRNA expression analysis in response to intermuscular bone development of blunt snout bream (Megalobrama amblycephala). *Sci. Rep.* **2016**, *6*, 31050. [CrossRef]
8. Sun, X.; Zhang, R.; Liu, M.; Chen, H.; Chen, L.; Luo, F.; Zhang, D.; Huang, J.; Li, F.; Ni, Z.; et al. Rmrp Mutation Disrupts Chondrogenesis and Bone Ossification in Zebrafish Model of Cartilage-Hair Hypoplasia via Enhanced Wnt/β-Catenin Signaling. *J. Bone Miner. Res.* **2019**, *34*, 2101–2116. [CrossRef]
9. Nunes, J.R.S.; Pértille, F.; Andrade, S.C.S.; Perazza, C.A.; Villela, P.M.S.; Almeida-Val, V.M.F.; Gao, Z.X.; Coutinho, L.L.; Hilsdorf, A.W.S. Genome-wide association study reveals genes associated with the absence of intermuscular bones in tambaqui (*Colossoma macropomum*). *Anim. Genet.* **2020**, *51*, 899–909. [CrossRef]
10. Yang, G.; Qin, Z.; Kou, H.; Liang, R.; Zhao, L.; Jiang, S.; Lin, L.; Zhang, K. A Comparative Genomic and Transcriptional Survey Providing Novel Insights into Bone Morphogenetic Protein 2 (bmp2) in Fishes. *Int. J. Mol. Sci.* **2019**, *20*, 6137. [CrossRef]
11. Zhong, Z.; Niu, P.; Wang, M.; Huang, G.; Xu, S.; Sun, Y.; Xu, X.; Hou, Y.; Sun, X.; Yan, Y.; et al. Targeted disruption of sp7 and myostatin with CRISPR-Cas9 results in severe bone defects and more muscular cells in common carp. *Sci. Rep.* **2016**, *6*, 22953. [CrossRef]
12. Nie, C.; Wan, S.; Chen, Y.; Zhu, D.; Wang, X.; Dong, X.; Gao, Z.-X. Loss of scleraxis leads to distinct reduction of mineralized intermuscular bone in zebrafish. *Aquac. Fish.* **2021**, *6*, 169–177. [CrossRef]
13. Kague, E.; Hughes, S.M.; Lawrence, E.A.; Cross, S.; Martin-Silverstone, E.; Hammond, C.L.; Hinits, Y. Scleraxis genes are required for normal musculoskeletal development and for rib growth and mineralization in zebrafish. *FASEB J.* **2019**, *33*, 9116–9130. [CrossRef]
14. Xiong, X.-M.; Robinson, N.A.; Zhou, J.-J.; Chen, Y.-L.; Wang, W.; Wang, X.-B.; Gao, Z.-X. Genetic parameter estimates for intermuscular bone in blunt snout bream (*Megalobrama amblycephala*) based on a microsatellite-based pedigree. *Aquaculture* **2019**, *502*, 371–377. [CrossRef]
15. Li, J.-T.; Wang, Q.; Huang Yang, M.-D.; Li, Q.-S.; Cui, M.-S.; Dong, Z.-J.; Wang, H.-W.; Yu, J.-H.; Zhao, Y.-J.; Yang, C.-R.; et al. Parallel subgenome structure and divergent expression evolution of allo-tetraploid common carp and goldfish. *Nat. Genet.* **2021**, *53*, 1493–1503. [CrossRef]
16. Yang, M.H.; Wang, Q.; Zhao, R.; Li, Q.S.; Cui, M.S.; Zhang, Y.; Li, J.T. *Cyprinus carpio* (common carp). *Trends Genet.* **2022**, *38*, 305–306. [CrossRef]
17. Tang, G.; Lv, W.; Sun, Z.; Cao, D.; Zheng, X.; Tong, G.; Wang, H.; Zhang, X.; Kuang, Y. Heritability and quantitative trait locus analyses of intermuscular bones in mirror carp (*Cyprinus carpio*). *Aquaculture* **2020**, *515*, 734601. [CrossRef]
18. Peng, J.; Zeng, D.; He, P.; Wei, P.; Hui, W.; Wu, T.; Zhuo, X.; Lin, Y. mRNA and microRNA transcriptomics analyses in intermuscular bones of two carp species, rice flower carp (*Cyprinus carpio* var. *Quanzhounensis*) and Jian carp (*Cyprinus carpio* var. *Jian*). *Comp. Biochem. Physiol. Part D Genom. Proteom.* **2019**, *30*, 71–80.

19. Xu, P.; Zhang, X.; Wang, X.; Li, J.; Liu, G.; Kuang, Y.; Xu, J.; Zheng, X.; Ren, L.; Wang, G.; et al. Genome sequence and genetic diversity of the common carp, Cyprinus carpio. *Nat. Genet.* **2014**, *46*, 1212–1219. [CrossRef]
20. Lv, Y.P.; Bao, B.L.; Jiang, Y.; Yang, L.L.; Li, J.L. Comparative analysis of intermuscular bones in lower teleosts. *J. Fish. China* **2007**, *31*, 661–668.
21. Bolger, A.M.; Lohse, M.; Usadel, B. Trimmomatic: A flexible trimmer for Illumina sequence data. *Bioinformatics* **2014**, *30*, 2114–2120. [CrossRef]
22. Li, H.; Durbin, R. Fast and accurate short read alignment with Burrows-Wheeler transform. *Bioinformatics* **2009**, *25*, 1754–1760. [CrossRef]
23. McKenna, A.; Hanna, M.; Banks, E.; Sivachenko, A.; Cibulskis, K.; Kernytsky, A.; Garimella, K.; Altshuler, D.; Gabriel, S.; Daly, M.; et al. The Genome Analysis Toolkit: A MapReduce framework for analyzing next-generation DNA sequencing data. *Genome Res.* **2010**, *20*, 1297–1303. [CrossRef]
24. Tribhuvan, K.U.; Kumar, K.; Sevanthi, A.M.; Gaikwad, K. MutMap: A versatile tool for identification of mutant loci and mapping of genes. *Indian J. Plant Physiol.* **2018**, *23*, 612–621. [CrossRef]
25. Wang, K.; Li, M.; Hakonarson, H. ANNOVAR: Functional annotation of genetic variants from high-throughput sequencing data. *Nucleic Acids Res.* **2010**, *38*, e164. [CrossRef]
26. Chen, C.; Chen, H.; Zhang, Y.; Thomas, H.R.; Frank, M.H.; He, Y.; Xia, R. TBtools: An Integrative Toolkit Developed for Interactive Analyses of Big Biological Data. *Mol. Plant* **2020**, *13*, 1194–1202. [CrossRef]
27. Weckx, S.; Del-Favero, J.; Rademakers, R.; Claes, L.; Cruts, M.; De Jonghe, P.; Van Broeckhoven, C.; De Rijk, P. novoSNP, a novel computational tool for sequence variation discovery. *Genome Res.* **2005**, *15*, 436–442. [CrossRef]
28. Zhang, Y.; Sun, X.Q.; Ye, Y.Q.; Wang, Q.; Li, Q.S.; Zhao, R.; Wang, H.W.; Li, J.T. Association between the Polymorphisms of fads2a and fads2b and Poly-Unsaturated Fatty Acids in Common Carp (*Cyprinus carpio*). *Animals* **2021**, *11*, 1780. [CrossRef]
29. Bradbury, P.J.; Zhang, Z.; Kroon, D.E.; Casstevens, T.M.; Ramdoss, Y.; Buckler, E.S. TASSEL: Software for association mapping of complex traits in diverse samples. *Bioinformatics* **2007**, *23*, 2633–2635. [CrossRef]
30. Alexander, D.H.; Novembre, J.; Lange, K. Fast model-based estimation of ancestry in unrelated individuals. *Genome Res.* **2009**, *19*, 1655–1664. [CrossRef]
31. Francis, R.M. pophelper: An R package and web app to analyse and visualize population structure. *Mol. Ecol. Resour.* **2017**, *17*, 27–32. [CrossRef]
32. Raymond, M.; Rousset, F. GENEPOP (Version 1.2): Population Genetics Software for Exact Tests and Ecumenicism. *J. Hered.* **1995**, *86*, 248–249. [CrossRef]
33. Feng, X.; Yu, X.; Tong, J. Novel Single Nucleotide Polymorphisms of the Insulin-Like Growth Factor-I Gene and Their Associations with Growth Traits in Common Carp (*Cyprinus carpio* L.). *Int. J. Mol. Sci.* **2014**, *15*, 22471–22482. [CrossRef] [PubMed]
34. Tang, Y.; Li, H.; Li, J.; Yu, F.; Yu, J. Association study between single nucleotide polymorphisms in leptin and growth traits in Cyprinus carpio var. Jian. *Genet. Mol. Res. GMR* **2016**, *15*, gmr.15037635. [CrossRef]
35. Wang, X.; Yu, X.; Tong, J. Molecular Characterization and Growth Association of Two Apolipoprotein A-Ib Genes in Common Carp (*Cyprinus carpio*). *Int. J. Mol. Sci.* **2016**, *17*, 1569. [CrossRef] [PubMed]
36. Sun, Y.; Yu, X.; Tong, J. Polymorphisms in Myostatin Gene and associations with growth traits in the common carp (*Cyprinus carpio* L.). *Int. J. Mol. Sci.* **2012**, *13*, 14956–14961. [CrossRef] [PubMed]
37. Knüppel, S.; Meidtner, K.; Arregui, M.; Holzhütter, H.G.; Boeing, H. Joint effect of unlinked genotypes: Application to type 2 diabetes in the EPIC-Potsdam case-cohort study. *Ann. Hum. Genet.* **2015**, *79*, 253–263. [CrossRef] [PubMed]
38. Wilkinson, G.N.; Rogers, C.E. Symbolic Description of Factorial Models for Analysis of Variance. *J. R. Stat. Soc. Ser. C* **1973**, *22*, 392–399. [CrossRef]
39. Della Ragione, F.; Takabayashi, K.; Mastropietro, S.; Mercurio, C.; Oliva, A.; Russo, G.L.; Della Pietra, V.; Borriello, A.; Nobori, T.; Carson, D.A.; et al. Purification and characterization of recombinant human 5′-methylthioadenosine phosphorylase: Definite identification of coding cDNA. *Biochem. Biophys. Res. Commun.* **1996**, *223*, 514–519. [CrossRef]
40. Peters, A.E.; Knöpper, K.; Grafen, A.; Kastenmüller, W. A multifunctional mouse model to study the role of Samd3. *Eur. J. Immunol.* **2021**, *52*, 328–337. [CrossRef]
41. Hagihara, M.; Endo, M.; Hata, K.; Higuchi, C.; Takaoka, K.; Yoshikawa, H.; Yamashita, T. Neogenin, a Receptor for Bone Morphogenetic Proteins. *J. Biol. Chem.* **2011**, *286*, 5157–5165. [CrossRef]
42. Havens, B.A.; Velonis, D.; Kronenberg, M.S.; Lichtler, A.C.; Oliver, B.; Mina, M. Roles of FGFR3 during morphogenesis of Meckel's cartilage and mandibular bones. *Dev. Biol.* **2008**, *316*, 336–349. [CrossRef]
43. Nakashima, K.; de Crombrugghe, B. Transcriptional mechanisms in osteoblast differentiation and bone formation. *Trends Genet.* **2003**, *19*, 458–466. [CrossRef]
44. Perrien, D.S.; Brown, E.C.; Aronson, J.; Skinner, R.A.; Montague, D.C.; Badger, T.M.; Lumpkin, C.K., Jr. Immunohistochemical study of osteopontin expression during distraction osteogenesis in the rat. *J. Histochem. Cytochem.* **2002**, *50*, 567–574. [CrossRef]
45. Su, S.; Dong, Z. Comparative expression analyses of bone morphogenetic protein 4 (BMP4) expressions in muscles of tilapia and common carp indicate that BMP4 plays a role in the intermuscular bone distribution in a dose-dependent manner. *Gene Expr. Patterns* **2018**, *27*, 106–113. [CrossRef]
46. Chinipardaz, Z.; Liu, M.; Graves, D.T.; Yang, S. Role of Primary Cilia in Bone and Cartilage. *J. Dent. Res.* **2021**, 220345211046606. [CrossRef]

47. Shi, W.; Ma, Z.; Zhang, G.; Wang, C.; Jiao, Z. Novel functions of the primary cilium in bone disease and cancer. *Cytoskeleton* **2019**, *76*, 233–242. [CrossRef]
48. Camacho-Vanegas, O.; Camacho, S.C.; Till, J.; Miranda-Lorenzo, I.; Terzo, E.; Ramirez, M.C.; Schramm, V.; Cordovano, G.; Watts, G.; Mehta, S.; et al. Primate Genome Gain and Loss: A Bone Dysplasia, Muscular Dystrophy, and Bone Cancer Syndrome Resulting from Mutated Retroviral-Derived MTAP Transcripts. *Am. J. Hum. Genet.* **2012**, *90*, 614–627. [CrossRef]
49. Topaz, O.; Indelman, M.; Chefetz, I.; Geiger, D.; Metzker, A.; Altschuler, Y.; Choder, M.; Bercovich, D.; Uitto, J.; Bergman, R.; et al. A Deleterious Mutation in SAMD9 Causes Normophosphatemic Familial Tumoral Calcinosis. *Am. J. Hum. Genet.* **2006**, *79*, 759–764. [CrossRef]
50. Cheng, F.; Sun, R.; Hou, X.; Zheng, H.; Zhang, F.; Zhang, Y.; Liu, B.; Liang, J.; Zhuang, M.; Liu, Y.; et al. Subgenome parallel selection is associated with morphotype diversification and convergent crop domestication in Brassica rapa and Brassica oleracea. *Nat. Genet.* **2016**, *48*, 1218–1224. [CrossRef]

MDPI
St. Alban-Anlage 66
4052 Basel
Switzerland
Tel. +41 61 683 77 34
Fax +41 61 302 89 18
www.mdpi.com

Biology Editorial Office
E-mail: biology@mdpi.com
www.mdpi.com/journal/biology

www.ingramcontent.com/pod-product-compliance
Lightning Source LLC
Chambersburg PA
CBHW061457220926
51850CB00004B/109
* 9 7 8 3 0 3 6 5 5 9 2 2 3 *